UNITS OF MEASUREMENT METRIC/ENGLISH CONVERSIONS

Length

1 meter = 39.4 inches = 3.28 feet = 1.09 yard

1 foot = 0.305 meters = 12 inches = 0.33 yard

1 inch = 2.54 centimeters

1 centimeter = 10 millimeters = 0.394 inch

1 millimeter = 0.001 meter = 0.01 centimeter =0.039 inch

1 fathom = 6 feet = 1.83 meters

1 rod = 16.5 feet = 5 meters

1 chain = 4 rods = 66 feet = 20 meters

1 furlong = 10 chains = 40 rods = 660 feet = 200 meters

1 kilometer = 1,000 meters = 0.621 miles =
    0.54 nautical miles

1 mile = 5,280 feet = 8 furlongs = 1.61 kilometers

1 nautical mile = 1.15 mile

Area

1 square centimeter = 0.155 square inch

1 square foot = 144 square inches = 929 square centimeters

1 square yard = 9 square feet = 0.836 square meters

1 square meter = 10.76 square feet = 1.196 square yards =
    1 million square millimeters

1 hectare = 10,000 square meters = 0.01 square kilometers =
    2.47 acres

1 acre = 43,560 square feet = 0.405 hectares

1 square kilometer = 100 hectares = 1 million square meters =
    0.386 square miles = 247 acres

1 square mile = 640 acres = 2.59 square kilometers

Volume

1 cubic centimeter = 1 milliliter = 0.001 liter

1 cubic meter = 1 million cubic centimeters =1,000 liters

1 cubic meter = 35.3 cubic feet = 1.307 cubic yards =
    264 US gallons

1 cubic yard = 27 cubic feet = 0.765 cubic meters =
    202 US gallons

1 cubic kilometer = 1 million cubic meters =0.24 cubic mile =
    264 billion gallons

1 cubic mile = 4.166 cubic kilometers

1 liter = 1,000 milliliters = 1.06 quarts = 0.265 US gallons =
    0.035 cubic feet

1 US gallon = 4 quarts = 3.79 liters = 231 cubic inches =
    0.83 imperial (British) gallons

1 quart = 2 pints = 4 cups = 0.94 liters

1 acre foot = 325,851 US gallons = 1,234,975 liters =
    1,234 cubic meters

1 barrel (of oil) = 42 US gallons = 159 liters

Mass

1 microgram = 0.001 milligram = 0.000001 gram

1 gram = 1,000 milligrams = 0.035 ounce

1 kilogram = 1,000 grams = 2.205 pounds

1 pound = 16 ounces = 454 grams

1 short ton = 2,000 pounds = 909 kilograms

1 metric ton = 1,000 kilograms = 2,200 pounds

Temperature

Celsius to Fahrenheit $°F = (°C \times 1.8) + 32$

Fahrenheit to Celsius $°C = (°F - 32) \div 1.8$

Energy and Power

1 erg = 1 dyne per square centimeter

1 joule = 10 million ergs

1 calorie = 4.184 joules

1 kilojoule = 1,000 joules =
    0.949 British Thermal Units (BTU)

1 megajoule = MJ = 1,000,000 joules

1 kilocalorie = 1,000 calories = 3.97 BTU =
    0.00116 kilowatt-hour

1 BTU = 0.293 watt-hour

1 kilowatt-hour = 1,000 watt-hours = 860 kilocalories =
    3,400 BTU

1 horsepower = 640 kilocalories

1 quad = 1 quadrillion kilojoules =
    2.93 trillion kilowatt-hours

Quantitative Prefixes

Large Numbers	Description	Small Numbers
exa 10^{18}	quintillion	alto 10^{-18}
peta 10^{15}	quadrillion	femto 10
tera 10^{12}	trillion	pico 10^{-1}
giga 10^{9}	billion	nano 10^{-}
mega 10^{6}	million	micro 10^{-6}
kilo 10^{3}	thousand	milli 10^{-3}

(e.g., a kilogram = 1,000 gm; a milligram = one-thousandth of a gram)

Robert K. Kaufmann
Boston University

Cutler J. Cleveland
Boston University

ENVIRONMENTAL
SCIENCE

Boston   Burr Ridge, IL   Dubuque, IA   New York   San Francisco   St. Louis
Bangkok   Bogotá   Caracas   Kuala Lumpur   Lisbon   London   Madrid   Mexico City
Milan   Montreal   New Delhi   Santiago   Seoul   Singapore   Sydney   Taipei   Toronto

TO RACHEL AND EVE

The world you get is the one your generation creates.
I hope it will be better than the one my generation leaves.
—*Robert K. Kaufmann*

TO ALL MY TEACHERS

. . . especially Charlie, Bob, Bruce, and Herman.
—*Cutler J. Cleveland*

ENVIRONMENTAL SCIENCE

1 2 3 4 5 6 7 8 9 0 DOW/DOW 0 9 8 7

ISBN 978–0–07–298429–3
MHID 0–07–298429–5

Publisher: *Thomas D. Timp/Margaret J. Kemp*
Senior Developmental Editor: *Joan M. Weber*
Senior Marketing Manager: *Tami Petsche*
Lead Project Manager: *Joyce M. Berendes*
Lead Production Supervisor: *Sandy Ludovissy*
Lead Media Project Manager: *Judi David*
Media Producer: *Daniel M. Wallace*
Senior Coordinator of Freelance Design: *Michelle D. Whitaker*
Cover/Interior Designer: *Jamie E. O'Neal*
Senior Photo Research Coordinator: *Lori Hancock*
Photo Research: *LouAnn K. Wilson*
Compositor: *Electronic Publishing Services Inc., NYC*
Typeface: *10/12 Palatino*
Printer: *R. R. Donnelley Willard, OH*

(USE) Cover Image: **Front cover image:** Anasazi Ruins Cliff Palace: © *George H. H. Huey / CORBIS*
Back cover images: (top): © *Albert Normandin / Masterfile;* (middle): © *Gary Gerovac / Masterfile;* (bottom): © *Royalty-Free/CORBIS*

The credits section for this book begins on page 538 and is considered an extension of the copyright page.

Library of Congress Cataloging-in-Publication Data

Kaufmann, Robert (Robert K.)
    Environmental science / Robert Kaufmann, Cutler Cleveland. – 1st ed.
      p. cm.
    Includes index.
    ISBN 978–0–07–298429–3 --- ISBN 0–07–298429–5 (acid-free paper)
    1. Environmental sciences--Textbooks. 2. Nature--Effect of human beings
      on--Textbooks. 3. Human beings--Effect of environment on--Textbooks. I. Cleveland,
      Cutler J. II. Title.
        GE105.K386 2008
        628--dc22   2006047312

www.mhhe.com

About the Authors

Robert K. Kaufmann is the director of and a full professor in the *Center for Energy & Environmental Studies* and the *Department of Geography and Environment* at Boston University. Before coming to the center, he was an economist at the WEFA Group and Chase Econometrics and a research scientist at Complex Systems Research Center at the University of New Hampshire. He received his PhD at the University of Pennsylvania in 1988 and a BS from Cornell University in 1979.

Professor Kaufmann teaches several undergraduate classes in energy and the environment, including Introduction to Environmental Science, Intermediate Environmental Science, and a course about environmental history. At the graduate level, he teaches classes in resource and environmental economics, ecological economics, and applied time series econometrics.

In addition to *Environmental Science,* Professor Kaufmann has written two other books, several book chapters, and more than sixty peer review papers on topics ranging from world oil markets, global climate change, and land use change to the global carbon cycle and ecological economics. Appearing in a variety of natural and social science journals, including *Science, Nature,* and *Proceedings of the National Academy of Sciences,* these papers have been cited more than one thousand times and have won awards from the International Association of Energy Economists, *Scientific American,* and the U.S. Wildlife Federation. Interviews with the author and his research results have appeared on the NBC *Nightly News* and the CBS *Nightly News* programs, as well as in *National Geographic, Reader's Digest,* and about a hundred newspapers, including *The Wall Street Journal, The New York Times,* and *The Financial Times.*

Professor Kaufmann's research efforts have been funded by approximately a million dollars in grants from institutions such as the National Science Foundation, NASA, the U.S. Environmental Protection Agency, and the Betty and Gordon Moore Foundation. In addition to doing consulting work for Nomura Securities, the European Central Bank, the World Bank, and the U.S. Department of Energy, Professor Kaufmann has also served as a panel member for the Intergovernmental Panel on Climate Change, the NASA Land Use/Land Cover Change Steering Committee, and the Project LINK Modeling Center, which maintains a global econometric model for the United Nations.

Professor Kaufmann is married to Shauna Tannebaum, and together they have two children, Rachel and Eve. In addition to "nachas from kinder," his happiness comes from racing his bicycle, looking for snakes and lizards while hiking, and fishing with his brother David in the fabulous Florida Keys.

Cutler J. Cleveland is currently a professor in the *Department of Geography and Environment* and the *Center for Energy and Environmental Studies* at Boston University. Professor Cleveland is editor-in-chief of the *Encyclopedia of Energy,* winner of an American Library Association award, editor of the *Dictionary of Energy,* and editor-in-chief of the journal *Ecological Economics.* Professor Cleveland is chairman of the Environmental Information Coalition, the governing body of the *Earth Portal.*

Professor Cleveland has been a consultant to numerous private and public organizations, including the Asian Development Bank, the United Nations Commission on Sustainable Development, Charles River Associates, the Technical Research Centre of Finland, and the U.S. Environmental Protection Agency. A member of the American Statistical Association's Committee on Energy Statistics, he is also a participant in the Stanford Energy Modeling Forum and part of an advisory group to the U.S. Department of Energy.

Professor Cleveland's research centers largely on oil supply, net energy analysis, natural resource scarcity, and the relations between energy use, natural resources, and economic systems. The National Science Foundation, the National Aeronautics and Space Administration, and the MacArthur Foundation have all supported Professor Cleveland's research. Professor Cleveland has published his findings in journals such as *Nature, Science, Ecological Modeling, Energy, The Energy Journal, The Annual Review of Energy, Resource and Energy Economics,* the *American Association of Petroleum Geologists Bulletin,* the *Canadian Journal of Forest Research,* and *Ecological Economics.*

Through the years, Professor Cleveland has received many awards. For his unique and innovative contributions to the field of energy economics, Professor Cleveland was given the Adelman-Frankel Award from the U.S. Association of Energy Economics. In addition to receiving publication awards from the International Association of Energy Economics and the National Wildlife Federation, he has also won teaching awards from the University of Illinois and Boston University's College of Arts and Sciences Honors Program.

Professor Cleveland is married to Karen Lefkowitz, and together they have two children, Molly and Sam. His interests include reading, bicycling, tennis, the Adirondack Mountains, and coaching his kids' soccer and lacrosse games.

Contents in Brief

PART ONE
INTRODUCTION TO ENVIRONMENTAL SCIENCE 1

Chapter 1   Environment and Society:
A Sustainable Partnership? 1

PART TWO
BASIC CONCEPTS IN ENVIRONMENTAL SCIENCE 19

Chapter 2   The Laws of Energy and Matter 19

Chapter 3   Systems: Why Are Environmental Problems So Difficult to Solve? 38

PART THREE
HOW THE NATURAL ENVIRONMENT WORKS 55

Chapter 4   The Physical Systems of Planet Earth: The Engine of Life 55

Chapter 5   The Flow of Energy in Biological Systems: Why Does It Matter? 75

Chapter 6   The Flow of Matter in the Environment: Why Does It Matter? 99

Chapter 7   Biomes: Where Do Plants and Animals Live? 123

Chapter 8   Succession: How Do Ecosystems Respond to Disturbance? 156

PART FOUR
HOW HUMAN SYSTEMS WORK 178

Chapter 9   Carrying Capacity: How Large a Population? 178

Chapter 10   An Ecological View of the Economy 202

Chapter 11   The Driving Forces of Environmental Change 224

PART FIVE
GLOBAL ENVIRONMENTAL CHALLENGES 246

Chapter 12   Biodiversity: Species and So Much More 246

Chapter 13   Global Climate Change: A Warming Planet 268

Chapter 14   A Reduction in Atmospheric Ozone: Let the Sunshine In 293

PART SIX
LIVING OFF THE LAND: SOILS, FOOD PRODUCTION, AND FORESTS 311

Chapter 15   Soil: A Potentially Sustainable Resource 311

Chapter 16   Agriculture: The Ecology of Growing Food 333

Chapter 17   Forests: So Much More Than Wood 354

PART SEVEN
AIR AND WATER RESOURCES 372

Chapter 18   Water Resources 372

Chapter 19   Air Pollution: Costs and Benefits of Clean Air 398

PART EIGHT
ENERGY AND MATERIALS 419

Chapter 20   Fossil Fuels: The Lifeblood of the Global Economy 419

Chapter 21   Nuclear Power 442

Chapter 22   Renewable Energy and Energy Efficiency 464

Chapter 23   Materials, Society, and the Environment 487

Chapter 24   A Sustainable Future: Will Business as Usual Get Us There? 509

Contents

Preface xiii
Guided Tour xx

PART ONE
AN INTRODUCTION TO ENVIRONMENTAL SCIENCE 1

Chapter 1 Environment and Society: A Sustainable Partnership? 1

Easter Island: The Island
    That Self-Destructed 2

The Principles of Sustainability 3
    *Principle 1: Sustainable Use of Natural Resources
        and Environmental Services 3*
    Principle 2: A Systems Perspective 3
    Principle 3: Equity and Fairness 4
    Principle 4: Incentives for Sustainable Behavior 5

CASE STUDY How Big Can Society Be? The Environmental
Plimsoll Line 5

Are We Headed in the Right Direction? 6
    *Violating Principle 1:
        Depletion and Degradation
        of Natural Resources and Environmental Services 6*
    *Violating Principle 2:
        Policies That Lack a Systems Perspective 9*
    *Violating Principle 3:
        Unequal Opportunities for Human Development 9*
    *Violating Principle 4:
        Actions Must Be Both Environmentally
        and Economically Sustainable 12*

POLICY IN ACTION The President's Council
on Sustainable Development: Blueprint for Sustainability
or Environmental Window Dressing? 14

Sustainability on a Personal Level:
    The Ecological Footprint 15

YOUR ECOLOGICAL FOOTPRINT The Footprint
of the *Moai* 16

SUMMARY OF KEY CONCEPTS 18
REVIEW QUESTIONS 18
KEY TERMS 18

PART TWO
BASIC CONCEPTS IN ENVIRONMENTAL SCIENCE 19

Chapter 2 The Laws of Energy and Matter 19

Lead: Industrial Marvel
    and Environmental Villain 20

POLICY IN ACTION Getting the Lead Out 21

Matter: Elements and Compounds 21
    Elements Essential for Life 22
    Elements in Earth's Crust 23

CASE STUDY The Pathways of Lead in Society
and the Environment 24

Atoms: The Building Blocks of Elements 24
    Subatomic Particles 24

Changes in Matter 25
    Chemical Reactions 25
    Nuclear Changes 25
    The Law of Conservation of Matter 27

The Importance of Energy 28
    Types of Energy 28

YOUR ECOLOGICAL FOOTPRINT
Lead and Your Listening Pleasure 29
    Energy and Work 30
    Energy and Power 32

The Laws of Thermodynamics 32
    The First and Second Laws of Thermodynamics 32
    Entropy as a Measure of the Quality of Energy and Matter 33

The Energy and Materials Balance
    of a Coal-Fired Electricity Plant 35

SUMMARY OF KEY CONCEPTS 37
REVIEW QUESTIONS 37
KEY TERMS 37

Chapter 3 Systems: Why Are Environmental
Problems So Difficult to Solve? 38

The Last Tree 39
What Are Systems? 39
    Predictable Behavior 39
    Do Systems Have Goals? 40

How Do Systems Generate Their Behaviors? 40
Using Energy to Generate Order 40
Homeostasis: Maintaining Order in the Face of Disturbance 41
Generating Homeostatic Behavior 42

CASE STUDY The Collapse of the Easter Island Civilization 45
Natural Selection 46

Why Are Systems So Hard to Manage? 47
Unpredictability 47
Complexity 47
Hierarchy 48
Time Lags 48

YOUR ECOLOGICAL FOOTPRINT The Direct and Indirect
Use of Natural Resources and Emission of Wastes 49

Distance Effects 50
Linear versus Nonlinear Relationships 50
The Scientific Method 50
A Reductionist versus Systems Perspective 51
Simulation Models 52

POLICY IN ACTION The IPCC: An Interdisciplinary Effort
to Make Climate Change Policy 53

SUMMARY OF KEY CONCEPTS 54
REVIEW QUESTIONS 54
KEY TERMS 54

PART THREE
HOW THE NATURAL
ENVIRONMENT WORKS 55

Chapter 4   The Physical Systems
of Planet Earth: The Engine
of Life 55

Small Changes, Big Effects 56
Energy from the Sun and
    Earth's Interior 56
Energy from the Sun 56
Solar Radiation Reaching Earth's Surface 56
Work Done by Solar Energy 57
Heat from Earth's Interior 57
Work Done by Heat from Earth's Interior 57

YOUR ECOLOGICAL FOOTPRINT How Much Sunlight
Do You Use? 58

A Simple Model of Physical Systems 59
Global Patterns of Atmospheric Circulation 59
Creating Gradients: Differences in the Intensity
    of Solar Radiation 59
Circulation Cells 61
Surface Winds 62
Global Patterns of Precipitation 62
The Environmental Services Provided by
    Atmospheric Circulation 63

Global Patterns of Oceanic Circulation 64
Physical Properties of Water 64
Patterns of Oceanic Circulation 65

El Niño: Illustrating the Environmental Services of Oceanic
    Circulation 67

CASE STUDY El Niño: A Link among Atmospheric, Oceanic,
and Crustal Circulation? 68

Movements in Earth's Crust 69
Plate Tectonics: Global Circulation of the Crust 69

POLICY IN ACTION Policy Responses to El Niño 70
The Rock Cycle: The Circulation of Materials within Earth's Crust 71
Natural Resources and Environmental Services from Movements
    in Earth's Crust 73

SUMMARY OF KEY CONCEPTS 74
REVIEW QUESTIONS 74
KEY TERMS 74

Chapter 5   The Flow of Energy in Biological
Systems: Why Does It Matter? 75

Smart, Fast Dinosaurs? 76
How Individual Organisms Use Energy 76
Obtaining Energy from the Environment 76
Maintenance: Endotherms versus Ectotherms 77

YOUR ECOLOGICAL FOOTPRINT Tracing Your Energy Flows
and Changes in Weight 80

Growth 82
Storage 82
Reproduction 84
Protection 86
Energy and Natural Selection 88

Energy Flows between Organisms 89
Food Webs and Chains: Who Eats Whom 89
How Many Predators? 91

CASE STUDY Will Catching Few Fish Restore the Marine
Food Web? 93

Concentrating Toxins through the Food Web 94

POLICY IN ACTION Can We Catch More Fish by Moving
Down the Food Chain? 96

SUMMARY OF KEY CONCEPTS 98
REVIEW QUESTIONS 98
KEY TERMS 98

Chapter 6   The Flow of Matter
in the Environment: Why Does It Matter? 99

Animal-Eating Plants 100
Matter: The Building Blocks of Life 100
Nutrient Capture by Autotrophs 100
Nutrient Capture by Heterotrophs 102

The Flow of Matter: Biogeochemical Cycles 102
Understanding Biogeochemical Cycles 102
The Carbon Cycle: The Master Cycle 104

YOUR ECOLOGICAL FOOTPRINT How Much Net Primary
Production Do You Use? 106

The Nitrogen Cycle: Keep Your Eye on Changes in Form 110

CASE STUDY Watching the Planet Breathe: The Wiggle
in the Mauna Loa Curve 110

The Phosphorus Cycle: Running Downhill 113
The Sulfur Cycle: A Gateway for Human Environmental Impacts 115
Interactions among the Cycles 117

Disrupting Biogeochemical Cycles:
Understanding Environmental Challenges 117

POLICY IN ACTION Is Dumping Iron into the Ocean
an Effective Policy to Slow Global Climate Change? 118

SUMMARY OF KEY CONCEPTS 121
REVIEW QUESTIONS 121
KEY TERMS 122

**Chapter 7   Biomes: Where Do Plants
and Animals Live? 123**

Introduction 124
Why Species Live Where They Do 124

Habitats and Niches 124
Niche and Adaptation: Linking Species to Their Environments 124

POLICY IN ACTION Climatological and Ecological Determinants
of Human Land Use 126

Environmental Gradients: Linking Adaptation to Place 127

The Types and Distribution of Terrestrial Biomes 130

CASE STUDY Keeping Track of Terrestrial Biomes—The Use
of Satellite Remote Sensing 131

YOUR ECOLOGICAL FOOTPRINT What Biome Do You
Live In? 134

Tropical Rain Forests 134
Tropical Dry Forests 136
Tropical Savannas 138
Deserts 139
Mediterranean Woodland and Scrubland 140
Temperate Grasslands 141
Temperate Forests 142
Boreal Forests 143
Tundra 144

The Types and Distribution of Aquatic Biomes 146

Rivers and Streams 147
Lakes 148
Estuaries 149
Temperate Coastal Seas 150
Tropical Coastal Seas—Coral Reefs 151
Upwellings 152
The Open Ocean 152

SUMMARY OF KEY CONCEPTS 155
REVIEW QUESTIONS 155
KEY TERMS 155

**Chapter 8   Succession: How Do Ecosystems
Respond to Disturbance? 156**

Introduction 157
Understanding Disturbances 157
What Is Succession? 159
The Pattern of Succession 160

The Macro and Micro Environments 160
How Does the Microenvironment Change through Succession? 160
How Do Changes in the Microenvironment Shape
*   Successional Communities? 161*
How Do Ecosystems Change through Succession? 161
How Does Succession Work? 163

Catastrophic Shifts in Ecosystems 164
Can Ecosystems Cope with Human Disturbances? 165

Can We Measure Ecosystem Health? 165
How Does Diversity Enhance Ecosystem Health? 167
Helping Ecosystems Heal: Ecological Restoration 169

CASE STUDY Disturbance in Aquatic Food Chains 170

YOUR ECOLOGICAL FOOTPRINT How Much Land
Do You Disturb? 172

POLICY IN ACTION Does Suppressing Small Fires Create
Large Fires? 174

SUMMARY OF KEY CONCEPTS 176
REVIEW QUESTIONS 177
KEY TERMS 177

PART FOUR
HOW HUMAN SYSTEMS
WORK 178

**Chapter 9   Carrying Capacity:
How Large
a Population? 178**

What Does the Collapse of Easter Island
Mean to You? 179
How Big a Population? 179

Population Growth 179
Limiting Factors 180

CASE STUDY How Much Harvest Is Sustainable? 182

Defining Carrying Capacity 183

The Maximum Number of Individuals 183
A Given Area 183
Maintained Indefinitely 184
Fluctuations around Carrying Capacity—
*   The Negative Feedback Loop 185*

POLICY IN ACTION Should We Use the Logistic Curve
to Manage Renewable Resources? 186

Understanding Human Population Growth 188

The Pattern of Births and Deaths 189
Age Structure and Population Momentum 189

Modifying the Idea of Carrying Capacity for People 191

The Malthusian Dilemma 191
The Maximum Number of Individuals
*   and the Demographic Transition 192*
A Given Area of the Environment 193
Maintained Indefinitely 194

YOUR ECOLOGICAL FOOTPRINT
Look for the Sustainability Label 195

Does Carrying Capacity Apply to People? 196
Identifying the Notion of Limits in Environmental Debates 196
Resource Pessimists 197
Resource Optimists 198
Clarifying Carrying Capacity for People 199

SUMMARY OF KEY CONCEPTS 201
REVIEW QUESTIONS 201
KEY TERMS 201

Chapter 10   An Ecological View of the Economy: The Four Steps of Economic Production 202
Introduction 203
Two Views of the Economy 203
The Economic System: Production and Consumption of Goods and Services 203
Linking Production and Consumption—The Circular Flow Model of the Economy 204
What's Missing? The Role of the Environment 204

Four Steps of the Economic Process: Links to the Environment 205
Step 1: Creating Natural Resources 205
Step 2: Providing a Habitable Environment 206
Step 3: The Production of Goods, Services, and Wastes 206
Step 4: Waste Assimilation 208

Economic Growth 209
Comparing Levels of Affluence 210
The Driving Forces behind Economic Growth and Rising Affluence 210
Technological Change and Labor Productivity 210

YOUR ECOLOGICAL FOOTPRINT Comparing National Ecological Footprints 212
Affluence and Materials Use 213
The Connection among Economic Growth, Energy, and Materials 214

The Economic Value of Environmental Goods and Services 214
Why Are Environmental Contributions Often Overlooked? 214
Valuing Environmental Goods and Services 215
Impacts of Environmental Degradation 216

CASE STUDY Can Human Ingenuity Substitute for a Degraded Environment? 219

Accounting for Environmental Degradation 220

POLICY IN ACTION Greening the GDP 222

SUMMARY OF KEY CONCEPTS 223
REVIEW QUESTIONS 223
KEY TERMS 223

Chapter 11   The Driving Forces of Environmental Change: Population, Affluence, and Technology 224
Introduction 225
The Root Causes of Environmental Impact 225
Population 225

YOUR ECOLOGICAL FOOTPRINT Personal Transportation 226
The Contribution of Population Growth to Environmental Change 228

Affluence 231
The Contribution of Affluence to Environmental Change 231
Poverty and Exposure to Environmental Health Risks 233

Technology 234
Technologies That Ease Environmental Problems: Fuel-Efficient Cars and Waste Recovery 234
Technologies That Worsen Environmental Problems: Feedlots 235

CASE STUDY Do More Efficient Automobiles Reduce Motor Gasoline Use? 235

Tying It All Together 236
How Societies Choose Technologies: Political–Economic Institutions 237
The Market 237
Choosing the "Right" Technology: Market Failures 238
Market Failures and the Environment 238

Designing Sustainable Institutions 240
Attitudes and Beliefs 240
Formulating Environmental Policy 241
Market-Based Incentives versus Command and Control 242

POLICY IN ACTION Reducing Motor Gasoline Consumption: CAFE Standards or Higher Prices? 244

SUMMARY OF KEY CONCEPTS 245
REVIEW QUESTIONS 245
KEY TERMS 245

PART FIVE
GLOBAL ENVIRONMENTAL CHALLENGES 246

Chapter 12   Biodiversity: Species and So Much More 246
The Biosphere Experiment 247
Defining Biodiversity 247
Patterns of and Mechanisms for Biodiversity 249

CASE STUDY Why So Many Fish Species? 250

The Importance of Biodiversity 254
Species Interactions and Ecosystem Function 254
Biodiversity as Insurance 255
Genetic Knowledge 255
Environmental Services 256

Why Is Biodiversity Declining? 256
The Rate of Extinction 256
Habitat Alteration 257
Introduction of Alien Species 258
Changes in Biogeochemical Cycles 259
Hunting and Harvesting 260

Preserving Biodiversity 260
Legal Protections 261
Preserving Species and Habitat 261

POLICY IN ACTION Preserving Biodiversity in the Face of Corruption 262
Market-Based Mechanisms 264

YOUR ECOLOGICAL FOOTPRINT
Biodiversity-Friendly Coffee 265

SUMMARY OF KEY CONCEPTS 266
REVIEW QUESTIONS 267
KEY TERMS 267

**Chapter 13   Global Climate Change:
A Warming Planet 268**

Climate Change and Norse Settlements of Greenland 269
Climate and Climate Change 269
   The Difference between Climate and Weather 269
   A Changing Climate 269
The Heat Balance of Planet Earth: The Cause
   of a Changing Climate 270
   How Much Energy Reaches the Earth's Surface? 270
   How Much Energy Escapes Back to Space? The Greenhouse Effect 271
Radiative Forcing and Human Activity 273
   Concentrations of Greenhouse Gases 273
   Changing Climate by Disrupting Global Biogeochemical Cycles 274

CASE STUDY Hemispheric Patterns in Temperature Change 274
   Will Emissions Grow? 276
Detecting Climate Change and Attributing It to Human
   Activity 277
   Detecting Climate Change 277

YOUR ECOLOGICAL FOOTPRINT How Much Carbon Dioxide
Do You Emit? 278
   Attributing Climate Change to Human Activity 280

POLICY IN ACTION Which Comes First—The Supply or Demand
for Energy-Efficient Capital? 282
   Why Are Many Skeptical? 283
How Will Human Activity Affect Climate? 284
   The Past as the Future 284
   Computers as Crystal Balls 285
   Why Does Temperature Rise? 285
   Can We Trust the Predictions? 286
The Impacts of Global Climate Change 286
   Can Biomes Move Faster Than Climate Changes? 286
   Will Food Supplies Decrease or Increase? 287
   Will We Drown under a Rising Sea? 288
   Will Large Number of Species Go Extinct? 288
Climate Change Policy 289
   What Should Be Done? 289
   No Silver Bullet 290
   The Kyoto Protocol: A First Step? 290

SUMMARY OF KEY CONCEPTS 291
REVIEW QUESTIONS 292
KEY TERMS 292

**Chapter 14   A Reduction in Atmospheric
Ozone: Let the Sunshine In 293**

An Environmental Success Story 294
The Atmosphere 294
   Components of the Atmosphere 294
   Layers of the Atmosphere 294
Stratospheric Ozone 295
   The Formation of Ozone 295
   The Distribution of Ozone 296
A Reduction in Stratospheric Ozone 296
Why Is Stratospheric Ozone Declining? 297
   The Halogen Depletion Hypothesis 298
   The Odd Nitrogen Hypothesis 299
   The Dynamic Uplift Hypothesis 300
   Which Hypothesis Is Correct? 300
   The Complete Explanation for the Reduction in Stratospheric Ozone 301
   What about a Reduction over the North Pole? 301
The Effects of Less Stratospheric Ozone 302
   Impact on Marine Ecosystems along Antarctica 303
   The Impact on Terrestrial Organisms (Including People) 303
Policies to Restore the Ozone Layer 304
   Policy Deadlock 304
   Breaking the Deadlock 305
   Institutional Determinants of Success 305

YOUR ECOLOGICAL FOOTPRINT Managing Your Fun in the
Sun on an Ozone-Depleted Planet 306

CASE STUDY The Link between Climate Change
and the Reduction of Stratospheric Ozone 308

POLICY IN ACTION Why Was the Solution to the Reduction
in Stratospheric Ozone Simple Relative to Global Climate Change? 309
   The Role of Technology 309

SUMMARY OF KEY CONCEPTS 310
REVIEW QUESTIONS 310
KEY TERMS 310

PART SIX
LIVING OFF THE LAND: SOILS,
FOOD PRODUCTION,
AND FORESTS 311

**Chapter 15   Soil: A Potentially
Sustainable Resource 311**

The Dust Bowl 312
Land Use, Soil, and Biological Activity 312
Soil Formation 313
   Soil Horizons 313
   Soil Formation 314
Soil Type 315
Soil Function 317
   Storing Water 317
   Storing Nutrients 319
Soil Erosion 320

YOUR ECOLOGICAL FOOTPRINT How Much Soil
Do You Erode? 322
   Impacts of Soil Erosion 324

CASE STUDY Where Has All The Soil Gone? 325

Conserving Soil 325

Using Soil Sustainably 327
Why Don't Farmers Use Optimal Soil Conservation? 327
Implementing Soil Erosion Policy—
Limits to Market-Based Mechanisms? 329

POLICY IN ACTION Contradictions in U.S. Soil Erosion Policy 330

SUMMARY OF KEY CONCEPTS 331
REVIEW QUESTIONS 331
KEY TERMS 331

Chapter 16   Agriculture: The Ecology
of Growing Food 333

The Land of Milk and Honey 334
A Brief History of Food Production 334
Hunting and Gathering versus Agriculture 334
Why Agriculture? 335

The Ecology and Economics of Agriculture 338
The Ecology of Agriculture 339
The Economics of Agriculture 339
Green Revolution Agriculture 341

The Benefits and Costs of the Green Revolution 343
Benefits of the Green Revolution 343
The Costs of the Green Revolution 344

YOUR ECOLOGICAL FOOTPRINT The Land Requirements
of Your Diet 346

CASE STUDY Agricultural Pollution via the Nitrogen Cycle 348

POLICY IN ACTION Reducing Agricultural Pollution
and Increasing Farmer Profit 349

The Future of Agriculture 350
Increasing Food Production 350
Can Farmers Reduce Material Inputs? 351

SUMMARY OF KEY CONCEPTS 352
REVIEW QUESTIONS 353
KEY TERMS 353

Chapter 17   Forests: So Much More
Than Wood 354

Deforestation in Seventeenth-Century England 355

How Quickly Are Forests Being Cut (and Regrowing)? 355
Causes for Deforestation 357
Forests to Agriculture 357
Timber 358

CASE STUDY Deforestation and Oil Prices 359

Property Rights and Fire 360
Mineral and Energy Production 360
Roads and Other Transportation Infrastructure 361

The Contribution of Forests to Human Well-Being 362
Direct Contributions 362
Indirect Contributions 363

YOUR ECOLOGICAL FOOTPRINT How Much Wood Do You
Use For Paper? 364

Are Rates of Deforestation Too High? 367

Comparing Direct and Indirect Contributions to
Economic Well-Being 367
Policies 367
Social Structure 368

How Can Deforestation Be Slowed? 368
Getting Prices Correct 368
Sustainable Forestry Practices 368
Debt for Nature Swaps 370

POLICY IN ACTION Sustainable Forestry Practices: What Price
and Who Pays? 370

SUMMARY OF KEY CONCEPTS 371
REVIEW QUESTIONS 371
KEY TERMS 371

PART SEVEN

AIR AND WATER RESOURCES 372

Chapter 18
Water Resources 372

The Legend of Ubar 373
The Hydrologic Cycle 373
Water Supply 376
Surface Water 376
Groundwater 376

CASE STUDY Climate Change and the U.S. Water Supply 378

Human Use of Water 380
Offstream Water Uses 380
Instream Uses 381

YOUR ECOLOGICAL FOOTPRINT How Much Water
Do You Use? 382

Threats to Sustainable Supplies of Clean Water 384
Diverting Surface Waters 384
Mining Groundwater 386
Domestic and Municipal Sewage 388
Industrial Water Pollutants 391
Agricultural Water Pollutants 391

Water and Conflict 392
Ensuring Access to a Sustainable Supply
of Clean Water 393
A Market for Water? 393
Increasing Efficiency 394
Controlling Water Pollution 395

POLICY IN ACTION Privatizing Water in Cochabomba, Bolivia 395

SUMMARY OF KEY CONCEPTS 397
REVIEW QUESTIONS 397
KEY TERMS 397

Chapter 19   Air Pollution: Costs and Benefits
of Clean Air 398

A Trip to Guangzhou, China 399
Pollutants 399

Carbon Monoxide 400
Particulate Matter 400

CASE STUDY A Link between Local Pollution
and Global Climate Change 402

Sulfur Dioxide 402
Nitrogen Oxides 405
Hydrocarbons 407

Concentrations 407

YOUR ECOLOGICAL FOOTPRINT How Much
Carbon Monoxide Do You Emit? 408

Vertical and Horizontal Mixing 410
Atmospheric Stability 410
Observed Concentrations 411

Why Have Emissions Declined? 412
Air Pollution as an Externality 412
Legislative Efforts to Internalize Air Pollution 412

The Optimal Level of Air Pollution 414
The Costs of Abatement Strategies 414
The Benefits of Cleaner Air 415

POLICY IN ACTION A Market for Sulfur Emissions 416

SUMMARY OF KEY CONCEPTS 418
REVIEW QUESTIONS 418
KEY TERMS 418

PART EIGHT
ENERGY AND MATERIALS 419

**Chapter 20   Fossil Fuels:
The Lifeblood
of the Global Economy 419**

Déjà-Vu All Over Again 420
The Past and Present
of Fossil Fuel Use 420

YOUR ECOLOGICAL FOOTPRINT How Much Energy
Do You Use? 422

The Formation of Fossil Fuels 424
Coal 424
Crude Oil and Natural Gas 425

Discovery, Extraction, and Processing 426
Coal 426
Crude Oil and Natural Gas 427

The Future for Oil and Natural Gas 430
How Much Oil and Natural Gas Remain? 430
When Will We Run Out of Oil? 432
Life after the Peak 433

POLICY IN ACTION The Energy Policy and Conservation Act
of 1975 435

What Do You Need to Know
about the World Oil Market? 435
Should the United States Reduce Its Dependence
on Imported Oil? 435

CASE STUDY Changes in OPEC Pricing Strategy 437

A Competitive Oil Market? 437
Why Are Oil Prices So High Now? 440

SUMMARY OF KEY CONCEPTS 441
REVIEW QUESTIONS 441
KEY TERMS 441

Chapter 21   Nuclear Power 442

Nuclear Power: A Faustian Bargain? 443
The Nature, Distribution, and Use of Uranium Resources 443
Uranium Resources and Production 444
The Promise and Current Status of Nuclear Power 445
The Nuclear Fuel Cycle 447
Inside a Nuclear Reactor 449
Nuclear Reactor Safety 450

The Disposal of Radioactive Wastes 451
The Nature and Classification of Radioactive Waste 451
The Long-Term Disposal of Radioactive Waste 452
The Yucca Mountain Site 452
Paying for Waste Disposal 453
Decommissioning 454

How Safe Is Nuclear Energy? 454
Three Mile Island 454

YOUR ECOLOGICAL FOOTPRINT How Much Nuclear Waste
Do You Generate? 456

Chernobyl 457

CASE STUDY The Chernobyl Disaster: Positive Feedback Run Amok 458
Calculating the Risk 458
Proliferation and Diversion 459

The Economics of Nuclear Power 460
Fusion 460

POLICY IN ACTION Should Taxpayers Subsidize Civilian
Nuclear Power? 461

A Nuclear Renaissance? 462

SUMMARY OF KEY CONCEPTS 462
REVIEW QUESTIONS 463
KEY TERMS 463

**Chapter 22   Renewable Energy
and Energy Efficiency 464**

Solar Energy: Back to the Future 465
The Quantity/Quality Paradox of Renewable Energy 465
The Direct Use of Solar Energy 467
The Solar Resource 467
Solar Thermal Collectors 467

Photovoltaic Energy 468
PV Technology 468
Applications of PV 470

Wind Energy 470
The Wind Energy Resource Base 470
Wind Turbine Technology 471
How Much Electricity Can Wind Provide? 471
Current Status of Wind Energy 471
Environmental and Siting Issues 472

Biomass Energy 472
Biomass Resource Base 473

Current Status of Biomass 473
Biomass Technology 473

CASE STUDY Does Ethanol Have a Positive Energy Balance? 474

Environmental Impacts of Biomass 475

YOUR ECOLOGICAL FOOTPRINT Your Renewable
Energy Footprint 475

Geothermal Energy 476
Geothermal Technology 476
Environmental Impacts of Geothermal Energy 476

Ocean Energy Systems 477
Ocean Thermal Energy 477
Tidal Energy 477
Wave Energy 479

Hydropower 479
Current Status of Hydropower 480
Environmental and Social Impacts of Hydropower 480

Hydrogen 480
Hydrogen Production, Storage, and Transport 480
Fuel Cell Applications 481
The Economic and Environmental Impacts of Hydrogen 481

Energy Efficiency 481

POLICY IN ACTION What Should the Role of Government
Be in Shaping Our Energy Future? 483

An Economic and Environmental Comparison of Solar
and Fossil Fuel Energy 484

SUMMARY OF KEY CONCEPTS 485
REVIEW QUESTIONS 486
KEY TERMS 486

Chapter 23 Materials, Society,
and the Environment 487

Materials: The Stuff of Life 488
The Materials Cycle 488
Mineral Formation, Occurrence, and Abundance 488
Formation and Occurrence 488

CASE STUDY Tracing the Flows of Arsenic 491

Classifying Mineral Resources 492
Abundance and Distribution 493
Mineral Exploration and Production 493

Materials and the Economy 495
Dematerialization 495
The Price of Materials 496

Energy and Resource Quality 498
The Fate of Materials 498
Material Wastes 499

YOUR ECOLOGICAL FOOTPRINT Recycling Batteries 500

Reducing Material Wastes 502
Source Reduction 502
Recycling 503

POLICY IN ACTION Pay-as-You-Throw Programs
for Municipal Solid Waste 504

There's Recyclable...And Then There's Recycled 506
Remanufacturing 506

The Benefits of Reducing Material Wastes 506

SUMMARY OF KEY CONCEPTS 507
REVIEW QUESTIONS 508
KEY TERMS 508

Chapter 24 A Sustainable Future: Will Business
as Usual Get Us There? 509

Behind the Headlines: Drilling for Oil in the Arctic National
Wildlife Refuge 510
Understanding Possible Solutions: The Parable
of the Plimsoll Line 511
Efficiency 512
Internalizing Externalities 512

YOUR ECOLOGICAL FOOTPRINT What Is Your
Overall Impact? 513

Working with the Market 514

CASE STUDY Recycling Environmental Tax Revenues 515

Eliminating Subsidies 516
Personal Choices 516
Scale: An Upper Limit on Size and Economic Well-Being? 517
The Economics of the Demographic Transition 518
*Increasing Efficiency versus Scale: The Environmental
Kuznets Curve 519*

POLICY IN ACTION When Is a Reduction Not a Reduction?
Scale versus Efficiency? 519

Living within Limits Imposed by Biogeochemical Cycles 521
*Operationalizing the Precautionary Principle in a World Filled
with Risk 522*

SUMMARY OF KEY CONCEPTS 523
REVIEW QUESTIONS 523
KEY TERMS 523

Glossary 524
Suggested Readings 536
Credits 538
Index 540

Preface

The recent rise in oil prices is a real-world energy and economic challenge we face today. Often the natural and social sciences are viewed by students as sterile and irrelevant to the real world. But the effects of the interaction between energy and economics are evident everywhere around us and thus constitute the underlying theme of this text.

A key propeller of our local and global economies and the driving force behind many governmental policies, energy is literally the powerhouse fueling both the problems and the progress of modern science. Not only will this text help students understand the issues faced by scientists grappling with the most critical challenge of our century—that is, the maintenance of an energy supply that will support the increasing population and the decreasing supply of fossil fuels—but it will make science relevant to students.

The twenty-four chapters in this text go well beyond simple descriptions of coal mines and solar cells. Our chapters integrate energy, economics, and policy to convey the serious nature of environmental challenges, such as rising oil prices, and to examine the root causes of these challenges and their wide-reaching effects. While doing this, we also emphasize our optimism regarding the resolution of these problems. Students will understand by the end of the course why oil's high costs extend far beyond paying more at the pump and how these costs affect our economic well-being and interactions with the environment.

WHY WRITE THIS BOOK?

Why write another text for a market that already has so many? In summary, our book is different. Many of the most popular traditional texts are, to varying degrees, "environmental encyclopedias," laden with facts about the environment and pollution but lacking comprehensive discussion that stresses interrelationships and linkages among problems and driving forces and solutions. As a result, these texts fail to teach students a coherent worldview that communicates how the environment affects their lives and how their actions affect the environment. Our frustration with this disparity drove us to write the first draft of this text more than ten years ago.

Starting with blank pages, we articulated our vision by crafting chapters and tools that we believed would facilitate learning and possibly kindle budding scientists. We used drafts in our own classes with hundreds of students, making multiple revisions as our vision clarified with each class taught. Through numerous reviews and developmental symposia, McGraw-Hill confirmed that environmental science instructors support a fresh approach that embraces energy and economic systems as key themes. The ensuing work coalesced around writing schedules, review panels, art and photo development, and careful selection of case studies and relevant examples. More than 100 environmental science instructors guided this text to improve its organization, accuracy, and currency. Every illustration was evaluated by a special panel of reviewers to ensure its accuracy and appropriateness. After having been class-tested, carefully appraised, and meticulously scrutinized, we hope to inspire confidence and enthusiasm for a stimulating new approach in environmental science texts.

HOW IS OUR BOOK DIFFERENT?

To fill the market gap left by current environmental texts, we have integrated information about the interaction between humans and their environment to develop a coherent body of knowledge instead of presenting an amalgamation of facts scattered throughout the book. Because teaching concepts and ideas can be a formidable challenge for instructors unfamiliar with the terminology used to discuss ecological and economic systems, we have introduced these concepts and terms in small doses throughout the text. Teaching ideas in conjunction with facts will expand the goals of the introductory environmental science course; and in addition to having a more complete understanding of existing environmental problems, students will obtain the tools necessary to make sense of new environmental problems.

In addition to employing an interdisciplinary approach, we aim to identify relevant environmental problems, understand the basis of those problems, and develop the tools necessary to solve those problems while concurrently establishing a framework that teachers can use and students can read as they would an introductory text in biology, chemistry, or economics.

An Interdisciplinary Perspective

We are scientists with formal training in both natural and social sciences. Based on this background, we view the world via the seamless flow of energy, materials, and information among physical, ecological, and human systems. This lens allows us to weave together an interdisciplinary story of the relationship between people and their environment. You will find the following overall themes throughout the text:

- **Energy and material flows:** In addition to introducing matter and energy, Chapters 2 and 3 discuss how the laws of conservation of matter and thermodynamics govern their transformations in the environment and society. Although other books have similar chapters, the emphasis on material and energy flows stops there. In our book, however, Chapters 2 and 3 lay the foundation for the chapters that follow. Chapter 4 describes how these principles govern the flow of energy and matter in physical systems and generate predictable movements in the atmosphere, oceans, and crust that are critical to ecological and economic systems. From there, we follow the flow of energy and materials in and among biological systems. Chapter 5 uses energy flows to explore the evolutionary strategies of individual organisms and entire food webs. Chapter 6 uses material flows to describe biogeochemical cycles and explore how human changes to these cycles cause environmental problems. The patterns described in Chapter 4 are combined with biological principles of energy flows from Chapter 5 to explain the distribution of individual organisms and entire biomes.

- **Ecology and economic systems:** We use the same energy and material flows to link human systems to physical and ecological systems. Chapter 9 describes the carrying capacity of nonhuman populations in relation to the availability of energy and materials. These principles are modified to explain arguments about limits on the size and well-being of human populations. In Chapter 10 we use material and energy flows to describe economic production and consumption as well as to illustrate how these flows cause environmental degradation. Consequently, our material and energy analysis of economic systems can be understood by environmental science students and instructors. Chapter 11, which describes how the severity of environmental impacts is driven by population, affluence, and technology, discusses how these impacts can be mitigated or exacerbated by social institutions that influence the choice of technology and implementation of policy. We then describe the economic principles that guide environmental policy without resorting to a host of alien-looking supply and demand curves.

By focusing on material and energy flows, we avoid economic theory and emphasize *economic systems*. Our emphasis on physical, chemical, and biological laws and principles that govern material and energy flows in human systems lets instructors teach their students about human systems without teaching "economics." We focus on how the economic system is linked to, and depends on, natural systems. Demonstrating and describing this linkage does not use standard economic tools such as marginal cost curves or production and utility isoquants. Instead we use the same biophysical analysis of energy and material flows in natural systems to link human economies to their environmental life support. As a result, instructors and students do not need any training in or prior knowledge of formal economics to use this text.

- **Costs and benefits of environmental policy:** In contrast to a simple list of environmental legislation, the interdisciplinary perspective prepares students to understand the costs and benefits of competing solutions to environmental challenges. This discussion builds on the principles in Chapters 10 and 11, which define the goals of environmental policy and the costs and benefits of various policy options. Chapter 13, a chapter about climate change, describes why zero carbon emissions is not a viable short-term policy goal because the cost of eliminating all uses of fossil fuels would be much larger than the damages avoided by halting emissions. Furthermore, we use the principles described in Chapter 11 to evaluate the effectiveness of various climate change options, such as carbon taxes, tradable permits, and standards. Finally, Chapter 13 describes issues of equity among developed and developing nations that increase the difficulty of international agreement on a climate change policy.

- **Systems perspective:** Chapter 3 explains what systems are and how feedback loops, stability, resistance, and resilience influence their behavior. Again, other books have a "systems chapter"; but the system perspective appears repeatedly throughout this text. For example, the ideas of resistance and resilience are discussed with regard to ecological succession in Chapter 8. One or more feedback diagrams appear in nearly all of the chapters after Chapter 3. The chapter about deforestation, for instance, uses a feedback diagram to illustrate how deforestation changes the physical, ecological, and economic environment in ways that either speed up or slow down future rates of deforestation.

- **Darwinian evolution and natural selection:** Although many books have a chapter about Darwinian evolution and natural selection, they frequently fail to portray these ideas in any meaningful way. As a result, Darwinian evolution often appears as a stand-alone theory. As environmental scientists, we know the opposite to be true: Darwinian evolution and natural selection explain material and energy flows in ecological systems and help explain the basis for and potential solutions to environmental challenges. Consistent with this integrated view, Darwinian evolution and natural selection are woven throughout the book. Natural selection is described in the chapter about systems theory. These principles are used in Chapter 5 to describe how the use of energy by individual organisms can be viewed as evolutionary strategies. Indeed, the section titled "Why Breathe

Oxygen" describes the evolutionary benefits of aerobic respiration. Similarly, adaptations of individual species and the "look" of entire biomes are explained in terms of evolutionary adaptations.

Focus on Underlying Environmental Causes

Although we applaud and encourage action at the individual level, this approach focuses on the *symptoms* of environmental deterioration rather than on the underlying *causes*. As Paul Ehrlich says, if the environmental agenda of every Earthling were recycling soda cans, stopping junk mail, screwing in energy-efficient lightbulbs, and using less water on lawns, Earth would still go down the drain. Our focus on the underlying causes surrounding environmental issues prepares students to understand the big picture.

The "Big Picture"

To understand the challenges that face them, students need the "big picture." As opposed to a simple recitation of facts, we emphasize how the world works and focus on the relationships between human systems and natural systems. We stress how the environment supports human well-being, shifting the emphasis from what we do *to* the planet to what it does *for* us and for other species.

Conceptual Illustration Program

As you flip through the pages of this book, you will notice that unlike other texts, ours has relatively few dazzling photographs of cute or creepy-crawly species that are on the brink of extinction; nor do we feature depressing photographs of pollution pouring into the air or water. Instead we rely on conceptual figures and graphs to tell the story. Captioned in a way to help "visual learners," our illustrations convey important ideas—such as the balance of material and energy flows and the ecological impact of ecosystem fragmentation. Important aspects of the links between humans and their environment are illustrated using graphs from scientific literature. Our diagrams show how natural systems are organized around the flow of matter and energy and how human alteration of these flows creates environmental challenges. Homeostatic mechanisms in natural ecosystems are illustrated by feedback loops, illustrating how human interventions can cause the system to re-equilibrate at another level or spin out of control.

DISTINCTIVE LEARNING TOOLS

The features of this text are unique and were carefully planned to enhance students' comprehension of environmental science. All chapters contain the following key components:

- **Student Learning Outcomes:** Each chapter begins with a list of Student Learning Outcomes (SLOs) that provide an immediate overview of the key concepts students are expected to learn. In addition, each of the *Your Ecological Footprint, Policy in Action*, and *Case Study* readings contains student learning outcomes that illustrate a comprehensive overview of the key concepts students are expected to gain from the featured material. SLOs express higher-level thinking skills that integrate the content and activities and lead to the development of a behavior, skill, or discrete usable knowledge upon completing the class.

- **Your Ecological Footprint readings:** *Your Ecological Footprint* is an empirical footprint exercise that describes and measures the amount of specific form(s) of environmental goods and services used to provide the students' lifestyle. Students are asked to audit various aspects of their daily lives, such as food, energy, or water consumption. Appropriate technical conversion factors are provided to make the simple calculations. Consistent with our interdisciplinary emphasis, the exercises build on each other to offer students information outlining how their lifestyle affects the environment, thus enabling them to independently analyze the cost and benefits of alternatives.

- **Policy in Action readings:** Concepts from environmental science can be used to guide social actions, particularly environmental policy. *Policy in Action* readings apply the systems perspective to decision making and thus help students understand how the links between natural and social sciences influence why some environmental policies are effective while others are not.

- **Case Studies:** Case studies in each chapter illustrate how the systems aspects of environmental challenges complicate possible solutions. For example, the Case Study in Chapter 18 discusses how climate change affects both precipitation and evaporation and subsequently outlines the difficulties in determining the effects of climate change on water supplies in the United States.

YOUR FEEDBACK IS WELCOME

We would appreciate hearing from both students and instructors about where and how we could improve this text. You, the users, are the real test of whether we have accomplished our goal of presenting environmental science as a big picture with an interdisciplinary systems approach. Please let us know what you think. We value your comments and suggestions. Please send your recommendations to the Earth and Environmental Science Division of McGraw-Hill Higher Education at 2460 Kerper Boulevard, Dubuque, IA 52001.

ACKNOWLEDGMENTS

As with most books, the research described here goes well beyond our own. This acknowledgment takes on added importance given the interdisciplinary nature of the book. We thank the many experts who helped us translate their original research results and expertise into readable paragraphs that are also technically correct:

Geoffrey A. Abers, *Boston University*

Richard Ahlstrom, *HRA, Inc., Conservation Archaeology*

Bruce T. Anderson, *Boston University*

Arthur Anker, *Smithsonian Tropical Research Institute, Panama City, Panama*

Maria del Carmen Vera-Diaz, *Amazon Institute for Environmental Research—Boston University*

Brynhildur (Binna) Davidsdottir, *University of Iceland*

Charles A.S. Hall, *SUNY—Syracuse*

Marc L. Imhoff, *Biospheric Sciences Branch, NASA's Goddard Space Flight Center*

Nancy Knowlton, *Scripps Institution of Oceanography— University of California, San Diego*

Daniel Nepstad, *Woods Hole Research Center*

Daniel Pauly, *Fisheries Centre, University of British Columbia*

Denis Poddoubtchenko, *Smithsonian Tropical Research Institute, Panama City, Panama*

F. Harvey Pough, *Rochester Institute of Technology*

J. Alan Pounds, *Monteverde Cloud Forest Preserve, Costa Rica*

Guido D. Salvucci, *Boston University*

Stephen Votier, *University of Plymouth*

Martin J. Wassen, *Copernicus Institute for Sustainable Development and Innovation—Utrecht University, The Netherlands*

Paul Zimansky, *Boston University*

We especially thank Charles A.S. Hall for putting us in touch with Marge Kemp, who recognized that we had something different to say about environmental science and education. The professional team involved in producing this text has vested many hours in making it the best environmental science text possible. The authors wish to express special thanks to McGraw-Hill for editorial support through Marge Kemp and Joan Weber; the marketing expertise of Tami Petsche; and the production team of Joyce Berendes, Sheila Frank, Michelle Whitaker, Lori Hancock, LouAnn Wilson, Sandy Ludovissy, Dan Wallace, and Judi David. Finally, we thank Laura Guild, who spent many hours computer editing various drafts of this book.

We rely on the massive feedback arranged by the publisher—through individual reviews, focus groups, art review panels, and accuracy checkers. These scholars and educators helped guide decisions about content, art, pedagogy, and all the other components that make this text a book:

David A. Francko, *Miami University of Ohio*

Walter A. Illman, *University of Iowa*

Andrew Lapinski, *Reading Area Community College*

The quality of the feedback we received from the art review panel members contributed significantly to this project. We thank these art review panel members:

David Aborn, *University of Tennessee—Chattanooga*

Robert Dill, *Bergen Community College*

Phillip B. Kneller, *Western Carolina University*

Henry M. Knizeski, *Mercy College*

Mark A. Smith, *Riverside Community College*

Michael Toscano, *San Joaquin Delta College*

We greatly appreciate the help of all the reviewers listed here—committed teachers who also have the same aims as the authors and editors:

David Aborn, *University of Tennessee—Chattanooga*

M. Stephen Ailstock, *Anne Arundel Community College*

Felix O. Akojie, *West Kentucky Community and Technical College*

Mark W. Anderson, *The University of Maine*

Roger Balm, *Rutgers University*

Jay L. Banner, *University of Texas—Austin*

Gary A. Beluzo, *Holyoke Community College*

William Berry, *University of California—Berkeley*

Donna Bivans, *Pitt Community College*

Jeffrey Braatne, *University of Idaho*

Hugh J. Brown, *Ball State University*

Christopher Cronan, *University of Maine*

Lynnette Danzl-Tauer, *Rock Valley College*

Paul G. Decelles, *Johnson County Community College*

Michael L. Denniston, *Georgia Perimeter College*

Robert Dill, *Bergen Community College*

Timothy J. Farnham, *University of Nevada—Las Vegas*

Dawn M. Ford, *University of Tennessee—Chattanooga*

David A. Francko, *Miami University of Ohio*

David L. Gorchov, *Miami University of Ohio*

Melissa Grigione, *University of South Florida*

Douglas J. Hallett, *Northern Arizona University*

Tracey Holloway, *University of Wisconsin—Madison*

Timothy V. Horger, *Illinois Valley Community College*

Debra Howell, *Chabot Community College*

Charles F. Ide, *Western Michigan University*

Walter A. Illman, *University of Iowa*

Zoghlul Kabir, *Rutgers University*

Phillip B. Kneller, *Western Carolina University*

Henry M. Knizeski, *Mercy College*

Nicole M. Korfanta, *University of Wyoming*

Andrew Lapinski, *Reading Area Community College*

Ernesto Lasso de la Vega, *Edison College*

Bobby Ann Lee, *West Kentucky Community and Technical College*

Les M. Lynn, *Bergen Community College*

Timothy F. Lyon, *Ball State University*

Blase Maffia, *University of Miami*

Matthew H. McConeghy, *Johnson & Wales University*

Mark A. McGinley, *Texas Tech University*

Chris Migliaccio, *Miami-Dade College*

Jerry T. Mitchell, *University of South Carolina—Columbia*

Royden Nakamura, *California Polytechnic State University— San Luis Obispo*

Michael J. Neilson, *University of Alabama—Birmingham*

Eugene Parker, *University of Maryland—Baltimore County*

Ervand M. Peterson, *Sonoma State University*

Julie Phillips, *De Anza College*

John Pleasants, *Iowa State University*

Thomas Pliske, *Florida International University*

Colin Polsky, *Clark University*

Jarka Popovicova, *Ball State University*

Lauren J. Preske, *University of Southern Indiana*

Pamela C. Rasmussen, *Michigan State University*

Seth R. Reice, *University of North Carolina—Chapel Hill*

Louis Scuderi, *University of New Mexico*

Cynthia L. Simon, *University of New England*

Daniel Sivek, *University of Wisconsin—Stevens Point*

Mark A. Smith, *Riverside Community College*

Barry D. Solomon, *Michigan Technological University*

James C. St. John, *Georgia Institute of Technology*

L. Harold Stevenson, *McNeese State University*

Jana H. Svec, *Moraine Valley Community College*

Anne Todd-Bockarie, *Philadelphia University*

Michael Toscano, *San Joaquin Delta College*

Jake F. Weltzin, *University of Tennessee—Knoxville*

Jeff White, *Lake Land College*

Ray E. Williams, *Rio Hondo College*

Jeffery S. Wooters, *Pensacola Junior College*

E. Lynn Zeigler, *Georgia Perimeter College*

Victoria Zusman, *Miami-Dade College*

Robert K. Kaufmann
Cutler J. Cleveland

TEACHING AND LEARNING SUPPLEMENTS

McGraw-Hill offers various tools and technology products to support *Environmental Science*. Students can order supplemental study materials by contacting their local bookstore or by calling 800-262-4729. Instructors can obtain teaching aids by calling the Customer Service Department at 800-338-3987, visiting the McGraw-Hill website at www.mhhe.com, or contacting their local McGraw-Hill sales representative.

Teaching Supplements for Instructors

 McGraw-Hill's **ARIS—Assessment, Review, and Instruction System** (www.mhhe.com/kaufmann1e) for *Environmental Science* is a complete online tutorial, electronic homework, and course management system, designed for greater ease of use than any other system available. Instructors can create and share course materials and assignments with colleagues with a few clicks of the mouse. All PowerPoint lectures, assignments, quizzes, tutorials, and interactives are directly tied to text-specific materials in *Environmental Science*; but instructors can also edit questions, import their own content, and create announcements and due dates for assignments. ARIS features automatic grading and reporting of easy-to-assign homework, quizzing, and testing. All student activity within McGraw-Hill's ARIS website is automatically recorded and available to the instructor through a fully integrated grade book that can be downloaded to Excel.

ARIS Presentation Center (found at www.mhhe.com/kaufmann1e)

Build instructional materials wherever, whenever, and however you want! ARIS Presentation Center is an online digital library containing assets such as photos, artwork, animations, PowerPoint presentations, and other media types that can be used to create customized lectures, visually enhanced tests and quizzes, compelling course websites, or attractive printed support materials.

Access to your book, access to all books! The Presentation Center library includes thousands of assets from many McGraw-Hill titles. This ever-growing resource gives instructors the power to utilize assets specific to an adopted textbook as well as content from all other books in the library.

Nothing could be easier! Accessed from the instructor side of your textbook's ARIS website, the Presentation Center's dynamic search engine allows you to explore by discipline, course, textbook chapter, asset type, or keyword. Simply browse, select, and download the files you need to build engaging course materials. All assets are copyright McGraw-Hill Higher Education but can be used by instructors for classroom purposes.

Instructors will find the following digital assets for *Environmental Science* at ARIS Presentation Center:

- **Color art:** Full-color digital files of *all* illustrations in the text can be readily incorporated into lecture presentations, exams, or custom classroom materials. These include all of the three-dimensional realistic art found in this edition, representing some of the most important concepts in environmental science.

- **Photos:** Digital files of photographs from the text can be reproduced for multiple classroom uses.

- **Additional photos:** Four hundred forty-four full-color bonus photographs are available in a separate file. These photos are searchable by content and will add interest and contextual support to your lectures.

- **Tables:** Every table that appears in the text is provided in electronic format.

- **Videos:** This special collection of eighty-four underwater video clips displays interesting habitats and behaviors for many animals in the ocean.

- **Animations:** Ninety-four full-color animations that illustrate many different concepts covered in the study of environmental science are available for use in creating classroom lectures, testing materials, or online course communication. The visual impact of motion will enhance classroom presentations and increase comprehension.

- **Active art:** These twenty-four special art pieces consist of key environmental science illustrations converted to a format that allows you to break down the art into core elements and then group the various pieces to create customized images. This is especially helpful with difficult concepts because they can be explained to students step by step.

- **Global base maps:** Twenty-two base maps for all world regions and major subregions are offered in four versions: black-and-white and full color, both with and without labels. These choices allow instructors the flexibility to plan class activities, quizzing opportunities, study tools, and PowerPoint enhancements.

- **PowerPoint lecture outlines:** Ready-made presentations that combine art, photos, and lecture notes are provided for each of the twenty-four chapters of the text. These outlines can be used as they are or tailored to reflect your preferred lecture topics and sequences.

- **PowerPoint slides:** For instructors who prefer to create their lectures from scratch, all illustrations, photos, and tables are preinserted by chapter into blank PowerPoint slides for convenience.

Earth and Environmental Science DVD by Discovery Channel Education (ISBN: 978-0-07-352541-9; MHID: 0-07-352541-3)

Begin your class with a quick peek at science in action. The exciting *new* DVD by Discovery Channel Education offers fifty short (three- to five-minute) videos on topics ranging from conservation to volcanoes. Search by topic and download into your PowerPoint lecture.

McGraw-Hill's Biology Digitized Videos (ISBN: 978-0-07-312155-0; MHID: 0-07-312155-X)

Licensed from some of the highest-quality life science video producers in the world, these brief video clips on DVD range in length from fifteen seconds to two minutes and cover all areas of general biology—from cells to ecosystems. Engaging and informative, McGraw-Hill's digitized biology videos will help capture students' interest while illustrating key biological concepts, applications, and processes.

Instructor's Testing Resource CD-ROM

This CD-ROM contains a wealth of cross-platform (Windows and Macintosh) resources for the instructor. Supplements featured on this CD-ROM include a computerized test bank, which utilizes EZ Test software to quickly create customized exams. This flexible and user-friendly program allows instructors to search for questions by topic, format, or difficulty level, and edit existing questions or add new ones. Multiple versions of tests can be created, and any test can be exported for use with course management systems such as WebCT®, Blackboard®, or PageOut. Word files of the test bank are included for instructors who prefer to work outside the test-generating software. Other assets on the Instructor's Testing and Resource CD-ROM are grouped within easy-to-use folders.

Transparencies (ISBN: 978-0-07331597-3; MHID: 0-07-331597-4)

A set of one hundred overhead transparencies includes key illustrations and tables from the text. The images are printed for great visibility and contrast, and labels are large and bold for clear projection.

eInstruction

This classroom performance system (CPS) utilizes wireless technology to bring interactivity into the classroom or lecture hall. Instructors and students receive immediate feedback through wireless response pads that are easy to use and engage students. eInstruction can assist instructors by

- Taking attendance.
- Administering quizzes and tests.
- Creating a lecture with intermittent questions.
- Using the CPS grade book to manage lectures and student comprehension.
- Integrating interactivity into PowerPoint presentations.

Contact your local McGraw-Hill sales representative for more information.

Course Delivery Systems

With help from WebCT®, Blackboard®, and other course management systems, professors can take complete control of their course content. Course cartridges containing website content, online testing, and powerful student tracking features are readily available for use within these platforms.

Learning Supplements for Students

ARIS (www.mhhe.com/kaufmann1e)

This site includes quizzes for each chapter, additional case studies, interactive base maps, and much more. Learn more about the exciting features provided for students through the *Environmental Science* website.

Interactive World Issues CD-ROM (ISBN: 978-0-07-255648-3; MHID: 0-07-255648-X)

This two-CD set for students explores controversial issues in water rights (Columbia River); migration (Mexico); urban spread (Chicago); population (China); and sustainability (South Africa).

Exploring Environmental Science with GIS by Stewart, Cunningham, Schneiderman, and Gold (ISBN: 978-0-07-297564-2; MHID: 0-07-297564-4)

This short book provides exercises for students and instructors who are new to GIS but are familiar with the Windows operating system. The exercises focus on improving analytical skills, understanding spatial relationships, and understanding the nature and structure of environmental data. Because the software used is distributed free of charge, this text is appropriate for courses and schools that are not yet ready to commit to the expense and time involved in acquiring other GIS packages.

Annual Editions: Environment 06/07 by Allen (ISBN: 978-0-07-351542-7; MHID: 0-07-351542-6)

This twenty-fifth edition is a compilation of current articles from the best of the public press. The selections explore the global environment, the world's population, energy, the biosphere, natural resources, and pollutions.

Taking Sides: Clashing Views on Environmental Issues, Eleventh Edition Expanded by Easton (ISBN: 978-0-07-351441-3; MHID: 0-07-351441-1)

This book represents the arguments of leading environmentalists, scientists, and policy makers. The issues reflect a variety of viewpoints and are staged as "pro" and "con" debates. Issues are organized around four core areas: general philosophical and political issues, the environment and technology, disposing of wastes, and the environment and the future.

Global Studies: The World at a Glance, Second Edition by Tessema (ISBN: 978-0-07-340408-0; MHID: 0-07-340408-X)

This book features a compilation of up-to-date data and accurate information about some of the important facts about the world we live in. Although it is nearly impossible to be an expert in all areas such as a nation's capital, type of government, currency, major languages, population, religions, political structure, climate, economics, and so forth, this book is intended to assist in understanding these essential facts in order to make useful applications.

Sources: Notable Selections in Environmental Studies, Second Edition by Goldfarb (ISBN: 978-0-07-303186-6; MHID: 0-07-303186-0)

This volume brings together primary source selections of enduring intellectual value—classic articles, book excerpts, and research studies—that have shaped environmental studies and our contemporary understanding of it. The book includes carefully edited selections from the works of the most distinguished environmental observers, past and present. Selections are organized topically around the following major areas of study: energy, environmental degradation, population issues and the environment, human health and the environment, and environment and society.

Student Atlas of Environmental Issues by Allen (ISBN: 978-0-69-736520-0; MHID: 0-69-736520-4)

This atlas is an invaluable pedagogical tool for exploring the human impact on the air, waters, biosphere, and land in every major world region. This informative resource provides a unique combination of maps and data to help students understand the dimensions of the world's environmental problems and the geographical basis of these problems.

Guided Tour
A Unique Integration of

Kaufmann/Cleveland's *Environmental Science* views the world via the seamless flow of energy, materials, and information among physical, ecological, and human systems. This lens weaves together an interdisciplinary story of the relationship between people and their environment.

Energy and Materials

Unlike other environmental science textbooks, Kaufmann/Cleveland introduce energy and material flows early and weaves these principles throughout the remaining chapters. Following these flows allows students to understand how their actions affect the environment and how the environment influences their economic well-being.

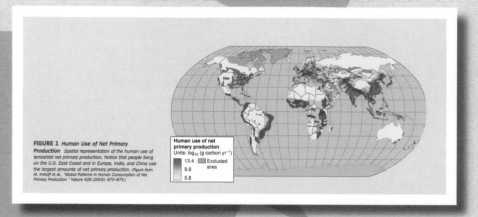

FIGURE 1 *Human Use of Net Primary Production* Spatial representation of the human use of terrestrial net primary production. Notice that people living on the U.S. East Coast and in Europe, India, and China use the largest amounts of net primary production. (Figure from M. Imhoff et al., "Global Patterns in Human Consumption of Net Primary Production." Nature 429 (2004): 870–873.)

Human use of net primary production
Units: $\log_{10}$ (g carbon yr^{-1})
13.4
9.8 — Excluded area
5.8

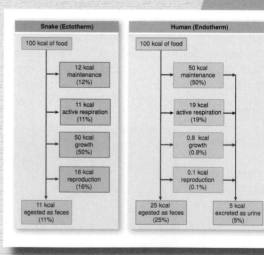

FIGURE 5.4 *Energy Allocation Strategies* Humans are endotherms, which means that we use about 50 percent of our energy for maintenance respiration, compared to 12 percent for snakes. This means that snakes have a lot more energy to devote to growth (50 percent) and reproduction (16 percent) compared to humans, which allocate less than 1 percent of their energy to growth and reproduction. (From C.A.S. Hall, C. Cleveland, and R. Kaufmann, Energy and Resource Quality: The Ecology of the Industrial Process. New York: John Wiley & Sons, 1986.)

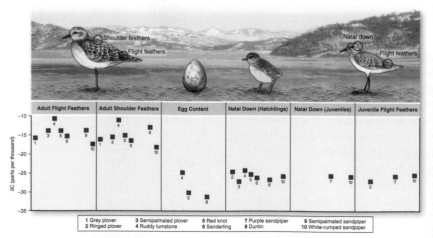

FIGURE 5.7 *Energy Used for Reproduction* The food eaten by birds in their winter homes has a higher ratio of carbon 13 to carbon 12 than the foods eaten in their summer grounds. High values for the ratio indicate that the energy used by adult birds to produce feathers comes from foods eaten in the winter grounds, whereas the energy used to produce eggs and raise offspring comes from food eaten in their summer breeding grounds. (From Klaassen et al., "Arctic Waders Are Not Capital Breeders." Nature, 413 (2001): 794.)

Energy, Economics, and Policy

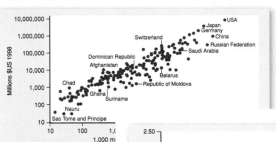

FIGURE 10.15 *International Comparison of Energy Use and GDP* Note that there is a strong correlation between energy use and the size of an economy. Data are for 2004.

Economics

By focusing on material and energy flows, Kaufmann/Cleveland emphasize the overall workings of the economy and avoid economic theory. This allows students to understand economic consequences of environmental challenges.

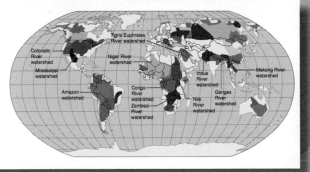

FIGURE 10.21 *Copper Depletion* The trend in ore grade for copper mined in the United States. (Source: Data from U.S. Geological Survey.)

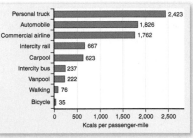

FIGURE 11.8 *Energy and Transportation* The energy intensity of various transportation modes. Source: Data from U.S. Department of Transportation.

FIGURE 18.18 *Potential for Conflict* Many of the world's major water sheds contain more than one nation. Under these conditions, upstream users control the quantity and quality of the water that is available to downstream users.

Policy

Rather than simply listing environmental legislation, Kaufmann/Cleveland weave the costs and benefits of environmental policies throughout so students can understand the complexity of environmental challenges and various solutions.

POLICY IN ACTION
Preserving Biodiversity in the Face of Corruption

In this chapter and many that follow, environmental policy is described in terms of what needs to be done. This simplification implies that decision makers develop a policy, it is implemented, and the policy either succeeds or fails. But this description overlooks an important determinant of success—**governance**, which is defined as the act of governing and exercising authority. Governance determines the degree to which the laws associated with environmental policy are enforced. As such, governance plays a critical role in the success or failure of environmental policy.

An important determinant of governance is **corruption**, which is defined as the unlawful use of public office for private gain. By outlawing certain actions, environmental policy often creates a black market, which generates a high price for an environmental good or service. For example, banning the sale of ivory raises its price, and prohibiting logging raises the price of wood. High prices generate enough money for lawbreakers to pay government officials to ignore illegal activities. For example, poachers or illegal loggers may pay park rangers to ignore activities that are forbidden by environmental policy. So corruption can reduce the effectiveness of even the best-conceived environmental policy.

We wish that our discussion of corruption could include the names of the guilty parties. But no one is willing to admit illegal actions. Instead political scientists try to estimate the level of corruption by surveying local officials and using the information to develop numerical indexes that measure the degree of corruption. Indexes

range between 1 and 10. A score of 1 indicates a high degree of corruption, while a score of 10 indicates a low degree of corruption.

A simple analysis shows a negative correlation between the corruption index and nations' species richness for birds and mammals (Figure 1). This negative correlation indicates that species richness tends to be greatest in the most corrupt nations. Similarly, biodiversity hot spots tend to be located in nations that are plagued by corruption.

The link between biodiversity and corruption is troubling because there are several reasons to believe that the government officials who enforce laws to preserve biodiversity are vulnerable to corruption. First, many of these government officials are poorly paid, which makes them vulnerable to bribes. Second, many conservation projects run for relatively short periods and are funded by organizations outside the host nations. The brief employment opportunities and the lack of close connections with the source of funds provide a rationale for corruption. Finally, it is difficult to measure the success of most conservation projects. Without such criteria, it is hard for funding organizations to determine whether their money is being spent effectively.

Corruption's influence on biodiversity has been tested by examining the link between corruption and the success of efforts to protect the remaining populations of African elephants and black rhinoceroses, which are protected by the Convention on International Trade in Endangered Species. To evaluate the convention's success, scientists measured

the percentages of change in the population of African elephants and black rhinoceroses between 1987 and 1994 in individual African nations and compared these changes to government efforts to save the species, such as national spending on protected areas (per square kilometer), per capita GDP, a measure of the human development index (a measure of the quality of life), and the index of governmental corruption.

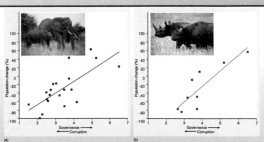

The results show that spending on protected areas, per capita GDP, and the human development index are not related to changes in the population size of African elephants and black rhinoceroses. Rather, changes in these populations are positively related to governance scores. Populations of African elephants and black rhinoceroses shrank in nations where corruption was high, and the populations grew in nations with relatively

little corruption (Figure 2). This result demonstrates the role of corruption. Shrinking populations in corrupt nations imply that poachers can convince government officials to ignore their illegal activities. On the other hand, the observation that populations grew in nations with less corruption implies that the Convention on International Trade in Endangered Species can be successful when local officials enforce the law.

ADDITIONAL READING
Smith, R.J, R.D.J. Moire, M.J. Wallop, A. Bainford, and N. Leader-Williams. "Governance and the Loss of Biodiversity." Nature 426 (2003): 67–70.

STUDENT LEARNING OUTCOME
• Students will be able to explain why and how corruption affects efforts to preserve biodiversity.

Active Learning with

Kaufmann/Cleveland's *Environmental Science* invites students to practice critical thinking and active learning throughout the text. By participating in different exercises and applications, students will learn how scientists approach problems and how they can actively apply new ideas and skills as well.

Your Ecological Footprint Readings

Students are asked to audit various aspects of their life, enabling them to analyze the costs and benefits of these choices.

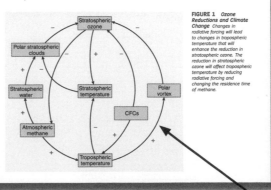

Your ECOLOGICAL footprint

How Much Wood Do You Use for Paper?

Much of the wood that is used in developed nations is converted to paper. In 2002 an average U.S. citizen used about 314 kg of paper. How much wood is required to supply this paper? Information given here will allow you to calculate the amount of wood used to produce the paper you consume.

It might seem that calculating this quantity should be simple. Some fact sheets report that approximately thirty-four trees are needed to make one metric ton of paper. This simple conversion is misleading because the amount of wood required depends on the type of tree used, the fraction of postconsumer fiber used, and the type of paper produced.

Much of the paper produced in the United States is made from wood fiber. The amount of fiber in a tree's wood varies by species. Hardwood species, such as northern red oak or the eastern cottonwood, tend to have more fiber than softwoods, such as jack pines or Douglas firs. These differences imply that more softwood is needed to produce a given quantity of paper.

Fibers from newly cut trees come from two sources: round wood and residues. Round wood includes whole trees. Normally the paper industry obtains fiber from small trees because large trees are more valuable as lumber. Residues are the materials that are left over from the normal operation of the timber industry, such as wood chips and scraps. Typically residues are burned or wasted if they are not used to make paper. But not all wood fibers have to come from newly cut trees. A third source of wood fibers is recovered paper, which includes old paper that is made available to the industry through recycling.

These three sources affect the amount of wood required to produce paper. The amount of new wood required to produce paper can be calculated from its content of virgin materials, which are fibers that come from trees. On the other hand, recovered paper does not require any newly cut wood. As indicated by the calculations that follow, increasing the fraction of fiber that comes from recovered paper reduces the amount of new wood required to produce paper.

Finally, the amount of wood needed to produce paper differs according to the type of paper. Papers vary in many aspects, including their lignin content. Lignin is a natural component of wood that binds the fibers. The presence of lignin enhances the printing properties of paper but reduces its strength and makes it more likely to discolor. Including lignin allows a large fraction of wood to be converted to paper. High-lignin papers include newsprint and office paper, such as copier paper, envelopes, and the like. Stronger papers, such as corrugated paper, are produced by lowering the lignin content. Removing lignin reduces the conversion efficiency of wood and increases the amount of wood needed to produce corrugated paper relative to newsprint.

Calculating Your Footprint

Each day you read newspapers, print material from the Web, and use paper for many other purposes. Here are three equations to let you calculate the quantity of wood required to produce three types of paper you consume.

Newsprint ___ kg wood/kg paper = 2.09 − 0. × % recycled

Office paper ___ kg wood/kg paper = 3.47 − × % recycled

Unbleached paperboard ___ kg wood/kg paper = 3 × % recycled

In these equations % recycled is the percentage of fiber that comes from recovered paper.

To help you convert kilograms of wood to lists the density of some softwoods produce paper. To illustrate, *Times*, which weighs about produced with 50 percent to produce the paper wood were cut from kg/4.52 kg/m³) of wood

Compile information includes paper used to produce the paper you use in printers, envelope. Paperboard is used in folded cartons store. In case you have trouble compiling in the United States in 2003 used about of office paper, and 148.2 kg of paperboard.

Interpreting Your Footprint

Suppose the paper you used was produced completely from virgin materials. In other words, 0 percent of its fiber came from how much could you reduce your use of wood if 40 percent used to produce the paper you consumed came from recovered.

The global economy produced about 16 kg of office paper, about newsprint, and about 32 kg of paperboard for every person on if we assume that all of this paper was produced from virgin materials not), by how much would you have to switch your paper usage paper to lower your use of wood so that it matched the global average.

ADDITIONAL READING

Bierman, C.J. *Handbook of Pulping and Papermaking*. San Diego: Academic Press, 1996.
The "Paper Calculator" at www.ofee.gov/recycled/descript.htm will all you to calculate many of the material and energy flows required to produce paper.

STUDENT LEARNING OUTCOME

• Students will be able to explain how using recycled paper slows the rate of deforestation.

	Density kg/m³		Density kg/m³
TABLE 1	**Wood Density**		
Softwoods		**Hardwoods**	
Jack pine	3.2	Eastern cottonwood	2.97
Douglas fir	3.61	Northern red oak	4.52

Calculating Your Footprint

Each day you read newspapers, print material from the Web, and use paper for many other purposes. Here are three equations to let you calculate the quantity of wood required to produce three types of paper:

Newsprint ___ kg wood/kg paper = 2.09 − 0.0209 × % recycled

Office paper ___ kg wood/kg paper = 3.47 − 0.0347 × % recycled

Unbleached paperboard ___ kg wood/kg paper = 3.04 − 0.034 × % recycled

In these equations % recycled is the percentage of fiber that comes from recovered paper.

To help you convert kilograms of wood to volume of wood, Table 1 lists the density of some softwoods and hardwoods that are used to produce paper. To illustrate, suppose you read the Sunday *New York Times*, which weighs about 1 kilogram. Assume that the newsprint is produced with 50 percent recovered fiber. The amount of wood required to produce the paper is 1.04 kg (2.09 − 0.0209 × 50). If this amount of wood were cut from a northern red oak, about 0.23 cubic meters (1.04 kg/4.52 kg/m³) of wood would be required.

Compile information about the amount of paper you use. Newsprint includes paper used to produce your newspaper; office paper includes the paper you use in printers, envelopes, letterhead, and printed forms. Paperboard is used in folded cartons that hold the things you buy in a store. In case you have trouble compiling your use, the average person in the United States in 2003 used about 35.4 kg of newsprint, 94.5 kg of office paper, and 148.2 kg of paperboard.

Interpreting Your Footprint

Suppose the paper you used was produced completely from virgin materials. In other words, 0 percent of its fiber came from recovered paper. By how much could you reduce your use of wood if 40 percent of the fiber used to produce the paper you consumed came from recovered paper?

The global economy produced about 16 kg of office paper, about 6 kg of newsprint, and about 32 kg of paperboard for every person on Earth in 2003. If we assume that all of this paper was produced from virgin material (it was not), by how much would you have to switch your paper usage to recycled paper to lower your use of wood so that it matched the global average?

Case Studies

Students are exposed to current, real-life environmental challenges through case study readings in every chapter.

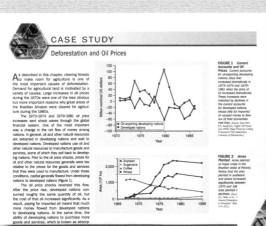

CASE STUDY

Deforestation and Oil Prices

As described in this chapter, clearing forests to make room for agriculture is one of the most important causes of deforestation. Demand for agricultural land is motivated by a variety of causes. Large increases in oil prices during the 1970s were one of the less obvious but more important reasons why great areas of the Brazilian Amazon were cleared for agriculture during the 1980s.

The 1973–1974 and 1979–1981 oil price increases sent shock waves through the global financial system. One of the most important was a change in the net flow of money among nations. In general, oil and other natural resources are extracted in developing nations and sold to developed nations. Developed nations use oil and other natural resources to manufacture goods and services, some of which they sell back to developing nations. Prior to the oil price shocks, prices for oil and other natural resources generally were low relative to the prices for the goods and services that they were used to manufacture. Under these conditions, capital generally flowed from developing nations to developed nations (Figure 1).

The oil price shocks reversed this flow. After the price rise, developed nations consumed roughly the same quantity of oil, but the cost of that oil increased significantly. As a result, paying for imported oil meant that much more money flowed from developed nations to developing nations. At the same time, the ability of developing nations to purchase more goods and services, which is known as absorp-

FIGURE 1 *Current Accounts and Oil Prices* Current accounts for oil-exporting developing nations (blue line) increased dramatically in 1973–1974 and 1979–1981 when the price of oil increased dramatically. These increases were matched by declines in the current accounts of developed nations, whose bills for imported oil caused money to flow out of their economies (red line). *(Source: Dean Parr R.K. Kaufmann, Higher Oil Prices Can OPEC Raise Prices by Cutting Production? PhD Dissertation University of Pennsylvania.)*

FIGURE 2 *Acres Planted* Acres planted of major crops in the Brazilian state of Parana. Notice that the area planted in soybeans and wheat increased significantly between 1970 and 198 area planted is declined. *(Seen D.L. Skole, et al., "Human Dimensions in Amazonia," Bios 314–322.)*

CASE STUDY

The Link between Climate Change and the Reduction of Stratospheric Ozone

Climate change and the reduction of stratospheric ozone often are presented as separate environmental challenges. In this book we discuss them in separate chapters. But contrary to this separation, the atmosphere is a dynamic system that includes many feedback loops. These loops imply that human-induced changes in climate and ozone cannot be separated. Anthropogenic climate change will postpone the date at which the reduction in stratospheric ozone is reversed; and the reduction in ozone may damp the ongoing increase in global temperature (Figure 1).

The stratospheric concentrations of chlorine and bromine probably peaked in 1997, but stratospheric ozone is expected to decline for many years. This ongoing reduction is due in part to global climate change. As described in Chapter 13, climate models forecast (with a high degree of certainty) that climate change will cool the stratosphere. A cooler stratosphere enhances the formation of polar stratospheric clouds. These clouds speed the chemical reactions that destroy ozone—so a cooler stratosphere will

FIGURE 1 *Ozone Reductions and Climate Change* Changes in radiative forcing will lead to changes in tropospheric temperature that will enhance the reduction in stratospheric ozone. The reduction in stratospheric ozone will affect tropospheric temperature by reducing radiative forcing and changing the residence time of methane.

Illustrated throughout the text, **feedback loops** show relationships between parts of a system and how they link and connect to each other.

Real-World Applications

Policy in Action Readings

Environmental science concepts can be used to develop policies. The **Policy in Action** readings help students understand why some environmental policies are effective, while others are not.

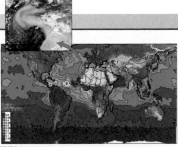

POLICY IN ACTION

Is Dumping Iron into the Ocean an Effective Policy to Slow Global Climate Change?

FIGURE 1 *The Biological Carbon Pump* Carbon dioxide that is taken up by marine organisms can either return to the ocean via respiration or sink toward the bottom. Carbon that sinks to the bottom causes additional quantities of carbon to flow into the ocean, which reduces the amount of carbon dioxide in the atmosphere.

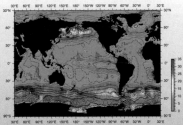

FIGURE 2 *High Nitrogen, Low Net Primary Production* Nitrogen often is a limiting nutrient in the ocean, but the highly colored areas along the Antarctic and the northern Pacific are areas where available forms of nitrogen are abundant but have relatively low levels of net primary production.

FIGURE 3 *Iron for the Ocean* Satellite measures indicate that winds pick up particles from the land surface and deposit them in the ocean or on other continents. An important source of such particles is northern Africa, where the lack of vegetation and fires can generate spectacular dust storms.

FIGURE 4 *Iron-Induced Diatom Blooms* As described in Chapter 7, the highest levels of net primary production occur along the shore, indicated by red, orange, and yellow. Away from the shore, areas of lower net primary production are indicated in blue. A small red area of high net primary production occurs where scientists added iron to the northeastern Pacific Ocean. These blooms consist of countless diatoms.

ADDITIONAL READING

Boyd, P. "Ironing Out Algal Issues in the Southern Ocean." *Science* 304 (2004): 396–397.

STUDENT LEARNING OUTCOME

• Students will be able to explain how changes in terrestrial ecosystems can affect distant marine ecosystems.

118

Student Learning Outcomes (SLOs)

Specifically written to develop higher level thinking skills, **Student Learning Outcomes (SLOs)** are a tool to assess a behavior, skill, or knowledge gained after studying the material. SLOs can be found at the beginning of each chapter, as well as each of the Policy in Action, Case Study, and Your Ecological Footprint readings.

THE LAWS OF ENERGY AND MATTER **2** CHAPTER

PART TWO Basic Concepts in Environmental Science

STUDENT LEARNING OUTCOMES

After reading this chapter, students will be able to

• Explain the importance of the law of conservation of matter for environmental science.

• Describe the limits that the laws of thermodynamics place on energy conversion.

• Distinguish the important differences between chemical, physical, and nuclear changes in matter.

• Provide examples of the entropy law in nature and in their everyday lives.

• Define and describe the major forms of energy they use in their everyday lives.

ENVIRONMENT AND SOCIETY

A Sustainable Partnership?

1

STUDENT LEARNING OUTCOMES

After reading this chapter, students will be able to

- Explain the mechanisms behind the collapse of Easter Island.
- Describe why equity and fairness are essential to sustainability.
- Identify the essential components of their ecological footprint.
- Explain how resource depletion and environmental degradation affect human well-being.
- Explain how subsidies and externalities contribute to environmental change.

Easter Island Moai *Easter Island is home to one of the most famous and well-studied examples of how a society can collapse due to mismanagement of its natural resources.*

EASTER ISLAND:
THE ISLAND THAT SELF-DESTRUCTED

Easter Island is the most isolated piece of inhabited land on the planet, lying in the Pacific Ocean some 3,747 kilometers (2,340 miles) from the coast of Chile, its nearest continental neighbor (Figure 1.1). Polynesians who made the harrowing journey by boat from islands hundreds of kilometers to the west settled the island sometime in the first centuries A.D. Most notable was the islanders' ingenious blend of technology, religion, and culture. Hundreds of statues called *moai* (mow-a) standing up to 10 meters (32.8 ft.) tall and weighing up to 85 metric tons (93.5 tons) were carved from quarries of volcanic rock, transported several kilometers, and raised to an upright position on platforms—all without the use of any metal tools, wheels, or a power source other than human labor (Figure 1.2). *Moai* probably were erected to honor high-ranking ancestors and to symbolize the connection between life and death. The statues also provided a peaceful outlet for rival groups on the island, who competed to build bigger and more ornate statues. For nearly 1,000 years the island was free of warfare and violent conflict.

The island was "discovered" by Dutch explorers on April 5, 1722, which happened to be Easter Sunday—hence its name. When the Dutch arrived, statue carving had ceased, and some statues were toppled and broken. By 1840 all the erected statues had been deliberately toppled by the islanders. Subsequent archaeological excavations revealed that spears, daggers, and other weapons, along with other evidence of conflict and warfare, first appeared when the statues were destroyed. There also was evidence of cannibalism.

What caused the Easter Island society to devolve from a cooperative, communal society to one riddled with strife and conflict? The answer appears to be a shortage of food caused by population growth and degradation of the island's environmental resources. Food scarcity generated competition and conflict among groups, which disrupted the communal system whereby the farmers and the people who fished for a living cooperated to feed the craftspeople.

Are the events of 300 years ago on a small remote island in the Pacific Ocean of any significance to civilization today? We believe they are. Easter Island had a finite amount of timber, agricultural, and fishery resources. The size of the Easter Island population clearly outstripped the ability of the island's resource base to support it (Figure 1.3).

FIGURE 1.2 *A Moai Statue* *Hundreds of statues called* moai *adorned the coast of Easter Island.*

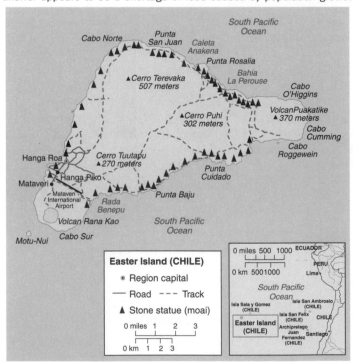

FIGURE 1.1 *Easter Island* *A map of Easter Island showing its location and the positioning of* moai *statues.* (Source: Data from the University of Texas Libraries.)

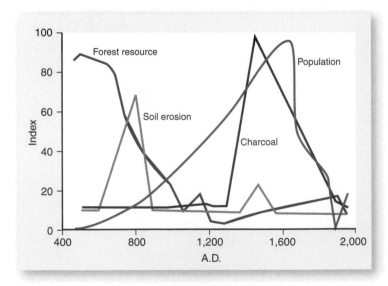

FIGURE 1.3 *A Pattern of Collapse* *Changes in population, deforestation, charcoal, and soil erosion on Easter Island.* (Source: Data from P. Bahn and J. Flenley, Easter Island, Earth Island, Thames & Hudson.)

Clearing forests enabled food production to expand, and with it population size. From the initial boatload of about 400 settlers, the population grew steadily to about 8,000 before collapsing abruptly around 1600. Even today the island is relatively barren grassland, which is littered with toppled statues and supports a tiny fraction of its former population.

We do not tell this story to predict a collapse of human civilization. Rather, we use the events on Easter Island to illustrate the meaning of sustainability and identify several principles that will guide our investigation of the relationship between people and their environment. It is an adventurous and far-reaching investigation, spanning fields such as biology, physics, history, demography, economics, and politics. We will discuss current problems and issues, journey back in time to earlier civilizations, and report on computer simulations for educated guesses about what the future holds. We will discover important differences between life in rich and poor nations and will discuss problems that affect the everyday lives of you and everyone on the planet. At the end of our investigation, we will not answer questions such as "Is Earth about to go the way of Easter Island?" or "How many people can Earth support?" But you will be prepared to answer a much more relevant question: Is your behavior, and the behavior of nations and the human population, moving away from or toward sustainability? ■

THE PRINCIPLES OF SUSTAINABILITY

Scientists debate which combination of events on the island were the most important and when they occurred, but one thing is clear: The islanders' society was not **sustainable.** As defined by the United Nations Commission on Sustainable Development, a sustainable society meets the needs of the present generation without compromising the ability of future generations to meet their own needs. This raises an important question about the Easter Islanders: What did they do that made their society unsustainable?

Principle 1: Sustainable Use of Natural Resources and Environmental Services

A **natural resource** is something we get from the environment to meet our biological and economic needs and wants. As such, natural resources are considered environmental goods that come from the lithosphere (the outer layers of Earth's crust) and the biosphere (the parts of Earth's surface and immediate atmosphere that are inhabited by living organisms).

Forests and their trees were among the most important natural resources used by the Easter Islanders. Pollen records show that a dense forest covered the island when the first settlers arrived. The settlers began to harvest trees for several purposes: to clear land for agriculture, to build canoes for fishing in coastal waters, to provide fuel for cooking, and to provide the timber for moving the *moai* from the quarries to various locations on the island. As population grew, so did the demand for wood. Pollen records also show a steady decline in forest cover until around 1400, when the entire island appears to have been deforested (Figure 1.3). To reach this point, the Easter Islanders mistakenly harvested their forests faster than the forests grew back.

Environmental services are natural processes that regulate conditions in the environment in ways that make the planet suitable for life. These services can take many forms.

Among the most important is **waste assimilation,** which is the ability of the environment to absorb, detoxify, and disperse wastes to make them less harmful. Wastes are generated in the process of producing, using, or disposing of goods or services. Wastes are the inevitable by-products of using energy and materials. **Pollution** is a waste that produces a physical, biological, or chemical change in air, water, soil, or food that potentially is harmful to humans or other living organisms.

Forests and their soils provided the most important environmental services on Easter Island. Forests help regulate the movement of water by pulling it from the soil and pumping it into the atmosphere. This water-pumping service helps maintain local and regional climate. Forests also protect the underlying soil. This soil was critical to food production. Without the protective cover of forests, soil on the island was washed out to sea, which undermined the production of food. Thus the importance of trees to the Easter Islanders went well beyond their use as timber.

The loss of forests, trees, and their soil as a source of natural resources and environmental services illustrate the first important **principle of sustainability:**

A sustainable society does not use natural resources or produce wastes faster than they are regenerated or assimilated by the environment.

Violating this principle contributed to the collapse of the Easter Island civilization because it used up the Island's supply of natural resources and environmental services.

Principle 2: A Systems Perspective

In hindsight, Easter Islanders probably did not fully understand or foresee the environmental implications of their actions. Many of their decisions had environmental consequences that ultimately were disastrous for their society. These mistakes highlight a second principle of sustainability:

Decisions that promote sustainability are consistent with the fact that human society is a system that is part of a larger system, the natural environment. Each of us has many connections to other parts of society and our environment. Each connection can have a positive or negative effect, so the net effect of human actions can differ from their original intent. The connections form feedback loops that can amplify or damp changes or disturbances to the environment. Sustainable societies must avoid actions that trigger feedback loops that amplify disturbances.

This principle has two important points. The first describes the need to understand society and its environment as an interconnected system. The second describes the need to understand how these connections can amplify or damp the effects of human actions.

A **system** is a collection of parts that generates a regular or predictable pattern. The parts of the Easter Island social system included the people, their tools, and their customs. Some people were farmers who grew food; some people were hunters who caught fish and seabirds; some people were artisans who carved the *moai;* other people were laborers who transported and erected the *moai.* Each group used a distinct set of tools such as terraced cultivation inside craters, wood canoes for fishing, or wooden rollers to move the *moai* (at least we think that is how they moved the *moai*). Social customs defined the connections among these groups. The farmers grew food that was distributed to the artisans, laborers, and hunters. The fishers provided meat to the farmers, artisans, and laborers. The construction of the *moai* provided cultural and religious benefits to the farmers and the hunters. These linkages constituted regular patterns of behavior.

The parts of the Easter Island environmental system included the soil, streams, topography, plants, animals, climate, and more. All of these parts were connected. The amount of soil affects tree growth. The presence of trees affects the local climate. The parts interacted with each other to generate a regular pattern. The climate was relatively stable; the trees and fish reproduced at a predictable rate; rainfall recharged springs at a predictable rate; and fertile soil was created and lost at a predictable rate. This regularity was vital to the islanders because it produced a relatively predictable and stable supply of food, wood, water, and environmental services.

Systems such as Easter Island have many connections, which make it difficult to determine how an action in one part of the system will affect another part of the system. For example, the Easter Islanders brought Polynesian rats as a source of food. But they ignored or were unaware of the fact that these rats ate the seeds of trees that provided wood, a critical natural resource. As the rat population increased, so did the number of rats available for humans to eat. But this increase was offset by the loss of trees caused in part by the rats. Thus introduction of the rat increased food supplies at first, but eventually reduced food supplies.

Such unintended consequences constitute the second principle for sustainability: A sustainable society must account for highly interconnected relationships with its environment and how these connections can cause decisions to succeed or fail. See *Case Study: How Big Can Society Be? The Environmental Plimsoll Line.* As will be described in more detail in Chapter 3, several aspects of the Easter Island lifestyle had unintended consequences that caused the society as a whole to collapse.

Principle 3: Equity and Fairness

The civilization on Easter Island flourished for centuries because groups cooperated. Agriculturists toiled in the fields to support the craftspeople. The craftspeople got fed and were engaged in an occupation that was highly regarded by society. The farmers and the rest of the population derived great religious and prestige value from the statues they commissioned. When food and timber became scarce, the difficulties associated with farming and fishing in a degraded environment meant that people were no longer willing to exchange food for *moai*, which led to the end of statue making and ultimately the craftspeople's way of life. This created a new group of people who now had to feed themselves.

To manage this difficulty, the islanders turned to the Birdman Cult. Following this tradition, there was a ritualistic competition for bird eggs, which were an important source of food. The winner of this contest gained control over much of the island's remaining resources. He and his extended family could demand tribute from the other Easter Islanders, the most common form of which was food. Obtaining this tribute was enforced by warriors—refusal to provide the tribute meant that one's house or agricultural fields could be burned. The Birdman Cult ensured that some groups were well fed, but it also meant that other groups did not have enough to eat and lived in fear of the current year's winner of the competition.

Given their weakened position, the losers of the competition had little incentive to cooperate with the winners to maintain the system. This breakdown in cooperation demonstrates the third principle of sustainability:

> The first two principles of sustainability must be meshed with the ethical and moral principles that govern fairness among nations, between genders, and among current and future generations.

If fairness is not present, the cooperation that supports the overall system will disappear, along with the benefits of that cooperation.

You can imagine a society that uses resources and environmental services in a sustainable way (principle 1) and avoids destabilizing its social–environmental system (principle 2) but is run by an elitist government with little

CASE STUDY

How Big Can Society Be? The Environmental Plimsoll Line

Whether it's a rubber raft in your backyard pool or an ocean cruise ship, anyone who has ever ridden in a boat knows the ultimate safety rule: Putting too much stuff on a boat causes it to sink! Cargo ships have lines on their hulls that show the maximum depth at which the ships can sit in the water. This mark is called the Plimsoll line, named after Samuel Plimsoll, a merchant and shipping reformer in the British Parliament in the late nineteenth century. Most nations engaged in international water transport now require all vessels to have a Plimsoll line.

Just as there is a maximum amount of cargo that can be loaded on a ship, there is a maximum load we can place on the environment through the extraction of resources and the discharge of wastes. This maximum can be envisioned as an environmental Plimsoll line: the maximum amount or rate of use of a resource or service at which it can sustained. As a ship is loaded, the Plimsoll line sinks toward the water level. As we use more of a particular resource or emit more waste, the planet sinks toward the environmental Plimsoll line. If the load becomes too great, the resource or service "sinks"—that is, it is depleted or degraded.

We demonstrate this with a simple example. Power plants in the Midwestern United States produce air pollutants that are transported to the Northeast, where they cause acid rain. Lakes can absorb or neutralize a certain amount of acidic precipitation without affecting their pH (a measure of water acidity). But as more acidic water is added to a lake, its ability to neutralize the acid is depleted, which can produce a sudden and dramatic change in water chemistry.

Emissions increased gradually through World War II, but the pH of northeastern lakes remained steady because the quantity of acid deposition was still within the lakes' neutralization capacity. After decades of neutralizing acid rain, northeastern lakes' pH suddenly plunged in the 1960s, killing many of the lakes' fish and plants. The sudden pH drop and the collapse of the fish population proved that sulfur dioxide emissions pushed the acid neutralization ability of many lakes beyond the Plimsoll line.

For most environmental goods and services, the Plimsoll line is not as clear. Nonetheless, a Plimsoll line does exist. We know this because Earth's ability to provide natural resources and environmental services is finite. The existence of the Plimsoll line is important because

the growth in population and living standards increases the overall size of society relative to the environment. This raises an important question: How big can society get before we breach the Plimsoll lines for key resources and environmental services? Unlike a line on the side of a ship, we can't say for sure how many people can live on this planet, how many resources they can use, or how much waste they can produce. Humans are adaptable. Facing a depleted resource, for example, we can develop technologies that use the resource more efficiently or use an alternative resource. Facing pollution, we can institute laws that restrict emissions. There also is uncertainty about how much depletion and degradation can occur before the environment changes in an irreversible and undesirable manner. Nonetheless, we always need to watch for evidence that we have overloaded Earth with people and their ecological footprints.

STUDENT LEARNING OUTCOME

- Students will be able to explain what it means for society to exceed an environmental limit or threshold.

---

respect for human life. History is full of authoritarian governments toppled by revolutions born of extremely unequal distribution of food, material wealth, health care, education, and political power. This principle, which clearly falls outside the fields of science, forces us to examine our values and ideals. The questions raised by such an examination are not easily answered because there is a wide range of opinions regarding what is "fair" or "just." Nevertheless, the issue is unavoidable.

You can think of the fairness principle from a systems perspective. If a large segment of a population believes it is not receiving its fair share, it may pressure those in power to redistribute wealth, political power, and the like. Those in power may respond by "tightening the screws" even more, resulting in even more unrest, and so forth. As history shows, this can produce a positive feedback loop that ultimately destabilizes the entire system.

Principle 4: Incentives for Sustainable Behavior

As the people of Easter Island cut forests at a rate that ultimately harmed their natural environment, no alarms went off telling them to stop or slow down. If anything, such nonsustainable behavior was rewarded. Converting forests to agricultural land faster than others allowed some groups to erect more *moai*, which increased their prestige relative to their neighbors. But this reward system accelerated the rate at which forests were cut, which led to the violation of sustainability principle 1.

This illustrates a fourth principle of sustainability:

Social incentives must reward those who act in a sustainable way and punish those who act in a nonsustainable manner.

If the social system rewards individuals who use natural resources or environmental services faster than they are replenished, who ignore the indirect effects of their actions, or who promote extreme inequality, society as a whole probably will collapse. Conversely, a society may be sustainable if its social systems punish such behaviors or reward people who limit the use of natural resources or environmental services, recognize the direct and indirect impacts of their actions, and lessen the inequality among social groups.

ARE WE HEADED IN THE RIGHT DIRECTION?

Understanding how violating the four principles of sustainability caused the Easter Island civilization to collapse forces us to ask some hard questions about our modern societies. Is the United States, Japan, Brazil, China, or any other nation moving toward or away from sustainability? Is the human population as a whole moving toward or away from sustainability? As a first step toward answering these questions, we compare the four principles of sustainability with actions in the United States and the rest of the world.

Violating Principle 1: Depletion and Degradation of Natural Resources and Environmental Services

Some of the most noticeable examples of nonsustainable human use of the environment can be seen in an examination of our use of natural resources. In many cases the rate at which we use natural resources exceeds the rate at which the environment supplies them (Table 1.1). By definition any use of a **nonrenewable resource** diminishes its future availability. But humans use or degrade potentially **renewable resources** such as soil, biodiversity, and forests faster than they are replenished by nature.

Two prominent examples illustrate this point. Earth's temperature is rising, which may foreshadow a rapid change in our planet's climate (Figure 1.4). Most scientists believe that this warming is caused in part by burning coal, oil, and natural gas, and by clearing and burning forests. If these activities continue at their present rates, scientists forecast that Earth's average temperature may increase by 1.4–5.8°C by the year 2100. This change may seem small and slow, but it is much faster than our climate has changed in the past 18,000 years. A rapid climate change would have serious environmental and social consequences. Rainfall patterns would change—perhaps significantly—thereby impairing food production in many regions. Sea level would rise, endangering billions of people who live in coastal regions. Extreme weather events such as hurricanes,

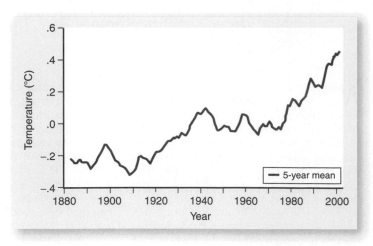

FIGURE 1.4 *Earth's Temperature* A 5 year running average of changes in the average temperature of Earth, 1880–2004. The vertical axis is the departure from the 1951–1980 average in °C. (Source: Data from Intergovernmental Panel on Climate Change.)

droughts, and heat waves may become more frequent. A warmer climate could also increase the spread of infectious diseases such as yellow fever.

A second global environmental alarm concerns the disappearance of species. This loss has been documented by the Global Biodiversity Assessment, which was sponsored by the United Nations Environment Programme and was the product of 1,500 scientists. Projections for the rate at which tropical forests are cut imply that 1–10 percent of all species will go extinct in the next quarter century. This rate of extinction would be approximately 1,000–10,0000 times faster than the average expected "background" extinction rate. Like global climate change, the elimination of species is caused by the actions of individuals, corporations, and governments. Clearing a forest in Brazil to grow soybeans or filling in a wetland in Oregon to build a shopping mall eliminates places where many nonhuman species live. The loss of species endangers our existence because we rely on other animals, plants, and microorganisms for food, materials, energy, and environmental services.

TABLE 1.1	The Rates of Use of Natural Resources Compared to Their Rates of Natural Renewal		
Resource	**Rate of Replenishment**	**Rate of Use or Degradation by Humans**	**Ratio of Use to Replenishment**
Biodiversity	20,000 years for evolution to create 20 species of mammals	20 species of mammals extinct in twentieth century	1,000:1
Crude oil	0.8 million barrels per year created by geologic processes	30 billion barrels per year used by the global economy	31,000:1
Tropical forests	1.0 million hectares of humid tropical forest regrowth per year	5.8 million hectares of humid tropical forest cut per year	5.8:1
Fertile soil	1 ton/hectare/year of new soil created	16 tons/hectare/year eroded from U.S. farmland per hectare per year	16:1

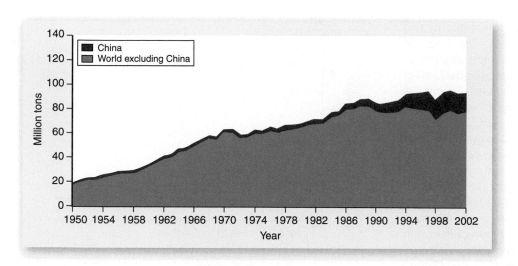

FIGURE 1.5 *Global Fish Catch* *Global fish catch, 1950–2002.* (Source: Data from United Nations Food and Agriculture Organization.)

Violations of sustainability principle 1 deplete natural resources and environmental services. **Resource depletion** occurs when we use natural resources or environmental services faster than the environment creates them. Depletion is illustrated by the problem of overfishing. Global fish harvesting has increased nearly fivefold in the last half century (Figure 1.5). Fish protein now accounts for 16 percent of global animal protein supply, and more than 10 million people worldwide work as fishers. Marine biologists estimate that the world's oceans can support a maximum of about 100 million tons of fish caught per year. We are now approaching that overall limit due to a huge increase in the number of fishing vessels and improvements in fishing technology. Many individual fish species have been relentlessly depleted. According to the United Nations Food and Agriculture Organization, about 70 percent of the world's marine fish stocks are heavily exploited, overexploited,

depleted, or slowly recovering. Just 10 percent of all large fish—both open-ocean species, including tuna, swordfish, and marlin, and the large groundfish such as cod, halibut, skates, and flounder—are left in the sea. Their depletion not only threatens the future of these fish species and the fishers who depend on them; it could also dramatically alter ocean ecosystems, with unknown global consequences.

Depletion usually follows a pattern described by the **best first principle**: Humans use the highest-quality sources of natural resources and environmental services first. As the high-quality sources of a resource are depleted, they are replaced by lower-quality sources. Low-quality sources require more effort to obtain than high-quality resources; therefore depletion makes it harder and harder to obtain resources. For example, fishery resources show a distinct spatial distribution (Figure 1.6). The largest fish populations are located in just a few coastal environments, which are targeted first by fishers. After

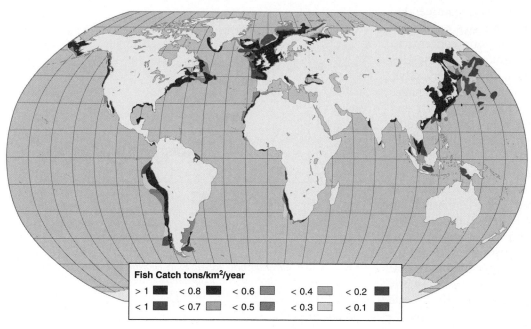

FIGURE 1.6 *The Geography of Global Fish Catch* *The location of marine fish catch, based on the average annual catch from 1950 to 2002.* (Source: Data from Sea Around Us Project.)

they are depleted, fishers move on to other lower-quality fish populations. For example, fishing fleets from New England used to fish exclusively on George's Bank, which was one of the world's most fertile fishing grounds, located in the Gulf of Maine 120 km off the coast of Massachusetts. Overfishing has decimated this fishery, and now vessels from New England travel as far as the Carolina coast to find enough fish to catch.

Why worry about depletion and degradation? The best first principle implies that depletion hurts the economy, the environment, and the health of humans and other species. For example, the depletion of fish has increased the sacrifices society must make to obtain a unit of this resource. This effect is measured by catch per unit effort, which is the quantity of fish caught divided by the fishing effort that produced it. The catch per unit effort for every fish and every region has plummeted in the last 40 years (Figure 1.7). The Japanese longline fishing industry illustrates the dramatic effects of depletion. This type of fishing consists of a main line, up to 100 kilometers long, and many shorter branch lines, each of which has a baited hook at its end. Longline fishing typically targets open-ocean fish such as tuna, swordfish, and billfishes. This fishing industry shows consistent and rapid declines in catch per effort, with catch rates falling from 6–12 down to 0.5–2 fish per 100 hooks, usually within the first ten years of exploitation.

Degradation of environmental services also exacts a significant toll on society, both directly and indirectly. At the dawn of the twentieth century the emissions of air pollutants were quite small. Pollutants were dispersed easily by the winds, so emissions had relatively little impact. As population, living standards, and energy use grew, emissions increased dramatically. In many regions emissions now exceed the ability of the atmosphere to dilute them to safe levels. This polluted environment imposes direct costs on human health, the economy, and the environment. More than 90 million people live in U.S. urban areas where air pollutants exceed the recommended safety standards. This pollution sickens and kills millions of people from asthma, allergies, bronchitis, lung cancer, and other health problems. These health problems cost the U.S. economy more than $30 billion per year in lost work days, trips to the hospital, and the like. Air pollution also damages trees and plants. Curtailed production of soybeans, corn, and other crops caused by their exposure to air pollution reduces farm income by more than $5 billion per year. Pollution damages forests, lakes, wetlands, and other ecosystems that provide important environmental services.

To cut these costs, environmental laws require firms to invest significant sums in devices that reduce emissions. In 1999 U.S. manufacturers spent about $150 billion on new capital equipment, of which $4.4 billion was spent on pollution control equipment. This $4 billion represents an opportunity cost—this money could have been invested in capital to increase the production of useful goods and services. These "lost goods and services" represent the indirect costs of environmental degradation.

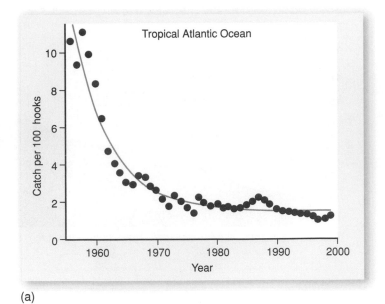

(a)

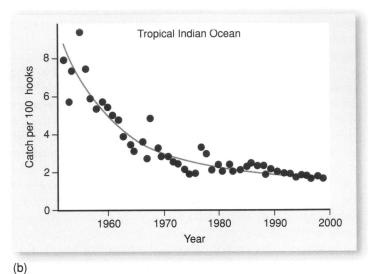

(b)

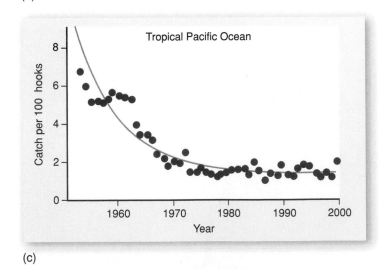

(c)

FIGURE 1.7 *Diminishing Returns in Global Fish Catch* *Changes in catch per unit effort for the Japanese longline fishing fleet in (a) tropical Atlantic Ocean, (b) tropical Indian Ocean, (c) tropical Pacific Ocean.*
(Source: Data from R. A. Myers and B. Worm, "Rapid Worldwide Depletion of Predatory Fish Communities," Nature 423: 280–283.)

Violating Principle 2: Policies That Lack a Systems Perspective

Why do people deplete natural resources and degrade environmental services? One reason is a lack of attention to principle 2—namely the recognition that human society is part of the environmental system and that this linkage can cause the effects of human actions to differ from their intent. People often make decisions based on their self-interest. Firms make decisions to maximize their profits, and individuals make decisions to maximize their economic well-being. But these decisions often trigger unforeseen feedback loops that produce unintended and unwanted outcomes. Unfortunately there is no shortage of examples.

Burning fossil fuels and smelting metals such as copper produce air pollutants such as sulfur dioxide. Before the 1970s pollutants usually were released to the atmosphere through a smokestack, and the winds dispersed them. Occasional weather conditions known as temperature inversions prevent winds from dispersing such pollutants. Under these conditions pollutants build to harmful and even lethal levels.

The U.S. Clean Air Act was passed in 1970 to improve the nation's air quality, particularly in densely populated areas. One inexpensive way to meet the act's requirements was to build much taller stacks that released pollutants above the temperature inversion layer. There were only two stacks taller than 150 meters prior to 1970. Between 1972 and 1978 industry built 178 stacks taller than 150 meters, and more are being built every year (Figure 1.8).

Taller stacks reduce pollution locally, but what goes up must come down. Taller stacks intensify air pollution downwind. In the 1970s scientists noticed fish populations in many lakes in New England and southern Canada shrinking significantly. The drop was caused by an increase in the acidity of the lake water. Scientists traced this problem to the sulfur dioxide released from coal-fired power plants hundreds of miles upwind in Ohio, Pennsylvania, Kentucky, Illinois, and Indiana. There tall smokestacks inject sulfur dioxide so high in the atmosphere that the prevailing winds carry it to New England and Canada. Along the way the sulfur reacts with water vapor to produce sulfuric acid. The acid eventually falls back to Earth as rain or snow and causes great environmental harm in lakes, rivers, and forests. Needless to say, people in Vermont and New Hampshire are not pleased that the "solution" to air pollution in the Midwest damages their forests and lakes. Similarly, the acidification of lakes in Sweden also is caused in part by pollutants emitted in Great Britain.

Violating Principle 3: Unequal Opportunities for Human Development

One of the clearest indications that the world is not moving toward sustainability is the huge difference among nations in the level of human development. The United Nations defines **human development** as the process of enlarging the range of people's choices by increasing their

FIGURE 1.8 *Smokestack* *Tall chimneys, such as this 364-meter chimney of a coal power plant located near Trbovlje, Slovenia, have long been used to disperse pollution beyond the local area.*

opportunities for education, health care, a clean environment, income, employment, and political freedom. These opportunities differ greatly among nations and define three broad categories: high development, medium development, and low development. People in nations with high levels of human development, such as Canada, France, and the United States, have high incomes, are highly educated, and live long lives. On the other hand, people in nations such as Kenya, Haiti, and Myanmar are poor, have little chance for educational advancement, and live much shorter lives.

Income inequality illustrates the stark difference between high and low levels of development. Individual income is measured by a nation's per capita gross domestic product (GDP). Per capita GDP is a dollar measure of all the goods and services produced in a country each year divided by that country's population size. In 2005 the United States, Canada, and many European nations had a per capita GDP that ranged from $25,000 to $50,000. Compare that to the $100–$1,000 range in nations with a low level of economic development such as Kenya, India, and Angola.

Similar differences exist for other aspects of human development such as education, literacy, and life expectancy (Table 1.2). In 2003 Canadians lived an average of 80 years; almost 100 percent of Canadian adults could read and write; nearly all Canadians had access to safe drinking water; and Canada's infant mortality rate was extremely low. Compare this to Kenya, where people lived an average of 47 years; 74 percent of adults could read and write; about 60 percent of the population had access to safe drinking water; and the infant mortality rate was 123 deaths per 1,000 live births.

Similar gaps exist among income classes of the same nation. Per capita GDP in the United States was about $38,000 in 2005, but 45 million people had no health care (almost 1 in every 6 people), and more than 10 percent of the population (36 million people) lived below the official poverty threshold. Per capita 2005 GDP in India was only $2,900, but people in the upper class had high incomes, owned elegant homes, drove expensive automobiles, and took luxurious vacations. The point is that the standard of living enjoyed by a rich Indian is closer to that of a rich American than it is to that of a poor Indian. Similarly, a poor American suffers many of the same problems as a poor Indian.

Income inequality in a country can be measured by income percentiles. For example, we can compare the total income of the top 20 percent (a "quintile") of a population to the income of the bottom 20 percent. If this ratio equals 1, income distribution is relatively equal. But in most nations this is not the case (Figure 1.9). The top quintile typically captures a far greater fraction of a nation's income than does the bottom quintile.

Sharp differences in levels of human development also exist between men and women. In almost every nation—including affluent nations such as the United States—women do not have equal access to income, education, land, credit, employment, and political power. This inequity exists despite the fact that in every society women have a profound effect on the well-being of their families, communities, economies, and local ecosystems. Women are almost always paid less than men, even for the same jobs. Globally about 70 percent of the world's poorest people are female, and their earnings average 25 percent lower than men. Women's education levels and literacy rates are far below those of men. An estimated 602 million women (33 percent of women) compared to 346 million men (19 percent of men) are illiterate. Girls and women also have higher rates of malnutrition than their male counterparts.

Clearly such sharp differences in economic development among nations, within nations, and between genders threaten the sustainability of the entire society. Are these differences increasing or decreasing? What is the trend in human development? There have been some spectacular success stories in nations with low and medium levels of development. In the last 30 years life expectancy has increased by more than 30 percent, and primary school enrollment has grown from 48 to 77 percent. Such gains have been especially strong in east and southeast Asia, where many governments such as those in China, Indonesia, Malaysia, and Singapore have attempted to couple economic growth with more equitable income distribution and improved health care, education, and employment for their citizens.

That's the good news. The bad news is that many people are worse off than they were ten or more years ago, and the divide between rich and poor continues to grow. The income gap between the fifth of the world's people living in the richest countries and the fifth in the poorest

TABLE 1.2	A Comparison of the State of Development in Canada and Kenya						
	GDP per Capita	Adult Literacy Rate	Population with Access to Clean Water	Population with Adequate Nutrition	HIV Prevalence (Ages 15–49)	Life Expectancy at Birth	Under-5 Mortality Rate
Canada	$31,000	100%	100%	100%	0.30%	80 years	6 per 1,000 live births
Kenya	$1,000	74%	62%	67%	7.0%	47 years	123 per 1,000 live births

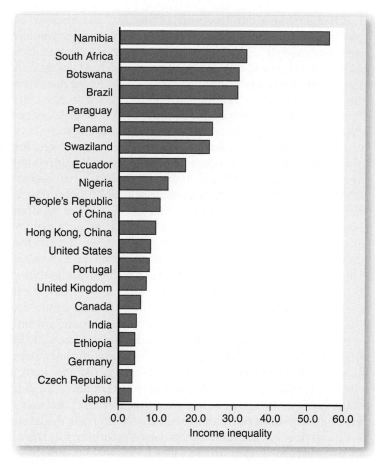

FIGURE 1.9 *Income Inequality* The ratio of the income of the top 20 percent of a population to the income of the bottom 20 percent. The richest 20 percent of the population in Namibia earns 50 times more than the bottom 20 percent. A lower value signifies greater equality in income distribution.

was 30 to 1 in 1960, 60 to 1 in 1990, and more than 70 to 1 today. More than 1.8 billion people living in eighty-nine nations are worse off economically than they were fifteen years ago. In nations such as Haiti, Sudan, Ghana, and Venezuela, per capita income is less today than in 1960. Unemployment hits young people the hardest in low-development nations, with rates reaching 30 percent in some cities in Kenya and Algeria (Figure 1.10). The United Nations "Poverty Clock" tracks what is perhaps the most disturbing indicator of human development: the number of people living in absolute poverty. In 2005 half the world's population—about 3 billion people—lived on less than $2 a day.

What do differences in development have to do with environmental change? Resource depletion and environmental degradation are correlated with living standards. Compared to people with low incomes, people with higher incomes drive more, spend more time in air-conditioned environments, eat in restaurants more often, live in bigger homes, take more extravagant vacations, and generally buy more goods and services. This translates directly into the use of more energy, materials, and environmental services. A comparison of India and the United States illustrates this point (Table 1.3). Compared to the average Indian, the average American uses much more energy, fresh water, and paper products and releases far more air pollution.

People in low-development nations also cause environmental damage, but their motivations are different. Their poverty and rapid population growth contribute to environmental degradation by limiting their choices. The poor are concerned with tonight's supper, not the world's future. Developing nations generally lack the capacity to build

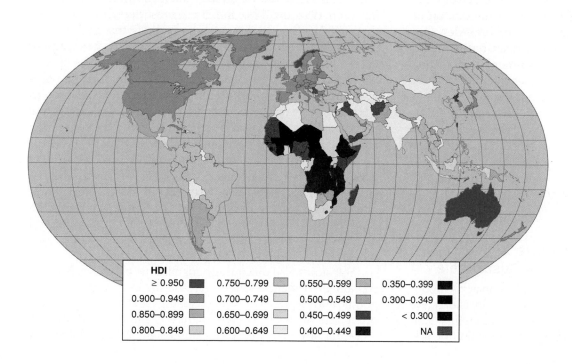

FIGURE 1.10 *Human Development* The pattern of human well-being as measured by the Human Development Index (HDI). The HDI is a measure of a country's income level, life expectancy, and quality of education. The HDI is a 0 to 1 scale. In 2003 Norway had the highest HDI (0.963) and Niger had the lowest (0.281). *(Source: Data from United Nations Development Programme.)*

HDI

≥ 0.950	0.750–0.799	0.550–0.599	0.350–0.399
0.900–0.949	0.700–0.749	0.500–0.549	0.300–0.349
0.850–0.899	0.650–0.699	0.450–0.499	< 0.300
0.800–0.849	0.600–0.649	0.400–0.449	NA

TABLE 1.3	A Comparison of Resource Use in the United States and India		
Resource	U.S. Consumption	Indian Consumption	Ratio of U.S. to Indian Consumption
Aluminum	4,137	420	33.7
Copper	2,057	157	44.8
Crude steel	93,325	20,300	15.7
Phosphate rock	40,177	2,381	57.6
Coal	672,036	184,992	12.4
Petroleum	666,032	53,294	42.7
Natural gas (terajoules)	21,387,719	397,250	183.9
Beef and veal (head)	35,989	11,758	10.5
Roundwood (1,000 cm³)	468,003	281,045	5.7
Pulpwood (1,000 cm³)	136,377	1,208	385.7

adequate sewer and water supply systems. But people still need water to survive, so they often use dangerously contaminated water. High rates of population growth magnify this problem because pollution intensifies as more people are forced to live in the same area.

These issues have been paramount at global environmental summits such as the Earth Summit in Rio de Janeiro in 1992 and the World Summit on Sustainable Development in Johannesburg in 2002. Low-development nations argue that they need economic assistance from the high-development nations to reduce environmental degradation. They want industrial nations to fund environmental protection by helping to pay for new technologies, enforcement of environmental treaties, and other measures that developing nations currently cannot afford. In essence, low-development nations assert that the only way they can reduce their impact on the environment is by transferring some of the wealth now enjoyed by people in high-development nations. See *Policy in Action: The President's Council on Sustainable Development: Blueprint for Sustainability or Environmental Window Dressing?*

Violating Principle 4: Actions Must Be Both Environmentally and Economically Sustainable

For some people and firms, ethical concerns about social responsibility and sustainability drive decisions. We applaud this effort, but let's be honest: Sustainability cannot depend solely on good will. Many decisions that people

and firms make about the production, use, and disposal of goods and services are financially motivated. Economists find that people spend money in a way that maximizes their economic well-being, while firms produce goods and services in a way that minimizes their costs. These decisions are based mainly on the price of goods and services.

In theory, the price of goods and services should reflect the degree to which their production, use, and disposal are sustainable. But for reasons we describe later, prices for many goods and services do not include their environmental impacts. This situation is called an **externality**: a cost associated with the production or consumption of a good that is not accounted for in the price of that good and that is borne by others in society. For example, when firms emit pollutants in the environment, they create an external cost: the damage the pollution inflicts on the economic and physical well-being of others. Externalities undermine sustainability because they make a good or service appear cheaper than it really is, leading to greater consumption than is desirable. Incorporating these external costs in prices is one important component of environmental policy. Including such information allows consumers to make decisions that are both economically sensible and environmentally sustainable.

Complicating such decisions, government policies sometimes distort prices in a way that makes it attractive for people to behave in a nonsustainable manner. A classic example of such perverse incentives is a **subsidy**: government-provided goods or services that would otherwise have to be purchased in the market, or special exemptions from standard required payments or regulations. Subsidies include direct payments from the government and tax breaks that target specific industries.

Governments around the world often subsidize the production of natural resources such as timber and energy. Between 1992 and 1997 the U.S. Forest Service spent more than $387 million building logging roads that are used by private timber firms (Figure 1.11). The U.S. government helps the energy industry by giving oil producers special tax breaks, maintaining waterways used by tankers, paying most of the cost of enriching the uranium fuel used in nuclear power reactors, and so on. No energy source is free of government support. Federal subsidies to the energy industry may run as high as $60 billion per year.

Subsidies often encourage people to produce or buy goods and services whose production, use, or disposal is not sustainable. The subsidies described in the previous paragraph speed timber harvests and accelerate the depletion of oil because they lower the cost of producing timber and oil. Lower costs generally lessen consumer prices, which increase consumption and therefore production.

Subsidies also discourage sustainable practices. Because sustainable practices often increase the cost of production, goods and services produced using sustainable practices often are more expensive than the same goods or services that are provided using nonsustainable practices. For

FIGURE 1.11 *Subsidies and the Environment* *Subsidies from the federal government to private companies, such as this logging road, encourage the depletion of natural resources.*

example, unless they are guided by a strong personal commitment to sustainability, many people would choose less expensive fish or lumber. Thus firms that harvest fish or timber in a sustainable manner cannot stay in business.

The tendency of most people to buy goods and services without regard to how they are produced creates a flip side to this principle of sustainability. Environmentally sustainable practices must also be economically sustainable. Firms that harvest trees, catch fish, or grow food in a way that is environmentally sustainable must be able to make a profit. Profit allows firms to stay in business. Without such financial rewards, we cannot expect firms to act in a way that promotes environmental sustainability if such effort harms their economic well-being. To remedy this situation, firms that produce goods or services in a sustainable manner should be rewarded financially, and individuals who purchase goods or services that are produced in a nonsustainable manner must be penalized (Figure 1.12).

An example of such an incentive is an **environmental performance bond**. This is a sum of money a firm must deposit with a government agency before it is granted a permit for an activity with the potential for significant environmental impact (such as road building, harvesting timber, or oil extraction). The bond is set at an amount equal to the best estimate of the worst potential future environmental damages. The bond is returned if the firm demonstrates that the anticipated damages did not (and will not) occur. Conversely, if damages occur, the funds are used to restore the site, and any remaining balance is returned. Thus the

bond provides an incentive for the firm to act in a sustainable manner. As we will describe, many firms produce fish, timber, food, and other resources in a manner that is sustainable for both the firms and the environment.

FIGURE 1.12 *Forest Certification* *The logo and initials "FSC" indicate that the Forest Stewardship Council (FSC) has certified that this timber has been harvested in a sustainable manner. The FSC seeks to balance the need for firms to operate profitably with the need to manage forests for long-term sustainability.*

POLICY IN ACTION

The President's Council on Sustainable Development: Blueprint for Sustainability or Environmental Window Dressing?

Our descendants may look back at June 13, 1992, as a defining moment in human history. On that date in Rio de Janeiro, Brazil, world leaders met to attempt to reverse the environmental and economic deterioration of the planet. This unique gathering, now known as the Earth Summit, drew more heads of state than any meeting in history. This summit produced a document called Agenda 21—a comprehensive road map for humanity to find its way toward sustainability.

Among its many provisions, Agenda 21 recommended that governments facilitate the development of national plans for sustainable development. In 1994 President Clinton established the President's Council on Sustainable Development, which was a panel of leaders from government, business, environmental, civil rights, labor, and Native American organizations. Over the next several years the council held dozens of open meetings with civic organizations, corporate boards of directors, local governments, environmental groups, civil rights organizations, and many other stakeholders in sustainable development. At the conclusion of these meetings and after its own deliberations, the council produced a report titled *Sustainable America: A New Consensus for Prosperity, Opportunity, and a Healthy Environment for the Future*. This report is centered on ten interdependent goals that flow from the council's belief that it is essential to seek economic prosperity, environmental protection, and social equity together. The report describes several major challenges facing the United States and how they can be addressed by changes in technologies, institutions, and lifestyles.

Cynics charge that the council's report is window dressing—yet another government report to gather dust on a shelf while politicians claim that real progress is being made. But there is reason to believe that this report represented a small but substantive step toward sustainability in the United States. The council included representatives from industry, government, and environmental groups who previously had strong disagreements; so it is remarkable what everyone agreed *should* happen. First, there was consensus that the United States has strong environmental laws such as the Clean Air Act and the Clean Water Act, which

National Goals toward Sustainable Development

Goal 1: Health and the Environment	Ensure that every person enjoys the benefits of clean air, clean water, and a healthful environment at home, at work, and at play.
Goal 2: Economic Prosperity	Sustain a healthy U.S. economy that grows sufficiently to create meaningful jobs, reduce poverty, and provide the opportunity for a high quality of life for all in an increasingly competitive world.
Goal 3: Equity	Ensure that all Americans are afforded justice and have the opportunity to achieve economic, environmental, and social well-being.
Goal 4: Conservation of Nature	Use, conserve, protect, and restore natural resources—land, air, water, and biodiversity—in ways that help ensure long-term social, economic, and environmental benefits for ourselves and future generations.
Goal 5: Stewardship	Create a widely held ethic of stewardship that strongly encourages individuals, institutions, and corporations to take full responsibility for the economic, environmental, and social consequences of their actions.
Goal 6: Sustainable Communities	Encourage people to work together to create healthy communities where natural and historic resources are preserved, jobs are available, sprawl is contained, neighborhoods are secure, education is lifelong, transportation and health care are accessible, and all citizens have opportunities to improve the quality of their lives.
Goal 7: Civic Engagement	Create full opportunity for citizens, businesses, and communities to participate in and influence the natural resource, environmental, and economic decisions that affect them.
Goal 8: Population	Move toward stabilization of the U.S. population.
Goal 9: International Responsibility	Take a leadership role in the development and implementation of global sustainable development policies, standards of conduct, and trade and foreign policies that further the achievement of sustainability.
Goal 10: Education	Ensure that all Americans have equal access to education and lifelong learning opportunities that will prepare them for meaningful work, a high quality of life, and an understanding of the concepts involved in sustainable development.

Source: From President's Council on Sustainable Development.

have produced significant improvements in air and water quality since the 1970s. Council members warned that these laws should not be weakened.

Second, there was consensus that market forces should be used in conjunction with environmental laws to promote sustainability. An example is environmental tax reform, which shifts the tax burden from income tax to taxes on pollution and resource depletion. For example, taxes on income could be reduced and the shortfall could be made up by taxes on fuels such as gasoline and coal, which are based on the amount of carbon dioxide and other pollutants they emit. Such a tax would encourage more efficient use of energy and promote the development of energy-efficient appliances, vehicles, and other machines. Another use of market forces would remove subsidies that currently promote nonsustainable use of the environment. For example, the federal government heavily subsidizes the development of water resources in the western United States. Therefore industry, households, and agriculture use water wastefully. Removing these subsidies would cause water prices to rise, thereby encouraging more sustainable use.

Third, the council asserted that the United States has both the reason and the responsibility to carry out global policies that support sustainable development. The future of the United States—its security, prosperity, and environment—is linked to the global environment. With the highest living standards in the world, the United States is the largest producer and consumer in history. With less than 5 percent of the world's population, the nation produces about a quarter of world GDP and consumes about a quarter of the world's resources. This has made the United States the world's largest producer of wastes and given us the incentive and capacity to create and use innovative technology to reduce pollution. Many nations seek to emulate the success of the U.S. system of environmental protection. In addition, problems such as climate change and biodiversity loss can be addressed only through global cooperation.

The council noted two important elements of building a strong U.S. engagement in the international environmental arena. The first is the commitment of financial resources that are needed by development assistance agencies such as the United Nations and the Global Environmental Facility. Matching our words with resources would demonstrate that we are serious about sustainability. Second, the United States should cooperate in key international environmental agreements. The council criticized the United States for not ratifying the U.N. Convention on Biological Diversity (the United States is the only major industrial nation that has not done so) even though ratification was supported by a broad cross section of U.S. industry and environmental groups. The United States also has not ratified the Kyoto Protocol on climate change—the international agreement to reduce the emissions of greenhouse gases. Again, the United States is the only industrial nation with significant emissions that has not ratified this agreement.

STUDENT LEARNING OUTCOME

- Students will be able to decide whether the President's Council on Sustainable Development is a serious attempt to promote sustainability.

SUSTAINABILITY ON A PERSONAL LEVEL: THE ECOLOGICAL FOOTPRINT

One of the biggest barriers to changes that would move society toward sustainability is the feeling that individuals can't make a difference. Many people look at environmental problems and wonder what they can do about global warming, the depletion of stratospheric ozone, or world population growth. Environmental problems seem so enormous that many people just shrug their shoulders and turn away. The feeling of being overwhelmed is reinforced by the dizzying range of opinions about the severity of environmental problems and the various "solutions" that are proposed. Some people think the global production of oil peaked yesterday, whereas others forecast that the peak will not come for another forty years.

What can one person do? A lot. Some of the most enduring and powerful images we carry through life are those of individuals who made a difference. Martin Luther King devoted and ultimately gave his life for the principles of equality and nonviolence. In his struggle against apartheid, Nelson Mandela suffered twenty-seven years in prison. A lone student, standing in the path of a tank in Tiananmen Square, China, was willing to die for freedom of expression.

How do your actions affect sustainability? The answer begins with recognizing that your contribution to resource depletion and environmental degradation is determined by decisions you make every day. The most important of these are decisions concerning what you buy. Different goods and services require different amounts and types of energy, materials, and environmental services. They also produce different types and amounts of pollution. Throughout this book we provide information that allows you to calculate how your purchases affect the environment. This impact can be understood using the concept of the **ecological footprint** (Figure 1.13). Your ecological footprint is equal to

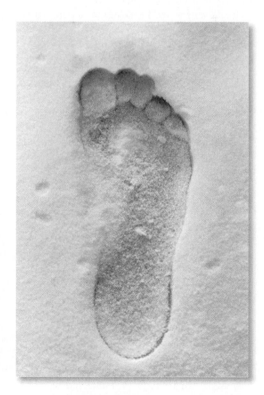

FIGURE 1.13
The Ecological Footprint

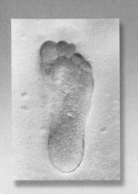

Your ECOLOGICAL *footprint*

The Footprint of the *Moai*

An ecological footprint can be used to assess the use of environmental goods and services that are associated with any human activity. The footprint of the *moai* on Easter Island quantifies how this religious and political institution affected the island's environment.

Measuring the footprint of a statue begins with asking what it takes to produce and transport a huge piece of stone weighing as much as 80 metric tons over rough terrain with nothing but human muscle power (Figure 1). The answer is a lot of energy! But in this case it wasn't fossil fuel energy, but the food energy consumed by the people who carved, moved, and erected the statues. Most of the food came from the land, so the footprint of the *moai* can be measured by the area of productive agricultural land that was used to support the human labor that carved, transported, and erected the statues. Let's look at the transportation part in more detail.

The statues were carved from a volcanic tuff in a quarry in the Rano Raraku crater, and they were transported to various sites on the island. A typical statue stands 4 meters tall and weights about 12.5 metric tons, although many statues are much smaller or larger. There are many theories about how the statues were moved from the quarry to their coastal sites. Archaeologist Jo Anne van Tilburg and her colleagues used a combination of archaeological and anthropological information about Polynesian culture and modern computer techniques to reconstruct a likely explanation. They suggest that most statues were moved with a system of wooden rollers. Palms and other tree species provided the wood for the track roller system. Various other plants were harvested to provide the cords and ropes used to lash and haul the statues. One concerted pull would have moved the statue about five meters. This pulling activity was followed by pauses to move the rollers, tighten the lashes, and so on.

Given the arduous and time-consuming nature of the task, van Tilburg and her colleagues asked how the islanders would choose a path from the quarry to the site where a statue was to be erected. They picked a likely typical transport path from the Rano Raraku quarry to a site on the eastern coast, a distance of about 10 kilometers. They used a three-dimensional computer map of the island's terrain to simulate the time and effort associated with alternative routes. The islanders probably picked a route that minimized total energy expenditure. Computer simulations suggested that the optimal route was the shortest one (10.1 km) because even though it required the most people (seventy), it took the shortest amount of time (about five days) and thus used the least energy.

The islanders lived in large extended families of forty-five to fifty people. Virtually every member provided some form of productive labor to the island economy. About eight males of appropriate age and strength per extended family would have been available to participate in the *moai* transport. Thus the seventy males who worked on the transport were supported by nearly nine extended families (between 391 and 435 people) who would have had to join forces to provide food for the workers as they hauled the statue over the 10 km path.

How much food was required? The average man who helped transport the *moai* required about 2,880 kcals of food energy per day. Of that, 500 to 600 kcals of protein would have been required to maintain muscle and other body tissue expended in the work task. Archeological data indicate the protein came primarily from rats, fish, and to a lesser

FIGURE 1 *Raising* **Moai** *Archaeologists reenacting the transportation and raising of a* moai *statue.*

extent chickens. The remaining 2,200 kcals were carbohydrates supplied by two staple island crops: 1,000 grams of sweet potato and about 500 grams of banana.

The seventy men moving the *moai* would have collectively required about 201,600 kcals per day. The average sweet potato yield was 7,200,000–14,820,000 kcals per hectare (3,000,000–6,000,000 kcals per acre), while banana cultivation yielded 6,916,000–13,832,000 kcals per hectare (2,800,000–5,600,000 calories per acre). Thus between 0.6 and 1.2 hectares (1.5–3.0 acres) of cultivated sweet potatoes and 0.65–1.3 hectares (1.6–3.2 acres) of cultivated bananas would have been required to support the workers. But cultivation on the island typically involved a fallow period in which a plot of land cultivated in one year would remain uncultivated the next year to allow soil fertility to recover. As a result, the total number of hectares required to produce a continuous crop would have to be roughly doubled to between 1.2 and 2.4 hectares (3–6 acres).

But the footprint of the *moai* did not stop there. The chief who had commissioned the transport task also needed a reserve supply of stored food to trade for the animal protein. In addition, the chief needed food of all types to host various social and ritual obligations inevitably associated with the project. Thus the chief required food to hire the carvers, feed the workers and their support network (extended families), trade for protein-rich foods, and "feed the gods" as well. Van Tilburg estimates that the total footprint of the *moai* transport task was about 20 hectares (50 acres) of productive land.

STUDENT LEARNING OUTCOME

- Students will be able to describe how it was possible to move massive statues long distances without the assistance of large animals or machines.

all the natural resources and environmental services used to produce your food, clothing, and shelter, as well as the other goods and services you use. Part of your footprint probably is obvious. You use some water every time you take a shower or cook pasta. You use electricity when you turn on the TV or your bedroom light, and you use gasoline when you drive a car. But your ecological footprint is far larger than you think. Some of your food was grown on eroding land with the use of pesticides and other persistent chemicals. Your use of electricity and gasoline increases the emission of carbon dioxide into the atmosphere and thereby contributes to global climate change. The newspaper you bought today may be made from paper produced from timber that was not harvested sustainably. See *Your Ecological Footprint: The Ecological Footprint of the* Moai.

Consider the footprints of alternative ways of getting to school or work. Let's assume that you live 10 kilometers from school or the office. If you drive a car, you would burn about a liter of gasoline, which represents about 8,000 kcals (Figure 1.14). Burning this gasoline emits many air pollutants. Your drive would generate a little more than two kilograms of carbon dioxide, which is an important greenhouse gas, a quarter kilogram of health-threatening carbon monoxide, and a few grams of smog-forming hydrocarbons and nitrogen oxides.

If you bicycle to work, you burn about 210 kcals of carbohydrate energy—about what's in a bowl of cereal. This is about forty times less energy than your trip by car, and you would burn no fossil fuels and release no harmful or toxic chemicals.

Each chapter in this book provides information to let you calculate how much of a particular resource you use or the quantity of a particular pollutant you emit. In the chapter about fossil fuels you will calculate how much oil, coal, and natural gas you use. In the chapter about global climate change you will calculate your emissions of carbon dioxide, which is one of the gases responsible for the ongoing rise in global temperature. We also provide information that shows how your ecological footprint is affected by your

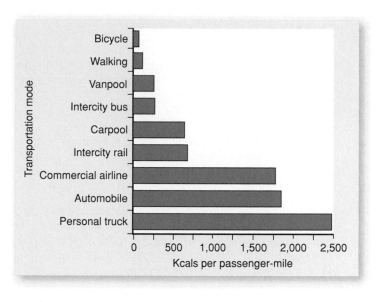

FIGURE 1.14 *Ecological Footprint of Transportation Modes* One component of your "transportation footprint" is the amount of energy used to move you a given distance.

decisions—for example, how much you could reduce your use of energy and emission of pollutants by riding your bike to the store rather than driving or eating a salad instead of a hamburger. By the end of the book you will know just how big your ecological footprint is, and you will be aware of a number of ways to reduce its size.

The purpose of the book's Your Ecological Footprint boxes is to illustrate the principles of sustainability in a context you can understand and relate to. You will be able to judge how sustainable your current lifestyle is—and more important, what changes you can make to reduce the size of your ecological footprint. The exercises will also prepare you to better understand environmental issues as they are discussed in the media and political arena.

SUMMARY OF KEY CONCEPTS

- A sustainable society meets the needs of the present generation without compromising the ability of future generations to meet their own needs.

- A sustainable society does not use natural resources or environmental services faster than the environment generates them. Natural resources include energy and materials we get from the environment to meet our biological and economic needs and wants. Environmental services are natural processes that regulate conditions in the environment and make the planet suitable for life.

- Society and the environment are interconnected systems—collections of parts that generate a regular or predictable pattern. The people on Easter Island were connected to each other through social relations; they were connected to their environment by their use of resources and environmental services.

- The best first principle means that as resources are depleted, society must use lower-quality resources. This depletion and substitution can have negative effects on the economy, the environment, and the health of humans and other species.

- Living standards are correlated with resource depletion and environmental degradation. Compared to people with low incomes, people with higher incomes use more energy, materials, and environmental services.

- Subsidies and externalities distort the prices that people pay for goods and services, often leading people to produce or buy goods and services whose production, use, or disposal is not sustainable.

- Your ecological footprint represents all the natural resources and environmental services used to feed, clothe, and shelter you, and to produce the goods and services you use. It can be used as a guide to compare alternative choices in how we consume goods and services.

REVIEW QUESTIONS

1. What were the principal ways in which the Easter Islanders depleted the resources of their environment?

2. What is the main difference between a natural resource such as crude oil and an environmental service such as a stable climate?

3. Go to the *Human Development Report* published by the United Nations Development Programme (http://hdr.undp.org/statistics/data/). Israel and Oman have roughly similar per capita GDP, but the former has a much higher score for the overall Human Development Index (HDI). Can you explain why?

4. Go to www.greenscissors.org/ and locate the section that reports the subsidies the federal government gives to the energy industry. What form of energy receives the largest handout?

5. Explain how government subsidies can accelerate depletion or pollution.

KEY TERMS

best first principle

ecological footprint

environmental degradation

environmental performance bond

environmental services

externality

human development

natural resource

nonrenewable resources

pollution

principle of sustainability

renewable resources

resource depletion

subsidy

sustainability

system

waste assimilation

THE LAWS OF ENERGY AND MATTER

2 CHAPTER

PART TWO Basic Concepts in Environmental Science

STUDENT LEARNING OUTCOMES

After reading this chapter, students will be able to

- Explain the importance of the law of conservation of matter for environmental science.
- Describe the limits that the laws of thermodynamics place on energy conversion.
- Distinguish the important differences between chemical, physical, and nuclear changes in matter.
- Provide examples of the entropy law in nature and in their everyday lives.
- Define and describe the major forms of energy they use in their everyday lives.

An image of the sun taken by a telescope on a NASA satellite *Energy from the sun drives nearly all natural processes, including life itself, and it supports human activity in countless ways.*

LEAD: INDUSTRIAL MARVEL AND ENVIRONMENTAL VILLAIN

Lead is a soft, extremely dense, bluish element found in Earth's crust. It is virtually indestructible because it resists corrosion and is noncombustible. Due to these unique physical properties, people for millennia have used lead in a wide variety of applications. More than 8,000 years ago inhabitants of North America used lead in body paint and ceremonial powders. In the days of the pharaohs Egyptians used lead to glaze pottery, to make solder, and to decorate ornamental objects. The Babylonians used lead sheets for garden flooring, for caulking, and for fastening bolts into masonry. Four thousand years ago the Chinese used lead to make coins, as did the ancient Greeks and Romans. The Romans also had fifteen different sizes of lead pipes to transport water across their vast empire in their legendary system of aqueducts (Figure 2.1). In fact, the Roman word for lead, *plumbum*, was used as the word for waterspouts; this is where our word *plumber* comes from. In the twentieth century lead was used in a variety of products, including paint, batteries, water pipes, ceramics, and gasoline.

Some of the chemical and physical properties that make lead an industrial marvel also make it a deadly toxin. Even very small amounts of lead can cause serious harm, particularly in children and pregnant women. Children poisoned by lead often suffer severe damage to the brain and nervous system, which may lower IQ and cause behavioral problems, impaired growth, hearing problems, and kidney damage. In adults lead causes high blood pressure, digestive problems, kidney damage, and nerve disorders. In large doses lead causes blindness, brain damage, and death.

Some of the obvious health effects of lead have been known for almost as long as people have used it. In 370 B.C. Hippocrates, founder of the field of medicine, associated severe attacks of abdominal pain with people who worked with lead. In the first century A.D. Dioscorides, founder of the field of pharmacology, noted that the ingestion of lead "causes oppression to the stomach, belly, and intestines by its severe pressure; it suppresses the urine, while the body swells and acquires an unsightly leaden hue." In 1773 Massachusetts Bay Colony passed an act that forbade lead in equipment used to distill rum after complaints "from North Carolina against New England Rum, that it poisoned their people, giving them dry bellyache, with a loss of the use of their limbs."

With the toxic effects of lead established for centuries, why is it still used widely today? There are several reasons, including uncertainty about how lead actually makes it into our bodies and about the subtler effects of lead on human health. See *Policy in Action: Getting the Lead Out*. The pathway from a ceramic mug made with lead to the human body is clear. But how does lead added to gasoline poison people? Tetraethyl lead ($Pb[C_2H_5]_4$) is produced in chemical factories and added to motor gasoline in petroleum refineries. When the gasoline is burned this lead is released to the atmosphere and deposited as small particles in soil, streets, buildings, and homes. People ingest the lead by inhaling the small particles or by ingesting them through food or direct contact with lead-tainted dust and dirt. In the case of tetraethyl lead, lead follows a complicated series of pathways before it affects human health.

The story of lead illustrates several physical and chemical laws that underlie environmental science. First, all materials go through numerous chemical and physical transformations—for example, mining lead, refining it, and then using it to make a battery. Sometimes materials change form when energy is burned, such as the conversion of tetraethyl lead in a liquid (gasoline) to a pollutant in the atmosphere. Second, lead—and every other material—is conserved. This means that no matter how many times lead materials are mined, melted, burned, boiled, bent, or "thrown away," the lead never disappears. This is true for every material people use. This chapter describes the physical and chemical laws that underlie environmental problems and environmental decision making. ■

FIGURE 2.1 *Lead in Ancient Rome* *The Roman water supply system was very advanced, using huge aqueducts to supply water for large cities. Elaborate systems of lead pipes moved water from one place to another using gravity.*

People have used lead and been aware of its toxic effects for nearly 8,000 years, yet its use and release to the environment continued to grow through the twentieth century. In the United States national policies to limit people's exposure to lead were not developed until the 1970s, by which time millions of people, particularly children, were affected by lead-related illnesses. The reasons for the slow response to the lead problem typify many environmental problems: uncertainty among scientists about the precise health affects of lead; vocal arguments from the lead industry that their products posed no significant threat to public health; government agencies that for too long bowed to industry pressure; and a public unable to sort out the conflicting arguments about lead as they were presented by the media.

By the 1920s some of the sources, damages, and health effects of lead were well documented in medical literature. Most of this information came in the form of case histories of individual children poisoned by lead. Not until the 1950s were large-scale efforts made to uncover the broader public health implications of lead. Notable among these was a 1951 study by the Baltimore Health Department that documented hundreds of reported cases of lead poisoning over a twenty-year period. The study was among the first to expose the extremely high risk faced by young children who ate paint chips or gnawed on windows covered with lead paint. The study formed the basis for 1951 legislation that prohibited lead paint on any interior surface. Unfortunately the law often was not followed and rarely was enforced. As a result, by the 1980s more than three-quarters of homes in Baltimore contained lead paint.

Public lead exposure escalated with the rapid increase in the use of leaded gasoline in the 1950s and 1960s. Much of the medical research on this source was funded by organizations that had a vested interest in selling lead-based products. These included General Motors, the Lead Industries Association (LIA), Standard Oil, and DuPont and Ethyl Corporation, makers of the gasoline additive tetraethyl lead. The LIA was particularly adept at cultivating relationships with organizations such as the American Public Health Association, the Sloan-Kettering Cancer Foundation, and scientists at Harvard Medical School and other universities. Not surprisingly, the results of the research sponsored by these groups claimed that lead posed no serious threat to public health. The results of industry-funded research influenced the federal government to take a do-nothing policy toward controlling people's exposure to lead. By today's standards, the alliance and collaboration between scientists, public health agencies, and the lead industry was highly inappropriate. It is a direct violation of the duties and responsibilities of individual scientists and public agencies to cooperate and collaborate with industries that have a vested interest in the results of research, at least without independent validation by other studies.

In the 1970s changing attitudes toward the conduct of public health research and toward environmental problems in general led to sweeping government regulation aimed at reducing the release of lead to the environment by the lead industry and lead products. The Clean Air Act placed strict limits on the concentration of lead in the atmosphere, which forced automakers and petroleum refiners to shift to unleaded gasoline.

The Clean Water Act strictly limited the concentration of lead in drinking water. This forced companies to reduce their release of lead to lakes, rivers, and groundwater, and forced municipalities to stop using lead in the repair and construction of pipes in public water supply systems. The Lead Poisoning Prevention Act (1978) set guidelines for allowable levels of lead in house paint, which forced manufacturers to reduce or eliminate leaded paints. Finally, the Occupational Health and Safety Administration issued regulations that limited the exposure of people to lead in workplaces.

These and other regulations have produced significant public health benefits. Leaded gasoline no longer is used except by some older farm machines; most paints now are lead-free; and factory workers face much lower exposure levels. But lead contamination will survive both us and future generations. Lead is long-lived in the environment and human tissue; the half-life of lead in the human body is twenty years. Soils remain contaminated with lead from decades of gasoline combustion, and millions of people live in older homes with lead-based paints. As a result, lead poisoning is one of the most common pediatric health problems in the United States. Three to four million children under age 6 have lead blood levels that exceed the federal health standard. This is far greater than the number of children affected by other common childhood diseases.

STUDENT LEARNING OUTCOME

- Students will be able to describe the events leading to the decision to ban leaded gasoline in the United States.

MATTER: ELEMENTS AND COMPOUNDS

Matter is the physical material of the universe: Matter is anything that has mass and takes up space. Everything that we feel and touch has mass, as do other things that are less obvious to our senses. A machine, the soil, a book, and trees are examples of matter. Although you can smell it but not see it, the natural gas that is burned in your stove is matter. Matter on Earth exists in an enormous variety of forms: living, nonliving; light, heavy; visible, invisible; liquid, solid; and so on. Yet all forms of matter are made of a few basic building blocks called **elements.** An element is a substance that cannot be broken down to other substances by ordinary

chemical means. Wood, which is not an element, can be burned to ashes and invisible gases; but gold, which is an element, cannot be broken down into anything else. Carbon (the main component of ashes), iron, gold, hydrogen (a lighter-than-air gas), and oxygen all are elements.

At this time scientists know of 116 different elements, 92 of which are found in the natural world (Figure 2.2). Some, like gold, silver, copper, and carbon, have been known for thousands of years. Others, such as meitnerium, darmstadtium, and ununquadium, have only recently been created by scientists. Each element is given a one- or two-letter symbol, often derived from

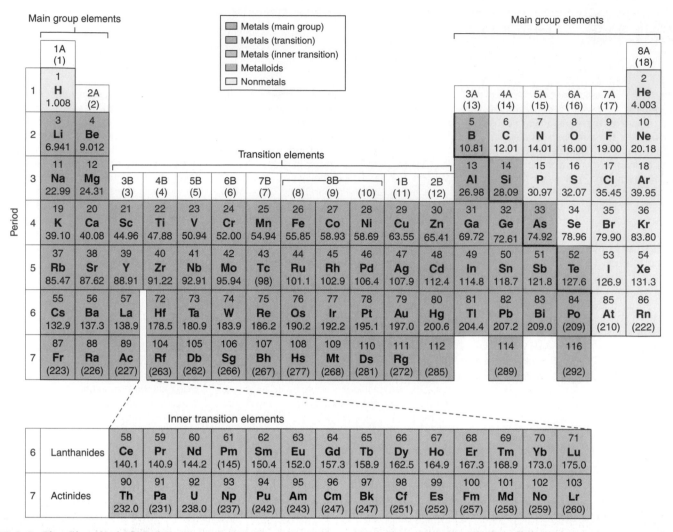

FIGURE 2.2 *The Chemical Elements* *The periodic table of the elements shows how scientists organize matter.*

the English name for the element, such as hydrogen (H) and carbon (C). Other symbols are derived from Latin or German names—for instance, the symbol for sodium is Na from the Latin *natrium*.

Elements combine in specific proportions to form **compounds**. Table salt, for example, is sodium chloride (NaCl), a compound that consists of the elements sodium (Na) and chlorine (Cl). Pure sodium is a metal, and pure chlorine is a poisonous gas that was used as a weapon during World War I. Chemically combined, however, they form an edible, nontoxic compound. This illustrates an important property of a compound: It often has properties quite different from those of its constituent elements.

Compounds often are abbreviated with chemical formulas that describe the fixed proportions of their elements. Numerical subscripts placed after the elements denote the ratios of the elements. Thus in carbon dioxide (CO_2) the ratio of carbon to oxygen always is 1:2, whereas in ammonia (NH_3) the ratio of nitrogen to hydrogen always is 1:3.

Elements Essential for Life

Chemical elements that are essential for life are called **nutrients**. About twenty-five of the ninety-two naturally occurring elements are used by plants and animals to grow and develop and thus are classified as nutrients. Of those twenty-five elements, carbon (C), oxygen (O), hydrogen (H), and nitrogen (N) make up 96 percent of most living organisms (Table 2.1). With sulfur (S) and phosphorus (P) these elements are called **macronutrients**. "Macro" describes the fact that organisms need these elements in relatively large amounts. Carbon is the principal macronutrient for organisms, typically comprising one-half to two-thirds of the dry weight of a plant or animal. Dry weight is the weight of something after all its water has been removed.

Organisms also need minute amounts of **trace elements** such as manganese (Mn), iodine (I), selenium (Se). Even though they are needed in very small amounts, trace elements are critical to the health of an organism. Some trace elements, such as iron (Fe), are required by all forms of life; only certain organisms require others. For example, vertebrates (animals

TABLE 2.1	Naturally Occurring Elements in the Human Body		
Symbol	Element	Atomic Number	Percent Weight of Human Body
O	Oxygen	8	65.0%
C	Carbon	6	18.5
H	Hydrogen	1	9.5
N	Nitrogen	7	3.5
Ca	Calcium	20	1.5
P	Phosphorus	15	1.0
K	Potassium	19	0.4
S	Sulfur	16	0.3
Na	Sodium	11	0.2
Cl	Chlorine	17	0.2
Mg	Magnesium	12	0.1

Trace elements (less than 0.01%): boron(B), chromium (Cr), cobalt (Co), copper (Cu), fluorine (F), iodine (I), iron (Fe), manganese (Mn), molybdenum (Mo), selenium (Se), silicon (Si), tin (Sn), vanadium (V), and zinc (Zn).

Source: Data from Campbell, *Biology, Third Edition*, Benjamin Cummings.

TABLE 2.2	Average Composition of Earth's Crust		
Symbol	Element	Atomic Number	Percent Weight of Earth's Crust
O	Oxygen	8	46.60%
Si	Silicon	14	27.72
Al	Aluminum	13	8.13
Fe	Iron	26	5.00
Ca	Calcium	20	3.63
Na	Sodium	11	2.83
K	Potassium	19	2.59
Mg	Magnesium	12	2.09
Tl	Titanium	22	0.44
H	Hydrogen	1	0.14
		Total	99.17
	All other elements		0.83
			100.00%

Source: Data from B. Mason, *Principles of Geochemistry*, Wiley.

with backbones) require the element iodine (I) to manufacture a hormone in the thyroid gland. A daily intake of only 0.15 milligrams of iodine maintains normal thyroid activity, but an iodine deficiency in the diet causes the thyroid to grow to an abnormal size, a condition known as goiter. Where it is available, iodized salt has reduced the incidence of goiter.

A detailed chemical analysis of your body, as well as the bodies of many other organisms on the planet, reveals elements that are not essential for growth and development and by this definition are not nutrients. These elements are taken up by organisms in the air they breathe, the water they consume, or the food they eat. See *Case Study: The Pathways of Lead in Society and the Environment.* As our story about lead indicates, some of these elements are toxic even at low concentrations. In addition to lead, metals such as cadmium (Cd) and aluminum (Al) are highly toxic, which is why their discharge into the environment is a public health concern. Other elements such as copper (Cu) and zinc (Zn) are essential trace nutrients for many organisms but can be toxic at higher concentrations. Copper mines and smelters release large quantities of copper to the air, water, and soil, damaging nearby vegetation and waterways.

Elements in Earth's Crust

The crust is Earth's outer layer. It contains all ninety-two naturally occurring elements, but most of them are extremely rare. In fact, just ten elements make up more than 99 percent of Earth's crust (Table 2.2). Many of the energy and material resources used by society are metal elements such as copper

(Cu), iron (Fe), uranium (U), and nickel (Ni); nonmetal elements such as phosphorus (P), sulfur (S), and silicon (Si); and compounds called hydrocarbon fuels such as oil, coal, and natural gas. The term hydrocarbon refers to the fact that these fuels are comprised principally of carbon and hydrogen.

Humans mine all ninety-two naturally occurring elements from Earth's crust. Many of the elements are familiar from everyday life; for example, iron is used to produce steel and aluminum is used to produce soft drink cans. Other elements are less obvious but just as common, as illustrated by the materials in a typical U.S. automobile (Figure 2.3). As you might expect, cars consist mostly of iron and aluminum. But they also contain zinc and cadmium (steel), nitrogen (plastic parts), silicon (ceramic electronic parts and glass windows), sulfur (rubber), and many other elements.

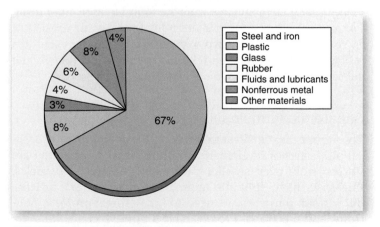

FIGURE 2.3 *Automobile Materials* The material composition of a typical US automobile.

CASE STUDY

The Pathways of Lead in Society and the Environment

The first step in understanding the human health and environmental effects of a toxic material such as lead is to identify its transformations as it goes from extraction to manufacturing, consumption, recycling, and ultimate disposal. This requires a systems perspective that views society as being connected to the environment by flows of energy and materials such as lead.

The release of lead to the environment occurs every time the metal is physically or chemically altered. In the extraction stage, lead-bearing rock is crushed and ground to separate the metal from the rock. In this process a lot of waste rock that contains small concentrations of lead enters the soil and nearby streams. Refineries melt and chemically treat the lead to further purify it, in the process releasing lead to the atmosphere and soil. The factories that use the refined lead to manufacture bat-

teries, leaded gasoline, paint, ammunition, and other products also release lead to the environment. About 65 percent of the lead in batteries is recycled each year; this figure is more than 80 percent in the United States due to lead battery recycling laws in many states. Note also that although leaded gasoline has essentially been banned in the United States, its continued use in other nations (Venezuela, Italy, France, Russia, Saudi Arabia, Nigeria, and China) releases huge quantities of lead to the environment.

Once in the environment, lead is ingested by people through four pathways: air, water, food, and soil/dust. For adults the main source of lead is food. Typically lead comes from soil particles that adhere to fruits and vegetables, as well as from tableware, lead-soldered food cans, lead crystal, lead wine seals, and so on. For children the main source of lead is soil or dust,

which they ingest from dirt on their hands, toys, windowsills, and other objects. Sources of lead in urban and suburban dust include lead-based paint and residues from leaded gasoline.

People are not the only organisms that ingest lead. Many waterfowl die of lead poisoning because they eat lead shot, perhaps mistaking it for seeds or small mollusks. Ingesting just one or two pellets can kill a bird. Lead from shot is passed up in the food chain when birds containing the shot are eaten by other animals. Studies show that waterfowl are the primary source of lead in the bald eagle's diet.

STUDENT LEARNING OUTCOME

- Students will be able to define the major flows of lead in the environment, as well as how those flows are affected by humans.

ATOMS: THE BUILDING BLOCKS OF ELEMENTS

The units of matter are called **atoms**. They are the smallest units of an element that can combine with other elements in a chemical reaction. Atoms are very tiny: More than a million atoms fit into the period at the end of this sentence. We represent atoms with the same abbreviations used for the elements made up of those atoms; thus O stands for both the element oxygen and a single atom of oxygen.

A **molecule** is an assembly of two or more tightly bound atoms. The package of atoms behaves as a single distinct object. Many elements in nature are found in molecular form. For example, oxygen as it is normally found in air consists of molecules that contain two oxygen atoms. We represent this molecular form of oxygen with the chemical formula O_2; the subscript tells us that two oxygen atoms are present.

Subatomic Particles

The atom is the smallest unit having the chemical and physical properties of its element; but these tiny bits of matter are formed from even smaller parts called subatomic particles. Physicists have split the atom into hundreds of particles, but for our purposes we need to know three: **protons, neutrons**, and **electrons**. Protons have a positive charge and are assigned a mass unit of one. Neutrons have no charge and

also have about one mass unit. Electrons have a negative charge and little mass. Protons and neutrons are packed tightly into the **nucleus** at the center of the atom. Electrons orbit the nucleus at nearly the speed of light (Figure 2.4).

The number of subatomic particles defines atoms of the various elements. All atoms of a particular element have the same number of protons in their nuclei. This quantity is termed the **atomic number** and is written as a subscript to the left of the element's symbol. The abbreviation $_6C$, for example, tells us that the element carbon has six protons in its nucleus. We can deduce the number of neutrons from a second quantity, the **atomic mass**, which is the sum of the number of protons plus neutrons in the nucleus of an atom. The atomic mass is written as a superscript to the left of the symbol of the element. For example, we use this shorthand to write an atom of carbon as $^{12}_{6}C$. A $^{12}_{6}C$ atom has six neutrons in its nucleus: the difference between its atomic mass (12) and its atomic number (6). An atom of iron, $^{56}_{26}Fe$, has 26 protons, 30 neutrons, and 26 electrons.

= Proton
= Neutron
= Electron

FIGURE 2.4 Helium Atom *A simplified view of a helium (He) atom that has two protons and two neutrons in the nucleus, and two electrons outside of the nucleus.*

Atoms of an element have the same number of protons (atomic number) but can vary by mass due to differences in the number of neutrons or electrons. Elements with the same atomic number but different atomic masses are called **isotopes**. Many elements in nature normally occur as a mixture of isotopes. For example, two common isotopes of uranium (atomic number 92) are

$$^{235}_{92}\text{U} \qquad \text{and} \qquad ^{238}_{92}\text{U}$$

Uranium-235   Uranium-238

Isotopes often are referred to by their atomic masses (the atomic number is dropped because it is the same for each isotope). Thus uranium-235 or U-235 is a common way to describe these two isotopes of uranium.

The discovery of isotopes by the English chemist Frederick Soddy in 1913 led to many fundamental advances in science and medicine. Medical procedures using isotopes help with the diagnosis or treatment of one out of every three hospitalized patients in the United States—procedures with a total estimated value of $10 billion. Isotopes are used daily in more than 36,000 diagnostic imaging procedures and in close to 100 million laboratory tests each year. These isotopes also play a vital role in the treatment of growing numbers of patients with cancer and other diseases. Physical scientists use isotopes in a range of applications, including the study of the history of climate change, deep ocean circulation, the exploration for minerals, nutrient cycling in forests, and the evolution of volcanic eruptions.

CHANGES IN MATTER

Matter can undergo three types of changes: physical, chemical, and nuclear. A **physical change** in matter is one in which a substance changes its physical form and appearance but not its chemical composition—for example, cutting steel changes a bar into two smaller pieces and many small shavings. The changes are cosmetic in the sense that the chemical composition of the steel is exactly as it was before. A physical change occurs when matter changes its state, such as when water turns to ice or steam. Again the chemical composition remains the same—water is H_2O regardless of whether it is a liquid, solid, or gas. These are the three states of matter.

Chemical Reactions

In **chemical changes** (also called **chemical reactions**) a substance is transformed into a different substance by changing its chemical composition. An example is the reaction between hydrogen and oxygen to form water:

$$2H_2 + O_2 \longrightarrow 2H_2O \qquad (2.1)$$

This reaction combines hydrogen and oxygen to form water. When we write a chemical reaction, we use an arrow to indicate the conversion of the starting materials, called **reactants**, to the **products**. The number 2 in front of the hydrogen means that the reaction starts with two molecules of hydrogen. Notice that all atoms in the reactants must be accounted for in the products. Matter is conserved in a chemical reaction: Reactions cannot create or destroy matter but can only rearrange it.

An important category of chemical reactions includes those carried out to obtain energy—in other words, to convert the energy stored in the reactants to a more usable form. Such reactants, called **fuels,** are substances that can be burned to produce heat. All common fuels (wood, coal, oil, natural gas) are carbon compounds that release their energy through **combustion**: the complete oxidation of a substance through the use of air or O_2.

The combustion of a fuel such as natural gas (CH_4) produces carbon dioxide, water, and energy:

$$CH_4 + 2O_2 \longrightarrow CO_2 + 2H_2O + 210.8 \text{ kcals} \qquad (2.2)$$

A kilocalorie is a unit of energy, and one kilocalorie (a kcal) is equal to the quantity of energy that will warm one kilogram of water by 1° Celsius. All carbon-based fuels undergo a similar chemical change when they are burned to release energy. Combustion has important biological, social, and environmental effects. Most organisms burn fuels like glucose ($C_6H_{12}O_6$) to provide heat for locomotion, growth, and development. People use the heat released from combustion to power cars, heat homes, and generate electricity. The release of CO_2 in this process is important because many scientists are concerned that the accumulation of CO_2 in Earth's atmosphere is changing Earth's climate.

Nuclear Changes

In the late nineteenth century many scientists believed that they understood thoroughly all the important properties of matter. But in 1898 the French scientist Henry Becquerel noticed that when rock containing the element uranium was placed next to photographic film, the film was exposed. Scientists weren't able to explain why something apparently emitted by the rock could cause such a phenomenon. That same year another French scientist, Marie Curie, identified and named the peculiar property of **radioactivity**. Radioactivity is the process in which some atoms naturally emit particles or rays with tremendous energies. More than fifty such **radioactive isotopes** or **radioisotopes** have been identified. All elements with atomic numbers greater than 83 are radioactive (Figure 2.2).

There are three types of radiation: alpha, beta, and gamma. Alpha radiation consists of particles with two protons and two neutrons. When a radioactive atom emits an alpha particle, its atomic number decreases by two units and its mass number decreases by four units. This process is known as **nuclear decay**. For example, the decay of plutonium into uranium is described as follows:

$$^{239}_{94}\text{Pu} \longrightarrow\; ^{235}_{92}\text{U} + ^{4}_{2}\text{Alpha particle} + \text{Energy} \qquad (2.3)$$

Beta particles are electrons emitted from a radioactive atom that move at very high speeds. Gamma rays have no charge or mass; they are high-energy X-rays that travel at the speed of light.

All radioactive elements have characteristic **half-lives**, which is the time it takes for the process of radioactive decay to convert half of the atoms of one element to atoms of another element. Each isotope has a characteristic half-life. For example, plutonium-235 has a half-life of 24,000 years. If we started with 10 grams of plutonium-235, only 5 grams would remain after 24,000 years. The half-lives of radioactive atoms range from millionths of a second to billions of years (Table 2.3).

Half-lives are not affected by physical treatment (heating or cooling) or by chemical reaction. This is why **radiocarbon dating** can be used as a reliable "archaeological clock" to determine the ages of shells, bones, and fossils. It also means that radioactive material produced in society, such as wastes from a nuclear power plant or a hospital, cannot be rendered harmless by chemical reaction or other practical treatment. All we can do is wait for these materials to decay. This means that special precautions must be used to isolate them and the damaging radiation they release, often for long periods of time (see Chapter 21).

Large amounts of energy are associated with nuclear changes in matter. Two changes that produce great quantities of energy are fission and fusion. **Fission** occurs when a heavy isotope such as uranium-235 splits into lighter isotopes (Figure 2.5(a)). When fission occurs, more neutrons and much energy are released. The neutrons that are released from the fission process are free to collide with other heavy nuclei, which release more neutrons and energy. The number of fission reactions and the energy released can escalate quickly and result in a violent explosion if the process is unchecked. Reactions that multiply in this way are called chain reactions (Figure 2.5(b)). This chain reaction powers the nuclear fission bomb. In nuclear fission electric power plants, the chain reaction is slowed to control the rate of fission. The energy is captured and converted to electricity.

Fusion occurs when the nuclei of two light elements are combined to form a heavier nucleus. In the process a small amount of matter is destroyed and a huge amount of energy is released. The fusion of two hydrogen nuclei to form one helium nuclei is the nuclear change that powers our sun and other stars (Figure 2.6). Fusion releases much more energy than fission, but the fusion of two hydrogen nuclei requires temperatures in excess of 100 million °C. This makes it difficult to harness fusion power on Earth. Scientists have been trying for years to sustain a fusion reaction in the laboratory. They have been unsuccessful thus far despite the massive sums of money spent on research and development. However, some optimists believe that fusion eventually will provide cheap, abundant energy. This issue is discussed in more detail in Chapter 21.

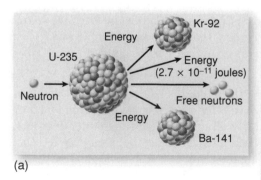

(a)

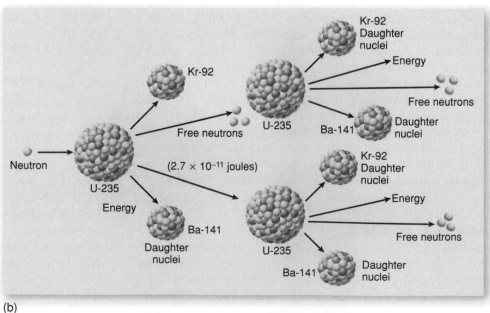

(b)

FIGURE 2.5 *Nuclear Fission* *(a) Fission of a uranium-235 nucleus by a neutron. (b) A nuclear chain reaction started by a single neutron that fissions one uranium-235 nucleus. Neutrons released from one fission event cause fission in nearby nuclei, releasing more neutrons and sustaining the chain reaction.*

TABLE 2.3	The Half-Lives of Some Radioactive Isotopes
Isotope	**Half-Life**
^{238}U	4.5×10^9 years (4.5 billion years)
^{40}P	1.3×10^9 years
^{239}Pu	24,000 years
^{14}C	5,730 years
^{137}Cs	30 years
^{230}U	20.8 days
^{222}Ac	5 seconds
^{212}Po	0.3 microseconds (0.3 one-thousandth of a second)

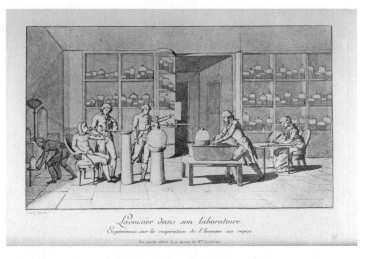

FIGURE 2.7 *Conservation of Matter* *Antoine Lavoisier conducts an experiment in this drawing made by his wife, who depicted herself at the table on the far right. The results of Lavoisier's experiment with the fermentation of wine demonstrated the law of conservation of matter. The form or quality of the matter changes in the course of the reaction, but the total quantity of matter stays the same.*

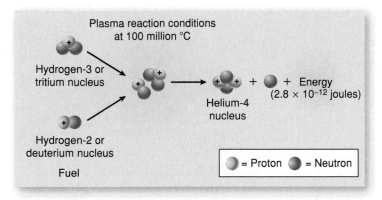

FIGURE 2.6 *Nuclear Fusion* *The nuclear fusion process involving two hydrogen atoms to force a helium atom.*

The Law of Conservation of Matter

In our discussion of chemical reactions, we stated that all atoms in the reactants must be accounted for in the products; matter is conserved in a chemical reaction. This property of matter was described in 1789 by the French chemist Antoine Lavoisier in experiments involving the fermentation of grapes into wine. Lavoisier noted that although the matter in the reactants (water, sugar, yeast) changed their chemical form, the total amount of matter did not change (Figure 2.7). The sugar, water, and yeast were changed into carbon dioxide, alcohol, and other products. But the total amount of each element at the beginning of the experiment was the same as the total amount at the end of the experiment. Lavoisier's work became the foundation for the **law of conservation of matter**: Matter is neither created nor destroyed in a physical or chemical transformation. All that changes is its form or quality. Scientists have found that the law of conservation of matter governs all physical and chemical changes of matter in the entire universe.

The law of conservation of matter has important implications for society and the environment. Goods and services do not appear out of thin air because people cannot create the materials from which they are derived. We extract elements from the environment and transform them chemically and physically into useful forms, but in doing so we do not create or destroy any matter. This means that the amount of each element in Earth's crust, oceans, or atmosphere is all we have to work with. This sets a broad but distinct limit on our ability to produce goods and services.

The law of conservation of matter also implies that all of the elements society extracts ultimately are returned to the air, water, or land after their use. This is especially important for elements that have harmful effects on the environment. See *Your Ecological Footprint: Lead and Your Listening Pleasure*. Consider the case of cadmium (Cd), a metal mined and refined for use as a pigment and stabilizer in plastics, in nickel-cadmium (Ni-Cd) batteries, and as protective plating for steel products. The chemical and physical transformations of cadmium by society release some of this metal to the air, water, and soil. Refining cadmium releases air pollution with small amounts of the metal; burning solid waste containing Ni-Cd batteries releases cadmium to the atmosphere; and landfills with paint and plastic leak cadmium to nearby streams. Cadmium that is inhaled or that makes its way into food and water can cause lung cancer and kidney damage.

Like lead, cadmium is extremely toxic even at low concentrations, so many nations have devised plans to reduce the release of cadmium to the environment. To develop these plans, we need to know how much cadmium is mined and where it goes after extraction. The conservation of matter means that we can track the flow of cadmium from the environment, through

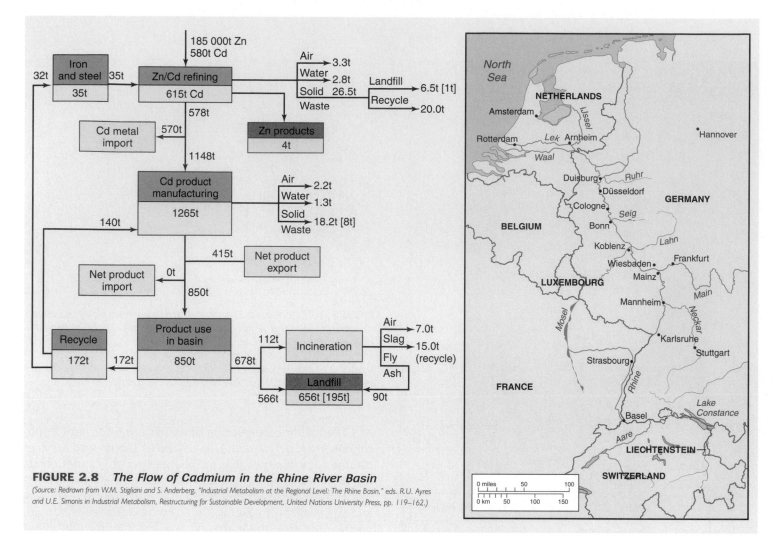

FIGURE 2.8   *The Flow of Cadmium in the Rhine River Basin*

(Source: Redrawn from W.M. Stigliani and S. Anderberg, "Industrial Metabolism at the Regional Level: The Rhine Basin," eds. R.U. Ayres and U.E. Simonis in Industrial Metabolism, Restructuring for Sustainable Development, United Nations University Press, pp. 119–162.)

society, and back to the environment. Chemists traced the movement of cadmium in the Rhine River basin in Europe (Figure 2.8). The Rhine flows north from Switzerland to the Netherlands where it empties into the North Sea, and it has tributaries in France, Germany, and Luxembourg. About 50 million people live in the river's 220,000 km² (85,000 mi²) drainage basin. The basin also is one of the most heavily industrialized regions in the world, being home to about 20 percent of the chemical industry in the Western Hemisphere.

In the early 1990s about 1,288 tons of cadmium were used in the basin each year to produce cadmium-based products. About 22 tons "leaked" to the environment from the factories manufacturing batteries, paint, plastic, and plating. Another 415 tons were exported from the basin in products shipped to other regions. Nearly half of the cadmium (566 tons) was tossed into landfills in the form of dead batteries, plastic containers, and old paint cans. About 172 tons of cadmium were recycled in the form of batteries and metal plating.

How do we know that the chemists did their homework correctly? The law of conservation of matter means that all the cadmium that is used within the basin must be accounted for. Tracing the flow of cadmium from mining to refining, manufacture, landfills, incineration, or recycling accounted for all 1,288 tons of cadmium in consumer products, landfills, the atmosphere, the soil, and the water. This analysis formed the basis for policies aimed at reducing the release of cadmium to the environment and recycling a greater fraction of cadmium use.

THE IMPORTANCE OF ENERGY

In about 340 B.C. Aristotle first used the term *energeia,* a word formed by combining two root forms meaning "at" and "work." He used the term to describe the operation or activity of anything. The modern term *energy* was derived from this concept of *energeia.* We now define **energy** as the ability to do work. Energy plays a critical role in society because it can be used to organize materials such as cadmium into useful goods, and it provides heat, light, and other useful services. In natural environments energy evaporates water, makes plants grow, and moves the large plates of Earth's crust.

Types of Energy

We live in the Boston area, and each month one of our families (Cutler's) gets three energy utility bills. One is from Boston Gas Company, the utility that sells natural gas;

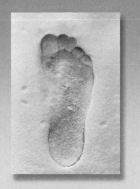

Your ECOLOGICAL *footprint*

Lead and Your Listening Pleasure

Every good or service requires different types and quantities of energy and materials to produce and hence releases different amounts of wastes to the environment. One of the most important determinants of your ecological footprint is the type of consumer goods you purchase. This point is illustrated by the use of lead. In 1996 about 1.5 billion kilograms of lead were consumed in the United States, about 5.8 kilograms per person per year. But some people use far more or far less than the general average due to the choices they make as consumers. Consider our choices in electronic devices. Many people own a Walkman(tm) or portable CD player that they listen to as they Rollerblade(tm), walk to work, or read the paper. Most such devices run on AA batteries. What initially appears as the cheapest option is to buy disposable lead-based alkaline batteries, which cost about 80 cents each. One AA alkaline battery weighs about 25 grams, of which about 40 percent, or about 10 grams, is lead. Let's assume your Walkman(tm) requires two AA batteries and that most of the time you listen to tapes. If you use your Walkman(tm) three to four times per week for a couple of hours each time, your batteries will last about 25 hours. Over a year you will buy and dispose of about 25 batteries, and in doing so release 250 grams of lead into landfills.

Another option is to buy rechargeable batteries, which initially are more expensive (about $1.25 each) and require the purchase of a recharging unit (as cheap as $10.00). But a rechargeable battery will last four to ten times longer than a disposable alkaline before it goes dead. Over the long run, rechargeable batteries end up being cheaper and less harmful to the environment than disposable batteries.

Rechargeable batteries aren't free of impacts. They are made from toxic materials, usually nickel and cadmium, and ultimately must be disposed of. But their longer life significantly reduces the release of harmful materials from their production, use, and disposal. All batteries, disposable or rechargeable, can be recycled. The benefits of rechargeable batteries can be enhanced by collecting old batteries and making a periodic visit to your local recycling center.

STUDENT LEARNING OUTCOME

- Students will be able to discuss the use of lead in regular and rechargeable batteries.

---

one is from Boston Edison, the regional utility that sells electricity; and one is from Arlington Fuel Oil Company, a local business that sells heating oil. Cutler's family also buys cylinders of liquefied propane gas (LPG). Natural gas is used to heat hot water and cook. Electricity is used to power appliances and lamps. Fuel oil is burned in a furnace to provide space heat. LPG powers the gas grill for summer barbecues.

This example illustrates that energy exists in many different forms. Yet all forms of energy can be expressed in heat equivalents—the quantity of energy they release when converted completely to heat. In this book energy is measured in kilocalories (kcals). For example, one gallon of gasoline contains about 31,254 kcals, one kilowatt-hour of electricity contains about 860 kcals, and one cubic foot of natural gas is equal to about 260 kcals. The conversion of energy to heat equivalents allows us to compare the quantities of different forms of energy.

Energy comes in many forms. **Electromagnetic radiation** or **radiant energy** is the energy carried by light. The energy generated by the sun is a form of radiant energy called solar energy. The radiant energy generated by the sun travels in the form of electromagnetic waves that differ in wavelength and energy content (Figure 2.9). Gamma rays, X-rays, ultraviolet waves, visible light waves, infrared waves, microwaves, TV waves, and radio waves are defined by their wavelengths. Some forms of radiant energy are familiar to us; others are not. The visible white light that reaches Earth can be passed through a prism to produce the full spectrum of colors from red to blue. That part of the electromagnetic spectrum is called the visible spectrum. But the warmth we feel from the sun is energy from parts of the electromagnetic spectrum that are not visible. The radiant energy that causes sunburn is from the ultraviolet part of the spectrum, which has shorter wavelengths than the visible portion. In general, the shorter the wavelength of a type of radiant energy, the higher its energy content.

The Earth intercepts about 1.3×10^{21} kcals of radiant energy each day. This amount is about 90,000 times the quantity of fossil fuel used by all human societies in 2006. In fact, the total amount of crude oil, natural gas, coal, and other fossil fuels in Earth's crust amounts to only about 11 days of sunshine! This illustrates an important aspect of solar energy: The total *quantity* is enormous, but it is spread out over a very large area. The amount that arrives at a given area of Earth's surface is very small. This presents a challenge to humans who want to use solar energy to run industrial economic systems. Elaborate and expensive collection systems often are required to collect and concentrate solar energy. We return to this important point in Chapter 22.

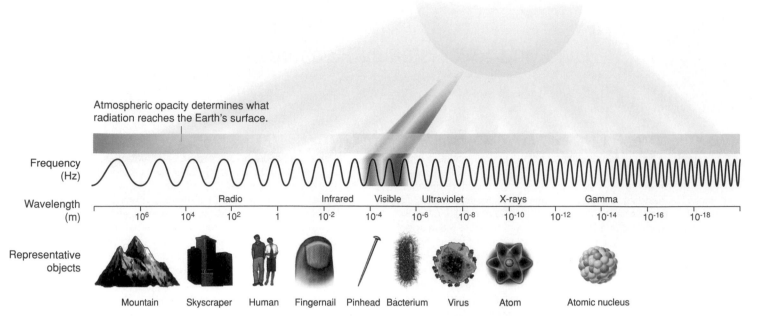

Atmospheric opacity determines what
radiation reaches the Earth's surface.

FIGURE 2.9 *The Electromagnetic spectrum* *The scale at the bottom indicates representative objects that are equivalent to the wavelength scale.*

Mechanical energy, the energy of the organized motion of matter, typically is the energy that drives the work done by machines. Mechanical energy includes potential and kinetic energy. **Kinetic energy** is the energy of motion. Examples of kinetic energy include flowing water, a car moving along a street, or a person walking up a flight of stairs. **Potential energy** is energy of position. Water that is stored behind a dam has gravitational potential energy. If the water is released from the dam, its potential energy is converted to kinetic energy in the form of flowing water.

Chemical energy is energy that is stored in the arrangement of elements, such as the energy stored in fossil fuels (like gasoline) or in carbohydrates (like simple sugars). The energy in gasoline and carbohydrates is stored in the chemical bonds between the atoms of the compounds. Combustion breaks these bonds and releases the energy. The energy stored in gasoline and carbohydrates is called chemical potential energy because it has the potential to do useful work in machines or in our bodies when released through combustion.

Nuclear energy is the energy that binds the protons and neutrons together in the nuclei of atoms. As we have seen, under certain conditions the binding energy can be released, and great amounts of heat energy are generated. This type of energy is used to generate electricity in nuclear power plants.

Electrical energy is the force of charged particles acting on one another. An electric current, for example, is caused by a flow of electric charges. In electrical wires and appliances, the flow of energy is caused by the back-and-forth flow of electrons.

Heat is the kinetic energy associated with the random motion of atoms and molecules. Temperature measures the average speed of atoms or molecules in a substance at a particular time. Heat is an important form of energy because all forms of energy can be expressed in their heat equivalent. For

example, gasoline is a form of chemical potential energy, but 1 liter (.26 gallons) of gasoline releases 31,254 kcals of heat when burned. Uranium is a form of nuclear fuel, but 1 kilogram (2.2 pounds) of uranium releases 4 billion kcals of heat when used as fuel in a nuclear power plant. Figure 2.10 gives the heat equivalents of some common forms of energy.

Energy and Work

Let's look more closely at what it means to say that energy is the "ability to do work." Scientists use a simple definition of work that is associated with basic forms of labor—a lift, a pull, or a push that moves an object. In physics work is defined as the force applied to an object times the distance over which the object is moved:

$$\text{Work} = \text{Force} \times \text{Distance} \qquad (2.4)$$

Pushing a book across a desk requires work because you must overcome friction. Work is a way of transferring energy to an object. When you push a marble across the floor, work is done and the marble gains kinetic energy. When an object is raised to a certain height, work is done and the object gains potential energy.

Work done by energy constantly shapes our physical and biological environment. One important example is how phase changes of matter—transitions between solid, liquid, and gaseous phases—often require large amounts of energy. The **heat of fusion** of water is the energy required to change a gram of water from a solid (ice) to the liquid state without changing its temperature. Water's heat of fusion is about 80 calories per gram. The **heat of**

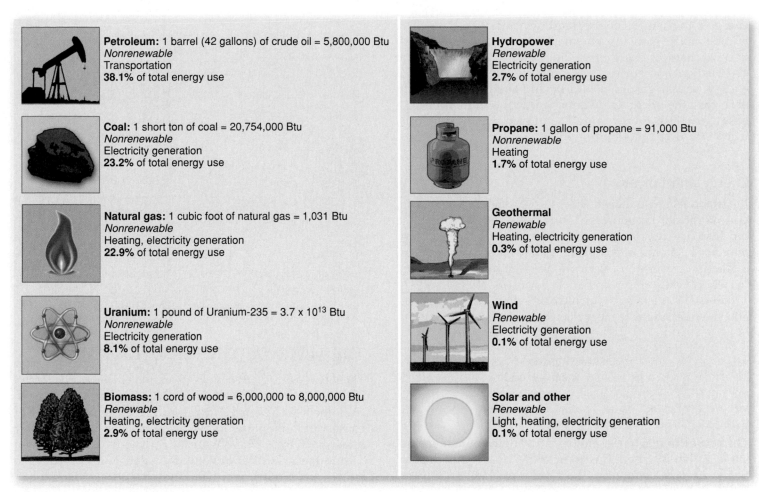

Petroleum: 1 barrel (42 gallons) of crude oil = 5,800,000 Btu
Nonrenewable
Transportation
38.1% of total energy use

Coal: 1 short ton of coal = 20,754,000 Btu
Nonrenewable
Electricity generation
23.2% of total energy use

Natural gas: 1 cubic foot of natural gas = 1,031 Btu
Nonrenewable
Heating, electricity generation
22.9% of total energy use

Uranium: 1 pound of Uranium-235 = 3.7×10^{13} Btu
Nonrenewable
Electricity generation
8.1% of total energy use

Biomass: 1 cord of wood = 6,000,000 to 8,000,000 Btu
Renewable
Heating, electricity generation
2.9% of total energy use

Hydropower
Renewable
Electricity generation
2.7% of total energy use

Propane: 1 gallon of propane = 91,000 Btu
Nonrenewable
Heating
1.7% of total energy use

Geothermal
Renewable
Heating, electricity generation
0.3% of total energy use

Wind
Renewable
Electricity generation
0.1% of total energy use

Solar and other
Renewable
Light, heating, electricity generation
0.1% of total energy use

FIGURE 2.10 *Forms of Energy* U.S. energy consumption by source, measured in heat equivalents.

vaporization of water is the energy required to change a gram of water into the gaseous state at the boiling point (100°C); this is equal to 539 calories per gram. These values are much higher than those for most other materials on Earth, which is one reason why water is unique among all substances. The high heat of vaporization of water makes it an effective coolant for the human body via evaporation of perspiration, extending the range of temperatures in which humans can exist. For the same reason water is used as a coolant in power generating facilities. As we will discuss in Chapter 4, the energy needed to drive phase changes in water explains why the oceans play a central role in determining the planet's climate.

Solar energy creates wind. Heating causes air to expand and become less dense. When this occurs cooler, heavier air presses inward and buoys the lighter, heated air, and a wind along the surface is created. This work done by solar energy drives global wind patterns because the equatorial regions are heated more intensely than higher latitudes. These mechanisms are also discussed in Chapter 4.

Society uses fossil fuels to do work in the manufacture of goods and services. Factories use oil, gas, and electric-ity to convert **raw materials** into consumer goods. We use energy in our homes to do various forms of work—running a dishwasher or a clothes dryer, for example.

To do work, energy must be converted from one form to another. An **energy converter** is a device that converts energy to work (Figure 2.11). Plants convert radiant

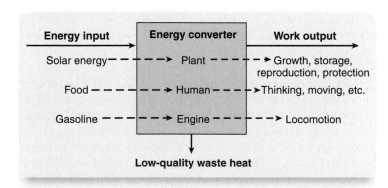

FIGURE 2.11 *Energy Conversion* An energy converter takes energy and uses it to do work. No energy converter can convert all of the energy input to work. Some of the energy input is converted to low quality waste heat.

energy into chemical energy (carbohydrates) and heat. The human body converts chemical energy (carbohydrates) into mechanical energy (walking) and heat. The internal combustion engine in a car converts chemical potential energy (motor gasoline) into kinetic energy and heat. In each case the energy converter (a plant, a human body, a car engine) use energy to do useful work (photosynthesis, locomotion, and motion).

Energy and Power

If a friend tells you that she hiked up a 5,000-foot mountain, you might not be impressed. But if she tells you that she did it in just twenty minutes, you would be amazed! You may not be aware of it, but the source of your amazement is the power output of your friend—she did a lot of work in a short time.

Power is the rate at which work is done, or, more generally, the rate at which energy is used:

$$\text{Power} = \frac{\begin{array}{c}\text{Quantity of work done}\\\text{(or quantity of energy used)}\end{array}}{\begin{array}{c}\text{Time required to do the work}\\\text{(or convert the energy)}\end{array}} \quad (2.5)$$

The power of machines, motors, and engines often is measured in horsepower. This term comes from preindustrial times when horses were the dominant source of work. Studies of work done by horses showed that the average horse can do work at a rate of 11 kcals per minute for sustained periods. This became the definition of 1 horsepower. In comparison, the average power of a new car sold in the United States in 2005 was 170 horsepower.

We often are impressed by the power of machines. The rate at which a fast train or a big earth-moving machine does work is impressive. These and other types of machines shape human society and our everyday lives. But the power of machines pales in comparison to some natural phenomena (Table 2.4). Volcanic eruptions, lightning, thunderstorms, and earthquakes are important because of their high power—they deliver enormous amounts of energy in a very short period. The power of these events enables them to shape the landscape of our planet and the living systems that it supports.

Fast and furious processes are not the only important work processes in nature. Many natural processes operate at a slow rate and therefore at low power. For example, a basic energy conversion in the environment is photosynthesis by plants—a process that converts solar radiation to chemical potential energy. Photosynthesis has a very small power output relative to a volcanic eruption or a speeding train, but it is the basis for nearly all life on Earth.

TABLE 2.4	Power of Various Events in Nature and Society		
Event		**Power**	**(Watts)**
Flight of hummingbird		10^{-1}	(.1)
CD player spinning U2's latest song		10^1	(10)
Running a 100-meter dash		10^3	(1000)
Intercity truck trip		10^5	(100,000)
Avalanche with 500-meter drop		10^7	(10,000,000)
Tornado		10^9	(1,000,000,000)
Lightning		10^{13}	(10,000,000,000,000)
Richter magnitude 8 earthquake		10^{15}	(1,000,000,000,000,000)

Source: Data from V. Smil, *General Energetics: Energy in the Biosphere and Civilization*, Wiley.

THE LAWS OF THERMODYNAMICS

The laws of thermodynamics (from the Greek words *therme* meaning "heat" and *dynamis* meaning "power") allow us to track the use of energy in work processes just as the law of conservation of matter allows us to track materials as they undergo change. The science of thermodynamics evolved during the Industrial Revolution, when humans began to burn fuels in powerful engines to produce goods and services. Thermodynamics can be used to understand every aspect of energy use in our daily lives, from the way we digest food to the way that our cars use gasoline.

The First and Second Laws of Thermodynamics

In 1865 the German physicist Rudolph Clausius investigated energy use by steam engines. The steam engine was a key driving force behind the Industrial Revolution because it was the first practical, affordable, and reliable machine that could convert the chemical energy in wood and fossil fuels to mechanical energy. A steam engine burns a fuel that contains chemical potential energy, such as wood or coal, in a boiler to convert water into steam. The engine then converts the steam to mechanical work, such as pumping water or propelling a train locomotive. In addition to mechanical work, the engine also generates waste heat—energy released from the boiler and other engine parts in the form of heat. The term "waste" refers to the fact that some of the chemical potential energy in the fuel is not converted to useful work, but instead is released to the general surroundings of the engine.

Clausius's experiments demonstrated two important laws of energy use. The **first law of thermodynamics** states that there is no increase or decrease in the quantity of energy in any energy conversion. The total energy input to an energy converter and the total energy output always are equal. In the case of the steam engine, the quantity of chemical potential energy in the fuel equals the sum of the mechanical energy and waste heat produced by the engine. This law applies to all energy conversions. For example, the amount of chemical potential energy in the food energy you consume is equal to the sum of the work done by your body (walking, thinking, and so on) plus the heat generated. Similarly, the amount of chemical potential energy in gasoline equals the sum of the work done by a car engine and its gadgets (such as the air conditioner and CD player) plus the waste heat generated.

The first law of thermodynamics means that the *quantity* of energy remains constant in every conversion process. But Clausius's experiments with steam engines revealed another fundamental law of energy conversion: The *quality* of energy also changes. The **second law of thermodynamics** states that *in all energy conversion processes energy loses its ability to do work and is degraded in quality.* The energy that the steam engine converts from chemical potential energy to waste heat has lost its ability to do useful work; it is degraded in quality.

Every energy conversion process can be described by its **efficiency**—the amount of useful energy or work output compared to the total energy input:

$$\text{Efficiency} = \frac{\text{kcals of work out}}{\text{kcals of total energy converted}} \quad (2.6)$$

Efficiency usually is expressed as a percentage. The second law of thermodynamics implies that no energy conversion process is 100 percent efficient because some useful energy always is lost as waste heat. The early steam engines Clausius studied were less than 1 percent efficient; subsequent improvements raised their efficiency to more than 20 percent by the twentieth century. A modern car engine converts 15 to 20 percent of its gasoline fuel to useful work (moving down the street). The remaining 75 percent is converted to waste heat. Humans convert food energy into useful work at about 20 percent efficiency, and convert the rest to heat. In fact, a human at rest gives off about the same amount of heat as a 100-watt lightbulb! The efficiencies of other important energy converters are shown in Table 2.5.

Entropy as a Measure of the Quality of Energy and Matter

After repeated experiments, Clausius observed that quantitative and qualitative changes in the form of the fuel could be described in terms of its degree of order or organization. The highly organized, useful chemical potential energy contained in wood or coal always was converted to a more random, disorganized state—waste heat (Figure 2.12). Clausius concluded that the actions of a steam engine could be described as the conversion of energy from an organized state to a disorganized state.

The principle that Clausius identified is called **entropy**. Entropy is the degree of order or organization in a system. Clausius's work applied the concept to energy, but since then scientists have found that the same principle also applies to all changes in materials. Matter and energy that are highly organized or highly ordered have low entropy. **Matter and energy that are highly disorganized or random have high entropy.**

Everyone has a sense of order and disorder. Books neatly arranged by subject on a shelf are orderly; books scattered through a room are disorderly. The cards in a new, unopened

TABLE 2.5	The Efficiencies of Different Energy Converters	
Type of Energy	**Energy Conversion**	**Efficiency Converter**
Plants	Electrical ⟶ Chemical	0.3–1%
Incandescent lamp	Electrical ⟶ Light	2–5%
Solar cell	Light ⟶ Electrical	2–30%
Steam locomotives	Chemical ⟶ Thermal ⟶ Mechanical	3–6%
Fluorescent lamp	Electrical ⟶ Light	10–12%
Automobile engine	Chemical ⟶ Thermal ⟶ Mechanical	15–20%
Nuclear power plant	Nuclear ⟶ Thermal ⟶ Mechanical ⟶ Electrical	30–35%
Home gas furnace	Chemical ⟶ Thermal	70–95%
Discharging a battery	Chemical ⟶ Electrical	72%
Electrical motor	Electrical ⟶ Mechanical	50–90%
Electrical generator	Chemical ⟶ Electrical	95–99%

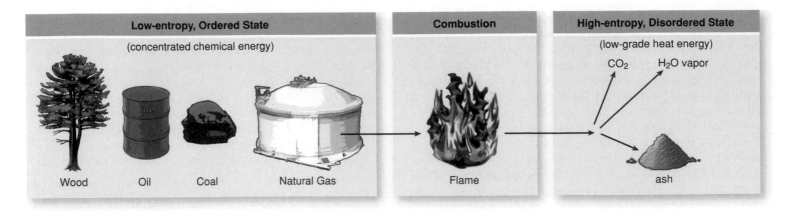

FIGURE 2.12 *The Entropy Law Applied to Energy Use* The combustion of energy converts the energy from a high-quality (low-entropy) state to a low-quality (high entropy) state. *(Source: Redrawn from R.S. Rouse and R.O. Smith,* Energy: Resource, Slave, Pollutant, *Macmillan, p. 277.)*

deck are ordered by number and suit; throwing the deck into the air produced a disorderly pile on the floor. A house is orderly; the lumber, nails, shingles, and windows piled in the work lot before construction are relatively disorderly.

The second law of thermodynamics also implies that orderly structures, patterns, and arrangements of energy and materials tend to drift toward disorder *by themselves.* This movement toward a greater state of entropy occurs without outside interference. Thus the tendency for energy and materials to move from an ordered, low-entropy state to a disordered, high-entropy state is a **spontaneous process.** An automobile engine spontaneously converts the low-entropy chemical potential energy of gasoline to forward motion and waste heat. Similarly, the low-entropy chemical potential energy of wood is spontaneously converted to heat in your fireplace.

The **entropy law** and the notion of spontaneous changes also apply to materials. If you place a single drop of ink in a glass of water, the ink will disperse spontaneously throughout the water (Figure 2.13). Eventually the ink will be evenly mixed and the water will take on a dark hue. If you leave a bicycle outside, it spontaneously crumbles into a pile of rust as the molecules of iron begin to flake off and fall to the ground or are dispersed by the wind. Junkyards and landfills prove the spontaneous tendency toward disorder described by the second law of thermodynamics. They are full of materials that have broken down or worn out due to the forces of entropy.

Disordered, high-entropy energy and materials do not organize *back* to a low-entropy, ordered state by themselves; human coaxing or some other outside intervention is required to achieve this. The movement toward a greater state of organization is called a **nonspontaneous process.** We know that the heat generated by the fireplace will not spontaneously reorganize itself into wood. The atoms of

ink will not reassemble themselves into a single drop, nor will a pile of rust reassemble itself back into a new bicycle. However, you could vacuum up the rust particles and transport them back to a factory where they could be melted, refined, and ultimately fashioned into a new bicycle. That process is nonspontaneous because it would not occur without a significant investment of time and energy.

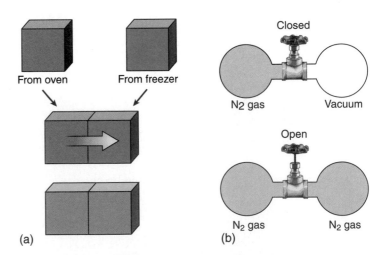

FIGURE 2.13 *Spontaneous Processes* (a) Suppose you have two cubes of metal, one in a red-hot oven, the other in a freezer. If you put the two blocks together, there is a spontaneous process that involves the flow of heat energy from the hot block to the cold block. (b) Suppose you have two glass containers connected by a valve. The left container holds nitrogen gas, while the right container has nothing (a vacuum). If you open the valve connecting the two containers, there is a spontaneous flow of nitrogen gas molecules from the left chamber to the empty right chamber.

THE ENERGY AND MATERIALS BALANCE OF A COAL-FIRED ELECTRICITY PLANT

The law of conservation of matter, the laws of thermodynamics, and the entropy law may seem far removed from the issue of sustainability; however, they provide a foundation for the study of all environmental problems. Pollution, resource depletion, and waste assimilation involve the flow of energy and materials in and between the environment and society. The depletion of natural resources and the release of pollution are caused by the extraction and processing of energy and materials to produce goods and services. These processes can be described by the physical laws discussed in this chapter because these laws apply to all transformations of energy and materials.

One method used to do this is called energy and materials balance. The "balance" is defined by the conservation of energy and matter—the quantity of energy and materials that enter a process must be equal to (be balanced by) an equivalent amount that leaves the process. The cadmium example demonstrates balance: The quantity of cadmium used in the Rhine basin to produce various goods is equal to the quantity of cadmium recycled, exported from the region, sent to landfills, or released to the air and water.

We know that energy and materials often are used together, so we can extend the balance concept to energy as well. The following example applies an energy and materials balance to a power plant that burns coal to generate electricity (Figure 2.14). The coal required to generate electricity contains various materials as well as chemical potential energy:

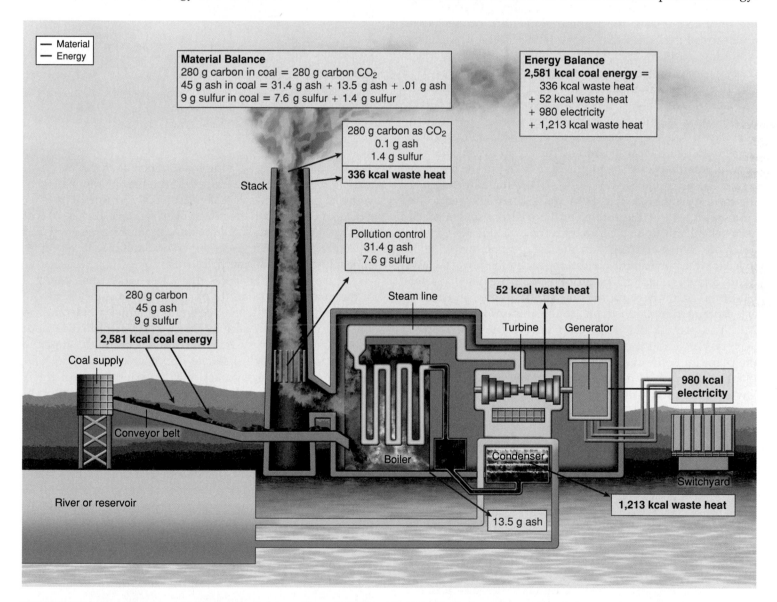

— Material
— Energy

Material Balance
280 g carbon in coal = 280 g carbon CO_2
45 g ash in coal = 31.4 g ash + 13.5 g ash + .01 g ash
9 g sulfur in coal = 7.6 g sulfur + 1.4 g sulfur

Energy Balance
2,581 kcal coal energy =
   336 kcal waste heat
 + 52 kcal waste heat
 + 980 electricity
 + 1,213 kcal waste heat

280 g carbon as CO_2
0.1 g ash
1.4 g sulfur
336 kcal waste heat

Stack

Pollution control
31.4 g ash
7.6 g sulfur

Steam line

52 kcal waste heat

Turbine   Generator

280 g carbon
45 g ash
9 g sulfur
2,581 kcal coal energy

Coal supply

Conveyor belt

Boiler

Condenser

980 kcal electricity

Switchyard

River or reservoir

13.5 g ash

1,213 kcal waste heat

FIGURE 2.14 *The Energy and Materials Balance for a Coal-Fired Power Plant* *The energy matter and energy conversions in this figure illustrate the conservation of matter and the laws of thermodynamics.* (Source: Redrawn from G.M. Masters, Introduction to Environmental Engineering and Science, Prentice-Hall.)

280 grams of carbon, 9 grams of sulfur, 45 grams of ash (various minerals that don't burn), and 2,581 kcals of chemical potential energy.

What happens to these materials and energy in the power plant? Let's track the energy first. The chemical potential energy in coal is converted three times, and each conversion is less than 100 percent efficient due to the second law of thermodynamics. First the chemical potential energy in coal is converted to heat in a boiler that converts water into steam. This conversion is 87 percent efficient: 13 percent of the energy released in combustion is vented to the environment through the smokestack. In the second energy conversion, the steam moves against the blades of a turbine and forces them to spin (like blowing on a pinwheel). The heat energy in the steam is converted to mechanical energy (spinning blades) with an efficiency of about 46 percent. In the final conversion, the turbine blades are connected to an electrical generator that converts the mechanical energy (spinning motion) to 1 kilowatt-hour of electricity. This energy conversion is about 95 percent efficient.

The materials in the coal also go through a series of transformations. When the coal is burned, some of the ash falls to the bottom of the furnace, similar to the ashes left in a fireplace after wood has been burned. The rest of the ash (so-called fly ash) and all the carbon and sulfur go up the smokestack. Special pollution control equipment installed in the stack can remove some of the ash and sulfur. The remaining sulfur (in the form of sulfur dioxide, SO_2) and ash are released to the atmosphere. All of the carbon in the coal is released to the atmosphere, mostly in the form of carbon dioxide (CO_2).

Note that the changes in the energy and materials in this example obey all the conditions set by the physical laws defined in this chapter. The process obeys the law of conservation of matter because the amount of carbon, sulfur, and ash entering the plant is equal to the amount leaving (Table 2.6). The process obeys the first law of thermodynamics because the amount of chemical potential energy entering the plant in the form of coal equals the amount leaving the plant in the form of electricity and waste heat. The second law of thermodynamics is met because every energy conversion is less than 100 percent efficient. The overall efficiency is 38 percent, meaning that 62 percent of the original energy in the coal is not converted to electricity. Finally, the process is consistent with the forces of entropy. High-quality chemical potential energy is converted to heat.

This example of energy and materials balance demonstrates a fundamental principle of environmental science: *Everything must go somewhere.* Literally we can never throw something

TABLE 2.6	Summary of Material and Energy Balance for a Coal-Fired Power Plant
Energy input to plant:	2,581 kcals of coal
Energy output from plant:	336 kcals of waste heat out the smokestack + 52 kcals of waste heat from turbine + 1,213 kcals of waste heat from condenser
Total energy output:	+ 980 kcals of electricity 2,581 kcals
Material input to plant:	280 g carbon 45 g ash 9 g sulfur
Material output from plant:	280 g carbon out the smokestack 13.5 g ash from furnace bottom + 31.4 g fly ash from particulate remover + 0.1 g fly ash out the smokestack 45.0 g ash 7.6 g sulfur from particulate remover + 1.4 g sulfur out the smokestack 9.0 g sulfur

Source: Data from G.M. Masters, *Introduction to Environmental Engineering and Science*, Prentice Hall.

"away" due to the laws of the conservation of energy and matter. Natural resources are extracted from the environment, converted to goods and services, and eventually returned as wastes, junk, and pollution. In the power plant example, the use of coal depletes coal resources; the bottom ash must be removed and hauled to a landfill; and the carbon and sulfur released to the atmosphere are serious air pollutants.

These simple principles that allow us to track changes in the quality and quantity of energy and materials are the basis for how we will analyze environmental problems. The strength of this approach is that it can be applied to both natural and human systems. Chapters 4 through 8 use the energy and materials balance approach to describe how organisms function, interact with one another, and interact with their physical environments to form the natural systems that support human life. Chapters 9 through 11 use the energy and materials balance approach to explore how society uses the environment to meet human needs, and in doing so, how society depletes natural resources and degrades the environment. By using this approach, we will demonstrate how studying the flow of energy and materials between the environment and society can be used to distinguish sustainable behaviors from those that are not.

SUMMARY OF KEY CONCEPTS

- All forms of matter are made of a few basic building blocks called elements—substances that cannot be broken down to other substances by ordinary chemical means. Elements combine in specific proportions to form compounds.

- Matter can undergo three types of changes: (1) physical, when a substance changes its physical form and appearance but not its chemical composition; (2) chemical, when a substance is transformed into a different substance by rearranging its chemical composition; and (3) nuclear, the process in which some atoms naturally emit particles or rays with tremendous energies.

- The law of conservation of matter states that matter is neither created nor destroyed in a physical or chemical transformation. This is a fundamental principle of environmental science because it means that all the materials that society extracts from the environment ultimately are returned to the environment as wastes.

- Energy is the ability to do work, whereas power is the rate at which energy is used per unit time. Energy converters are devices that convert energy into useful work.

- The first law of thermodynamics states that energy is conserved; the second law of thermodynamics states that no energy conversion process is 100 percent efficient.

- Entropy is the degree of order or organization in a system. Orderly structures, patterns, and arrangements of energy and materials tend to drift toward disorder *by themselves*. This tendency for energy and materials to move from an ordered, low-entropy state to a disordered, high-entropy state is a spontaneous process.

REVIEW QUESTIONS

1. On your next trip to the supermarket or convenience store, choose one of the food items you purchase and look at the nutrition label. What essential trace nutrients does it contain?

2. How much do you pay for the energy used to power your computer? Check the nameplate on the back of the machine. Usually the energy rating is in amps and volts, not watts. Use this equation to convert to watts: (Amps × Volts) = Watts. Calculate your cost per hour of operation like this: Cost per hour = Watts ÷ 1,000 = Kilowatts × approximately $0.09 per kilowatt-hour. Multiply this number by the number of hours per week or month that you use the computer.

3. Distinguish between energy and power. What is the most powerful energy conversion device you used today?

4. Go to www.ndted.org/EducationResources/CommunityCollege/Radiography/Physics/carbondating.htm. Explain how carbon-14 dating is used to calculate the age of fossils. What other uses does radiocarbon dating have?

KEY TERMS

atomic mass	fission	nutrient
atomic number	fuel	physical change
atoms	fusion	potential energy
chemical change	half-lives	power
chemical energy	heat	products
chemical reaction	heat of fusion	protons
combustion	heat of vaporization	radiant energy
compound	isotopes	radioactive isotopes
efficiency	kinetic energy	radioactivity
electrical energy	law of conservation of matter	radiocarbon dating
electromagnetic radiation	macronutrients	radioisotopes
electrons	mechanical energy	raw materials
elements	molecules	reactants
energy	neutrons	second law of thermodynamics
energy converter	nonspontaneous process	spontaneous process
entropy	nuclear decay	trace elements
entropy law	nuclear energy	
first law of thermodynamics	nucleus	

3 SYSTEMS

Why Are Environmental Problems So Difficult to Solve?

Treeless Easter Island *Although Easter island was once covered by forests, it now has few trees. And the trees that it does have were brought to the island after the arrival of European explorers in the late 1770s.*

STUDENT LEARNING OUTCOMES

After reading this chapter, students will be able to

- Explain how systems are able to generate predictable behaviors and display homeostasis.

- Explain how natural selection can generate traits that allow organisms to thrive in particular environments.

- Compare and contrast linear and nonlinear growth and their effects on society's ability to manage the environment.

- Identify the characteristics of systems that make it difficult to solve environmental challenges, and discuss why potential solutions sometimes make the problem worse.

- Explain how the scientific method can be used to evaluate competing explanations for observable phenomena.

A wide-angle photo of a treeless Easter Island reminds us of a story that we read to our children, *The Lorax* by Dr. Seuss (Theodor Seuss Geisel). The book tells the story of the Once-ler, an ambitious character who develops a way to convert truffula trees into thneeds, which are goods that can satisfy many sorts of human needs and wants. To expand his business, the Once-ler sets up factories and increases the rate at which he cuts truffula trees. The environmental impacts of these actions are witnessed by the Lorax, who warns the Once-ler about their effects. Ignoring the Lorax's concerns, the Once-ler continues to cut truffula trees. About two-thirds of the way through the story, Dr Seuss writes,

> And at that very moment, we heard a loud whack!
> From outside in the fields came a sickening smack
> of an axe on a tree. Then we heard the tree fall.
> *The very last Truffula Tree of them all!*
> No more trees. No more Thneeds. No more work to be done.

That same sickening smack must have reverberated across Easter Island. By the time Jacob Roggeveen arrived on Easter Island in 1722, no trees or bushes higher than 3 meters (9.8 feet) were present. This barren landscape contrasts starkly with the forests that awaited the first human inhabitants. Analyses of charcoal burned by the islanders and pollen in mud cores dug from the island's swamps show that up to twenty-one species of trees populated forests that covered a significant portion of Easter Island. These same methods indicate that the forest was largely eliminated between the 1400s and 1600s.

How could the Easter Islanders have cut down the last of their trees? The islanders knew that trees were critical to their way of life. Why did the Easter Island civilization fail to recognize the ecological changes that threatened its sustainability, and why did its members fail to change their lifestyle to avert the collapse?

One cause for these failures is associated with a simple truth—systems are difficult to understand and manage. Although the links between Easter Islanders and their environment seem simple relative to industrial societies, the relationships were complex. The environmental system on Easter Island was more fragile than those on the islands their ancestors inhabited. The relatively cold, dry climate slowed the rate at which trees regrew. This effect was exacerbated by the lack of nutrients in the soil. There were no sharp rocks just below the surface to warn the islanders that their soil was eroding. Variations in rainfall from one year to the next made it hard to tell whether several years of poor harvests represented a run of bad luck or an ongoing decline in agricultural productivity. There was a long lag time between the cause of a problem (deforestation) and its ultimate effect (decreased agricultural productivity). This lag made it hard for the Easter Islanders to judge whether their actions made the situation better or worse. Without a Lorax to lecture them about their unsustainable ways, the Easter Island civilization collapsed. ■

WHAT ARE SYSTEMS?

A **system** is a collection of parts, which are known as storages and flows, that interact with each other to generate regular or predictable patterns or behaviors. A **storage** is a system part where energy or materials stay for an extended period. Clouds are a storage of water in the climate system. Fats are an energy storage in living organisms. **Flows** are movements of energy or materials between storages. Rain is a flow of water from clouds to the soil. Predation, the act of one animal eating another, is a flow of energy from prey to predator.

Predictable Behavior

Systems have flows and storages that change over time and space in an orderly fashion. Order is a relative concept that can be understood by defining its opposite—disorder. Disorder is synonymous with randomness, the lack of a regular or meaningful pattern. A random sequence has no information and cannot be used to predict the future. For example, the outcome of a coin toss is random: There is a 50 percent chance that the coin will land heads up and a 50 percent chance that the coin will land tails up. If you flip a coin many times and record the sequence of heads and tails, the sequence will show no regular pattern (Figure 3.1). Without a pattern, it is not possible to forecast whether the next flip will result in a head or tail. Should a pattern appear, you would suspect that the coin is "fixed."

Systems exhibit nonrandom or regular patterns of behavior. For example, the climate system generates a regular pattern of temperature and precipitation over space and time. For reasons described in Chapter 4, large amounts of precipitation fall near the equator and small amounts of precipitation fall at locations 30° north and 30° south of the equator. A regular pattern of past behavior allows you to make an "educated guess" about future outcomes. An "educated guess" means your prediction is more likely to be correct than a random guess. Continuing with the previous example, understanding this behavior allows you to make

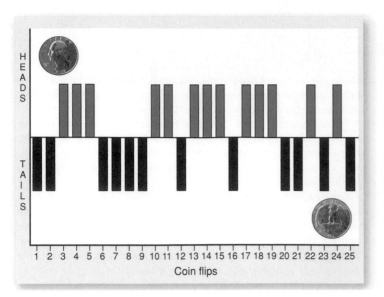

FIGURE 3.1 *Random Pattern The sequence of heads and tails from a coin toss shows no pattern. Regardless of how much you know about previous tosses, the odds on the next toss are 50 percent heads and 50 percent tails.*

an "educated guess" that Puerto Ayacucho (5° south of the equator) receives more precipitation annually than Alice Springs (25° south of the equator).

Do Systems Have Goals?

The defining characteristic of a system is its regularity and predictability. By itself, this predictability or regularity cannot be judged as good or bad—it just is. The pattern of temperature and rainfall generated by the climate system is not good or bad. There have been many patterns of temperature and precipitation during Earth's history. Some climates were cooler and wetter than today, whereas others were hotter and drier.

Other regular, predictable behaviors can be judged by objective criteria. Charles Darwin was the first to describe an objective way of evaluating behaviors of biological systems. According to Darwin's brilliant insight, biological behaviors and traits can be judged by the number of offspring that an individual leaves in the next generation, which is known as **fitness.** A behavior increases fitness if it improves an individual's chance of leaving offspring in the next generation. A behavior lowers fitness if it reduces an individual's chance of leaving offspring in the next generation. For example, some animals show parenting behavior, whereas others do not. Parenting can increase fitness by protecting offspring from potential predators and teaching offspring how to get food. But parenting may reduce fitness by increasing an individual's exposure to predators or by taking time away from finding another mate. These trade-offs can be evaluated by answering the following question: Does parenting increase the number of offspring

in the next generation left by an individual who parents relative to the number of offspring left by an individual who does not parent?

Throughout this book we will evaluate behaviors by human systems relative to their sustainability. Environmental impacts are often determined by economic systems. The purpose of an **economic system** is to produce and distribute goods and services that people associate with material well-being. Economic behaviors are judged by their efficiency, which refers to getting the most out of the resources used. Economic systems are said to be efficient when nobody can be made better off without making somebody else worse off. But economic efficiency should not be equated with another principle of sustainability, fairness. Economic changes often produce losers as well as winners, but efficiency does not indicate whether such changes are economically good or bad.

HOW DO SYSTEMS GENERATE THEIR BEHAVIORS?

A casual glance around the environment finds a world filled with order and predictability. Do order and predictability violate the laws of thermodynamics and the principle of entropy? The answer is no. Storages and flows operate in a manner that is consistent with the laws of thermodynamics.

Using Energy to Generate Order

Why do storages and flows change in an orderly fashion over time and space? Ironically, the answer is based on the second law of thermodynamics. As described in Chapter 2, the second law states that energy and materials flow spontaneously from low entropy to high entropy. Such differences in entropy are known as a **gradient.** For example, differences in the concentration of sodium ions constitute a gradient that causes a spontaneous flow from areas of high concentration to areas of low concentration (Figure 3.2).

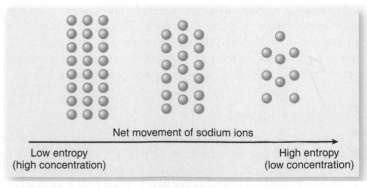

Net movement of sodium ions

Low entropy
(high concentration)

High entropy
(low concentration)

FIGURE 3.2 *A Gradient of Sodium Ions There is a high concentration of sodium ions (Na^+) on the left and a low concentration on the right. This gradient in concentrations causes sodium ions to flow spontaneously from areas of high concentration (low entropy) to areas of low concentration (high entropy).*

Gradients do not appear spontaneously. Energy is required to create a gradient. We illustrate this concept by explaining how nerve cells send signals to and from the brain (Figure 3.3). The nerve cell uses energy to power "sodium pumps" that move sodium ions from an area of low concentration (inside the cell) to an area of higher concentration (outside the cell).

Systems also contain spontaneous flows. Because these movements are consistent with the laws of thermodynamics, no energy is required. When a nerve cell "fires," it opens channels that allow the sodium ions to flow spontaneously from an area of high concentration to an area of low concentration. The spontaneous flow allows the nervous system to transmit information to and from the brain very quickly.

The patterns generated by spontaneous and nonspontaneous flows of materials and energy ultimately depend on new supplies of energy to the system. Because energy cannot be recycled (see Chapter 2), all systems need a source of new energy to do work. Without this input, work will stop and the orderly pattern will cease.

Homeostasis: Maintaining Order in the Face of Disturbance

The world in which many systems operate changes constantly. Temperature increases or decreases; the supply of natural resources grows or shrinks. Each of these changes can disturb a system by altering a storage, flow, or both. Understanding how a system reacts to a disturbance is critical to understanding how it operates.

Some systems can maintain their behavior when disturbed, an ability that is termed **homeostasis.** Homeostasis is measured by the system's ability to maintain a storage or flow, which is termed the **set point.** The set point for human body temperature is about 37°C (98.6°F). This is said to be your "normal" temperature.

There are many ways to characterize a system's homeostatic ability. **Stability** refers to a system's ability to return to a set point. An unstable system cannot return to its set point after a disturbance. An unstable system can be envisioned as a ball on a flat plane (Figure 3.4a). The initial location for the ball is the set point. The slightest disturbance causes the ball to roll away from the set point. Because the surface is flat, the ball does not return to its set point.

Homeostasis creates stability, which allows a system to return to its set point. For a stable system, the set point can be envisioned as a ball at the bottom of a depression (Figure 3.4b). A disturbance causes the ball to move away from the bottom, but the sides of the depression move the ball back toward the set point. Human body temperature is part of a stable system. When your body temperature drops below 37°C (98.6°F), shivering generates additional heat. When your body temperature rises above 37°C, sweating dissipates excess heat via evaporation.

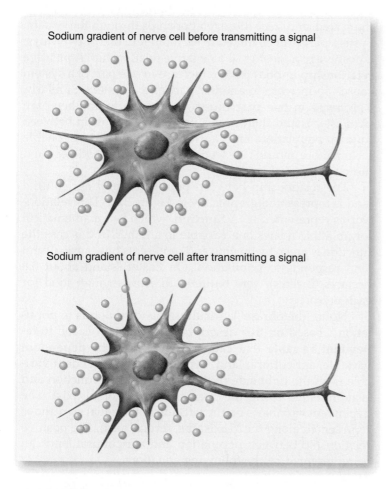

FIGURE 3.3 *Nerve Cells* Nerve cells create a gradient by using energy to pump sodium ions out of the cell. When the nerve transmits a signal, channels in the cell membrane open and allow the ions to flow spontaneously back into the cell. This flow eliminates the gradient.

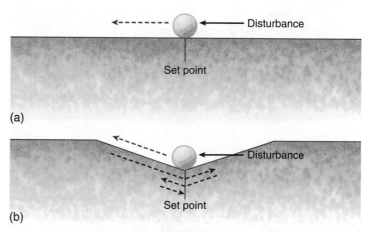

FIGURE 3.4 *(a) An Unstable System* The ball does not return to its set point following a disturbance that moves the ball to the left or the right. *(b) A Stable System* The ball will return to its set point after a disturbance that moves the ball to the left or the right. The dotted blue line represents the initial movement due to the disturbance. The red dotted line represents movements back to the set point due to homeostasis.

The ability of a system to withstand a disturbance is termed **resistance.** A resistant system shows a relatively small change when "hit" by a relatively large disturbance. After the ball is "hit," the steep depression in Figure 3.5a prevents the ball from rolling far from its set point. On the other hand, the shallow sides in Figure 3.5b imply that even a small disturbance will move the ball far from its set point.

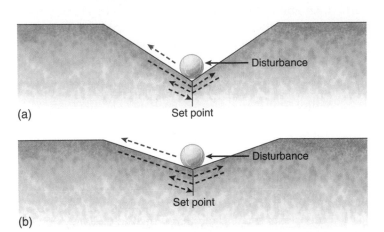

FIGURE 3.5 *(a) A Resistant System* The steep sides imply that the ball will move only a short distance from its set point following a disturbance, which is given by the size of the arrow. *(b) A Less Resistant System* The shallow sides imply that the same disturbance will move the ball much farther from its set point. The dotted blue line represents the initial movement due to the disturbance. The red dotted line represents movements back to the set point due to homeostasis.

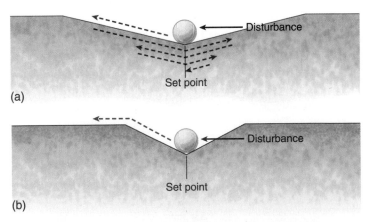

FIGURE 3.6 *(a) A Resilient System* The wide bowl returns the ball to its set point even after the ball has moved far from its set point. *(b) A Less Resilient System* The narrow bowl returns the ball to its set point only if the disturbance moves the ball a short distance from its set point. If the disturbance moves the ball far from its set point, the ball will jump out of the bowl. The flat surface will not move the ball back to its set point. The dotted blue line represents the initial movement due to the disturbance. The red dotted line represents movements back to the set point due to homeostasis.

Such a system is said to have little resistance. Your body is fairly resistant with regard to its internal temperature. Even naked, you can maintain an internal temperature of 37°C in air that ranges from 16°C to 49°C (61°F–120°F).

The ability of a system to return to its set point is described by **resilience.** A resilient system can return to its set point even when moved far from its set point. The system represented in Figure 3.6a is resilient because the ball will return to its set point even when moved far from that set point. Conversely, the system represented in Figure 3.6b is not resilient because the ball will not return to its set point once it moves a short distance from its set point. Your body is not very resilient with regard to internal temperature. If your internal temperature drops below 31°C (87.8°F) or rises above 41°C (105.8°F), your temperature probably will not return to 37°C (98.6°C). Instead you probably will die from either hypothermia or hyperthermia.

Generating Homeostatic Behavior

To understand how some systems can generate homeostatic behavior, we must classify the relationship between parts of the system and trace these relationships through the system. System parts can be linked to each other in one of two ways: a positive relationship or a negative relationship. A **positive relationship** implies that an increase in one part of a system causes an increase in another part of the system. Similarly, a decrease in one part causes a decrease in another part. On Easter Island there was a positive relationship between human population and agricultural land. More people meant more mouths to feed. Growing this food required more agricultural land.

The relationship between population and agricultural land is represented graphically in Figure 3.7a. This relationship is represented by a **function,** which is a mathematical formula that relates one variable to another. In this case the function is given by the line and represents how agricultural land responds to population. On Easter Island about 0.8 hectares (2 acres) were required to grow enough food for each person.

Notice the phrase "agricultural land responds to population." Based on this description, population is the **independent variable** that causes a change in the **dependent variable,** agricultural land. Typically the independent variable is on the right side of the equal sign in a function and is represented along the X-axis (the horizontal axis). The dependent variable is on the left side of the equal sign and is represented along the Y-axis (the vertical axis). The positive relationship between population and agricultural land can be represented by the following equation:

(Coefficient)

Agricultural Land = 0.8 × Population   (3.1)

(Dependent variable)   (Independent variable)

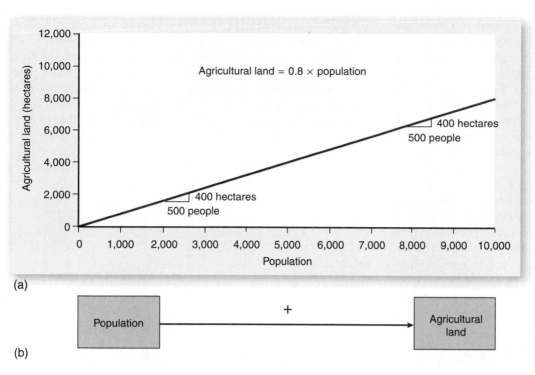

(a)

(b)

FIGURE 3.7 *(a) Positive Relationship The positive relationship between population and agricultural land on Easter Island. The functional relationship between the independent variable (population) and agricultural land (dependent variable) is given by Equation 3.1. The relationship is linear: An increase in population of 500 people increases agricultural area by 400 hectares.* *(b) Systems Diagram The positive relationship between human population and agricultural land as represented in a systems diagram.*

The other type of relationship, a negative relationship, implies that a change in one part of a system causes the part to which it is linked to move in the opposite direction. If two parts of a system have a negative relationship, an increase in one part of the system will cause a decrease in the other part, and vice versa. On Easter Island there was a negative relationship between the human population and forests. As the population grew, the forests shrank because people cut the trees for use as firewood and canoes (Figure 3.8a).

Positive and negative relationships are summarized with either a (+) or (−). The line that connects two parts of a system that have a positive relationship is designated with a (+) (Figure 3.7b) while the line that connects two parts of a system that have a negative relationship is designated with a (−) (Figure 3.8b). This simplification is critical. Systems often have many parts that are connected in many ways. By following the plus and minus signs through the system, you can understand how a system responds to a disturbance.

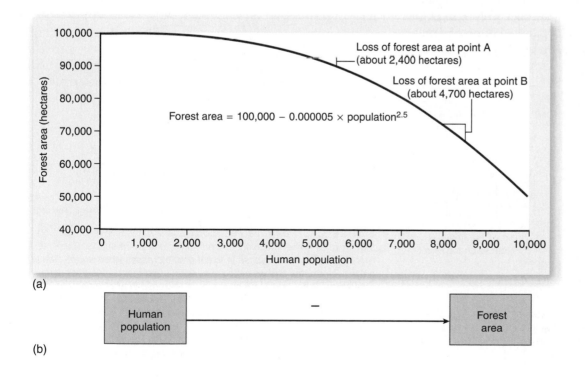

(a)

(b)

FIGURE 3.8 *(a) Negative Relationship The negative relationship between the human population and forested area on Easter Island. A hypothetical functional relationship between the independent variable (population) and forested area (dependent variable) is given by the following equation: Forested area = 100,000 − 0.000005 × Population$^{2.5}$. This relationship is nonlinear. An increase in 500 people at a population of 5,000 people reduces forest area by about 2,400 hectares, but an increase of 500 people at a population of 8,000 people reduces forest area by about 4,700 hectares.* *(b) Systems Diagram The negative relationship between human population and forested area as represented in a systems diagram.*

The effects of a disturbance on a system can be evaluated using the notion of a **feedback loop.** A feedback loop includes the linkages that move through the system and ultimately connect back to it. For example, if A is linked to B, B is linked to C, C is linked to D, and D is linked to A, then A, B, C, and D constitute a feedback loop (Figure 3.9).

The effect of a disturbance on a system depends on whether the feedback loop is positive or negative. A **positive feedback loop** does not change the effect of the disturbance after one complete loop. The loop A → B → C → D is a positive feedback loop if a disturbance that has a positive effect on A still has a positive effect on A after the disturbance comes back to A from D. On Easter Island a positive feedback loop linked the human population, agricultural land, and food supply (Figure 3.10). Suppose that the human population, agricultural land, and food production are at **equilibrium,** which is a state in which there is no net change. Suppose a disturbance (perhaps the arrival of new immigrants) increases the size of the population. The increase in population increases agricultural area, which is represented with a (+) on line 1. The increase in agricultural area increases food supply, which is represented by a (+) on line 2. The increase in food supply increases population, which is represented by a (+) on line 3. This feedback loop is positive because the initial disturbance, an increase in population, moves through the system and returns in a way that increases the population. Similarly, a disturbance that reduces the human population will move through the system and come back in a way that further reduces the human population.

On the other hand, a **negative feedback loop** changes the effect of the disturbance after one complete loop. The loop A → B → C → D is a negative feedback loop if a disturbance that has a positive effect on A has a negative effect on A after the disturbance comes back to A from D. On Easter Island a negative feedback loop linked human population, forest area, and wood supply (Figure 3.11). Suppose that

population, forest area, and wood supply are in equilibrium. That equilibrium is disrupted by a disturbance, such as an increase in rainfall and temperature that increases forest area. The increase in forest area increases wood supply. The positive relationship between forested area and wood supply is indicated by a (+) in line 1. The increased supply of wood increases the availability of firewood and the number of trees

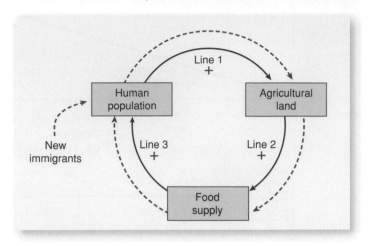

FIGURE 3.10 *Positive Feedback Loop* A positive feedback loop that includes population, agricultural land, and food production. The (+) on line 1 indicates that an increase in population increases demand for agricultural land. The (+) on line 2 indicates that an increase in agricultural land increases food supply. The (+) on line 3 indicates that an increase in food supply increases population. As a result, a disturbance that increases the human population moves through the system and comes back to the human population in a way that increases the human population further. Dotted lines trace the effect of new arrivals through the system; blue dotted lines represent an increase.

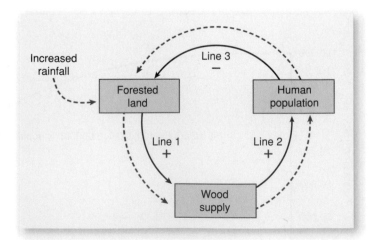

FIGURE 3.11 *Negative Feedback Loop* A negative feedback loop that includes human population, forest area, and wood supply. The (+) on line 1 shows that an increase in forest area increases wood supply. The (+) on line 2 indicates that an increase in wood supply increases the human population. The (−) on line 3 says that an increase in wood supply reduces forest area. As a result, a disturbance that increases forest growth moves through the system and then comes back to the forest in a way that reduces forested area back toward its value before the disturbance. Dotted lines trace the effect of increased rainfall through the system; blue dotted lines represent an increase; red dotted lines represent a decrease.

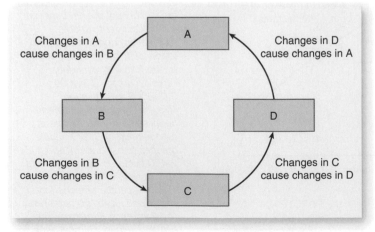

FIGURE 3.9 *Feedback Loop* The relationships among A, B, C, and D loop back to themselves because A is linked to B, B is linked to C, C is linked to D, and D is linked back to A. Together A, B, C, and D form a feedback loop.

CASE STUDY

The Collapse of the Easter Island Civilization

Chapter 1 introduced the idea of sustainability by recounting the rise and fall of the Easter Island civilization. Although the story contains many characters, such as people, rats, forests, and sea birds, the plot is relatively simple. In the opening act population grows slowly but steadily over time (Figure 1.3 on page 2). In the second act population collapses back toward its original levels. These events can be explained using a systems perspective (Figure 3.13). A positive feedback loop encouraged the population to grow. This growth also was part of a positive feedback loop that caused and exacerbated the environmental problems of soil erosion and the loss of forest area. These problems were parts of several negative feedback loops that caused the human population to shrink. This decline was amplified by the rats, which were supposed to increase food supply but eventually reduced food supplies.

In the opening act population growth is encouraged by a positive feedback loop. Population growth led to an increase in agricultural land. The increase in agricultural land increased food supply. The increase in food supply allowed the population to grow, which further increased agricultural land, and so on.

Population growth and the resultant increase in agricultural land reduced forest area and increased soil erosion. The severity of these problems increased over time due to positive feedback loops. Agricultural land is more vulnerable to soil erosion than forested areas. For reasons described in Chapter 15, erosion reduced the depth of soil from which crops could obtain water and nutrients. As a result, crops grown in shallow soils yielded less food. The reduction in food supply increased the demand for agricultural land. This last link completes the positive feedback loop. Each movement through the loop reduced soil depth and increased agricultural land.

Much of the land used for agriculture on Easter Island originally was covered by forests. As soil erosion increased the demand for agricultural land, it also reduced forest area. The reduction in forested area was reinforced by population growth. As the population grew, it used more wood for cooking and the construction of fishing canoes. Eventually use of wood exceeded the rate at which forests produced new trees. Beyond this threshold, the growing population reduced the area covered by forests. Eventually most of the forest was lost.

Soil erosion and the loss of forests initiated negative feedback loops that moved the population back toward its original level. The loss of forest area made it difficult to create new agricultural land. Without the ability to create new agricultural land, the negative effects of soil erosion on food supply predominated. The reduction in food supply decreased the population. The loss of forests also reduced the availability of wood for cooking and the construction of canoes. Without canoes, the islanders could no longer catch fish. The loss of fish reduced their food supply, which reduced population. This negative effect was reinforced by the loss of wood for cooking and warmth.

The negative feedback loops may have stabilized population at much higher levels had the Easter Islanders not made the mistake of bringing in rats. When people arrived, there were no permanent terrestrial vertebrates (animals with a spine). Introducing rats provided a critical source of protein, but the rats escaped human control and established wild populations. For reasons described in Chapter 12, the introduction of new species often disrupts the existing environment. These disruptions changed the environment in ways that made it less hospitable for humans.

In the wild, rats had two important food supplies: palm nuts, which were the seeds for the palm trees, and the eggs of sea birds that

nested on the island. Because the rats had no native predators or diseases, their population increased rapidly. As their population grew, so did their consumption of bird eggs. The rats and the people ate so many eggs that the nesting population of birds declined over time. This reduced human food supplies and amplified the loss of food caused by soil erosion. The rats also ate many palm nuts, which reduced the ability of the forest to regrow. The lack of forest regrowth quickened the loss of forests associated with their conversion to agricultural land and wood harvests. This amplified the negative feedback loop that reduced population growth.

Toward the end, the combination of the negative feedback loops and the actions of the rats caused the population to decline sharply. By the time Europeans reached Easter Island, the population had declined significantly relative to its peak. This reduction undermined the social system so much that the Islanders were no longer making *moai*. Although people still lived on Easter Island, the system that made Easter Island one of the most intriguing populations had disappeared.

ADDITIONAL READING

Bahn, P., and J. Flenley. *Easter Island, Earth Island.* London: Thames and Hudson, 1992.

Van Tilburg, J. A. *Easter Island.* Washington, DC: Smithsonian Institution Press, 1994.

Redman, C. L. *Human Impact on Ancient Environments.* Tucson: University of Arizona Press, 1999.

STUDENT LEARNING OUTCOME

- Students will be able to explain how positive feedback loops allowed the human population on Easter Island to grow and how negative feedback loops caused it to shrink back toward its original level.

that can be made into fishing canoes. These changes have a positive effect on the size of the human population, which is indicated by the (+) on line 2. As the population increases, people cut more trees. This cutting has a negative effect on forest area, which is indicated by the (–) on line 3. The reduction in forest area moves the forest (and eventually the human population) back toward the original equilibrium.

Positive and negative feedback loops determine whether a system is capable of homeostasis. Unstable systems are characterized by positive feedback loops. These loops amplify the effects of a disturbance so that it moves the system further and further from its initial values or set points.

The positive feedback loop that links population, agricultural land, and food production encouraged the Easter Island population to grow steadily for several centuries.

Systems that are capable of homeostasis are characterized by negative feedback loops. These loops offset the effects of a disturbance so that the system tends to return to its original values or set points. The negative feedback loop that included population, forest area, and wood supply helped cause the Easter Island population to crash back toward its original value. (A more detailed description of the collapse and feedback loops is given in "Case Study: The Collapse of the Easter Island Civilization.")

Natural Selection

Positive and negative feedback also generate **natural selection,** which is the process by which individuals' inherited needs and abilities are more or less closely matched to resources available in their environment, giving those with greater "fitness" a better chance of survival and reproduction. Natural selection makes use of the fact that the offspring of many organisms do not look or behave exactly like their parents. Furthermore, individual offspring do not look or behave exactly like their siblings. This variation is the raw material of natural selection.

The amount of variation present depends in part on how an organism reproduces. Organisms that reproduce asexually generate relatively little variation. This variation is largely generated by the small changes that occur every time genetic material is copied. These changes are only a small component of the great variation that is present in organisms that reproduce sexually. This process generates offspring that get half of their genetic material from each parent. These combinations are unique: No individual looks exactly like its parents or siblings (other than identical twins).

Natural selection focuses on the differences in traits that are critical to survival, such as the abilities to obtain food and escape from predators. Whether a trait helps or hurts an individual's probability of obtaining food or escaping from a predator depends on the environment. An individual's environment is defined by both the physi-

cal setting and the other species that are present. Because the climate and species present vary, traits that help an individual obtain food in one environment may diminish its chance of obtaining food in another. For example, camouflage may help an individual survive in an environment where predators use sight to find their prey; but it would be of little help in an environment where predators use smell to find their prey.

Being able to find food and escape predators influences an individual's probability of survival, but the ability to leave offspring is the ultimate criterion for natural selection. Traits that increase an individual's probability of leaving offspring will be more common in the following generation because there will be fewer offspring from individuals with traits that diminish the probability of leaving offspring. This increases the number of individuals in the next generation with traits most favorable to survival and reproduction.

The role of variation and natural selection is illustrated in Figure 3.12. Suppose a population of brown lizards is in equilibrium with its generally brown environment but suddenly finds itself living in a green environment. This is a challenging situation—assume that these lizards escape predators by blending in with the background. The offspring of the original population have a range of colors that is shown in the box labeled Generation 2. Some of the offspring are a darker shade of brown, while

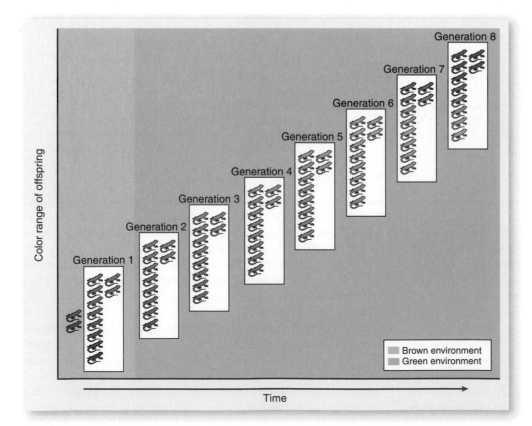

FIGURE 3.12 *Natural Selection*
The change in color in a hypothetical lizard population that is driven by natural selection. In each generation, young lizards display a range of colors from brown to green. Because green lizards blend in with their background better than brown lizards, natural selection increases the probability that the greenest of the offspring will live long enough to reproduce and leave offspring in the next generation. This process is repeated in each generation so that the lizard population changes mostly from brown individuals to green individuals.

others are a lighter shade of brown. Of these, the light brown individuals have the greatest chance of evading predators, and therefore they are most likely to produce a second generation.

These lizards also produce offspring with a range of colors. Again, some offspring are darker than their parents, while others are lighter. So long as the environment favors lizards that can best blend in with the green background, natural selection makes it more likely that the lighter brown individuals will produce the next generation. If these lighter lizards are genetically capable of producing green individuals, the environment remains green, and blending in remains a predominant determinant of fitness, eventually the lizard population will consist mostly of green individuals.

WHY ARE SYSTEMS SO HARD TO MANAGE?

Understanding what systems are and how they behave is critical to our ability to develop sustainable living habits. As described in later chapters, economic activities disturb the environment in many ways that cause society to use natural resources or environmental services faster than they are generated (in violation of sustainability principle #1). Because such behavior seems nonsensical, it is natural to ask why societies behave unsustainably.

Unpredictability

Systems may have regular, predictable, or purposeful behaviors, but these behaviors do not imply a predetermined outcome. We do not know exactly how many systems will behave. The residents of Easter Island could not know exactly how much food they would grow because rainfall varied from year to year. This lack of complete predictability implies that system behavior contains an element of uncertainty, which scientists call **stochastic** behavior. In any given year the actual rainfall on Easter Island equaled average rainfall plus or minus a stochastic element.

The importance of the stochastic element can be evaluated by comparing the observed values to the average. Are the observations clustered near the average, or are they dispersed widely around the average? **Variance** measures the degree of dispersion or scatter. If the observations are clustered close to the average, variance is small. Conversely, variance is large if the observations are dispersed widely around the average.

A large variance implies that the value will be relatively unpredictable from one year to the next, and extreme events will be relatively common. If rainfall has a large variance, in many years rainfall will be much less than average (droughts), and in other years rainfall will be much greater than average (floods).

Unpredictability and extreme events make it difficult to manage environmental systems. As will be discussed throughout the book, the environment varies from one year to the next, and this unpredictability makes it difficult to determine whether environmental changes are due to natural variation or human activities. For example, some people still cling to the highly unlikely notion that the increase in global temperature is due entirely to natural variability.

The element of chance forces people to make decisions without complete information, which is termed **risk management.** Farmers cope with the unpredictable element of precipitation by planting a crop that will do well given average conditions. For other years, the farmer makes contingency plans. When rainfall is significantly less than average, a farmer could use an irrigation system that would bring water from a nearby river. But building an irrigation system is expensive. Should the farmer build such a system?

Risk management dictates that the answer depends in part on the variance of precipitation. If the variance is small, there will be relatively few years in which rainfall will be significantly less than average, so an expensive irrigation system will be used rarely. Under these conditions the farmer would be economically better off not building the irrigation system and suffering the economic losses in the few years when rainfall is insufficient. On the other hand, a large variance implies that there will be many years in which rainfall is insufficient. Under these conditions the irrigation system will be used frequently; therefore it makes economic sense for the farmer to build the irrigation system.

Complexity

Complexity refers to the number of storages and flows and the number and strength of feedback loops in a system. Increasing any of these attributes increases complexity. Complexity makes it more difficult to understand how a system will respond to disturbance. This lack of understanding lies at the heart of many environmental problems. If we represent the positive and negative feedback loops on Easter Island, the complexity of Figure 3.13 indicates why it was difficult for Easter Islanders to determine whether they were violating the principles of sustainability.

Complexity also depends on the presence and strength of positive and negative feedback loops. Many systems have both positive and negative feedback loops; the effect of a disturbance depends on which feedback loop is stronger. For example, the Easter Island system had a positive feedback loop that included human population, agricultural land, and food supply. But this loop also was linked to a negative feedback loop that included population, agricultural land, forest area, and wood supply. These loops imply that converting forested areas to agricultural land had both a positive and negative effect on population. It is not possible to know which effect predominates simply by looking at Figure 3.13.

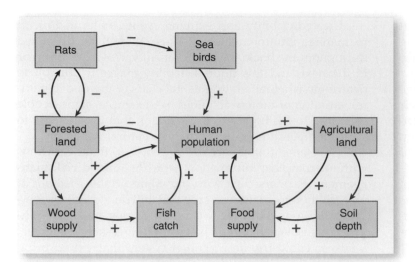

FIGURE 3.13 *Easter Island System* *The human population on Easter Island was linked to the soil, trees, fish, birds, and rats through a series of positive and negative feedback loops. In the short run, positive feedback loops cause the population to grow. In the longer term, negative feedback loops predominated, and the human population was moved back toward its original level.*

(Source: Redrawn from Bahn and Flenley, Easter Island, Earth Island. London: Thames and Hudson, 1992.)

Equally troubling, complexity makes it difficult to know whether efforts to move society toward sustainability will solve the problem, have no effect, or make it worse. Indeed, complexity often frustrates the intent of environmental policy so that a policy aimed at solving a problem may generate a new problem. These confounding effects are illustrated throughout this book in the Case Studies, including this chapter's description of the collapse of the Easter Island civilization.

Hierarchy

Systems often are part of a larger system. A system that is part of a larger system is said to be a **subsystem.** Systems and their subsystems form a hierarchy, which is a group or "ladder" of systems arranged according to their function. A system includes all the rungs below it, each of which can be considered a subsystem.

The organization of the natural world can be viewed as a hierarchy of systems. Atoms are ordered into complex biological molecules such as proteins. The molecules of life are organized into minute structures called organelles, such as mitochondria, which are the components of cells. Similar cells are grouped into tissues, and specific arrangements of tissues form organs. Organs such as the heart, lungs, and blood vessels are organized into organ systems such as the circulatory system. Organ systems are organized to form individual organisms. At even higher levels of organization, individuals are organized to form populations, communities, and ecosystems.

Human systems, such as economic systems, also are arranged as hierarchies. Individuals sell their labor. Groups of laborers are organized into an assembly line, which produces a good such as a computer. Assembly lines are organized into factories, which purchase the raw materials for the computer and ship the finished goods. Factories are part of a corporation that may produce several different types of goods and services. Corporations that produce similar types of goods or services form a sector, such as the electronics sector. Sectors interact with each other (see *Your Ecological Footprint: The Direct and Indirect Use of Natural Resources and Emission of Wastes*) and consumers to form a national economy. National economies trade with one another to form a global marketplace.

Hierarchy makes it difficult to understand and manage systems because behavior at one hierarchical level cannot be predicted from behavior at lower (or higher) hierarchical levels. The people of Easter Island lived in several villages, each of which was a subsystem of the total population. Each village specialized in a different activity (fishing, agriculture, and so on). The viability of these specializations changed as the Easter Island system changed, so some villages prospered while other villages suffered. This created conflict among villages that accelerated the collapse of the Easter Island civilization. These subsystem dynamics are not shown in Figure 3.13, which represents the human population as a single storage in the Easter Island system.

Time Lags

A **time lag** refers to the period that lapses between a cause and its effect. Systems are made up of linkages that display lags varying from very short to very long. For example, a reduction in topsoil can have an almost immediate effect on crop yield. On the other hand, there may have been a long lag between the size of the rat population and the amount of forest area. Rats eat seeds that grow into trees, so there can be a twenty-year lag between a rat eating a seed and a noticeable loss in the availability of firewood.

Long lag times make it difficult to identify and solve environmental problems because they make it hard to con-

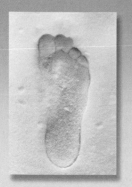

Your ECOLOGICAL *footprint*

The Direct and Indirect Use of Natural Resources and Emission of Wastes

Starting with Chapter 4, Your Ecological Footprint boxes will allow you to calculate your use of natural resources or environmental services or your emissions of waste. To do so, the Ecological Footprints will provide information about the amounts of natural resources, environmental services, or emission of wastes that are associated with a particular activity. These numbers are termed intensities. For example, the Ecological Footprint in Chapter 20 provides energy intensities (the amount of energy associated with a dollar's worth of various activities), whereas Chapter 13 provides carbon intensities (the amount of carbon dioxide that is emitted by a dollar's worth of various activities).

This Ecological Footprint describes the methods that are used to calculate some of these intensities. To be complete, the calculation includes both the direct and indirect uses or emissions. Direct uses or emissions often are easy to measure. For example, it is relatively easy to measure the amount of fuel you use to drive your car, heat or cool your home, and cook your food. Similarly, it is easy to measure the amount of carbon dioxide you release when you drive your car, heat or cool your home, and cook your food.

On the other hand, indirect uses are more difficult to measure because they occur elsewhere in the economy. Indirect uses or emissions often are associated with the production, transport, or disposal of a good or service. For example, how much energy was required to grow and transport the food you eat or the CD you listen to? Similarly, how much fertilizer contaminated the waterways in the process of growing your food? How much air pollution was generated by the factory that made your CD?

Although they seem small, indirect uses or emissions should not be ignored. They can be equal to or greater than direct uses or emissions. As such, intensities that include only direct uses or emissions would understate the environmental impact of your activities.

To include both the direct and indirect uses and emissions, the intensities in some of the Ecological Footprints are calculated using either life cycle analysis or input–output tables. As described in Chapter 23, life cycle analysis is used to calculate the flows of energy and materials that are associated with a specific good or service.

Input–output tables are used to track the flow of a natural resource or the emission of a waste through the entire economy. To do so, input–output tables divide purchases between intermediate and final demand sectors. Final demand sectors buy goods and services from the intermediate sector but do not resell these purchases. For example, you buy goods (such as cars and CDs) and services but do not sell them (people sell their used cars, but the value of these sales is not included in the total measure of economic activity).

Intermediate sectors buy goods and services from one another, but these purchases are designed to produce a good or service that will be resold. These purchases are termed interindustry flows. Interindustry flows of the automobile sector include the purchase of steel from the steel sector, the purchase of glass from the glass sector, and the purchase of leather from the leather sector.

The interindustry flows can be used to calculate how many cents' worth of goods and services from one sector are needed to produce a dollar's worth of output. For example, the interindustry flows associated with the automobile sector can be used to calculate how many cents' worth of rubber, glass, and leather are needed to produce a dollar's worth of car. The steel, glass, and leather sectors all purchase goods and services from other sectors; so the value of the steel purchased by the automobile sector includes all the materials purchased by the steel industry. Input–output tables trace these flows and can be used to calculate intensities that include both direct and indirect purchases.

These monetary intensities can be converted to intensities measured in physical units. Physical intensities measure the physical quantities of materials used (or emissions of wastes) that are used both directly and indirectly to produce a dollar's worth of material. For example, input–output tables can be used to calculate the quantity of energy that is used to produce a dollar's worth of automobiles, the kilograms of water that are used to produce a dollar's worth of steel, or the kilograms of carbon dioxide that are emitted to produce a dollar's worth of paper. These physical intensities are consistent with the second law of thermodynamics: All of the material that goes into the economy is contained (embodied) in the goods and services produced by the economy and its emission of wastes. No material is created or destroyed. As such, intensities measured in physical units allow you to calculate all the natural resources, environmental services, or wastes that are associated with the goods and services you purchase.

ADDITIONAL READING

Miller, R.E., and P.D. Blair. Input–Output Analysis: Foundations and Extensions. Englewood Cliffs, NJ: Prentice Hall, 1985.

STUDENT LEARNING OUTCOME

- Students will be able to explain why looking at only their direct purchases of energy and materials will underestimate the true size of their environmental footprint.

nect cause and effect. The Easter Islanders probably did not make the connection between the size of the rat population and the rate at which their forests were disappearing. This hid one part of a potential solution to their problem.

Equally frustrating, long lags diminish the effectiveness of environmental policy because the problem may continue long after the cause has been removed or reduced. As described in Chapter 14, several international treaties have stopped the production and trade of chlorofluorocarbons (CFCs), the chemicals that deplete ozone in the upper layers of the atmosphere. But the concentration of ozone will continue to decline because global warming in the lower layers of the atmosphere will cool the next layer up, which will accelerate the rate at which ozone molecules are converted to oxygen molecules.

Distance Effects

Distance effects refer to the separation in space between a cause and effect. This geographical separation makes it difficult to recognize how our everyday activities cause environmental problems. Many of us live in urban or suburban environments, so we do not see how extracting natural resources from the environment and disposing of wastes generate environmental problems (recognizing this connection is one reason for the exercises associated with Your Ecological Footprint). As a result, it appears as though our everyday activities have no environmental impact. Few of us think about the pollution associated with a coal-fired electricity plant when we turn on a light because most of us live far from such a plant. Similarly, few of us think about the environmental impacts associated with clear-cutting (cutting all trees in a particular area) when we use a piece of paper because most of us live far from the forest that is being logged. Without this firsthand experience, it is more difficult to be concerned about these problems. And without this concern, it can be difficult to generate the consensus that is needed to create an environmentally sustainable resource policy.

Even when such a consensus is present, distance effects can make it difficult to implement environmental policy. The cause and effect of many environmental problems often are separated by political boundaries. These boundaries mark the reach of law, so people who live on the side of the border where the environmental problems occur cannot restrict the activities of people who live on the other side of the border, where the problem is generated. For example, much of the sulfur that causes acid deposition in Canada is emitted by power plants in the United States.

Linear versus Nonlinear Relationships

The components of a system are linked by linear and nonlinear functions. Linear functions are represented with a straight line. For example, the line in Figure 3.7a represents the linear relationship between population and agricultural land on Easter Island. Assuming that each hectare of land is equally productive, the amount of agricultural land needed for each additional person remains the same, regardless of the size of the population.

This constancy disappears in nonlinear relationships, which are represented with a function other than a line. For example, Figure 3.8a uses a power function to represent the relationship between the human population and forests. At low population levels such as point A, an increase of 500 people has a relatively small effect on forest area—about 2,400 hectares (5,298 acres). Small populations use relatively little wood. This wood may be produced by the natural growth of the forest so that there is little net change in forest area. As the population grows to point B, so too does its consumption of wood. As people harvest trees faster than they grow, the forest shrinks. This shrinkage reduces the rate at which the forest produces wood, which accelerates the loss of forest area associated with a given level of harvest. As a result, an increase in 500 people at point B reduces forest area by about 4,700 hectares (11,609 acres), which is nearly double the loss that occurs at point A.

Nonlinear relationships make it difficult to manage environmental problems. Most of us tend to see the world through "linear glasses." This assumption allows us to believe that environmental problems will evolve gradually in proportion to the cause of the problem. But nonlinear relationships imply that the severity of an environmental problem can increase suddenly even if the cause of the problem increases gradually. For example, Easter Islanders may have ignored their effect on the forest when their population was small. Even as the population grew, they may have underestimated their effect because they assumed the effect would grow in proportion to the population.

Understanding Environmental Problems   Systems are difficult to understand and manage, and there are many examples of environmental policies that fail. But these difficulties and failures do not mean that society cannot develop policies that move society toward sustainability. There are also many examples of management practices and policies that have ameliorated environmental problems. These success stories are distinguished from the failures by the way in which scientists approach the problem and the way in which society designs policy.

The Scientific Method

Over the last 300 years scientists have expanded our understanding of how the world works by using the scientific method. Following this method, progress is made by developing and testing hypotheses. People outside of science often criticize this method by saying "that's just a hypothesis." By saying so, they imply that a hypothesis is wrong. After all, *hypothesis* is just a fancy word for an idea about how the world works. And as we all know, everyone has his or her own pet ideas about how the world works.

But the scientific method is much more than giving everyone the freedom to think up his or her own explanation. The scientific method is a set of rules that describes the way in which scientists state their hypothesis. Most important, the scientific method specifies a set of rules for deciding which hypothesis is "better." If everyone plays by these rules, scientists can develop a hypothesis about how the world "really works."

Scientific inquiry begins with the formulation of a hypothesis: an explanation that accounts for a set of facts that can be tested (Figure 3.14). For example, it is a fact that Earth is home to many different species of plants and animals. Darwin's theory of evolution began with a hypothesis about how that diversity was generated.

Another criticism of the scientific method is that scientists assume their hypothesis is correct. For example, evolutionary biologists sometimes are criticized because they assume that the theory of natural selection is correct. In fact, this assumption is critical for the process of testing. To test a hypothesis, scientists assume their hypothesis is correct and think of things they would expect to see in the real world if that were the case. Returning to Darwin's theory, if species change through an evolutionary process, there should be intermediate forms of life. For example, if birds evolved from reptiles, there should be some organisms that have characteristics of both reptiles and birds.

To be considered a scientific hypothesis, the hypothesis must offer some prediction that can be tested against facts, which are observations of the real world. Testing a hypothesis is the purpose of experiments. Scientists do **experiments,** which are a set of actions and observations to verify or falsify a hypothesis or research a causal relationship between phenomena.

Scientific testing is different from the tests you are accustomed to taking. Hypotheses are not right or wrong. Rather, scientists say that a hypothesis is consistent or inconsistent with the facts. The difference is more than a matter of semantics. A hypothesis cannot be proved correct. If a hypothesis is consistent with experimental facts, the hypothesis is said to be **validated.** Once validated, the hypothesis is retained,

and scientists think of new ways to test it. A successful hypothesis is always being tested. At best a hypothesis that has been validated by many experiments can be considered a theory. But theories, too, are tested constantly. Darwin's theory still is the subject of considerable testing.

On the other hand, a hypothesis can be proved incorrect. Often the experimental results are not consistent with a hypothesis. The scientific method dictates that experimental observations or facts must be believed and that the hypothesis must, in such a case, be disbelieved. The hypothesis is then said to be **invalidated** or **falsified.** Once invalidated, the hypothesis must be changed to explain the facts. If it cannot be changed to account for the facts, the hypothesis is discarded. To date, no experiment has invalidated Darwin's theory of evolution in a scientifically meaningful fashion.

A Reductionist versus Systems Perspective

The scientific method has clarified the cause for many environmental problems, such as the cause of reductions in ozone in the upper layers of the atmosphere. Despite these successes, society's attempts to deal with the environment often misidentify the cause of a problem, develop policy that does not solve a problem, or develop policy that makes a problem worse. Why such a poor record for a species that is capable of space travel?

The poor record is caused by a basic difference between space travel and environmental problem solving. Space travel and many other projects can be understood using a reductionist approach. A **reductionist approach** is based on the premise that the best way to learn about something is to break it into parts and study the parts separately. This approach has been quite successful, particularly in the natural sciences (biology, chemistry, physics). This approach is reflected by the departmental setup of your college or university, where classes are offered by departments that are organized around specific bodies of thought called disciplines (chemistry, biology, economics, anthropology).

The limits of a reductionist approach were first recognized by Frederick von Bertlanfy, a German scientist working in the 1920s who realized that many things in nature and human society cannot be explained by dissecting them into their individual parts. Understanding these topics requires a different perspective, one that emphasizes patterns and connections. This perspective now is known as **general systems theory,** which is basically concerned with problems of relationships, structures, and interdependence rather than with the constant attributes of an object.

General systems theory connects information from many different disciplines and is especially useful for analyzing environmental issues and sustainability. Sustainability is defined by the connections between society and its environment, therefore, dissecting these connections into separate parts does not generate the requisite information. For

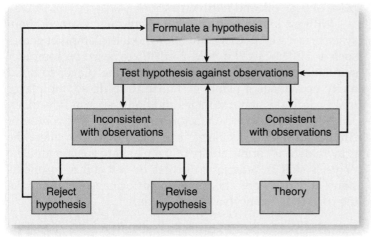

FIGURE 3.14 *The Scientific Method* *The scientific method starts with the formulation of a hypothesis. The predictions generated by the hypothesis are compared against observations. If the observations are inconsistent with the hypothesis, the hypothesis is modified or discarded. If repeated comparisons are consistent with observations, the hypothesis becomes a theory.*

example, the Easter Island civilization was not unsustainable in itself. Rather, its members' actions were not sustainable in relation to the environment in which they lived.

As described in the remainder of this book, an **integrated systems approach**—one that uses information from many disciplines—is needed to understand and solve specific environmental problems and generate general policy that moves society toward sustainability. This approach is needed because such issues require information from many disciplines. For example, treating the climate as a sustainable source of environmental services requires that information about the physical aspects of the climate system be integrated with the economic aspects of activities that change global climate. (See *Policy in Action: The IPCC: An Interdisciplinary Effort to Make Climate Change Policy.*)

Simulation Models

Including information from many disciplines sounds like a good idea, but it can be very difficult to implement. At first glance economics seems to have little in common with chemistry, which seems to have little in common with anthropology, which has little in common with physics. To bridge these differences, environmental scientists use the universal language of mathematics. Environmental scientists use mathematics to represent what economics, chemistry, anthropology, physics, and other disciplines say about a particular problem. By formalizing these various insights into mathematical equations and linking these equations together, environmental scientists create mathematical representations that simulate the behavior of environmental systems. These mathematical representations are known as **simulation models.**

An integrated systems approach is needed to make a simulation model. To build the model, scientists must choose which storages and flows to include. This decision is based on whether a storage or flow affects the behavior under investigation without regard to discipline. For example, a model that represents global climate change has to include the system components that influence temperature and precipitation (an issue for climatologists). These variables depend in part on the amount of greenhouse gases such as carbon dioxide and methane stored in the atmosphere (an issue for atmospheric chemists). The flow of these gases to the atmosphere depends on biological activity (an issue for biologists) and economic activity (an issue for economists). Clearly, studying climate change cannot stop at boundaries of climatology.

Building a simulation model also forces scientists to quantify the nature (positive or negative) and strength (weak or strong) of relationships. For example, the amount of carbon dioxide emitted by burning coal, crude oil, and natural gas is positively related to economic activity (increases in economic activity increase energy use) and is negatively related to technical change (new energy-efficient technologies reduce emissions). In addition, model building forces scientists to quantify relationships. As economic activity increases by $1, by how much do emissions of carbon dioxide increase?

Much like a scientific hypothesis, a simulation model is validated by testing it against observations. If the model includes the relevant components and their relationships are represented by equations that have the correct sign and the correct strength, the model will be able to simulate real-world behavior. If the simulated behaviors are sufficiently close to reality, the model is said to be validated. Once the model passes this test, it is used to understand system behavior and the effectiveness of policy by playing "what if" games. These "what if" games are known as **scenario analysis.** A scenario is simply a set of values for disturbances or relationships between variables. For example, once a climate model is validated, it can be used to consider what happens to global temperature if faster economic growth accelerates the use of oil and ultimately the emission of carbon dioxide (the gas thought primarily responsible for global climate change). Knowing this effect before temperature increases allows decision makers to determine whether policy aimed at slowing oil use is needed. The model also could be used to evaluate the effectiveness of alternative policies. For example, the model could be used to compare the effectiveness of increasing the energy efficiency of new cars to the effectiveness of imposing taxes that raise the price of coal, oil, and natural gas.

Without knowing it, you probably have learned about systems using computer simulation models. Computer games such as SIMM-City™ or SIM-Earth™ are based on computer simulation models. These games allow the player to build a city or manage a planet sustainably. To do so, the player makes decisions about storages and flows, such as how many roads to build, how to run factories, and so on. The computer simulation model calculates how these decisions affect the growth of a city or the ability of the planet to sustain life. The player's skill is determined by his or her ability to understand how decisions affect system performance and to make the correct decisions in a timely fashion.

POLICY IN ACTION

The IPCC: An Interdisciplinary Effort to Make Climate Change Policy

To address some of the difficulties that are associated with understanding and managing the threat posed by global climate change, the World Meteorological Organization and the United Nations Environment Programme established the Intergovernmental Panel on Climate Change (IPCC) in 1988. The Intergovernmental Panel on Climate Change includes thousands of the world's leading scientists. Their goal is to assess the most up-to-date scientific, technical, and socioeconomic research in the field of climate change. This goal is not much different than the charge given to hundreds of scientific bodies that have previously investigated environmental problems. Yet the Intergovernmental Panel on Climate Change is different. It represents an important first step in an interdisciplinary scientific investigation that will lay the groundwork for policies to manage the threat to human and natural systems posed by global climate change.

The interdisciplinary nature of global climate change is reflected by the panel's structure. The Intergovernmental Panel on Climate Change is organized into three working groups. Working Group I concentrates on the climate system; Working Group II concentrates on impacts and response options; and Working Group III concentrates on economic and social dimensions. These divisions reflect aspects of global climate change rather than the discipline of scientists who study climate change. To understand the climate system, Working Group I includes physical and social scientists. Climatologists analyze how the climate system may respond to changes in the atmospheric concentration of greenhouse gases. Economists forecast the rate at which economic activities emit greenhouse gases. To understand how these emissions accumulate in the atmosphere, Working Group I also includes ecologists, oceanographers, and atmospheric chemists.

Once invited to participate, experts are expected to abandon their individual focus. Rather, they are expected to build their knowledge into a more complete understanding about a particular aspect of climate change. The working groups describe this synthesis in a series of assessment reports. The first assessment report was published in 1990; a second set was published in 1995; and a third group of reports was published in 2001. A fourth set of assessments will be published soon.

As implied by their title (such as *Climate Change 2001*), these reports are supposed to assess the existing knowledge regarding climate change. To do so, they attempt to rank the certainty of knowledge regarding climate change. These rankings include categories such as "we are virtually certain that" or "it is our judgment that." As science progresses, more aspects of the problem can be moved toward the "we are virtually certain" category. For example, the 1990 assessment report stated that "global mean temperature has increased by 0.3–0.6°C over the past 100 years," but the scientists could not say whether this increase was caused by natural variation or the emission of greenhouse gases associated with economic activity. Five years later, the second assessment report confirmed the change in temperature and added "the balance of evidence suggests that there is a discernible human influence on global climate." In 2001 the report stated "there is new and stronger evidence that most of the warming observed over the last 50 years is attributable to human activities."

Establishing an interdisciplinary consensus about what is known about global climate change is an important first step in a negotiation process that may ultimately result in an international treaty. Global climate change affects the entire planet. Yet the severity of this threat can be assessed by only a few nations that have a large, well-funded scientific establishment. Differences between developed and developing nations in their ability to assess the threat makes it difficult to generate an international agreement about what, if anything, should be done about global climate change. Nations that cannot assess the threat worry that they will be taken advantage of by nations with a more thorough understanding of the problem.

The knowledge assembled by the Intergovernmental Panel on Climate Change also allows the negotiations to proceed to policy formation. Without scientific consensus, international negotiations probably would bog down over disagreements about what is known about climate change. Nations that want to do little or nothing about climate change would use scientific uncertainty to argue that climate change is not a real problem. Conversely, nations that want strict measures would overstate the threat. The panel's emphasis on scientific consensus allows the negotiators to focus on the question of what should be done. This is the crux of the problem. The impacts of global climate change and policies to ameliorate its impact will differ among nations. For example, nations that emit large amounts of greenhouse gases may argue for less stringent policy than nations that contribute little to global climate change because a strict policy may have a large negative effect on the more polluting economies. Such disagreement is legitimate; these differences can be narrowed only if negotiators focus on the real sources of disagreement.

TO GET THE LATEST OFF THE WEB
The official IPCC Web site: www.ipcc.ch/

STUDENT LEARNING OUTCOME
- Students will be able to explain how the Intergovernmental Panel on Climate Change may be able to facilitate an international agreement about climate change.

SUMMARY OF KEY CONCEPTS

- Systems are a collection of storages and flows that interact with each other in ways that are consistent with the laws of thermodynamics to generate regular or predictable behavior. Stable systems can return to a set point following a disturbance. This ability is termed homeostasis. Homeostasis can be evaluated by resistance, which measures the degree to which a system changes in response to a disturbance, and resilience, which measures the degree to which a system can return to a set point following a disturbance.

- The relationship between parts of a system can be described by positive or negative relationships and feedback loops. Positive feedback loops amplify the effects of a disturbance and tend to destabilize a system. Negative feedback loops offset the effects of a disturbance and tend to stabilize a system.

- Environmental systems are difficult to manage due to the variability of individual storages and flows, the complexity of linkages, geographic and temporal separations between cause and effect, and nonlinear relationships between variables.

- The scientific method allows us to evaluate competing hypotheses about how the world works. Reductionism has been a powerful tool in many disciplines, but many environmental problems cannot be understood by breaking the system into individual parts and studying them in isolation. Instead many environmental problems can be understood using an integrated systems approach in which information from many disciplines is used to explore the linkages among system parts. This approach is formalized in simulation models, in which mathematical equations are used to simulate the system and run scenarios.

REVIEW QUESTIONS

1. Explain the following contradiction: The second law of thermodynamics states that the universe as a whole tends toward a greater state of disorder, but the environment is filled with systems that display predictable behavior.

2. Natural selection has been described as being blind. How can a random process generate traits that seem precisely designed for life in a given environment?

3. Formulate a general rule to determine whether a feedback loop is positive or negative. *Hint:* The rule should be based on the number of relationships and the sign (positive or negative) of those relationships.

4. Characterize the relationships in the flow diagram for Easter Island (Figure 3.13) with respect to time lags. Describe the role of time lags in the collapse of the Easter Island civilization.

5. During the first five years of the twenty-first century, oil prices rose from about $20 per barrel to well over $70 per barrel. Do you think a reductionist or integrated systems approach could best explain this price increase?

KEY TERMS

complexity	homeostasis	scenario analysis
dependent variable	independent variable	set point
economic system	integrated systems approach	simulation models
equilibrium	invalidated	stability
experiments	natural selection	stochastic
falsified	negative feedback loop	storage
feedback loop	positive feedback loop	subsystem
fitness	positive relationship	system
flows	reductionist approach	time lag
function	resilience	validated
general systems theory	resistance	variance
gradient	risk management	

THE PHYSICAL SYSTEMS OF PLANET EARTH

The Engine of Life

4

STUDENT LEARNING OUTCOMES

After reading this chapter, students will be able to

- Explain how a convection cell can explain patterns of circulation in the atmosphere, ocean, and Earth's surface.

- Explain how gradients in incoming solar energy generate predictable patterns of temperature, precipitation, and winds.

- Describe the factors that cause ocean water to circulate horizontally and vertically.

- Describe how energy from Earth's interior causes matter to flow from the interior to the surface and back again to the interior.

- Compare and contrast changes in atmospheric and oceanic circulation that are associated with El Niño events.

Frozen New York *The movie The Day After Tomorrow portrays a cold future caused by a change in ocean circulation in the Atlantic Ocean. Although scientists view this as a very unlikely worst-case scenario that would occur over decades, not weeks, changes in this ocean circulation pattern have generated global climate changes during the last 18,000 years.*

The Statue of Liberty looking out on a newly frozen New York City Harbor is one of the most memorable scenes in the film *The Day after Tomorrow* (opening photograph). In this movie the planet undergoes a dramatic change in climate. Within weeks the planet changes from the world we know to one in which giant snow storms hit much of the Northern Hemisphere, tornadoes strike Los Angeles, and grapefruit-sized hail falls on Tokyo. As a result, large areas of the United States, Canada, and Europe are rendered uninhabitable. U.S. citizens flee to Mexico, reversing the usual direction of illegal immigration.

Financially the film was a modest success—it took in nearly $200 million. One reason for its popularity was the special effects. Another reason was that like all good thrillers, the film included just enough "facts" for many viewers to suspend disbelief—despite warnings from scientists that such rapid climate changes are highly unlikely, even in the worst cases.

How did the filmmakers generate just enough believability to pack the theaters? They blamed the sudden climate changes on a shift in the flow of seawater between the ocean surface and ocean depths.

This change stopped the Gulf Stream, which is a surface current that brings warm water from the tropics along the eastern coast of the United States and on toward western Europe. Without this warm water, air temperatures in this region would cool. But even if the Gulf Stream stopped, significant cooling would occur over decades, not weeks.

So why did the audience believe that a change in ocean circulation could generate a dramatic change in climate? Some may have heard TV weather forecasters blame bad weather on El Niño events, which are circulation changes in the Pacific Ocean. Others may have known that Earth's climate is generated by a complex system, in which small changes in one part reverberate throughout the planet—the so-called butterfly effect, named for some people's claim that a butterfly flapping its wings in the Amazon changes the weather in New York City. Regardless of the reason, the audience was basically correct—energy from the sun and the Earth's core drives a climate system that creates predictable patterns in the circulation of Earth's atmosphere, ocean, and land surfaces that support life on this planet. ∎

ENERGY FROM THE SUN AND EARTH'S INTERIOR

Energy from the Sun

Einstein's famous equation $E = mc^2$ (the amount of energy created equals the amount of matter destroyed times the speed of light squared) describes the sun's thermal nuclear fusion reaction that converts hydrogen atoms to helium and energy. This reaction occurs in the sun's interior, where about 4 million metric tons (4.41 million tons) of hydrogen (matter) are destroyed per second to create about 3.9×10^{26} watts (energy). This tremendous release of energy raises the temperature of the sun's interior to about 10 million°C (18 million°F). Most of this heat is absorbed by the outer layers of the sun, which are known as the **photosphere.** The temperature of the photosphere varies between 5,300°C and 8,600°C (9,572–15,512°F). In general, scientists assume that the photosphere radiates energy to Earth at about 6,400°C (11,552°F). Telling you this temperature seems like a bit of trivia now, but Chapter 13 describes its importance to global climate change.

The energy emitted by the sun becomes more diffuse as it travels through space. After traveling 150 million kilometers (93 million miles), the amount of solar radiation that reaches the upper layers of the Earth's atmosphere is equivalent to 1.97 calories per square centimeter (12.71 kcal/inch²) per minute. This quantity is known as the **solar constant.**

Solar Radiation Reaching Earth's Surface

There is quite a difference between the amount of solar radiation that reaches the upper layers of Earth's atmosphere and the quantity that reaches Earth's surface (Figure 4.1). As solar radiation enters the atmosphere, it collides with gas molecules, dust, and other particles. These collisions cause it to **scatter** in various directions. Some of the radiation is turned back toward space in a process called **reflection.** Reflection in the atmosphere accounts for about 6 percent of the incoming radiation. The percentage reflected increases after large volcanic eruptions, which eject large numbers of particles into the atmosphere, creating more opportunities for sunlight to bounce back.

Clouds also reflect solar radiation. On a totally overcast day clouds reflect 60–70 percent of incoming solar energy. When averaged over the year and across the globe, clouds reflect approximately 21 percent of the incoming solar radiation. We say approximately because this global fraction has not been measured accurately, yet it is critical to arguments about global climate change. As you might expect, any change in cloudiness could have a dramatic effect on global climate.

Finally, some of the energy that reaches Earth's surface is reflected back to space. The amount reflected depends on the type of surface. For example, forests reflect much less solar radiation than snow. When averaged over the year and the entire planet, Earth's surface reflects about 4 percent of incoming solar radiation. Together these three mechanisms

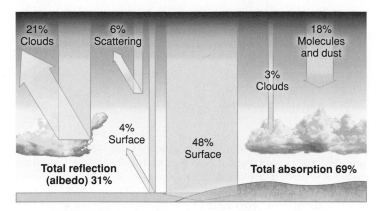

FIGURE 4.1 *Incoming Solar Energy* *The fate of incoming sunlight among reflection and absorption in the atmosphere, land surface, and bodies of water.*

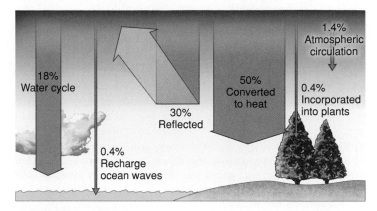

FIGURE 4.2 *Work Done by Solar Energy* *Solar radiation heats the atmosphere and ocean, drives the water cycle and wind patterns, and recharges ocean waves. A small fraction (0.4%) is converted to biological energy by green plants via photosynthesis, forming the base of food chains and webs that are discussed in Chapter 5.* (Redrawn from V. Smil, General Energetics: Energy in the Biosphere and Civilization. New York: John Wiley & Sons, 1991.)

reflect about 31 percent of the incoming solar radiation back to space. The global total for the percentage of incoming solar radiation that is reflected back to space is known as **albedo.**

Energy that is not reflected back to space is absorbed. Molecules, dust, and clouds in the atmosphere absorb about 21 percent of the incoming radiation. Molecules and dust absorb about 18 percent and clouds absorb about 3 percent. Another 48 percent is absorbed by Earth's land and water surfaces. Together Earth's surface and its atmosphere absorb about 69 percent of the radiation that reaches the outer layers of Earth's atmosphere. Consistent with the laws of thermodynamics, the percentages absorbed and reflected sum to 100 percent, which accounts for all incoming solar radiation.

Work Done by Solar Energy

The 69 percent of the incoming solar energy that is absorbed drives the physical systems of planet Earth (Figure 4.2). About one-half of the solar energy is converted directly to heat. This heat influences the temperature of the atmosphere. About one-fifth of incoming solar energy is used to drive the water cycle. About 2 percent of the solar energy drives atmospheric circulation. Smaller quantities (less than 1 percent) of the energy are used to power ocean waves. Finally, a tiny fraction (about 0.4 percent) is used directly by plants and phytoplankton (single-celled algae). *Your Ecological Footprint: How Much Sunlight Do You Use?* allows you to calculate the amount of sunlight used to provide some of the goods you consume every day.

Heat from Earth's Interior

The other source of energy to Earth's surface comes from its interior. Heat comes from several sources. A small portion is heat left over from the formation of the planet. Most of this heat, however, is associated with the radioactive decay of elements, especially uranium, thorium, and potassium

(remember from Chapter 2 that radioactive decay converts matter to energy). Together these sources raise the temperature to between 1,200°C (2,192°F) and 1,400°C (2,552°F) at about 100 kilometers (62 miles) below the surface. Most scientists believe that the temperature at Earth's center is about 4,300°C (7,772°F), but this value could be 50 percent higher or lower.

Luckily for us, Earth's surface temperature is much lower. This difference sets up a gradient that causes heat to flow from the hot interior toward the cooler surface. This flow is about 1.5×10^{-6} calories per square centimeter (9.68×10^{-6} cal/inch^2) per second. These measurements are fairly accurate despite the uncertainty about the temperature of Earth's interior.

If you add the flow of heat from Earth's interior over the globe for an entire year, the total is about 2×10^{17} kcal. This total is large: It is greater than the global use of coal, oil, natural gas, and electricity generated from nuclear and hydropower plants. Nonetheless, this large total is much less than 1 percent of the energy that arrives from the sun. So incoming solar radiation is the main source of energy for the physical systems of planet Earth.

Work Done by Heat from Earth's Interior

Although this flow is tiny compared to the sun's contribution, the sun's energy affects only the upper few meters of Earth's surface. Below the surface the main source of energy is heat from Earth's interior. Some of this energy is used to do work, such as forming new crust and pushing the continents apart (see the section on movements in Earth's crust). Heat from Earth's interior also melts the existing crust and heats ocean water.

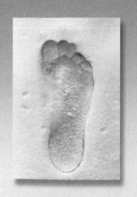

Your ECOLOGICAL *footprint*

How Much Sunlight Do You Use?

The physical systems of planet Earth use solar energy to generate the natural resources and environmental services you consume. This box provides information that allows you to calculate the quantity of solar energy that is used to provide the food you eat, the clean water you drink, and the paper you use.

The link between solar energy and food supply is fairly direct. Plants use solar energy to combine molecules that contain a single atom of carbon (carbon dioxide, CO_2) into molecules that contain six molecules of carbon (glucose, $C_6H_{12}O_6$). Glucose stores some of this solar energy in a form you can use.

How much solar energy is stored in your food? Only about 0.3 percent of the solar energy that hits a leaf is stored in glucose (photosynthesis has an efficiency of about 0.3 percent). This efficiency implies that a plant must capture about 300 kcal of solar energy to store 1 kcal of energy. Of this 1 kcal, about 0.4 kcal are converted to plant material (40 percent efficient). And only about 10 percent of a plant is edible (10 percent efficient). Because all these conversions are relatively inefficient, the average kcal of plant foods such as red peppers or tomatoes represents about 7,500 kcal of solar energy. This energy total increases by a factor of 10, to 75,000 kcal of solar energy, for a kcal of meat. For reasons described in the next chapter, animals convert only about 10 percent of plant food to edible meat.

Solar energy also is required to produce drinkable (potable) water. Due to the forces of gravity, water eventually makes its way to the ocean. There it mixes with sodium (and other minerals), and the high concentration of these minerals makes seawater undrinkable. Seawater is made drinkable via the work done by solar energy. This energy separates water from the minerals and moves the water over land and to a reservoir near you.

Evaporation is a very energy-intensive process: As described on page 64 of this chapter, 540 kilocalories are required to convert a kilogram of liquid water to a kilogram of water vapor, which is a gas. In addition, great amounts of solar energy are required to lift the water so it forms a cloud and to move the cloud horizontally (paths 1 and 2 of the convection cell in Figure 4.3).

The solar energy used to provide potable water can be calculated by dividing the amount of solar energy used to drive the hydrological cycle over a year by the annual rate of precipitation globally. This quotient implies that 750 kcal of solar energy are required to generate a kilogram of precipitation.

Solar energy also is used to produce paper. Paper is made from wood created by trees. This woody material is generated by the same process that generates edible energy. As with plants, only about 0.3 percent of the solar energy that hits a tree's leaf is stored in glucose (0.3 percent efficient). Most of this energy (50–70 percent) is used to produce roots. Only about 5 percent is used to make plant material that can be used to make paper. This implies that about 6,000 kcal of solar energy are needed to make about 1 kcal of plant material. On average, dry wood contains about 4,500 kcal per kilogram. This implies that about 27 million kcal of solar energy are required to make a kilogram of material that can be converted to paper.

The papermaking process also is relatively inefficient. The Kraft process is used to make most (about 80 percent) of the paper in the United States. In this process dry wood is soaked and heated in a liquid that consists mainly of sodium. Between 40 and 65 percent of the dry wood becomes usable pulp. These efficiencies imply that about 2 kilograms of dry wood are needed to make 1 kilogram of paper. All together, these inefficiencies imply that about 54 million kilocalories of solar energy are required to produce a kilogram of paper.

Calculating Your Footprint

How many kilocalories do you eat per day? Most dieticians suggest that people eat about 2,500 kcal per day. (On food containers, Calories with a capital C is equivalent to a kilocalorie.) The Food and Agriculture Organization (part of the United Nations) reports that the average U.S. citizen used 4,821 kcal per day in 2003 (this included food that was wasted). In the next chapter you will determine the energy content of your food. Multiply this sum by 7,500. This represents the amount of solar energy used to grow your food.

_____ kcal sunlight = 4,821 food kcal/day × 7,500 kcal/kcal

How much water do you drink? The British Dietetic Association suggests that a healthy adult should drink about 2.5 liters of water per day. If you assume that 750 kcal of solar energy are required to generate a kilogram of precipitation (1 liter of water weighs 1 kilogram), how much solar energy is required to generate 2.5 liters of drinking water?

_____ kcal sunlight = 2.5 liters/day × 750 kcal/liter

The average U.S. citizen used 314 kilograms of paper per year in 2004. This translates to about 0.86 kg/day. If you assume that 54 million kilocalories of solar energy are required to generate a kilogram of paper, how much solar energy is required to grow the paper you use every day?

_____ kcal sunlight = 0.86 kg/day × 5.4×10^7 kcal/kg

The total amount of solar energy you use for food, water, and paper is the sum of these three uses (your total should be about 101.3 million kcal per day).

Interpreting Your Footprint

What area of Earth's surface is required to capture the sunlight used to grow your food, generate your drinking water, and grow the trees used to make your paper? If we assume that the solar constant at the equator is 28,320 kcal/m²/day and that Earth's albedo is about 31 percent, you can use the following formula to calculate the area used:

_____ m²/day = [1.01×10^8 kcal/day/28,320 kcal/m²/day] × 1/0.69

As will be described in Chapter 11, the average person living in the United States uses more energy and materials than the average person on the planet. The Food and Agriculture Organization reports that the average person on Earth uses 3,272 kcal per day and uses 0.15 kilograms of paper per day. How much more sunlight do you use than the average person on Earth? How much more land is required to capture this sunlight?

STUDENT LEARNING OUTCOME

- Students will be able to explain why the land they use extends well beyond the house or apartment they live in.

A SIMPLE MODEL OF PHYSICAL SYSTEMS

Earth's atmosphere, oceans, and land surface are in constant motion. We perceive these motions as winds, ocean currents, volcanoes, and earthquakes. This motion is not spontaneous, and much of the motion is not random. Rather, the atmosphere, oceans, and land surface move in a fairly regular pattern that is powered by solar energy and heat from Earth's interior. These patterns are generated by gradients associated with geographic variations in the rate at which solar energy and heat from Earth's interior warm air, evaporate water, and move the land surface. The link between these geographic variations and the predictable patterns can be described by the operation of a simple engine known as a **convection cell.**

Despite its fancy name, you can understand the operation of a convection cell by following the flow of water in a coffeepot that is being heated on a stove (Figure 4.3). Movements in a convection cell are powered by the concentrated application of energy. For the coffeepot in Figure 4.3, the concentrated application of energy occurs at point (a), where the flame from the stove heats the water. Here the stove warms the water directly above the flame. This warming reduces the density of the water. As the water becomes less dense, it rises along path 1.

When the hot water reaches the top of the pot, it moves horizontally along path 2 in Figure 4.3. The flow of water along path 2 is powered by the upward flow of water along path 1 and the downward flow of water along path 3. The heat from the stove causes water to rise steadily from point (a). As the water rises, it pushes the water ahead of it along path 2. At point (c) water falls back toward point (d). This sinking motion pulls water toward point (c). As a result, water flows spontaneously along path 2 from point (b) to point (c).

Water cools as it moves along the surface of the coffeepot. This cooling increases its density. By the time the water moving along path 2 reaches point (c), it is more dense than the water below it. Without support from the underlying water, the water at point (c) spontaneously sinks in the same way that a bowling ball perched on top of a straw will fall. The water sinks (and the bowling ball falls) when the potential energy is converted to kinetic energy. As such, the sinking along path 3 is spontaneous.

Sinking water at point (d) creates a gradient that completes the convection cell. Movement along path 3 causes water molecules to "pile up" and increase the pressure of water at point (d). Meanwhile the water molecules that rise rapidly over the flame generate open space at point (a). This open space reduces pressure at point (a). The difference in pressure between points (d) and (a) creates a gradient that causes water to flow spontaneously from high pressure at point (d) to low pressure at point (a).

The flow along path 4 completes the circulation generated by the convection cell. The predictable flows along paths 1, 2, 3, and 4 continue so long as the flame heats the water at point (a). As we describe next, the flows in this simple convection cell can be used to understand how solar energy and heat from Earth's interior drive the circulation of Earth's atmosphere, oceans, and land surfaces.

GLOBAL PATTERNS OF ATMOSPHERIC CIRCULATION

Creating Gradients: Differences in the Intensity of Solar Radiation

Early explorers, such as Columbus and Magellan, were able to navigate sailing ships based on a detailed knowledge of wind speeds and directions in different regions of the world. The predictable pattern of wind speeds and directions is generated by an energy gradient that is associated with geographic differences in the rate at which solar energy reaches Earth's surface. Because Earth is a sphere, the amount of energy that hits its surface varies from place to place (Figure 4.4). The greatest amount of solar radiation is intercepted by surfaces that are perpendicular to the sun. The geographic location of these surfaces changes during the year. In mid-June, areas located 23.5° north of the equator (Tropic of Cancer) are perpendicular to the sun; areas 23.5° south of the equator (Tropic of Capricorn) are perpendicular in mid-December. (The equator is 0° latitude, and the north and south poles are 90° north and south latitude, respectively.) The location that is perpendicular to the sun changes between these two latitudes over the year because Earth is tilted on its axis.

This tilting and the curvature of Earth's surface reduce the amount of solar radiation that hits Earth's surface at higher latitudes. Near the north pole, the amount of

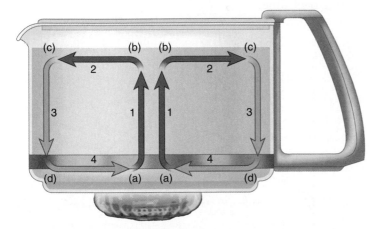

FIGURE 4.3 *Convection Cell Circulation of water in a coffeepot is driven by convection cells. The application of heat from the stove drives water from point (a) to point (b) along path 1. The constant flow of water along path 1 and the descent of water along path 3 cause water to flow from point (b) to point (c) along path 2. The potential energy of the dense water at point (c) is converted to kinetic energy as the water moves from point (c) to point (d) along path 3. Water flows from an area of high density at point (d) to an area of low density at point (a) along path 4.*

FIGURE 4.4 *Geographical Variation in the Intensity of Sunlight* The intensity decreases as one moves away from the equator.

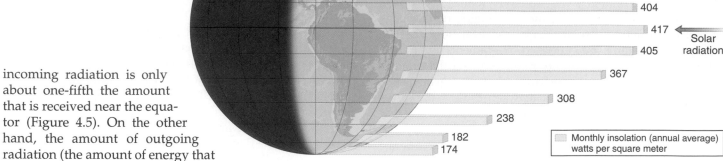

173
186
237
308
367
404
417 ← Solar radiation
405
367
308
238
182
174

☐ Monthly insolation (annual average) watts per square meter

incoming radiation is only about one-fifth the amount that is received near the equator (Figure 4.5). On the other hand, the amount of outgoing radiation (the amount of energy that returns to space from the Earth's atmosphere) does not vary greatly by latitude. The differences between incoming and outgoing radiation create areas near the equator that have a net surplus of energy and areas toward the poles that have a net loss of energy.

The difference between incoming and outgoing radiation determines an area's **radiation balance.** There is a net gain in energy where incoming radiation exceeds outgoing radiation. This excess energy makes the areas near the equator relatively warm. Areas near the poles tend to be relatively cool because there is a net loss in energy—outgoing radiation exceeds incoming radiation. By themselves, these differences imply that temperature would rise year after year near the equator and decline near the poles.

Even if you have never been to the equator or the poles, you know that these regions do not continuously warm or cool. Instead temperatures remain relatively constant over

time because heat is transferred between regions. The energy surplus and deficit set up a gradient that causes energy to flow from low latitudes (warm areas with an energy surplus) to high latitudes (cool areas with an energy deficit) (Figure 4.6). This transfer generates predictable flows of air and water. In the Northern Hemisphere, ocean water warms near the equator and flows northward. As the water flows north, it releases heat to the atmosphere. As described in this chapter's opening story, the eastern coast of the United States and the western coast of Europe are warmed by heat released from the Gulf Stream. These and other transfers do not eliminate the differences in temperature among regions (areas near the equator are still warmer than areas near the poles), but they prevent regions of Earth from warming and cooling over time.

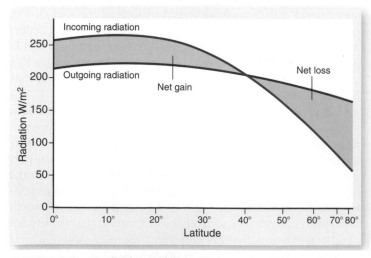

FIGURE 4.5 *Radiation Balance* The radiation balance of Earth by latitude. Incoming radiation exceeds outgoing radiation between the equator and about 40° north or south. At higher latitudes outgoing radiation exceeds incoming radiation.

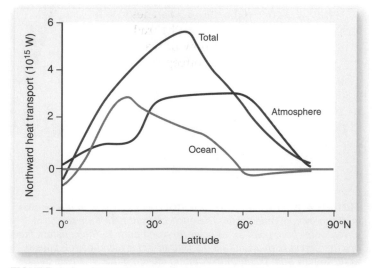

FIGURE 4.6 *Heat Transport* The quantity of heat carried north by the ocean, atmosphere, and total circulation. Positive values indicate heat moving north; negative values near the equator and near 60°N indicate that ocean heat at these latitudes tends to move south.

Circulation Cells

You can make an educated guess about the directions of the prevailing winds based on a location's latitude. The predictability is generated by latitudinal differences in net gain or loss of energy that set up temperature and pressure gradients. These gradients cause air to travel from areas of high pressure to areas of low pressure in movements that we call **winds.**

Near the equator, the solar energy surplus increases air temperature. This warm air becomes less dense and thus tends to rise (Figure 4.7). Rising air masses near the equator correspond to the water that rises in the coffeepot along path 1 in Figure 4.3. The rising air leaves behind a region of low pressure known as the **equatorial low.**

After rising vertically some 20 kilometers (12.4 miles) above Earth's surface, the warm air tends to move horizontally toward the poles. This flow is powered by the rising air beneath it. The high-altitude flow of air away from the equator is equivalent to the horizontal movement of water across the top of the coffeepot along path 2 in Figure 4.3.

As the air moves toward the poles, it cools and becomes more dense. This high, heavy air contains significant amounts of potential energy that is converted to kinetic energy at about 30° north and south of the equator. Here the high-altitude air becomes more dense than the air below it and sinks toward the surface. This downward flow of air corresponds to the sinking coffeepot water along path 3 in Figure 4.3. As the air sinks, it "piles up" at the surface and creates a region of high pressure that is known as a **subtropical high.**

The gradient in pressure between a subtropical high and an equatorial low causes air to move from the subtropical high to the equatorial low. These ground-level winds, known as the **trade winds,** correspond to the flow of water along the bottom of the coffeepot via path 4 in Figure 4.3.

These four flows constitute a regular pattern of circulation. There are six such circulation cells, three on each side of the equator (Figure 4.8). The two circulation cells on either side of the equator are known as **Hadley cells.** There are other areas of low pressure at about 60° north and south of the equator.

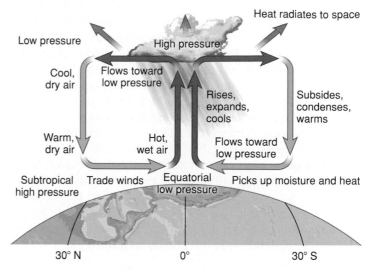

FIGURE 4.7 *Atmospheric Convection Cell* *Warm air rises at the equator. As it rises, it cools and generates large amounts of precipitation. Air cools as it moves away from the equator. At about 30° north or south the dense air falls toward Earth's surface. As it does, it warms and tends to reduce precipitation. Once this air reaches the surface, pressure differences cause air to move along the surface toward the equator.*

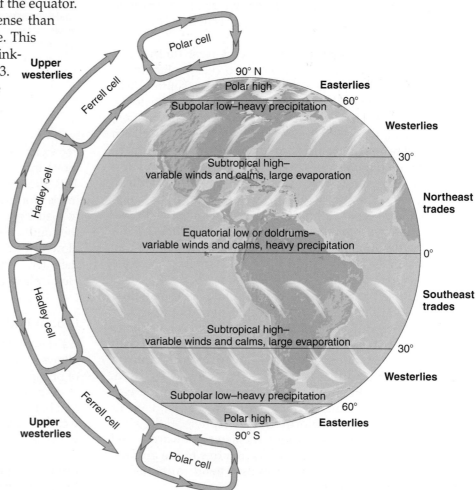

FIGURE 4.8 *Atmospheric Circulation* *Three cells on each side of the equator control vertical and horizontal air flows. Surface winds in the Northern Hemisphere bend toward their right due to the Coriolis effect; that same effect bends surface winds to their left in the Southern Hemisphere.*

These zones of low pressure, known as the **polar fronts,** are associated with two more sets of circulation cells. One set lies between 30° and 60° north and south of the equator, and each is known as a **Ferrell cell.** Finally, a third set of cells, each of which is known as a **polar cell,** cause air to circulate between 60° and 90° north and south of the equator.

The three circulation cells on each side of the equator move great quantities of heat from the equator toward the poles. This long-distance transport is possible because the cells are extremely leaky. Not all of the air that descends at the subtropical high blows back toward the equator. Some of the air and its heat are captured by the midlatitude Ferrell cells, which move the air and its heat toward the subpolar low at ground level. This flow is mainly responsible for the large quantity of heat that is transported toward the pole by the atmosphere between 30° and 60° (Figure 4.6).

Surface Winds

Each of the three circulation cells is associated with surface winds. The Hadley cells generate the trade winds, which move toward the equator. The low-pressure polar fronts create surface winds that move from the poles toward the polar front (known as the polar **easterlies**) and from the subtropical highs toward the polar fronts (known as the midlatitude **westerlies**). These names refer to the directions from which the winds come.

You may be surprised by the east–west direction of these winds—the cross sections of the cells in Figure 4.8 show that the surface winds move north or south. These north–south winds take on an east–west direction due to the **Coriolis effect.** The Coriolis effect makes objects in motion, such as an air mass, bend toward their right in the Northern Hemisphere and bend toward their left in the Southern Hemisphere (Figure 4.9). This bending is generated by differences in the rate at which Earth rotates. Earth rotates fastest at the equator and slows toward the poles. The rotation speeds differ because all parts of the planet must make one full rotation per day, but the distance traveled differs by latitude. The distance around Earth is greatest at the equator, so locations on the equator must travel the farthest, and go the fastest, to complete one revolution in a day.

The speed at which the planet rotates is transmitted to objects on the planet. An object at the equator moves to the east at the same rate as the equator. Air or water that starts at the equator retains this easterly speed as it moves north or south. As it moves away from the equator, the air passes land that is moving east at a slower speed. This difference in speed makes the air move to the east relative to the ground, which represents a bend in air flow to the right in the Northern Hemisphere. If the air is moving south from the equator, its faster eastward speed still makes it move toward the east relative to the ground. This eastward direction appears as a bend to the left in the Southern Hemisphere.

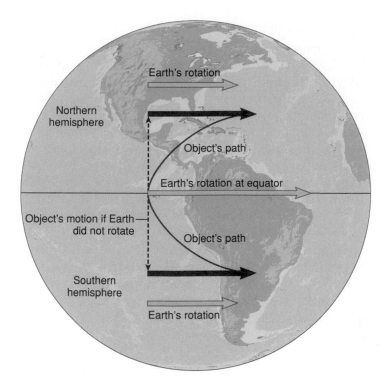

FIGURE 4.9   The Coriolis Effect *The speed of Earth's rotation is greatest at the equator and slows with latitude, as indicated by the green arrows. The difference between the green arrows is given by the black arrows that point east. This easterly direction is transmitted to objects that move north or south from the equator. As a result, objects moving north in the Northern Hemisphere bend toward their right (red line), and objects moving south in the Southern Hemisphere bend toward their left (red line).*

The bending caused by the Coriolis effect generates the east–west component of the ground-level winds in Figure 4.8. For example, the trade winds in the Northern Hemisphere move from the subtropical high toward the equatorial low. This direction makes the trade winds blow toward the south in the Northern Hemisphere. As they move south, they move to their right, which means they blow from the northeast to the southwest. In the Southern Hemisphere the trade winds blow toward the north. As they move north, they move to their left, which means that trade winds in the Southern Hemisphere blow from the southeast to the northwest.

Global Patterns of Precipitation

As described in Chapter 3, knowing a location's latitude allows you to make an educated guess about its annual rate of precipitation. This predictability is generated by the global pattern of atmospheric circulation. Movements of air along and between circulation cells influence the global location of wet and dry areas. At a local level, air movements interact with the land surface to determine the rate of precipitation.

The ability of air to hold water changes as it moves through a circulation cell. The moisture content of air tends to increase as it moves along the ground. In this portion of the cell, air picks up water that evaporates from the surface of oceans, soils, and vegetation. The northeast trade winds, for example, pick up large quantities of water as they blow across the warm portions of the Atlantic Ocean.

Air cools as it rises in the upward portion of the convection cell. Cool air holds less water than warm air. This causes the water to **condense** (change from a gas to a liquid) and precipitate as rain or snow. This effect generates large amounts of rain or snowfall near the equator (such as at the top of Mt. Kilimanjaro in Tanzania) and the polar fronts. For example, some of the water from the Atlantic Ocean that is picked up by the northeast trade winds falls as rain at Puerto Ayacucho (5° north), Venezuela.

Air absorbs little water as it moves parallel to Earth's surface along the upper arm of the circulation cell. This implies that the air is relatively dry as it enters the descending portion of the circulation cell. As the air descends, its temperature and pressure rise. These changes increase the air's ability to hold water, which means that this air is not likely to generate significant quantities of precipitation. As a result, many of the world's deserts are found under subtropical highs at about 30° north and 30° south. This effect is partially responsible for the lack of rain in areas such as Alice Springs, Australia, and the Sahara Desert in northern Africa.

Not all of the world's deserts are associated with the sinking arms of circulation cells. Local geography also creates many deserts. The Gobi Desert is located 40° north of the equator in the middle of the Asian continent. This area is a desert because it is far from the ocean. By the time surface winds reach the middle of Asia, they already have lost much of their water.

Other deserts are formed downwind from mountain ranges. As air passes over a mountain range, the air's temperature and pressure change in ways that generate **orographic precipitation** (Figure 4.10). Air is forced to rise as it flows over mountains. The increase in elevation cools the air, reduces its ability to hold water, and increases precipitation on the windward side of the mountain. The causes

for these high rates of precipitation mimic those associated with the rising arm of a circulation cell. As air sinks on the leeward side of the mountain, the dry air warms, which increases its ability to hold water and hence limits condensation. Consequently there is relatively little rainfall on the leeward side of a mountain range. The causes for these low rates of precipitation mimic those associated with the sinking arm of a circulation cell. This orographic effect creates deserts such as the Mojave Desert, which is located on the leeward side of the San Gabriel and San Bernardino mountains that run parallel to the California coast.

The different processes that create deserts imply that not all deserts are alike. Deserts at the subtropical highs tend to be relatively warm. But deserts that form far from an ocean often have a continental climate, which may mean very warm summers and very cold winters.

The Environmental Services Provided by Atmospheric Circulation

The global pattern of atmospheric circulation supports human well-being both directly and indirectly. Later chapters describe how the work done by air movements produces clean air and drinkable water. These services are vital, but so too is the predictability of climate.

To illustrate the importance of predictability, consider three possible patterns for atmospheric circulation. One possibility is no circulation. Air would move by **advection** (the motion of individual molecules), which is a slow process. Under these conditions the latitudinal temperature gradient would be much greater, making equatorial regions much warmer and polar regions much cooler than they are.

Another possibility is a completely random flow of air. Imagine what Earth would look like if the atmosphere circulated in a random pattern. The direction of winds would change unpredictably. Under these conditions an area could be hot during some periods and cold during others. Similarly, an area could be a desert one year and receive

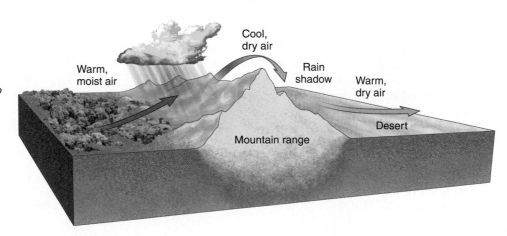

FIGURE 4.10 *Orographic Precipitation* Air cools as it rises over a mountain. As it cools, the air holds less water; this enhances precipitation on the windward side of the mountain. On the leeward side the sinking air warms, which increases its ability to hold moisture and decreases precipitation.

large amounts of precipitation the next. These ever-changing unpredictable conditions would challenge the survival of local plants and animals.

Now consider the benefits associated with the fairly predictable patterns described previously. Winds tend to warm some areas and cool others. Similarly, areas near the equator and areas near the polar fronts tend to have more precipitation than areas near the subtropical highs. And even when these patterns change, the change tends to be fairly predictable. In the Boston area we start our bicycle trips by riding toward the southwest in the summer and to the northwest in the winter. That way the wind will be at our backs on the way home.

Predictable patterns of temperature and precipitation allow plants, animals, and human societies to adapt to the range of conditions in each area. As described in Chapter 7, plant and animal species have evolved complex adaptations to cope with the local patterns of temperature and precipitation. However complex these adaptations are, they are small relative to the flexibility that would be needed to survive in environments where temperature and precipitation change randomly.

Societies also adapt to their climates. People who live in dry areas often invest large sums to build dams that capture and store water when it is available. People who live in wet areas may build dikes and levees to prevent frequent rains from washing away their homes. Life would be a lot more difficult (and expensive!) if local conditions alternated randomly between floods and droughts.

GLOBAL PATTERNS OF OCEANIC CIRCULATION

Physical Properties of Water

"Water is the basis for life" is a cliché. But this statement is true. Water is the primary component of living organisms, but its importance goes well beyond this. The physical properties of water play a critical role in the environmental services that determine our planet's ability to support life.

Water damps large swings in temperature. This ability is determined by its specific heat and its ability to circulate as a liquid. **Specific heat** is the amount of heat energy required to raise the temperature of a material of a particular mass. As described in Chapter 2, 1 kilocalorie is required to raise the temperature of 1 kilogram of water 1°C. The specific heat of water is 1.0. Most other substances on Earth have a much smaller specific heat (rock has a specific heat of about 0.2), which implies that water heats up and cools down less rapidly than most other materials. This is one reason why many people spend their summer vacation at the shore: Coastal areas have cooler summers (and warmer winters) than interior portions of the continent (the so-called continental climate).

The effect of the high specific heat of water is enhanced by the great amounts of energy that are required to change water from one state to another. Evaporating a kilogram of water (changing its physical state from a liquid to a gas) requires 540 kcal. As a result, evaporating water dissipates lots of heat; this is how sweating cools us. Similarly, melting a kilogram of ice at 0°C requires 80 kcal to convert it to a liquid at 0°C.

Finally, oceans heat up and cool down less rapidly than land because land is a solid. The sun's rays heat the first couple of meters of land, but they penetrate water far beyond the first couple of meters. Furthermore, the energy absorbed in the upper layers can be "mixed" into the deeper layers, which makes the ocean a much larger reservoir for heat than the land surface.

But there are limits on this mixing that are determined by another important physical property of water—the relationship between water's temperature and its density. Unlike most substances, the density of water does not increase steadily with declining temperatures. Water is most dense at about 4°C (39.2°F). Above and below this temperature, water becomes less dense. Ice at 0°C (32°F) floats on top of liquid water, which is why you can skate on a frozen lake. If the density of water increased as it cooled below 4°C, lakes would freeze from the bottom up rather than from the top down.

In most streams, lakes, and oceans, surface waters are warmer than deeper layers. The change in temperature with depth is known as the **temperature profile.** The temperature profile is important because water becomes less dense as temperature warms beyond 4°C. Differences in density and temperature create layers in aquatic environments. A layer of less dense warm water floats on top of a layer of denser cold water (Figure 4.11). Differences in density also slow the

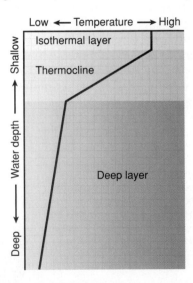

FIGURE 4.11 *Layers in Water Bodies The red line indicates that temperature generally decreases with depth. This means that less dense warm waters float atop more dense cold waters. These layers are separated by the thermocline, which is the portion of the water column where the temperature and density change rapidly.*

rate at which water in the layers mixes. The **thermocline** is the portion of the water column where temperature changes very rapidly. This change in temperature and density acts as a barrier that makes it difficult for cold bottom waters to mix with warm surface waters. As described in Chapter 7, this barrier plays an important role in aquatic ecosystems.

This layering is reinforced by the rate at which water absorbs sunlight. Pure (distilled) water absorbs about 53 percent of the incoming solar radiation in the first meter. This energy is then converted to heat. Another 53 percent of

the remaining radiation is absorbed by the next meter, and so on. The exponential decline in sunlight makes ocean bottoms dark as well as cold.

Patterns of Oceanic Circulation

The pattern of surface winds (Figure 4.12a) looks very similar to the pattern of surface ocean currents (Figure 4.12b). This is no coincidence. Surface ocean currents are generated by wind patterns. Given information about a location's latitude and

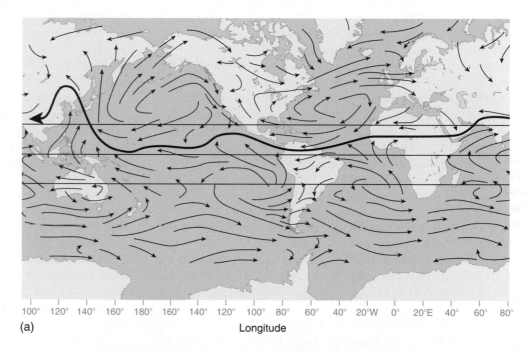

(a)   Longitude

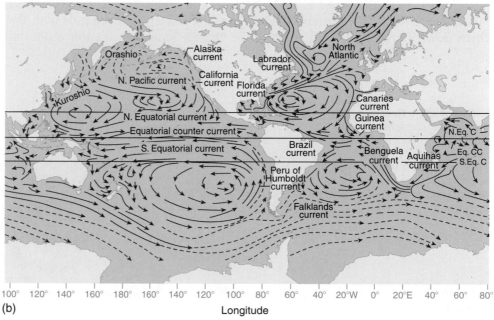

(b)   Longitude

FIGURE 4.12   *Oceanic and Atmospheric Circulation (a) Surface winds tend to blow from the east along the equator and between 30° north and south and blow from the west between 30° and 60° north and south of the equator. (b) These surface winds generate circular patterns in surface ocean circulation. Water flows west along the equator and north and south when it bumps against land at the western end of the ocean basin. As it reaches 30° north and south, changes in surface winds blow the water back across the ocean. At the eastern end of the ocean basin, the water is again deflected north and south.*

coastal position (eastern or western boundary of an ocean basin), you can make an educated guess about the direction of surface ocean currents.

As winds blow along the ocean surface, some of the air's movement is transferred to the water. As this movement is transferred to deeper layers in the ocean, its direction changes due to **Ekman transport.** Ekman transport describes a process that causes surface ocean currents (first 100 meters—328 feet) to move 90° to the right of the surface winds in the Northern Hemisphere and 90° to the left of the surface winds in the Southern Hemisphere. In short, the atmosphere and ocean can be viewed as a series of layers, with the atmosphere being the least dense and the deep ocean layers being the most dense. Surface winds drag the surface ocean layer, and the surface ocean layer drags the one below it, and so on. At each transfer between layers, the movement slows and the Coriolis effect bends the movement to the right (Northern Hemisphere) or left (Southern Hemisphere) relative to the layer above it. When all of these changes are added up, the overall ocean current moves 90° to the left or right of the surface winds.

The combination of surface winds and Ekman transport generates the general pattern of surface ocean circulation. At the equator surface waters tend to move west. South of the equator surface currents have a southward component, while north of the equator currents have a northward component. This tends to move surface waters away from the equator, which makes the surface layer at the equator relatively shallow. The shallow surface ocean layer at the equator is critical to El Niño events, which are described in the next section.

The surface waters continue west until they reach the western boundary of the ocean basin (the eastern shore of a continent). Here the water is deflected north or south. This deflection generates currents such as the Gulf Stream, which flows along the eastern coast of the United States,

and the Kuroshio Current, which flows along the eastern coast of Japan. As these waters move north and south, they transfer heat to higher latitudes. Warm water in the Gulf Stream is one reason for the relatively mild climate of the Mid-Atlantic states (New York, New Jersey).

These currents flow north and south along the western boundaries of the ocean basin until they reach latitudes between 30° and 60°. Here the dominant winds, the westerlies, blow from west to east, causing surface waters to flow east until they reach the eastern boundary of an ocean basin (the western shore of a continent). The currents again are deflected north and south, which causes cold surface water to flow back toward the equator along the western coasts of continents. Examples of these equator-bound flows include the Alaska and California currents. Because these currents originate in cold Arctic water, the water moving south past San Francisco (38° north) is colder than the water flowing north past New York City (41° north).

Together these individual currents generate a circular motion known as a **gyre.** In general, surface currents tend to move west along the equator, toward the poles along the western boundary of ocean basins, east in high latitudes, and back toward the equator along the eastern boundary of ocean basins.

In addition to this circular flow of surface waters, ocean waters also follow a predictable path between the surface and deep layers. These flows are known as **thermohaline circulation.** This name is associated with the circulation's drivers—the direction and strength of deep ocean currents are driven by differences in temperature and salinity. Thermohaline circulation moves water from the surface to the bottom, around the world, and back to the surface (Figure 4.13).

We start our trip through thermohaline circulation as surface currents in the Atlantic Ocean move water toward the pole with the power equivalent of 100 Amazon Rivers. As the water moves toward the pole, it loses heat, which

FIGURE 4.13 *Thermohaline Circulation* Changes in temperature and salinity drive vertical and horizontal movements in ocean waters. Cold dense surface waters sink in high latitudes in the north Atlantic and along the Antarctic. Deep water flows back toward the surface in various areas in the Indian and Pacific Ocean.

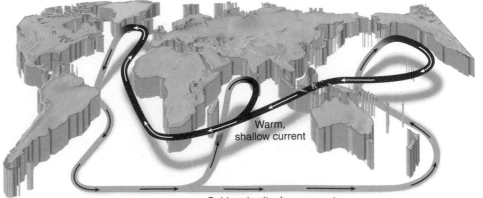

Warm, shallow current

Cold and salty deep current

warms the eastern U.S. coast and western Europe. Changes in the strength of thermohaline circulation over the last 18,000 years have strongly affected climate: Slowdowns are associated with cooling.

By giving up its heat, the water cools, which increases its density. This reduces the differences in density along the water column (from top to bottom). Without much difference in density, cold water sinks to depths of 1,000 meters (3,281 feet) off the coast of Greenland. From here the deep water moves south across the equator and continues until it reaches the Antarctic, where it joins water that sinks along the Antarctic coast. Together these waters form Antarctic bottom water, which flows eastward along the Antarctic. As the water reaches the Pacific Ocean, some of the deep water is deflected north by the Pacific plate (see the next section for a description of Earth's crustal plates).

Surface waters constantly join deep water circulation; therefore, an equal amount of deep water must return to the surface. Most of the deep water that returns reaches the surface via the slow process of diffusion. This flow occurs throughout the ocean basin.

Huge quantities of deep ocean water rise back to the surface in a few select areas. These areas are known as **upwellings.** Many of these upwellings are created off the western shores of continents (the eastern boundary of an ocean basin) by the interaction of winds, surface currents, and the shoreline at about 30° north or south. In this zone the combination of trade winds and Ekman transport causes surface waters to flow away from the shoreline. You can envision this as wind "scraping" away the top layer and piling it on the far side of the ocean basin. Because the wind prevents surface water from returning and evening out the differences in sea level, deeper water takes the place of the surface water. These movements generate massive upwellings off the western coasts of Africa and South America. As the deep water rises, it brings many nutrients, which support high rates of biological productivity. The economic importance of these flows is described next.

El Niño: Illustrating the Environmental Services of Oceanic Circulation

Oceanic circulation provides several important environmental services. By redistributing heat, oceanic circulation is an important determinant of climate. Equally important, oceanic circulation constantly redistributes materials that are critical to life. Without this work, the ocean would support much less life than it does currently.

These environmental services are highlighted by the economic and biological impacts of an **El Niño,** which is a change in ocean circulation off the coast of Peru that occurs every three to seven years. In most years southeast trade winds blow from the subtropical high off the coast of Peru to the Indonesian equatorial low, which is normally just north of Australia (Figure 4.14a). These winds blow surface waters across the Pacific Ocean so that the

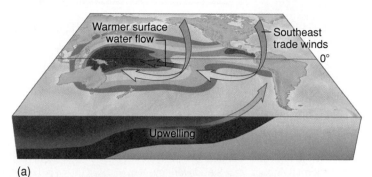

(a)

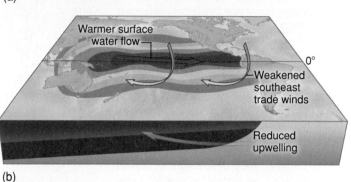

(b)

FIGURE 4.14 *El Niño Events* (a) The pattern in atmospheric and oceanic circulation during a non–El Niño year. At this time, the southeast trade winds blow across the Pacific to the Indonesia low, where the water that evaporates from the ocean precipitates along the rising arm of the Hadley cell. The strong trade winds blow surface waters away from the coast of Peru, which raises sea level in the western Pacific and allows deep water to rise to the surface off the coast of South America. (b) During an El Niño event, the southeast trade winds weaken. This reduces rainfall in southeast Asia and Australia and allows warm water from the western Pacific to flow east. This prevents deep water from rising to the surface off the coast of Peru.

sea level in the western Pacific is about 1 meter higher than the sea level in the eastern Pacific. This difference in sea level remains so long as the trade winds blow "normally." Years in which this pattern is especially strong are termed *La Niña* events.

As the trade winds blow across the warm Pacific waters, they pick up large amounts of moisture. This moisture is the source for the rain that nourishes tropical rain forests in Indonesia and wheat production in Australia, as well as the monsoon rains that support rice production in India. At the same time the trade winds and the slope in sea level allow deep water to move to the surface off the coast of Peru. This creates a strong upwelling that supports a large population of anchovy fish.

During an El Niño event this pattern of atmospheric and oceanic circulation changes (Figure 4.14b) due to a weaker than normal zone of high pressure as measured at Tahiti. Scientists have not yet identified the disturbance that initiates an El Niño event. (See *Case Study: El Niño: A Link among Atmospheric, Oceanic, and Crustal Circulation?*) Regardless of the initial cause, the changes that lead to an El Niño event

CASE STUDY

El Niño: A Link among Atmospheric, Oceanic, and Crustal Circulation?

Most scientists believe that El Niño events (discussed in this chapter) are generated by interactions between atmospheric and oceanic circulation. According to this hypothesis, El Niño events can be triggered by a large thermal input that raises sea surface temperatures. These thermal inputs can be driven by the normal pattern of oceanic circulation. Nonlinear and stochastic elements in the pattern of oceanic circulation cause variations in the rate at which heat accumulates and dissipates in various portions of the ocean. These variations can warm portions of the sea surface and trigger an El Niño event.

A geophysicist, Daniel A. Walker, hypothesizes that a different sequence of events produces an El Niño event. Walker says that the thermal input to the oceans comes from Earth's interior. This hypothesis is based on an intriguing correlation of seismic activity under the eastern portion of the Pacific Ocean near Easter Island and the onset of El Niño events. This correlation can be used to illustrate possible linkages among the physical systems of planet Earth and the difference between statistical correlation and physical causation.

The East Pacific Rise is located west of Easter Island. Along this rise, tectonic plates move 160–170 mm (6.3–6.7 inches) per year. This rate is one of the most rapid in the world. As a result, seismic activity along the East Pacific Rise has been studied extensively for more than thirty years. During this period scientists have tracked the number of earthquakes and the amount of energy they release.

While plotting these data, Walker noticed an interesting pattern: Months with the greatest number of earthquakes or months with earthquakes that release the greatest amount of energy precede the onset of El Niño events (Figure 1). This hints at a relationship between seismic activity and El Niño events. A simple statistical analysis suggests that if the seismic events and the El Niño events occurred randomly, the probability of finding a sequence in which periods of heightened seismic activity precede El Niño events would be 1 out of 313. Such long odds imply that the sequence probably is not generated by random chance. Rather, there may be a physical connection between seismic activity and El Niño events.

But statistical correlation should not be confused with physical causation. Correlation

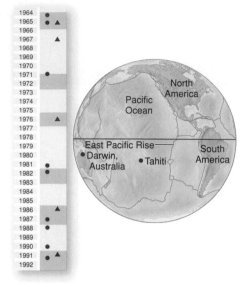

FIGURE 1 *Correlation between El Niño Events and Seismic Activity* The timing of El Niño events is indicated by the shaded years. Notice that the shaded areas are correlated with months in which there are earthquakes along the East Pacific Rise with a magnitude of eight or more (triangles). A similar correlation exists between El Niño events and months with the greatest release of seismic energy along the East Pacific Rise (circles). (Source: Redrawn from a figure in The New York Times.)

indicates that the timing of the two events is similar but *does not mean* that the two events are connected. For example, some Wall Street analysts have noticed a correlation between the winner of the Superbowl and the performance of the stock market. When the Superbowl champions are a team from the old American Football League, stocks usually decline over the year; stocks rise over the year if the winner is a team from the old National Football League. Clearly the victor does not determine the performance of the stock market. So this relationship is simply a correlation. To show causation, scientists must describe a physical mechanism that connects events.

Scientists have come up with several explanations for how seismic activity may trigger El Niño events. These hypotheses are based on the relationship between air temperature and pressure. In general, warm air has less pressure than cold air, so hypotheses seek to explain how seismic activity along the East Pacific Rise could affect air temperature.

One explanation postulates that seismic activity is associated with volcanic activity on the seafloor that releases large amounts of heat. If this heat were to reach the ocean surface, it could raise sea surface temperature, which would raise air surface temperature, which would reduce atmospheric pressure. Many scientists are skeptical of this hypothesis because the thermocline probably would prevent the heated water from reaching the surface.

Another explanation focuses on changes in the rate at which surface waters absorb sunlight. Scientists find that water from ocean vents has a higher concentration of nutrients, which increases the local rate of photosynthesis. If seismic activity increases the concentration of materials critical to sea life and if those materials reach the surface, the resultant increase in biological activity could increase the amount of sunlight that is absorbed by the upper layers of the ocean. This could warm the sea surface, increase air temperature, and reduce atmospheric pressure. Another explanation hypothesizes that seismic activity along the East Pacific Rise alters circulation in the surface and intermediate layers of the eastern Pacific in a way that increases surface temperature and reduces pressure.

Most scientists disagree with the explanations for the statistical correlation between seismic activity and the onset of El Niño events offered by Walker and his colleagues. Their hypothesis may be validated by further research. But for now, the statistical relationship between El Niño events and seismic activity is nothing more than an interesting correlation.

ADDITIONAL READING

Walker, D. "More Evidence Indicates Link between El Niños and Seismicity." *Eos* 76 (1995): 33–36.

Walker, D. "Seismicity of the Pacific Ocean Rise: Correlations with the Southern Oscillation Index?" *Eos* 69 (1988): 857–867.

STUDENT LEARNING OUTCOME

- Students will be able to explain how differences between correlations and causation affect our ability to predict what will happen in the future.

are linked via a positive feedback loop, which reinforces the initial disturbance so that the zone of high pressure off the coast of Peru becomes progressively weaker.

As the subtropical high becomes weaker, so too does the pressure gradient between the coast of Peru and the Indonesian low (Figure 4.15). The weaker gradient slows the trade winds. The change in the strength of the trade winds reduces the amount of moisture that moves toward the Indonesian equatorial low. As a result the monsoons fail in India and Australia has a drought. Conversely, rainfall tends to increase over portions of the United States, the Pacific coast of Australia, and northwestern Peru. The weakened trade winds cannot maintain the difference in sea level, so warm surface water comes rushing back across the Pacific Ocean via the relatively shallow surface layer along the equator. This movement slows the upwelling off the coast of Peru.

These changes in atmospheric and oceanic circulation disrupt environmental services, which affects biological and human systems throughout the world. The El Niño of 1982–1983 caused a drought that reduced agricultural production in Brazil from 18 percent of GNP in 1978 to 8 percent of GNP in 1983. Without the upwelling off the coast of Peru, there is a big reduction in the anchovy population. Without as many anchovy to catch, fishermen are put out of business, and this loss of income ripples through the Peruvian economy. The El Niño of 1982–1983 caused about $2 billion in damages to the Peruvian economy. These economic losses are not inevitable—advances in our ability to predict El Niño events and clever policy can ameliorate the impacts (see *Policy in Action: Policy Responses to El Niño*).

MOVEMENTS IN EARTH'S CRUST

Other than during an earthquake, you cannot feel Earth's crust moving—although with advanced GPS (global positioning satellites) equipment you can measure these movements. Despite this subtlety the slow but continuous motion

of Earth's crust plays an important role in supporting life. But unlike the solar-driven services provided by the atmosphere and ocean, environmental services provided by Earth's crust are powered by heat from Earth's interior.

Plate Tectonics: Global Circulation of the Crust

The idea that the continents fit together like a giant jigsaw puzzle has been around for a long time. In 1596 a Dutch mapmaker, Abraham Ortelius, wrote that the eastern coast of South America "fits into" the western coast of Africa. In 1912 this notion was formalized into the scientific theory of *continental drift* by the German meteorologist Alfred Wegener. Continental drift hypothesizes that the continents were part of a continuous landmass a long time ago and since have moved to their present positions. Wegener gets credit for developing the hypothesis because he proposed tests that went beyond the jigsaw puzzle analogy. If the continents were joined, the same plants and animals would have lived in the west coast of South America and the east coast of Africa. Wegener pointed out that the fossilized plants and animals found in the eastern coast of South America were related to those found in the western coast of Africa. If the continents were located at different latitudes and longitudes long ago, their climates would have been different from today's and so would the types of plants and animals present. Consistent with this hypothesis, the fossilized remains of tropical plants are found in Antarctica, which is too cold to support any plants, let alone tropical plants. Unfortunately for Wegener, his hypothesis did not explain how the continents could move.

This final piece of the puzzle came together by the 1960s, when better maps of the ocean floor and earthquake epicenters showed that Earth is not a ball of solid material. Rather, Earth consists of separate plates that can move. Each **plate** is a large rigid slab of solid rock. The movements of these plates and a host of related phenomena, such as earthquakes and volcanoes, are explained by the theory of plate tectonics.

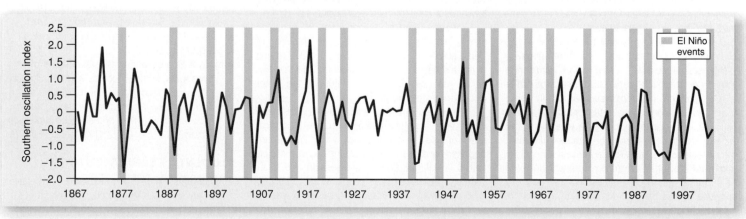

FIGURE 4.15 *Measuring El Niño Events* The southern oscillation index (red line) measures the difference in air pressure at Tahiti and Darwin, Australia. A large negative value indicates an El Niño event, such as the severe El Niño event of 1982–1983.

POLICY IN ACTION

Policy Responses to El Niño

The El Niño event of 1982–1983 was the most economically destructive on record. Damage to infrastructure and reductions in agricultural and industrial production cost the global economy nearly $13 billion. Such losses raise the question of whether decision makers can develop policies to lessen the impact of future El Niño events. The answer seems to be yes, based on the experiences of Peru and Brazil.

The success of policy in Peru and Brazil is based on an interdisciplinary understanding of the natural and social science aspects of El Niño events. The first step in developing successful policy is recognizing El Niño for what it truly is—a global phenomenon that links atmospheric and oceanic systems (Figure 1). Until the early 1970s most people thought that El Niño was a local phenomenon that affected only the Peruvian coast. The El Niño events of 1972–1973 and 1976–1977 changed that perception because their climatic and economic impacts were felt worldwide. This recognition led to interdisciplinary scientific efforts to understand the causes and effects of El Niño. This research was organized by the National Climate Program and the World Climate Research Program in the United States. At the international level, South American nations along the eastern shore of the Pacific formed the Regional Programme for the Study of the El Niño Phenomena. After the 1982–1983 event, these efforts evolved into a global program known as the Tropical Ocean and Global Atmosphere.

These programs generated considerable information about the physical and social aspects of El Niño events. One of the most important accomplishments was the development of statistical models that allowed climate scientists to forecast El Niño events several months in advance. Early warning makes it easier for policy makers to minimize the impacts of El Niño events.

But early warning is only the first step in the chain of successful policy. Forecasts of upcoming El Niño events are of little use if nobody outside the government believes them. Government credibility is a big problem in nations such as Brazil, where a majority of people believe that natural resource policy is designed to benefit the wealthy at the expense of the poor. To overcome these difficulties, the governor of Ceará, a region in the northeastern portion of Brazil that is prone to drought during El Niño events, launched a grassroots campaign to convince people that the forecasts of an impending drought issued by the local water authority were based on science.

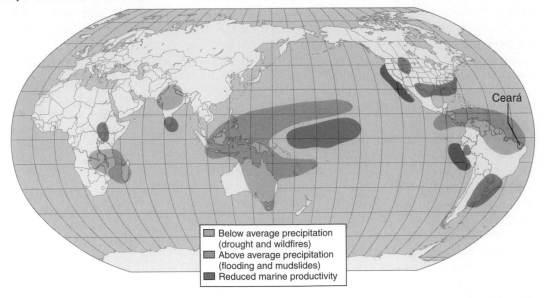

FIGURE 1 *Climatic and Biotic Effects of El Niño Events* *El Niño events are associated with droughts in some areas, such as Africa and southeast Asia (red areas), and floods in other areas, such as southern areas of the United States, and changes in fisheries off the western coasts of North and South America. (Figure from NOAA.)*

Once decision makers have information about an impending El Niño event and have the people's attention, policy must encourage actions that reduce the impacts. One possible array of actions is illustrated by Brazil's response to the warning that there would be an El Niño in 1991–1992 (the forecast was correct). Through the government agricultural agencies, farmers were advised that they should plant their seeds relatively early in the growing season. In addition, farmers were advised to plant strains of crops (cultivars) that mature in a relatively short period and are more resistant to drought (have a higher wilting point—see Chapter 15). Although these alternative cultivars produce less food under good conditions than the cultivars usually planted, the alternatives perform better under the drought conditions that are associated with an El Niño event. To ease the financial difficulties associated with these changes, the government provided low-interest loans through local banks.

Another suite of actions involved managing water use. To increase supply, the governor of Ceará accelerated the construction of a $13 million dam on the Pacajus River. To slow the rate at which water was used (mostly for irrigation), the water agency restricted water use.

How successful were these policies? Restricting water use extended supply. At normal rates of use the water supply would have been depleted by December 1992; instead the supply lasted until April 1993. This extension and other policies ameliorated the effect of the 1991–1992 El Niño on agriculture in Brazil. In 1992 rainfall was 23 percent below normal and agricultural production declined by about 20 percent. This reduction was much smaller than the 85 percent decline in agriculture that accompanied the 30 percent reduction in rainfall during the 1987 El Niño. The main difference between these two El Niño events was that there were no policies to counteract the 1987 El Niño. Clearly the ability to forecast El Niño events, the ability to convince citizens that the events are coming, and the technologies to offset the effects of drought make it possible to develop policy that can offset short-term climate fluctuations. Whether these policies hold any lessons for efforts to deal with long-term changes in climate is explored in Chapter 13.

ADDITIONAL READING

Golnaraghi, M., and R. Kaul. "The Science of Policymaking—Responding to ENSO." *Environment* 37, no. 1 (1995): 16–44.

Lemos, M. "A Tale of Two Policies: The Politics of Climate Forecasting and Drought Relief in Ceará, Brazil." *Policy Sciences* 36, no. 2 (2003): 101–123.

Patt, A., P. Suarez, and C. Gwata. "Effects of Seasonal Climate Forecasts and Participatory Workshops among Subsistence Farmers in Zimbabwe." *Proceedings of the National Academy of Sciences of the United States of America* 102, no. 35 (2005): 12623–12628.

STUDENT LEARNING OUTCOME

- Students will be able to explain why effective policy to lessen the effects of environmental changes must go beyond the ability to predict those effects.

The theory of plate tectonics is based on the notion that Earth consists of layers (Figure 4.16). These layers are defined by the transfer of heat and motion. The **asthenosphere,** which is the upper part of the mantle, has a consistency that is somewhere between a liquid and a solid. This allows it to move large quantities of heat from Earth's center. This energy is then transferred to the **lithosphere,** which is the outermost layer of crust and uppermost mantle. The lithosphere is not a continuous layer that encircles the planet. Rather, the lithosphere consists of about a dozen major plates (Figure 4.17). Continents ride on top of these plates.

The movements of these plates can be understood using the notion of the convection cell from Figure 4.3. The rising portion of the convection cell pushes material from the asthenosphere between the plates at midocean ridges (Figure 4.18(a)). These movements are equivalent to path 1 in Figure 4.3.

The crust also moves horizontally, parallel to Earth's surface. The creation of new crust pushes the plates apart. This lateral motion of the plates, which produces continental drift, corresponds to movements along path 2 in Figure 4.3.

The speed at which a plate travels depends on the rate at which the convection cell builds new crust. Crust moves about 2 centimeters (0.8 inch) per year in the North Atlantic; movements in the Pacific usually are greater, up to 16 centimeters (6.3 inches) per year. The slow but steady movement greatly shifts the location of crust over millions of years. The eastern coast of South America "fits" into the western portion of Africa because about 180 million years ago these two continents were part of the same landmass (Figure 4.19).

The amount of new crust formed at the ridges is roughly balanced by the amount of crust destroyed. Crust is destroyed in ocean trenches. These trenches occur at the edges of the plates opposite to the ocean ridges. At these edges plates "bump" into one another. On continents this bumping causes both edges to buckle. In oceans one plate slips below the other (Figure 4.18(b)). As the old crust slips below the adjoining plate, the crust heats up and drops into the mantle. This corresponds to the descending movement along path 3 in Figure 4.3.

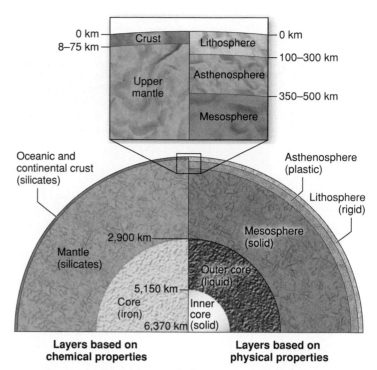

FIGURE 4.16   *Earth Layers*   *The layers of Earth as defined by physical properties of heat transfer and motion, such as the lithosphere, asthenosphere, and mesosphere, or layers as defined by chemical properties, which include the crust, mantle, and core.*

The Rock Cycle: The Circulation of Materials within Earth's Crust

The movement of material between the crust and the mantle is only one of the material cycles that occur at Earth's surface. Another cycle involves the building up and wearing down of land surfaces. In this process the rocks that make up Earth's surface change in a process known as the **rock cycle** (Figure 4.20).

In the constructive phase of the cycle, rocks are built primarily from molten material that originates in the upper mantle and the crust. Rocks that are formed from this molten

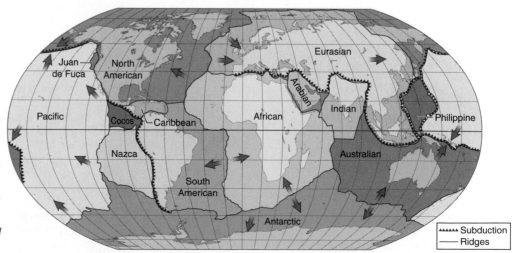

FIGURE 4.17   *Tectonic Plates*   *Earth's outer layer consists of a series of plates, which move in the directions indicated by the arrows. New crust forms at midocean ridges, which is why arrows near these ridges point away from each other. Crust is destroyed in subduction zones, where one plate moves under another—here the arrows point toward each other.*

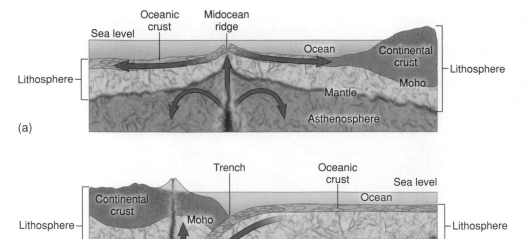

(a)

(b)

FIGURE 4.18 *Creation and Destruction of Crust* (a) The creation of new crust occurs at midocean ridges, where molten material from Earth's interior reaches the surface. (b) The destruction of crust occurs in midocean trenches, where one plate drops below another.

material are known as **igneous rocks,** which means "born of fire." As implied by their name, the formation of igneous rock is driven by heat from Earth's interior. Examples of igneous rocks include basalt and pumice.

Over time rocks break down from exposure to sunlight, water, and intervals of heating and cooling. These processes, known as **weathering,** break rocks into relatively fine particles called **sediment.** Particles are carried away from their parent material by wind or water in a process known as **erosion.** Because the movements of

wind and water are fairly regular, erosion often deposits the particles in a regular pattern. If the pattern persists for a long period, particles accumulate atop one another in a series of layers.

As the particles accumulate, upper layers exert pressure on the lower layers. This pressure squeezes (compaction) and cements (cementation) the particles together as **sedimentary rocks** in a process known as **consolidation.** Small particles are converted to shale, and larger particles are converted to sandstone. Over long periods thick layers

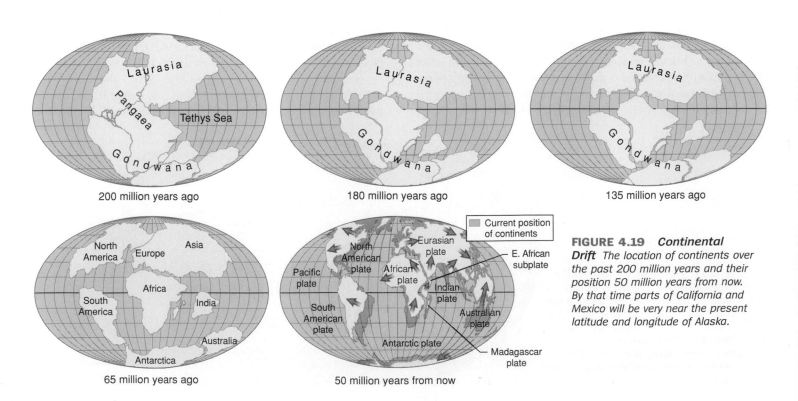

200 million years ago

180 million years ago

135 million years ago

65 million years ago

50 million years from now

FIGURE 4.19 *Continental Drift* The location of continents over the past 200 million years and their position 50 million years from now. By that time parts of California and Mexico will be very near the present latitude and longitude of Alaska.

FIGURE 4.20 *The Rock Cycle*
Igneous rocks are formed from molten material that originates in Earth's interior. Over time, igneous rocks may be broken into smaller particles and deposited. These particles are eventually converted to sedimentary rocks. If sedimentary or igneous rocks are subjected to sufficient heat and pressure, they are converted to metamorphic rocks. These rocks can be melted back to the precursors of igneous rocks or exposed to the surface, both of which can start the rock cycle anew.

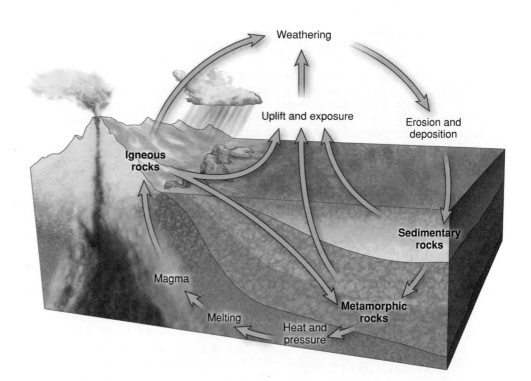

of sedimentary rock may form. For example, sedimentary rock nearly 10 kilometers (6.2 miles) thick lies at the bottom of some areas of the Gulf of Mexico, where the Mississippi River dropped much of the sediments it carried from interior portions of the North American continent.

If the burial of sedimentary or igneous rocks subjects them to extended periods of high temperature and pressure, they may be changed to **metamorphic rocks.** Continuing with the previous example, sandstone may be changed to quartzite and shale may be converted to slate. If the temperatures and pressures are high enough, the metamorphic rock may melt back to the precursors of igneous rocks. Similarly, once exposed to the surface, metamorphic rocks may weather. These two processes start the rock cycle anew.

Natural Resources and Environmental Services from Movements in Earth's Crust

Some of these environmental services are fairly obvious. The people of Iceland warm almost all of their homes and buildings with heat energy from the rising arm of a convective heat cell (Iceland sits nearly atop the Mid-Atlantic Ridge). This reduces the amount of coal, oil, or natural gas they need to import.

The motion of Earth's surface also is critical to the evolution of life. Plate tectonics separated the continents more than 100 million years ago. On each continent evolution proceeded separately, which increased the diversity of living organisms. There are many more families of mammals than reptiles in part because mammals evolved after the continents had separated. As described in Chapter 12, biological diversity generates a wide variety of environmental services.

Heat from Earth's interior also plays a critical role in the chemical content of ocean water. Cold ocean water sinks into cracks in the ocean floor near midocean ridges. Here the water is heated, which spurs a series of chemical reactions that change the composition of seawater. For example, these reactions remove magnesium and increase the concentration of calcium and silicon. Silicon is an important component of diatoms: microscopic sea organisms that are an important source of food for many creatures. As the seawater is heated further, it is ejected back into the water column. This circulation allows the ridges to process significant quantities of seawater over time. A quantity of water equal to the entire ocean is processed every 8 million years. Although this seems like a long time to people, it is a short time when compared to the geological history of the planet.

Heat from Earth's interior also creates mineral resources that are directly useful to humans. If metals and other minerals were scattered randomly throughout Earth's crust, they would be too diffuse to mine economically. But heat from Earth's interior and the motion of Earth's crust creates areas where metals and minerals are concentrated severalfold relative to their abundance in an average piece of crust. This makes it less expensive for humans to mine these resources.

The motion of Earth's crust and the rock cycle help form fossil fuels (coal, oil, and natural gas), which are the most important source of energy for industrial societies. Sedimentation of eroded particles buried organic material deep in the crust, away from the oxygen that would have allowed it to decay completely. The pressure of the overlying sediments and heat from Earth's interior refined the material and concentrated its energy content.

SUMMARY OF KEY CONCEPTS

- Latitudinal differences in the rate at which solar energy strikes Earth's surface create temperature gradients that support three circulation cells on each side of the equator. These circulation cells generate predictable patterns in temperature (temperature decreases with latitude), wind direction (toward the equator between the equator and 30° and between 60° and the poles, but away from the equator between 30° and 60°), and precipitation (precipitation is greatest at the equator and 60°).

- The physical properties of water are critical to life on Earth. Its specific heat is greater than most other materials, which slows changes in temperature. Water also requires large amounts of energy to change among its three states. This too slows changes in temperature. Water is most dense at 4°C (39.2°F), and as its temperature moves from this value, its density changes significantly. Vertical temperature differences stratify water bodies into different layers, which makes it difficult for materials in one layer to reach another layer.

- Oceans circulate both horizontally and vertically. Horizontal circulation along the surface is driven by surface winds, Ekman transport, and the location of continents. These factors cause ocean surface waters to move in a circular pattern. Thermohaline circulation is driven by temperature, winds, and salinity so that surface waters sink in the North Atlantic and along Antarctica. Bottom waters return to the surface via diffusion throughout the ocean and in concentrated flows known as upwellings.

- The materials that make up Earth's surface move through a cycle. Heat from Earth's interior forces new material to the surface, where it cools and adds to the surface material. This addition pushes the plates away from these areas. Surface material is destroyed when it drops back toward Earth's interior. These forces change the latitude and longitude of continents over time.

- The rocks that make up Earth's surface also move through a cycle. Surface rocks are weathered into smaller particles, which are carried away by wind and water and may be deposited in a fairly predictable pattern. Over time the particles are converted back into rocks.

REVIEW QUESTIONS

1. What are the sources of the energy that emanates from Earth's interior and the sun? What are their relative magnitudes?

2. Explain the circulation of Earth's atmosphere, oceans, and surface materials in terms of the convection cell shown in Figure 4.3.

3. If Earth rotated to the west, how would the pattern of surface winds and surface ocean currents change relative to their current directions?

4. What would happen to the pattern of deep water circulation in a lake if water were completely transparent to solar energy (that is, if all of the light that struck the lake surface was absorbed by the lake bottom)?

5. As will be explained in Chapter 20, oil is derived from living organisms that lived millions of years ago. Much of the world's oil supply is found in the Persian Gulf, where there is little biological activity to serve as a source of oil. How can you explain this seeming contradiction?

KEY TERMS

advection	Hadley cell	sediment
albedo	igneous rocks	solar constant
asthenosphere	lithosphere	specific heat
condense	metamorphic rocks	subtropical high
consolidation	orographic precipitation	temperature profile
convection cell	photosphere	thermocline
Coriolis effect	plate	thermohaline circulation
easterlies	polar cell	trade winds
Ekman transport	polar front	upwellings
El Niño	radiation balance	weathering
equatorial low	reflection	westerlies
erosion	rock cycle	winds
Ferrell cell	scatter	
gyre	sedimentary rocks	

THE FLOW OF ENERGY IN BIOLOGICAL SYSTEMS

Why Does It Matter?

5

CHAPTER

STUDENT LEARNING OUTCOMES

After reading this chapter, students will be able to

- Explain how evolutionary strategies can be described in relation to the allocation of energy among six uses.

- Compare and contrast the costs and benefits of endothermy versus ectothermy.

- Compare and contrast alternative strategies for the timing and quantity of energy allocated toward reproduction.

- Explain what determines the total biomass of organisms living in a given area and the number of trophic positions present.

- Explain why the concentration of toxic materials in living organisms is many times greater than the concentration of those materials in the physical environment.

- Explain how the availability of food changes with the trophic position.

Fast Dinosaurs *The view of dinosaurs has changed based on how scientists think dinosaurs used energy.*

The three *Jurassic Park* movies portray dinosaurs as quick and clever. In one memorable scene a *Tyrannosaurus rex* nearly chases down a speeding jeep (opening photograph). In another scene two *Velocoraptors* figure out how to open a door and hunt for two children as a well-coordinated team. These capabilities contrast with previous representations of dinosaurs as slow, dimwitted creatures.

Aside from improvements in moviemaking technology, the *Jurassic Park* trilogy portrays dinosaurs in a new and somewhat controversial manner. This new perspective depends largely on a change in how scientists think dinosaurs used energy. Many scientists now think that dinosaurs used energy rapidly, even at rest. This ability may have allowed dinosaurs to move and behave more like present-day mammals than present-day reptiles.

Evidence for this view comes from the study of energy flows in individual dinosaurs and entire ecosystems. Baby dinosaurs were tiny relative to their parents. A baby duckbill dinosaur weighed just 170 grams (6 oz), which was only 1/16,000 of an adult's weight. To become an adult within a reasonable period, a baby duckbill dinosaur had to grow very quickly. This implies that the baby duckbill could consume huge amounts of energy and rapidly convert it to body structures. This would have been possible only if the baby duckbill dinosaur was very active.

Additional evidence for high rates of activity comes from studies of energy flow among dinosaurs. If dinosaurs used energy rapidly while at rest, relatively little of their food energy could have been used to increase their body weight. As a result, many plant-eating dinosaurs would be needed for each meat-eating dinosaur. This notion is consistent with the ratio of bones from plant-eating dinosaurs found by paleontologists relative to the number of bones from meat-eating dinosaurs. Comparisons to the modern environment show that the ratio of bones from meat-eating dinosaurs relative to plant-eating dinosaurs is closer to the ratio of present-day mammalian meat eaters to plant eaters rather than the ratio of present-day reptilian meat eaters to plant eaters. Regardless of whether this new view of dinosaurs is correct, the fact that the hypothesis is tested by analyzing how dinosaurs used energy indicates that energy flows play a critical role in past and present biological systems. ■

HOW INDIVIDUAL ORGANISMS USE ENERGY

Plants and animals display a wide variety of behaviors. Some animals migrate long distances while others hibernate. Some trees drop their leaves in the fall while others remain green year-round. These behaviors can be understood as a use of energy for one or more of the following six purposes: maintenance, growth, storage, reproduction, protection, and obtaining more energy (Figure 5.1). All plants and animals use energy for these purposes, yet the way in which they do so is unique to each plant or animal. For example, some bat species solve the problem of obtaining food in the winter by hibernating, whereas some birds solve the same problem by migrating around the globe. Understanding the energy flows that are associated with the six purposes will allow you to make sense of behaviors that might otherwise seem counterintuitive.

Obtaining Energy from the Environment

Day-to-day survival by most living organisms depends on their ability to obtain energy from the environment. This need is based on the second law of thermodynamics, which states that the ability of energy to do useful work decreases as it is used to do work. As living organisms use their energy for maintenance, growth, reproduction, and protection, they must get new supplies of energy from their environment.

Organisms have evolved two general strategies for obtaining energy from the environment: using inorganic forms of energy such as sunlight to generate organic forms of energy such as glucose, and obtaining organic forms of energy directly from other organisms. Organisms that convert inorganic forms of energy to organic forms of energy

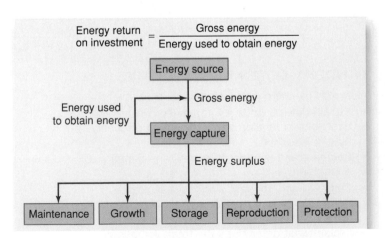

FIGURE 5.1 *How Living Organisms Use Energy* The productivity of energy capture can be measured by energy return on investment, which measures the amount of energy obtained from the environment (gross energy) relative to the amount of energy used to obtain it. The energy surplus is allocated among maintenance, growth, storage, reproduction, and protection. This allocation constitutes an evolutionary strategy that influences the fitness of individual organisms.

are called **autotrophs,** which is Greek for "self feeders." Organisms that get their energy from other organisms are called **heterotrophs,** which is Greek for "other feeders."

The term *autotroph* is slightly misleading. The first law of thermodynamics states that energy cannot be created; therefore, autotrophs do not really *create* their own food. Rather, autotrophs use inorganic forms of energy to power chemical reactions that store energy in the bonds of organic molecules. Many autotrophs generate organic forms of energy via photosynthesis. The general formula for **photosynthesis** is given by Equation 5.1:

$$6CO_2 + 6H_2O \xrightarrow{\text{Solar energy}} C_6H_{12}O_6 + 6O_2 \qquad (5.1)$$

Solar energy is critical to photosynthesis. Carbon dioxide (CO_2) and water (H_2O) do not combine spontaneously to form the sugar glucose ($C_6H_{12}O_6$). For example, seltzer contains water and carbon dioxide (the bubbles), but seltzer will not generate its own sugar (such a process would save drink companies the cost of purchasing sugar). Rather, photosynthesis uses solar energy to break the bond between the hydrogen and oxygen atoms in the water molecule and to incorporate the hydrogen atoms in the carbon dioxide molecule to form glucose. This process converts some of the kinetic energy of solar energy to potential energy that is stored in the bonds of the glucose molecule. As a work process, photosynthesis is relatively inefficient. Less than 1.0 percent of the solar energy is converted to potential energy in the glucose molecule.

Following another strategy, heterotrophs obtain energy-containing molecules by eating other organisms. This process is much more efficient than photosynthesis. But this high efficiency does not come easily. No organism volunteers to be eaten by another organism. Animals and even plants try to protect themselves from those who would make them into a meal (see the section about protection). As a result, heterotrophs must work to obtain food.

The effort required to obtain food can be evaluated with the concept of energy return on investment. The energy return on investment for obtaining energy measures the amount of energy obtained divided by the amount of energy used to obtain it (Figure 5.1). Generally, the energy return on investment for obtaining food energy must be greater than 1. That is, the amount of food energy obtained must be greater than the amount of energy used to obtain it. The difference between the amount of energy obtained and the energy used to obtain it is the energy surplus. This surplus is critical because it represents the amount of energy that is available for maintenance, growth, reproduction, storage, and protection. No energy will be available for these uses if the strategy for obtaining food consistently has an energy return on investment that is less than or equal to 1. Under these conditions, the organism will die after it has depleted its storage of energy.

The effect of energy return on investment on the viability of alternative strategies for obtaining food can be illustrated by the types of foods eaten by one family of lizards (Figure 5.2). Small (under 100 grams—3.5 oz) species of lizards in the iguanid family generally eat other animals (carnivorous), whereas larger species of lizards (over 100 grams) generally eat plants (herbivorous). In many cases large lizard species are not herbivorous throughout their life. Immature green iguanas may eat many insects, but adults are largely herbivorous.

This change in diet with size is influenced by the type of animals that lizards can catch and still generate surplus energy. Lizards use energy slowly and tire quickly (see the section about maintenance for more details). As a result, lizards can catch only animals that use energy in the same way, such as insects. Lizards cannot chase down birds and mammals, which use energy quickly and have considerable endurance. Under these conditions, catching insects has an energy return on investment greater than 1 for small lizards because the amount of food energy in the insect is large relative to the energy used by the lizard to catch the insect. But the energy return on investment for eating insects is small for large lizards because the amount of food energy captured is small relative to the amount of energy used by the lizard to catch the insect. The energy return on investment for catching birds and small mammals would be significantly greater than 1 because the amount of food energy captured would be large relative to the amount of energy used to catch the bird or mammal. But most large lizards cannot catch small birds or mammals, so they are mostly herbivorous.

Some large lizards are carnivorous, and these exceptions highlight the importance of energy return on investment. Large lizards that eat birds and mammals do not chase them down like some mammalian predators, such as lions. Instead large lizards that eat birds and mammals are wait and ambush predators. These lizards hide or camouflage themselves, stay still, and make a quick lunge or a very short run when potential prey comes within range. This hunting style is used by the world's largest lizard, the Komodo dragon (Figure 5.3), which can be 3 meters long (10 feet) and weigh 160 kilograms (353 lb). This lizard lies in ambush and uses its claws or jaws to attack a wild pig or deer that wanders within range. These attacks often do not kill the prey. Instead a host of bacteria live on the dragon's claws and teeth. A scratch or a bite will initiate a fatal infection. Should the prey escape the initial attack, the dragon does not give chase. Instead the dragon relies on its keen sense of smell to locate its prey after it dies.

Maintenance: Endotherms versus Ectotherms

The second law of thermodynamics states that systems must use energy to maintain order against the constant tendency toward disorder. The use of energy to maintain order in a living system is termed **maintenance respiration.** Maintenance respiration takes several forms. Many

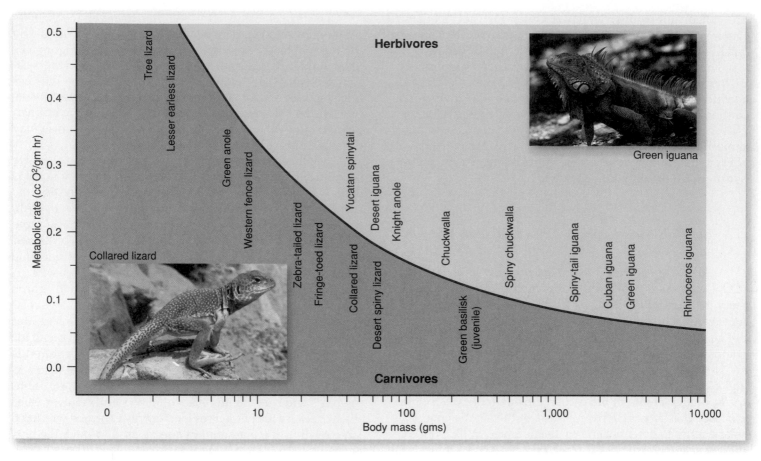

FIGURE 5.2 *Lizard Size and Diet* Large lizards such as the green iguana in the upper right corner generally are herbivorous (lizard species above the line), whereas smaller lizards, such as the collared lizard in the lower left corner, generally are carnivorous.
(From F. Harvey Pough, "Lizard Energetics and Diet." Ecology 54, no. 4 (1973): 837–844.)

organisms use some of their energy to remove and detoxify wastes. This prevents wastes from accumulating and disrupting bodily functions. The human body uses significant amounts of energy to power the kidneys, which remove wastes from the bloodstream, and to power the liver, which breaks down wastes in the bloodstream. Another form of maintenance respiration includes the production of red blood cells. As red blood cells age, their metabolic abilities wear down. Consequently your body always is scavenging old red blood cells (the liver scavenges about 3 million cells per second) and producing new cells.

To us, the most interesting component of maintenance respiration is the rate at which the body's metabolic machinery runs and its effect on an organism's ability to manage its body temperature. Animals have evolved two strategies for managing body temperature. Some species use energy at such high rates that they can use waste heat to raise and maintain their body temperature well above the ambient environment. These animals are known as **endotherms.** Endotherms include birds, mammals, and some very active fish such as

the Atlantic bluefin tuna. You too are an endotherm; to calculate your use of energy, see *Your Ecological Footprint: Tracing Your Energy Flows and Changes in Weight.*

Another group of animals uses energy relatively slowly, and their waste heat is not sufficient to raise their body temperature above ambient temperature. Instead these animals obtain their body heat from the environment. These animals are known as **ectotherms.** Ectotherms can raise their body temperature above ambient temperature by sitting in the sun; but once the sun goes down, their body temperature is the same as the environment. Ectotherms include most fish, insects, amphibians, and present-day reptiles. Many people think of these animals as being "cold-blooded," but this is not necessarily the case.

The differences between endotherms and ectotherms affect how they allocate energy (Figure 5.4). Endotherms can raise their body temperature because their **basal metabolic rate,** the rate at which an organism uses energy while at rest, is about 10 times greater than the basal metabolic rate of an ectotherm. Put another way, an endotherm uses a large fraction of its energy

FIGURE 5.3 *The Komodo Dragon* *The world's largest living lizard is the Komodo dragon, which lives on several small Indonesian islands and eats wild boar and deer.*

budget "idling" at a high rate. As a result, endotherms use a relatively large fraction of their energy budget for maintenance respiration. On the other hand, ectotherms "idle" at a lower rate than endotherms and therefore use a smaller fraction of their energy budget for maintenance respiration. As a result, ectotherms can allocate a larger fraction of their surplus energy to other uses, such as reproduction.

Given these differences, why are some animals endothermic while other animals are ectothermic? At first glance it seems terribly inefficient to use so much energy for maintenance respiration. This energy could be used for growth, reproduction, or protection. What are the benefits of being an endotherm?

The ability of endotherms to maintain body temperatures warmer than the ambient environment carries several advantages. Most organisms have a range of internal body temperatures at which their bodies can best carry out the physiological processes necessary for life. Outside this range, the body works less well. Beyond some extreme the body may not work at all, and the organism dies.

The human body regulates its body temperature within a range of about 1°C (1.8°F) at an average of about 37°C (98.6°F). Using the negative feedback loops described in Chapter 3, your body can return its temperature to 37°C so long as it remains between 34°C (93.2°F) and 41°C (105.8°F). Internal body temperatures beyond this range can be fatal because the body no longer can return to its set point. Death occurs within a short period at body temperatures above 43°C (109.4°F) and below 28°C (82.4°F).

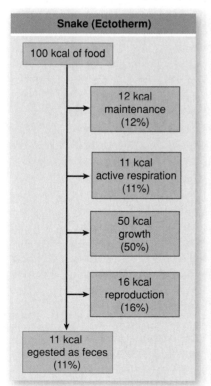

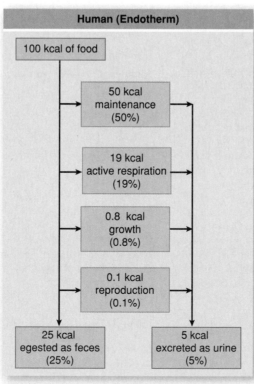

FIGURE 5.4 *Energy Allocation Strategies* *Humans are endotherms, which means that we use about 50 percent of our energy for maintenance respiration, compared to 12 percent for snakes. This means that snakes have a lot more energy to devote to growth (50 percent) and reproduction (16 percent) compared to humans, which allocate less than 1 percent of their energy to growth and reproduction.*
(From C.A.S. Hall, C. Cleveland, and R. Kaufmann, Energy and Resource Quality: The Ecology of the Industrial Process. New York: John Wiley & Sons, 1986.)

Your ECOLOGICAL *footprint*

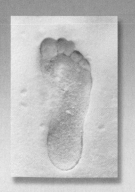

Tracing Your Energy Flows and Changes in Weight

Over the last decade U.S. citizens have become increasingly concerned about their weight. The fraction of the U.S. population that is classified as overweight or obese has increased. These terms are defined relative to an index that accounts for both weight and height called *body mass index* (BMI—a person's body weight in kilograms divided by the square of his or her height in meters). Individuals with a BMI of 25–29.9 are considered overweight; people with a BMI of 30 or greater are considered obese.

There are many hypotheses about the general increase in weight. Some argue about the role of advertising; others say that a sedentary lifestyle is responsible. Whatever the ultimate cause, we can see the immediate cause by tracing the intake and use of energy.

The minimum amount of energy required by a resting human is described by *basal metabolic rate*—the energy used for maintenance. Basal metabolic rate varies by gender, weight, and age. You can calculate your basal metabolic rate using one of the following equations:

$$\text{Male basal metabolic rate} = 66.5 + 13.75 \times (wt) + 5.003 \times (ht) - 6.775 \times (age) \qquad (1)$$

$$\text{Female basal metabolic rate} = 655.1 + 9.563 \times (wt) + 1.85 \times (ht) - 4.676 \times (age) \qquad (2)$$

Here *wt* is your weight (kilograms), *ht* is your height (centimeters), and *age* is your age in years. These equations were developed by Harris and Benedict in 1919 and measure the kilocalories used per day. Despite their age, the equations still are considered accurate.

Humans are endotherms, so basal metabolic rate also depends on environmental temperature. Scientists find that the human body uses the least energy for thermoregulation (regulating body temperature) at about 26°C (78.8°F). Temperatures above or below that minimum increase energy use through sweating or shivering, respectively.

In addition to general body shape, energy requirements are determined by other uses. A pregnant woman uses about 10 percent more energy than her nonpregnant counterpart. A person fighting an infection with a fever uses more energy than when healthy. Furthermore, not all weight is equal. In general, a kilogram of muscle burns more energy than a kilogram of fat. So all else being equal, a person who uses a lot of energy for storage burns energy more slowly (per unit body weight).

Most people are not at rest all day. Exercise increases the amount of energy used by the body. The more vigorous the exercise, the more energy is required. Table 1 lists the energy burned to sustain a half hour of various activities. The energy cost of these activities varies by height, weight, and gender, but adding this information would greatly increase the complexity of the calculation.

TABLE 1 Energy Cost of Everyday Activities

Activity	Kcal per 30 Minutes of Activity
Walking (3 mph)	113
Climbing stairs	308
Bowling	103
Basketball	274
Aerobics class (high impact)	240
Home activities (light tasks)	86
Home activities (moderate tasks)	120
Home activities (vigorous tasks)	137
Lawn mowing (riding)	86
Lawn mowing (power)	188
Lawn mowing (push)	206
Ping-Pong	137
Mopping	120
Running (8-minute miles)	440
Bicycling	211

Calculating Your Footprint

You can use the information given to approximate your daily change in weight. Use the equations in this box to calculate your basal metabolic rate. Increase this total by the amount of energy you use for the activities described in Table 1. Next determine your energy intake: Use Table 2 to link the foods you eat to their kilocalorie equivalent. Remember to correct your energy intake for the fraction of energy that is lost as feces (see Figure 5.4).

Use the following equations to calculate your energy consumption. (You may have to change units because most food categories have more than one food type.) If you can't find a food item in Table 2, many food containers show the energy content of the food inside.

TABLE 2	Energy Content for Selected Food Types		
Food Item	**Energy Content**	**Food Item**	**Energy Content**
Beverages		Milk products and eggs	
Beer	141 kcal/12 oz	Whole milk	208 kcal/cup
Wine	74 kcal/3.5 fl oz	Semiskim milk	179 kcal/cup
Coffee	15.8 kcal/6 oz	Cheese	114 kcal/oz
Tea	2 kcal/fl oz	Eggs	75 kcal/egg
Fats		Cereals, sugar, potatoes, vegetables, and fruits	
Fat for frying	124 kcal/teaspoon	Flour	419 kcal/cup
Margarine	815 kcal/stick	Sugar	774 kcal/cup
Meat		Potato	220 kcal/potato
Beef	218 kcal/3 oz	Vegetables (average)	50–80 kcal/100 grams
Pork	204 kcal/3 oz	Fruits	35–120 kcal/100 grams
Minced meat	86 kcal/slice		
Sausage	144 kcal/link		

_____ kcal (beverages) = _____ oz/day × _____ kcal/oz

_____ kcal (fats) = _____ teaspoon/day × 124 kcal/teaspoon

_____ kcal (margarine) = _____ stick/day × 815 kcal/stick

_____ kcal (meat) = _____ oz/day × _____ kcal/oz

_____ kcal (dairy and eggs) = _____ cups/day × _____ kcal/cup

_____ kcal (plant material) = _____ gram/day × _____ kcal/day

Total Energy Intake _____ kcal

Energy used = Basal metabolic rate [from Equation 1 or 2] + energy used in everyday activities [from Table 1]

Interpreting Your Footprint

Your change in weight can be calculated as the difference between your energy intake and your use of energy (1 kg of body weight is equal to about 7,700 kcal).

_____ Weight change kg/day = (_____ kcal eaten − _____ kcal used)/7,700 kcal/kg

These numbers vary but will be similar from day to day, so multiply the difference by 7 to determine your weekly balance. If you take in more energy than you consume, a positive number indicates that you will gain weight. Conversely, if you take in less energy than you use, a negative number indicates that you will lose weight. Use the number calculated to determine whether you are gaining weight, losing weight, or remaining at about the same weight.

If you are gaining weight, determine the weekly reduction in food intake that would balance your energy intake and energy use. If you don't like dieting, determine the increase in weekly activity that would raise your energy use to the point at which it equals your energy intake.

STUDENT LEARNING OUTCOME

- Students will be able to describe some of the reasons that obesity is becoming more common in the United States and what can be done to reverse this trend.

Ectotherms tend to be less sensitive to changes in their body temperature. The desert iguana, a lizard that lives in the deserts of Arizona and California, performs best when its body temperature is about 38°C (100.4°F—warmer than "warm-blooded" mammals like yourself). But this lizard can be active at body temperatures between 27°C (80.6°F) and 41°C (105.8°F), which is termed its **active range.** To maintain this range, the lizard moves into the shade when its body temperature rises close to 41°C and basks in the sun when it drops back toward 27°C. Beyond this active range, body functions degrade and the lizard becomes inactive. At body temperatures above 50.5°C (122.9°F), the desert iguana dies within a short period. At temperatures below 8°C (46.4°F), the desert iguana is functionally dead—it cannot move to eat or avoid being eaten.

The relationship between body temperature and physiological capability determines when an organism can be active. Endotherms can be active day or night, winter or summer. This flexibility is possible because they can maintain their body temperature in a changing environment. On the other hand, ectotherms can be active only when environmental conditions allow them to maintain a suitable body temperature. Desert iguanas can be active for only a portion of the day during the spring, summer, or fall when they can keep their body temperatures between 27°C and 41°C. During the hottest months these lizards are active only in the early morning or late afternoon. During the other portions of the day or the year, the lizards remain in their burrows because they cannot raise (or lower) their body temperature to levels within the active range.

Extending the period that an organism can be active is an important benefit for endothermic animals. Endotherms can use this extra time to collect food, which gives them more energy to use for growth, storage, reproduction, and protection. Equally important, the same metabolic machinery that gives them a high basal metabolic rate also allows them to use energy quickly for other purposes, such as running. This gives endotherms more speed and endurance than ectotherms, which is why ectothermic lizards cannot chase down endothermic mammals.

But there are no absolute advantages to endothermy versus ectothermy. Rather, the advantages and disadvantages depend on the physical environment in which an organism lives and the organisms with which it interacts. That is why we see both endothermic and ectothermic organisms. Can you guess where ectotherms predominate? The answer will be discussed in Chapter 7.

Growth

Growth is the process by which structural components of the body, such as bones and muscles, get bigger and stronger. Generally these processes are known as **maturation,** a growth process in which juveniles increase in size and change in form to the point at which they are capable of

reproduction. At this point energy is redirected from growth to other uses such as reproduction (see the section about reproduction).

Some organisms retain their general body shape (morphology) as they mature. But many species change their form dramatically as they mature—a process known as **metamorphosis.** Insects (the most numerous class of animals) go through metamorphosis. Those of you who raised fruit flies in a genetics laboratory will remember that fruit flies start life as little wormlike creatures known as larvae. After eating nonstop for about six days, the larvae wall themselves up in tiny sacks known as pupae. After four more days they emerge as flies. As adult flies they have about four days to find mates and reproduce before they die. Because the flies are ectotherms, the length of these stages depends on ambient temperature.

Insects are not the only organisms that go through such dramatic changes. Many mollusks, such as clams and oysters, start life as free-swimming larvae without shells. Frogs also change through their life cycle. Tadpoles hatch from eggs and spend many weeks eating nonstop. Eventually the tadpoles grow hind and front legs, absorb their tails, replace their gills with lungs, and leave the water as frogs.

Why do some organisms go through such dramatic changes? There is no definite answer, but energy flows may provide a clue. Some body forms may be specialized to take advantage of food sources that are available temporarily. Fruit flies spend their larval stage crawling through rotting fruit, something that would be impossible for the mature fly. Similarly, tadpoles can feed on the huge increase in algae that is associated with the mixing of the water column in spring (see Chapter 7).

An alternative explanation for metamorphosis may be mobility. Many aquatic organisms that are relatively immobile as adults pass through a larval stage in which they are relatively mobile. For example, mussels, clams, corals, and barnacles have larval stages that move with ocean currents (Figure 5.5). When they reach a spot that may be hospitable, the larvae can settle to the bottom and metamorphose to the adult form. Consistent with the importance of dispersal, lace coral can revert to its larval stage and reenter the water column if the location where it settles does not suit its needs.

Storage

Most of us eat three (or more) meals per day. Such regularity is the exception rather than the rule. Many plants and animals live in areas where the availability of energy is irregular. Environmental conditions, such as cold temperatures, may prevent plants from carrying out photosynthesis year-round. Similarly, encounters between predator and prey depend in part on random chance, which implies that many predators do not eat daily. To ensure access to energy during periods when plants and animals cannot obtain new supplies of energy, plants and animals store energy.

(a)

(b)

FIGURE 5.5   *The Influence of Body Form on Mobility*   *The larval form of a barnacle (a) moves with the currents until it finds an open spot on a hard substrate. There it settles and metamorphosizes into its adult form (b), which does not move.*

The need for storage differs among organisms and depends in part on the environment. Some animals live in environments where cold weather or the lack of precipitation makes it difficult to obtain energy for long periods. In response, some animals reduce energy use by slowing their basal metabolic rate. **Hibernation** is a state in which the metabolic rate slows by as much as 99 percent. Nonetheless, hibernating animals still need energy. This energy is obtained from energy that is stored when food is available. For example, hibernating bats use energy that they stored during the summer.

In other environments predators may be active year-round, but they may not be successful every day. To tide themselves over between successful hunts, these predators store lesser quantities of energy. For example, lions eat every few days because their hunts are successful only about one time out of six. Storage gives lions energy between meals.

Plants also need to store energy. Sugar maple trees that live in temperate climates are dormant during the winter. Sugar maples store excess energy generated during the summer in their roots. This storage is complemented with the energy used to make the trees' leaves. Before maple trees drop their leaves, they break down the chlorophyll (the molecule that captures the sunlight used to power photosynthesis) and remove its constituents along with energy. Removing this matter and energy exposes the brilliant colors of fall foliage. When spring comes and it is time to regenerate their leaves, sugar maples move this energy and matter back to their branches from their roots as sap. People tap into this sap to make maple syrup, which is a high-energy food.

Other plants also store energy in their roots. Potato and carrot plants store a significant fraction of the energy they generate during the growing season in tubers known as potatoes and tap roots known as carrots. High concentrations of energy make these roots ideal food for humans and other animals.

The ways in which plants and animals store energy is shaped by natural selection. Most animals store a significant fraction of their energy as fat, whereas plants store a significant fraction of their energy as carbohydrates. Why does natural selection favor the use of fat in animals? The body has relatively few uses for fat. It cannot use fat to make muscles or bone. Storing energy as fat has some real disadvantages—some components of fats cannot be respired aerobically, and fats are relatively difficult to move in body fluids. (Can you guess why?) Animals that live in cold or aquatic environments use fat for insulation or buoyancy, but even animals that live in warm environments store energy as fat.

Natural selection favors fat as a storage molecule in animals because of the unique properties of fats. One gram of fat can store about twice the amount of energy that can be stored in a gram of protein or starch. This high storage capacity is critical to animals that must move about. An animal's ability to catch food or run away from predators, which increases its likelihood of leaving offspring in the next generation, is enhanced by storing the most energy in the least amount of weight. The advantages of increased mobility associated with fats outweigh the disadvantages associated with storing energy as fat. These disadvantages outweigh the benefits for immobile plants.

The need to move about also influences where animals store fat within their bodies. Mobility is maximized if fat is stored near the center of gravity. Your center of gravity is near the bellybutton. Consistent with the pressures to store energy and maintain mobility, people tend to put on weight at their hips and thighs.

Reproduction

All organisms die eventually—some sooner and others later. Some insects live for only days or weeks. Other animals live for years or decades. Some trees live for centuries or even millennia. This period seems long relative to the human life span, but even thousands of years is short compared to the billions of years that living organisms have existed on Earth. Clearly the success of living organisms is based on their ability to produce new individuals rather than their ability to live a long time. In your more philosophical moments, you may be tempted to ask why natural selection has favored reproduction over long life. It is not possible to answer this question definitively, but natural selection and energy flows provide a possible answer.

Natural selection favors strategies that allow individuals to contribute the greatest number of offspring to the next generation. To do so, living organisms use a significant fraction of their energy to produce offspring and, in some cases, to care for offspring in ways that increase the likelihood that the offspring will survive and produce more offspring. Allocating energy to produce and care for offspring is termed **reproduction.**

Using energy for reproduction reduces the amount of energy that is available for other uses such as maintenance. Natural selection may favor individuals that use their energy to produce offspring now instead of trying to repair damage to their own bodies even though not repairing the damage may increase the likelihood that the individual will die and therefore not reproduce in the future. Similarly, natural selection may favor the design of body parts (teeth, lungs, heart) that function just long enough to get the individual through its reproductive years (the time during which it produces offspring). As the individual ages beyond its reproductive years, the parts may begin to fail so that there is a loss of function. Such losses decrease the probability of survival and reproduction. Together theses changes are known as **senescence.**

You can almost think of this as planned obsolescence—bodies are designed to last so long as an individual has a high probability of producing offspring. Evidence for such obsolescence comes from the teeth and reproductive ages of red deer. Females have a reproductive life span of about fourteen years, whereas males have a reproductive life span of about six years. This short time is due to competition among males for mates. To "bulk up" for this competition, young males grow much faster than females (Figure 5.6a). To support these high rates of growth, males consume large amounts of food, which wears down their teeth much faster than those of females (Figure 5.6b). This wear reduces the ability of older males to obtain food. A relatively poor diet tends to speed senescence, so males tend to live about five fewer years than females. Because males are unlikely to mate in their later years, natural selection has not favored changes that would increase the durability of their teeth.

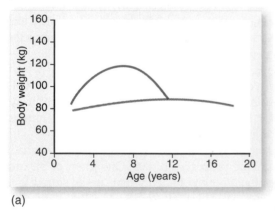

(a)

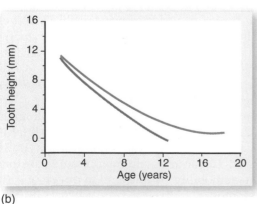

(b)

Male red deer

Female red deer

FIGURE 5.6 *Live Fast, Die Young*
(a) Male red deer get bigger faster (green line) than females (brown line). (b) As a result, the height of the mandibular first molar for males (green line) wears out faster than females (brown line). This reduction leads to malnutrition, which reduces the life expectancy of male deer. Because old males rarely breed, natural selection has not favored changes that would increase the durability of male teeth. (Data from J. Carranza, S. Alarcos, C.B. Sanchez-Prieto, J. Valencia, and C. Mateoa, "Disposable-Soma Senescence Mediated by Sexual Selection in an Ungulate." Nature 432 (2004): 215–218.)

Energy allocated toward reproduction is used in many ways. For example, much of the energy used by seedlings to produce their first leaves comes from the plants that produced the seeds. Similarly, the energy in a chicken's egg is used by the single fertilized cell to produce a chicken. These concentrated storages of energy make seeds and eggs a popular component of many animals' diet.

Because natural selection determines success based on the number of offspring left in the next generation, some species invest large amounts of energy in reproduction. To get this energy, some bird species, such as the grey plover, travel thousands of kilometers to their summer breeding grounds in the Arctic. These travels seem directed by the availability of large amounts of energy. By analyzing the ratio of carbon isotopes ($C^{13}:C^{12}$), scientists can identify the food that is used by the parents to produce their feathers and eggs. The ratios indicate that the parents use foods from their winter homes to produce their feathers, but they use energy from their Arctic breeding grounds to produce their eggs (Figure 5.7). This implies that the large amounts of energy that are available in the summer breeding grounds compensate for the energy used to get there.

The energy allocated toward reproduction does not stop at birth. Many species care for their offspring, which is termed **parenting;** this too requires significant amounts of energy. For example, a female Pacific gray whale may lose 25 percent of her weight while nursing. Similarly, humans invest large amounts of energy in parenting. Humans cannot take care of themselves until they are about 8–10 years old, during which time parents must provide food, shelter, and clothing. In developed nations this period of dependence has been elongated to the age of 18 or 21. During this period parents invest considerable time, money, and effort. The U.S. government estimates that raising a child born in 1999 costs about $240,000.

Natural selection favors strategies that maximize the number of offspring that live long enough to reproduce (not just the number of offspring produced). To be successful, plants and animals face important trade-offs regarding the amount of energy invested in reproduction, the timing of that investment, and how that investment is allocated among offspring. A plant or animal can invest a large or small fraction of its energy budget in offspring. Similarly, it can reproduce relatively early in its life cycle or wait. And a plant or animal can use its reproductive energy to produce a few large offspring or many smaller offspring.

FIGURE 5.7 *Energy Used for Reproduction* The food eaten by birds in their winter homes has a higher ratio of carbon 13 to carbon 12 than the foods eaten in their summer grounds. High values for the ratio indicate that the energy used by adult birds to produce feathers comes from foods eaten in the winter grounds, whereas the energy used to produce eggs and raise offspring comes from food eaten in their summer breeding grounds. (From Klaassen et al., "Arctic Waders Are Not Capital Breeders." Nature, 413 (2001): 794).

Each of these strategies is mutually exclusive. For example, the strategy used to allocate energy among offspring varies between two extremes. At one extreme energy can be used to produce many small offspring. Producing many offspring increases the chances that some will live long enough to reproduce. But producing small offspring reduces the probability of survival for the individuals. At the other extreme, energy can be used to produce a few large offspring. Producing a few large offspring increases the probability that individual offspring will survive. But producing just a few offspring reduces the number that may live long enough to reproduce. The trade-off between the number and size of offspring is similar to the parable about the dangers of putting all your eggs in one basket. If you put all your eggs in one basket and nothing happens to the basket, the decision seems wise. On the other hand, if the basket drops and all the eggs break, the decision seems foolish.

The success of a reproductive strategy depends on the environment. In environments with relatively little competition, natural selection may favor organisms that allocate a relatively large fraction of their energy toward reproduction, make that investment relatively early in their life cycle, and use it to produce many small offspring. Such strategies are sometimes said to be **r selected** because they enable these organisms to increase their numbers rapidly when environmental conditions permit. The term *r selected* comes from an equation used to describe population growth (discussed in Chapter 9); *r* represents the maximum rate of growth possible.

In crowded environments, where there is considerable competition, natural selection may favor strategies that allocate a relatively small fraction of the energy budget toward reproduction, make that investment relatively late in the life cycle, and use it to produce a few relatively large offspring. Organisms that pursue such strategies are sometimes said to be **K selected.** The term *K* refers to the maximum number of individuals a unit of landscape can support sustainably.

As with other evolutionary strategies, neither K selection nor r selection is inherently better. Rather, reproductive strategies are an adaptation to a particular environment and especially the nature of competition and disturbances in those environments (as discussed in Chapter 8). In environments where there are many predators and parenting is difficult, natural selection may favor the production of many small offspring. For example, a female lobster lays 10,000–80,000 eggs and may hatch about 10,000 larvae, but only about ten of these larvae become adults. In environments with fewer predators and where parenting is more effective, natural selection may favor the production of a few larger offspring. For example, African elephants produce only one very large baby at a time.

Protection

Heterotrophy has proved to be a very common way of "making a living" because plants and animals are a concentrated source of energy. To avoid being an easy target, most plants and animals have evolved ways of protecting themselves. Strategies used by animals are fairly obvious and run the gamut from sharp teeth, horns, and claws, which can be used to fend off predators, to speed, which can be used to escape from predators, to camouflage, which can be used to hide from predators.

But what about plants? At first glance plants seem helpless. Their need for sunlight forces them into the open (they can't hide under rocks). The thin leaves used to collect solar energy and gases (carbon dioxide and oxygen) make it easy for animals to consume significant chunks of the leaves. And the extensive root system that is needed to collect water and materials makes it impossible for plants to run away. Given this vulnerability, you may wonder why herbivores haven't consumed all of the plants.

One reason is that carnivores control the population of herbivores. But this is just a small part of the answer. Careful analyses show that many plants protect themselves from herbivores by using a significant fraction of their energy budget to fill their leaves, stems, and roots with chemicals that discourage animals from eating them by tasting bad or by slowing the herbivores' growth rate. Should a wild radish plant be chewed, the insect's saliva stimulates the production of mustard oils and glycosides that discourage insects from eating the plant. Such reactions reduce the amount of damage done by insect herbivores and increase plants' evolutionary success. Plants that increase their defenses in response to an insect attack early in the growing season produce more seeds than plants that do not increase their defenses.

Many of the plant species that we treasure as spices and herbs get their flavor from protective chemicals. Protective chemicals are responsible for the flavors associated with mustard, garlic, and onions. Many of these spices are especially popular in tropical climates (Figure 5.8a). Whether cooks know it or not, the spices' defensive chemicals slow the rate at which food spoils by discouraging the growth of bacteria (Figure 5.8b). The understanding that native plants have a wide array of protective chemicals that may be of use to people has led to the drive to save Earth's rain forests (discussed further in Chapter 12).

Animals too have evolved chemicals that make them taste bad or poisonous to potential predators. For example, birds that eat monarch butterflies spit them out, and animals that eat poison arrow frogs die shortly thereafter. Other examples of poisonous or distasteful organisms include the puffer fish. The liver of the puffer fish has a special toxin that paralyzes the nervous system. This "fugu fish" is a delicacy in Japan and is prepared so that just enough poison remains to make your lips tingle.

In an interesting twist of evolution, the animals with the worst taste sometimes are among the most beautiful (Figure 5.9). This beauty is a warning. Poisonous chemicals

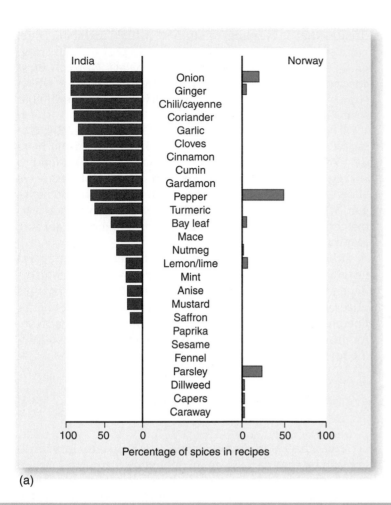

(a)

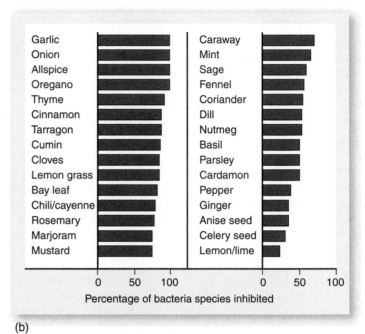

FIGURE 5.8 *Spicy Food* (a) Cooks in India add more spices to their meals than cooks in Norway. In general, the hottest spices are used in India. (b) These spices contain protective chemicals that slow or stop the growth of bacteria. As such, they slow spoilage in hot environments. *(From The New York Times.)*

(b)

(a)

(b)

FIGURE 5.9 *Beautiful but Dangerous!* Bright colors warn potential predators to avoid (a) poison arrow dart frogs and (b) monarch butterfly.

are an effective means of protection only if potential predators know to stay away. A poison dart frog will not live to reproduce if the predator dies *after* it has eaten the frog. To help predators avoid such mistakes, animals protected with poisons have evolved bright colors, especially red, black, or yellow. Most predators will not eat animals with these colors, whether they are poisonous or not!

Some species without protective chemicals sometimes have the warning colors of animals with protective chemicals, which is termed **mimicry.** The nonpoisonous scarlet kingsnake is red, black, and white like the poisonous coral snake (Figure 5.10). Such mimics get a free ride on the energy used for protection by the coral snake: They do not invest energy in defensive chemicals (and hence have energy to use for other purposes), but they are protected from predators because they look like species that have defensive chemicals. However, natural selection limits the level of protection enjoyed by these free riders. If the numbers of mimics grows large relative to the population of animals with defensive chemicals, the mimic strategy will become less effective because predators will not learn to avoid either the mimic or the animals with defensive chemicals.

Energy and Natural Selection

As described in the preceding sections, there are many ways in which an organism can obtain and use energy. Organisms do not develop these strategies deliberately. No species decides that it is going to support a high metabolic rate, use a significant fraction of its energy to produce offspring, or fill its tissues with poisonous chemicals. Nor can most species make the detailed calculations that indicate which strategy is best. How does natural selection "choose" what seems to be the ideal strategy?

The role of natural selection can be illustrated by asking why animals breathe oxygen. Living organisms have developed several biochemical pathways to convert food to energy, which is termed **respiration.** One set of pathways allows organisms to convert food to energy without oxygen, which is termed **anaerobic respiration.** Another uses oxygen to convert food to energy, which is termed **aerobic respiration.**

Most of the organisms you are familiar with depend heavily on aerobic respiration. Why do so many organisms depend on aerobic respiration? We can see the advantages of aerobic respiration by calculating the quantity of energy liberated relative to the anaerobic pathway. One pathway for anaerobic respiration can be summarized by the following formula:

$$C_6H_{12}O_6 \longrightarrow 2C_3H_6O_3 + Energy \qquad (5.2)$$

Here the by-product is lactate, $C_3H_6O_3$. Aerobic respiration can be summarized by the following formula:

(a)   (b)

FIGURE 5.10   *Mimicry   Similarity of colors may cause potential predators to confuse the nonpoisonous scarlet kingsnake (a) with the poisonous coral snake (b), and this confusion may give the nonpoisonous scarlet kingsnake a chance to escape.*

$$C_6H_{12}O_6 + 6O_2 \longrightarrow 6CO_2 + 6H_2O + \text{Energy} \qquad (5.3)$$

The critical difference between the two forms of respiration concerns the amount of energy obtained from a molecule of glucose. An organism that respires glucose via the anaerobic pathway (Equation 5.2) obtains 47 units of energy, whereas the aerobic pathway generates 686 units of energy. Given the same amount of food, the aerobic organism has 15 times more energy. This difference allows aerobic organisms to allocate more energy to maintenance, growth, reproduction, protection, and obtaining more food. Based on this advantage, organisms that depend mainly on aerobic respiration have been able to outcompete organisms that are capable of anaerobic respiration in environments where oxygen is present.

ENERGY FLOWS BETWEEN ORGANISMS

Natural selection favors strategies that work best for the individual. But the evolution of these strategies also influences other organisms. Some uses of energy, such as maintenance, do not increase the structure of an organism or its offspring, whereas other uses, such as growth, storage, or reproduction, increase the structure of the organism and its offspring. This difference is critical to heterotrophs, which view that individual and its offspring as a potential source of food. This implies that the strategies governing the allocation of energy among maintenance and growth, storage, and reproduction influence the food supply for predators. And the availability of food influences the strategies of both the predator and its prey.

Food Webs and Chains: Who Eats Whom

An organism's strategy for obtaining new supplies of energy defines its role in the environment. This role is defined by the flow of energy between organisms. These flows link organisms to one another through **food chains** and **food webs.** Put simply, food chains and webs show who eats whom. Food chains and webs differ in complexity: Food webs portray a more complicated and realistic picture. Nonetheless, this section emphasizes food chains because their simplicity allows us to illustrate some of the basic principles of energy transfer between organisms.

All food chains and webs start with autotrophs. On land, photosynthetic green plants are the dominant autotrophs (Figure 5.11a). In lakes, streams, and oceans, tiny single-cell photosynthetic algae called **phytoplankton** are the dominant autotrophs (Figure 5.11b). The role of autotrophs is dictated by the second law of thermodynamics. Because no energy conversion is 100 percent efficient, some energy is lost as animals eat plants or other animals. And because energy cannot be recycled, some energy is lost from

the food chain as a plant or animal uses energy for any of the six purposes that are described in the first section of this chapter. These losses are replaced by autotrophs, which convert inorganic forms of energy to organic forms. Without this steady input of new low-entropy energy, the food chain would collapse.

The rate at which autotrophs convert inorganic forms of energy to organic forms of energy is known as **gross primary production.** Gross primary production measures the amount of organic energy generated by autotrophs per unit area per unit time. For photosynthetic autotrophs, gross primary production measures the rate at which green plants or phytoplankton use solar energy to produce sugar from carbon dioxide and water per unit area.

Autotrophs use a significant fraction of the energy captured by photosynthesis for maintenance. Subtracting maintenance from gross primary productivity yields **net primary production.** Autotrophs use net primary production for growth, storage, or reproduction. These uses increase the size and numbers of autotrophs, which are measured by **biomass.** Biomass describes the mass of a species or group of species.

By definition, net primary production determines the amount of energy that is available to the rest of the food chain. Energy used to produce new structure (such as leaves), used for storage (taproots), or used for reproduction (seeds) can be eaten by other organisms. From an energy perspective, herbivores eat the net primary production of autotrophs. This net primary production is transferred along the food chain when herbivores are eaten by carnivores. This energy may be transferred again when carnivores are eaten by other carnivores. Ultimately net primary production is transferred to the organisms that eat dead plants and animals.

The position along the food chain at which an organism obtains energy is termed its **trophic position.** Animals in the second trophic position eat autotrophs and are known as **grazers** or **primary consumers.** On land insects are among the most important primary consumers (Figure 5.11a). In aquatic environments, such as lakes, streams, and oceans, small multicellular organisms known as **zooplankton** are among the most important primary consumers (Figure 5.11b).

Grazers or primary consumers are eaten by animals in the third trophic position, which are known as **carnivores** or **secondary consumers.** On land small mammals, reptiles, or amphibians often fill the role of secondary consumers. In aquatic environments small fish often serve as secondary consumers. The process of animals eating other animals can continue for several trophic positions. On land and in water there are often third- and fourth-level consumers. For example, small carnivores (such as frogs) often are eaten by medium-sized carnivores (such as snakes), which are eaten by larger carnivores (such as hawks). Rarely are there carnivores in the fifth or sixth trophic position, for reasons that are described in the next section.

FIGURE 5.11   *Food Chains* (a) A simple terrestrial food chain for a desert ecosystem in which grasshoppers are the primary grazers, which are eaten by coachwhip snakes (secondary consumers), which are eaten by great horned owls, which are tertiary consumers. (b) A simple aquatic food chain in which phytoplankton are the primary producers, which are eaten by zooplankton (primary consumers), which are eaten by silversides (secondary consumers), which are eaten by bluefish (tertiary consumers), which are eaten by ospreys (fourth-level consumers). Neither chain shows detritivores.

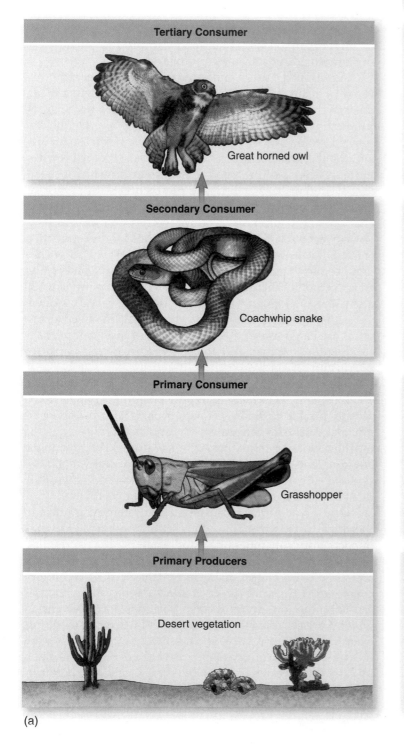

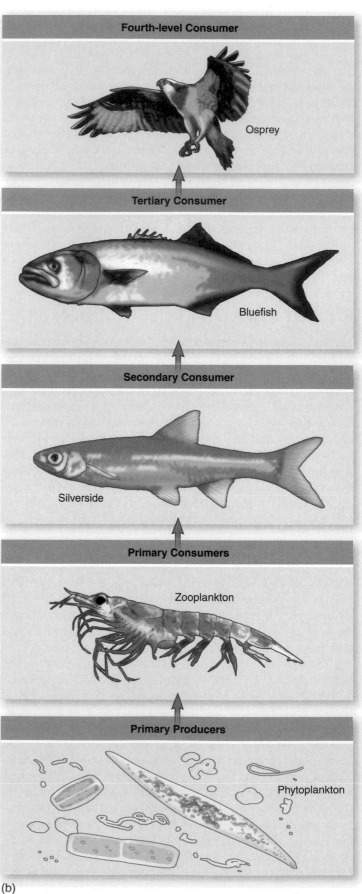

(a)

(b)

Finally, decomposers play an important role in the food chain. The organisms that die at every stage of the food chain are called **detritus** and are eaten by organisms known as **decomposers** or **detritivores.** Decomposers come in all shapes and sizes. Bacteria, fungi, lobsters, earthworms, and vultures all are decomposers. Some anthropologists think that early humans ate the carcasses of dead animals.

Earthworms are very different from vultures, and these differences reflect their role in decomposition. In the first step, large **scavengers** (animals that eat dead animals) such as vultures consume most of a dead animal. Once the carcass is exposed by scavengers, smaller organisms such as maggots gain access. Finally, the organic matter that remains is decomposed by bacteria and fungi. On land bacteria generally consume remnants of animals, whereas fungi generally consume the remnants of plants. Decomposers are distinguished by their use of exoenzymes, which break down organic material before it is consumed.

Decomposers extend the food chain and, in some cases, increase the availability of food energy (but not the total amount). The grass *Spartina* is abundant in tidal marshes. This grass has a high rate of net primary production, but only a small fraction of this production can be eaten directly because the grass is hard to digest. Nonetheless, the grass becomes an important source of energy to the food chain as detritus. After the plant dies, microbial decomposers increase its digestibility by other organisms. As organisms eat the dead plant, they also eat the microbial decomposers, and these decomposers improve the nutritive value of the marsh grass. Detrital food chains are important in many environments, such as salt marshes, grasslands, small streams, and even the open ocean (see Chapter 7).

You probably eat several products of decomposer food chains. The bacteriology department at Cornell University requires that all food brought to its parties be microbially mediated. On first thought this might seem repulsive; but the parties include lots of delicious breads, crackers, cheeses, sour cream, and of course wines, beers, and other beverages—all generated with the help of microbes, which do their job by partially decomposing what we feed them!

The notion that most of us participate in the decomposer food chain implies that food chains are more complicated than those pictured in Figures 5.11(a) and 5.11(b). Most organisms eat from several trophic positions. Some organisms (**omnivores**) eat both plants and animals. Other organisms eat only other animals, but they eat animals from several different trophic positions. For example, coachwhip snakes eat both insects, which are herbivores, and small lizards, which are carnivores.

The variety of eating habits is expressed by the notion of a food web. A food web is similar to a food chain but has many more lines that connect organisms at various trophic positions (Figure 5.12). Under these conditions it is more difficult to describe species' trophic positions. The coachwhip snake that eats both insects and lizards would be assigned to a trophic level between 3 and 4. If the snake obtains 50 percent of its energy from herbivorous insects and 50 percent of its energy from carnivorous lizards, the coachwhip snake would be assigned a trophic position of 3.5. This complexity extends to human management—for example, connections among species in food webs makes it difficult to manage fisheries. (See *Case Study: Will Catching Fewer Fish Restore the Marine Food Web?*)

How Many Predators?

The shape and length of a food chain is determined by the rate of net primary production and the rate at which energy is lost from the chain. Net primary production determines the amount of energy that starts through the food chain. Energy is lost as it is transferred from one trophic position to the next because animals cannot digest all of the plant or animal material they eat. Instead, much of the energy passes through the animal and is egested as fecal material. Digestive efficiency depends in part on the type of food eaten. Plants generally are less digestible than animals.

Energy also is lost at each trophic position as the plant or animal uses energy. Animals use a significant fraction of their energy budget for maintenance. This use dissipates energy without increasing biomass. As a result, an organism's biomass increases by an amount that is significantly less than the amount of energy it absorbs from food. The rate at which heterotrophs create new biomass per unit area in a given time period is termed **secondary productivity** and is roughly analogous to the difference between gross and net primary productivity in autotrophs.

The amount of energy at each trophic position is represented with an **energy pyramid** (Figure 5.13). These diagrams are shaped like pyramids because the amount of energy declines at higher trophic positions. The rate of decline is measured by **ecological efficiency,** which is the percentage of energy from one trophic level that is incorporated in the next level. Ecological efficiency generally ranges between 1 and 10 percent. This is why you multiplied the amount of solar energy used to produce a kilogram of plant food by a factor of 10 to determine the amount of solar energy in a kilogram of meat (the conversion of plant material to meat is only about 10 percent efficient) in Your Ecological Footprint in Chapter 4.

Two aspects of the energy pyramid are important to understanding the types and quantities of plants and animals in an area: the rate at which the pyramid narrows and the number of trophic positions or stories in the pyramid. The rate at which the pyramid narrows depends on the secondary productivity of the species present (Figure 5.13). Pyramids dominated by endotherms generally narrow more rapidly than pyramids dominated by ectotherms (ectotherms have a lower basal metabolic rate and so convert a greater fraction of the food they eat to biomass). For example, the secondary efficiency of kangaroo rats is about

FIGURE 5.12 *Food Webs* *A more complete view of the feeding relations in a desert ecosystem such as the Sonoran Desert, much of which lies within the state of Arizona in the southwestern United States. Notice that many of the same organisms in Figure 5.11 are present here, but the food web contains many more species and connections between species. Notice also that we have omitted detritivores.*

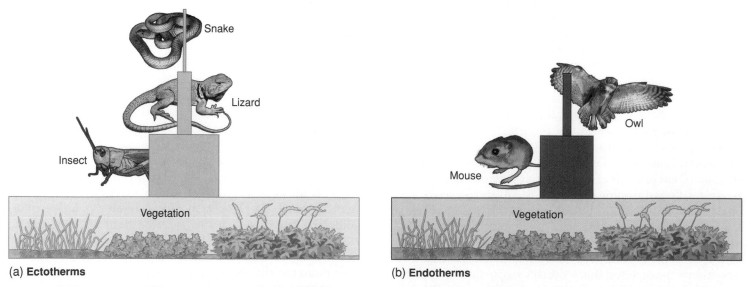

(a) **Ectotherms**

(b) **Endotherms**

FIGURE 5.13 *Biomass Pyramids* *The biomass pyramids for ecosystems dominated by ectotherms and endotherms. The biomass pyramid for endotherms narrows more rapidly than for ectotherms because endotherms use more energy for maintenance than ectotherms. As a result, biomass pyramids for ecosystems dominated by endotherms often have fewer levels than pyramids for ecosystems that are dominated by ectotherms.*

CASE STUDY

Will Catching Fewer Fish Restore the Marine Food Web?

Humans now are an important part of marine food chains. To feed our ever-increasing demand for protein, we trawl the oceans, catching large numbers of fish. Such catches are not sustainable in many parts of the world. To prevent a collapse in fish populations, many nations are implementing policies that reduce the number of fish caught. But have human interventions undermined resilience so much that reducing the number of fish caught will not restore the food chain?

In the North Sea fishing has changed the food web in ways that extend well beyond the fish we eat. The link to other species is apparent to anyone who has fished in the ocean from a boat. As the fish are filleted and the remnants are tossed overboard, birds mob the boat, competing for the discards.

This scene occurs on a much larger scale around commercial fishing boats. These boats discard a significant fraction of the fish that they catch. Globally 25–30 million metric tons (27.5–33 million tons) of fish are discarded each year. Fishers discard some of their catch for a variety of reasons, including legal limits on the minimum landing size. In the North Sea there are minimum sizes for commercial fish species such as haddock or whiting. Undersized fish cannot be sold legally. To prevent such fish from taking up valuable storage space, the catch is sorted as it is brought onboard. "Keepers" are separated from the undersized fish, which are then "thrown back." The idea was to return small fish to the sea alive, but most of the fish that are thrown back are already dead.

These fish are eaten by large seabirds such as the great skua. The great skua is larger than most species and so is able to outcompete smaller species for the discards. Discards have been a boon to the great skua—its population increased with the number of fish discarded. Furthermore, the skua changed their diet in response to the availability of discards. When the quantity of whiting discarded by fishers decreased between 1986 and 2002, the proportion of these fish in the skua's diet decreased. This void was filled in part by haddock discards, the availability of which increased between 1986 and 2002.

Over the last forty years the quantity of discards in the North Sea fishery has declined faster than fish stocks. This loss of food has put increasing pressure on the enlarged skua population. As top-level predators, skua also eat other foods, such as sand eels. But the population of sand eels also is declining. For reasons not well understood, the spawning population of sand eels in 2000 was the lowest value ever observed.

Due to declining supplies of its two favorite foods, the great skua may turn its attention to seabirds. Such a change in diet could have a significant effect on bird populations. For example, about 5,000 skua live on Foula (one of the Shetland Islands). These birds eat a total of about 4.4×10^7 kcal per year. If the fraction supplied by small seabirds increased by 5 percent, this would represent about 2.22×10^6 kcal. Assuming that the average energy content of birds is 2.61 kcal per gram, this would represent about 850 kg of bird biomass. This is equivalent to about 2,100 black-legged kittiwakes or Atlantic puffins.

It is difficult to determine how increased predation would affect these other bird populations. The potential prey species have relatively long lives, so if the skua eat adult birds, this loss could reduce the population significantly. Alternatively, if the skua eat eggs or young birds, the change in diet may have relatively little effect on the population of the skua's potential prey. Given these effects, it is possible that the change in the skua's diet could significantly reduce the population of its prey—and perhaps eventually that of the skua.

The potential for such changes tests the resilience of the marine systems. If changes in the marine food web have gone too far, simply reducing the number of fish caught will not return the marine food web to its starting point. As a result, society may have to design fishing policy that is more sophisticated than simply limiting catch.

ADDITIONAL READING

Votier, S.C., et al. "Changes in Fisheries Discard Rates and Seabird Communities." *Nature* 427 (2004): 727–730.

STUDENT LEARNING OUTCOME

- Students will be able to explain why some environmental changes cannot be reversed by simply stopping the activity that caused the change.

---

1 percent; the secondary efficiency of the side-blotched lizard is about 20 percent; and the efficiency of some salamanders can be over 90 percent.

The notion of ecological efficiency is used to test hypotheses about how dinosaurs used energy. Paleontologists have tried to reconstruct energy pyramids for dinosaur food chains from fossils. Calculations based on the size and numbers of bones indicate that the ratio of predators to prey for dinosaur food chains is low, just like the ratio of predators to prey in food chains dominated by modern endothermic mammals (Figure 5.14). This similarity is an important component of the hypothesis that dinosaurs were more like modern mammals than modern reptiles.

The number of trophic positions or stories in an energy pyramid is determined by the width of the first story and the rate at which the pyramid narrows. All energy that flows through the pyramid ultimately is derived from the first story. Everything else being equal, a pyramid with a broad base can support more trophic positions.

The number of trophic positions also is influenced by the rate at which the pyramid narrows. As energy is used and lost between stories, less energy remains to support the next level. Most pyramids extend four or five stories at most. After this many transfers and uses, not enough energy remains to support another level. Everything else being equal, a pyramid that narrows rapidly (has a lower ecological efficiency) can support fewer trophic positions.

Energy pyramids can help us understand some important aspects of animal behavior and evaluate some important claims regarding solutions to environmental issues.

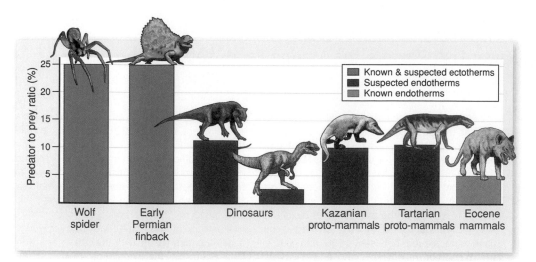

FIGURE 5.14 *Predator–Prey Ratios The ratio of predators to prey for wolf spiders (ectotherms), early reptiles such as finbacks (ectotherms), prehistoric mammals (endotherms), and dinosaurs. Notice that the predator–prey ratio for dinosaurs is closer to that of endothermic mammals than ectothermic spiders or finbacks. This implies that dinosaurs may have used a large portion of their energy budget for maintenance respiration.* (Redrawn from R. Bakker, Dinosaur Heresies. New York: William Morrow and Company, 1986.)

Blue whales are the largest living creatures on Earth, yet their food consists of tiny shrimplike creatures called krill. This food chain may seem puzzling—how can such a large organism obtain enough energy by eating a very small organism?

The answer is determined by the principles of energy pyramids. To survive, a blue whale needs large amounts of food (up to 3.6 metric tons, or 4 tons, per day). Such large amounts of food are needed because of their huge size (100 metric tons—110 tons) and their need to generate large amounts of body heat. The internal temperature of a blue whale is about 37°C (98.6°F), which is about the same as that or humans; yet blue whales live in water that is barely above freezing. Tremendous amounts of energy are available only at the lower positions of the food chain. That is, the biomass of krill is sufficiently large and abundant to supply the needs of the blue whale. As the whale swims, it eats nearly constantly by sucking in huge amounts of water and filtering out the krill.

The blue whale cannot eat "higher" on the food chain because there are not enough medium-sized organisms (third- and fourth-level heterotrophs) to support the blue whales' energy needs. Because these bigger fish are less abundant than krill, the whale simply cannot catch these medium-sized organisms fast enough to satisfy its energy needs. The same principles influence the diet of very large land animals. The largest land animals such as the African elephant (5.5–6.5 metric tons or 6.1–7.2 tons), the white rhinoceros (2 metric tons or 2.2 tons), and the hippopotamus (up to 3.5 metric tons or 3.9 tons) are vegetarians. These vegetarians are large relative to top-level predators such as lions (up to 400 kg or 880 lb) and Siberian tigers (200–300 kg or 440–660 lb).

Energy pyramids also can be used to evaluate the claim by vegetarians that society can increase its food supply by not eating meat. This argument is partially correct. Following the flow of energy from plants to cows to people, it is clear that the amount of energy that is available to people increases if people eat primary producers (plants) instead of primary consumers (cows) (Figure 5.15).

The correctness of this argument depends on where the cows graze. A vegetarian diet increases the human food supply only if the cows graze in areas that could otherwise be used to grow crops. If crops cannot be grown in the area, eating meat may increase the food supply. Many areas that are too arid to support crops can support sparse growths of grass. Cows, sheep, and goats can digest this grass and convert it to meat or dairy products. Under these conditions, eating meat increases the human food supply if the number of livestock allowed to graze arid lands is kept below quantities that inflict permanent damage. Similar arguments are made about fishing: More fish are available at lower trophic positions, but this strategy may not be effective for increasing the number of fish caught. (See *Policy in Action: Can We Catch More Fish by Moving Down the Food Chain?*)

Concentrating Toxins through the Food Web

In the early 1960s scientists noticed that the population of many raptors (birds of prey), such as bald eagles, ospreys, and peregrine falcons, were shrinking rapidly. One reason for the decline was a lack of breeding success. That is, the birds were not raising young. Further investigation revealed that the lack of success started early in the process—the eggs' shells were too thin, and the parents crushed them during incubation.

Research showed that pesticides were interfering with the birds' metabolism of calcium, which is a critical component of eggshells. Such interference required very high levels of pesticides. Indeed, the levels of pesticides in the birds' tissues were many times greater than the concentration of pesticides in the soil or water. How could their pesticide levels become so high?

The answer was found by following the concentration of pesticides and other toxic materials through the food chain. The concentration of pesticides and other toxic materials in each trophic position increases by a factor of approximately 10 relative to the position just below it. Top-level carnivores such as raptors suffer concentrations of pesticides and other toxic materials thousands or mil-

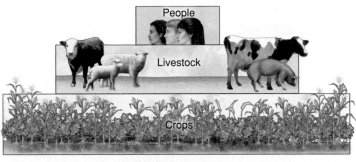

Vegetarians

Meat-based diet

FIGURE 5.15 *Vegetarians* *Vegetarians eat from the first level of the biomass pyramid, which includes plant-based foods such as wheat, corn, and soybeans. This first level has more biomass and hence represents a greater food supply than the second trophic level, which includes animals such as cows, pigs, and chickens. A vegetarian diet does not increase food supply in areas where the soil or climate will not support crops.*

lions of times greater than their concentration in the soil or water. The increasing concentration of pesticides or other toxic materials in living organisms via the food chain is known as **biomagnification.**

If you read the material about secondary efficiency carefully, you may already have an inkling about the mechanisms for biomagnification. As described on page 91, ecological efficiency varies from 1 to 10 percent. This range of efficiencies is the inverse of the range for increased toxin concentrations—a tenfold to hundredfold increase from one position to the next.

The relationship between ecological efficiency and biomagnification is based on a simple fact: Pesticides and toxic materials that accumulate via biomagnification cannot be metabolized. That is, individuals cannot use them as a source of energy. For example, the pesticide DDT can be metabolized to another compound DDE, but DDE cannot be broken down further. As the organism takes in food and uses its energy, carbon dioxide and water are returned to the atmosphere (Equation 5.3), but the pesticide remains. As a result, the pesticide accumulates in the organism's tissues over its lifetime. (Can you guess the type of material it is stored in?)

But an organism lives only so long. As a result, the concentration of pesticides or toxic materials can increase only so much. How does the concentration increase so dramatically in the bodies of top-level carnivores?

The process of biomagnification is amplified by the flow of energy between trophic positions. Organisms at each trophic position eat food that has concentrated the pesticide or toxic material over the lifetime of organisms in the trophic position on which they feed. As a result, organisms at each trophic position eat food that has a concentration of pesticides or toxic materials greater than the food eaten by organisms at the trophic level just below it. For example, the osprey eats large bluefish that have concentrations of pesticides up to ten times larger than the small fish, Atlantic silversides, that are eaten by the bluefish. As a result, the

accumulation of pesticides or toxic materials continues beyond the lifespan of an individual organism.

We can illustrate this effect by following the flow of energy and pesticides through a couple of positions of the food chain given in Figure 5.16. Suppose that small fish such as Atlantic silversides eat zooplankton that have about 0.4 parts per million (ppm) of the pesticide DDE. Over their lifetime, the Atlantic silversides accumulate the DDE so that its concentration in their tissue increases about tenfold to 0.23 ppm. Larger fish such as

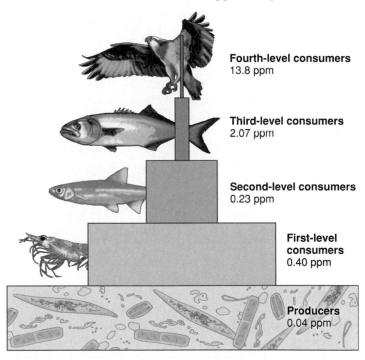

Fourth-level consumers
13.8 ppm

Third-level consumers
2.07 ppm

Second-level consumers
0.23 ppm

First-level consumers
0.40 ppm

Producers
0.04 ppm

FIGURE 5.16 *Biomagnification* *The concentration of DDE increases from 0.04 ppm in primary producers to 13.8 ppm as it moves up the food chain. These large increases are possible because the concentration increases by about a factor of 10 at each step. For example, the concentration increases about ninefold (2.07 ppm/0.23 ppm) from silversides (second-level consumers) to bluefish (third-level consumers).*

Can We Catch More Fish by Moving Down the Food Chain?

How can the world feed a growing number of hungry people? In the 1960s one answer seemed to be the vast supply of fish in the world's oceans. Based on this notion, several nations increased the number of boats and people in their fishing fleets. This extra fishing effort caught significantly more fish. Between 1960 and 1980 the biomass harvested by the world's fisheries increased by 5–7 percent per year. Since 1980 this increase slowed and now has disappeared: Total catch has not increased significantly since the 1990s. As the world's population continues to increase, policy makers need to know whether the number of fish caught can resume its upward trend. We can begin to answer this question by using trophic webs to understand how we increased catch in the past and why we probably cannot increase catch significantly in the future.

Marine food webs consist of four or five trophic positions. The top of the food web is occupied by marine mammals and by large piscivorous (fish-eating) fish such as the north Atlantic swordfish and the yellow fin tuna. These often are the most expensive fish in the market. Have you ever wondered why? One important reason is their relative scarcity. Although these fish can be very large, there is relatively little tuna or swordfish biomass in the world's oceans. This scarcity is dictated by the energy pyramid. As the top-level carnivores, their biomass is small relative to creatures at the lower positions of the pyramid.

Most people prefer the taste of tuna or swordfish to that of mackerel or herring. Nonetheless, the catch of these smaller fish grew relative to the catch of tuna and swordfish over the last few decades. The reason is simple. Fish like herring and mackerel are at a lower trophic position, so there are more fish to catch. In other words, the world has fished "down the trophic web" to increase the biomass of fish caught.

The tendency to fish down the food web is illustrated by Figure 1. The average trophic position of fish caught has declined from about 3.4 in the early 1950s to about 3.0 in the middle 1990s. It did not decline steadily because of changes in the anchovy fishery. The average trophic position dropped dramatically in the 1960s and early 1970s because of a dramatic increase in the catch of Peruvian anchovy, which occupies a relatively low trophic position of about 2.2. The Peruvian anchovy fishery collapsed (due to overfishing and an El Niño event—see the Policy in Action box in Chapter 4) in 1972–1973, and this caused the average trophic position of the global fisheries catch to increase. Since the middle 1970s the average trophic position has declined steadily.

Has the strategy of fishing down marine food webs been successful? The answer is mixed (Figure 2). In some regions, such as the northeast Atlantic (2b) and the Mediterranean (2d), the strategy generally has been successful. The Mediterranean catch increased between 1950 and 1994 as the mean trophic position decreased. In the southeast Pacific (2c), the largest catch corresponds to an intermediate trophic position. And in the northwest Atlantic (2a), the relationship is opposite to that expected: The largest catch occurs at the highest trophic levels.

One reason that catch did not increase at lower trophic positions is a lower limit on the size of fish that can be caught and sold. Fish at the lower positions of the food web tend to be smaller than fish at the upper positions. At

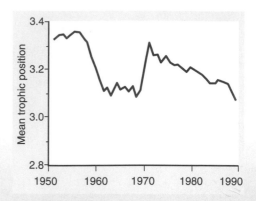

FIGURE 1 *Mean Trophic Position The average trophic position of fish caught by global marine fisheries.* (From Pauly et al., "Fishing Down Marine Food Webs." Science 279 (1998): 860–863).

some point fish are too small or their population is too diffuse to support a commercial fishery. For example, the biomass of lanternfish is large compared to other fish currently caught; but their population is diffuse, so harvesting these fish is too expensive relative to the number of fish caught.

Finally, the interconnectedness of the food web also may prevent an increase in catch at lower trophic positions. North Sea fishing fleets recently have turned to Norwegian pout as a commercial fish. This fish is important to the diet of cod and saithe, which are important commercial species in the area. By catching pout, humans reduce the food supply for other commercial species. In addition, pout also eat large quantities of krill. The krill eat copepods, which are an important source of food for the fish used by people. By reducing the pout popu-

bluefish eat the Atlantic silversides and again concentrate the DDE tenfold to about 2 ppm. Ospreys eat these bluefish and accumulate the DDE in their tissue tenfold to nearly 14 ppm. Based on these three exchanges, the concentration of pesticide increases by about a thousandfold.

The process of biomagnification makes it difficult to reduce the level of pesticides in top-level carnivores. Pesticides and toxic materials remain in the environment long after their production is banned. For example, the biomagnification of DDE and the decline of the raptor population continued

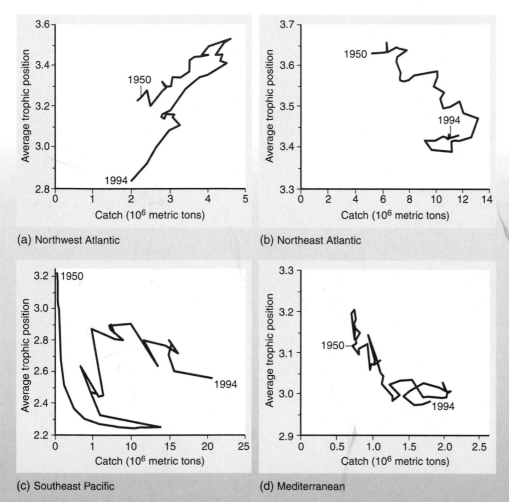

FIGURE 2 *Catch and Average Trophic Position* Biomass pyramids indicate that it should be possible to catch more fish at lower trophic positions. This relationship would be indicated by a general movement down and to the right. This relationship is consistent with data in (b) the northeast Atlantic and (d) the Mediterranean but is inconsistent with the relationship in (a) the northwest Atlantic and (c) the southeast Pacific.

lation, there are more krill, fewer copepods, and less food for commercially important fish. This too reduces the catch of fish as people fish down the food web.

Experience with marine fisheries indicates that moving down the food web can increase the number of fish that can be caught. But this strategy cannot infinitely increase food supply. Fish at the lower trophic positions tend to be small and have a low population density, both of which discourage the formation of an economically viable fishery. In addition, fishing at lower positions increases negative interactions with other fish populations. These factors probably will prevent further increases in fish catch from the world's oceans.

ADDITIONAL READING

Pauly, D., V. Christensen, J. Dalsgaard, R. Froese, and F. Torres. "Fishing Down Marine Food Webs." *Science* 279 (1998): 860–863.

Caddy, J.F., J. Csirke, S.M. Garcia, and R.J.R. Grainger. "How Pervasive Is 'Fishing Down Marine Food Webs'?" *Science* 282 (1998): 1383–1383a.

Pauly, D., R. Froese, and V. Christensen. "*How Pervasive Is 'Fishing Down Marine Food Webs'?*" (Response.) *Science* 282 (1998): 1383a–1383b.

STUDENT LEARNING OUTCOME

- Students will be able to explain why the availability of food for people is not determined solely by the trophic level from which the food is obtained.

even after the application of DDT stopped in 1972. During this period, several captive breeding programs at universities such as Cornell maintained a population of peregrine falcons. At first captive birds were reintroduced to the environment, but the presence of DDT in the food chain made breeding difficult. By the 1990s the concentration of DDT in the food chain declined significantly, and these birds started to establish wild breeding populations. Now breeding pairs of peregrine falcons have established themselves throughout the U.S. East Coast, including downtown Boston and New York.

SUMMARY OF KEY CONCEPTS

- Living organisms use energy for six purposes: maintenance, growth, storage, reproduction, protection, and obtaining more energy from the environment. The way in which energy is used varies among individuals and can be viewed as an evolutionary strategy. The success or failure of evolutionary strategies for obtaining energy and allocating it among the five other uses depends on the environment and the other organisms in that environment. Strategies that are successful in one environment may not work in other environments.

- The viability of strategies for obtaining energy from the environment can be evaluated with the concept of energy return on investment. This ratio measures the amount of energy obtained relative to the amount of energy used to obtain it. In general, EROI must be greater than 1; otherwise the organism will not have any energy for the other five functions.

- As energy and matter pass from autotrophs to heterotrophs and from heterotrophs to other heterotrophs, organisms become linked to one another. These linkages are known as food chains and food webs. All food webs start with autotrophs. Energy then moves through grazers and consumers. All food chains end with detritivores. The location of an organism in this transfer of energy is known as its trophic position.

- Gross primary production measures the amount of organic energy generated by autotrophs per unit area per unit time, and net primary production represents the amount of energy that is converted to structures that can be consumed by heterotrophs. This energy is lost from the food chain as organisms use energy and as energy is transferred between organisms. The length of the food chain is positively related to net primary production and ecological efficiency.

- The concentration of pesticides and other toxic materials increases with the trophic position. Individual organisms concentrate these materials in their bodies as they burn energy. This increased concentration is passed to the next trophic level. Because this process is repeated at each trophic position, top-level predators may have concentrations of toxic materials that are thousands or millions of times greater than their environment.

REVIEW QUESTIONS

1. In the early 1900s some scientists argued that the order of living organisms violated the second law of thermodynamics. Evaluate this claim.

2. Some species migrate long distances to their breeding grounds. How could natural selection favor such behavior?

3. Based on the discussion in the Policy in Action box, draw a food web for the North Sea fishery. Describe how harvesting pout affects the availability of cod and other commercially important fish species.

4. Suppose you have two energy pyramids, both with five trophic positions. One pyramid is dominated by ectotherms; the other is dominated by endotherms. Assuming that the environmental concentration of pesticides was the same for both pyramids, in which pyramid would the top-level predator have the greatest concentration of pesticides?

KEY TERMS

active range	energy pyramid	parenting
aerobic respiration	food chain	photosynthesis
anaerobic respiration	food web	phytoplankton
autotrophs	grazers	primary consumers
basal metabolic rate	gross primary production	r selected
biomagnification	heterotrophs	reproduction
biomass	hibernation	respiration
carnivores	K selected	scavengers
decomposers	maintenance respiration	secondary consumers
detritivores	maturation	secondary productivity
detritus	metamorphosis	senescence
ecological efficiency	mimicry	trophic position
ectotherms	net primary production	zooplankton
endotherms	omnivores	

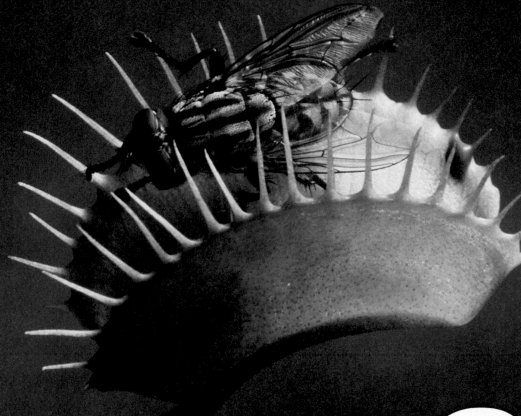

THE FLOW OF MATTER IN THE ENVIRONMENT

Why Does It Matter?

6

CHAPTER

STUDENT LEARNING OUTCOMES

After reading this chapter, students will be able to

- Describe how the availability of matter limits the growth of autotrophs and heterotrophs.
- Compare and contrast food webs and biogeochemical cycles.
- Describe how changes in one cycle affect storages and flows in another cycle.
- Explain how biogeochemical cycles can be used to describe many human impacts on the environment.
- Explain how the rates of nonspontaneous flows and residence times affect the rates at which atoms of carbon, nitrogen, and phosphorus can complete one turn through their respective cycles.

Venus flytrap *The Venus flytrap grows in nutrient poor soils. To obtain nutrients, it uses specialized leaves, traps, to obtain nutrients from insects.*

A fly unknowingly steps on a tiny hairlike structure; instantly the Venus flytrap encloses its victim (opening photograph). A few days later the trap opens and a mummified fly falls out. Digestive juices have allowed the plant to absorb much of the fly's soft body parts.

The Venus flytrap does not "eat" insects in the usual sense. The plant derives little or no energy from the fly. Nor does it have much need for energy from the fly: The plant has plenty of green leaves with which it can generate its own food. What does the Venus flytrap obtain from its victims?

Insects are a significant source of nutrients for the Venus flytrap. The Venus flytrap and other insect-eating plants tend to live in soils that are deficient in many nutrients such as nitrogen. Poor soil makes it difficult for the roots of the Venus flytrap to get enough nutrients. Instead this plant has evolved specialized leaves that can trap and absorb nutrients from its insect victims.

The Venus flytrap goes to such lengths because nutrients are critical to plant growth. A plant uses nutrients to build leaves, stems, roots, and the rest of its infrastructure. Indeed, nutrients are important to all plants and animals. Without these building blocks of living tissue, the energy generated via photosynthesis is of little use to the Venus flytrap.

In many ways the flow of matter is tightly linked to the flow of energy. When a Komodo dragon eats a deer, it obtains both energy and nutrients. Yet there are some important differences. Matter can be recycled, whereas energy cannot. As a result, we will look at nutrient cycles instead of one-way food chains. Nutrients are used to build different structures, so we will look at cycles for each type of nutrient. ∎

MATTER: THE BUILDING BLOCKS OF LIFE

Organic matter is defined as molecules that contain carbon. Carbon, one of the most important building blocks of living organisms, is usually combined with hydrogen and oxygen. For example, these elements are the sole constituents of the glucose and lactate molecules that are described in Chapter 5. **Carbohydrates** and **fats** are organic molecules that contain only carbon, hydrogen, and oxygen.

But living organisms are made up of elements other than carbon, hydrogen, and oxygen. If your body were to be separated into its constituent elements, significant amounts of nitrogen, potassium, and calcium would be present. A very careful look would also reveal small amounts of trace elements such as iron and sulfur.

Although present in small amounts, these other elements are crucial to living organisms. For example, nitrogen is an important component of **proteins,** which are structurally complex, three-dimensional molecules that are made up of amino acids. Proteins make up most of the working parts of living tissues. For example, proteins are critical to the function of muscle.

The matter required by living organisms is called **nutrients.** Nutrients include all elements that are needed by a plant or animal to grow and function. Usually the term nutrient refers to nitrogen, phosphorus, sulfur, and potassium. Less often the term refers to elements such as calcium, iron, cobalt, sodium, and molybdenum.

Organisms have developed two strategies for obtaining nutrients. Most autotrophs get their nutrients from the environment in an inorganic form. On the other hand, most heterotrophs get their nutrients from plants and other animals in an organic form. Like the strategies used to obtain energy, the need to obtain nutrients influences the shape and behavior of individual organisms.

Nutrient Capture by Autotrophs

Except for the Venus flytrap and other insect-eating plants, most autotrophs get their nutrients from the environment in an inorganic form. For example, most plants use nitrogen in the form of nitrate (NO_3) and phosphorus in the form of phosphate (PO_4). These forms of nitrogen and phosphorus are said to be **available** because plants can use them immediately without converting them to another form. On the other hand, most plants cannot use nitrogen in the form of molecular nitrogen (N_2). This form of nitrogen is said to be **unavailable.** A plant could suffer from a shortage of nitrogen even though the atmosphere is mainly (78 percent) molecular nitrogen.

Vascular plants, which are plants that have a well-developed system of conducting tissue to transport water, mineral salts, and sugars, absorb nutrients that are dissolved in water from the soil through their roots. Nutrients are transported to the other parts of the plant along with water. Plants do not have a circulatory system with a pump (the heart) like many animals. Instead water evaporates from leaves through a process known as **transpiration** (Figure 6.1). As water evaporates, it "pulls" water up from the roots as if the water molecules were linked in an invisible chain. Plants transpire 300–1,000 kilograms (662–2,205 lb) of water to make 1 kilogram of dry plant material. Relatively little of this water is combined with carbon dioxide to make glucose via photosynthesis.

Plants require many different nutrients and use them in specific combinations. For example, the chemical makeup of phytoplankton has a relatively constant ratio of nitrogen to phosphorus at about 16 units of nitrogen for every 1 unit of phosphorus. This 16:1 proportion, known as the **Redfield ratio,** is not constant—it varies between phytoplankton and plants and among various parts of a plant.

You can think of the proportion of nutrients as a recipe for making living tissue. Relatively little variation is allowed. One missing ingredient will slow plant growth significantly. The nutrient that is in least supply relative to the quantity required by the recipe is termed the **limiting nutrient.** The importance of the limiting nutrient is described by **Liebig's law of the minimum.** This law states that the growth rate of plants often is determined by the nutrient that is least abundant or least available relative to the needs of the plant.

Liebig's law can be illustrated by the following example (Table 6.1). Suppose the production of 1 unit of plant biomass requires 100 units of carbon, 16 units of nitrogen, and 1 unit of phosphorus. Suppose that a plant has 300 units of carbon, 32 units of nitrogen, and 5 units of phosphorus. How many units of biomass can the plant build from this supply of nutrients? The plant has enough carbon to make 3 units of biomass (300/100), enough nitrogen to make 2 units of biomass (32/16), and enough phosphorus to make 5 units of biomass (5/1). The plant can make 2 units of biomass because there are only 32 units of nitrogen. In this example nitrogen is the limiting nutrient. Notice that phosphorus is not the limiting nutrient, even though there are only 5 units available. Put simply, the limiting nutrient is not always the nutrient that is present in the smallest quantity.

Farmers use Liebig's law to increase the amount of food grown per hectare. Farmers test their soil for the abundance of nutrients and compare it to the quantities required by their crops. Farmers can speed the growth of their crops by adding fertilizers that contain the limiting nutrients. Continuing with the previous example, adding nitrogen to the soil would overcome the limit on growth imposed by the availability of nitrogen. On the other hand, fertilizing the soil with phosphorus would waste the farmers' money because it would not alleviate the limit on plant growth that is imposed by the supply of nitrogen.

FIGURE 6.1 *Transpiration* Water molecules evaporate from leaves, and as they do, they pull water molecules through the tree and from the soil. This water carries nutrients that are needed by the plants.

TABLE 6.1	The Limiting Nutrient		
	Carbon	Nitrogen	Phosphorus
Supply	300	32	5
Redfield Ratio	100	16	1
Units of Biomass Possible	3	2	5

Nutrient Capture by Heterotrophs

When animals eat plants or other animals, they obtain nutrients as well as energy. Because they come from living organisms, these nutrients usually are in an organic form. For example, most animals get much of their nitrogen in the form of proteins.

Animals obtain nutrients when they consume plants or animals, but the nutrients are not available until they cross to the interior portion of the body, a process called **absorption.** This occurs in the intestine or stomach in many animals, such as people. To be absorbed, organic materials are not broken into individual elements. Rather, complex forms of organic molecules are broken into smaller building blocks in a process known as **digestion.** Once digested and absorbed, these building blocks can be reassembled into proteins or carbohydrates.

Like plants, animals need nutrients in specific forms. Some forms cannot be digested. For example, you could eat a stomach full of grass, but you would not be able to digest the carbohydrates because your stomach does not produce enzymes that can break down the cell walls of the grasses. Indeed, our inability to digest celery is one reason that it is such a good diet food. Its energy return on investment is less than one.

Similarly, many heterotrophs cannot make all of the building blocks from individual elements, so they must have them ready-made in their diet. The human body consists of millions of different proteins that are made from only twenty types of amino acids. Your body can make ten of these amino acids if it has all the ingredients. That is, your body knows the recipe for these ten amino acids, such as cysteine. These ten amino acids are known as **nonessential** amino acids. On the other hand, your body does not know the recipe for the other ten amino acids, such as tryptophan. Even if it has all the ingredients, it cannot make them. Instead you must obtain these ten amino acids, known as **essential** amino acids, directly from your food.

The concept of limiting nutrient usually is not applied to heterotrophs. Nonetheless, heterotrophs need to consume a variety of nutrients to remain healthy. This variety is termed a balanced diet. Even foods that are good for you do not constitute a complete diet. For example, pasta is a healthful food. It consists mainly of carbohydrates and is low in fat. Nonetheless, you could not exist for long on a diet of only pasta because it has no protein—and without protein your body would run short on nitrogen and essential amino acids. This shortage would limit your ability to build and repair proteins. As a result, your body's ability to function would slowly degrade. **Kwashiorkor** is the medical name given to the human malady brought on by a lack of protein. This disease is relatively rare in affluent nations but is common in poorer nations. As you might expect, children are most susceptible to kwashiorkor. Without nitrogen and essential amino acids, children cannot build the proteins that are required by a developing body.

THE FLOW OF MATTER: BIOGEOCHEMICAL CYCLES

For all you know, you may have just inhaled the same molecule of carbon dioxide that Abraham Lincoln exhaled while stating "Four score and seven years ago. . . ." Similarly, the nitrogen atoms in your brain may have been part of the jaw muscles of a *Tyrannosaurus rex* about 70 million years ago. These recombinations are possible because matter is not used up as it passes from one organism to the next. Unlike energy, matter is used over and over again. The potential for recycling is critical because, for all practical purposes, Earth does not get any new supplies of matter as it does for energy.

The ways in which matter is recycled are known as **biogeochemical cycles.** The term biogeochemical is an agglomeration of three words: biological, geological, and chemical. Biogeochemical cycles represent movements among the biological, geological, and chemical systems of Earth. These cycles are like a road system. The carbon cycle includes the many routes by which a carbon atom may have traveled from Lincoln's lips to yours, and the nitrogen cycle includes the routes by which a nitrogen atom may have traveled from the jaw of *Tyrannosaurus rex* to your brain. For example, after leaving Abraham Lincoln's lips, an atom of carbon may spend some time in the atmosphere. From there a plant may have incorporated the carbon atom into its structure via photosynthesis. The plant may have been eaten by a cow, and the carbon may have been incorporated into its milk. This milk was converted to yogurt via the detritivore food chain, which respired the carbon atom back into the air in the yogurt container; then you opened the container and inhaled the carbon atom.

Understanding Biogeochemical Cycles

Before we discuss biogeochemical cycles in detail, let's explain the rules that govern these cycles. Pictorial representations of biogeochemical cycles are easy to understand if you remember that the figure is like a road map that shows the flow of matter from one location to the next. Locations in a biogeochemical cycle are termed **storages.** A storage is a location where matter stops along its journey. As such, a storage is represented as a place that stores an element such as carbon or nitrogen (Figure 6.2). There are five basic storages in the environment: oceans (and other water bodies), the atmosphere, Earth's crust, soil, and **biota,** which are all living organisms. For example, about 78 percent of the atmosphere consists of nitrogen in the form of molecular nitrogen (N_2). This makes the atmosphere a large storage for nitrogen. Nitrogen also is an important component of living tissue, so biota are also a storage for nitrogen.

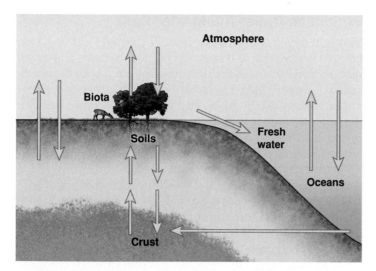

FIGURE 6.2   *Biogeochemical Cycles* *Elements flow between four main storages: the atmosphere, crust, biota, and oceans. Storages are in Bold, and flows are represented by arrows.*

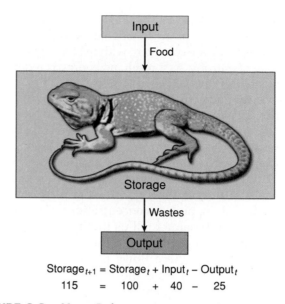

$$Storage_{t+1} = Storage_t + Input_t - Output_t$$
$$115 \quad = \quad 100 \;\; + \;\; 40 \;\; - \;\; 25$$

FIGURE 6.3   *Mass Balance* *The amount of matter in a storage in the next period (Storage$_{t+1}$) is equal to the quantity in the storage in the current period (Storage$_t$) plus the amount of matter that flows into the storage (Input$_t$) minus the amount of matter that flows out of the storage (Output$_t$).*

Matter always is moving into and out of storages. A movement from one storage to another is called a **flow**. A flow measures the quantity of matter that moves from one storage to another per unit of time. For example, photosynthesis causes carbon to flow from the atmosphere to the biota. Gross primary production (discussed in Chapter 5), which in the Florida Everglades is about 20 grams of carbon per square meter per year, is the annual flow of carbon from the atmosphere to the biota.

The relationship between flows and storages is governed by the law of conservation of matter (Figure 6.3), which states that matter cannot be created or destroyed. This implies that the amount of matter in a storage is determined by the rate at which matter flows into and out of that storage. For example, a lizard (a biotic storage for carbon) that starts the year with 100 units of carbon will end the year with 115 units of carbon if the lizard absorbs 40 units of carbon from its food and exhales 25 units of carbon via aerobic respiration (100 + 40 − 25 = 115).

The size of a flow relative to the size of a storage determines **residence time**—the time that an atom spends in a storage. For example, the average molecule of carbon dioxide spends about five years in the atmosphere but about 3,000 years in the deep layers of the ocean.

The laws of thermodynamics define two types of flows: spontaneous and nonspontaneous. **Spontaneous flows** are consistent with the tendency of matter to move from low to high entropy—that is, to become more disorganized. You can identify spontaneous flows as movements from areas of high concentration to areas of low concentration. For example, the combustion of fossil fuel emits large amounts of carbon dioxide into the atmosphere. As it does, the concentration of carbon dioxide in the atmosphere increases relative to the concentration of carbon dioxide in the ocean (Figure 6.4). The difference

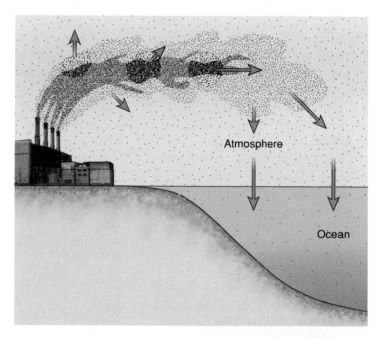

FIGURE 6.4   *Spontaneous Flows* *Spontaneous flows cause matter to flow from areas of high concentration to areas of low concentration. In the carbon cycle, emissions from fossil fuel combustion set up a gradient in the concentration of carbon dioxide. Fossil fuel combustion emissions have a relatively high concentration of carbon dioxide, which is indicated by the dark shading. This increases the concentration of carbon dioxide in the atmosphere relative to its concentration in the ocean, which is indicated by the lighter shading. These differences cause carbon dioxide to flow from areas of high concentration to areas of low concentration, as indicated by the arrows, such as those that point from the atmosphere to ocean.*

in concentration causes carbon dioxide to flow spontaneously from the atmosphere to the ocean. Spontaneous flows require no work because they are consistent with the tendency toward a greater state of entropy. As a result, no energy is required to drive spontaneous flows between storages.

Nonspontaneous flows are the opposite of spontaneous flows. Nonspontaneous flows oppose the general tendency for matter to become more disordered. Nonspontaneous flows move matter from areas of low concentration to areas of high concentration. For example, trees pull carbon from the atmosphere, where it is found in low concentrations, and incorporate it in their structure, where it is found in higher concentrations (Figure 6.5). The laws of thermodynamics tell us that movements from low concentrations to high concentrations do not occur by themselves. Rather, these flows must be powered by energy. To move carbon from the atmosphere, trees use solar energy to drive photosynthesis.

Every biogeochemical cycle must have at least one nonspontaneous flow. You can understand this by thinking of spontaneous flows as moving matter "downhill" while nonspontaneous flows move matter "uphill." If all the flows in a biogeochemical cycle were spontaneous, the matter in a particular cycle would move downhill and accumulate in the "lowest" storage. This accumulation would be a dead end that eventually would stop the cycle. Such a dead end is prevented by nonspontaneous flows, which move matter back "uphill" and thereby keep matter moving through the cycle.

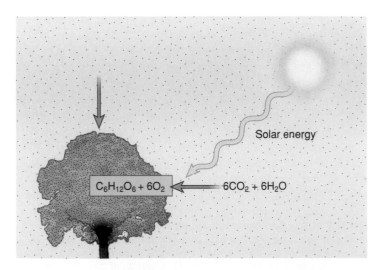

FIGURE 6.5 *Nonspontaneous Flows Nonspontaneous flows cause matter to flow from areas of low concentration to areas of high concentration. In the carbon cycle, photosynthesis causes carbon dioxide to flow from low concentrations in the atmosphere (light shading) into the biota, which have a higher carbon concentration (dark shading). The work associated with this nonspontaneous flow is powered by the solar energy that drives photosynthesis.*

The Carbon Cycle: The Master Cycle

We might never guess that carbon is essential to life because it makes up less than .004 percent of Earth's crust. This small fraction puts carbon in fourteenth place among all elements. Such an obscure element hardly seems to merit the attention of ecologists, chemists, physicists, and economists. But by mapping the carbon cycle, scientists have learned that carbon is critical to the flow of biological energy, the flow of industrial energy, and the regulation of Earth's climate.

Carbon is found in six major storages: the atmosphere, biota, soils, oceans, crust, and fossil fuels (Figure 6.6). These storages are connected by eight flows: photosynthesis, respiration, deforestation, fossil fuel combustion, solution and dissolution in the ocean, sedimentation, and death. As with other biogeochemical cycles, there is no start and no end to the carbon cycle; nonetheless, the easiest place for us to start the journey through the carbon cycle is the nonspontaneous flow of carbon from the atmosphere to the biota. Each year photosynthesis by terrestrial plants moves about 120 petagrams (1 petagram = 10^{15} grams = 10^{12} kilograms, or 132 billion tons) of carbon from the atmosphere to the biota. Once in the biotic storage, the carbon dioxide is used to build organic molecules such as carbohydrates, proteins, and fats.

If you could look inside the box labeled biota, you would see these organic molecules moving between producers and consumers. These names and the general pattern of movement should be familiar—they are the food chains and webs described in Chapter 5. Indeed, food chains and webs, which describe the flow of carbon within the biotic storage, are an important component of the carbon cycle.

The principles described in Chapter 5 also help us to understand what happens to carbon in the biotic storage. To calculate your use of this carbon, see Your Ecological Footprint: How Much Net Primary Production Do You Use? The biomass pyramid (Figure 5.13 on page 92) describes how biomass declines from one trophic position to the next. We explained that energy's ability to do work is lost as it is used. But carbon is matter, and matter cannot be used up or disappear. So what happens to the carbon? As the energy is used, the organic molecules are respired aerobically, and they are broken into carbon dioxide and water (see Equation 5.3 in Chapter 5, page 89) and flow to the atmosphere. Each year respiration by organisms other than detritivores returns to the atmosphere about half (60 petagrams, or 66.2 billion tons) of the carbon dioxide that is absorbed by photosynthesis.

Another portion of the carbon that flows from the atmosphere to the biota becomes part of the detritus food chain. Partially decomposed organic matter becomes part of the soil carbon storage. Eventually the organic material in the soil is decomposed into its constituents, water and carbon dioxide, which return to the atmosphere. This flow

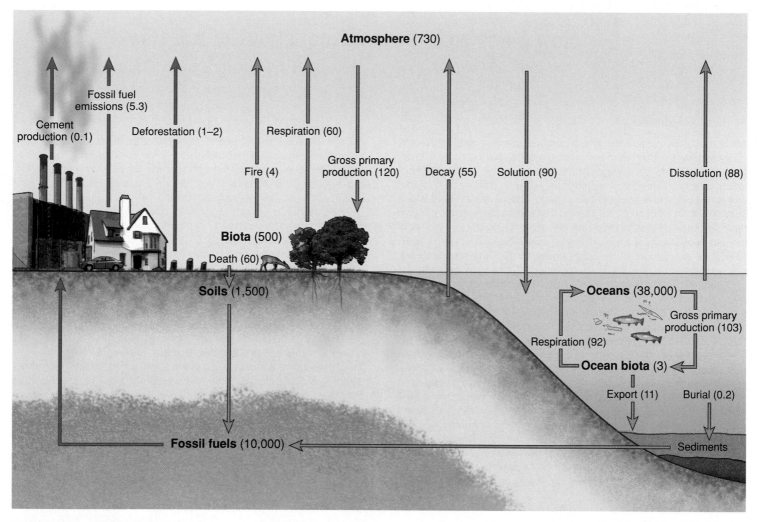

FIGURE 6.6 *Carbon Cycle* *Storages are measured in petagrams, which are quadrillion (10^{15}) grams. Flows are measured in petagrams per year. Flows associated with human activity are in red.*

of carbon is known as decay and accounts for about 55 petagrams (60.6 billion tons). Together with respiration, these flows account for *most but not all* of the carbon removed from the atmosphere by photosynthesis.

We emphasize "most but not all" because each year very small amounts of carbon escape respiration and decay when they become trapped in environments without oxygen. Without oxygen, decomposition is incomplete. Although only a tiny portion of organic material escapes decomposition, accumulation over millions of years has stored large amounts of carbon as fossil fuels. Scientists estimate that about 10,000 petagrams (11,025 billion tons) of carbon are stored as coal, oil, natural gas, oil shale, and tar sands.

Until recently the flow of carbon from fossil fuels to the atmosphere was minuscule—nearly zero. The fossil fuel storage represented a dead end for the carbon cycle. But the Industrial Revolution increased the use of coal, oil, and natural gas. Burning fossil fuels completes the process

of breakdown back to carbon dioxide and water. In 2003 humans burned about 5.2 billion metric tons (5.7 billion short tons) of coal, 29.2 billion barrels of oil, and 95.5 trillion cubic feet of natural gas, which caused about 6.9 petagrams (7.6 billion tons) of carbon to flow from the fossil fuel storage to the atmosphere. Notice that increases in economic activity have increased fossil fuel emissions relative to the value in Figure 6.6, which is based on storages and flows in the 1990s compiled by the Intergovernmental Panel on Climate Change.

The combustion of fossil fuels is not the only flow in the carbon cycle affected by economic activity. Prior to the expansion of human civilization, the amount of carbon stored in biota changed very slowly from year to year because the amount taken up through photosynthesis was nearly equal to the amount emitted through respiration and decomposition. But human activity has disturbed the biotic storage. Over the last several hundred years humans

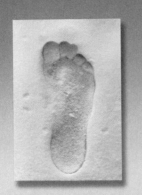

Your ECOLOGICAL *footprint*

How Much Net Primary Production Do You Use?

The role of humans in the global carbon cycle can be quantified by measuring the fraction of net primary production (NPP) that we use. NPP is a critical factor because it represents the total amount of biological energy available to heterotrophs. As such, the fraction of NPP consumed by humans conveys two aspects of society's ecological footprint. The finite supply of NPP implies that the fraction used by people represents the degree to which we are approaching the upper limit of the supply of biological energy. Second, as we increase our use of NPP, we reduce the amount of biological energy available to other species, which reduces the planet's ability to support these other species.

Determining the fraction of NPP consumed by humans requires two difficult calculations: the NPP generated by the entire planet and the amount used by humans. Scientists calculate NPP by combining climate data (such as temperature and precipitation) and remotely sensed data (data measured by satellites) with computer simulation models of biological processes. The satellite data are used to estimate autotrophic biomass. The model combines this information with temperature and precipitation data, which largely determine the rate at which autotrophs photosynthesize and respire. The balance between these two is NPP. Using these methods, scientists estimate that terrestrial plants generate about 57 petagrams (62.8 billion tons) of carbon annually.

Humans use NPP in many ways. Food is the most obvious. All food is derived directly (like vegetables) or indirectly (like meat) from NPP. Humans also use NPP indirectly in the form of paper, fiber, and wood for energy or construction.

At first glance it seems easy to use their chemical composition to convert a kilogram of beef or paper to kilograms of carbon. But this simple approach would understate human use of net primary production. Much of what humans use is above-ground biomass. But this is often supported by a nearly equal amount of below-

ground biomass. Similarly, economic production processes often discard a significant amount of the original material. For example, not all harvested timber is converted to wood that can be used for construction, and the remaining portions are sometimes wasted rather than being used for other purposes (such as paper).

The degree to which these indirect uses are included allows scientists to generate three estimates for the fraction of NPP used by society. At the low end, humans use about 3 percent of the NPP generated by terrestrial and aquatic ecosystems. This 3 percent represents only the organic material used directly by people or their domesticated animals.

A more inclusive estimate says that people use about 19 percent of the planet's NPP. This estimate includes both the NPP used directly by humans and the NPP lost as humans replace natural ecosystems with their own land uses. For example, as humans replace forests with croplands or pasturelands, the amount of NPP generated by these lands is reduced. This lost NPP is included in the 19 percent.

An even more inclusive estimate asserts that people use about 39 percent of the planet's NPP. This percentage includes the uses just described plus the secondary effects of human activity. Human activity erodes soils and emits wastes into the atmosphere. As described in

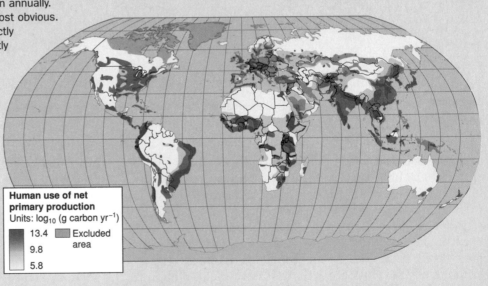

FIGURE 1 *Human Use of Net Primary Production* Spatial representation of the human use of terrestrial net primary production. Notice that people living on the U.S. East Coast and in Europe, India, and China use the largest amounts of net primary production. (Figure from M. Imhoff et al., "Global Patterns in Human Consumption of Net Primary Production." Nature 429 (2004): 870–873.)

Human use of net primary production
Units: $\log_{10}$ (g carbon yr^{-1})

- 13.4
- 9.8
- 5.8
- Excluded area

have reduced the area covered by forests, a process known as **deforestation.** By reducing the number of trees through burning and other uses, deforestation reduces the amount of carbon stored in the biota. Most of this carbon flows to the atmosphere. In the 1990s deforestation and other changes in land use caused 1–2 petagrams (1.1–2.2 billion

tons) of carbon to flow from the biota to the atmosphere annually.

The other important set of flows moves carbon from the atmosphere to the ocean and from the ocean to the atmosphere. For a long time these two flows were approximately equal. This balance was created and maintained by

Chapters 15 and 19, respectively, these actions degrade environmental services. This degradation includes a reduction in NPP.

There is no "correct" answer regarding the fraction of NPP consumed by society. Including the indirect uses reduces the accuracy of the estimate. Nonetheless, it is clear that excluding these indirect uses significantly understates the effect of humans on the environment.

Scientists have calculated the quantity of NPP used by people in various locations (Figure 1). In general, people with a high material standard of living use more NPP than people with a lower material standard of living. The data used to generate this map are reproduced in Table 1. You can use this information to calculate your use of NPP. Sum your use of these products over a year, multiply by the conversion factor, and sum this number over all products. In Chapter 8 you will use this information to calculate the amount of land from which you draw these goods.

Calculating Your Footprint

We recognize that it may be difficult to fill in all of the blanks here. You should have compiled data about your diet in Chapter 5's Your Ecological Footprint (page 81). We will explore your use of paper and wood in Chapter 17's Your Ecological Footprint (page 318). You can either read ahead and use that information here, or use the average values for U.S. citizens, which are given in parentheses.

NPP consumed as plant-based food $= \dfrac{(3,774) \text{ kcal/day} \times 0.36 \text{ gram/kcal}}{\times 365 \text{ days/year}}$

NPP consumed as meat $= \underline{(124)} \text{ kg/year} \times 9,295 \text{ grams/kg}$

NPP consumed as milk $= \underline{(115.8)} \text{ kg/year} \times 501 \text{ gram/kg*}$

NPP consumed as eggs $= \underline{(14.6)} \text{ kg/year} \times 3,630 \text{ gram/kg**}$

NPP consumed as paper $= \underline{(314)} \text{ kg/year} \times 991 \text{ gram/kg}$

NPP consumed as wood $= \underline{(1.06)} \text{ m}^3\text{/year} \times 2,483 \text{ gram/m}^3$

* Milk weighs about 3.92 kg per gallon.
** An egg weighs about 56.7 grams (2 oz).

Interpreting Your Footprint

Ecologists estimate that net primary production by Earth's terrestrial ecosystems is about 57 petagrams of carbon (quadrillion (10^{15} grams)). If all 6.48

TABLE 1 — The Grams of Carbon of NPP Used to Produce Various Products

Product	NPP Equivalent
Plant-based foods	0.36 grams carbon per kcal
Meat	9,295.2 grams carbon per kg
Milk	500.8 grams carbon per kg
Eggs	3,630.2 grams carbon per kg
Paper	990.8 grams carbon per kg
Wood (fuel)	2,483.4 grams carbon per m³

Data from calculations by Imhoff et al., 2004.

billion (10^9) people used net primary production at the rate you just calculated, what fraction of global net primary production would people use?

This fraction is probably much larger than the 7 percent of terrestrial net primary production used directly by the human population that is estimated by ecologists. Why such a difference? As will be explained in Chapter 11, the average person living in the United States uses much more energy and material than the average denizen of the planet. To give you an idea of this difference, the average person living in 2002 used about 2,804 kcal of plant-based food per day, 39 kg of meat per year, 44 kg of milk per year, 8.4 kg of eggs per year, about 56 kg of paper per year, and about 0.51 m³ of wood. Use these numbers to calculate the quantity of net primary production used by the average person. Do these calculations move you closer to the 7 percent estimate?

ADDITIONAL READING

Imhoff, M.L., L. Buonoa, T. Ricketts, C. Loucks, Robert Harriss, and W.T. Lawrence. "Global Patterns in Human Consumption of Net Primary Production." *Nature* 429 (2004): 870–873.

Vitusek, P.M., P.R. Ehrlich, A.H. Ehrlich, and P.A. Matson. "Human Appropriation of the Products of Photosynthesis." *Bioscience* 36 (1986): 368–373.

the spontaneous flow of carbon from the storage of high concentration to the storage with the lower concentration. These movements created an equilibrium between the amounts of carbon in the atmosphere and ocean.

This equilibrium has been disrupted by the combustion of fossil fuels and deforestation. These two flows add carbon to the atmosphere, which increases the concentration of carbon in the atmosphere relative to the ocean. The greater atmospheric concentration of carbon causes carbon to flow spontaneously from the atmosphere to the ocean. The size of this flow is limited by a negative feedback loop, termed the Revelle factor. As carbon dioxide

FIGURE 6.7 *Unknown Carbon Sink The difference between the change in carbon stored in the atmosphere and flows into and out of the atmosphere. Negative values indicate that there is less carbon in the atmosphere than indicated by estimates for flows into and out of the atmosphere.* (Data from Richard Houghton, personal communication.)

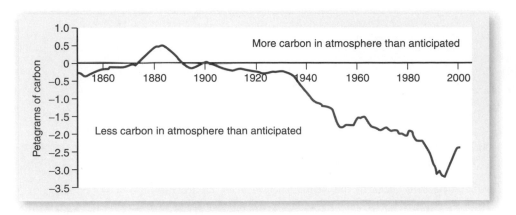

dissolves in the ocean, it reduces the ocean's pH (makes it more acidic). The lower pH slows the rate at which carbon dioxide dissolves in the ocean. Currently the flow of carbon from the atmosphere to the ocean is about 2 petagrams (2.2 billion tons) greater than the flow of carbon from the ocean to the atmosphere.

Despite the scientific certainty that the global carbon cycle is governed by the laws of conservation, scientists cannot "balance" the storages and flows. That is, summing the best estimates for the flows of carbon to and from the atmosphere indicates that there is less carbon in the atmosphere than expected. During the 1990s the atmosphere was missing about 3 petagrams (3.3 billion tons) of carbon per year (Figure 6.7). This missing carbon is associated with an **unknown carbon sink.**

The unknown carbon sink is either an unknown mechanism that removes carbon from the atmosphere or a known mechanism that removes carbon faster than estimated by scientists. There are several hypotheses concerning the unknown carbon sink. Many are based on negative feedback loops that include the atmospheric concentration of carbon dioxide (Figure 6.8). One hypothesis is that the increasing concentration of carbon dioxide in the atmosphere increases net primary production, and this speeds the rate at which carbon is pulled from the atmosphere. Experiments show that plants grow faster at higher concentrations of carbon dioxide, but it is not clear whether this increase is significant in the real world. If the growth of plants is not limited by the availability of carbon in the atmosphere, increasing its concentration will not increase plant growth. On the other hand, the mechanism may be boosted by human activities that increase the availability of nitrogen to plants (see the next section about the nitrogen cycle).

Another hypothesis for the unknown carbon sink focuses on climate. The increasing concentration of carbon dioxide in the atmosphere is partially responsible for the global increase in temperature (discussed in Chapter 13). As the world gets warmer, this could enhance plant growth, which would speed the rate at which plants remove carbon

from the atmosphere. Satellite data show that Earth's surface has become greener over the last twenty years (Figure 6.9). These changes are positively related to temperature increases.

Alternatively, higher temperatures could accelerate the rate of decay. Decay frees nutrients from organic material. If these nutrients are limiting in a Liebigian sense, the greater supply could accelerate net primary production and therefore speed the rate at which carbon dioxide is removed from

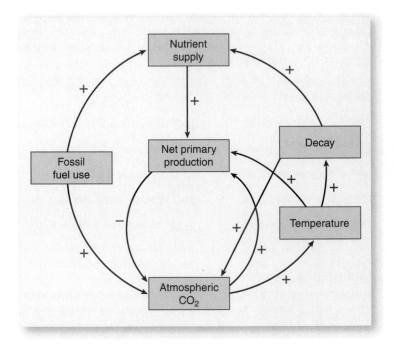

FIGURE 6.8 *Feedback Loops and the Unknown Carbon Sink Changes in the atmospheric concentration of carbon dioxide change the flow of carbon to and from the atmosphere. The unknown carbon sink may be associated with negative feedback loops that change carbon flows to and from the atmosphere, moving the atmospheric concentration of carbon dioxide back toward its initial value. This effect is opposed by a positive feedback loop that includes atmospheric CO_2, temperature, and decay and reinforces the ongoing increase in atmospheric CO_2.*

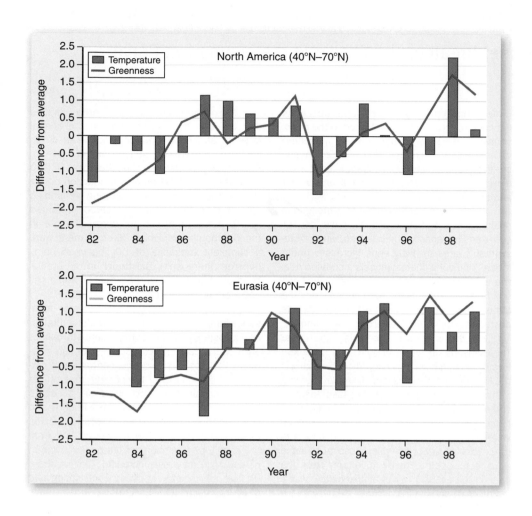

FIGURE 6.9 *Surface Temperature and Plants* *There is a positive relationship between temperature and satellite measures of surface vegetation: Warm years are associated with greater quantities of vegetation.* (Redrawn from L. Zhou, C.J. Tucker, R.K. Kaufmann, D. Slayback, N.V. Shabanov, and R.B. Myneni, "Variations in Northern Vegetation Activity Inferred from Satellite Data of Vegetation Index during 1981 to 1999." Journal of Geophysical Research 106 (2001): 20069–20083.)

the atmosphere. But if these nutrients were not limiting, accelerated rates of decay would increase the flow of carbon dioxide to the atmosphere.

Even though scientists cannot balance the global carbon cycle, it is clear that the amount of carbon entering the atmosphere is greater than the amount of carbon leaving the atmosphere. Over the last forty-five years the amount of carbon stored in the atmosphere has increased, which we see as a significant increase in the atmospheric concentration of carbon dioxide (Figure 6.10). Between 1959 and 2004 the concentration of carbon in the atmosphere increased from about 317 parts per million (ppm) to 377 ppm. This increase is important because the amount of carbon dioxide in the atmosphere influences the amount of heat retained, which alters global climate (see Chapter 13). Notice too that the increase in Figure 6.10 is not steady. Within each year the concentration of carbon dioxide rises and falls. This intra-annual cycle allows us to watch the planet "breathe." See *Case Study: Watching the Planet Breathe.*

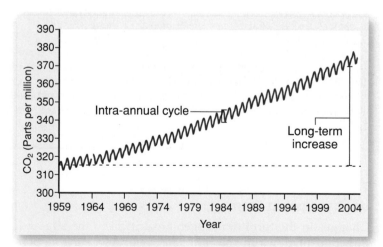

FIGURE 6.10 *Atmospheric Carbon Dioxide* *The atmospheric concentration of carbon dioxide at the observatory at Mauna Loa, Hawaii. The increase over time is associated with the combustion of fossil fuels and deforestation. The intra-annual cycle is associated with seasonal patterns of net primary production and respiration in the Northern Hemisphere.*

CASE STUDY

Watching the Planet Breathe: The Wiggle in the Mauna Loa Curve

Since 1958 the concentration of carbon dioxide in the atmosphere has been measured daily at Mauna Loa, a high mountain in Hawaii. This air is relatively free from local pollutants and so is thought to be representative of air in the Northern Hemisphere. As shown in Figure 6.10 on page 109, the measurements show two movements. Since 1958 there has been a general increase in the atmospheric concentration of carbon dioxide. This increase is associated with a variety of human activities that are described in this chapter and Chapter 13. The data also show an annual cycle. Each year the concentration of carbon dioxide rises and falls. What is responsible for these annual changes?

These changes reflect seasonal differences in carbon flows between the atmosphere and the biota. As described on page 104, two flows connect the atmosphere and biota. Photosynthesis moves carbon from the atmosphere to the biota. Respiration moves carbon from the biota to the atmosphere. To generate an annual cycle, photosynthesis must vary relative to respiration in a regular pattern. For some portion of the year photosynthesis must exceed respiration, producing a net flow of carbon from the atmosphere to the biota so that the atmospheric concentration of carbon dioxide declines. At other times of the year respiration must exceed photosynthesis, producing a net flow of carbon from the biota to the atmosphere and increasing the atmospheric concentration of carbon dioxide. What causes these regular changes in photosynthesis and respiration?

Many plants in temperate latitudes are either annuals or deciduous. For reasons described in Chapter 7, deciduous trees drop their leaves in the fall or during the dry season. Similarly, annual plants die during the fall and winter. The dead leaves and plants decay. As they decay, their carbon is returned to the atmosphere. The flow of carbon to the atmosphere via respiration is greater than photosynthesis, so the atmospheric concentration of carbon dioxide increases through the fall and winter and into early spring.

In late spring and summer this process is reversed. The seeds of annuals grow, deciduous trees regenerate their leaves, and photosynthetic rates in evergreens increase. Under these conditions photosynthesis increases relative to respiration, and there is a net flow of carbon from the atmosphere to the biota. Thus the atmospheric concentration of carbon dioxide decreases during the late spring and summer.

But this still does not explain the cycle. The seasons in the Northern and Southern Hemispheres are out of phase: When it is spring in the Northern Hemisphere it is fall in the Southern Hemisphere and vice versa. Because there is some exchange of atmospheric CO_2 between hemispheres, why doesn't the spring–summer increase in photosynthesis in one hemisphere cancel the fall–winter increase in respiration in the other hemisphere? Equally important, does the peak in the annual cycle of atmospheric concentration of carbon dioxide occur in the early spring of the Northern Hemisphere or in the early spring of the Southern Hemisphere?

To answer these questions, you need to remember that Mauna Loa is in the Northern Hemisphere and so largely reflects conditions there. In addition, there are important differences between hemispheres regarding the ratio of land to water. The Northern Hemisphere has more land than the Southern Hemisphere. This difference is even greater in the temperate latitudes—most of the land in the Southern Hemisphere is in the tropics.

The difference in land area implies that the trough (the low point) in the annual cycle of atmospheric carbon dioxide is determined by the growing season in the Northern Hemisphere. Because of the differences in land area, the rate of net primary production during the Northern Hemisphere growing season is greater than the amount of carbon returned to the atmosphere via the decay of biota in the Southern Hemisphere. Even if there were a complete exchange of CO_2 between hemispheres (there isn't), the trough in the annual cycle of atmospheric carbon dioxide would still occur during September.

Scientists use this annual cycle to measure changes in biota in temperate latitudes. As explained in this chapter, scientists cannot balance the carbon cycle. Some scientists postulate that the increased concentration of carbon dioxide and higher temperatures stimulate photosynthesis, and this increase accounts for some of the "missing carbon." If increased concentrations of atmospheric carbon dioxide increase the biota in the temperate latitudes, the amplitude (the change from peak to trough) of the cycle may also increase. That is, there may be more biota regrowing each spring and decaying each fall. Several scientists have looked for a change in the amplitude of the cycle, but these analyses do not provide conclusive evidence for an increase in Northern Hemisphere biota.

STUDENT LEARNING OUTCOME

- Students will be able to explain how changes in the atmospheric concentration of carbon dioxide can be used to identify changes in terrestrial vegetation.

The Nitrogen Cycle: Keep Your Eye on Changes in Form

Changes in the form of nitrogen are critical to the size and direction of flows in the nitrogen cycle. As in the carbon cycle, living organisms build nitrogen into many forms. But unlike the carbon cycle, in which most of the carbon outside living organisms is found in a single inorganic form (carbon dioxide), there are many forms of inorganic nitrogen. These inorganic forms include ammonium (NH_4^+), ammonia (NH_3), nitrite (NO_2^-), and nitrate (NO_3^-). Differences among these forms are important because each organism requires nitrogen in a particular form. Other forms are unavailable and therefore are unusable.

Most of the planet's nitrogen is found in the crust, where it is unavailable to any living organisms. We will ignore this storage. Instead our voyage through the nitrogen cycle starts in the atmosphere (Figure 6.11). The atmosphere stores almost 4 billion teragrams (a teragram is 10^{12} grams or a billion kilograms—4.4×10^{15} tons) of nitrogen, nearly all of which is molecular nitrogen N_2. This form, which

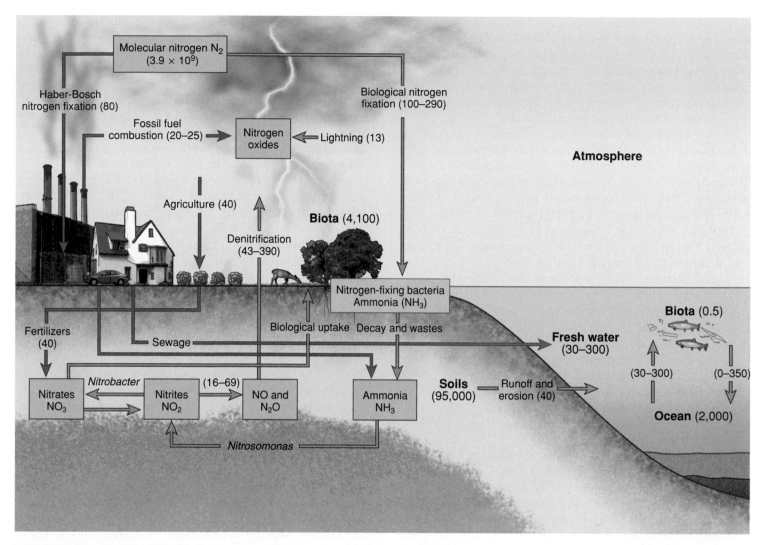

FIGURE 6.11   *The Nitrogen Cycle* *Storages are measured in teragrams, which are 10^12 grams. Flows are measured in teragrams per year. Flows associated with human activity are shown in red. Notice that the amount of nitrogen fixed by human activity is about the same as the amount fixed by biological activity. Lack of numbers indicates great uncertainty about its value.*

makes up about 78 percent of the atmosphere, consists of two nitrogen atoms that are bonded together tightly. To break this bond, large amounts of energy are needed.

To enter the biotic and soil storages, molecular nitrogen must be broken apart and recombined with other elements. This conversion is called **nitrogen fixation.** The flow of nitrogen from the atmosphere to the biotic and soil storages is not spontaneous and is driven mostly by two work processes: lightning strikes and nitrogen-fixing organisms. Lightning releases heat as it passes through the atmosphere, and this heat converts molecular nitrogen to nitrogen oxides, such as nitrogen oxide (NO). Earth's surface is hit by lightning about one hundred times each second. Despite this frequency, lightning strikes fix only about 13 teragrams (14.3 million tons) of nitrogen per year.

Living organisms also fix nitrogen. In the soil nitrogen-fixing bacteria such as *Rhizobium* convert molecular nitrogen to ammonia (NH_3), which is quickly converted to ammonium (NH_4^+). In the oceans bacteria such as cyanobacteria convert nitrogen to ammonia. These bacteria can fix nitrogen because they have an enzyme, nitrogenase, that splits molecular nitrogen. The ability to fix nitrogen makes these organisms critical to the formation of new soils, to the replenishment of soils that have lost much of their nitrogen to fires and erosion, and to agriculture and forestry, where humans remove large amounts of nitrogen along with crops and timber.

Nitrogen fixation is a relatively slow, energy-intensive process. Because of this limit, nitrogen-fixing bacteria often are at a competitive disadvantage relative to many other organisms. As a result, many nitrogen-fixing bacteria live cooperatively with other organisms, usually plants, in an arrangement known as **symbiosis.** The nitrogen fixers provide nitrogen in a form that is available to the plants while the plants provide water and energy-containing carbon compounds to the nitrogen fixers. Scientists recently found that if the nitrogen fixers "hold out" on their host plants, the host plants will stop supplying them with energy-containing carbon compounds.

The symbiotic relationship with nitrogen fixers is essential to plants known as **legumes,** such as soybeans. Legumes have special nodules on their roots that house nitrogen-fixing bacteria. This association gives legumes access to nitrogen, which allows these plants to have a higher protein content than nonlegumes. This high protein content makes legumes an important component of a vegetarian diet. By eating a diet that includes a variety of legumes, vegetarians can obtain all of the essential amino acids.

In addition to nitrogen fixation, the death and subsequent decay of living organisms also generate ammonia. After death, detritivores convert organic forms of nitrogen such as proteins to inorganic forms such as ammonia through a process known as **mineralization.**

Ammonia often is converted to nitrate, in a process known as **nitrification,** in two steps. In the first step ammonia is converted to nitrite (NO_2^-). This conversion is done by bacteria known as *Nitrosomonas.* In the second step nitrite is converted to nitrate (NO_3^-), a conversion done mainly by bacteria known as *Nitrobacter.*

Once converted to nitrate, the nitrogen moves in one of two general directions. Many plants can take up nitrate directly. Absorbed from the soil, the nitrogen is converted and incorporated into the plant's tissues in various forms. From there the nitrogen travels through the food chain. Like carbon, nitrogen is lost from one trophic position to the next. This lost nitrogen returns to the soil as dead material or as waste products. In either case the nitrogen can reenter the nitrification process. Recycling these forms of nitrogen closes the biological portion of the cycle.

Alternatively, nitrate can be converted back to molecular nitrogen in a process known as **denitrification.** In the first stage the nitrate is converted to nitrite. The nitrite is converted to nitric oxide (NO). This nitric oxide is then converted to nitrous oxide (N_2O), which is converted to molecular nitrogen. Denitrification usually is performed by bacteria such as *Pseudomonas.* These bacteria and many of the others responsible for denitrification often are anaerobic and live in **anoxic** environments (areas without oxygen).

Human activities have created new storages and flows in the nitrogen cycle. Humans use nitrogen for several purposes, two of the most important of which are the production of fertilizers and gunpowder. As described in Chapter 16, nitrogenous fertilizers helped boost crop yields in the Green Revolution. Nitrogen, in the form of ammonium nitrate, also is a necessary component of gunpowder. (The bomb that destroyed the federal building in Oklahoma City in 1995 was made from fertilizer and gasoline.)

Since the Chinese invented gunpowder about a thousand years ago, society has manufactured gunpowder and fertilizer from biological storages of nitrogen such as bird droppings. In 1909 Fritz Haber and Carl Bosch developed a way to "fix" nitrogen from the atmosphere using natural gas and lots of energy. Because of the huge growth in the demand for both fertilizers and gunpowder, industry now uses the Haber-Bosch process to fix about 80 teragrams (88.2 million tons) of nitrogen annually, which is slightly less than the total amount of nitrogen fixed by biological activity.

In many areas human storages and flows of nitrogen now are larger than natural flows. City sewage contains large amounts of nitrogen. Similarly, feedlots for domesticated animals also generate significant amounts of nitrogenous waste. A feedlot with 10,000 pigs produces an amount of nitrogenous waste equal to that from a city of 18,000 people. These highly concentrated flows of nitrogen can be dangerous to human health. As the nitrogenous wastes are converted from ammonia to nitrate, nitrates may accumulate in local waterways. High concentrations of nitrate can be toxic, especially to babies, who often have large populations of bacteria living in their stomachs that convert nitrate to nitrite. Nitrite reacts with hemoglobin to reduce the blood's ability to transport oxygen. This causes **methehemoglobinemia,** also known as "blue baby syndrome."

Natural selection affects the form of nitrogenous wastes. Some nitrogenous wastes, such as ammonia, are toxic and so must be highly diluted. Only organisms with constant access to water, such as fish, excrete ammonia. Other forms of nitrogenous wastes, such as urea and uric acid, are less toxic and are excreted by terrestrial organisms such as humans. Making urea or uric acid requires additional amounts of energy relative to ammonia, but such investments help terrestrial organisms face the challenges of life without a constant supply of water. This is especially true for animals that live in the desert and cannot afford to lose large amounts of water with their nitrogenous wastes.

Humans also increase the flow of nitrogen oxides to the atmosphere by planting crops and burning fossil fuels. Worldwide, farmers now plant more legumes than ever. This increases global nitrogen fixation by about 40 teragrams (44.1 million tons). Another 20–25 teragrams (22–27.6 million tons) are added annually by burning fossil fuels. The process of combustion heats air to very high temperatures (usually above 1,100°C–2,012°F). This heat splits atmospheric nitrogen and allows it to recombine with oxygen to form nitrogen oxides (NO_x). As described in Chapter 19, these oxides are air pollutants that lead to several environmental problems such as acid deposition and photochemical smog. In other areas nitrogen oxides reach the soil (nitrogen deposition), often via precipitation, and enhance plant growth. Indeed, nitrogen deposition may alleviate limits on the supply of nitrogen and allow the increased concentration of atmospheric CO_2 to enhance net primary production. Nitrogen deposition may be responsible for the unknown carbon sink as represented by the link between fossil fuel use, nutrient supply, net primary production, and atmospheric CO_2 in Figure 6.8.

The Phosphorus Cycle: Running Downhill

The phosphorus cycle is different from the carbon and nitrogen cycles because the atmosphere stores little phosphorus. Instead phosphorus tends to move "downhill" from the biota, to soils, into streams that make their way to the ocean, and eventually to ocean sediments. The cycle is completed after 10–100 million years when Earth's "heat engine" moves ocean sediments back to the surface, where the phosphorus can be taken up by the biota.

Our voyage through the phosphorus cycle starts with Earth's crust (Figure 6.12). Although phosphorus makes up only a tiny portion of Earth's crust (0.1 percent), some minerals have a high concentration of phosphorus, such as apatite (as much as 36 percent). The forces of geological uplift and soil erosion expose these minerals to sunlight,

heating, cooling, and precipitation. These forces, along with biological activity, free phosphorus from storage in Earth's crust and start its journey through the cycle.

Once freed, phosphorus can take many forms. Much of it tends to combine with metals, generating materials that do not dissolve in water, or become attached to the surface of other materials. In these forms inorganic phosphorus is relatively immobile, which means little is washed into streams. Equally important, most of these forms are unavailable to biological organisms. As a result, phosphorus often is a limiting nutrient, especially in aquatic environments.

Some phosphorus is converted to phosphate ions (PO_4^{3-}), the form used by most plants. Once absorbed through the roots, phosphorus is used to build important compounds, such as proteins and DNA. In addition, phosphorus is used to build adenosine triphosphate (ATP), which transports energy around a cell.

After being incorporated, phosphorus travels with carbon and nitrogen through the food chain. At each trophic level, phosphorus returns to the soil through metabolic waste products, death, and decomposition. The amount of

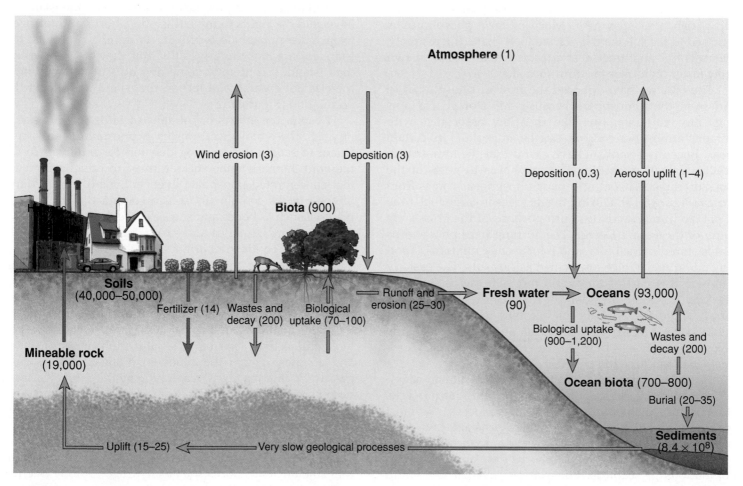

FIGURE 6.12   *The Phosphorus Cycle* Storages are measured in metric tons, flows are measured in metric tons per year. Flows associated with human activity are in red.

phosphorus in the soil is much larger than the amount in the biota. Nonetheless, the flow from the soil to the biota is approximately equal to the opposite flow.

In spite of this balance, there is a slow but steady loss of phosphorus from the land. Each year a small fraction of phosphorus is washed into rivers and streams. From here the phosphorus makes its way to the ocean, where it dissolves in seawater. Algae take up some of the phosphorus, and this starts its movement through the oceanic food chain. As with the terrestrial food chain, some is lost from the biotic storage as phosphorus moves from one trophic level to the next. Some of this phosphorus returns to the food chain, but a small amount sinks to the bottom and becomes part of the ocean sediments. Over millions of years this slow movement from land to sea has created a huge storage of phosphorus at the bottom of the ocean.

Once phosphorus reaches the ocean, little returns directly to land. In some areas fish and birds return significant quantities of phosphorus to the land. For example, salmon spend much of their adult lives swimming and eating in the ocean. As they do, they incorporate phosphorus into their bodies. This phosphorus returns with the salmon as they migrate upstream. During their upstream journey the salmon are eaten by animals; those that survive the trip perish after spawning. Much of their phosphorus is returned to the soil via the detrital food chain. Interestingly, scientists find that trees near streams where salmon spawn grow faster than trees far from such sites.

Thus far we have ignored the flow of phosphorus to and from the atmosphere because this storage contains only tiny amounts. But like the flow associated with salmon, atmospheric flows can be important in certain areas. Recently scientists have found that the atmosphere provides small quantities of phosphorus to parts of the Amazon forest. This atmospheric phosphorus comes from Africa (Figure 6.13). During the dry season large fires (due in part to human activity) burn portions of the grasslands south of the Sahara Desert. The burning frees phosphorus that is stored in the biota, and it forms tiny mixtures of solids and liquids, which are known as **aerosols.** The easterly winds in this part of the world move the aerosols across the Atlantic Ocean and deposit them in the Amazon rain forest. There the phosphorus becomes incorporated into the food chain. Currently scientists are not sure how the flow of phosphorus from Africa affects net primary production in the Amazon basin.

As with the carbon and nitrogen cycles, human activities are increasingly important in the phosphorus cycle. Humans use phosphorus for several purposes, the most important of which is fertilizer. Phosphorus also is used in laundry detergents, but this use has been reduced.

Humans obtain some phosphorus from areas with high populations of seabirds. For example, Nauru is a south Pacific island that supported a high population of seabirds

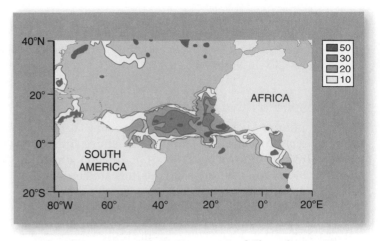

FIGURE 6.13   *Atmospheric Transport of Phosphorus The movement of phosphorus-containing dust particles from Africa to South America.* (From R.M. Swap, Gastang, S. Greco, R. Talbot, and P. Kallberg, "Saharan Dust in the Amazon Basin." Tellus 44B (1992): 133–149.)

for thousands of years. These birds spent most of their time catching fish from the sea and excreting wastes on the island. During this time the wastes, known as guano, accumulated. During the last hundred years the islanders have sold the rights to mine this phosphorus and nitrogen to produce phosphate fertilizer. As mining proceeds, the islanders are becoming wealthy. But they also are losing their island. It is literally being dug up and sold, and now there is not enough soil left on the island to support the population (Figure 6.14).

Despite the efforts of millions of birds for thousands of years, these biological sources of phosphorus are insufficient to satisfy the growing demand for fertilizer. To fill the void, humans mine rocks formed in ocean sediments that concentrated phosphate over millions of years. Many of these rocks contain phosphate combined with organic material such as bones and shells. These deposits, generally found in North America and Africa, have become available as a result of continental uplift.

Regardless of the source, human use of phosphorus tends to create the same problem. Phosphorus is the limiting nutrient in many environments. Phosphates used by people find their way into the environment and relax limits that once controlled biological activity. Rapid increases in biological activity can be especially problematic in lakes and streams, where the biomass of algae can increase greatly. During the night, respiration by the extra algae and decomposition of dead algae can suck all the oxygen out of the water. This oxygen depletion can kill the fish and other aerobic organisms in the water. And if this were not bad enough, the dead algae, fish, and other organisms are decomposed by anaerobic organisms, which often respire foul-smelling substances such as hydrogen sulfide (see the section about the sulfur cycle).

FIGURE 6.14   *Disappearing Island*   *Most of the soil on the island of Nauru has been removed by the process of mining for phosphorus and nitrogen.*

The process in which the addition of phosphorus (or some other limiting nutrient such as nitrogen) causes a boom and subsequent bust in biological activity is known as **eutrophication.** Eutrophication was a relatively common problem in the 1950s and 1960s when laundry detergents contained more phosphate. For example, eutrophication was primarily responsible for the pollution of Lake Erie. People would wash their clothes, and the sewage system would return the wash water to lakes and streams. The problem has been reduced greatly because laws passed in 1972 required that sewage be treated before being returned to waterways (see Chapter 18). In addition, concern by consumers encouraged manufacturers to make detergents that contain less phosphates. Eutrophication still occurs but now is often caused by runoff from agricultural land that has been fertilized and from feedlots that contain many domesticated animals.

The Sulfur Cycle: A Gateway for Human Environmental Impacts

Generally sulfur comprises about 0.25 percent of living organisms. As such, sulfur is an important, albeit small, component of living organisms. Unlike nitrogen and phosphorus, sulfur usually is abundant relative to biological needs. As a result, rarely is sulfur a limiting nutrient. Instead sulfur is important because many human activities increase storages and flows of sulfur. For example, forests require 1–5 kilograms of sulfur per hectare (5.4–27.2 lb/acre) per year; in polluted areas human activities add 10–100 kilograms per hectare per year (54.3–543.4 lb/acre).

Sulfur is found in five storages that are connected by several flows (Figure 6.15). Flows usually are very small relative to storages, except for the atmosphere. For this storage, annual flows from volcanic activity and the formation of aerosols from sea spray are larger than the storage. This implies that the residence time of sulfur in the atmosphere is very short. Indeed, sulfur aerosols from sea spray remain in the atmosphere for a couple of weeks, and sulfur from volcanic eruptions may last a few years at most. Similarly, soils store relatively little sulfur. Instead the vast majority of sulfur is stored in oceans and ocean sediments. In sediments sulfur is found in two forms, sulfate (SO^{2-}_4) and (S^{2-}).

These two forms of sulfur are linked to each other through biological activity in much the same way that carbon dioxide and oxygen are linked to each other. In anaerobic environments sulfur-reducing bacteria use sulfate ions to derive energy from organic molecules in roughly the same way that aerobic organisms use oxygen to derive energy from organic molecules (see Equation 5.3). On net

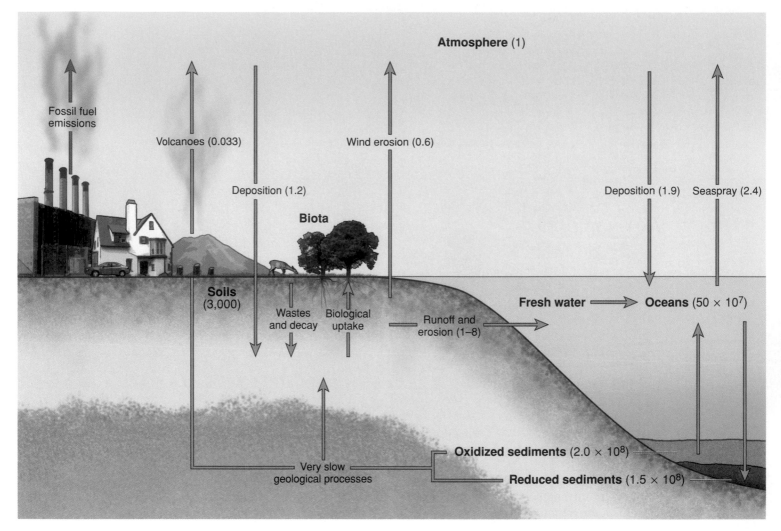

FIGURE 6.15   *The Sulfur Cycle   Storages are measured in 10^{12} moles of sulfur and flows are measured in 10^{12} moles per year. Flows associated with human activity are in red.*

these sulfur-reducing bacteria convert sulfates and organic molecules to energy, carbon dioxide, water, and hydrogen sulfide (H_2S). This gas smells like rotten eggs and is responsible for the foul smell associated with tidal marshes.

Some of this hydrogen sulfide is converted back to sulfates by purple and green sulfur bacteria. These bacteria use sunlight to drive a reaction in which hydrogen sulfide, carbon dioxide, and water are combined to generate organic molecules, sulfates, and hydrogen ions. In some ways this reaction is analogous to photosynthesis, in which green plants use sunlight, carbon dioxide, and water to generate organic molecules. Prior to the evolution of green plants, the interaction between sulfur-reducing bacteria and purple and green sulfur bacteria was analogous to the modern interaction between green plants and animals. But the presence of green plants and the evolution of a high-oxygen atmosphere has relegated this interac-

tion to a sidebar of biochemistry. Instead the presence of atmospheric oxygen means that most hydrogen sulfide is oxidized directly back to sulfates.

The two common forms of sulfur rarely are found by themselves. Instead they are usually combined with other elements. S^{2-} combines with iron to form a mineral called pyrite (FeS_2), which is also known as "fool's gold." Pyrite can be formed by colorless bacteria that combine sulfates and oxides of iron in ocean sediments. Under other conditions oceanic sulfates crystallize with calcium to form gypsum ($CaSO_4 \cdot 2H_2O$). Both pyrite and gypsum are relatively stable in the ocean, and this stability can sequester (tie up) sulfur for long periods. Eventually geological uplift returns the pyrite and gypsum to the surface, where weathering releases the sulfur as sulfates. The weathering of pyrite and gypsum, along with flows from the atmosphere, generates the abundant supplies of sulfur that rarely limit biological activity.

Much of our interest in the sulfur cycle arises from the environmental impacts generated by human activities that alter flows and storages. Human activity increases the rate at which sulfates weather from pyrite. Pyrite often is associated with coal and metal sulfides such as copper. Mining these materials generates large amounts of waste materials, which are known as **tailings.** Tailings often have a relatively high concentration of pyrite. Bringing these tailings to the surface allows them to break down, which releases their sulfates. These sulfates combine with water to produce sulfuric acid. Such waters, known as acid mine drainage, seep into ground and surface waters.

Humans also increase the flow of sulfates to the atmosphere. Coal contains relatively large quantities of sulfur, up to 5 percent, which varies by coal type. When coal is burned without scrubbers, its sulfur is released into the atmosphere. There the sulfur can form aerosol particles that reflect sunlight and affect the heat balance of the planet (see Chapter 13). In addition, these aerosol particles can combine with water, form sulfuric acid, and return to the surface as acid deposition (discussed in Chapter 19). This acidification makes soil and freshwater ecosystems less hospitable to life.

Interactions among the Cycles

So far we have discussed the carbon, nitrogen, phosphorus, and sulfur cycles as if they function separately. But this is not the case. Biogeochemical cycles interact in ways that slow or accelerate flows and build or deplete storages. As a result, changes in the carbon cycle may affect the nitrogen cycle, and vice versa. Understanding these linkages is critical to understanding how humans cause environmental challenges, as well as their potential solutions. See *Policy in Action: Is Dumping Iron into the Ocean an Effective Policy to Slow Global Climate Change?*

We can understand the interaction among cycles by returning to Liebig's law. As discussed previously, biological tissues are made of carbon, nitrogen, phosphorus, sulfur, and other elements. These elements often flow together as they move from one storage to the next. The combination of elements differs slightly among storages. These differences can be expressed as the ratio of carbon to some other nutrient, which is termed the carbon:nutrient (C:Nt) ratio (Nt stands for nutrient). For example, the carbon to nitrogen ratio in plants may be significantly higher than it is in animals.

Because of these differences, the limiting factor may change as biological material moves from one storage to the next. Under these conditions, nutrients that no longer are limiting may be returned to the environment. Conversely, nutrients may be removed from the environment as they become limiting factors. Such changes may alleviate nutrient limits on some other biological process and thereby

accelerate a flow in another cycle. Conversely, the changes may limit some other biological process, thereby slowing a flow in some other cycle.

Interactions among cycles are especially critical in forests because much of their biomass consists of trees, and a significant portion of this biomass is consumed through the detritivore food chain. Trees are unique because their C:Nt ratios differ greatly by part (leaf, stem, root, trunk, bark) and age. For example, the carbon to nitrogen ratio in leaves is about one-fifth that of branches. That is, a leaf has about five times more nitrogen per unit biomass than a branch. Similarly, old trees have a much higher C:Nt ratio than young trees. These differences have an important effect on the rate at which carbon returns to the atmosphere. Most of the carbon in a tree returns to the atmosphere via the detritivore food chain (only a small portion of living tree biomass can be consumed). Detritivores generally have a lower C:Nt ratio than the tree materials. Even leaves can have a carbon to nitrogen ratio that is five to ten times higher than the microbes that consume them.

These differences imply that the rate at which the decomposer populations grow is limited by nitrogen. When sufficient quantities of available nitrogen are present, the decomposer community can grow quickly and consume large amounts of dead material. This tends to accelerate the flow of carbon from the biota to the atmosphere. Conversely, the flow of carbon from the biota to the atmosphere is slowed by a lack of available nitrogen. Thus storages of available nitrogen affect the rate at which carbon flows from dead biota to the atmosphere.

DISRUPTING BIOGEOCHEMICAL CYCLES: UNDERSTANDING ENVIRONMENTAL CHALLENGES

Changes in the storages and flows of biogeochemical cycles can help illustrate how human activities affect the environment. From a global perspective, humans have played a minuscule role in biogeochemical cycles for most of our 4 million year history. During this period evolutionary change linked storages and flows to produce predictable biogeochemical cycles. As a result, the sizes of most storages and flows did not change much from year to year. This generated a fairly predictable supply of natural resources and environmental services for humans and other species.

Over the last two centuries, however, humans have increased the types and quantities of natural resources they use and the pollutants they emit. These changes threaten to disrupt the ability of biogeochemical cycles to provide natural resources and environmental services for all species. And the threat to these resources and services lies at the heart of many environmental challenges.

Is Dumping Iron into the Ocean an Effective Policy to Slow Global Climate Change?

Several years ago scientists performed an experiment designed to slow global climate change. Moving back and forth across patches of ocean in northern and southern areas of the Pacific, scientists dumped large quantities of iron into the ocean and measured subsequent rates of net primary production and the rates at which carbon sank toward the bottom. The results were a qualified success. Adding iron increased net primary production, but the increase did not increase the quantity of carbon moved to the deep layers of the ocean, which would keep the greater primary production from slowing the accumulation of carbon in the atmosphere.

Despite its limited success, the experiment was a clever attempt to manipulate biogeochemical cycles for policy purposes. To understand just how clever the idea was, we need to answer three questions: How will increasing net primary production slow global climate change? How will adding iron increase the rate of net primary production? Why conduct the experiment in the Pacific?

Net primary production by phytoplankton moves carbon dioxide that is dissolved in ocean water into the biota. This flow of carbon reduces the concentration of carbon dioxide in seawater and creates a gradient that causes carbon dioxide to flow spontaneously from the atmosphere to the ocean. The size of this gradient and the quantity of carbon that ultimately flows from the atmosphere to the ocean depends on the flow of carbon from the biota that sinks toward the ocean sediments

relative to the flow of carbon from the biota to the seawater due to respiration (Figure 1). Increasing the flow of carbon that sinks to the ocean bottom relative to the quantity that is respired back to surface ocean water increases the amount of carbon that is removed from the atmosphere. And a reduction in the quantity of carbon dioxide that is stored in the atmosphere may slow global climate change. As described in Chapter 13, nearly all scientists agree that increases in the atmospheric concentration of carbon dioxide are partially responsible for the increase in global temperature over the last 150 years.

The flow of carbon dioxide to ocean sediments relative to the flow that returns carbon dioxide to surface waters depends in part on the availability of nutrients. In areas of the ocean with low nutrient availability, phytoplankton grow slowly. Under these conditions, very small organisms known as picophytoplankton capture a large fraction of organic wastes. Because of their small size, picophytoplankton do not sink. As a result, little organic carbon is moved to ocean sediments. Much of it returns to the surface waters as carbon dioxide.

The balance between flows changes in areas with abundant nutrients. In areas with abundant

nutrients, phytoplankton absorb and store large quantities of nutrients. This allows them to reproduce rapidly, generating a "bloom" of phytoplankton. Such blooms are possible because the generation time of phytoplankton is much shorter than the generation time of their main predator, zooplankton. Phytoplankton are much larger than picophytoplankton, so a significant fraction of the phytoplankton sinks toward the ocean sediments during a bloom. In addition, zooplankton eat the phytoplankton and excrete fecal material, which then sinks. Both flows reduce the fraction of net primary production that returns to surface waters as carbon dioxide.

The population of phytoplankton is constrained by some limiting nutrient, but which nutrient? The average composition of marine organic material is 106 C:16 N:1 P and is relatively constant. Early analyses by Alfred Redfield (of Redfield ratio fame on page 101) indicated that phosphorus limited growth in the oceans. But further research has shown that the productivity of phytoplankton, in both coastal areas and open oceans, usually is limited by the availability of fixed inorganic nitrogen such as nitrate, nitrite, and ammonium. Despite its general importance, scientists noticed areas with abundant supplies

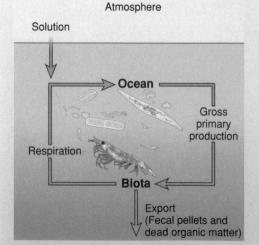

FIGURE 1 *The Biological Carbon Pump* Carbon dioxide that is taken up by marine organisms can either return to the ocean via respiration or sink toward the bottom. Carbon that sinks to the bottom causes additional quantities of carbon to flow into the ocean, which reduces the amount of carbon dioxide in the atmosphere.

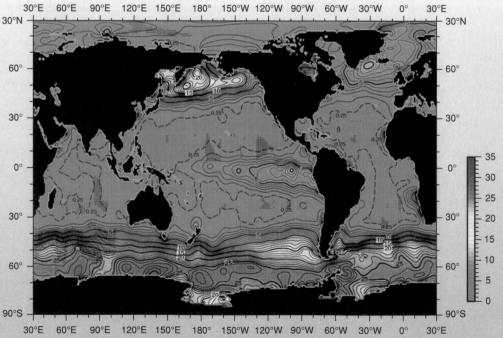

FIGURE 2 *High Nitrogen, Low Net Primary Production* Nitrogen often is a limiting nutrient in the ocean, but the highly colored areas along the Antarctic and the northern Pacific are areas where available forms of nitrogen are abundant but have relatively low levels of net primary production.

of fixed inorganic nitrogen relative to the population of phytoplankton—the so-called high-nitrate low-chlorophyll areas (Figure 2).

Then scientist Jon Martin hypothesized that biological activity in these areas is limited by iron. At first glance the importance of iron seems unlikely. Iron is the fourth most abundant element in Earth's crust (see Table 2.2 on page 23). As such, it should also be readily available in seawater. But significant areas of the ocean have a relatively low iron concentration. Most of the ocean's iron is carried from land via wind. Winds pick up small particles that contain iron, carry these particles over long distances, and deposit them in the ocean (Figure 3). Because there is relatively little land in the Southern Hemisphere, especially its southern part, the area from which winds in the Southern Hemisphere can pick up iron is limited. Thus the availability of iron is especially low.

If iron is the limiting nutrient, adding iron could spur phytoplankton blooms that ultimately move carbon from the atmosphere to the ocean bottom. The first parts of the experiments were a success. Adding iron generated phytoplankton blooms in both northern and southern areas of the Pacific (Figure 4). The persistence of the blooms varied between regions. The blooms slowed within days in the northern Pacific. Here much of the bloom was populated by diatoms, which secrete coverings using silica. As the diatoms took up the limited supply of silica, silica became the limiting nutrient. In the southern Pacific, where there are

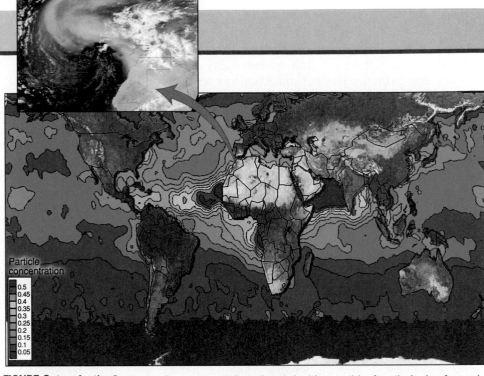

FIGURE 3 *Iron for the Ocean* Satellite measures indicate that winds pick up particles from the land surface and deposit them in the ocean or on other continents. An important source of such particles is northern Africa, where the lack of vegetation and fires can generate spectacular dust storms.

Particle concentration
0.5
0.45
0.4
0.35
0.3
0.25
0.2
0.15
0.1
0.05

abundant supplies of silicic acid (a form of silica that is available to diatoms), the bloom continued for an extended period.

But even this extended period was not enough to increase the rate at which carbon sank to the ocean bottom. In both locations, measurements showed that only a small fraction of the carbon taken up by the phytoplankton sank. Furthermore, the actual fractions were smaller than the values used by modelers to argue for the importance of the experiments.

ADDITIONAL READING

Boyd, P. "Ironing Out Algal Issues in the Southern Ocean." *Science* 304 (2004): 396–397.

STUDENT LEARNING OUTCOME

- Students will be able to explain how changes in terrestrial ecosystems can affect distant marine ecosystems.

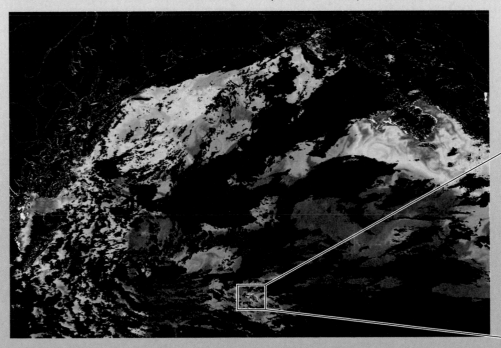

FIGURE 4 *Iron-Induced Diatom Blooms* As described in Chapter 7, the highest levels of net primary production occur along the shore, indicated by red, orange, and yellow. Away from the shore, areas of lower net primary production are indicated in blue. A small red area of high net primary production occurs where scientists added iron to the northeastern Pacific Ocean. These blooms consist of countless diatoms.

Recall from Chapter 1 that sustainability principle 1 requires that humans use natural resources and environmental services no faster than the environment supplies them. Sustainable use consists of extracting materials and energies from storages at the same rate at which biogeochemical cycles replenish those storages. For example, timber harvests are sustainable when the rate at which trees are cut equals the rate at which net primary production increases the carbon stored in trees. Nonsustainable use consists of extracting materials and energies from storages at rates greater than the rates at which biogeochemical cycles replenish their storage. For example, the carbon cycle adds about 0.8 million barrels of usable crude oil to the fossil fuel storage each year. Humans extract about 25 billion barrels of crude oil annually. As a result, the amount of carbon stored as crude oil shrinks by 24.9992 billion barrels per year. This reduction implies that it is only a matter of time (about 100 years) until there is little carbon stored as crude oil.

Shrinking storages cause environmental problems in many ways. Most directly, smaller storages reduce the availability of natural resources. This reduction is known as resource depletion. As society extracts oil, less remains for use in the future. If the activities described in Chapter 17 persist, the area covered by tropical rain forests will continue to shrink. Deforestation and oil shortages are two important environmental problems that are discussed in Parts Five through Eight.

We can use the same principles to understand environmental problems caused by increased waste production. Sustainable use of waste assimilation services requires that wastes be produced no faster than environmental services can assimilate them. Waste emissions increase the inflows to a storage. Rarely do the outflows increase in proportion to the inflows. As a result, the amount of material in the storage increases. The accumulation of potentially harmful materials in the atmosphere and waterways causes air and water pollution. As mentioned previously, humans now fix slightly less nitrogen than the nonhuman component of the nitrogen cycle. A significant fraction of this nitrogen is applied to agricultural fields in the American Midwest. As farmers increase their nitrogen application, greater amounts end up in the Mississippi River and eventually the Gulf of Mexico. There the extra nitrogen (and phosphorus) fertilizers increase net primary production. As these algae die, they sink to the

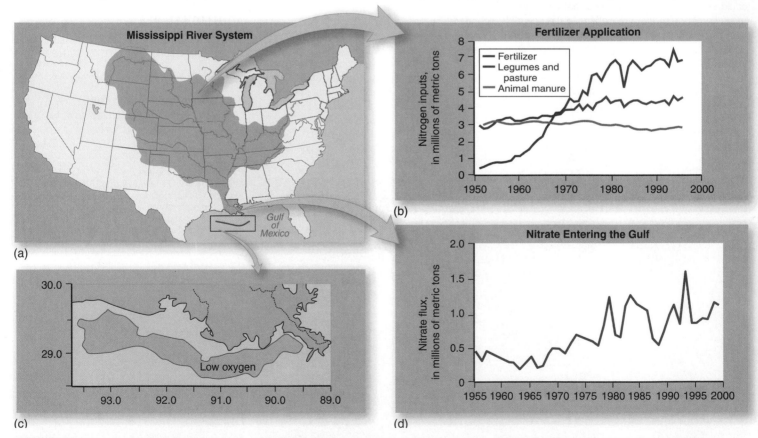

FIGURE 6.16 *Low-Oxygen Zones* (a) The green areas show the portion of the Midwest from which the Mississippi River draws its water and nutrients. (b) Over the last fifty years the amount of fertilizer added to these regions has increased significantly. (c) As a result, the amount of nitrogen in the Mississippi River water that enters the Gulf of Mexico has increased. (d) This increase generated large areas where the oxygen levels *are reduced significantly.* (b) (Redrawn from D.A. Goolsby and W.A. Battaglin, Nitrogen in the Mississippi Basin: Estimating Sources and Predicting Flux to the Gulf of Mexico, USGS Fact Sheet 135-00(2000)). (c) (Figure redrawn from G.F. McIsaac, M.B. David, G.Z. Gertner, and D.A. Goolsby, "Nitrate Flux in the Mississippi River." Nature 414 (2001): 166–167.) (d) (Redrawn from N.N. Rabalais, R.E. Turner, et al., Characterization of Hypoxia: Topic 1—Report for the Integrated Assessment of Hypoxia in the Gulf of Mexico. NOAA Coastal Ocean Program Decision Analysis Series no. 15. Silver Spring, MD: NOAA Coastal Ocean Program, 1999.)

bottom, where they decay. More organic material uses more oxygen, so oxygen levels are lowered significantly in much of the Gulf of Mexico (Figure 6.16). These zones can no longer support as many fish or the people who catch and eat them.

Increasing the size of some storages can create myriad environmental problems. Some materials are directly hazardous to human health. For example, increasing the storage of sulfur in the atmosphere causes respiratory problems. When the wind stopped diluting sulfur emissions in London during December 1952, the tripled concentration of sulfur dioxide contributed to a threefold increase in the death rate. Such health effects will be discussed in greater detail in Chapter 19.

The accumulation of pollutants in storages also can injure or kill species that provide environmental services. Once in the atmosphere, sulfur quickly is converted to forms that increase the acidity of precipitation. Consistent with the flow of the sulfur cycle, sulfur is transferred to soils and waterways. As a result these storages become acidified, and this can reduce or, in extreme cases, stop all environmental services. For example, many lakes in the Adirondack Mountains of New York State are too acidic to support living organisms; they no longer recycle nitrogenous wastes or provide recreational fishing opportunities.

SUMMARY OF KEY CONCEPTS

- Living organisms are made up of carbon, hydrogen, oxygen, and smaller amounts of other elements that are termed nutrients. Some of these nutrients are in forms that are usable by living organisms (available), while other forms are unavailable. The ratio of elements that are used to make living organisms is relatively fixed, which implies that the availability of a single element can limit the rate at which plants and animals can create new biomass.

- Matter flows between living organisms and the environment through myriad pathways known as biogeochemical cycles. These cycles consist of four storages: the crust, biota, atmosphere, and oceans. Movements between storages are known as flows, which can be either spontaneous or nonspontaneous. All biogeochemical cycles contain at least one nonspontaneous flow. Storages and flows are governed by the laws of conservation and thermodynamics.

- The carbon cycle is viewed as the master cycle on Earth. The most important nonspontaneous flow is photosynthesis. Humans have changed the carbon cycle by burning fossil fuels and altering land use. Due to these changes, the atmospheric concentration of carbon dioxide has increased over time. Scientists cannot balance the flow of carbon into and out of the atmosphere.

- Flows in the global nitrogen cycle are determined by the many forms of nitrogen. The most important nonspontaneous flow is nitrogen fixation, the process by which atmospheric nitrogen, which is unavailable to most living organisms, is converted to forms that are available to the biota. Using the Haber-Bosch process, humans now fix slightly less nitrogen than natural processes, which leads to eutrophication.

- A complete turn through the phosphorus cycle is very long because it depends in part on movements in Earth's crust, which are the main nonspontaneous flow. Humans have accelerated the cycle by mining phosphorus. The extra phosphorus has increased biological activity, especially in aquatic environments. This too has led to eutrophication.

- So much sulfur flows through the environment that this element usually does not limit biological activity. Humans have changed the storages and flows of the sulfur cycle by mining and burning coal. These changes tend to cool the planet and increase the acidity of precipitation and waterways.

- The effects of human activities on the environment can be understood in terms of biogeochemical cycles. Humans change the storages and flows of cycles, and this disrupts their normal function. The normal functioning of biogeochemical cycles provides the natural resources and environmental services on which economic well-being depends, so disrupting storages and flows usually reduces economic well-being.

REVIEW QUESTIONS

1. How do the law of conservation and the laws of thermodynamics make biogeochemical cycles similar to and different from food chains?

2. Vegetarians often are told to eat a combination of legumes (soybeans or peas), whole grains (rice or corn), and seeds or nuts (sesame or sunflower seeds). What is the reason for such a recommendation?

3. Suppose that the Redfield ratio is 100:16:1 (C:N:P). If a plant has 300 units of carbon, 50 units of nitrogen, and 7 units of phosphorous, which is the limiting nutrient?

4. The Haber-Bosch process uses large amounts of natural gas. Explain why this is so.

5. Explain the similarity and differences between nitrogen fixation and photosynthesis.

KEY TERMS

absorption

aerosols

anoxic

available

biogeochemical cycles

biota

carbohydrates

deforestation

denitrification

digestion

essential

eutrophication

fats

flow

kwashiorkor

legumes

Liebig's law of the minimum

limiting nutrient

methehemoglobinemia

mineralization

nitrification

nitrogen fixation

nonessential

nonspontaneous flow

nutrients

proteins

Redfield ratio

residence time

spontaneous flow

storages

symbiosis

tailings

transpiration

unavailable

unknown carbon sink

BIOMES

Where Do Plants and Animals Live?

STUDENT LEARNING OUTCOMES

After reading this chapter, students will be able to

- Explain how specializations limit many species to a relatively limited geographical range.
- Describe how the presence or absence of other species influences a species' range.
- Describe how many species can coexist within a limited geographical range.
- Explain how temperature, water, and light influence the shape and longevity of leaves.
- Explain the separation between light and nutrients in aquatic biomes and how this separation, or lack thereof, affects net primary production in various aquatic biomes.

Similar Environments, Similar Looks
The Texas Horned Lizard found in the US Southwest and (insert) the Mountain Devil from Australia.

INTRODUCTION

A "living pincushion" (opening photograph, (a))—that is how people describe the Texas horned lizard (sometimes incorrectly called "horned toad"). Horned lizards live in arid regions of the American Southwest. Horned lizards have flat round bodies, much like pancakes. Their tails, legs, bodies, and heads are covered with spines. Their diet consists almost entirely of ants.

Mountain devils live in arid regions of Australia. Their bodies also look like pancakes and are covered by spines (opening photograph, (b)). Their diet consists almost entirely of ants. To the untrained eye the horned lizard and the mountain devil are nearly indistinguishable. But these two species, which live on opposite sides of the planet, are not related. They arrived at their body shape and diet independently.

Are these similarities a coincidence? No. They were generated by natural selection. Plants and animals that live in similar environments often face similar challenges and develop similar solutions to such challenges. Both horned lizards and mountain devils need to find food and avoid predators in arid regions where there is relatively little plant cover. Ants are common in arid environments, so they provide a reliable source of food. But ants are difficult to digest, so both lizard species need large digestive systems and thus have evolved a rounded body shape. This round body shape makes it difficult for the lizards to outrun predators. Instead their spiny appearance and color allow them to blend in with the background and may protect them from predators.

The similarity between the horned lizard and mountain devil is known as **convergent evolution,** which is the evolution of similar characteristics in unrelated species due to similar environmental stresses. Convergent evolution is possible because natural selection favors behavioral or physiological traits, known as **adaptations,** that allow a plant or animal to thrive in a particular environment. As a result, most species live only in certain regions. In any given area, the group of interacting species is known as a **community.** A community and the physical environment in which the community lives are called an **ecosystem.** The *Dictionary of Biology* defines a **biome** as a major regional community of plants and animals with similar life forms and environmental conditions. This largest geographical biotic unit is named after the dominant type of life form, such as tropical rain forest, grassland, or coral reef. ■

WHY SPECIES LIVE WHERE THEY DO

Habitats and Niches

Where does a species live? At first that seems a simple question. It is easy to list the geographical regions where a plant or animal species can be found. Texas horned lizards are found from Kansas to Texas and west to southeast Arizona. Isolated populations also exist in Louisiana. The geographical locations and environmental conditions where a plant or animal lives are called a species' **habitat.**

But the concept of habitat does not fully answer the question of where an organism lives. Why does the Texas horned lizard live in Kansas and Texas, and why are there isolated populations in Louisiana? To answer this question we need to understand the concept of a **niche,** which includes the totality of a species' environmental requirements. These requirements include how and where an organism gets its energy and nutrients and the ways it interacts with other species. For example, the Texas horned lizard lives only in areas where there are large ant populations. Being an ectotherm, it also must live in climates where it can maintain its active body temperature for a significant portion of the year.

But a species' niche extends beyond its direct requirements. Plants and animals also change the environment in ways that affect the habitat of other species. The American alligator digs holes in the banks of lakes and rivers. In addition to providing refuge for the alligator, these holes store water during the dry season, providing habitat for smaller animals such as turtles. Thus the alligator's niche includes the function of making water available for other species.

The concept of niche is the basis for understanding where species live. The **competitive exclusion principle** states that no two species can share the same exact niche indefinitely unless other factors limit the density of the better competitor. Over time, natural selection favors changes that reduce the degree to which species share the same niche. For example, plant communities in Alaska known as tussocks usually are limited by nitrogen, yet many species coexist within a small area. Shouldn't the species that is best able to obtain nitrogen succeed while the others disappear? In Alaska natural selection has generated a different solution: Species have evolved different strategies for obtaining nitrogen. Each species gets its nitrogen in a different form, during a different month, or from a different soil depth (Figure 7.1). For example, the Bigelow sedge gets most of its nitrogen as nitrate during June, whereas the dwarf birch gets most of its nitrogen as ammonium during August. These strategies allow the species to split the niche and thereby coexist.

Niche and Adaptation: Linking Species to Their Environments

Adaptations are characteristics that affect an individual's chance to survive and produce offspring. Some adaptations are well suited for a fairly wide geographic habitat;

others work best in a relatively narrow range of conditions. Such adaptations are **specializations.** For a description of human adaptations to our environment, see *Policy in Action: Climatological and Ecological Determinants of Human Land Use.*

The living world is filled with adaptations. Things you might overlook often are a response to some aspect of the environment. Many of you probably have not thought about the shape of a leaf. Myriad shapes are possible: long or short, thick or thin, smooth or rough, and so on. Why do leaves look as they do? A leaf's shape is no accident. Natural selection favors a shape that allows each plant to thrive in its environment. Plants that live in areas with lots of water have thin leaves with lots of surface area. A large surface area allows the leaf to capture significant amounts of sunlight, which is needed to power photosynthesis. But a thin leaf transpires much water, so the plant must have access to sufficient water. The banana tree, which lives in tropical rain forests, has giant leaves (Figure 7.2(a)). On the other hand, plants that live in areas with little water often have thick stubby leaves with relatively little surface area. This shape slows the rate at which water is lost from the leaf. The creosote bush, which lives in the southwestern U.S. deserts, has very small leaves (Figure 7.2(b)).

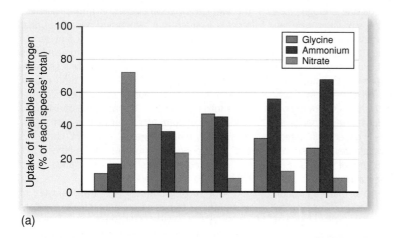

(a)

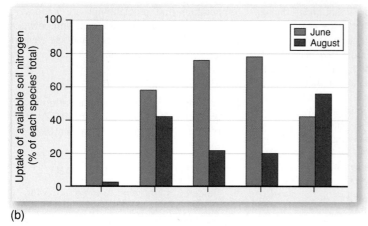

(b)

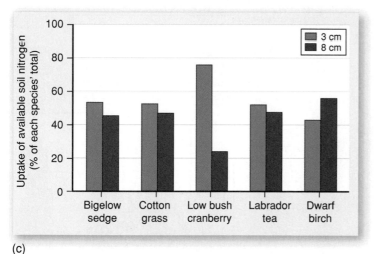

(c)

FIGURE 7.1 *Splitting the Niche* The source of nitrogen for five plant species in the Alaskan tussock. Species differ in (a) the form of nitrogen uptake (b), the month of uptake, and (c) the soil depth.
(Source: Data from R.B. McKane et al., "Resource-Based Niches Provide a Basis for Plant Species Diversity and Dominance in Arctic Tundra." Nature 415: 68–71.)

FIGURE 7.2 (a)
Leaves That Fit the Environment
(a) The giant leaves of the banana plant, which lives in the rain forests of Costa Rica.
(b) A leaf of the creosote bush, which lives in the deserts of the U.S. Southwest.

(b)

POLICY IN ACTION

Climatological and Ecological Determinants of Human Land Use

The location of terrestrial biomes is determined by solar radiation, temperature, and precipitation. These factors also affect how people use land. Wheat fields often are located on what were once temperate grasslands, while rubber plantations lie atop what were once tropical rain forests. In short, human patterns of land use are tuned to climatic and ecological patterns.

The climatic and ecological patterns that determine land use can sometimes be subtle. As shown in Figure 7.12 on page 138, the tropical savanna runs east-west across the African continent just below the Saharan Desert. Although this swath of land is classified as a single biome, there are striking differences in how people use it. Agricultural societies live in western Africa (such as Niger). Here farmers grow millet (a type of grain) in an area that receives 500–600 mm of precipitation per year. Nomadic herders (people who herd livestock over wide areas) live in eastern Africa (such as Kenya). But this area also receives 500–600 mm of precipitation per year. Why the dramatic difference in lifestyle?

One reason might be the seasonal variations in annual precipitation. In western Africa there is a single rainy season from July to September when western Africa receives most of its annual precipitation. This is followed by a nine-month dry season. Eastern Africa has two dry seasons, each about three months. These dry seasons are split by two rainy seasons, one from February to May and another from October to December.

Differences in the seasonal pattern of precipitation affect the land's ability to grow crops and support livestock. In western Africa the single rainy season delivers relatively large quantities of water to the soil over a short period, during which the soil has enough water to support crops. These crops must mature rapidly. As the soil dries, the rains will not return for another nine months. Thus the single intense rainy season can support crops that mature in a single, relatively short growing season.

Any single rainy season in eastern Africa delivers about half the water that is delivered by the one rainy season in western Africa. After either of the shorter rainy seasons, there is not enough water in the soil to support the complete development of a crop. Instead the soil will run out of water and the plants will die before the next rain comes.

But the very conditions that make agriculture difficult in eastern Africa make herding livestock possible. Because the two dry seasons in eastern Africa are relatively short, soil moisture usually does not drop to the low levels that prevail during the dry season in western Africa. The availability of water during the dry season supports shrubs with relatively deep roots. These roots allow plants to be relatively productive year-round, which makes them a good source of food for livestock. Herding is more difficult in western Africa because the dryer conditions make it difficult to graze livestock.

The effect of seasonal differences in the availability of water on crops versus herding is reinforced by geographic differences in the variability of precipitation between years. Although east and west Africa receive the same amount of precipitation annually, the variation around that average differs significantly between the two regions. The variance of precipitation (see Chapter 3) is much greater in western Africa, where annual precipitation may be significantly less than or greater than the average for several years in a row. Historical records indicate that alternating periods of drier and wetter than normal years may last ten to fifteen years. On the other hand, precipitation in eastern Africa shows significantly less variation. Years that are drier or wetter than normal occur, but rarely do four or more drier or wetter years follow each other. The absence of extended droughts in eastern Africa makes it relatively easy to maintain livestock herds. Livestock can "wait out" short periods of drought. On the other hand, the extended droughts of western Africa would wreak havoc on livestock herds.

The striking difference in the lifestyles of west African farmers and east African herders confirms the effect of climatic and ecological patterns on human patterns of land use. Average temperature and precipitation are important; but variance, both within the year and between years, also influences how humans use environmental goods and services.

ADDITIONAL READING

Ellis, J., and K.A. Galvin. "Climate Patterns and Land Use Practices in the Dry Zones of Africa." *Bioscience* 44, no. 5 (1994): 340–349.

STUDENT LEARNING OUTCOME

- Students will be able to explain how local climate affects the strategies that people use to obtain food in sub-Saharan Africa.

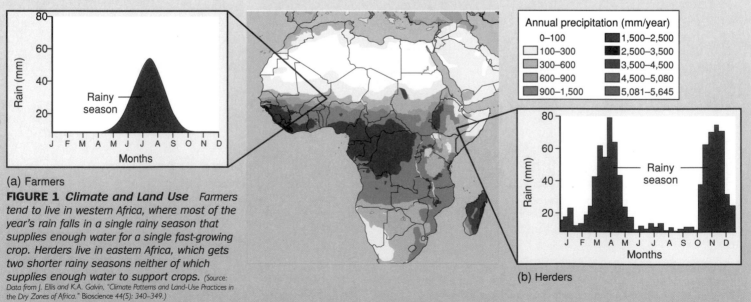

(a) Farmers

FIGURE 1 *Climate and Land Use* *Farmers tend to live in western Africa, where most of the year's rain falls in a single rainy season that supplies enough water for a single fast-growing crop. Herders live in eastern Africa, which gets two shorter rainy seasons neither of which supplies enough water to support crops.* (Source: Data from J. Ellis and K.A. Galvin, "Climate Patterns and Land-Use Practices in the Dry Zones of Africa." Bioscience 44(5): 340–349.)

Annual precipitation (mm/year)
0–100
100–300
300–600
600–900
900–1,500
1,500–2,500
2,500–3,500
3,500–4,500
4,500–5,080
5,081–5,645

(b) Herders

Animals also have adaptations. Most frogs rapidly lose water across their skin, so they live near water. Some frogs, such as the waxy tree frog, extend their niche to include relatively dry areas. The waxy tree frog prospers in these areas by secreting a waxy substance that it spreads constantly across its body to slow the loss of water.

Environmental Gradients: Linking Adaptation to Place

We can understand the relationships among niche, adaptation, and geographical region by thinking of the world as a collage of environmental conditions such as precipitation, tempera-

ture, and sunlight. Each condition varies over the planet. Recall from Chapter 4 that large amounts of precipitation fall near the equator and at 60° N or S, while dry areas tend to occur at 30° north and south. Temperature tends to decrease with latitude.

Changes in conditions from one region to the next are called **environmental gradients** (these differ from thermodynamic gradients, which are based on differences in entropy). In some regions areas of high precipitation gradually blend into areas of low precipitation and then back again to high precipitation (Figure 7.3(a)). Similar gradients can be drawn for other environmental conditions such as temperature (Figure 7.3(b)). Together, gradients separate the planet into regions. Some regions are cold and wet;

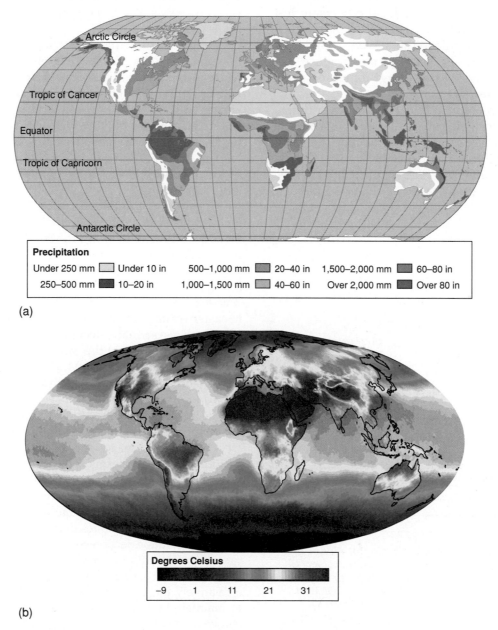

(a)

(b)

FIGURE 7.3   *Environmental Gradients*   (a) Global annual precipitation and (b) temperature in July 2003 as measured by satellites.

others are cold and dry; others are hot and wet or hot and dry. If we consider other environmental conditions such as the availability of solar energy or soil type, the number of combinations increases geometrically.

Adaptations to environmental conditions determine a species' geographic distribution. Specializations allow a species to prosper within a relatively narrow range of conditions. Beyond this range, a species' specializations are poorly suited for environmental conditions, which reduces the probability that the members of a species can survive and reproduce. As a result, species rarely are found beyond regions where their adaptations work well. For example, the horned lizard is found only in areas with large ant populations, and the Venus flytrap is found only in wetlands (sections of land that are covered by water during some portion of the year). The area along a gradient where an individual can obtain just enough energy and materials to survive is called the **survival range** (Figure 7.4). For a population to survive, individuals must obtain enough energy and materials to reproduce. The area where individuals have enough energy for reproduction defines a species' **reproductive range.**

The effect of precipitation gradients on the distribution of plants is illustrated by a hike up Mt. Lemmon just outside Tucson, Arizona (Figure 7.5). Starting at the base, the creosote bush is among the first plant species you see. As you climb higher, you encounter the catclaw acacia and later the Arizona rosewood. At the top you enter a forest of ponderosa pines. This order is generated by a gradient in soil moisture. Mt. Lemmon is located in the Sonoran Desert, so soils

tend to be dry at the mountain base. Soil moisture tends to increase at higher elevations due to orographic precipitation and lower temperatures, which slow evaporation.

The effect of soil moisture on net photosynthesis varies among plant species (Figure 7.6). The ponderosa pine is very sensitive to soil moisture. Once soil moisture drops below −15 bars (a measure of soil moisture), net photosynthesis drops nearly to zero. Because of its sensitivity to soil moisture, the ponderosa pine is found only at the top of Mt. Lemmon. The catclaw acacia can tolerate relatively low levels of soil moisture. Even when soil moisture drops below −30 bars, the catclaw acacia can maintain its maximum rate of net photosynthesis—so the catclaw acacia can live well down from the mountaintop. Finally, the creosote bush can maintain positive rates of net photosynthesis at very low levels of soil moisture, so it can inhabit the very dry lowlands.

A similar principle limits the reproductive range of the little brown bat. Its range is associated with the energy required for hibernation. This quantity is determined by temperature and length of hibernation. Due to its small size and its need to fly to capture food, a little brown bat can store 1.5–3.5 grams of fat, which allows the bat to hibernate for four to seven months. This period imposes a northern limit on the bats' range—beyond this limit, the time that bats cannot catch insects is too long relative to the bats' ability to store energy.

In addition to climate, range can be determined by interactions with other species. If you drive east on U.S. Interstates 10 and 20 along the southern tier of the United States you will cross through eight states (Figure 7.7). As you do so, you will encounter different numbers of lizard species. There are many lizard species in California, Arizona, and New Mexico, but there are far fewer species in eastern states such as Mississippi, Alabama, and Georgia. Why such a systematic decline?

Temperature is not the cause because average temperatures in Alabama and New Mexico are fairly similar. On the other hand, there is a strong gradient in precipitation from the arid portions of the Southwest to the moist areas of the Southeast. This gradient seems to influence the number of lizard species: Most live in arid regions.

One explanation for this distribution is potential competition from endotherms. The low basal metabolic rate of ectotherms is a useful adaptation for animals that live in hot dry environments where net primary production is relatively low. These areas do not suit endotherms because their high basal metabolic rate implies that they need a lot of food.

This situation is reversed in the eastern states, where greater precipitation increases net primary production, raising the amount of energy available. Endotherms can capture a relatively large share of this energy because their ability to maintain body temperature lengthens the time that they can be active. Fewer species of lizards may be found in the Southeast because ectotherms find it difficult to compete

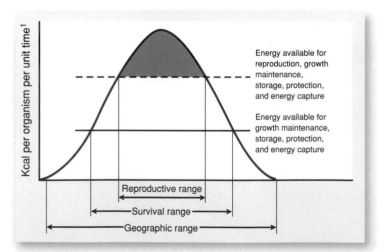

FIGURE 7.4 *Adaptation and the Distribution of Plants and Animals* *The survival range includes all areas where an individual's adaptations enable it to obtain energy and materials for its basic needs. In other areas, these same adaptations allow the individual to obtain extra energy and materials, which allow it to reproduce. This smaller area, where most of the population lives, is termed the reproductive range.* (Source: Redrawn from C.A.S. Hall et al., "The Distribution and Abundance of Organisms as a Consequence of Energy Balances Along Multiple Environmental Gradients." Oikos 65: 377–390.)

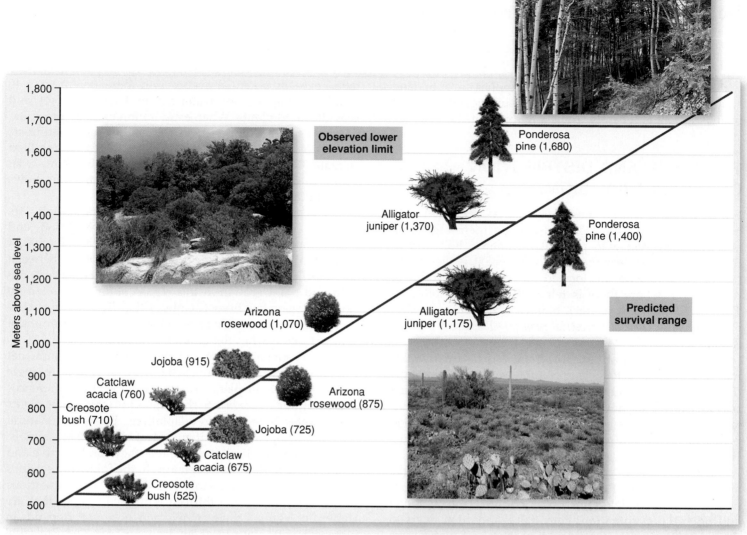

FIGURE 7.5 *Elevational Distribution of Plants* The distribution of plants as one travels up Mt. Lemmon, just outside Tucson, Arizona. One starts in the desert and ends in a pine forest. The predicted survival range corresponds to the lowest elevation at which a plant can generate enough energy to cover the requirements for survival. The observed lower elevation corresponds to the lowest elevation at which plants are found. This elevation tends to be higher because competition from other species keeps plants from living at the lower edge of their survival range. (Source: Redrawn from J.A. Bunce et al., "Role of Annual Leaf Carbon Balance in the Elevational Distribution of Plant Species Along an Elevational Gradient." Botanical Gazette 140: 288–294.)

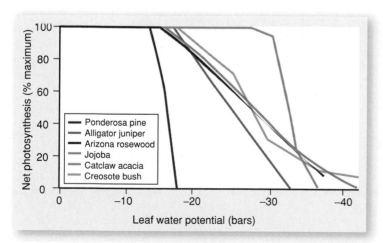

FIGURE 7.6 *Moisture and Photosynthesis* The relationship between the availability of water and the maximum rate of photosynthesis for species that live in the Santa Catalina Mountains, Arizona. Plants adapted to arid environments can maintain photosynthetic rates near their maximum even as moisture drops. (Source: Modified from J.A. Bunce et al., "Role of Annual Leaf Carbon Balance in the Elevational Distribution of Plant Species along an Elevational Gradient," Botanical Gazette 140: 288–294.)

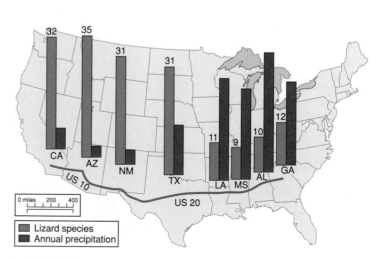

FIGURE 7.7 *Where to Find Lizards* As you drive east, the number of lizard species you find declines. On the other hand, annual precipitation increases. (Source: Data from F.H. Pough.)

with endotherms where food energy supplies are abundant. Neither endothermy or ectothermy is superior; the success of an adaptation depends on the environment.

THE TYPES AND DISTRIBUTION OF TERRESTRIAL BIOMES

From space, terrestrial biomes appear as broad latitudinal bands. Biomes are distinguished by their vegetation and are associated with particular climates. Rain forests tend to be found near the equator where it is warm and rainy, while deserts are located 30° north and south of the equator, where there is relatively little precipitation. The use of satellites to track biomes is described in *Case Study: Keeping Track of Terrestrial Biomes—The Use of Satellite Remote Sensing.*

The distribution of terrestrial biomes is determined primarily by three factors: temperature, water, and sunlight (Figure 7.8). Temperature affects the length of the **growing season,** which is the number of consecutive days during which temperature remains above 0°C (32°F). During this period plants must generate enough energy to reproduce or store enough energy to survive until the next growing season.

The distribution of biomes also is determined by the availability of water. Plants get water from soil. The amount of water in soil is determined by the rates of precipitation and **potential evaporation.** Potential evaporation refers to the amount of water that would evaporate if water were available (that is, if the soil never dried out). When precipitation exceeds potential evaporation (hereafter referred to simply as evaporation), water accumulates in the soil and is available to plants. When evaporation exceeds precipitation, soil water supplies dwindle. Under these conditions, plants slow the rate at which they lose water by closing their stomata. But closing their stomata prevents the plants from taking in fresh supplies of carbon dioxide, and this slows their photosynthesis.

Finally, the availability of light may limit plant growth. The amount of light available is determined by latitude and local climate. Because of the tilt of Earth's axis, the greatest amount of sunlight reaches Earth's atmosphere between 23.5° north and 23.5° south (see Chapter 4). However, sunlight and plant growth in this region often are limited by clouds. Remember from Chapter 4 that high rates of convection generate many clouds, which reduce the amount of sunlight that reaches the ground.

In any given area, one of these factors may limit net primary production in much the same way that a single nutrient may limit plant growth. Geographical differences in the limiting factor create challenges for which local plants and animals have evolved adaptations. For plants, natural selection has generated different leaf shapes. For animals, local climate may limit the time that they can be active and influence their allocation of energy. In the following sections we use these limiting factors and the relevant adaptations to describe nine of the thirteen terrestrial biomes that are shown in Figure 7.9. To understand how you depend on biomes beyond your local environment, see *Your Ecological Footprint: What Biome Do You Live In?*

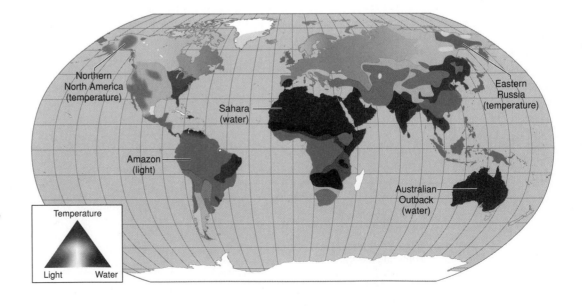

FIGURE 7.8 *Limiting Factors Geographical differences in the factors that limit net primary production. Notice that water limits net primary production in deserts, such as the Sahara and the Australian outback. Temperature limits net primary production in forested areas of northern North America and eastern Russia. Finally, clouds reduce the amount of light that reaches the surface near the equator, which limits net primary production by tropical rain forests, such as the Amazon.* (Source: Redrawn from Nemani et al., "Climate-Driven Increases in Global Terrestrial Net Primary Production from 1982 to 1999," Science 300: 1560–1563.)

CASE STUDY

Keeping Track of Terrestrial Biomes—The Use of Satellite Remote Sensing

Scientists have up-to-date information about the status of terrestrial biomes for the entire globe. They can map the extent of each biome. They can tell when biomes "green up" or "brown out." They can estimate biomass at each location along with the rate of net primary production. But how do scientists generate this information? The planet is too big to visit each location. Rather, this information is generated from data that are collected by satellites. Instruments on the satellites do not take photographs in the traditional sense. Instead they measure the amount of light that is reflected by Earth's surface, and this information is processed in a way that allows scientists to diagnose Earth's status.

Most of the data used to track terrestrial biomes are collected by NOAA (National Oceanic and Atmospheric Administration) satellites. As these satellites orbit Earth, they collect information about each location. These locations, called pixels, vary in size from 1 square meter to 64 square kilometers. The size of the pixel is determined by the instrument that collects the information.

For each pixel, an onboard optical instrument such as the Advanced Very High Resolution Radiometer (AVHRR) measures the amount of sunlight that is reflected at various wavelengths, which are termed channels. Channel 1 measures reflected light with wavelengths between 0.58 and 0.68 μm (micrometers—millionths of a meter). Channel 2 measures reflected light with wavelengths between 0.725 and 1.1 μm.

We emphasize channel 1 and channel 2 because scientists use them to calculate the normalized difference vegetation index (NDVI) with the following formula:

$$NDVI = \frac{(Channel\ 2 - Channel\ 1)}{(Channel\ 2 + Channel\ 1)}$$

The value of NDVI can be used to determine whether a pixel is covered by water, snow, bare soil, or vegetation. Areas covered with water or snow reflect lots of light in channel 1 but relatively little light in channel 2 (Figure 1). As a result NDVI is negative for areas covered by water or snow. Rocks and bare soil reflect about the same amount of light in channels 1 and 2, so the value of NDVI for these areas is approximately zero. Vegetated areas reflect a large amount of light in channel 2 but little light

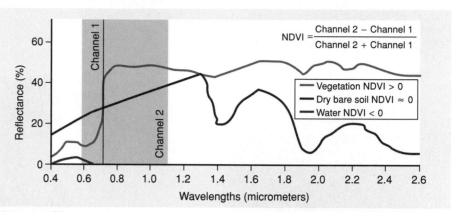

FIGURE 1 *Satellite Measurements* Instruments on satellites measure the fractions of light reflected at various wavelengths. The fractions reflected differ greatly among bare soil, vegetation, and water. Shaded areas represent the reflectances that are measured in channel 1 and channel 2 by the Advanced Very High Resolution Radiometer. (*Source: Data from Lillesand and Keifer.*)

in channel 1. So vegetated areas have relatively large NDVI values.

Once scientists separate vegetated areas from snow, water, rock, and bare soil, they need to determine the type of biomes present. Again NDVI generates useful information. NDVI values depend on the amount and structure of the surface vegetation. Biomes with lots of biomass and complex structure, such as tropical rain forests, have larger NDVI values than biomes with simpler structure and less biomass, such as temperate grasslands.

Unfortunately for scientists, there are no sharp boundaries for NDVI values that can be used to distinguish biomes. To assign a pixel to a biome, scientists also use seasonal variations in NDVI. For example, the NDVI of a tropical dry forest will decline during the dry season, whereas the NDVI of a tropical rain forest will display smaller fluctuations. Based on these differences, scientists can use the seasonal changes in NDVI, along with their absolute levels, to determine the type of biome present at any location.

Even with this additional information, scientists cannot assign pixels to biomes with absolute certainty. As a result, maps that show the location of biomes contain errors. For example, some pixels that are labeled "tropical rain forest" are not covered by tropical rain forests. To measure these errors, scientists "ground truth" their maps. They go to various parts of the planet and compare the biomes present with the biomes indicated by their satellite-based maps. Comparisons show that the maps' accuracy varies by biome. Scientists can

identify open scrublands (like chaparral) with great accuracy (above 90 percent). This implies that 90 percent of the areas mapped as scrublands are scrublands when scientists visit those locations. But scientists are less able to identify deciduous needleleaf forests (a type of temperate forest): The accuracy of such classifications is about 60 percent. All told, scientists estimate that the accuracy of their maps is between 80 and 90 percent. Although not perfect, the information generated from satellites lets scientists gather much more information about terrestrial biomes than if they had to frequently visit the entire planet's surface.

ADDITIONAL READING

Defries, R.S, M. Hansen, J.R.G. Townshend, and R. Sohlberg. "Global Land Cover Classifications at 8 km Spatial Resolution: The Use of Training Data Derived from LANDSAT Imagery in Decision Tree Classifiers." *International Journal of Remote Sensing* 19 (1998): 3141–3168.

Lillesand, T.M., and R.W. Keifer. *Remote Sensing and Interpretation.* New York: John Wiley & Sons, 1994.

STUDENT LEARNING OUTCOME

- Students will be able to explain why NDVI values for vegetated surfaces are positive, values for bare soil are about zero, and values for water are negative.

FIGURE 7.9 *Terrestrial Biomes* *The location of thirteen terrestrial biomes, showing the climatic factors that determine the location and adaptations of local plants and animals. These adaptations are illustrated by the shape and longevity of leaves, which reflect the availability of water, temperature, the length of the growing season, the availability of sunlight, and grazing by herbivores. In boreal forests, short growing seasons favor trees that retain their leaves. The longer growing seasons in temperate forests allow enough time for trees to drop their leaves in the fall and regrow them in the spring. In grasslands and savannas, grazing by herbivores has favored adaptations that allow grasses to grow from areas near the ground, which are less likely to be eaten. In deserts of North America and Africa, the lack of water has favored sharp leaves that we recognize as spines on cactus and euphorbs respectively. In tropical dry forests, many trees drop their leaves during the dry season and regrow them during the rainy season. In tropical rain forests, the struggle for light favors large thin leaves, which dissipate heat and lose lots of water.* (Source: WWF Ecoregions.)

● *Namibia Desert,* The Republic of Namibia, Africa

● **Desert and Dry Shrublands** *Mojave Desert,* Nevada, USA

● **Boreal Forest** *Northern Canada Shield Taiga,* Manitoba, Canada

●**Tropical and Subtropical Seasonal Forest** *Tumbes-Piura Dry Forest,* Ecuador, South America

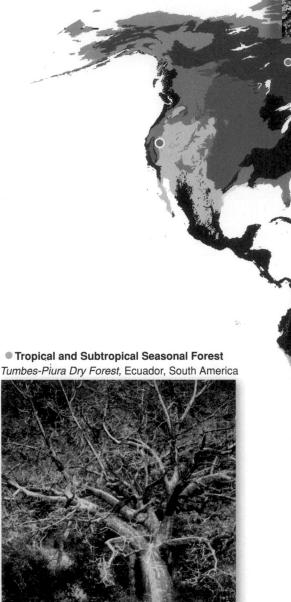

■	Tropical rain forest, subtropical moist forest
■	Tropical and subtropical seasonal forests
■	Tropical grasslands and savannas
■	Deserts and dry shrublands
■	Temperate rain forest
■	Temperate coniferous forests
■	Temperate broadleaf and mixed forests
■	Mediterranean woodlands and scrub
■	Temperate grasslands and savannas
■	Boreal forests
■	Tundra
■	Rock and ice
■	Montane grasslands and shrublands

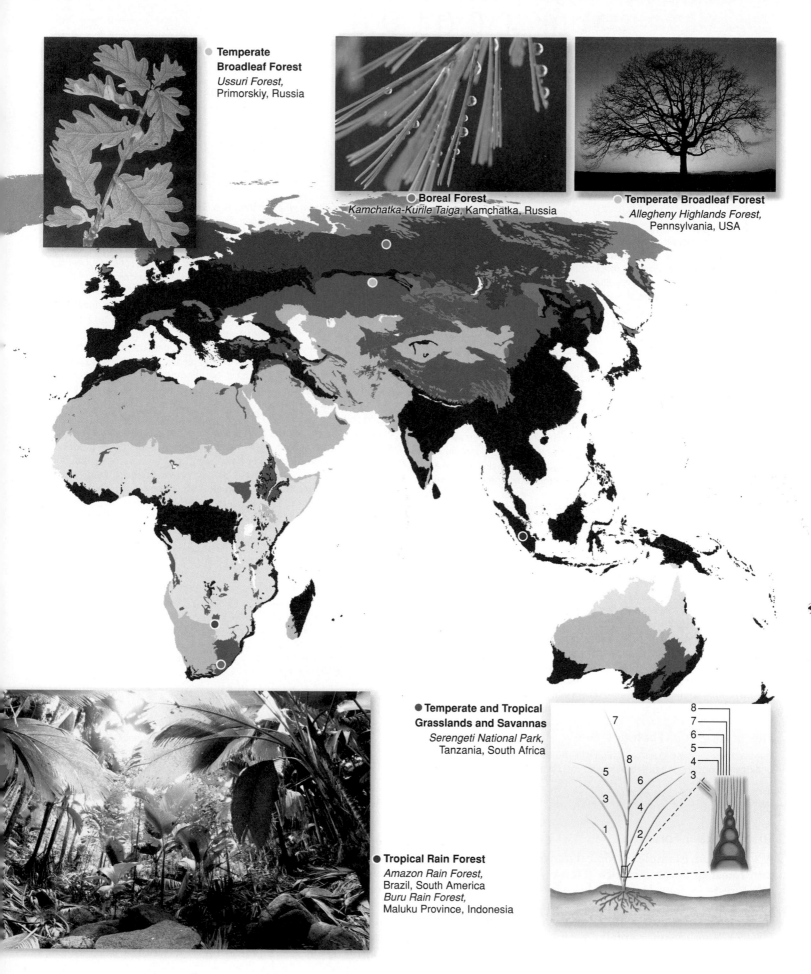

Temperate Broadleaf Forest
Ussuri Forest,
Primorskiy, Russia

Boreal Forest
Kamchatka-Kurile Taiga, Kamchatka, Russia

Temperate Broadleaf Forest
Allegheny Highlands Forest,
Pennsylvania, USA

Temperate and Tropical Grasslands and Savannas
Serengeti National Park,
Tanzania, South Africa

Tropical Rain Forest
Amazon Rain Forest,
Brazil, South America
Buru Rain Forest,
Maluku Province, Indonesia

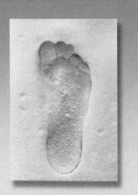

Your ECOLOGICAL *footprint*

What Biome Do You Live In?

At first this seems to be a relatively simple question to answer. Figure 7.9 on pages 132–133 maps the global distribution of biomes. To determine which biome you live in, just locate your area on the map and match the color to the legend. Those of us living in Boston are in the temperate broadleaf and mixed forest biome.

But this simple lesson in reading a map understates the geographic extent of your footprint. In reality, you consume environmental goods and services that come from a wide variety of biomes that are distributed across the planet. To calculate your use of net primary production (Chapter 6), you examined three types of environmental goods: foods derived from plants, foods derived from livestock, and materials derived from trees. Not all of these materials are produced in your local biome. For example, those of you in the American Southwest use paper from trees grown in temperate conifer forests or deciduous and mixed forests. Here in Boston most of the grains we eat come from plants grown in what were once temperate grasslands.

Table 1 identifies the biomes from which environmental goods are harvested or, in some cases, the biomes in which the species originated. For example, much of the corn and wheat that is grown in the United States and worldwide comes from agricultural fields that were once covered by temperate grasslands. For other environmental goods, supplies are obtained from a wide variety of biomes. For example, potatoes are grown in every U.S. state. The biome listed for potatoes is montane grasslands and scrubland—the plant is thought to have originated in highlands of South America, most likely in the area now bordering Lake Titicaca.

Calculating Your Footprint

To help identify the biome you live in, match the environmental goods that you consume to their biomes. Make a list of the biomes from which you obtain environmental goods. How many biomes supply you with environmental goods? This number is a more accurate representation of your footprint.

To demonstrate the importance of this expanded notion of your footprint, think about the following questions: How many of the environmental goods you use come from biomes other than the one in which you live? How would your lifestyle change if you could consume only products that came from the biome in which you live?

Interpreting Your Footprint

The answers to these questions highlight the benefits of trade, which go well beyond obtaining items that cannot be obtained locally. Many of the benefits come from what economists call comparative advantage, which is trade based on what a region does best. To illustrate, let's examine a hypothetical example in which corn and wood are traded between Iowa farmers and forest owners in Massachusetts. The soils and climate of temperate grasslands allow Iowa farmers to grow corn at lower cost than a landowner in Massachusetts, who would have to chop down a local temperate broadleaf forest and replace it with a cornfield. On the other hand, landowners of temperate broadleaf forests in Massachusetts can harvest trees at lower cost than an Iowa farmer who tries to grow a forest on what was once temperate grassland.

Suppose the Massachusetts landowner can harvest 10 trees per hectare and grow only 25 bushels of corn per hectare. In Iowa growing 100 bushels of corn requires 1 hectare, but growing 10 trees would require 4 hectares of irrigated land in Iowa. (Why must the land be irrigated?)

To illustrate the benefits of trade, suppose that trees and corn are priced so that the cost of 100 bushels equals the cost of 10 trees. Under these conditions the Iowa farmer can grow one extra hectare of corn and receive 10 trees, which would have required 4 hectares to produce in Iowa. These 4 hectares could have been used to produce 400 bushels of corn. So the Iowa farmer gets 10 trees for 100 bushels of corn instead of the 400 bushels the farmer would have given up to grow the trees in Iowa. Conversely, the Massachusetts landowner gets 100 bushels of corn for 10 trees instead of the 4 hectares (and 40 trees) that would have been needed to grow that much corn in Massachusetts. In summary, both the Iowa farmer and Massachusetts landowner are better off via trade than they would have been if their footprints were restricted to their local biomes.

This example ignores the environmental impacts of corn production, tree harvests, and transportation of traded goods. In some cases these environmental costs can be large and must be subtracted from the benefits of trade. If the environmental impacts are large, trade can harm one or both parties. This may be the case in some trade between developing nations that export natural resources to developed nations such as the United States.

STUDENT LEARNING OUTCOME

- Students will be able to explain how their environmental footprint extends beyond their local biomes and how this increases their material standard of living.

Tropical Rain Forests

A well-known example of tropical rain forests is the Amazon rain forest in South America (Figure 7.10). Other tropical rain forests are found in Australia, Africa, and Asia. On each continent tropical rain forests are located within 10° of the equator, where they are supported by warm temperatures (25–27°C) and heavy rainfall (2–4 meters per year). Notice that we said warm, not hot. Contrary to popular belief, tropical rain forests are not hot because much heat is dissipated by evaporating water. Despite high rates of evaporation, rainfall exceeds evaporation during all months, which creates abundant supplies of water.

The warm temperatures and heavy rainfall give tropical rain forest plants their shape. Individual leaves tend to be large and thin (Figure 7.9). Big leaves increase the amount of light captured, compensating for cloudiness and competi-

TABLE 1 Home Biomes for Environmental Goods

Environmental Goods	Biomes		
Beverages		**Meats and Fish**	
Beer	Temperate grasslands	Beef	Temperate grasslands, deserts and dry scrublands
Wine	Mediterranean woodlands		
Coffee	Tropical rain forest	Tuna	Open ocean
Tea	Montane grasslands and scrublands	Whitefish	Temperate coastal seas
Grains		Shrimp	Tropical and temperate coastal seas
Wheat	Temperate grasslands	Clams, mussels, and oysters	Estuaries, tropical and temperate coastal seas
Corn	Temperate grasslands		
Rice	Temperate grasslands and wetlands	Salmon	Temperate coastal seas
Soybeans	Temperate and tropical grasslands	Anchovies and sardines	Upwellings
Fruits and Vegetables		**Paper, Wood, and Fiber**	
Sugarcane	Tropical rain forests	Paper	Temperate broadleaf and mixed forests, temperate coniferous forests
Sugar beets	Temperate grasslands		
Citrus	Tropical dry forest, Mediterranean woodlands	Lumber	Temperate broadleaf and mixed forests, temperate coniferous forests, tropical and subtropical seasonal forests
Lettuce	Temperate grasslands, Mediterranean woodlands		
Tomatoes	Deserts and dry scrublands (origin)	Inexpensive wood products	Temperate broadleaf and mixed forests, temperate coniferous forests
Potatoes	Montane grassland (origin)		
		Expensive wood products	Tropical rain forests
		Cotton	Tropical grasslands and savannas (irrigated warm deserts)

tion from other plants. Thin leaves can better dissipate heat generated by absorbing large amounts of direct sunlight. These benefits impose costs—large thin leaves tend to transpire lots of water. But in the tropical rain forest water is abundant year-round.

Competition for light gives tropical rain forests their characteristic three-dimensional shape. To capture light trees grow tall, and the tree leaves mingle to form a closed canopy. This canopy provides habitat for a wide variety of animal life. Some animals, such as birds, monkeys, and sloths, rarely come to the ground. The closed canopy also means that relatively little light reaches the forest floor, which restricts plant growth. As a result, it is relatively easy to walk through a tropical rain forest—in contrast to the common notion that these forests are dark, impenetrable places that require a swing of the machete for every step forward. This notion

Amazon rain forest

Amazon rain forest

Belem, Brazil 26.7°C 2,438 mm

Kisangani, Zaire 25.3°C 1,760 mm

Kuala Lumpur, Malaysia 27.5°C 2,685 mm

Amazon rain forest

Tropic of Cancer

Equator

Tropic of Capricorn

■ Wet season   ■ Dry season   ■ Mean annual temperature   □ Mean annual precipitation

FIGURE 7.10 *Tropical Rain Forests* *Tropical rain forests are located near the equator. These forests appear to be impenetrable from the banks of a river but are actually relatively open.*

may come from glimpses of rain forests from rivers or roads. Edges of tropical rain forests look dense because roads and rivers create gaps that allow light to reach the surface, where it can support plants living close to the ground. But such gaps are the exception, not the rule.

Abundant sunshine, copious precipitation, and warm temperatures create ideal growing conditions. As a result, tropical rain forests can be described by a number of superlatives. The rate of net primary production is among the highest. The number of plant and animal species in tropical rain forests also is among the highest. Up to 300 species of trees may live in a single hectare of a tropical rain forest—compared to a few dozen species per hectare in midlatitude forests. The many species are connected in surprising ways. Many tree species depend on a particular bee, bat, or bird for pollination. The seeds of some trees must pass through the guts of seed eaters before they **germinate,** which is the process by which a seed starts to grow and develop. In addition, many trees depend on fruit-eating birds (such as parrots, toucans,

and hornbills) to spread their seeds. Without these pollinators and seed eaters, the trees cannot reproduce or disperse; and without the trees, the forest cannot survive.

Tropical Dry Forests

A striking example of a tropical dry forest is located on the island of Madagascar. Other tropical dry forests are found in South America, Africa, Australia, and Asia (Figure 7.11). On each continent they are located between 10° and 25° north or south of the equator, where they are supported by hot temperatures (25–35°C) and abundant rainfall (1–2 meters per year). Although abundant, the rainfall is highly seasonal. Much of the rainfall occurs during a six- or seven-month "rainy season" that usually includes June in the Northern Hemisphere and December in the Southern Hemisphere. Can you explain this timing? (Think about the tilt of Earth's axis and the mechanism that generates high rates of precipitation at the equator.) During this rainy season, precipitation is greater than evapora-

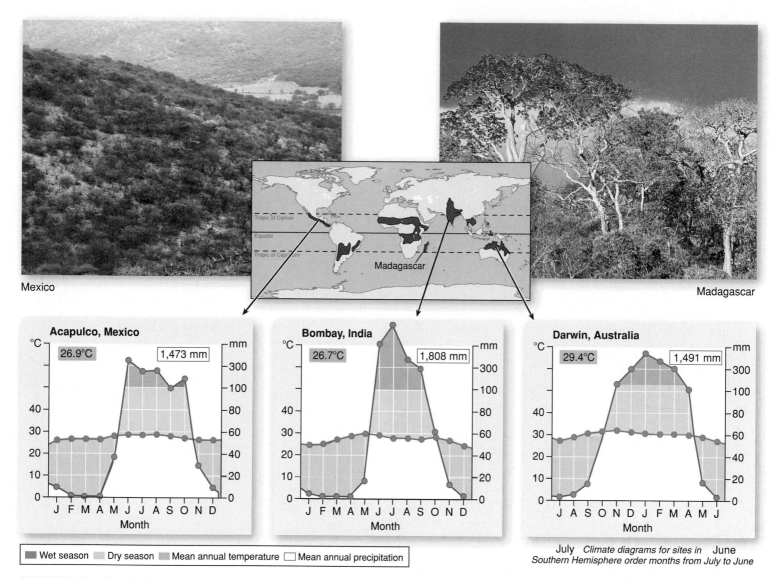

FIGURE 7.11 *Tropical Dry Forests Tropical dry forests are located beyond tropical rain forests just north and south of the equator. The presence of a dry season favors adaptations that allow trees to lose their leaves in the dry season and regrow them during the rainy season.*

tion, and water is abundant. During the dry season, evaporation exceeds precipitation, and water is relatively scarce. It is during the dry season that temperatures can exceed 30°C.

The shapes and lifespans of leaves are adapted to cope with the dry season. At the start of the dry season the soil begins to dry, leaves close their stomata to conserve water, and photosynthesis slows. As the soil dries further, some tree species change the angle of their leaves from horizontal, which maximizes the capture of sunlight, to vertical, which reduces the sunlight (and heat) captured by each leaf. Cooler leaves lose less water. Some leaves also curl up to decrease the amount of sunlight absorbed. These behaviors are reversible. After it rains, leaves resume their usual shapes and rates of photosynthesis. Other tree species in tropical dry forests may drop their leaves and regrow them in the rainy season.

Seasonality also affects the three-dimensional structure of tropical dry forests. In addition to having lower rates of net primary production and less biomass than tropical rain forests, tropical dry forests tend to be shorter—the height of trees often

is positively related to rainfall. Wet/dry seasonality induces trees in tropical dry forests to allocate their biomass differently than trees in tropical rain forests. In tropical rain forests light often is the limiting factor, so natural selection favors trees that grow tall. Water often is the limiting factor in tropical dry forests, so natural selection favors trees that use a larger fraction of their energy to create roots that seek and store water rather than leaves, which lose water through transpiration.

The seasonal availability of water also influences the timing of animal life. Many animals time the birth or hatching of their offspring to coincide with the seasonal availability of food, which is determined by the availability of water. Other animals avoid water shortages by migrating to wetter areas, such as lakes, or to tropical rain forests. Contrary to the common perception that animals hibernate only in cold climates, at least one tropical primate, the fat-tailed dwarf lemur, avoids the dry season by hibernating for up to seven months. During this hibernation, the lemur slows its basal metabolic rate and allows its body temperature to fluctuate with environmental conditions.

Venezuela

Cheetah

FIGURE 7.12 *Tropical Savannas* *Tropical savannas have few trees due to the lack of rainfall. Large areas of grass provide food for many grazers and predators, such as cheetahs, that eat them.*

Tropical Savannas

A famous tropical savanna is Kruger Park in South Africa, which is home to elephants, lions, zebras, and other species (Figure 7.12). Other tropical savannas are found in South America, Africa, Australia, and Asia. On each continent tropical savannas are located between 10° and 20° north and south. Here tropical savannas are supported by warm temperatures (25–35°C) and lesser amounts of rainfall (usually less than 1 meter per year). The seasonality of this rainfall is more extreme than tropical dry forests: The rainy season is shorter and the dry season is longer.

The relatively small amount of rain and the long dry season give savannas their general appearance—they are mostly grass with widely scattered trees. This spacing gives tree roots room to collect water from a relatively large area. But the spacing among trees is not necessarily random. The location of trees is determined by myriad factors such as grazing and even the location of old termite nests. Termites tend to concentrate nitrogen, which often is the limiting nutrient in terrestrial ecosystems.

In tropical savannas the predominant plant species are grasses. Grasses "green up" during the growing season and "brown" during the dry season. Browning is caused by drying, which generates the other main challenge to plants— fire. At the start of the rainy season lightning strikes may ignite large fires. These fires burn the tops of the grasses, but the grasses can regrow quickly. Unlike many plants, grasses do not grow from the top. Rather, they grow from areas at or below the soil surface, where they are protected from fire (Figure 7.9). Fires give savannas their open parklike appearance by eliminating entanglements of brush.

Savanna vegetation is also shaped by grazers. Some scientists argue that grazing stimulates some plants to increase their photosynthetic rates and the rates at which they absorb nutrients. As a result, areas that are grazed moderately may have higher rates of net primary production and higher nutrient contents than areas that are not grazed. For example, the net primary production of the shortgrass plains in the Serengeti (Kenya) may be greatest when wildebeests eat about 25 percent of the grass.

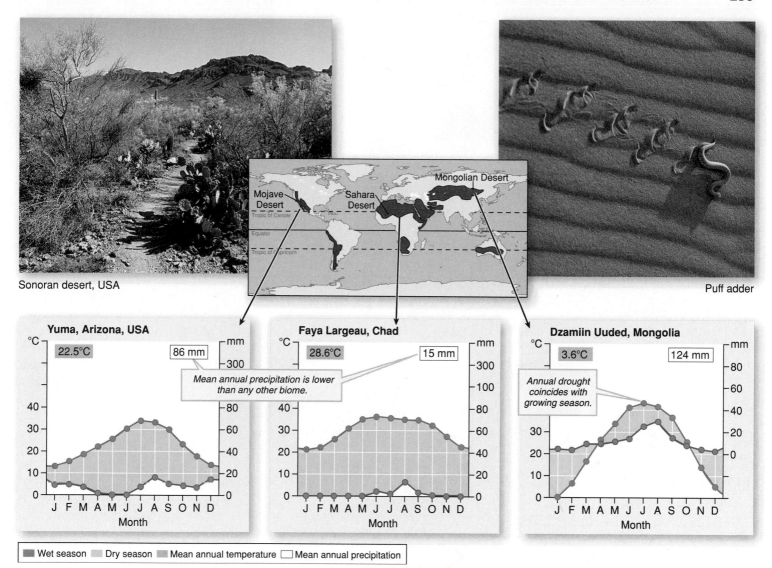

Sonoran desert, USA

Puff adder

FIGURE 7.13   *Deserts* *Many deserts are located at about 30°N and 30°S, where air from the Hadley cells sinks. Desert plants and animals, such as the Puff adder have many adaptations to the lack of moisture and shifting sands.*

Animals also are influenced by the grass and seasons. The abundance of relatively edible grass (compared to trees) supports large populations of mammalian grazers, such as antelope and wildebeest. Many of these grazers have long faces. This shape creates room for the teeth that are needed to grind grass while keeping the animals' eyes and ears on the lookout for predators. Well-known savanna predators include big cats such as lions, leopards, and cheetahs. In addition to these hazards, grazers are challenged by the dry season, which forces many species to undertake long migrations in search of food.

Finally, savannas have relatively large detritivore food chains. Termites are conspicuous—their mounds can be 6 meters high and house several million individuals. Other detritivores include dung beetles, which feed on the droppings of the mammalian herbivores. These beetles collect the dung and roll it into large balls. They then lay eggs in the balls and bury them. After hatching, the larvae feed on the dung.

Deserts

Familiar deserts include the Saharan Desert, which covers several nations in northern Africa, and Death Valley in the Mojave Desert (Figure 7.13). Other deserts are found on every continent other than Antarctica. On each continent, many are located at about 30° north and south of the equator. This location is consistent with their definition as places where potential evaporation exceeds annual precipitation. No single precipitation rate can be used to define a desert. For example, the Sonoran Desert in Arizona may receive up to 300 mm per year, which is similar to the amount received by some tropical savannas. Nor can deserts be defined by temperature. Summer temperatures in the desert of central Mongolia rarely exceed 20°C. (Can you explain the location of this desert?)

The lack of water keeps desert vegetation to a minimum. At one extreme is the Saharan Desert, where huge areas of sand dunes and rocks lack vegetation. But most deserts (including the Sahara) support some vegetation. When people think of desert vegetation, they envision cacti and the similar euphorbs

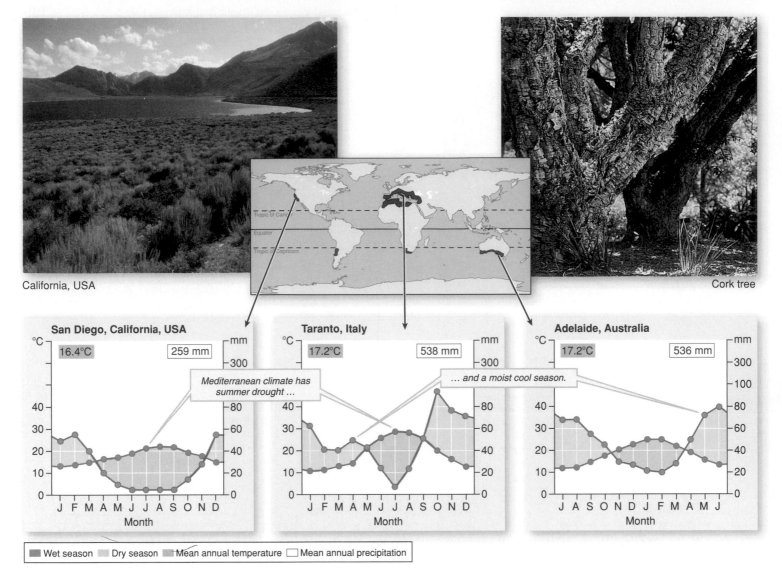

California, USA

Cork tree

San Diego, California, USA
16.4°C
259 mm

Mediterranean climate has summer drought ...

Taranto, Italy
17.2°C
538 mm

... and a moist cool season.

Adelaide, Australia
17.2°C
536 mm

■ Wet season　■ Dry season　■ Mean annual temperature　□ Mean annual precipitation

FIGURE 7.14　*Mediterranean Woodland and Shrubland* Mediterranean woodland and shrubland areas tend to have warm, dry summers, which encourage fires. As a result, local plants have many adaptations for frequent fires, including thick bark.

in African deserts (Figure 7.9). Cacti look very different from most other plants, which reflects selective pressures to conserve and store water. Most cacti are fairly round, reducing the surface area across which the plant may lose water. Leaves lose a lot of water, and most cacti do not have leaves. Instead natural selection has favored the conversion of leaves into spines, which protect the plants and their fluids from herbivores.

Although evaporation exceeds precipitation over the span of an entire year, in some deserts precipitation exceeds evaporation for weeks or months. This short period of water availability favors adaptations that allow desert plants and animals to complete their life cycles quickly. Annual plants may grow, flower, and die in just a few weeks, which makes the desert bloom seemingly overnight. Spadefoot toads are active for only a couple of weeks each year. The toads spend most of the year buried underground. They quickly come to the surface when it rains, then eat and lay eggs in temporary ponds. The tadpoles have just a few weeks to mature and store enough food until they bury themselves to wait for next year's rain.

Some desert animals such as the kangaroo rat are able to go their entire lives without drinking water. Instead these animals obtain water directly from their food (most plant and animal food consists largely of preformed water) and by converting their food to energy (metabolic water). For example, metabolizing 1 gram of fat generates about 1 gram of water. Metabolic water is the basis for camels' desert success. There is no water in camels' humps. Rather the humps contain fat, which the camels convert to energy and water when needed.

Mediterranean Woodland and Scrubland

In North America the most recognized Mediterranean woodland and scrublands include the semiwooded areas around Los Angeles that are known locally as chaparral (Figure 7.14). In other areas this biome is known as matoral (Spain), fynbos (South America), mallee (Australia), and garrigue (eastern Mediterranean). Mediterranean biomes are found on every continent (except Antarctica), usually

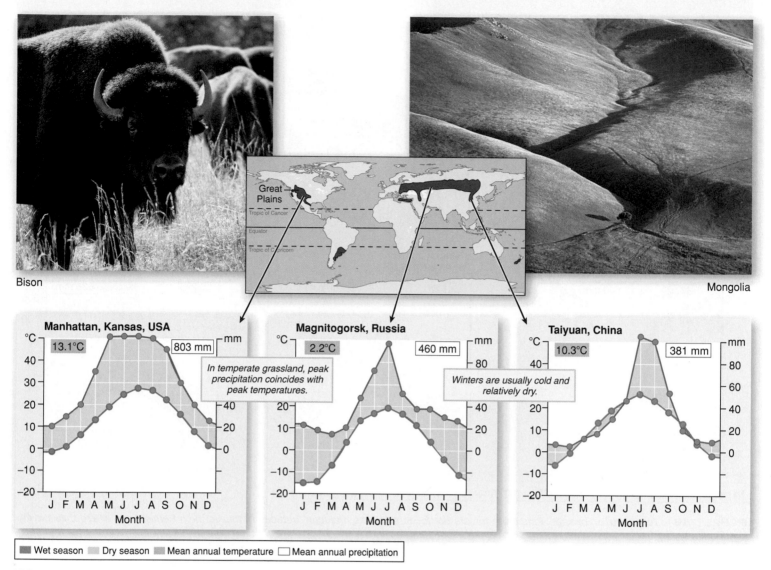

Bison

Mongolia

Great Plains

Tropic of Cancer

Equator

Tropic of Capricorn

Manhattan, Kansas, USA

°C | 13.1°C | 803 mm | mm

In temperate grassland, peak precipitation coincides with peak temperatures.

J F M A M J J A S O N D
Month

Magnitogorsk, Russia

°C | 2.2°C | 460 mm | mm

Winters are usually cold and relatively dry.

J F M A M J J A S O N D
Month

Taiyuan, China

°C | 10.3°C | 381 mm | mm

J F M A M J J A S O N D
Month

■ Wet season ■ Dry season ■ Mean annual temperature □ Mean annual precipitation

FIGURE 7.15 *Temperate Grasslands* *In North America temperate grasslands include tall prairie grasses and many bison. The vegetation in some areas of the Mongolian steppe tends to be shorter and sparser.*

between 30° and 40° north or south of the equator, but they can extend farther north or south. At these latitudes winters are fairly mild, and summers often are hot. These biomes receive 250 to 600 mm of precipitation per year. Little of this precipitation falls during the summer, so evaporation generally exceeds precipitation during the hottest months.

The quantity and timing of precipitation shapes these biomes. Yearly precipitation is sufficient to support a limited number of trees. The lack of precipitation during the summer makes this biome vulnerable to fire (see Chapter 8's Policy in Action on pages 174–175); fire shapes many of the local plants and animals. Because temperatures remain above 0°C (32°F) year-round, most trees are evergreen. But they have small leaves to cope with the dry season. In addition, trees often have thick bark, which helps them survive fires.

Shrubs often pursue another strategy. Some shrubs such as chamise burn easily but can sprout quickly following a fire. Indeed, many plants native to Mediterranean biome areas that people use as herbs, such as rosemary, thyme, sage, and oregano, contain aromatic chemicals that enhance

burning. Other plant species grow and reproduce when water is available and die back during summer.

The combination of climate, fires, and plants' adaptations to fires shapes the animal community. The hot dry conditions force many local animals such as snakes, rabbits, and foxes to spend part of the day underground. Their burrows also serve as sanctuaries for less mobile species during fires. More mobile species escape fires by running. Larger grazers, such as deer, must often travel long distances to find enough food because many plants have tough leaves with a waxy covering to reduce water loss. The presence of large grazers also supports a small population of larger predators, such as mountain lions and wolves.

Temperate Grasslands

The Great Plains of North America are a well-known example of temperate grasslands (Figure 7.15). Others are found in South America, Europe, and Asia. On each continent, temperate grasslands tend to be located in the middle

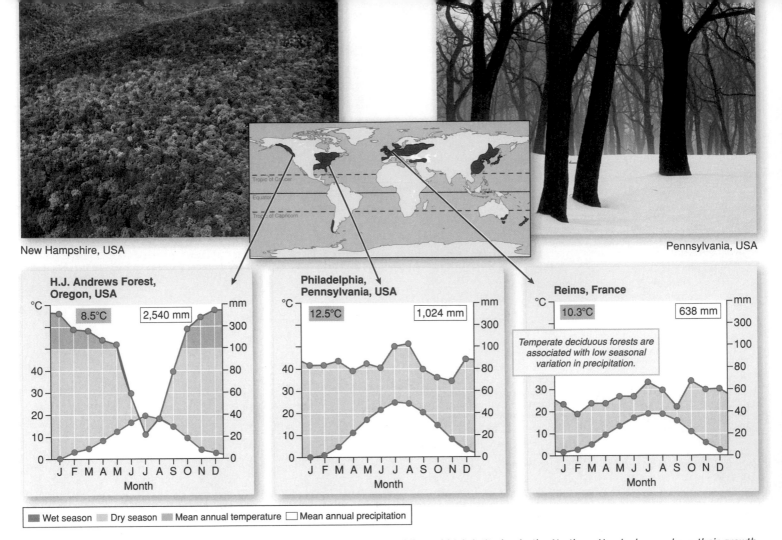

New Hampshire, USA

Pennsylvania, USA

H.J. Andrews Forest, Oregon, USA
8.5°C   2,540 mm

Philadelphia, Pennsylvania, USA
12.5°C   1,024 mm

Reims, France
10.3°C   638 mm

Temperate deciduous forests are associated with low seasonal variation in precipitation.

■ Wet season ■ Dry season ■ Mean annual temperature □ Mean annual precipitation

FIGURE 7.16 *Temperate Forests Temperate forests are found at middle and high latitudes in the Northern Hemisphere, where their growth is often limited by the length of the growing season. In response, many temperate forests are characterized by brilliant colors in the fall and bare trees in the winter.*

of the continent and hence sport a continental climate—hot summers and cold winters. Long cold winters imply that the growing season is relatively short. The growing season coincides with most of the year's precipitation, which usually totals between 300 and 1,000 mm per year.

This range in precipitation supports a variety of grass species. In North America precipitation increases as we move east from the Rocky Mountains. In the driest areas grasslands are known as shortgrass **prairies** (prairie is the North American term for grassland). As we move farther east, the grass becomes taller due to the presence of bluestem and Indian grasses. These species can be 3 meters high, hence the name tallgrass prairie.

Fire also shapes temperate grasslands; it spreads quickly due to dry grass and strong winds. Like their counterparts on the tropical savanna, grasses survive fires by growing from areas at or below the soil surface (Figure 7.9). This allows them to survive and thrive in conditions where fire is frequent. Indeed, some grasses like big bluestem grow taller and produce more biomass after a fire.

The rich soil and lush grass support huge herds of grazers—such as bison in North America, which once numbered in the millions. Together with other herbivores, these

animals consume up to 50 percent of net primary production in grasslands, compared to 5–10 percent in most other biomes. Again, grasses survive this onslaught by growing from places beyond the grazers' reach.

Adaptations that help grasses withstand the effects of grazing force grazers to modify their eating habits. Before the arrival of Europeans, bison and prairie dogs dominated the North American prairie. Prairie dogs live underground in large groups that are known as towns. The high density of prairie dogs exerted heavy pressure on local plants, but in general this pressure seems to have had a positive effect. Grazing by prairie dogs increases the nitrogen content and digestibility of the local plants while leaving total biomass unchanged. Bison seem to recognize these improvements and prefer to graze near prairie dog towns.

Temperate Forests

Familiar temperate forests include those in northern California that are home to the giant sequoias (Figure 7.16). Many of you probably live in the temperate forest areas that cover much of the East Coast of North America west to the Mississippi River. Other temperate forests are found in Europe, Asia, and Australia. Consistent with their ability to support trees, these areas receive 650–3,000 mm of precipitation per year. Most temperate forests are located between 40° and 50° north and south, though some extend to 30°.

Temperate forests have a relatively short growing season. This brevity is partially responsible for one of the most important properties of many temperate forests: **deciduousness,** which refers to the dropping of leaves. Each year deciduous trees drop their leaves (usually, but not always, in the autumn) and grow new leaves the following spring. As a result, deciduous forests are green for only part of the year.

At first glance shedding leaves seems wasteful. But the selective pressures that favor deciduousness can be understood by analyzing the costs and benefits of leaves. There are many costs associated with shedding leaves. Plants use considerable amounts of energy and nutrients to build leaves, only to lose them when the leaves are shed. A tree without leaves cannot photosynthesize and so cannot generate new supplies of energy.

But keeping leaves also imposes costs. Leaves provide energy to a tree only when environmental conditions permit, but they require energy for maintenance at all times. Trees lose water from their leaves. Trees also lose energy and nutrients when leaves are eaten by herbivores. Finally, the photosynthetic rate of a leaf tends to decline as it gets older.

In temperate forests, the costs of keeping a leaf for the next growing season may be greater than the benefits of keeping it. The importance of this balance is indicated by adaptations that reduce the cost of losing old leaves and building new leaves. Just before dropping their leaves, trees withdraw about half of the leaves' nitrogen and a third of their phosphorus. The maximum photosynthetic rate for the leaves of deciduous trees tends to be somewhat higher than evergreen trees, whereas the amount of defensive chemicals in deciduous leaves tends to be lower than the amount in evergreen leaves. The higher rate of photosynthesis and the lower cost of construction shorten the **payback period** for deciduous tree leaves relative to evergreen tree leaves. (The payback period is the time it takes a leaf to capture an amount of energy equivalent to the amount of energy used to make the leaf.) For example, the leaves of the deciduous Japanese elm require 9–15 days to generate the amount of energy required to produce them compared to the 30 days required for the leaves of the Japanese evergreen *Shiia seiboldi* (there is no common English name). Because of this short payback period, deciduous trees can afford to drop their leaves.

Animals of the temperate forest also must cope with the relatively short growing season. As with deciduous trees, natural selection favors adaptations that allow animals to ride out the cold winter. For example, many animals (like bats and bears) hibernate during the winter. During hibernation animals slow their basal metabolic rates and live off the energy they stored during the growing season. Other animals migrate to areas where more food is available. Many bird species that live in temperate forests during the spring, summer, and fall fly toward the equator (or to "summer" on the other side of the equator) for the winter.

Deciduousness also creates opportunities for the detritivore food chain. Although trees withdraw many nutrients from leaves before they drop, each year's crop of dead leaves (along with the branches and dead trees) provides much organic material for the detritivore food chain. Some of these detritivores are well known, such as earthworms and insects. Even more important are the many fungi that return nutrients to the soil. Without these detritivores forest growth would slow considerably because the dead leaves and trees would "lock up" limiting nutrients such as nitrogen.

Boreal Forests

Boreal forests grow at high latitudes (50° to 65° north) where few people live (Figure 7.17). Boreal forests exist in North America, Asia, and Europe as a more or less continuous belt around the Northern Hemisphere, where the growing season is usually shorter than six months. These forests receive 200–600 mm of precipitation per year, which is about the same as temperate grasslands. Nonetheless, there is enough water to support trees because the cold temperatures allow precipitation to exceed evaporation for much of the year. In addition, boreal forests are marked by climatic extremes. There may be a 100°C temperature difference between the warmest and coldest days of the year. Similar variations occur between years. The average January temperature in Anchorage, Alaska, ranged from −35°C to −7.7°C between 1971 and 1986.

The shape and activity of boreal forests are controlled by solar radiation, air temperature, and the availability of soil water. The maximum day length varies between 16 and 24 hours, and many tree species that live in the boreal forest require long periods of sunlight to trigger their growing season. Natural selection favors day length (rather than temperature) to control growing cycles because day length is a more reliable indicator than temperature, which has a large variance. Because the sun is never really overhead, tree canopies are shaped to absorb light at low angles. Although moisture usually is plentiful, roots cannot absorb water from frozen soil, so trees in boreal forests often suffer from water shortages.

The extreme environmental conditions in boreal forests and the requisite specializations reverse the costs and benefits of dropping and regrowing leaves discussed for temperate forests. Many boreal tree species, such as spruce, are evergreens. The short growing season makes it difficult to grow new leaves each year: By the time they appear, the growing season could be half over. Instead leaves may have

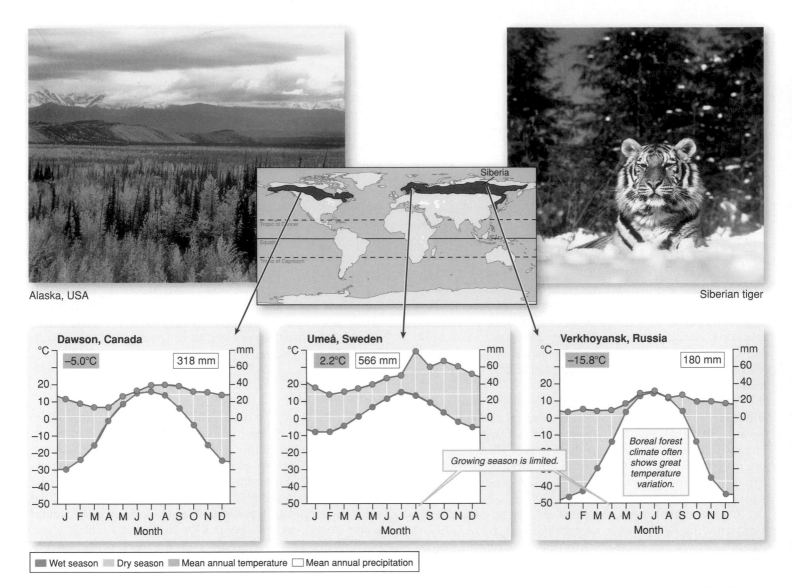

FIGURE 7.17 *Boreal Forests* *Boreal forests are found at high latitudes in the Northern Hemisphere, where the growing season is short. With little time to grow new leaves, many boreal forests are dominated by trees that are green year-round. They are also home to large predators, such as the Siberian tiger.*

very long life spans—in some cases 25–30 years. Keeping leaves over the winter requires little energy because cold temperatures reduce maintenance respiration. The cold temperatures also slow decomposition, which reduces the availability of nutrients in the soil. This scarcity increases the cost of gathering nutrients to grow a new leaf relative to the cost of keeping the old leaf. Finally, the harsh climate favors specializations for water transport. These specializations increase the cost of building new leaves, and along with the relatively short growing season, elongate the payback period for leaves. This long payback period implies that boreal trees cannot afford to drop their leaves.

There may be few famous boreal forests, but their animal inhabitants are well known. They include caribou, reindeer, moose, wolves, Siberian tigers, grizzly bears, and many birds. Birds such as the white-throated sparrow tend to migrate to the boreal forest in the spring, raise their young, and fly south in the fall. Many local mammals also migrate

or hibernate. Those that are active year-round, such as the caribou or moose, are often well insulated with fat, fur, or feathers. In addition, the fur or feathers of some mammals (snowshoe hare) and birds (ptarmigan) turn white during the winter for camouflage. Such animals often have feet and body shapes adapted for snow. As implied by its name, the snowshoe hare has broad feet that help keep it on top of the snow as it moves (sinking into the snow greatly increases the energy cost of moving).

Tundra

In the United States the most talked-about area of tundra probably is the north slope of Alaska, where plants and animals share the landscape with oil-producing facilities (Figure 7.18). Beyond Alaska, the tundra is a thin belt that stretches across North America, Europe, and Asia, most of which is north of the Arctic Circle. Precipitation ranges from

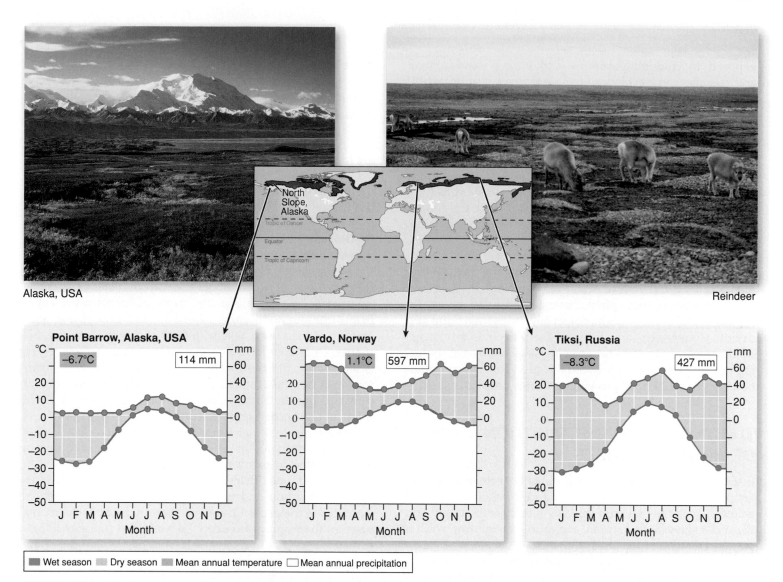

FIGURE 7.18 *Tundra Tundra is found at the northern edge of the Northern hemisphere. The vegetation tends to be short, which keeps it out of the wind and buried in snow during winter. In summer these plants are grazed by large herbivores such as reindeer.*

200 to 600 mm per year, which is similar to the boreal forest. However, the tundra's temperature range is less extreme than the boreal forest. (Can you explain why? *Hint:* Look at the map in Figure 7.9.) Nonetheless, the growing season is shorter than that of the boreal forest.

The tundra is covered with low-lying plants such as mosses, lichens, grasses, sedges, and members of the heath family. These plants are covered by snow much of the year, which allows them to avoid extreme winter temperatures. There are no trees because of the short growing season and the permafrost that is just below the soil surface in many areas of the tundra. **Permafrost** is any soil material that remains below 0°C (32°F) for two or more years. Roots cannot extract frozen water from the soil, so this layer of permafrost keeps the **rooting depth,** which is the vertical distance from the soil surface that contains 95 percent of a plant's roots, too shallow for trees.

Because of the cold temperatures, many ecologists have formulated "laws" that describe how the cold climate shapes local animal life. One law states that animals living at high latitudes have smaller surface areas than their lower-latitude relatives. For example, the ears of the arctic fox are smaller than the ears of the red fox, which lives in boreal forests, which are shorter than the ears of the fennec, which lives in African deserts. Shorter ears reduce the surface area across which the arctic fox can lose body heat. Similarly, many tundra mosses and lichens grow in rounded clumps that reduce wind exposure.

The need to conserve heat or avoid cold is not the only factor that influences the shape of tundra plants and animals. A larger body fortifies organisms against the cold, but the advantages of large size must be weighed against several disadvantages. Small size is an advantage for some plants and animals such as lemmings (mouselike rodents) because it allows them to spend the winter under the snow, which protects them against temperature extremes. On the other hand, larger animals such as northern reindeer remain exposed to cold temperatures and strong winds. Small animals also can develop and mature more rapidly, which allows them to complete their life cycles within the short growing season. Tundra birds such as the ptarmigan and Bewick's swan are smaller than their southern cousins.

Contrary to popular belief, cold winter temperatures are not the most important factor that determines species' ability to inhabit the tundra. Rather, the limiting factor often is length of the growing season, which can be as short as forty days. The large whopper swan cannot live in the tundra because the growing season is too short to raise its young. To survive and reproduce during the short growing seasons, species have adaptations that either speed their development so they can mature in one growing season or slow development so they can spread their development over several growing seasons. Some plants have gained an advantage by forsaking bud scales, which protect new buds from wind and drying. Conversely, some insects have gained an advantage by developing over several growing seasons. By not maturing in a single season, they avoid the effects of adverse conditions in years when the growing season is unusually short.

THE TYPES AND DISTRIBUTION OF AQUATIC BIOMES

Just as the global patterns of solar radiation, precipitation, and temperature influence the structure and location of terrestrial biomes, aquatic biomes are shaped by salinity, temperature, and the availability of light, oxygen, and nutrients. **Salinity,** which is the concentration of salt, defines two classes of aquatic biomes: freshwater biomes and saltwater biomes. The concentration of salt in freshwater is less than that of most living cells, so freshwater multicellular organisms tend to lose water and therefore must have ways to maintain it. (Can you explain why? Think back to our discussion of gradients in Chapter 3.) Maintaining water is less difficult for marine organisms because the concentration of salt in seawater is about the same as that in cells.

In both fresh and salt water, the most productive biomes are located where there are ample amounts of light, oxygen, and nutrients. These factors rarely are found together. As described in Chapter 4, water absorbs solar radiation, so light is most available near the water surface. The depth to which light can support positive rates of net primary production is termed the **euphotic zone.** The euphotic zone may be only a meter deep in estuaries where particles and living organisms cloud the water, or it may be 120 meters deep in the open ocean where the water is clear.

Complicating aquatic life, most nutrients are located below the euphotic zone. Nutrients fall downward with waste materials such as fecal pellets and with dead organic matter. Once they fall below the euphotic zone, their return is slowed by the thermocline (see Chapter 4). This layering separates the availability of light (near the surface) and nutrients (on the bottom), thus limiting the productivity of many aquatic ecosystems (Figure 7.19).

High rates of net primary production occur where nutrients return to the euphotic zone. This can occur in several ways. In many temperate lakes nutrients mix into the euphotic zone during the spring and fall due to changes in solar radiation and winds. These factors cannot return nutrients to the euphotic zone in most marine environments because the water is too deep. Instead most marine biological activity occurs in relatively small areas where the ocean is shallow, where winds or water currents bring bottom water to the surface, or where rivers meet the ocean.

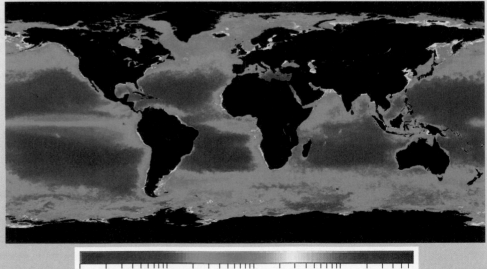

FIGURE 7.19 *Aquatic Net Primary Production* Aquatic net primary production (as proxied by chlorophyll concentration) in aquatic ecosystems.

Chlorophyll *a* concentration (mg/m³)

Rivers and Streams

River and stream biomes are called **lotic** systems, which is the Greek word for "flowing." Lotic biomes are found on every continent except Antarctica. Well-known examples include the world's great rivers, such as the Mississippi and Amazon rivers (Figure 7.20). Because rivers and streams move precipitation to the ocean (Chapter 18), almost all areas have at least a small stream.

Looking at a stream, you may think it ends at its banks; but scientists have determined that the hydrological and biological boundaries of these waterways extend dozens of meters downward and up to several kilometers laterally. The **riparian zone** is the transition zone between the waterway and the surrounding terrestrial environment. Below the flowing waters, the soil is saturated in the **phreatic zone.** Even these areas contain many living organisms, such as bacteria. The lack of clear boundaries implies that lotic biomes are best understood as land and water systems.

This land-water linkage is crucial because the terrestrial environment influences streams' water chemistry and nutrient supply. Scientists demonstrated this link by clearing all the vegetation around Hubbard Brook in the White Mountains of New Hampshire. Removing the vegetation changed the chemistry of the stream and the organisms in it. A more detailed description of this experiment can be found in Chapter 8.

The kinetic energy of flowing water shapes its inhabitants. Plants and bottom-dwelling animals that live in lotic systems have adaptations that allow them to avoid being swept away with currents. Many animals are small, which allows them to hide among rocks. Other animals have a low profile that keeps them out of the fastest currents, and some build small houses from sand grains that they glue to large rocks. Some creatures have developed ways to use the flowing water. For example, the net spinning caddis fly, which is common in temperate trout streams, depends on flowing water to bring its food.

(b)

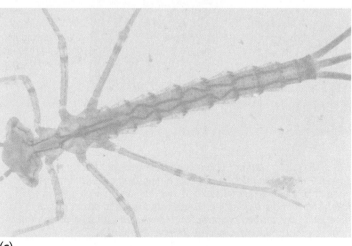
(c)

(a)

FIGURE 7.20 *Rivers and Streams* (a) Large rivers such as the Amazon provide critical habitat to both aquatic and terrestrial organisms, and people often use them as a source of water and transportation. (b) Smaller, rapidly flowing waters of streams are populated by organisms, such as insect larva, (c) with a flattened shape that allows them to avoid being swept away.

Flowing water also influences the food web. Unlike other systems where photosynthesis is the most important source of energy, lotic food webs often depend heavily on leaves and insects that fall into the river. Detritus in the form of dead leaves is especially important. When a leaf falls into a small river, it sinks to the bottom and is colonized by fungi and bacteria. Many of these bacteria fix nitrogen, and this increases the nutritional value of the decaying leaf for the specialized insects that feed on it.

Lakes

Lake ecosystems are called **lentic,** which is the Greek word for "layer." Limnologists (scientists who study lakes) believe that there are more lakes now than at any time in Earth's recent past. Most lakes are temporary because rivers tend to wear down the barriers at the lake outlet. Thus special geological processes are required to form lakes. Some of the processes that form lakes include glacial action (the Finger Lakes in New York State); crustal upwarping or bending (Lake Titicaca in Peru and Bolivia); volcanic action that either blocks river movement (Lake Atitlan in Guatemala) or forms a crater (Crater Lake in Oregon); crustal block slippage (Lake Tanganyika in Kenya, Lake Baikal in Russia); and animal activity by beavers and humans (beaver lakes, Lake Mead in Nevada and Arizona).

Lakes often are divided horizontally and vertically (Figure 7.21). The **littoral zone** includes shallow waters that are near the shore. Here the combination of terrestrial plants such as cattails and aquatic algae generates high rates of net primary production relative to the other parts of the lake. This energy supports a relatively large array of species that live on both the land (dragonflies, turtles, ducks) and in the water (fish, crayfish, frogs). Beyond the near shore is the **limnetic zone.** In the upper layer, which is known as **epilimnion,** sunlight powers high rates of net primary production when nutrients are available. Their availability varies seasonally with the strength of the thermocline. When the **thermocline** disappears in the spring and fall, winds can mix deep water and their nutrients back to the surface. This mixing is blocked during the summer, when the thermocline strengthens. The **hypolimnion** includes the deep waters, which tend to be dark and cold. As a result net primary production tends to be low. Most of the species that live in the hypolimnion are heterotrophs or detritivores.

The depth of the hypolimnion influences the rate of net primary production. Shallow lakes tend to be more productive because the nutrients that fall to the bottom can mix quickly back to the surface. In deeper lakes, nutrients tend to be lost for at least one growing season. You can make an educated guess about a lake's rate of net primary produc-

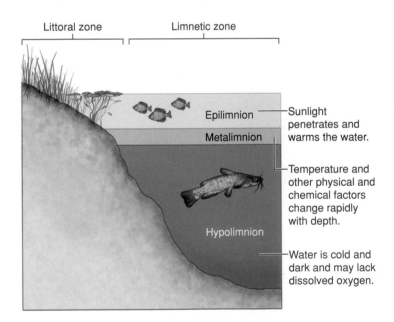

Littoral zone   Limnetic zone

Epilimnion — Sunlight penetrates and warms the water.

Metalimnion

— Temperature and other physical and chemical factors change rapidly with depth.

Hypolimnion

— Water is cold and dark and may lack dissolved oxygen.

Oligotrophic

Eutrophic

FIGURE 7.21 *Lakes Lakes can be divided into lateral and vertical zones. The size and depth of these zones, along with the nutrient content of local terrestrial systems and human activity, influence a lake's net primary production. Lakes with low rates of net primary production are termed oligotrophic (top). Lakes with high rates of net primary production are termed eutrophic (bottom).*

tion based on its color—blue or green. Blue lakes generally have small populations of phytoplankton and therefore have relatively low rates of net primary production. Such lakes are termed **oligotrophic. Eutrophic** lakes have high rates of net primary production. Their large populations of phytoplankton make the water appear green.

Estuaries

Rivers that flow into the ocean create a highly productive biome known as an **estuary,** which is a semi-enclosed coastal body of water that has a free connection with the open ocean, and within which seawater is measurably diluted with freshwater derived from land drainage (Figure 7.22). Well-known estuaries include the San Francisco and Chesapeake bays in the United States and the Pearl River estuary in China.

The combination of freshwater flowing into salt water and the tidal cycle create three biological habitats: salt marshes, mudflats, and channels. **Salt marshes** are farthest from the sea and represent the transition from land. These areas often can be recognized by tall salt grasses, which thrive because they can secrete excess quantities of salt. As described in Chapter 5, dead grasses are an important source of energy via the detritivore food chain.

As implied by their name, **mudflats** are large areas of mud that are exposed to the air during low tide. The cycling between land and aquatic environments is a challenge for most organisms—dry to wet, freshwater to salt water, high oxygen to low oxygen, and warm to cold. Because relatively few species can cope with these ranges, mudflats have relatively few species but often house many individuals of those species. The mudflats (and salt marshes) connect to the ocean via **channels.** Channels always contain water. This allows them to serve as nurseries for many commercial species of fish (such as sole) and shellfish (such as crabs).

Although relatively few species can tolerate the wide range of conditions in an estuary, these areas are among the most productive of all biomes. Estuaries tend to be relatively shallow. As a result, the euphotic zone extends to the bottom, where there is a high concentration of nutrients. The concentration of nutrients is especially high in estuaries because rivers bring new supplies that are derived from soil and dead organic material, which wash into the river. For example, the nutrients that wash into the estuaries at the end of the Mississippi River are collected from the entire middle portion of the United States (Figure 6.16 on page 120).

The mixing of salt and fresh water also maintains the concentration of nutrients in the euphotic zone. In many estuaries in temperate climates, a layer of less dense freshwater "floats" on top of a dense layer of salt water. This bottom layer forms a wedge of salt water that moves up and down with the tide, scraping nutrients from the bottom as it moves. Near its forward edge the salt wedge mixes with the freshwater, and this returns nutrients to the surface.

FIGURE 7.22 *Estuaries Estuaries are located where rivers meet the sea. Zones in estuaries are defined by ebb and flow of tides. These flows influence the salinity, temperature, and depth of the water, which influence the types of organisms that live in an estuary and the zones they live in.*
(Source: Redrawn from J.L. Sumich and J.F. Morrissey, Introduction to the Biology of Marine Life, Eighth Edition, Jones & Bartlett.)

Temperate Coastal Seas

For many readers, temperate coastal seas are the aquatic biome that lies just beyond their local shore (Figure 7.23). Familiar examples include Jones Beach, which is just outside New York City, and the rocky coastline of Maine. In this biome many animals are **benthic,** which means they live on or near the seafloor. Here the tides and the nature of the sea bottom influence the species that live in a particular area.

Rocky bottoms give animals a semipermanent place to call home, whereas animals living on a soft bottom must cope with constant shifts. Rocky bottoms tend to support **epifauna**—animals that are attached to or move about the surface. In this zone many species tend to be **filter feeders:** They obtain food from passing water. Filter feeders include mussels and barnacles. Soft bottoms tend to be populated by **infauna**—animals that live within the bottom material. This environment is more conducive to scavengers, such as small shrimp and polychaete worms; other inhabitants include filter feeders, such as tube worms, which build tubelike homes in soft bottoms.

Differences in the animals that live in areas with hard and soft bottoms can be seen most easily in the **intertidal zone**—the area between land that is wetted by the high tide but always covered by the low tide. The upper intertidal zone is never completely underwater—it is wet during very high tides or splashes from large waves. Thus relatively few aquatic species live here, regardless of whether it is a rocky coast or a sandy beach. More species are present in the middle intertidal zone, where individuals must survive both periods in which they are exposed to the atmosphere and periods when they are submerged. A rocky bottom gives these species something to hold onto as the water ebbs and flows. You may recognize these areas as rocks covered with barnacles, mussels, starfish, and algae. The intertidal zone on a sandy beach seems relatively uninhabited. To find the inhabitants you must dig into the sand, which houses worms and sand crabs. Closer to the water, in the lower intertidal zone, the number of species increases even further. On rocky shores you can find sea anemones, brittle stars, and sea cucumbers. On sandy beaches the animals that inhabit the lower intertidal zone still are found below the surface. These species include sand dollars and soft-shelled clams known to New Englanders as "steamahs."

Adaptations that prevent these creatures from being swept away by the tides also limit the mobility of infauna and epifauna; nonetheless, many of these species are extremely widespread. This creates a seeming contradiction—how do such slow-moving organisms extend their range? Many benthic animals start life as tiny free-floating

FIGURE 7.23 *Temperate Coastal Seas* (a) Animals that live in the intertidal zone with a hard bottom must resist drying during low tides and must be able to withstand being submerged during high tide. (b) In areas with a soft bottom, such as beaches, organisms such as clams often live within the sand.

(a)

(b)

larvae that look nothing like the adults. They may remain in this form for weeks or months as they are carried with the currents. When they find a hospitable spot, they metamorphose into their adult forms (see Chapter 5). This mobility comes at a cost—the larvae are very vulnerable to predation (remember from Chapter 5 that much less than 1 percent of lobster larvae become adults). This disadvantage seems to be offset by the advantages of mobility. For example, some mussel species, which barely move as adults, are found throughout the Atlantic and Pacific oceans.

Tropical Coastal Seas—Coral Reefs

Like terrestrial biomes, tropical coastal seas tend to have more species than their high-latitude counterparts. Perhaps the most diverse and well-known examples of this biome are coral reefs. The most famous of these may be the Great Barrier Reef, which stretches for 2,000 kilometers along the eastern shore of Australia (Figure 7.24). Other reefs are found worldwide wherever conditions permit. These conditions include warm water (above 20°C) and shallow depth (less than 50 meters).

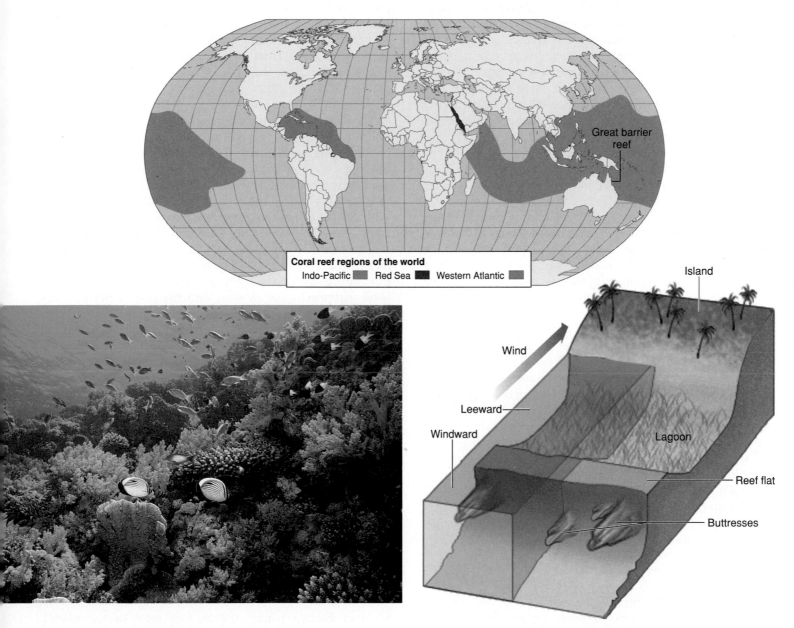

FIGURE 7.24   *Coral Reefs* *Coral reefs are found in warm oceans. The reef creates different zones that host different organisms. The reef flat is home to many species, with fewer numbers in the lagoon, such as the giant clam.* (Sources: Map redrawn from Reef Relief, www.reefrelief.org; line art drawn from J.L. Sumich and J.F. Morrissey, *Introduction to the Biology of Marine Life, Eighth Edition*, Jones & Bartlett.)

Large reefs are built by small animals known as stony corals. These animals live as a colony; individual units are known as polyps. Each polyp has tentacles that can capture small organisms and move them to a mouth in the polyp's center. Individual polyps sit in a small cup of calcium carbonate, which they make themselves. Together these calcium carbonate cups are the coral's hard skeleton. This skeleton is covered by a thin layer of tissue, which connects individual polyps. Corals grow upward by about 1 mm/year and grow outward by about 8 mm/year. At these rates it takes a long time to form large coral reefs.

The corals and their hard skeletons are critical to this biome. The skeletons provide a hard substrate, which often is a limiting factor in tropical marine environments. As such, coral reefs provide habitat and a place of anchor for many species. Furthermore, reefs generate diverse habitats. The windward side of the reef provides a transition from the deepwater habitat to the reef itself. Here breaking waves create a splash zone that poses difficulties for the corals but creates many holes and narrow spaces that help hide small fish from predatory larger fish. Beyond this zone is the reef flat, where the water is fairly shallow; in some places it just covers the coral at low tide. Many coral species find these conditions ideal, and this portion of the reef houses the most species, including spectacular fish that often find their way to home aquariums. Beyond the reef flat is a lagoon, where the water deepens and there is little wave action. The lagoon often has a mix of corals and sandy bottoms. Without the hard substrate and the three-dimensional habitat provided by the corals, the number of species declines. The most famous lagoon animal may be the giant clam, which can be over a meter long.

So far we have talked about the diversity of fish and other heterotrophs. But the water is sparkling clear, which implies that the population of phytoplankton is relatively small (remember that oligotrophic lakes are clear). Without a large population of autotrophs, the energy source for coral reefs was a mystery until scientists realized why corals need sunlight (hence their need for relatively shallow water). The light is used by single-celled algae that live in the corals' tissue. These algae are present in large numbers and provide enough energy to satisfy the daily requirements of many coral species. In return, the corals provide the algae with shelter, carbon dioxide for photosynthesis, nitrogen, and phosphorus. (Where do you think the carbon dioxide, nitrogen, and phosphorus come from?)

Upwellings

Isolated areas of highly productive aquatic biomes are located where nutrient-rich bottom waters mix into the euphotic zone. Such areas are known as **upwellings** (Figure 7.25). Upwellings are created in three general ways. A unique type of upwelling occurs along the equator in the Pacific Ocean. Here Ekman transport (see Chapter 4) moves surface water north and south away from the equator, which allows deeper water to take its place.

Another upwelling occurs along the Antarctic continent, where cold water sinks along the coast and warmer surface waters remain offshore. Between these two areas, deep water comes to the surface. This upwelling supports a highly productive food chain. At the top are the world's largest animals, baleen whales. These whales filter their food, krill, from seawater in huge quantities (see Chapter 5). These krill are so abundant that they also support large populations of seabirds, who fly long distances to the Antarctic to reproduce.

Coastal upwellings are found where winds blow surface waters away from coasts. This allows deep, nutrient-rich waters to take their place. Consistent with the global patterns of atmospheric and oceanic circulation described in Chapter 4, these upwellings usually are located on the eastern boundaries of ocean basins off the coasts of California, Peru, northern Africa, and southern Africa. The upwelling along the eastern boundary of the Pacific ocean (the west coast of South America) is among the most productive regions of the world's oceans. The productivity's reliance on winds is illustrated by the changes that occur off the coast of South America during an El Niño event. In most years there are about 2 micromoles (a millionth of a mole) of nitrate per unit of water, and the phytoplankton fix about 219 milligrams of carbon per cubic meter per day. But during the severe El Niño event of 1982–1983 nitrate levels dropped close to zero, and the rate of net primary productivity dropped by about 95 percent to about 10 milligrams of carbon per cubic meter per day. This loss of net primary productivity reduced the population and density of other organisms via the food web. Some species such as hake moved to deeper waters. Sardines shifted their range south so that few fish were caught off the coast of Ecuador but more were caught off the coast of Chile. In addition, each sardine had less body weight and a smaller oil content (fish store energy in oils), which indicated that they were not eating as well as in other years. The anchovy fishery was affected most severely by El Niño. In 1983 the catch was only a hundredth of that caught ten years earlier. Once the El Niño condition abated and net primary production returned to higher rates, so did the anchovy population. Indeed, anchovies may be adapted to the boom and bust cycles of coastal upwellings.

The Open Ocean

Most of Earth's water, and in fact most of the planet, appears blue (instead of greenish blue). This blue is sunlight that is reflected back to the surface from the depths. As in

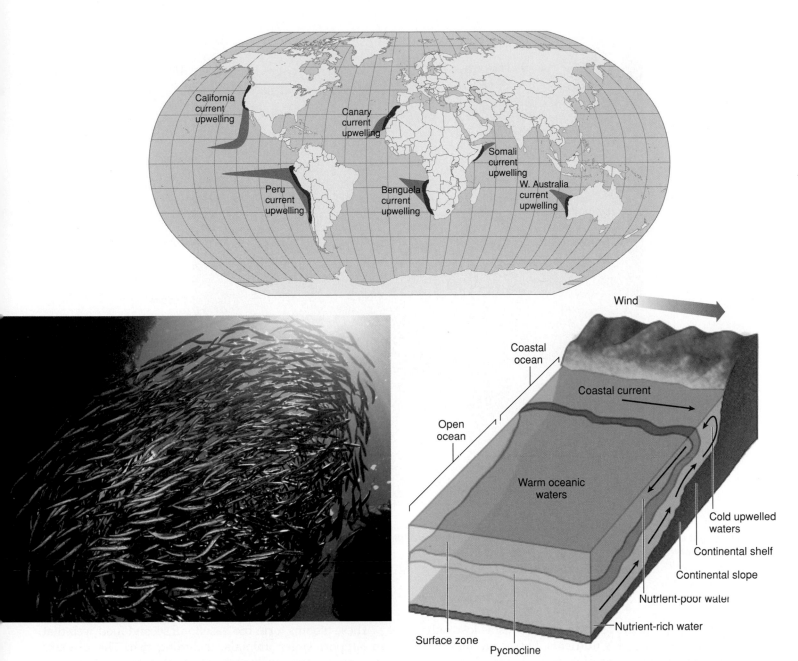

FIGURE 7.25 *Upwellings Upwellings tends to occur on the western coasts of continents. Here winds blow surface waters away from the shore, which allows nutrient-rich bottom waters to move toward the surface. The combination of light and nutrients supports high rates of net primary production, which move through the food chain to support large fish populations.* (Sources: Redrawn from J.L. Sumich and J.F. Morrissey, Introduction to the Biology of Marine Life, Eighth Edition, Jones & Bartlett.)

oligotrophic lakes, light penetrates the open ocean to great depths because phytoplankton populations are small. This implies that most of the world's oceans have low rates of net primary production.

Life in the open ocean is separated vertically (Figure 7.26). Most **pelagic** (open water) species live in the **epipelagic** zone, which roughly coincides with the euphotic zone. Below the **mesopelagic** zone extends up to 1,000 meters toward the bottom. Here the light is very dim. Bottom-dwelling species

live in almost complete darkness: The only light comes from living organisms, which use light to recognize other members of their own species and to lure potential prey. The deepest layer of the ocean is known as the **hadal** zone.

The pelagic zone hosts two fairly separate food webs. One web consists mainly of very small phytoplankton (termed picoplankton), which are less than 2 microns in size, and small protozoa that feed on the plankton. This web accounts for a significant fraction of the total net primary

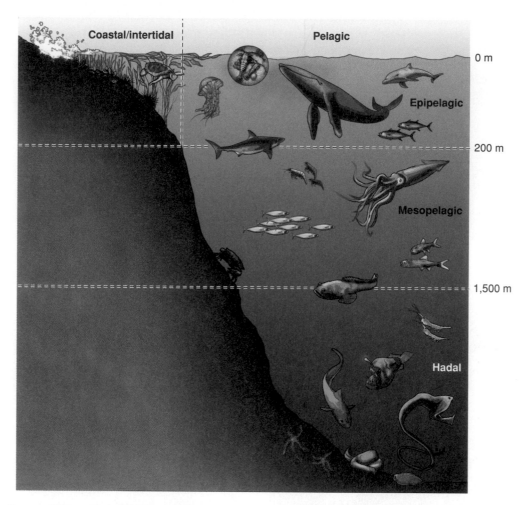

FIGURE 7.26 *Open Ocean Differences in light, temperature, and pressure shape the organisms that live in various ocean layers. The availability of light in the euphotic zone means that energy for food webs in the upper layers comes from photosynthesis. In the hadal zone detritus from the upper layers is an important source of energy for food webs.*

production in open oceans because the organisms lose only a small fraction of the web's nutrients. Losses are minimal because the small organisms sink very slowly—less than 0.1–0.2 meters per day. Despite the high rates of production relative to the other web, the small size of the autotrophs prevent them from supporting larger forms of life such as fish. Too many trophic positions would be needed between the tiny picoplankton and larger fish.

Scientists are just now starting to understand a second food web, although they do not agree on its size and importance. This second food web is based on phytoplankton that are one to three orders of magnitude (10 to 1,000 times) larger than the picoplankton. Larger plankton are present in small numbers in surface waters throughout the world's oceans. Most of the time nutrients limit their net primary production. But sometimes nutrients mix into the euphotic zone, and when they do, these larger primary producers bloom (increase their population greatly) for short periods.

These blooms form the base for a second food web that can support larger animals, including fish. The primary consumers in this web include zooplankton that migrate vertically each day. During the daylight hours, migratory zooplankton reside just below the epipelagic zone. In the mesopelagic zone the dim light hides them from potential predators. As the sun sets, migratory zooplankton move up to the epipelagic zone, where they feed on plankton. As the sun rises, they move back toward the mesopelagic zone and restart the cycle.

Blooms that support the zooplankton and larger animals last for only short periods. Because of their larger size and their tendency to form clumps, phytoplankton sink much faster (10–20 meters per day) than picoplankton. As they sink, they take the nutrients out of the euphotic zone, and this ends the bloom. Without nutrients, phytoplankton go into a dormant state and may become active only when they return to an area of sufficient light and nutrients.

SUMMARY OF KEY CONCEPTS

- Adaptations are behaviors or capabilities that allow species to thrive in some areas and fail in others. The area where a species lives is its habitat, and its role in that habitat is its niche. Two species generally cannot share the same niche indefinitely. Natural selection favors adaptations that allow species to avoid competition by changing their niche relative to similar species.

- A species' range is determined by the area where it can obtain enough energy for growth, storage, maintenance, protection, and reproduction. Its ability to obtain this energy is determined by how well its adaptations suit a given climate and the other species that live in the area. Because climate and species vary over the globe, most species live in small areas relative to the entire terrestrial or aquatic environment.

- On land, the geographic distribution of plants is determined by the availability of light, water, and temperature. The availability of light is determined primarily by latitude and secondarily by the structure of the local plant community. The availability of water is determined by the balance between precipitation and potential evaporation. This balance is correlated with latitude. Finally, temperature has its greatest effect by determining the length of the growing season. This too generally varies by latitude.

- The shape and life span of leaves varies among biomes. Leaves are large in areas where water is abundant and small where water is scarce. Leaves have a relatively long life span where climate conditions permit positive rates of net primary production year-round or where the growing season is too short for leaves to regrow. Deciduous trees are found in areas where short growing seasons or long dry seasons limit the period for positive rates of net primary production.

- The activity of terrestrial animals also is shaped by the length of the growing season. Where the growing season extends throughout the year, many animals are active year-round. In areas where water or temperature limits the growing season, animals often avoid these limits by hibernating or migrating.

- Aquatic biomes are shaped by salinity, temperature, and the availability of light, nutrients, and oxygen. The physical characteristics of water separate light, which is most abundant at the surface, and nutrients, which are most abundant at the bottom. This spatial separation is reinforced by density differences among top and bottom waters. The most productive aquatic ecosystems are located where the water is shallow, where winds or water currents bring bottom waters to the surface, or where rivers meet the sea. Animal life in aquatic biomes is most abundant where a solid substrate provides a place of anchorage and rocks or corals create vertical structure.

REVIEW QUESTIONS

1. Without the effects of disease or disturbance, why is it impossible for two species to share the same exact niche for extended periods? Why might natural selection favor changes that reduce interactions between species rather than a battle to the finish in which one species becomes extinct?

2. Explain how temperature and the availability of light and water affect the shapes and life spans of leaves.

3. Explain why light and nutrients often are separated in aquatic biomes. Explain the mechanisms that create areas where light and nutrients are abundant in the same layer of the water column.

4. Explain how the combination of climatic conditions and the distribution of endotherms affects the range of ectotherms in the United States.

5. Why is a solid bottom important in aquatic biomes?

KEY TERMS

adaptations

benthic

biome

channel

community

competitive exclusion principle

convergent evolution

deciduousness

ecosystem

epifauna

epilimnion

epipelagic

estuary

euphotic zone

eutrophic

environmental gradients

filter feeders

germinate

growing season

habitat

hadal

hypolimnion

infauna

intertidal zone

lentic

limnetic zone

littoral zone

lotic

mesopelagic

mudflat

niche

oligotrophic

payback period

pelagic

permafrost

phreatic zone

potential evaporation

prairies

reproductive range

riparian zone

rooting depth

salinity

salt marsh

specializations

survival range

thermocline

upwellings

8 SUCCESSION

How Do Ecosystems Respond to Disturbance?

STUDENT LEARNING OUTCOMES

After reading this chapter, students will be able to

- Describe the characteristics of a disturbance that influence their impact on the environment and living organisms.
- Describe the mechanisms that cause plant and animal communities to change over time following a disturbance.
- Describe why ecosystems do not always return to their original state following a disturbance.
- Describe ways in which scientists evaluate how human activity is stressing ecosystems.
- Describe the ways in which people try to restore ecosystems to their original state following human disturbance.

Before and After *A view of Mount St. Helens on May 17, 1980, which was one day before the eruption, and May 18, 1980 which was the day that the volcano erupted.*

single prey species will vary strongly with the population of its main prey species. Population changes from one trophic position to the next are termed **trophic cascades.** An ecosystem's stability also is enhanced if a prey species is eaten by several predators. Under these conditions there will be a relatively small change in the prey population should something happen to one of its predators. Conversely, the prey population will vary directly with the predator population if it interacts strongly with that predator.

This source of diversity is illustrated by food webs in temperate estuaries. Here Deborah Finke and Robert Denno set up food webs that had a single autotroph (marsh-grass), a single herbivore (planthoppers), and one or more predators—three spider species and the mirid bug (Figure 8.15). Less diverse food webs displayed various types of trophic cascades relative to more diverse systems. For example, webs that included one predator, the mirid bug, had fewer planthoppers and more plant biomass than webs that also included the three spider species. These results indicate that increasing the diversity of predators diminishes the impact of predation on herbivores and damps the cascading effects on marshgrass, which is the base of the food web. For a description of how food chain diversity is affected by disturbances, see *Case Study: Disturbance in Aquatic Food Chains.*

Helping Ecosystems Heal: Ecological Restoration

As we have just described, human activities can reduce ecological resistance and resilience. Furthermore, even healthy ecosystems tend to recover slowly relative to human wants. This has spawned a new science called **ecological restoration,** which is the process of reestablishing to the extent possible the structure, function, and integrity of indigenous ecosystems and sustaining the habitats they provide. To calculate your effect on natural ecosystems, see *Your Ecological Footprint: How Much Land Do You Disturb?*

The definition of ecological restoration is deceptively simple. Those restoring an ecosystem must know what it looked like before human disturbance. This is easier said than done! Some of the controversy about forest management in the American Southwest centers on what the forests looked like before Europeans arrived. There is some evidence that tree density was much lower than today (Figure 8.16). Reducing tree density so the forests look like those that prevailed before the arrival of European settlers is one rationale for government policy to increase harvests.

Equally important, in what ways should a restored ecosystem look like the original? Should we focus on restoring the original species, or should we focus on restoring the ecosystem's capabilities? Scientists also need criteria to measure their success. They know that a restored ecosystem will never be the same as the original ecosystem, but how close is close enough? Many argue that restoring a forest of Ponderosa pines must also restore the original pattern of fire occurrence. This pattern is thought to have included

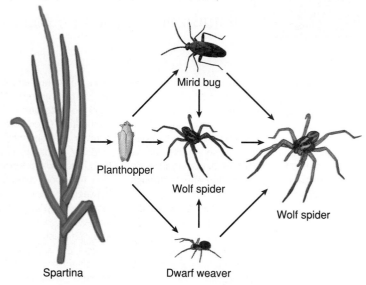

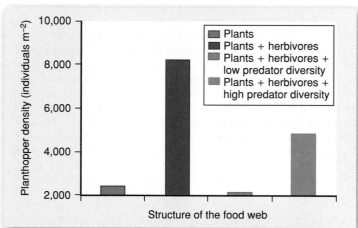

FIGURE 8.15 *Trophic Cascades Scientists manipulated salt marshes to set up food webs with either a few predator species (low predator diversity) or several predator species (high predator diversity). Food webs with high predator diversity had more planthoppers than food webs that had low predator diversity* (Source: Redrawn from D.L. Finke and R.F. Denno, "Predator Diversity Dampens Trophic Cascades." Nature 429: 407–410.)

frequent, low-intensity fires that had relatively little effect on healthy adult trees but killed many young trees. For example, there is some evidence that only one to four trees were established per hectare per decade in the three centuries before the 1870s. Since then the rate has increased to hundreds or thousands of trees, which may have increased the density of current forests. To evaluate the effect of fire suppression, see *Policy in Action: Does Suppressing Small Fires Create Large Fires?*

Consistent with the differences between primary and secondary succession, the first step in ecological restoration often includes efforts to reestablish healthy soil. For example, introducing nitrogen-fixing plants such as peas and clover is the first step in converting land covered by coal wastes to pastureland. In areas contaminated by

CASE STUDY

Disturbance in Aquatic Food Chains

Much of this chapter describes disturbance and succession in terrestrial ecosystems. But disturbances also initiate succession in aquatic systems. As on land, the effects of disturbances on aquatic ecosystems can be quite complex. To understand these effects, ecologists quantify the effect of a disturbance on individual species and trace the effects through the food chain.

In the river systems of northern California, the most important primary producers often are single-celled algae that adhere to rocks and other parts of the river bottom. These algae are consumed by two types of grazers: those that are susceptible to predation and those that are resistant to predation. Grazers that are resistant to predators include the larvae of the caddisfly, which are too large for most predators to eat. Furthermore, they protect themselves with strong protective cases by "gluing" together small stones. Other grazers such as the larvae of midges or mayflies are smaller and have no protective cases, so they are vulnerable to predation by fish.

These four groups are linked via a food chain that can be simplified as follows (Figure 1). Algal biomass depends on the amount of light, which increases photosynthetic rates and therefore increases algal biomass and the population of grazers, both resistant and nonresistant, that eat algae and therefore reduce algal biomass. The population of nonresistant grazers is related positively to algal biomass but is related negatively to predator biomass (pred-ators eat nonresistant grazers). The population of resistant grazers also is related positively to algal biomass but is not related to the predator population (predators don't eat a significant number of resistant grazers). Finally, the population of predators is positively related to their food, nonresistant grazers.

The food chain describes the direct links between species, but a computer model of this food chain reveals some indirect links. There is an indirect link between the population of resistant and nonresistant grazers. Because both sets of grazers eat algae, a population increase in one set of grazers implies a population decrease in the other set of grazers. An increase in the population of resistant grazers increases the quantity of algae they consume, which reduces the food supply for the nonresistant grazers.

The competition for food between resistant and nonresistant grazers also generates a negative relationship between the populations of resistant grazers and predators. An increase in the population of resistant grazers reduces the population of nonresistant grazers. A reduction in the population of nonresistant grazers reduces the predator population. Conversely, increasing the predator population reduces the population of nonresistant grazers, which increases the population of resistant grazers.

These indirect relationships imply that any disturbance that affects the population of resistant grazers also will affect the predator population. Ecologists find that floods reduce the population of the resistant grazers relative to the population of nonresistant grazers. The vulnerability of caddisfly to floods is caused by the same factors that make it resistant to predation. The heavy protective cases restrict caddisfly larvae to the river bottom. The river bottom is a dangerous place during a flood because rushing water rolls small stones along the river bottom, and these stones can crush the caddisfly larvae. Their large size, which protects caddisfly larvae from predation, also prevents them from hiding in small spaces where they could avoid the rolling stones. As a result, floods kill more resistant grazers than nonresistant grazers.

These links imply that floods should increase the predator population by reducing the population of resistant grazers, whereas a lack of floods should reduce the predator population by allowing the population of resistant grazers to increase relative to nonresistant grazers. These predictions are confirmed by experimental manipulations of streams. Experimenters find that adding resistant grazers, caddisfly larvae, to streams reduces algal biomass by 83 percent, reduces nonresistant grazer populations of midges and mayflies by 56 percent, and reduces predator populations by 23 percent. Similarly, streams that are managed to control floods have smaller populations of algae, nonresistant grazers, and predators than streams that are not managed for flood control.

FIGURE 8.16 *Tree Density, Then and Now* The density of forests on Mt. Trumbull, Arizona in 1870 (a painting by H.H. Nichols) and a photograph of that same area in 1994–1995. Notice that the forest is now more dense than 100 years ago.

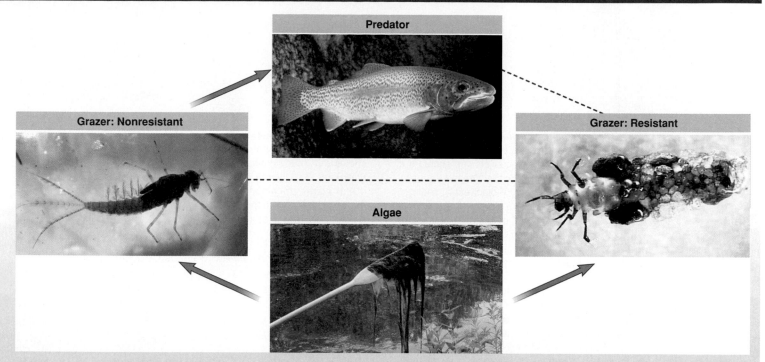

FIGURE 1 *Lotic Food Chain A simplified food chain from a lotic system in northern California. Direct flow of energy between species is shown by arrows. Dotted lines indicate indirect links between species. As humans stopped floods, the population of resistant grazers increased. This allowed them to eat more algae, which reduced food for nonresistant grazers. As their population shrank, so did the population of predators such as rainbow trout.* (Source: Modified from J.T. Wooten et al., "Effects of Disturbance on River Food Webs," Science 273: 1558–1561.)

The indirect effects of disturbances on predator populations has important implications for efforts to restore salmon populations. To date, managers have concentrated on policies that affect the fish directly, such as opening dams that block migration. A complementary strategy may focus on flood control. By suppressing disturbances, flood control may reduce the salmon's food supply. By allowing floods, managers may be able to restore food supply and ultimately the fish population.

ADDITIONAL READING

Wooton, J.T., M.S. Parker, and M.E. Power. "Effects of Disturbance on River Food Webs." *Science* 273 (1996): 1558–1561.

STUDENT LEARNING OUTCOME

- Students will be able to describe how disturbances help maintain populations of competing species.

heavy metals such as nickel, the first step often involves plants such as pennycress that can accumulate heavy metals. The plants are then removed, which reduces the concentration of toxic materials.

Next scientists attempt to reestablish the indigenous ecosystem using principles that are based on ecological succession. For example, succession was relatively ineffective in reestablishing the climax community in areas of Venezuela that were disturbed by building roads. Scientists found that the disturbed areas were colonized by a fungus that inhibited succession. To get around this problem, scientists inoculated the area with the "right" type of fungi, and this accelerated succession. Similarly, scientists were able to accelerate the rate of revegetation on the Fresh Kills landfill on Staten Island (where New York City disposes of much of its solid waste) by planting a limited number of trees. Although these trees did not spread, they facilitated

succession by providing roosting sites for birds. The birds carried seeds from native trees onto the landfill, which accelerated succession.

Marine biologists use the same principles to restore coral reefs by transplanting coral from healthy reefs. This process has its drawbacks—the transplanted coral has a high mortality rate, and the process stresses the healthy reefs from which the coral is removed. To avoid these problems marine biologists are experimenting with techniques to farm coral for transplantation. To date the results have been encouraging—in one case scientists found that the lagoon portion of the reef could be used to grow loosely scattered corals. Furthermore, when transplanted, these corals had high survival rates and helped restore some aspects of the reef.

The most ambitious restoration project involves Everglades National Park at the southern tip of Florida.

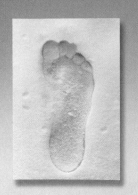

Your ECOLOGICAL *footprint*

How Much Land Do You Disturb?

Human activities disturb the environment in many ways. Our actions alter biogeochemical cycles, emit air and water pollutants, and convert natural ecosystems to human uses. These conversions generate a variety of environmental challenges. As described in the fifth and sixth sections of this book, these challenges include loss of biodiversity, changes in global climate, loss of forests, and soil erosion. Here we provide information that will allow you to calculate the amount of land that you disturb and therefore the size of your footprint on natural ecosystems.

To understand the impact of your footprint, we need to differentiate land use and land cover. Land cover describes the community that occupies a given landscape. Land cover includes both natural and human systems. Examples of natural land covers include forests and grasslands, whereas human land covers include urban areas such as houses and roads. Land use refers to how a natural or human land cover is used. Wildlife reserves, timber harvesting, mining, and agriculture are types of land use. Land use does not map directly to land cover, and vice versa. For example, forest land cover can be used as a wildlife refuge or as a source of timber. Conversely, agricultural land use can include wheat fields that look like grasslands or tree plantations that look like forests.

The difference between land use and land cover is critical because human activities convert natural land covers to particular land uses, but many land uses are not sustainable. For example, forests (land cover) are used as a source for timber (land use). But after several decades most of the economically valuable trees have been harvested, and logging may no longer economically viable. Similarly, grasslands (land cover) are converted to agriculture (land use); but continuous agriculture increases soil erosion and reduces fertility, and this reduces the economic viability of using that landscape as farmland.

The limited life span of many human land uses implies that (1) human activity continuously converts natural land covers to human land uses and (2) human activity creates a new category of land cover—disturbed land. Disturbed land includes landscapes that are abandoned because they cannot be used for their original land uses. As such these lands are left to succession, which may or may not be able to regenerate the communities that made up the original land covers.

Calculating Your Footprint

Here we focus on the area of natural land covers that are converted to human land uses to supply the goods you consume. You can approximate one component of this disturbance from your use of net primary production and the rate of net primary production of natural land covers. Table 1 in Chapter 6's Your Ecological Footprint box (page 107) allowed you to calculate the amount of net primary production required to generate your use of plant-based food, meat, eggs, milk, paper, and wood. Table 1 in Chapter 7's Your Ecological Footprint box (page 135) allowed you to identify the biomes from which many of your environmental goods originate. Table 1 here lists the average annual rate of net production in these biomes.

Together this information will let you calculate the area needed to generate a particular environmental good. For example, Chapter 6's Your Ecological Footprint indicated that the average person in the United States consumes about 3,774 kcal of plant-based foods per day. If we assume that most of these foods are grains, Table 1 below shows that many grains are grown on land that was once temperate grassland. This biome has a rate of net primary production of about 408.6 grams of carbon per square meter per year. This would imply that your plant-based foods require an area of about 1,213 m² per year, which is calculated as follows:

$$1{,}214 \text{ m}^2/\text{year} = [3{,}774 \text{ kcal/day} \times 365 \text{ days/year} \times 0.36 \text{ grams carbon/kcal}]/408.6 \text{ grams carbon/m}^2/\text{year}$$

You should repeat this calculation with the other environmental goods listed in Table 1 from Chapter 6's Your Ecological Footprint. Be careful—the use of other environmental goods is listed on an annual basis, so you don't need to include the 365 days per year. When you are done, sum the values.

Interpreting Your Footprint

As will be described in Chapter 11, the average person living in the United States consumes more environmental goods than the average world citizen. Repeat the calculations with the global averages listed in Chapter 6's Your Ecological Footprint. How much larger is your environmental footprint than the global average?

The land area you just calculated probably understates the size of your footprint. If the land used to grow your food, paper, and wood is managed sustainably, your footprint equals the area you calculated. If the land is not managed sustainably, the total understates your foot-

TABLE 1	Average Annual Rate of Net Production in Biomes
Biome	**Net Primary Production (grams carbon per square meter per year)**
Tropical forest	1,014.92
Temperate forests	701.92
Boreal forests	211.68
Tropical savannas and grasslands	663.26
Temperate grasslands	408.63
Deserts	78.56
Tundra	97.27
Wetlands	1,228.57

Source: Data from I.C. Prentice et al., "The Carbon Cycle and Atmospheric Carbon Dioxide," eds. J.T. Houghton et al., in *Climate Change 2001: The Scientific Basis.* Cambridge University Press.

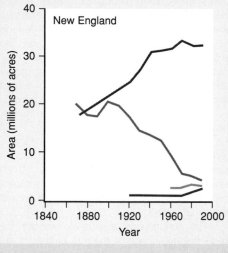

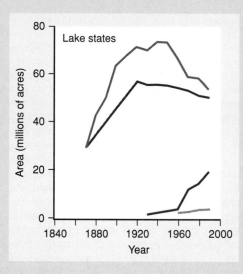

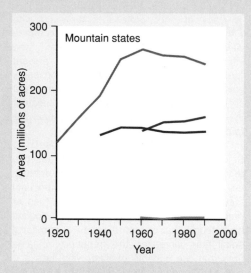

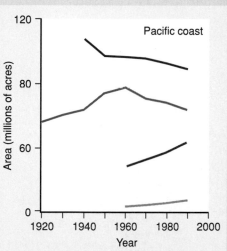

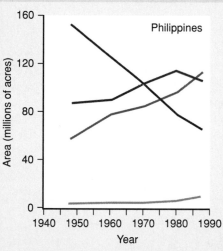

FIGURE 1 *Abandoned Land* The area of abandoned land (blue line) has been increasing in the United States and the Philippines. In some places abandoned land is roughly similar to the area in farmland (green line), forest areas (red line), or urban areas (orange line). *(Source: Data from A.P. Dobson et al., "Hopes for the Future: Restoration Ecology and Conservation Biology." Science 277: 515–522.)*

print. Suppose that production per unit land area declines 1 percent per year. By how much will your footprint grow over the next decade?

If the land used to grow your food, paper, and wood is managed poorly, some areas may eventually be abandoned. Evidence indicates that in many sections of the United States areas of degraded and abandoned land have increased over time (Figure 1). In the Pacific Coast region the area of degraded land is approaching the area occupied by active farms. In the Philippines the areas of degraded land is about the same as the area occupied by agriculture. And both of these categories are greater than the remaining area of forested land.

ADDITIONAL READING

Dobson, A.P., A.D. Bradshaw, and J.M. Walker. "Hopes for the Future: Restoration Ecology and Conservation Biology." *Science* 277 (1997): 515–522.

STUDENT LEARNING OUTCOME

- Students will be able to explain how the use of natural resources and land management determine the area of land used to provide food, paper, and wood.

POLICY IN ACTION

Does Suppressing Small Fires Create Large Fires?

Many of your professors grew up with government-sponsored television commercials that featured "Smokey the Bear," who informed listeners that "Only you can prevent forest fires." Consistent with that message, government managers would suppress (put out) forest fires as soon as they were detected. By doing so managers hoped to preserve the forest, save lives, and protect property.

But many of you have never seen a Smokey the Bear commercial. His absence reflects a change in policy. Government managers now allow some forest fires to burn until they reach a critical size or threaten lives or property. Why the change in policy? Ecologists have found that in some cases suppressing small fires increases the frequency of large fires that are capable of destroying lives and property. Feedback loops associated with fire

suppression work against the actual goals of fire suppression.

The counterintuitive effects of fire suppression may be illustrated by the size and frequency of wildfires in southern California and the northern portion of Baja California, which is part of Mexico. These regions have different policies regarding fires. Since 1892 California has actively suppressed fires. However, there has been little or no effort to suppress fires in Baja California.

Both southern California and Baja California have three types of ecosystems that are likely to burn: grasslands, coastal sage scrub, and chaparral. The effect of suppression varies by ecosystem. The likelihood and size of fires in grasslands and coastal sage scrub are determined mostly by precipitation during the growing season. Greater precipitation increase net primary production. High levels of production

create more biomass that can serve as fuel for a fire during the dry season, when the vegetation dries and dies. Because fires depend on events in that year's growing season, the effects of fire suppression do not carry over from one year to the next; so fire suppression policy has little effect on the frequency or size of fires in grasslands and coastal sage scrub.

The probability of fire in the chaparral ecosystem is determined largely by its age. Until the age of twenty years, chaparral stands contain relatively little dead material. This fraction increases with age, eventually reaching 50–70 percent. Dead chaparral biomass is an ideal fuel that increases the likelihood that a fire will start and spread.

Fire suppression may have reduced the frequency but increased the size of fires in the chaparral ecosystem. The frequency of fires is

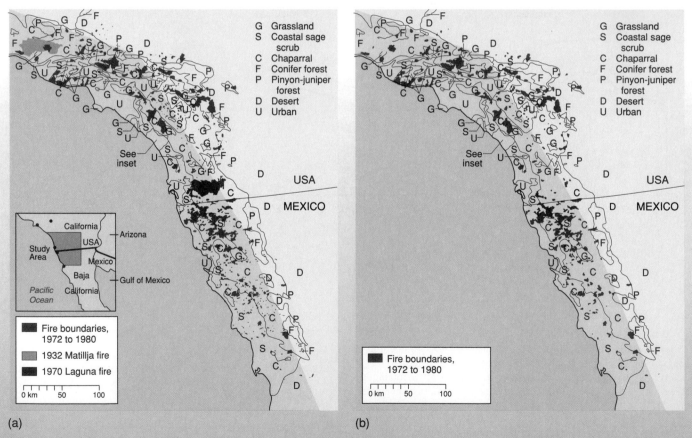

(a)   (b)

FIGURE 1 *Fire Suppression and Fire Size* (a) Some scientists argue that fire suppression in southern California reduced the frequency but increased the size of fires relative to Baja California in Mexico, where there is no fire suppression. (b) Other scientists argue that there is no difference in the size and frequency of fires when fires that occurred before 1972 are removed. (Source: (a) Data from R.A. Minnich, "Fire Mosaics in Southern California and Northern Baja California," Science 219: 1287–1294; (b) Data from J.E. Keeley and C.J. Fotheringham, "Historic Fire Regime in Southern California Shrublands," Conservation Biology 15(6): 1535–1548.)

reduced by suppressing fires shortly after they start. But suppressing fires allows trees to live longer than they would without intervention. This creates large areas occupied by older stands of chaparral, which are highly flammable. Under these conditions a fire can burn wide areas.

Without fire suppression, the many fires that ignite each year create a mosaic of chaparral stands of various ages, and this spatial mosaic tends to prevent larger fires. As a stand ages and the availability of fuel increases, the likelihood of fire increases. When a fire occurs, it destroys the older stands but is extinguished when it reaches younger, less flammable stands. Under these conditions a fire cannot spread. Fire perpetuates the mosaic by serving as a disturbance that reinitiates succession. After the fire, a relatively nonflammable stand of young plants grows in place of the burned stand. This younger stand acts as a firebreak when the neighboring stand of chaparral, which extinguished the first fire, is old enough to burn.

The effect of fire suppression on the size and frequency of fires between 1972 and 1980 can be seen in Figure 1(a). There were many small fires in the chaparral ecosystems south of the border, where there is no attempt to suppress fires. North of the border there were fewer fires, but individual fires were much larger. Skeptics argue that this pattern is spurious—if the large fires that occurred before 1972 are removed, the size and frequency of fires are similar (Figure 1(b)). Skeptics also assert that any differences in the frequency and size of fires are associated with differences in climate, land use, and population densities. In response, those who argue for the importance of fire suppression point to differences in stand ages north and south of the U.S.–Mexico border (Figure 2) and to the fact that this border is perpendicular to the differences in climate. Finally, they say that their hypothesis about the effect of fire suppression can account for all types of fires, whereas the skeptics' hypothesis can account for only fires associated with Santa Ana winds. (Remember the criteria for accepting and rejecting scientific hypotheses that are described in Chapter 3.)

If those who argue for the relevance of fire suppression are correct, forest managers in southern California may need to change their strategy to preserve the forest and protect lives and property. Managers may have to set fires that recreate the mosaic of different-aged stands. To do so, fires must be

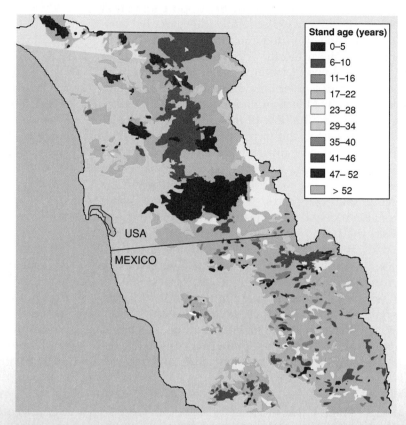

FIGURE 2 *Fire and Age Chaparral stands in southern California occur in large blocks that are the same age. In Baja California same-aged stands are smaller and mixed together. This difference is presented as evidence that fire suppression in southern California creates large stands of old chaparral that are highly flammable. (Source: Data from R.A. Minnich, "An Integrated Model of Two Fire Regimes." Conservation Biology 15(6): 1549–1553.)*

sufficiently large—about 1,000–2,000 hectares (2,471–4,942 acres), which is consistent with the median size of fires in Baja California. The rate at which they set fires must be increased to a level consistent with fires in Baja California, which burn 50,000–100,000 hectares (123,550–247,106 acres) per year. Finally, the fires should be set in early summer rather than the fall, when the Santa Ana winds can drive fires out of control.

ADDITIONAL READING

Keeley, J.E., and C.J. Fotheringham. "Historic Fire Regime in Southern California Shrublands." *Conservation Biology* 15, no. 6 (2001): 1535–1548.

Minnich, R.A. "Fire Mosaics in Southern California and Northern Baja California." *Science* 219 (1983): 1287–1294.

Minnich, R.A. "An Integrated Model of Two Fire Regimes." *Conservation Biology* 15, no. 6 (2001): 1549–1553.

STUDENT LEARNING OUTCOME

- Students will be able to describe how efforts to reduce the frequency of natural disturbances create the potential for larger disturbances.

Originally the Everglades was a huge wetland that allowed freshwater from Lake Okeechobee to flow slowly to the Bay of Florida (Figure 8.17). This flow generated a series of parallel ridges and sloughs, each about 500 meters (1,640 ft) wide. The ridges were high enough to support terrestrial ecosystems based on sawgrass, while the sloughs were filled with water and supported aquatic biomes. As human populations in southern Florida increased and agriculture expanded, people diverted water, which reduced the flow through the Everglades. Over time the sloughs filled in, and this eliminated much of the Everglades' aquatic ecosystems. Diversion also reduced the flow of freshwater to the Bay of Florida, which increased the bay's salinity. Together with the increasing amounts of nitrogen drained from agricultural systems, these factors are reducing the population of sea grasses, which are the base for the bay's food web.

To restore the Everglades, the South Florida Water Management District and the U.S. Army Corps of Engineers have devised the Comprehensive Everglades Restoration Plan. The plan, which will cost tens of billions of dollars, will increase the flow of water through the Everglades by removing hundreds of kilometers of canals and levees. The water will be rerouted through an expanded version of the Everglades, which will include an additional 24,000 hectares (59,280 acres) of farmland that will be purchased from private owners. The plan also includes the construction of a wastewater treatment plant that will remove

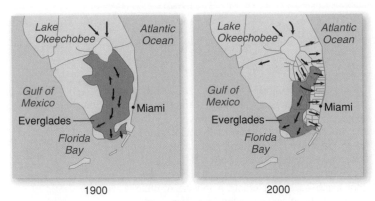

1900   2000

FIGURE 8.17   *Water Flow, Then and Now*   In 1900 water *from Lake Okeechobee flowed slowly to the Bay of Florida through the Everglades. In 2000 much of this water flows through diversions that bring water to people living on the eastern coast of Florida. As a result, less water flows through the Everglades, which has allowed the original channels, known as sloughs, to fill in.* (Source: Data from M. Schrope, "Save Our Swamp." Nature 409: 128–130.)

nitrogen and other nutrients. To prevent flooding, the government will drill hundreds of wells that will be used to store water during heavy rains. This last component has provoked considerable scientific skepticism because nothing on this scale has been done before. Despite the need to modify this and other parts of the plan, the restoration of the Everglades is moving forward because it is supported by a scientific and political consensus.

SUMMARY OF KEY CONCEPTS

- Ecosystems are not constant—they are affected constantly by disturbances. Disturbances vary in frequency and spatial scale, and their product determines the rate at which an ecosystem is turned over. Disturbance frequency and the variability around this frequency influence whether plants and animals have adaptations to cope with disturbances.

- Succession is the process by which ecosystems respond to disturbances. Succession can be fairly predictable because the microenvironment tends to change in an orderly process. These changes are important because they tend to determine the adaptations that are suited for an environment. In general, species that inhabit early successional communities tend to grow quickly and allocate much of their energy toward reproduction. Species that inhabit later successional communities tend to grow slowly and allocate less of their energy to reproduction. Changes in adaptations can affect the functioning of the entire ecosystem. In general, early successional communities tend to have higher rates of net primary production, be less diverse, and lose more nutrients than later successional communities.

- Succession is driven by facilitation, tolerance, and inhibition. Through facilitation, species change the microenvironment in ways that make it less hospitable for themselves or more hospitable for the species that will replace them. Some species have adaptations that allow

them to tolerate conditions that are unfavorable for other species; this generates successional changes between communities. Finally, some species create conditions that inhibit other species, which tends to stabilize a community and thereby slow succession.

- In some cases succession is replaced by catastrophic shifts between ecosystems. Each of the ecosystems is stable and so can persist for extended periods due to positive feedback loops. But some disturbances trigger positive feedback loops that establish an alternative ecosystem, which also is stable. Changes between these states are not symmetric—reversing conditions that led to the switch often will not change the ecosystem back to its original state.

- An ecosystem's health is determined largely by its ability to resist and recover from disturbances. Both traits are determined in part by diversity. Diversity of functional groups increases the ability of ecosystems to recover from disturbances, while diverse food webs and the prevalence of weak interactions increase an ecosystem's ability to resist disturbances.

- To help ecosystems recover from human disturbances, humans supplement succession through ecological restoration. These processes focus on reestablishing the soil and plant and animal communities. Ecological restoration usually works in conjunction with succession rather than replacing it.

REVIEW QUESTIONS

1. Explain the conditions under which an ecosystem with a high rate of disturbance may have a low turnover rate.

2. Explain why there is a nonlinear relationship between the spatial scale of disturbances and the time required for recovery.

3. Suppose that two species differ greatly in their generation time—twenty minutes for species A and twenty years for species B. Which species is more likely to develop adaptations to a disturbance that has a frequency of about fifty years?

4. Categorize the following species as members of early successional communities or late successional communities: roaches, dandelions, Siberian tigers, and oak trees. Explain your choices.

5. Explain the differences among facilitation, tolerance, and inhibition. Give at least one example of each.

6. Explain the difference between ecological stability, resistance, and resilience. How are they affected by diversity?

KEY TERMS

allelopathy

annuals

climax community

competition

disturbances

ecological resistance

ecological restoration

ecological stability

ecosystem health

facilitation

frequency

functional group

inhibition

light compensation point

light saturation point

macroenvironment

microenvironment

perennials

pioneer community

predictability

primary succession

redundancy

resilience

secondary succession

spatial scale

strong interaction

succession

tolerance model

trophic cascades

turnover rate

weak interaction

9 CARRYING CAPACITY

How Large a Population?

Signs of Distress *When Jacob Roggeveen arrived at Easter island in 1722, he did not mention that any of the moai were toppled. When Captain Cook arrived in 1774, he mentioned that some of the moai had been toppled from their platform. Analysts speculate that these toppled maoi indicate some form of social distress on the island.*

STUDENT LEARNING OUTCOMES

After reading this chapter, students will be able to

- Compare and contrast the effects of density-dependent and density-independent factors on the size of plant and animal populations.

- Describe the factors that determine carrying capacities for plant and animal populations.

- Explain how sanitary, economic, and social conditions affect the rate of population growth.

- Explain why the concept of carrying capacity for plants and animals can't be applied directly to human populations.

- Compare and contrast the notion of environmental limits as seen by resource optimists and resource pessimists.

The collapse of the Easter Island civilization that is described in Chapters 1 and 3 contains a moral, but the moral depends on who tells the story. To some people the toppled *moai* symbolize the dire consequences that befall human populations that grow beyond their environment's ability to support them. To others the toppled *moai* represent a failure to develop a technology or social order that could ensure survival on a tiny isolated island.

Choosing between interpretations depends in part on whether you believe that the biological concept of carrying capacity can be used to analyze human populations. **Carrying capacity** is the maximum number of individuals of a population that can be maintained indefinitely by the environmental goods and services of a given area of the environment without depleting the environment's ability to produce those resources or generate those services. If applied to the entire human population, the notion of carrying capacity implies that there is an upper limit to the number of people who can live sustainably on Earth.

If you believe that the biological concept of carrying capacity applies to people, the collapse of the civilization on Easter Island may foreshadow what is to come for the human population. According to this interpretation, the Easter Island civilization collapsed because it depleted natural resources and triggered a series of positive feedback

loops. There is no doubt that the human population dwindled and the supply of natural resources (trees, fertile soil, seabirds) shrank during the 1,300 years between the arrival of the Polynesians (A.D. 300–400) and the arrival of Europeans in 1722. The rate at which these natural resources disappeared was accelerated by a series of mistakes, like the introduction of the Polynesian rat, and rituals (the construction of statues) whose effect could not be stopped easily once started. As we describe in the fifth section of this book, the global human population is depleting many of its natural resources and environmental services and may have triggered some feedback loops that will be difficult to stop.

Some people believe that carrying capacity does not apply to people because human ingenuity and social institutions differentiate people from all other species. People are not like rabbits, increasing their population without regard to the supply of natural resources. According to this view, the Easter Island civilization collapsed because it could not avoid the consequences of population growth and resource depletion. Even after most of the resources were depleted, the civilization might have persisted if it had developed a social structure that favored cooperative management. Instead it made matters worse by adopting rituals that intensified competition for the few resources that remained. ■

HOW BIG A POPULATION?

Population Growth

Before we can explore whether there is an upper limit on population, we need to understand how a population grows toward this limit. A **population** is a group of individuals of the same species that inhabits a specific area. The rate at which the number of these individuals changes from one period to the next is termed **population growth.** By definition, population growth is equal to the number of births minus the number of deaths. This relationship can be expressed as follows:

$$\frac{dN}{dt} = \text{Birthrate} - \text{Death Rate} \qquad (9.1)$$

in which N is population size and the term $\frac{dN}{dt}$ represents the change in population in a given period (in "math speak" it says the change in population per change in time). Because

many animals and plants do not have live offspring (they lay eggs or produce seeds), you should interpret the term **birthrate** to mean the number of new individuals that are produced by a population in a period. **Death rate** refers to the number of individuals that die in a given period. A population shrinks when the birthrate is less than the death rate. Conversely, when the birthrate is greater than the death rate, a population grows.

The role of birthrate in Equation 9.1 implies that most populations grow in a nonlinear fashion. Because birthrate often depends on population size, large populations generally have greater birthrates than small populations. As such, large populations have the potential to grow by a greater number than small populations.

The positive relationship between population size and birthrate defines two types of population growth: geometric and exponential (Figure 9.1). **Geometric population growth** occurs in populations that produce a single batch of offspring in a year. Under these conditions, population

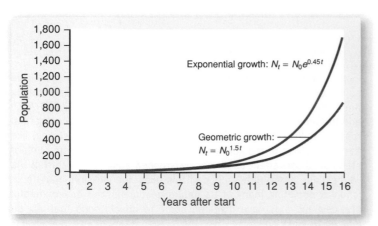

FIGURE 9.1 *Geometric and Exponential Population Growth*
Both geometric and exponential population growth are nonlinear because the change in population from one period to the next depends on the size of the population. Despite the example given, exponential growth is not always greater than geometric growth. In these examples N is the size of the population and t is the years after the start.

increases by a constant ratio from one generation to the next. Many insect populations, which produce a single batch of offspring per year, grow geometrically.

Other populations produce offspring throughout the year. Continuous population growth is known as **exponential population growth.** Exponential population growth can be represented by the following equation:

$$\frac{dN}{dt} = r_{max}N \qquad (9.2)$$

in which r_{max} is the **maximum intrinsic growth rate.** This is defined as the maximum per capita rate at which a population can grow under ideal conditions. For example, if Equation 9.2 has a yearly frequency, a value of 2.0 would indicate that a single individual could produce two new individuals per year. As you might expect, many species have values for r_{max} that are significantly greater than 1. Chapter 5 describes the so-called r selected species, which have very high values for r_{max}. Examples of r selected species include roaches and dandelions.

A population that grows exponentially can be described by the following equation:

$$N_t = N_0 e^{r_{max}t} \qquad (9.3)$$

in which N_0 is the starting population (at time zero), N_t is the population t periods after the start, and e is the base of the natural logarithm (about 2.72). If a population starts with four individuals ($N_0 = 4$) and has an r_{max} of 2.0, the population would be nearly 30 within two years ($N_t = 4 \times 2.72^2$).

Within a decade, which is not very long from a population perspective, the population could reach 2 billion (you should verify this number using Equation 9.3).

As indicated by Figure 9.1, both geometric and exponential population growth can generate very large populations within a relatively short period. Do populations really grow this quickly? Yes, under some conditions. About 9,500 years ago melting glaciers created vast areas of open space in the British Isles. This space was colonized in part by the scots pine; the scots pine population grew exponentially for about 500 years. Similarly, the whooping crane population has been growing exponentially since the 1940s, when efforts were made to protect both the birds and their summer and winter habitats.

These examples of exponential growth may seem obscure—because they are! Exponential growth is possible only under ideal conditions. Ideal conditions include low **population density,** which describes the number of individuals per unit area, and abundant supplies of the factors that limit growth. Such conditions are present for short periods. These periods include the colonization of new areas, a sudden increase in food supply, or the implementation of a new protection regime.

Limiting Factors

Beyond these conditions, populations usually grow more slowly than their maximum intrinsic rate of growth. And many populations do not grow at all because of some **limiting factor.** A limiting factor is the factor that is in least supply relative to the needs of a population. As such, a limiting factor is similar to a limiting nutrient. Scientists divide limiting factors into two categories: density-independent factors and density-dependent factors. **Density-independent factors** are factors that affect population growth but are not related to the size of the population. Density-independent factors often are associated with weather. Chapter 8 described how storms topple trees that live in climax communities, opening a gap that creates an opportunity for population growth by species that are associated with early successional communities. Storms are not related to population size: The size of a population does not increase or decrease the likelihood of a storm. Instead the frequency and severity of hurricanes or other density-independent factors are determined by chance—storms are stochastic events.

The other category of limiting factors is **density-dependent factors.** This category includes factors that are related to population size. The size of a population affects the rate at which it grows or shrinks via factors such as predation, disease, and reproduction. A high population density increases the death rate due to disease because disease spreads more rapidly when individuals are crowded together. Similarly, a high population density fills the places where individuals can hide and so increases the death rate due to predation. Density-dependent factors tend to slow population growth as the population becomes larger. Conversely, density-dependent factors tend to enhance population growth when population density declines. These effects imply that there is a negative feedback loop between population density and the rate of population growth (Figure 9.2).

The effect of density-dependent factors on the rate of population growth can be represented by the following equation:

$$\frac{dN}{dt} = r_{max}N\left(\frac{K - N}{K}\right) \tag{9.4}$$

in which K is carrying capacity. Equation 9.4 has a name—the **logistic equation.** The logistic equation states that the change in population is determined by two factors: (1) the maximum potential growth rate, which is represented by the term $(r_{max} \times N)$; and (2) the importance of density-dependent factors, which is represented by $\left(\frac{K-N}{K}\right)$. The first term indicates that everything else being equal, a large population has the *potential* to add more new individuals than a small population. But everything is not equal. As population grows, density-dependent factors become more important. Their importance is represented by the difference between carrying capacity (K) and population size (N). As population N approaches carrying capacity K, the term $\left(\frac{K-N}{K}\right)$ shrinks, which slows the rate of population growth. The slowdown in population growth that occurs as a population approaches its carrying capacity is known as **environmental resistance.** (This should not be confused with the term *ecological resistance*, described in Chapter 8, or the general notion of resistance, which is the ability of a system to withstand a disturbance.)

The effect of population size and density-dependent factors on population size and rate of growth can be described by two curves. The S-shaped logistic curve in Figure 9.3(a) describes the *size* of a population (N) over time. The *growth* of a population over time is described by the bell-shaped curve in Figure 9.3(b). This curve is used to manage the sustainable harvest of biological resources such as fish. See *Case Study: How Much Harvest Is Sustainable?*

The relationship between these curves can be illustrated as follows. Suppose we start with a small population of frogs living in a pond, which is indicated by the low level of the S-shaped curve in section A of Figure 9.3(a). When the number

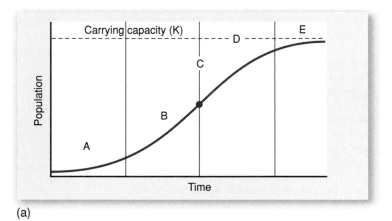

(a)

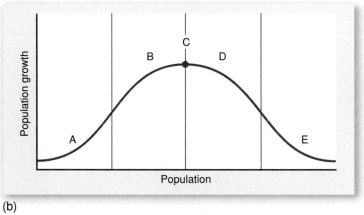

(b)

FIGURE 9.3 *(a)* **The S-shaped Logistic Curve** *This curve traces population size over time. Sections A, B, D, E, and point C identify changes in the importance of potential population growth relative to density-dependent factors. Density-dependent factors are least important in section A, but the increase in population is small because population is small. As the population expands (section B), density-dependent factors become more important, but the increase in population is larger because the population is bigger. At point C the increase in population is greatest. Beyond this point population continues to grow—but at slower rates because the importance of density-dependent factors increases faster than the size of the population. (b)* **The Bell-Shaped Curve** *This curve traces the rate of population growth over time. Population grows most slowly at small (section A) and large (section E) populations and grows most rapidly at intermediate populations (point C). For those of you who have taken calculus, the bell-shaped curve is the first derivative of the logistic curve and C is the inflection point.*

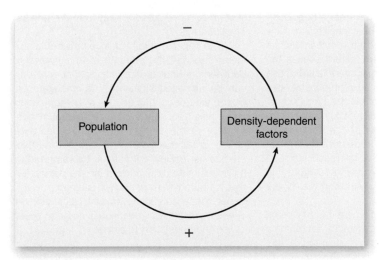

FIGURE 9.2 *Negative Feedback Loops: Population Growth and Density-Dependent Factors* *The negative feedback loop that includes population and density-dependent factors tends to stabilize population size at some set point. As population exceeds this set point, the increase in density-dependent factors tends to move population back to its original size. Conversely, reductions in population below the set point reduce density-dependent factors in a way that tends to increase population back toward the set point.*

CASE STUDY

How Much Harvest Is Sustainable?

Renewable resources such as food, wood, and fiber are generated by plant and animal populations. These populations can provide a sustainable supply *if* they are managed properly. Managers have two goals. Sustainability requires that the population not shrink over time. Within this limit managers want to produce the greatest amount of resource possible. These two goals imply that managers strive to steady the population at the level that generates the greatest supply year after year. This rate of production is known as the maximum sustainable yield.

A systems perspective can be used to understand how populations can be managed for maximum sustainable yield. To maintain a particular population size, the amount of resource harvested must equal the rate at which it is produced. For a fishery (a fishery is both the fish population and the people and equipment that harvest the fish) in which density-dependent factors predominate, the rate at which the fish population grows is given by a bell-shaped curve like Figure 9.3(b). At carrying capacity population does not grow. Without growth even a small harvest will reduce the size of the population. Ironically the negative feedback loop that includes population and density-dependent factors implies that sustainable harvests can be increased by reducing the population. As shown in Figure 1, a reduction in fish biomass from 900 kg to 800 kg reduces the importance of density-dependent factors, and this increases population growth. This allows sustainable harvest to increase from 75 to 80 kg.

There is a limit to which reducing population size increases the sustainable harvest. A population with a biomass of 600 kg corresponds to the peak in the bell-shaped curve, 87 kg of new fish biomass. This peak value of 87 kg is the maximum sustainable yield. To obtain the maximum sustainable yield, managers must maintain 600 kg of fish. Populations slightly greater than or less than this value grow more slowly than 87 kg. As a result, harvesting 87 kg when the population is greater or less than 600 kg is not sustainable. This means that managers must know the ideal population biomass and be able to maintain it.

In reality, managers do not know the population size or the rate of growth that corresponds to the maximum sustainable yield. Without accurate information about these variables, what will happen if managers harvest too

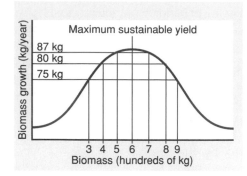

FIGURE 1 *Sustainable Harvests The relationship between fish biomass and sustainable harvest for a hypothetical population. Horizontal lines refer to the sustainable level of harvest at any population. For example, a population with a biomass of 800 kg generates 80 kg of fish that can be harvested sustainably. If population is managed to keep its biomass above 600 kg, the population can return to its managed level if managers make mistakes. But if the population is managed to keep biomass below 600 kg, the population can collapse if managers make mistakes.*

many or too few fish? Will these mistakes be amplified or corrected? In other words, is the population that they are managing stable?

If biomass exceeds the level that generates the maximum sustainable yield and if management mistakes are not too serious, a negative feedback loop will correct mistakes. To illustrate, suppose managers continually harvest 80 kg of fish each year from a population of 800 kg—the fishery is in equilibrium. Now suppose a temporary change in ocean circulation (a density-independent factor) reduces the population to 700 kg. If managers fail to notice this decline, what happens to the population?

A population with a biomass of 700 kg grows by 85 kg per year (Figure 1). If managers continue to harvest 80 kg of fish, the population will grow to 705 kg. If managers continue to harvest 80 kg per year, the population will continue to grow until it reaches 800 kg. At this level harvest will equal growth, and the fishery will be back to equilibrium.

Conversely, what happens to the fish population if a temporary increase in net primary production increases the population to 900 kg? The population increase boosts the importance of density-dependent factors, which slow the rate of growth. A population of 900 kg grows by 75 kg per year (Figure 1). If managers harvest 80 kg, the fish population will shrink by 5 kg per year. If managers continue to harvest 80 kg,

the population will continue to shrink until the population reaches 800 kg. At that level harvest will equal the growth rate, and the fishery will be back to equilibrium. As such, maintaining population above the level that generates the maximum sustainable yield ensures that small mistakes are self-correcting.

This caveat is critical. If the population is managed at a level equal to or below the population that generates maximum sustainable yield, small mistakes can cause the fishery to collapse. Suppose the fish population is managed to harvest 80 kg per year from a population of 400 kg. Although the harvest is sustainable, the population is less than the level that generates maximum sustainable yield, which is 600 kg.

What happens if a temporary change in ocean circulation reduces the fish population biomass to 300 kg? If managers fail to notice the decline in population, they will continue to harvest 80 kg, but a 300 kg population generates only 75 kg of fish. As a result, the population shrinks to 295 kg. This reduces the rate at which the population produces new fish, and the harvested fish biomass continues to exceed the biomass of new fish produced. This causes the population to shrink year after year. Eventually the population will collapse and the resource may be lost. In this case the equilibrium of 400 kg population and 80 kg harvest is unstable—a slight disturbance causes the system to move steadily away from the original equilibrium.

The negative feedback loop between population size and the strength of density-dependent factors implies that biological populations can be managed sustainably. If managers err on the side of caution and keep the population above the level that generates maximum sustainable yield, the negative feedback loop will correct small management errors. But the fishery has no resilience if fish are harvested from a population that is smaller than the population that generates maximum sustainable yield. Under these conditions any miscalculation or change in density-independent factors could cause the population to collapse.

STUDENT LEARNING OUTCOME

* Students will be able to explain how population size affects the sustainable level of harvest and the stability of this harvest.

of frogs (N) is small relative to carrying capacity (K), density-dependent factors allow growth rates close to the maximum. Rapid growth rates are possible because there are plenty of places to hide, there is plenty of food to eat, and the pond's water is clean. Even though the term ($\frac{K-N}{K}$) is nearly 1.0, the maximum rate of population growth is relatively small because the number of frogs capable of reproduction ($r_{max} \times N$) is small. The small increase in population is represented by the low level of the bell-shaped curve in section A of Figure 9.3(b).

As the frog population grows, density-dependent factors become less conducive to growth (section B). That is, the term ($\frac{K-N}{K}$) shrinks, but at the same time the larger population increases reproduction—the importance of ($r_{max} \times N$) grows. The increase in reproduction more than offsets the effects of density-dependent factors. As a result, the population grows at ever-increasing rates in section B. This continues until point C, which represents the fastest rate of population growth. This point corresponds to the steepest slope on the S-shaped curve in Figure 9.3(a). For those of you taking calculus, this is the inflection point—the point at which population growth changes from growth at ever-faster rates to growth at ever-slower rates. Beyond point C the balance between the reproduction and density-dependent factors reverses. In section D density-dependent factors have a larger effect on population growth than the increasing size of the population. As a result, population growth slows. Nonetheless, population continues to grow, although at an ever-slower rate (Figure 9.3(a)).

As indicated in section E, eventually population growth stops (Figure 9.3(a)). In section E population remains relatively constant, as indicated by the flat portion of the logistic curve in Figure 9.3(a). The flat portion corresponds to the carrying capacity (K) of that population.

DEFINING CARRYING CAPACITY

As defined previously, carrying capacity is the *maximum number* of individuals of a population that can be *maintained indefinitely* by the environmental goods and services that are generated by a *given area* of the environment. Each of these italicized phrases is described here to clarify the meaning of carrying capacity.

The Maximum Number of Individuals

The *maximum number* of individuals is determined by the balance between the amount of environmental goods and services that are required by each individual and the quantity of these goods and services that are provided by the environment. Dividing the availability of environmental goods and services by the quantity required by each individual indicates the maximum number of individuals that can be supported.

This quotient is illustrated with a simple example. Think about the factors that determine the maximum number of frogs that can live in a pond. Suppose the area of the pond is 240 square meters (2,582 ft²), and each female frog needs 2 square meters (21.5 ft²) to lay her eggs. This implies that the pond can support 120 female frogs, or 240 frogs in total if we assume that half the population is female (Table 9.1).

The pond is more than just a place to lay eggs, and carrying capacity may be less than 240 frogs if other environmental goods or services are less abundant than the space required to lay eggs. Continuing with this example, assume that the pond provides flies for the frogs to eat, space for the frogs to hide from predators, and clean water (Figure 9.4). If each frog needs 2 square meters to hide, the pond can support 120 frogs. If each frog needs 100 liters (264 gal) of water to dilute its waste, a 10,000 liter (2,642 gal) pond can support 100 frogs. In this case the pond's ability to dilute the frogs' wastes determines the maximum number of frogs. As such, the waste-processing ability of the pond is the limiting factor.

A Given Area

A *given area* of the environment is critical to the notion of carrying capacity because the availability of environmental goods and services varies with the type of ecosystem. As described in Chapter 7, reproductive range depends on factors such as soil moisture, net primary production, the length of winter, and competition from other species. As a result, the carrying capacity for a particular population depends on conditions that are unique to a local ecosystem.

TABLE 9.1	The Pond's Carrying Capacity for Frogs			
	Pond's Environmental Goods and Services			
	Hiding Places	Waste Processing	Egg-Laying Spots	Food
Pond Supply	240 m²	10,000 liters	240 m²	500,000 flies per year
Requirement per Frog	2 m²	100 liters	2 m² per mated pair	1,000 flies per year
Sustainable Frog Population	120 frogs	**100 frogs**	120 mated pairs	500 frogs

Value in bold indicates carrying capacity as determined by the most limiting environmental good or service.

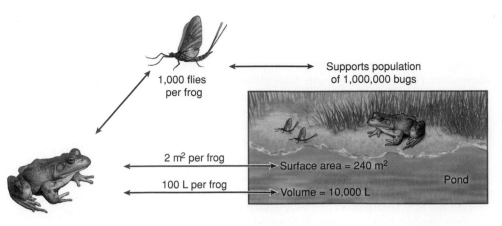

FIGURE 9.4 *Frog Carrying Capacity*
Carrying capacity is determined by the ratio of the frogs' needs relative to the pond's ability to provide those needs. The pond has an area of 240 m², holds 10,000 liters of water, and supports a population of 1 million flies. Each frog needs 1 m² to lay eggs and 100 liters to process its wastes, and eats 1,000 flies per year. By comparing these needs with the pond's ability to provide them in Table 9.1, we can identify the factor that determines the pond's carrying capacity for frogs.

We can illustrate the importance of the local environment by comparing the ability of two 10,000 liter (2,642 gal) ponds to support frogs. Suppose one pond is deep with a small surface area while the other pond is shallow with a large surface area (Figure 9.5). If the ponds' abilities to dilute wastes depend on their oxygen content, the shallow pond will be able to support more frogs because its large surface area allows oxygen to mix throughout the water column. (Why might waste dilution depend on oxygen content? Review the nitrogen cycle in Chapter 6.) Even though the ponds have the same amount of water, the shallow pond has a higher carrying capacity for frogs than the deep pond.

Maintained Indefinitely

Maintained indefinitely refers to the ability of an ecosystem to provide the same quantity and quality of environmental goods and services over time. For most of us, indefinitely means a very long time. When we speak of populations, indefinitely means many, many generations. In order for a population to be within its carrying capacity, the ecosystem must support the same number of individuals generation after generation. If the carrying capacity of a pond is 100 frogs, the pond must be able to support 100 frogs for hundreds or thousands of generations (or as long as the pond lasts).

There is a simple rule to determine whether a population can exist indefinitely: A population cannot use environmental goods or services faster than the environment provides them. Use of environmental goods or services within this limit ensures that the environment can sustain the population indefinitely. On the other hand, a population that uses environmental goods or services faster than they are generated by the environment is not sustainable. Nonsustainable use cannot be maintained indefinitely because it reduces a finite supply or degrades the environment's ability to generate those goods or services in the future. These changes reduce the maximum number of individuals that can be supported by a unit of the environment.

Returning to our previous example, suppose that food is the limiting factor in a pond and that each frog eats 1,000 flies per year. Furthermore, assume that each fly produces an average of 0.5 new flies per year (that is, the fly population increases geometrically with a ratio of 1.5). If there are 1 million flies in the pond, how many frogs can live in the

pond indefinitely? Your first guess may be 1,000 frogs. After all, there are 1 million flies and each frog eats 1,000 flies. One million flies divided by 1,000 flies per frog is 1,000 frogs.

But can 1,000 frogs live in the pond indefinitely? The answer is no. Put simply, 1,000 frogs will eat flies faster than the 1 million flies replace themselves. In the first year the 1 million flies will replace themselves and add an additional

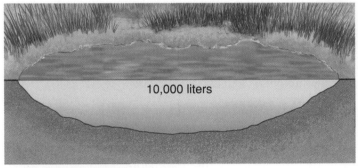

Shallow pond with large surface area

Deep pond with little surface area

FIGURE 9.5 *Not All Ponds Are Equal* *Ponds that have the same volume can have different abilities to process wastes generated by frogs. Less oxygen diffuses into the deep pond with little surface area relative to the shallow pond with a large surface area. This will allow the shallow pond to process more wastes than the deep pond.*

0.5 million flies. Under these conditions, the fly population is 1.5 million. But each frog eats 1,000 flies, so the 1,000 frogs will eat 1 million flies. As a result, 500,000 flies will remain at the end of the first year. The next year the 500,000 flies will produce 750,000 flies (500,000 × 1.5 = 750,000). But the 1,000 frogs will eat all of these flies and still be hungry! There will be no flies left, and the frogs will have nothing to eat in the second year. The frog population will collapse because they ate flies faster than the flies reproduced. Clearly the pond cannot support 1,000 frogs indefinitely.

The simple rule for sustainability requires that the frog population not eat flies faster than they replace themselves. Of the 1.5 million flies at the end of the first year, 1 million of those flies simply replace the previous generation. The remaining 0.5 million flies represent a potentially sustainable food source. Frogs can eat these 0.5 million flies without reducing the original population of 1 million flies. To calculate the number of frogs that can live in the pond indefinitely, divide the 0.5 million flies by 1,000 flies per frog. This calculation shows that the pond's fly population could support 500 frogs indefinitely.

Fluctuations around Carrying Capacity—The Negative Feedback Loop

The logistic curve is a very simple representation of population growth. According to the S-shaped curve in Figure 9.3(a), a population stops growing when it reaches carrying capacity. In reality, populations rarely remain at their carrying capacity. Rather, they tend to fluctuate around their carrying capacity (Figure 9.6). These fluctuations are generated by a negative feedback loop that includes time lags in

the relationship between the size of the population and the density-dependent factors that influence the rate of population growth (Figure 9.2).

Plant and animal populations tend to grow when the number of individuals is below the carrying capacity. When the frog population is below carrying capacity, there are extra supplies of food, there are plenty of places to hide, and there is relatively little waste to foul the water. Continuing with our example of the frog pond, suppose the frog population is 75, which is well below the carrying capacity of 100 frogs. When there are only 75 frogs, the pond can dilute and detoxify wastes faster than the frogs excrete wastes. These conditions ensure high water quality that allows the population to grow toward its carrying capacity (section A in Figure 9.7(b)).

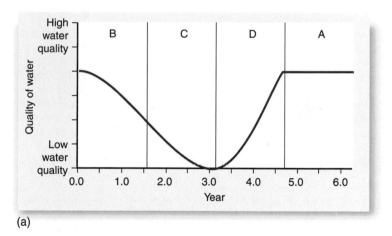

(a)

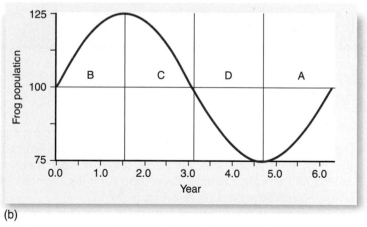

(b)

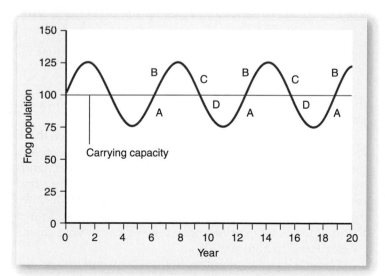

FIGURE 9.6   *Population Fluctuations* *Populations tend to fluctuate around their carrying capacity rather than reaching their carrying capacity and remaining there. The frog population is below carrying capacity in sections A and D. Population is greater than carrying capacity (overshoot) in sections B and C.*

FIGURE 9.7   *Population and Water Quality* *(a) Water quality starts to decline in section B as the frog population expands beyond its carrying capacity (b). Eventually (section C) the decline in water quality causes the population to decline. Population continues to decline below carrying capacity (section D) because water quality is low. Water quality starts to improve in section D because the population, which is below carrying capacity, generates wastes less rapidly than the pond can process them. This allows the pond to process wastes that accumulated when population was greater than carrying capacity in sections B and C. The resultant increase in water quality allows the population to grow (section A), which starts the cycle anew.*

POLICY IN ACTION

Should We Use the Logistic Curve to Manage Renewable Resources?

As described in this chapter's Case Study, using the logistic curve to manage renewable resources allows society to have its cake and eat it too. Up to the point of maximum sustainable yield, managers can increase sustainable harvest by increasing the amount of resource harvested. But is this really possible? Unfortunately there is relatively little evidence that the logistic curve accurately describes the growth of wild populations. As a result, we may be mismanaging some of our most important potentially renewable resources.

The most infamous application of the logistic curve is its use to explain changes in the deer population of the Kaibab Plateau in New Mexico. Between 1905 and 1937 the park supervisor and visitors estimated the size of the deer population. In the park supervisor's original form, the curve traced by these data does not really look like a logistic curve. Since the original publication, other authors have replotted the data in ways that made it appear that the data are consistent with a logistic curve. Over time the original graph was forgotten, and the retouched graphs became "proof" for the operation of the logistic curve. In reality scientific research has identified very few wild populations that behave according to the logistic curve.

The lack of evidence is troubling because the logistic curve is used to manage many renewable resources, especially fisheries. Using a logistic curve to manage fisheries was pioneered by Dr. William Ricker. He developed the Ricker curve, which shows a bell-shaped relationship between population and the number of juvenile fish that reach maturity (termed *recruitment*). This relationship implies that managers can increase sustainable harvest by reducing the fish population. This notion was accepted readily because it seems consistent with fish biology. For example, large salmon populations reduce the number of hatchlings because females who spawn late in the season destroy the nests of fish that have spawned earlier. Similarly, large salmon populations increase the number of big fish that cannibalize small fish.

But is there any evidence that fisheries can be managed reliably using the Ricker curve? Fishery biologists have produced a plethora of graphs that purport to show a bell-shaped relationship between population and recruitment. But do the bell-shaped curves really represent the data?

From a statistical perspective, the answer seems to be no. Look closely at the right sides of the bell-shaped curves in Figure 1. In most cases there are few data points. These observations are missing because fishing reduced the salmon population before managers started to measure population and recruitment. As a result, there are few examples to support the theory that recruitment declines as population nears its carrying capacity. In many cases a horizontal line would fit the data as well as or better than the bell-shaped curve. In the case of sockeye salmon, the data for Karluk seem nearly spherical—that is, the points look like a blast from a shotgun. In this case there may be no relationship between population and recruitment.

There are several reasons why the relationship between populations and recruitment may not look like a bell-shaped curve. Density-dependent factors affect population via several mechanisms, not all of which tend to depress growth when a population gets big. Big salmon stocks may increase population growth. As described in Chapter 6, salmon that return to their spawning rivers carry large quantities of nutrients (especially phosphorus). These nutrients are released into the water when the salmon die after spawning. Higher nutrient concentrations increase the river's net primary production. Higher rates of net primary production increase the food available for the young salmon. This positive feedback loop suggests that a hefty salmon population increases recruitment rather than decreasing it as the Ricker curve suggests.

Alternatively, the potential for a bell-shaped relationship between population and recruitment may be disrupted by density-independent factors. This may be the case for the recruitment of cod on George's Bank—a shallow part of the Atlantic Ocean off the New England coast where cod lay their free-floating eggs. The rate at which the hatchlings make it to reproductive

As the frog population grows and reaches its carrying capacity, population growth does not halt abruptly. Instead the frog population may continue to grow and thereby exceed its carrying capacity. Exceeding carrying capacity is termed **overshoot** (section B in Figure 9.7(b)).

Populations may overshoot their carrying capacity because of time lags in the relationship between population size and density-dependent factors. Continuing with our example, suppose the pond's population grows beyond 100 frogs. Beyond this number, the frogs produce waste faster than the pond can dilute and detoxify them. This causes wastes to accumulate in the pond (section B in Figure 9.7(a)). At first the decline in water quality may have little effect on the success of reproduction, so the population continues to grow beyond its carrying capacity.

Overshoot is temporary because the negative feedback loop between population and water quality will eventually reduce the frog population. As the frog population expands beyond 100, the frogs excrete more wastes than the pond can dilute and detoxify. As a result, water quality declines ever more rapidly. Eventually the water becomes so fouled that the population shrinks back toward its carrying capacity (section C in Figure 9.7(b)).

The decline in population does not end when the population drops back to its carrying capacity. Instead the population drops below its carrying capacity (section D in

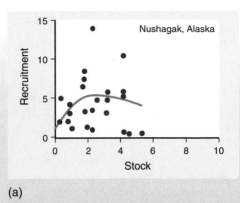

(a)

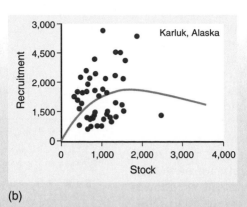

(b)

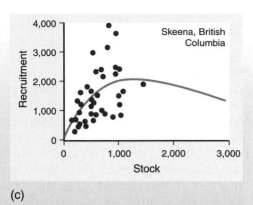

(c)

FIGURE 1 *Ricker Curves Ricker curves fit to real observations for sockeye salmon fisheries in (a) Nushagak, Alaska, (b) Karluk, Alaska, and (c) the Skeena region of British Columbia that appear in the scientific literature. The solid lines represent the bell-shaped relationships between the sizes of the salmon populations and recruitment (the number of juvenile fish that reach maturity). Notice that the observations, which are given by the circles, do not seem to fit the solid lines.* (Source: Data from C.A.S Hall, "An Assessment of Several of the Historically Most Influential Theoretical Models Used in Ecology and of the Data Provided in Their Support." Ecological Modeling 43(1–2): 5–31.)

age depends on winds. In some years swirling winds keep the eggs and the young cod on George's Bank. As described in Chapter 7, shallow waters have higher rates of net primary production, and the availability of food increases the fraction of the young cod that reach maturity. In other years winds blow the eggs and the young cod into deeper water. There lower rates of net primary production generate less food for the young cod, and a smaller fraction reach maturity. Because the wind (density-independent factor) is so important to the success of the young cod, this effect may overwhelm any bell-shaped relationship between population and recruitment.

Despite the appeal described in the Case Study, using the logistic curve to manage renewable resources may cause society to mismanage some of its most important resources. Even if we could measure population and recruitment accurately, a bell-shaped curve may not represent the relationship between these variables correctly. If there is no bell-shaped relationship, we may be incorrectly reducing the size of stocks in the vain hope that these reductions will increase sustainable harvests.

ADDITIONAL READING

Hall, C.A.S. "An Assessment of Several of the Historically Most Influential Theoretical Models Used in Ecology and the Data Provided in Their Support." *Ecological Modeling* 43, nos. 1–2 (1988): 5–31.

STUDENT LEARNING OUTCOME

- Students will be able to describe why the logistic curve may not be the most effective way to manage fisheries.

Figure 9.7(b)). Because the frog population exceeded 100 individuals during the period that corresponds to sections B and C, wastes have accumulated in the pond's water. These accumulated wastes can be eliminated only after the frog population has dropped below 100 frogs. Until these accumulated wastes can be processed, poor water quality continues to reduce the frog population in section D. Eventually the pond will process the accumulated wastes, and water quality will return to normal. At this point the frog population will grow and start the cycle again (section A in Figure 9.7(b)).

The size and frequency of such fluctuations may be large and unpredictable in real populations. The longer the time lag, the greater the fluctuation. The size and regularity of the fluctuations also depend on the relative importance of density-dependent and density-independent factors. See *Policy in Action: Should We Use the Logistic Curve to Manage Renewable Resources?* If density-dependent factors predominate, the fluctuations may be fairly predictable. On the other hand, if density-independent factors are more important, fluctuations may be irregular and can be dramatic. For example, an especially cold winter may kill a significant number of frogs regardless of the size of the frog population. Indeed, if the density-independent factors are frequent and relatively severe, a population may not reach its carrying capacity.

UNDERSTANDING HUMAN POPULATION GROWTH

The first modern attempts to count the human population were made near the end of the seventeenth century. Those early censuses put the human population at about 680 million (Figure 9.8). Getting there, the human population grew very slowly for hundreds of millennia. But since then the human population has grown at an extraordinary rate. Between 1700 and 1750, the human population grew at an annual rate of 0.25 percent, which corresponds to a **doubling time** of 280 years. By 1850 the growth rate doubled to 0.55 percent per year, and by the 1960s the growth rate reached a peak of 2.06 percent per year. At that rate the population doubles every 34 years. Since then the annual growth rate has declined to about 1.2 percent. Despite this slowdown the human population is expected to grow through most of the next century before it levels off between 9 and 10 billion people, although considerable uncertainty is associated with this forecast (Figure 9.9).

At the simplest level, the human population grows for the same reason that a frog population grows—the number of births exceeds the number of deaths. Human births are measured by the **crude birthrate,** which is the number of births per 1,000 people per year. Human deaths are measured by the **crude death rate,** which is the number of deaths per 1,000 people per year. *Crude* refers to the fact that birth and deaths are not adjusted for the number of women in their reproductive years or the age structure of the population. In 2004 the crude birthrate for the human popula-

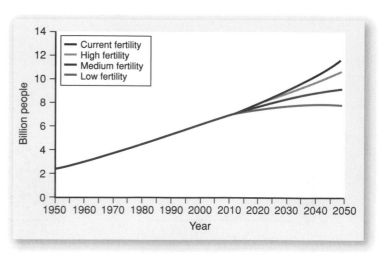

FIGURE 9.9 *Future of the Human Population* Projections *for the size of the human population depend on total fertility rates. Population grows fastest if fertility rates remain at their current level. Many analysts project that fertility rates will decline. The more that fertility rates decline, the smaller the population will be. If fertility rates decline to the lowest levels, populations will decline in the second half of the twenty-first century.* (Source: Data from the United Nations, World Population Prospects.)

tion was 21 per 1,000 and the crude death rate was 9 per 1,000. This means that the population grew at a rate of 12 people per 1,000, which is equivalent to a 1.2 percent annual increase. The annual increase in population is simply population size multiplied by the annual growth rate. In 2004 the world population was 6.4 billion, so the annual increase was 77 million people (0.012 × 6.4 billion = 77 million).

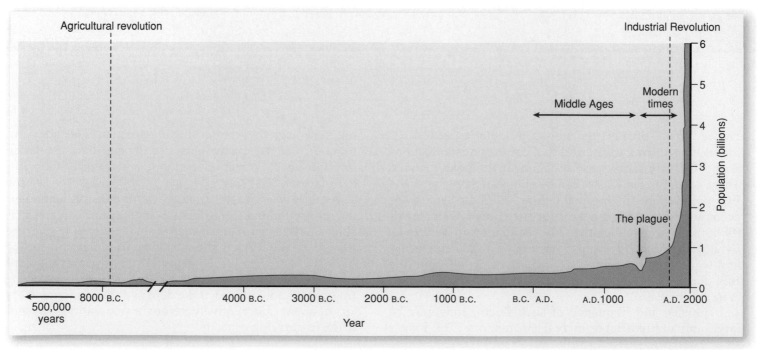

FIGURE 9.8 *Human Population Growth* For much of human history population grew very slowly. Starting with the Industrial Revolution, the human population has grown dramatically.

Crude death rates and crude birthrates are simple ratios, but they provide some powerful insights into the causes of population growth over the last two centuries. Crude birthrates have fallen over the last two centuries, but population has grown because crude death rates have fallen even faster. Thus to understand changes in the human population, we need to understand the factors that influence crude death rates and crude birthrates.

The Pattern of Births and Deaths

For the world as a whole, as well as for individual nations, the increase in the human population has been due primarily to lower death rates. These reductions are associated with changes that many of us take for granted. Public sanitation systems that provide clean water reduce crude death rates significantly. In nations without adequate sanitation systems, waterborne diseases such as typhoid fever, cholera, and dysentery kill more than 400 children *every hour*. (As described in Figure 18.16 on page 366, waterborne diseases are responsible for a significant fraction of all diseases in many developing nations.) Crude death rates also are reduced by the availability of adequate nutrition, clothing, and shelter. In populations without these basic necessities, millions of children under the age of 5 die each year due to illnesses caused by insufficient calories or protein in their diets. Finally, simple medical advances, such as the availability of antibiotics, reduce the number of people who die from epidemics and plagues. Modern medical advances such as heart bypass surgery and cancer treatments have a relatively small effect on the crude death rate and, ultimately, growth rates because many of the people saved by these advances will not produce any more children.

Births decline more slowly than deaths because births are influenced by a complex set of economic, cultural, political, religious, and demographic factors. The crude birthrate is influenced by the fraction of the population that is made up of women in their childbearing years (defined by many population scientists as ages 15–44) and the average number of children each of those women will bear over her lifetime. This average is known as the **total fertility rate** (TFR). **Replacement level fertility** is the total fertility rate at which a population remains constant. At this level each couple will bear the number of children needed to replace themselves. Replacement fertility must be greater than 2 because some children die before they reach reproductive age and some women have no children. In nations with low rates of infant mortality such as the United States, replacement level fertility is about 2.1. In nations with higher rates of infant mortality such as Uganda, replacement level fertility can be as high as 2.5.

The average total fertility rate for the world was about 2.7 in 2004. This means that today's 15-year-old girl is expected to have fewer than three children during her lifetime. Many factors influence the number of children a woman will have. Fertility rates tend to be lower in nations

where women have access to safe, reliable methods of birth control and information about how to use them. Fertility rates also tend to be lower where both men and women have access to education and employment opportunities, and where the costs of raising and educating children are high. Fertility rates tend to be higher in nations with high rates of infant mortality and where women marry and begin having children when they are young. Finally, a host of religious and cultural traditions influence family size.

The importance of these factors is mirrored by international differences in total fertility rates (Figure 9.10). In nations such as the United States and Germany, high income, access to contraception, and relatively equal opportunities for women favor low fertility rates. In nations such as India and Kenya, poverty, high infant mortality rates, lack of access to contraceptives, and the lack of educational and economic opportunities for women lead to high total fertility rates. Regardless of differences among nations, total fertility rates tend to decline over time. The current global total fertility rate of 2.7 is just over half its value of 5.0 in 1950.

Age Structure and Population Momentum

Population growth also depends on the number of women in their reproductive years. The distribution of women and men among age groups is illustrated by **age classes** and **age structure.** Age classes are the number of males and females in an age group. Age classes generally are divided into three

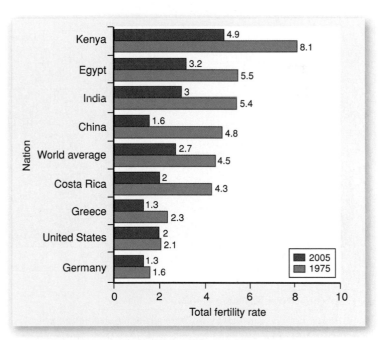

FIGURE 9.10   *Total Fertility Rates* Fertility rates vary among nations and over time. For most nations total fertility rates declined between 1975 and 2005. In general, total fertility rates in low-income nations such as Kenya are greater than total fertility rates in high-income nations such as Germany. (Source: Data from 2005 World Population Data Sheet, Population Reference Bureau.)

groups: pre-reproductive (ages 0–14 years), reproductive (15–44 years), and post-reproductive (ages 45 and older). Diagrams that represent the number of males and females in various age groups are known as **population histograms** (Figure 9.11).

The distribution of a population among these age classes determines its age structure. Populations in nations such as Mexico have large population fractions younger than 15. This distribution generates a histogram that has a pyramid shape (Figure 9.12(a)). As people in the bottom of the pyramid move into their reproductive years, the crude birthrate may rise significantly. This built-in potential for growth is called **population momentum.** On the other hand, the population of Denmark is spread relatively evenly throughout the age

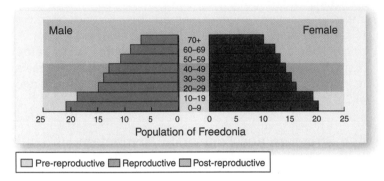

FIGURE 9.11 *Population Histogram* *A population histogram for the hypothetical nation of Freedonia. Red bars represent the number of females in the age group given by the numbers on the far side of the bars. Blue bars represent the number of males in the same age group. For example, the second blue bar from the bottom shows that there are just fewer than 20 million males between the ages of 10 and 19.*

classes (Figure 9.12(b)). Populations that display this rectangular shape have little tendency to grow. At the other extreme are histograms that look like inverted pyramids (Figure 9.12(c)). For European nations such as Italy, the number of people moving into their reproductive years is less than the number aging beyond their reproductive years. This tends to reduce the number of births, increase the number of deaths, and cause populations to shrink.

Age structure and population momentum also provide important insights into the socioeconomic conditions of the past and present. Over a decade of very high total fertility rates in the United States (and many European nations) at the end of World War II produced a large cohort of children known as the **baby boomers.** The baby boomers, born between 1946 and 1964, have generated a bulge in the age structure pyramid for the United States that has influenced culture at every stage. When the baby boomers reached school age (Figure 9.13(a)), more schools had to be built; but years later many of these schools closed due to declining enrollments. Baby boomers now make up nearly half of the U.S. adult population; they dominate the job market; they buy most of the goods and services; and they cast the most votes in elections (Figure 9.13(b)).

The United States will face some unprecedented challenges as the baby boomers reach retirement. This will be caused by changes in the **age–dependency ratio,** which is the number of people under the age of 15 plus the number of people over 65 divided by the number of people between those ages. This ratio is important because people between 15 and 65 hold most of the jobs, earn most of the income, and therefore support the young and the elderly. As the baby boomers age, the age–dependency

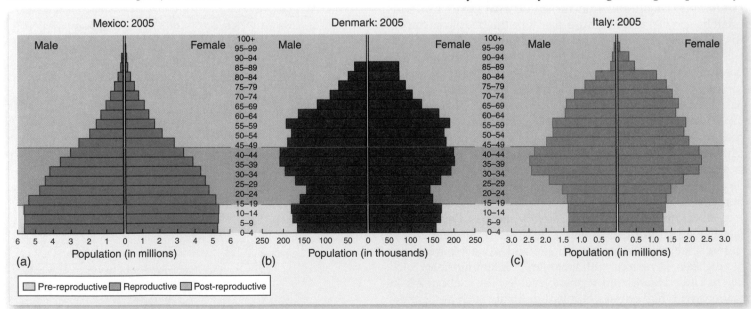

FIGURE 9.12 *Histograms and Population Growth* *(a) The population histogram for Mexico is shaped like a pyramid. This shape implies that the population will grow rapidly as the large number of children reaching their reproductive years exceeds the number of adults that age beyond their reproductive years. (b) Little or no growth is expected for nations that have a rectangular histogram, such as Denmark. Here the number of children that reach their reproductive years equals the number of people who age beyond their reproductive years. (c) Populations that have histograms shaped like an inverted pyramid, such as Italy, are likely to decline because the number of children aging into their reproductive years is less than the number of people aging beyond their reproductive years.* (Source: Data from U.S. Census Bureau.)

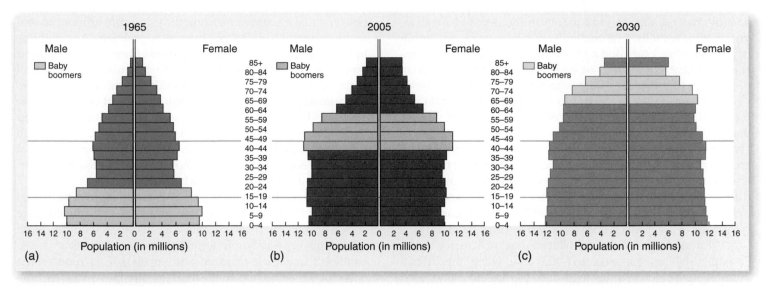

FIGURE 9.13   *U.S. Baby Boom* *Baby boomers include people born in the U.S. between 1946 and 1964. (a) By 1965 this huge group of children enlarged the base of the U.S. population histogram. (b) Forty years later (2005) they create a bulge of middle-aged people who are at the peak of their earning power. (c) By 2030 baby boomers will generate a bulge of older people who will collect Social Security and require considerable medical care. (Source: Data from U.S. Census Bureau.)*

ratio for the U.S. population will increase. Fewer people will work to support a large group of retired baby boomers. This will be especially problematic for the Social Security trust fund. The baby boomers have been paying into the fund to support their parents; but when they retire, this large bulge of older Americans will draw their checks from a relatively small group of workers (namely you, their children) who will have to pay relatively high taxes. The same problem may exist for health care. By the year 2030 more than 20 percent of the population will be over 65, increasing the demand for health care (Figure 9.13(c)). The additional cost of health care will fall on a smaller proportion of the population who will be working and paying taxes.

MODIFYING THE IDEA OF CARRYING CAPACITY FOR PEOPLE

Can we use the concept of carrying capacity to calculate an upper limit on the size of the human population? At first glance it would appear so. Just as we did for frogs, we could identify and measure all the environmental goods and services used by people and compare them to the rate at which the environment provides them. However, such calculations would be misleading because the relationship between people and the environment is far more complex than it is for frogs and other nonhuman populations. But this hasn't stopped people from trying.

The Malthusian Dilemma

The hypothesis that carrying capacity can be used to quantify the limits on human population was made famous by the economist-philosopher Thomas Malthus. In 1787 Malthus wrote an *Essay on Population* in which he described how the environment limits the size of human populations. Malthus's argument is simple: Like other animals, human populations grow exponentially (Figure 9.14). But food production, the factor that ultimately limits the size of the human population, grows linearly.

To Malthus exponential growth implied that population and food production were on a collision course because of a simple mathematical truth: Exponential growth eventually catches up to and surpasses linear growth (Figure 9.14). This implies that the human population will eventually exceed food production. At relatively low population densities food production is greater than the demand for food. Extra food allows the population to grow. Over time food production grows

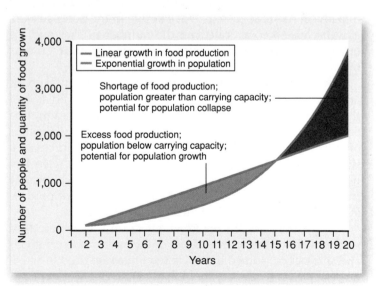

FIGURE 9.14   *Malthusian Collapse* *According to Malthus, the human population grows exponentially while food production increases linearly. Exponential growth by the human population will always catch up to and exceed linear growth in food production. When this happens, the population exceeds food supply, and the population must shrink back toward carrying capacity.*

linearly, but population (and therefore the demand for food) grows exponentially. Eventually the two lines cross, and population can exceed the food supply. When this occurs, Malthus predicted that famine and disease would reduce the human population to levels consistent with the food supply.

Malthus applied his theory to the English population and predicted that it was on the verge of collapse. Malthus thought that the English population was growing faster than its ability to grow corn. He argued that England should slow its population growth through voluntary means, such as postponing marriage. If the English failed to adopt voluntary means, Malthus warned that involuntary means would slow growth. These means included disease, starvation, and general economic misery—all symptoms of a society that has exceeded its carrying capacity.

Malthus's prediction of collapse has not materialized. Perhaps Malthus miscalculated the time when the human population would catch up to and exceed the supply of food (or some other environmental good or service). Alternatively, Malthus may have been wrong because people are so different than plants and animals that the concept of carrying capacity cannot be applied to human populations. We can understand some of these differences by returning to three phrases in the definition of carrying capacity.

The Maximum Number of Individuals and the Demographic Transition

Calculating the maximum size of nonhuman populations is relatively simple because each member of a plant or animal population requires about the same amount of environmental goods and services. Equally important, this quantity remains the same over long periods. We can calculate the maximum number of frogs that can live in the pond sustainably because we can calculate the number of flies and the quantity of wastes excreted by a frog each year. Furthermore, these requirements do not change much over time.

In one important sense, people use environmental goods and services just like frogs. Remember, the second law of thermodynamics states that matter and energy cannot be created or destroyed. Because of this limit, society depends on the environment to provide the natural resources that we use for food, shelter, and clothing. Similarly, society depends on environmental services to dilute and detoxify the wastes that humans generate in the process of producing and using that food, shelter, and clothing.

But people use the environment to do much more than feed and clothe themselves. Most of the environmental goods and services consumed in the United States are used in activities that are not necessary for biological survival. For example, most of the coal burned in the United States is used to generate electricity, but only a small portion of this electricity is used for biological survival, such as cooking and heating homes. Instead most of the electricity is used to produce goods and services such as cars and CD players—things that make life more enjoyable but are not necessary for survival. As a result, it does not make sense to measure the minimum amount of coal that is required to support a human population. You might be unhappy if the use of coal and other environmental goods and services were restricted to the amounts required for biological survival. Thus it is not possible to calculate an upper limit on the human population from a minimum set of environmental goods and services. And even if we could, this number may not be socially relevant. Few of us strive for a world in which the planet is filled with people who live at the limit of physical survival. People do live under these circumstances, but generally they are not satisfied by such an existence.

Indeed, there is evidence that feedback loops slow population growth when people have access to good health care and a modicum of economic well-being. As crude death rates drop, especially infant mortality rates, parents are more confident that their children will survive. This reduces the need for large families, so the total fertility rate drops. Income gains slow total fertility rates via several mechanisms. As economic opportunities increase for adults, the income earned by children becomes less important to a family's well-being. Children thus become more of a financial liability than an asset, which encourages parents to have fewer children. These same opportunities create options for women to work outside the household, which also reduce total fertility rates. Rising incomes also diminish crude death rates, and hence fertility rates, by improving basic health care and nutrition.

How important are these and other negative feedback loops? The historical experience of industrial nations indicates that these feedback effects slow population growth considerably (Figure 9.15(a)). For most of human history, which corresponds to Phase I in Figure 9.15(a), high death rates and high birthrates combined to generate slow rates of population growth. Starting 200 years ago in what are now developed nations, improvements in public sanitation and food production reduced crude death rates (Phase II in Figure 9.15(a)). At the same time, high total fertility rates and pyramidal age structures produced high rates of population growth similar to those in developing nations today. But income gains and lower infant mortality rates helped reduce birthrates during Phase III in Figure 9.15(a) to close the gap between crude birthrates and crude death rates. Eventually birthrates drop to equal death rates, and the population stops growing (Phase IV in Figure 9.15(a)). This pattern of dropping death rates and later birthrates is called the **demographic transition.** Now that they have passed through this transition, most developed nations have low birthrates, low death rates, and low or negative rates of population growth (Figure 9.15(c)).

Many developing nations, such as Mexico, are just beginning their demographic transition (Figure 9.15(b)). Death rates already have dropped, and birthrates are dropping. But how much further and how fast they will drop is an open question. A relatively short lag between the drop in death rates and birthrates will stabilize the human population at 9 to 10 billion. But if a lack of natural resources, environmental degradation, or other forces prevent the income gains that slow birthrates, population may reach much higher levels.

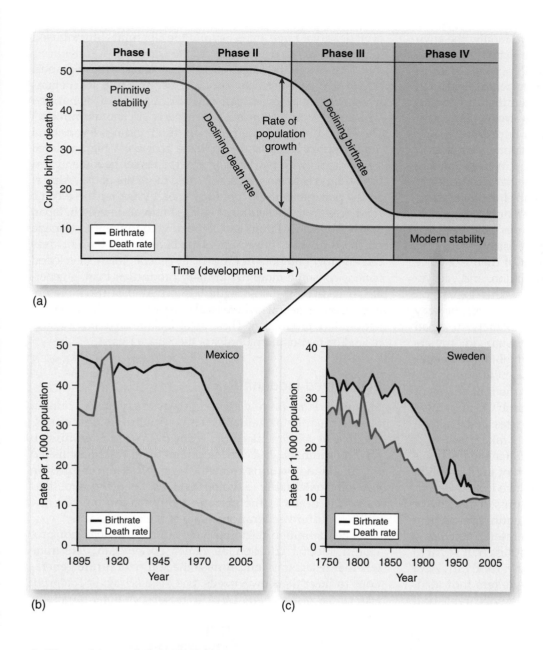

(a)

(b)

(c)

FIGURE 9.15   *The Demographic Transition* (a) The demographic transition occurs in four phases. In the first phase crude birthrates and crude death rates are high but about equal. As a result, population grows slowly. In Phase II access to clean water reduces crude death rates while crude birthrates remain high. As a result population grows rapidly. Population growth slows in Phase III, when reductions in total fertility rates lower the crude birthrate. Eventually crude birthrates and crude death rates equalize in Phase IV, which generates the very slow population growth rates that characterize developed nations. Consistent with these changes, developing nations such as Mexico (b) are in Phase III of the transition while developed nations such as Sweden (c) have completed the transition and are in Phase IV.

A Given Area of the Environment

The quantities and types of resources and services that are available to a population of plants or animals from a given area of the environment are determined by the environment in which the plants or animals live. In general, plants and animals can't significantly increase the types or quantities of environmental goods or services that are available to them. Frogs can't change the rate at which the pond processes its wastes or provides flies. As a result, the maximum number of frogs (or any plant or animal) that can live in a given area of the environment is determined by the characteristics of the local environment.

But the local environment does not determine the quantities and types of environmental goods and services that are available to the human population. People *can* modify their environment to increase the quantities and types of goods and services provided. In addition, people can move

environmental goods and services over long distances. Both of these strategies can increase carrying capacity.

People differ from plants and animals because they can increase carrying capacity by modifying the environment to increase the supply of environmental goods and services. Consider food production. Until about 12,000 years ago people obtained their food by **hunting** (capturing wild animals) and **gathering** (collecting edible plants). The quantity of food available to hunter–gatherers is determined by the rate at which the local ecosystem produces edible plants and animals. The quantity of food produced by the unaltered local ecosystem imposes an upper limit on population size in that area. As a result, the physical and biological characteristics of the local ecosystem determine the carrying capacity of a hunter–gatherer population just as the physical and biological characteristics of a pond determine the carrying capacity for frogs.

About 12,000 years ago people started to modify the environment through **agriculture:** replacing the natural ecosystem with one that supports the plants and animals of people's choosing. By converting grasslands and forests to wheat fields and pastures, people increase the amount of food that can be produced in a given area. These increases raise carrying capacity. Even the simplest agricultural ecosystem can support ten to fifty times more people than the same area of a natural ecosystem.

Humans also increase their carrying capacity by expanding the area from which they obtain environmental resources and services. Referring back to the pond example, the geographic extent of a frog's existence is defined by the size of the pond: 240 square meters and 10,000 liters of water. Frogs cannot enlarge the pond. The geographic area of the frogs' existence is fixed.

People expand the geographic extent of their existence by using natural resources, waste assimilation, and ecosystem services from distant environments. This expansion includes the use of neighboring environments all the way to environments from distant parts of the planet. An urban center such as New York City covers 780 square kilometers (301 mi²), but it uses a much larger area to obtain resources and assimilate wastes. For example, the 4.5 billion liters (1.2 billion gal) of sewage produced daily by the inhabitants of New York City eventually reach the Atlantic Ocean, where currents, tides, and waves dilute the wastes and bacteria that detoxify the wastes. These processes take place over thousands of square kilometers of water that are many times the size of New York City. Similarly, much of the 12 million kg (26.4 million lb) of solid waste produced each day in New York City is transported to landfills outside the city or is burned in incinerators. Using Earth's entire atmosphere to dilute the emissions from local incinerators immensely expands the area that generates the environmental services used by the 7.4 million inhabitants of New York City. As a 1991 strike by private building service workers indicated, life in New York City can quickly become unbearable and even deadly if huge areas beyond the city limits are not used to produce resources and assimilate wastes.

Nations use the same strategy to expand the geographic extent of their existence. This is especially true for nations whose citizens enjoy a high material standard of living. The Japanese economy often is described as an "economic miracle." The island of Japan is densely populated and has few natural resources. Nonetheless, the Japanese generate a high material standard of living by drawing environmental goods and services from ecosystems scattered across the entire planet. Almost all the paper and wood products used by the Japanese come from trees grown in southeast Asia. Almost all oil used in Japan comes from the Middle East. Similarly, consumers in Canada purchase bananas grown in Colombia because banana trees cannot grow outdoors in Canada. In general, many developed nations consume more net primary production than is generated by local ecosystems (Figure 9.16). This deficit also is present in developing nations where local ecosystems have low rates of net primary production. These areas include an east-west strip of land on the southern border of the Sahara Desert.

Maintained Indefinitely

Maintained indefinitely is another way of saying that human actions must be sustainable. The official definition of sustainable development comes from the World Commission on Environment and Development that was established by the United Nations General Assembly. The commission defined **sustainable development** as "development that meets the needs of the present without compromising the ability of future generations to meet their own needs."

This definition is deceptively simple—it does not provide a specific criterion by which society can determine whether its actions compromise the ability of future generations to meet their own needs. See *Your Ecological Footprint: Look for the Sustainability Label*. To make such determinations,

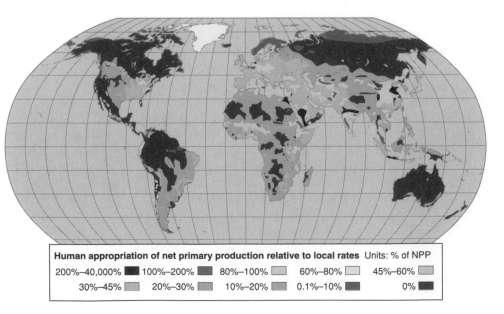

FIGURE 9.16 *Trading Net Primary Production* Many populations depend on environmental goods that are generated by net primary production beyond their local environments. As a result, people living in some areas, such as the eastern U.S. coast and western Europe, consume more than 100 percent of net primary production of their local ecosystem. More than 100 percent of net primary production also is consumed in areas where population density is high, such as eastern China, or arid areas such as the Arabian Peninsula. (Source: Data from M.L. Imhoff et al., "Global Patterns in Human Consumption of Net Primary Production." Nature 429: 870–873.)

Human appropriation of net primary production relative to local rates	Units: % of NPP

200%–40,000%	100%–200%	80%–100%	60%–80%	45%–60%
30%–45%	20%–30%	10%–20%	0.1%–10%	0%

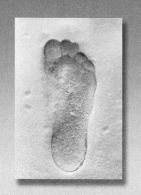

Your ECOLOGICAL *footprint*

Look for the Sustainability Label

Calculating whether your actions are sustainable is tricky. It may be possible for one person to consume environmental goods and services at a high rate, but what if everyone consumed them at the same rate? Equally complex, how can you, as an individual, determine whether your actions are sustainable? Assembling the relevant information is nearly impossible. To illustrate, suppose someone asked whether your weekly tuna sandwich is sustainable. Some of the information you would need to know includes where your tuna came from (a can is not the answer) and how many other people are eating sandwiches with tuna caught from the same fishery. Currently this information is not printed on the label.

Such information soon may be provided by the Marine Stewardship Council. The Marine Stewardship Council is a nonprofit, nongovernmental organization that was set up in 1996 by the World Wildlife Fund and Unilever, a multinational corporation that produces many goods, including processed foods made from fish, such as fish sticks. The goal of the Marine Stewardship Council is to provide information that allows consumers to purchase fish that are harvested in a sustainable manner. This information will appear as an "eco-label" that signifies that the seafood was harvested in a sustainable manner.

Decisions about whether a fishery is being managed in a sustainable fashion are based on three criteria:

Criterion 1: A fishery must be conducted in a manner that does not lead to overfishing or depletion of the exploited populations. For populations that are depleted, the fishery must be conducted in a manner that demonstrably leads to their recovery.

Criterion 2: Fishing operations should allow for the maintenance of the structure, productivity, function, and diversity of the ecosystem (including the habitat and associated dependent and ecologically related species) on which the fishery depends.

Criterion 3: The fishery is subject to an effective management system that respects local, national, and international laws and standards and incorporates institutional and operational frameworks that require use of the resource to be responsible and sustainable.

These criteria are related to the expanded definition of carrying capacity in this chapter and the principles of sustainability from Chapter 1. The first criterion reads like the definition that is applied to nonhuman populations—resources cannot be used faster than they are replenished. The second criterion recognizes that people can modify the environment in a way that alters its carrying capacity—in this case, the methods used to catch the fish cannot disrupt the habitat and diversity (both functional diversity and the strength of linkages in the food chain, as discussed in Chapter 8) in a way that reduces the carrying capacity of the local environment. Criterion 3 recognizes that carrying capacity also has a cultural component—remember that sustainability principle 3 in Chapter 1 states that sustainable use of resources must promote equity. A fishery cannot be sustainable if it is managed in a way that benefits only a portion of society.

As of April 2004 the Marine Stewardship Council had certified ten fisheries as sustainable. These fisheries include an Alaska salmon fishery, the rock lobster fisheries in western Australia and Baja California, and the Thames Blackwater herring fishery. Certification means that the fish caught from these fisheries carry the Marine Stewardship Council eco-label. As of April 2004 this eco-label appeared on 195 products sold in seventeen countries.

How does the Marine Stewardship Council eco-label help consumers and producers? For producers, certification carries several potential benefits. After the Thames Blackwater herring fishery was certified, the price of its fish rose by nearly 50 percent. This price jump is essential to the financial health of the fishery—remember from sustainability principle 4 in Chapter 1 that both the biology and economics must be sustainable. For the Alaskan salmon fishery, the eco-label may allow sellers to differentiate wild fish from farmed fish. This separation may help sales: Over the past two decades the price and catch of Alaskan salmon declined due mostly to the availability of less expensive farmed fish. Many people feel that wild fish taste better than farmed fish, and some people are concerned about the environmental effects of fish farming. If the eco-label gives consumers confidence that the fish they are buying are wild, they may be willing to pay more.

The eco-label also may increase market access. Many developed nations have laws that require goods and services to be produced in environmentally friendly ways. These laws serve two purposes—to protect the environment and to protect domestic producers against less expensive imports. Often these laws are used to prohibit sales from developing nations. This excuse for banning imports may be thwarted by an objective measure of sustainable production. That is, a developing nation can use the eco-label to argue against any efforts to exclude its fish from markets in developed nations.

Eco-labels also benefit consumers and the environment. Labels allow consumers to determine whether their fish have been caught in a sustainable manner. This information is important to some consumers—sales of Marine Stewardship Council labeled fish and fish products are growing between 10 and 20 percent per year. And as retailers try to market Marine Stewardship Council fish, their efforts increase consumer awareness about the sustainability of current fishing practices. For example, one U.S. supermarket chain sponsored a marketing program titled "Fish for Our Future" in which it highlighted Alaskan salmon fishing practices. This effort probably reached more people than the total number of students taking environmental science classes in U.S. colleges and universities.

Ultimately, the degree to which your weekly tuna sandwich is sustainable depends on how the fish are caught. You, your children, and your grandchildren may be able to eat a weekly tuna sandwich if you are willing to spend a bit more for fish that are certified by the Marine Stewardship Council. Almost by definition, fish caught in a sustainable manner are more expensive because fewer fish are caught and the techniques are designed to reduce environmental impacts. Whether these benefits justify the higher price is up to you.

ADDITIONAL READING

Roheim, C.A. "Early Indications of Market Impacts from the Marine Stewardship Council's Eco-Labeling of Seafood." Mimeo.

STUDENT LEARNING OUTCOME

- Students will be able to explain how sustainable management goes beyond simply harvesting a resource in a manner consistent with the rate at which the environment generates new supplies.

economists and ecologists can use two criteria. One is based on the notion that humans are no different from plant or animals. This perspective lies behind the notion of **strong sustainability,** which states that for human actions to be sustainable, they cannot degrade the environment or use environmental goods or services faster than they are generated by the environment.

If the economy was managed according to the notion of strong sustainability, the use of nearly all nonrenewable resources such as oil would be forbidden. Because new supplies of oil are generated very slowly, only very low rates of use would satisfy the criteria of strong sustainability. Geologists estimate that about 6 million barrels of oil are formed each year. If this oil were divided evenly among the planet's 6 billion people, that would leave you with about .001 barrels, which is about 1.5 liters (0.4 gal). This quantity would not power many of the material comforts you take for granted (the average U.S. resident consumed about 26 barrels of oil in 2004).

Similarly, managing the economy according to the notion of strong sustainability would forbid any action that would degrade the environment. But almost all activities degrade the environment to some degree. Even "environmentally friendly" agricultural techniques accelerate soil erosion relative to the rate that it occurs in natural ecosystems. Similarly, even the best water treatment plants do not remove all pollutants before the wastewater is returned to the environment.

These difficulties have spurred an alternative definition for sustainability. This alternative recognizes that human well-being comes from three sources: goods and services that are produced by the environment, goods and services that are produced by economic forms of capital (such as machines), and goods and services that are produced by social institutions (such as education and justice). These three sources of goods and services are somewhat interchangeable. That is, if human activities degrade the environment's ability to produce goods or services, it may be possible to offset these losses by increasing the amount of goods and services produced by economic capital or social institutions. For example, human activities diminish the ocean's ability to produce fish. In some places people use economic capital and social institutions to build and operate fish farms. Fish from these farms make up for some of the reduction in fish generated by the ocean. Similarly, agriculture has accelerated the rate of soil erosion. In some places, like the United States, farmers have offset the loss of soil by increasing the use of fertilizers.

The potential for substitution is used to define **weak sustainability.** Weak sustainability states that an action that degrades the environment or uses a natural resource or a waste-processing service faster than it is generated by the environment can be sustainable if these losses are offset by an increase in another source of well-being—either economic capital or social institutions. Returning to the preceding examples, actions that reduce the ocean's ability to produce fish are sustainable if these losses are offset by increased production by fish farms. Low rates of soil erosion may be sustainable if increasing fertilizer use offsets the loss in production due to soil erosion.

Weak sustainability recognizes that human well-being does not depend solely on the environment, as is the case for nonhuman species. But it recognizes that there is a limit on the ability to replace environmental goods and services with either economic capital or social institutions. For example, some fish farms are located in areas that previously supported large populations of wild fish; and fish farms have been criticized for their potential environmental damages. Similarly, fertilizers cannot eliminate a plant's need for soil. Because environmental goods and services complement economic capital and social institutions, the concept of weak sustainability does not exempt people from the basic principles of sustainability.

DOES CARRYING CAPACITY APPLY TO PEOPLE?

Clearly humans are different from other species on Earth. Because we do not seek to maximize our population, because we can modify the environment to increase the types and quantities of environmental goods and services, and because we do not depend solely on our immediate environment for our economic well-being, the concept of carrying capacity cannot be applied directly to humans. To some people these modifications imply that the notion of carrying capacity has no relevance to human populations. To others the notion of carrying capacity lies behind the collapse of the Easter Island civilization and may portend a similar fate for the entire human population. As described next, the notion of carrying capacity lies just below the surface of many environmental debates. Highlighting its role will allow us to define ways that carrying capacity can be used to evaluate whether our actions are leading us to the same fate that befell the inhabitants of Easter Island.

Identifying the Notion of Limits in Environmental Debates

Politicians rarely argue about whether the voting public has exceeded its carrying capacity. Nonetheless, debate about whether there is a carrying capacity for people lies at the heart of many environmental controversies. Even a casual reading of the daily newspaper shows that nearly all stories about the environment have one or more of the following threads: Are there too many people, do our activities reduce the production of environmental goods and services, and are our actions sustainable? The answers to these questions depend on whether you believe the environment imposes limits on people.

The role of limits in environmental debates is illustrated by the political debate about global climate change. Ultimately the economic debate about what, if anything, should be done about global climate change depends on the answer to the following question: Are there limits to the quantity of carbon dioxide that humans can emit by burning fossil fuels and still depend on the atmosphere to provide a favorable climate now and in the future? If there is no limit, there is no need to curtail human activities that emit carbon dioxide. On the other hand, if the rising concentration of carbon dioxide will change Earth's climate in a way that makes life more difficult, humans may need to limit activities that emit carbon dioxide.

Whether there are limits on the size of the human population and people's behavior depends on the strength of the relationship between economic well-being and the environment. Does environmental degradation reduce economic well-being? Or can humans maintain their standard of living by replacing environmental goods and services with economic capital and social institutions? If humans use their ingenuity to replace environmental goods or services that are degraded or destroyed by environmental degradation, carrying capacity may not apply to people. For example, if people can invent adaptations to a changing climate (and don't care about other species), there may be no need to limit carbon dioxide emissions. Similarly, if modern agricultural technology can compensate for the negative effects of soil erosion, there is no need to worry about activities that cause erosion.

But what if human knowledge and technologies cannot compensate for the degradation and loss of the environmental goods or services? What if humans decide that they do care about other species? Under these conditions society must take actions to preserve important environmental goods and services. If humans cannot adapt to a changing climate, society must limit the emission of carbon dioxide. Similarly, if agricultural technology cannot offset the negative effects of soil erosion, farmers must find ways to reduce soil erosion.

These possible outcomes are summarized by two contrasting views about the relationship between people and the environment. At one end of the spectrum are the so-called resource pessimists, who hypothesize that there is a strong link between humans and their environment and that this strong link implies environmental limits. Based on this hypothesis, resource pessimists argue that the finite supply of environmental goods and services puts an upper limit on the size and material well-being of human populations. According to resource pessimists, we are approaching or already have exceeded these limits. At the other end of the spectrum are the so-called resource optimists, who hypothesize that human ingenuity always will develop new ways of using environmental goods and services or replacing them with economic capital or social institutions. The lack of environmental limits implies that the concept of carrying capacity does not apply to human populations.

Resource Pessimists

Although English society did not collapse as Malthus predicted, neo-Malthusians express similar ideas (Figure 9.17). Population is growing exponentially in developing nations. The material standard of living is growing exponentially in industrial societies. Resource pessimists acknowledge that the biological definition of carrying capacity cannot be applied directly to people, but they maintain that exponential growth will eventually outrun the environment's ability to provide goods and services. Thus resource pessimists envision the same collision that Malthus predicted for nineteenth-century England, but this time at a global scale.

The economic systems of today are far more complex than those of preindustrial England; therefore, resource pessimists cannot identify a single resource such as corn that will limit population or economic well-being. But this hasn't stopped them from trying. Resource pessimists use computer simulation models to analyze the relationships among population growth, economic growth, and the environment. The first of these analyses was done by Donella Meadows and her colleagues. In 1972 they published *The Limits to Growth*, a study that used computer models to represent historic trends in population growth, pollution, and the depletion of natural resources. Their results were rather gloomy. They forecast that the size of world population and its economic well-being would increase until 2000. After that, both population and standard of living would decline along with life expectancy. These reductions would be caused by too many people demanding too many resources and generating too much pollution.

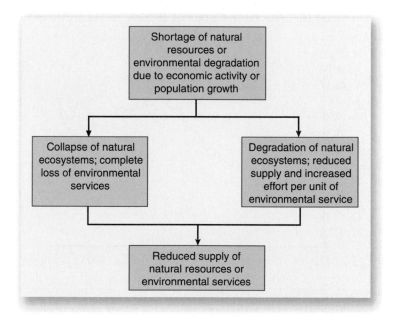

FIGURE 9.17 *Resource Pessimists Resource pessimists argue that resource depletion or environmental degradation will reduce the size of the human population and/or lower its material standard of living.*

In 1991 Meadows and her colleagues updated their original analysis with *Beyond the Limits*. In this study the authors compared their original predictions with two decades of additional information. The authors noted that people adapted to several serious environmental problems. The responses to environmental problems included better technology such as improved energy efficiency, reduced population growth in some nations, and social policy such as the Montreal Protocol to reduce ozone depletion. In spite of these successes, they emphasized that populations and standards of living continue to grow exponentially and that many environmental challenges are becoming more severe. As a result, Meadows and other resource pessimists conclude that society must act quickly and forcefully to avoid a Malthusian population collapse.

There is some support for the arguments of Meadows and her colleagues. In the early 1960s some scientists and economists argued that the vast supply of fish in the world's oceans could feed a growing number of hungry people. If possible, this would reduce the rate at which natural ecosystems were being converted to agricultural land. The number of fishing boats and crews grew exponentially, as did the technology to locate fish. The number of fish caught increased rapidly through the mid-1980s (Figure 9.18). This increase was possible because the number of fish produced by the marine ecosystem exceeded people's ability to harvest them.

But the number of fish harvested cannot grow exponentially forever because there is an upper limit to the rate at which marine ecosystems produce fish. Biologists estimate that the ocean can produce a maximum of about 80–100 million metric tons (88–110 million tons) of fish each year. Since the world first caught 85 million metric tons (94 million tons) in 1990, the number of fishing boats and crews continued to increase. Despite this increased effort, the total number of fish caught has declined. Clearly human technology has collided with the ocean's ability to provide fish.

During the 1950s many scientists and economists hailed nuclear power as the energy source of the future. They argued that it would provide an infinite, nonpolluting, and inexpensive source of energy. Some predicted that nuclear power would be too cheap to meter (installing meters would be more expensive than simply sending everyone the same low monthly bill). These claims appeared to be prescient in the early stages of the nuclear power industry. During the 1960s the number of nuclear power plants in the United States and the amount of electricity they generated grew exponentially. This success led to Project Independence, which was a plan by the U.S. government to reduce U.S. oil imports by building 2,000 nuclear power plants by 2000.

Since the mid-1970s the record of the nuclear power industry has contradicted all the early promises. The promise of an infinite supply was limited by worries that the United States was running out of inexpensive sources of uranium, the fuel used to run the plants. The promise of a nonpolluting source of energy was limited by the inability to develop an acceptable plan to dispose of radioactive wastes. The promise of an inexpensive source of energy was limited by the high cost of building new plants. Cost overruns increased the cost of power plants nearly tenfold relative to the original estimates and raised the price of electricity to consumers. Growth in nuclear power plants came to a sudden halt. Since 1978 no new nuclear power plants have been ordered, and dozens were canceled. As of 2000 there were only 104 nuclear power plants operating in the United States—a far cry from the grandiose plans of Project Independence.

Resource Optimists

Most resource optimists do not deny that environmental problems exist, but they disagree with resource pessimists about their severity. Resource optimists accept Malthus's statement that populations can grow exponentially; they disagree with his contention that the supply of environmental goods and services grows linearly. Resource optimists assert that human knowledge and technical capabilities grow exponentially. They note that improved technology allowed corn production to increase exponentially, which undermined Malthus's predictions for England.

According to this perspective, people avoid a Malthusian disaster thanks to our ingenuity. Ingenuity takes several forms (Figure 9.19). As the supply of a resource dwindles, its price often rises. This provides an incentive to find new supplies. If these supplies are too difficult to locate, ingenuity may develop a replacement. If a replacement cannot be found, the increasing price may spur the development of technology that increases efficiency (using less of a resource to do the same task). Increasing efficiency also can alleviate concerns about pollutants by reducing the amount of emis-

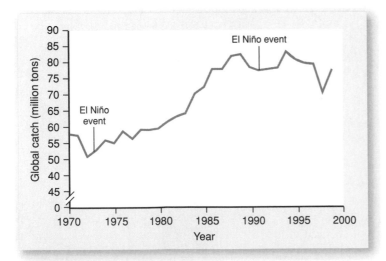

FIGURE 9.18 *Fish Catch Despite ever-increasing efforts to catch more fish, the annual catch of the world's ocean fishery has leveled off since the mid-1980s. Resource pessimists argue that this reflects a limit on the ocean's ability to provide food for a growing human population.* (Source: Data from R. Watson and D. Pauly, "Systematic Distortions in World Fisheries Catch Trends," Nature 414. 534–536.)

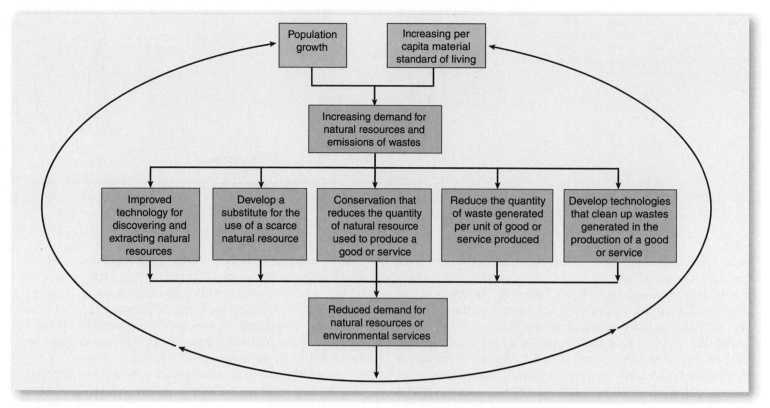

FIGURE 9.19   *Resource Optimists* *Resource optimists recognize that population growth and economic activity may degrade the environment, but these changes do not limit the size or the economic well-being of human populations. Environmental degradation and resource shortages change technology, develop substitutes, increase efficiency, reduce emissions, or develop cleanup technologies that reduce environmental impacts to allow population and economic well-being to continue growing.*

sions generated by an activity. If such reductions are too difficult, ingenuity may spur the innovation of techniques that can clean up the wastes. These capabilities allow society to avoid the type of collapse envisioned by resource pessimists.

There is some evidence for the ability of ingenuity to solve environmental challenges. Examples include society's response to the oil crisis and the reduction of ozone in the upper layers of the atmosphere. Since the 1930s chemical compounds known as chlorofluorocarbons (CFCs) have leaked from air conditioners and refrigerators and made their way to the upper layers of the atmosphere. There the CFCs destroy ozone molecules. When scientists finally recognized the cause of this loss of ozone, the public pressured industry and government to reduce the production of these chemicals. As will be described in Chapter 14, industry developed replacements for CFCs that allow people to enjoy a cold beverage in a cool room on a hot day. As a result, the amount of CFCs leaking into the atmosphere will slow dramatically. Over the next several decades this reduction gradually will reverse the depletion of ozone.

In other instances the market stimulates the changes needed. In the late 1970s and early 1980s a perceived shortage in oil supplies triggered negative feedback loops that alleviated the crisis, at least in the short term. As the supply of crude oil declined, the price of oil increased. Higher

prices made oil production more profitable. To earn these profits, firms developed new technologies that allowed them to drill in places that were previously too expensive, such as the deep waters off the coast of Mexico and in the North Sea. There they found new oil fields that now supply significant quantities of oil. At the same time higher prices reduced the profits of firms and the disposable income of consumers who used oil. This gave firms and individuals an incentive to conserve energy. The combination of increased supply and lower demand allowed oil prices to collapse in 1986. Although Figure 9.20 shows that oil prices have risen since 2002, the price of oil remains below that of the early 1980s (after adjustment for inflation).

Clarifying Carrying Capacity for People

As with many debates, the truth about whether there is a carrying capacity for people lies between the extremes posed by resource pessimists and resource optimists. Their respective arguments are instructive because they identify how the environment limits the size and affluence of the human population and how these bounds change constantly.

The laws of conservation of energy and matter dictate that all materials used and all wastes generated must come from and return to the environment. As such, humans are

FIGURE 9.20 *Oil Prices* The price of oil corrected for the effects of inflation. Even the recent price increase has not pushed prices back to where they were in the late 1970s and early 1980s.

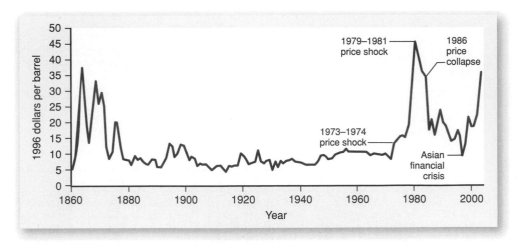

completely dependent on the environment. On the other hand, ingenuity allows us to change the types and quantities of environmental goods and services we use and the types and quantities of wastes we return to the environment. This flexibility, along with the notion of weak sustainability, implies that the degree to which our numbers and economic well-being are linked to the environment is determined ultimately by time. Ingenuity may allow us to change the types of goods and services we depend on; the critical question is whether we have enough time to develop an alternative before a shortage can impose limits in a Malthusian sense. Given enough time, humans probably could replace nearly any environmental good or service. On the other hand, some replacements could take a long time, and before a solution arrives, the resulting shortage could shrink the population and reduce its well-being.

"Timing is everything" is a common cliché. Yet its importance is illustrated by how resource optimists and pessimists tell the story of the "wood crisis" that occurred in England during the sixteenth century. At the start of the sixteenth century England was a relatively poor nation with few domestic industries. By midcentury England increased its production of industrial goods such as iron and glass, in large part by using wood and charcoal that were obtained from local forests. As a result forested areas in England shrank, and the price of fuel increased severalfold.

Both resource optimists and pessimists agree on these facts; but as with Easter Island, they differ in deriving the story's moral. Resource optimists emphasize the response to higher wood prices. Higher prices prompted several social responses to ease the shortage. As shown in Figure 9.19, higher prices prompted alternative supplies. Some of this additional wood came from colonial New England. In addition to religious freedom, much of England's interest in North America centered on abundant supplies of natural resources such as wood, which was shipped back to England in special "mast ships" (ships designed to transport tall trees for use as masts on large sailing ships). In addition, high wood prices prompted innovations designed to reduce wood use. In 1709 Abraham Darby developed a way to produce iron using

abundant supplies of coal. Although it was abundant, coal could not be used before 1709 to make iron because coal contained impurities such as sulfur that ruined the final product. With Darby's technique and other technological innovations, coal became the dominant source of energy, and this powered the start of the Industrial Revolution, which increased both the population and its standard of living.

Resource pessimists agree on the outcome, but they point out that the timing of Darby's innovation created several large bumps in the road to the Industrial Revolution. By 1709 several generations of people living in England had endured high fuel prices and low supplies. While these conditions did not constitute a Malthusian collapse, they did cause economic and social disruption. The reduced supply and high price of wood cut the production of industrial goods and thereby reduced living standards. High prices and cold winters (the sixteenth century fell within the "little ice age," which was an extended period of cold—see Chapter 13) made it difficult for many people to keep warm. These reductions in living standards created social unrest that focused on access to forested lands. Unhappiness about sales of forests to foreign buyers contributed to a civil war in which the British monarchy was temporarily driven from the throne. So the mechanisms described by resource optimists alleviated the wood shortage in the long term, but in the short term the wood crisis limited the size and economic well-being of the English population.

Try to keep the importance of time in mind as you read about the environmental challenges that are described in Chapters 12–23. In the long term human ingenuity probably will develop solutions to these challenges. But the timing of these solutions is critical. For example, there are energy sources other than oil. But how long will it take to develop these alternatives and substitute them for oil? How long would it take to convert every gas station so you could recharge your car's battery or replace its natural gas canister? Similarly, how long would it take to retrain every mechanic to fix these new automobile engines, and how long would it take for everyone to be able to buy new cars?

Such long lead times appear daunting; but what if policy ignores the problem until it is at hand? Under these conditions the availability of environmental goods and services will limit the size and well-being of the human population during the long times required to develop and implement workable solutions. Indeed, some of these limits may occur during your lifetime. On the other hand, such limits probably will not become apparent if environmental challenges are addressed well before they become limiting. Such foresight is the ultimate goal of environmental policy—to develop solutions for problems that never occur.

SUMMARY OF KEY CONCEPTS

- Population growth is determined by the difference between the birthrate and the death rate. Both of these rates are determined by density-independent and density-dependent factors. When density-dependent factors are unimportant, populations have the potential to grow geometrically or exponentially. As populations expand, density-dependent factors become more important, which slows population growth. The point at which population growth stops due to density-dependent factors is termed carrying capacity.

- Real-world populations seldom stop growing immediately upon reaching their carrying capacity. Time lags in the relationship between population growth and density-dependent factors allow many real-world populations to fluctuate around their carrying capacity.

- Carrying capacity is defined by the interactions between a nonhuman population and its environment. The concept of carrying capacity must be modified if applied to people, who can change the types and quantities of goods and services they use, modify the environment to increase the types and quantities of environmental goods and services available, obtain the goods and services from distant environments, and replace environmental goods and services with economic capital and social institutions.

- Views about whether there is a carrying capacity for people vary between two extremes. Resource pessimists argue that exponential growth eventually outruns the environment's ability to provide goods and services, which will cause the human population to shrink and standards of living to decline. Resource optimists argue that ingenuity allows people to alleviate limits associated with environmental goods and services, which permits the human population to expand and its material standard of living to increase steadily.

- Given enough time, people probably could replace nearly any environmental good or service. But developing replacements could take a long time. Before a solution arrives, a reduction in supply could shrink the population and reduce its well-being. Such limits probably will not materialize if environmental challenges are addressed well before they become limiting.

REVIEW QUESTIONS

1. Which type of species is more likely to grow exponentially—a species that lives in an early successional community or a species that lives in a late successional community?

2. Which type of fish population is easier to manage—a population that is limited by density-dependent factors or a population that is limited by density-independent factors?

3. Population is supposed to stop growing when it reaches its carrying capacity. But populations in the real world often exceed their carrying capacity. Explain how this is possible.

4. Explain how the notion of carrying capacity that is used to analyze nonhuman populations must be modified to analyze human populations.

5. Pick an environmental challenge. Explain how resource pessimists see that challenge affecting human population size and material standards of living. Explain how resource optimists see a solution to the challenge. Explain how the timing of the solution may affect its success.

KEY TERMS

age classes	density-dependent factors	overshoot
age structure	density-independent factors	population
age–dependency ratio	doubling time	population density
agriculture	environmental resistance	population growth
baby boomers	exponential population growth	population histograms
birthrate	gathering	population momentum
carrying capacity	geometric population growth	replacement level fertility
crude birthrate	hunting	strong sustainability
crude death rate	limiting factor	sustainable development
death rate	logistic equation	total fertility rate
demographic transition	maximum intrinsic growth rate	weak sustainability

10 AN ECOLOGICAL VIEW OF THE ECONOMY

The Four Steps of Economic Production

STUDENT LEARNING OUTCOMES

After reading this chapter, students will be able to

- Identify the key differences between the circular flow and ecological views of the economy.

- Define the four steps of the economic process.

- Describe the connections among economic growth, energy, and materials use.

- Understand the reasons behind international differences in affluence.

- Explain why green accounting is important for sustainability.

The New York Stock Exchange *When people think of the "economy," they usually associate it with monetary measures of our well-being, such as those produced every day in financial markets. This chapter will explore the environmental basis of economic development that is rooted in our use of natural resources and ecosystem services.*

INTRODUCTION

It is less than two hours until the State of the Union address, and the president's political advisers pace nervously. It is an election year, and an important ingredient of a successful reelection campaign is a healthy, growing economy. They are waiting for a report on the rate of growth in the **gross national product** (GNP), which is the most widely used indicator of a society's economic well-being. A growing GNP means that new jobs are being created and that there are more goods and services for people to buy. A growing GNP gives a president the political courage to ask, "Are you better off than you were four years ago?" (Figure 10.1).

The fax machine beeps, and the advisers sigh in relief—the economy grew at a respectable 4 percent annual rate. The president can now claim that his or her economic policies are working. This means that you will have a good job when you graduate that will allow you to raise your standard of living.

The success of the U.S. economy goes beyond your hard work and the president's policies. To produce 11.7 trillion dollars' worth of GNP in 2004, the U.S. economy burned nearly 100×10^{15} Btus of oil, natural gas, coal, and other energy; used 140 million metric tons of copper, lead, and other metals; used 170 million metric tons of wood products; and consumed 630 million metric tons of food. These activities eroded at least 20 million metric tons of soil, emitted over 5,900 million metric tons of CO_2, created more than 214 million metric tons of municipal solid waste, and generated 1.5 million metric tons of hazardous chemical waste.

The use of energy and materials and the emission of wastes deplete energy reserves, change global climate, cut forests, erode cropland, deplete fisheries, and reduce biological diversity. These impacts imply that we need a bigger picture of the economy. A *complete* measure of economic activity would include these unwanted side effects in addition to the goods and services produced.

In this chapter we expand the traditional view of the economy to include the role of the environment. From start to finish, the environment generates natural resources and dilutes and detoxifies pollutants

FIGURE 10.1 *Economic Growth* *A large and growing gross national product (GNP) is considered a sign of a healthy economy.*

in ways that boost economic well-being. These environmental goods and services are diminished by the challenges described in Parts Five through Eight (Chapters 12–23), such as the loss of biodiversity, climate change, and soil erosion. Understanding these links is critical: Environmental challenges are not simply a matter of preservation, but threaten your job and your economic standard of living. ■

TWO VIEWS OF THE ECONOMY

The Economic System: Production and Consumption of Goods and Services

An economy is a system in which goods and services are produced and distributed among people to satisfy needs and wants. **Goods** are things that people find useful such as cars, computers, and chocolate cakes. **Services** are flows of benefits over time, such as watching a movie, being advised by a financial planner, being treated for a sprained ankle in a hospital, or being educated at a university.

Economies have two functions: the production and the consumption of goods and services. **Production** is the transformation of inputs into goods and services. Goods and services are produced by businesses that are known as **firms.** Firms produce goods and services with **labor** and **capital.** Labor refers to efforts by workers. Capital refers to machinery, factories, equipment, tools, and transportation and communication networks. Together capital and labor are known as **factors of production.**

Many combinations of labor and capital can be used to produce a good or service, and these combinations are called **technology.** Think of technology as a recipe that firms use to produce a good or service. For example, steel can be produced from an open-hearth furnace or an electric arc furnace, and a car can be made of steel or aluminum and plastic. The technology used depends on the quantities of factors used and their respective prices. In general, firms choose the technology that allows them to produce the goods or services at least cost. Because the prices of labor and capital change constantly and because scientists, engineers, and entrepreneurs develop new technologies, the technology used to produce a good or service changes constantly.

Goods and services are produced for a reason: Firms can make a profit by selling them to people. The satisfaction that people get from these goods and services is known as **utility.** Economists assume that people spend their income on the combination of goods and services that generates the largest amount of utility. This combination is different for each person. Some people prefer rock-and-roll, whereas others like jazz; some want to wear the latest fashions or drive the newest cars, whereas others wear five-year-old dungarees or ride their bikes. The process of buying music or pants (or any good or service) is termed **consumption.**

Linking Production and Consumption—The Circular Flow Model of the Economy

Production and consumption are linked via a circular flow (Figure 10.2). Firms use factors of production to produce goods and services. Firms pay wages for labor and profits for capital. Households (the consumers) purchase the goods and services with cash and credit payments, which are known as **personal consumption expenditures.** Governments receive

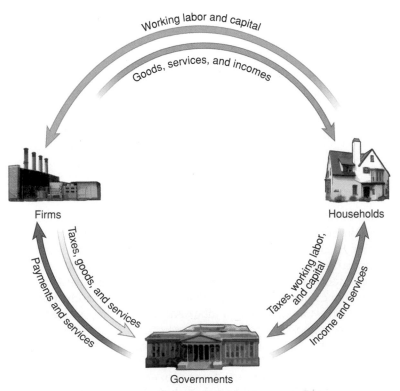

FIGURE 10.2 *Circular Flow Model of the Economy* In this model the economy is depicted as the exchange of money among firms, households, and governments.

money in the form of taxes and provide goods and services such as highways, national parks, and Social Security.

These flows can be used to measure economic activity. **Gross national product** is the value of all the goods and services that are sold to households. As such GNP measures the amount of goods and services produced by the citizens of a nation in a year regardless of where these citizens reside. **Gross domestic product** (GDP) measures the goods and services produced in a nation regardless of the citizenship of the people or firms in that nation. So the value of cars produced in the United States by Japanese firms appears in Japanese GNP and U.S. GDP.

On the other side of the cycle, payments by firms to households as wages and profits represent the income that households can use to buy goods or services. The flow of money from firms to households is **gross national income** (GNI). Because the money associated with income and purchases flows in a cycle, GNP and GNI are roughly equal.

What's Missing? The Role of the Environment

We have just described the conventional view of the economic process. By conventional we mean it is prescribed nearly everywhere—from most high school social studies classes, to economics departments in most universities, to the President's Council of Economic Advisers, to the heads of powerful international institutions such as the World Bank. Despite its almost universal acceptance, the circular flow model of the economy contains a fundamental error: It is inconsistent with the basic physical laws of the natural world that were described in Chapter 2.

Remember the notion of entropy from Chapter 2? Entropy decreases with economic production and consumption. Cars are highly organized relative to their parts, and they do not assemble themselves. Nor do economic goods and services come from thin air. Cars are made of iron, aluminum, plastic, and other materials. A doctor's office or university classroom is made of steel, concrete, wood, and other materials. This leads to a simple yet powerful observation: Economic production is a work process that uses energy to upgrade natural resources into useful goods and services.

This simple truth allows us to take a larger view of the economic system (Figure 10.3). The laws of conservation of matter and thermodynamics instruct us that energy and matter are neither created nor destroyed. All energy and materials come from the environment. All the wastes generated by the use of energy and materials ultimately return to the environment. Thus the economic system is a subsystem of the environmental system. Recognizing this hierarchy expands the simple notion of economic production to give four steps. Ignoring the interdependence between the economy and the environment obscures the fact that our economic well-being depends on a healthy and thriving natural world. Thus environmental challenges that degrade the environment threaten our economic well-being.

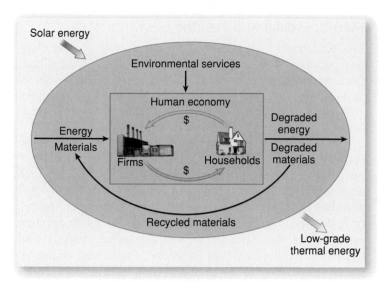

FIGURE 10.3 *Ecological Model of the Economy* In this model the economy is connected to the environment by flows of energy, materials, and environmental services.

FIGURE 10.4 *Natural Resource Extraction* The Bingham Canyon copper mine in Utah is the largest human-made excavation on Earth.

FOUR STEPS OF THE ECONOMIC PROCESS: LINKS TO THE ENVIRONMENT

Step 1: Creating Natural Resources

In the first step of the economic process the planet's biogeochemical cycles produce natural resources. Recall that a natural resource is something that exists in nature that can be used by humans at current economic, technological, social, cultural, and institutional conditions. We say these natural resources are "produced" because natural resources are highly concentrated collections of energy and materials relative to other sources that we do not use. If you dig a hole in your backyard, after a few meters you will hit bedrock—Earth's crust. If you chip off a chunk, you will find tiny amounts of economically useful elements such as gold, copper, lead, and phosphorus. In fact, you would find minute amounts of nearly *all* ninety-two naturally occurring elements. The average concentration of an element in the crust is called its **crustal abundance.** Most elements' crustal abundance is only a few grams per metric ton.

Such low concentrations imply that crustal abundance will not support an economically viable copper mine in your backyard or the vast majority of other locations. Instead copper is produced in a small handful of mines located in the southwestern United States, Chile, and Peru (Figure 10.4). These are areas where biogeochemical cycles have concentrated copper several times greater than its crustal abundance and moved it closer to the surface. Rocks that contain high concentrations of metals and minerals metals are called **ores.** A kilogram of copper ore has ten to one hundred times more copper than the average rock.

Renewable natural resources also are characterized by a high degree of organization. Fish live throughout the world's oceans, but fishing vessels do not randomly trawl the open ocean. For reasons explained in Chapters 4 and 7, the open ocean is a biotic desert. Instead most fishing occurs near coasts and in zones of upwelling where oceanic circulation, wind patterns, and river runoff concentrate nutrients to support high rates of net primary production and large fish populations. These regions thus support a rich food chain where the concentration of fish can be 66,000 times greater than that in the open ocean (Table 10.1). The vast majority of fish caught each year are taken from a small handful of coastal fisheries.

The high degree of organization or concentration applies to all other economically useful forms of energy and materials. As described in Chapters 15 and 16, most of the world's agricultural output comes from regions where biogeochemical cycles have produced soil that is far richer in nutrients, organic material, and other biologically important attributes than most of the soil on the planet. Most of the world's timber is harvested from forest ecosystems that produce dense accumulations of stored

TABLE 10.1	Productivity of Ocean Ecosystems		
Ecosystem Type	% of Ocean Area*	Average Net Primary Production gCalories/m²/yr	Fish Production per Unit Area per Year (Relative to the Open Ocean)
Open ocean	92	57	1
Continental shelf	7.4	162	660
Upwelling zones	0.1	225	66,000

*Other ecosystems not listed include algal beds and reefs and estuaries.

Source: Data from J.H. Ryther, "Photosynthesis and Fish Production in the Sea," *Science* 166: 72–76; data from P. Colinveaux, *Ecology 2*, Wiley.

carbon (wood), and most of the world's drinking water comes from lakes and reservoirs where the planet's geomorphology stores great quantities of freshwater before it returns to the sea.

The high concentration of energy and materials in natural resources does not occur spontaneously. Consistent with the laws of thermodynamics, the concentration of materials and fuels into natural resources requires work done by the environment. Recall from Chapter 3 that energy is required to produce a gradient, which is nothing more than a difference in the concentration of economically useful ores relative to their crustal abundance. This means that energy is used to produce the high concentrations of fish, metals, and other energy and materials that form our natural resource base.

Every natural resource therefore has an **environmental energy cost:** the energy required to create a natural resource of a given concentration. Recall from Chapter 4 that there are two sources of environmental energy: solar energy and heat from Earth's interior. These energies power environmental processes such as atmospheric and oceanic circulation, the rock cycle, net primary production, and other processes that create and sustain natural resources.

The amount of energy used by the environment to create natural resources depends on their degree of concentration. More environmental energy is used to concentrate a resource further from its average concentration. Higher-grade metal ores have much higher concentrations than their crustal abundance and thus require more environmental energy to create than lower-grade metal ores (Figure 10.5). For example, the environment uses three times more energy to concentrate iron in ores that are nearly 30 percent iron relative to ores that are 5 percent iron.

Step 2: Providing a Habitable Environment

The economy's dependence on natural resources is obvious because we see and use them in our everyday lives. Furthermore, they are readily measured in both physical and dollar terms. We know how much coal, copper, timber, and fish are extracted each year and what the average price is per ton of coal, fish, or other natural resource.

But not everything that sustains economic production comes from a sawmill, oil refinery, farm, or mine. Less obvious, but equally important, are the many ways in which ecosystems make economic production—and life itself—possible. Ecosystem functions refer to the habitat, biological, or system properties or processes of ecosystems. **Ecosystem services** are the values that people derive, directly and indirectly, from ecosystem functions. Prominent examples include regulation of the chemical composition of the atmosphere, climate control, nutrient recycling, soil formation, pollination, habitat for biodiversity, and regulation of the hydrologic cycle (Table 10.2). For example, if all vegetation were removed, Earth's average temperature would rise by about 8°C (14.4°F) and precipitation would decline by about 50 percent.

Step 3: The Production of Goods, Services, and Wastes

There is a wise saying in cooking: If you want to bake a cake, you need to break some eggs. Eggs are an indispensable ingredient in many recipes. The same principle applies to economic systems: If you want to produce goods and services, you have to use energy and materials and generate some wastes. The laws of conservation of matter and thermodynamics dictate these requirements.

The United States used nearly 100 quadrillion (10^{15}) Btu in 2005 (Figure 10.6). One quadrillion Btu are called a "quad" of energy. The 100 quads of energy use are equivalent to about 2,800 gallons of gasoline per person per year and represent about 23 percent of global energy use.

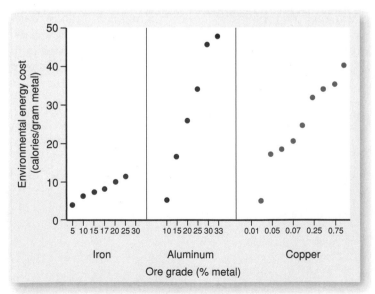

FIGURE 10.5 *Environmental Energy Cost* The quantity of environmental energy used to produce various ore grades of iron, aluminum, and copper. (Data from M.J. Lavine and T.J. Butler, Use of Embodied Energy Values to Price Environmental Factors: Examining the Embodied Energy/Dollar Relationship. Center for Environmental Research, Cornell University.)

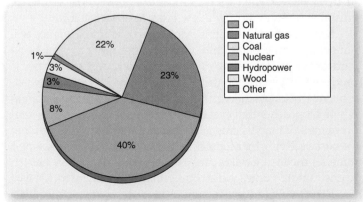

FIGURE 10.6 *U.S. Energy Use* Energy use in the United States in 2005. (Data from U.S. Department of Energy.)

TABLE 10.2	Ecosystem Functions and Services	
Ecosystem Service	**Ecosystem Functions**	**Examples**
Gas regulation	Regulation of atmospheric chemical composition.	CO_2/O_2 balance, O_3 for UVB protection, and SO_x levels.
Climate regulation	Regulation of global temperature, precipitation, and other biologically mediated climatic processes at global or local levels.	Greenhouse gas regulation, DMS production affecting cloud formation.
Disturbance regulation	Capacitance, damping and integrity of ecosystem response to environmental fluctuations.	Storm protection, flood control, drought recovery, and other aspects of habitat response to environmental variability mainly controlled by vegetation structure.
Water regulation	Regulation of hydrological flows.	Provisioning of water for agricultural (such as irrigation) or industrial (such as milling) processes or transportation.
Water supply	Storage and retention of water.	Provisioning of water by watersheds, reservoirs, and aquifers.
Erosion control and sediment retention	Retention of soil within an ecosystem.	Prevention of loss of soil by wind, runoff, or other removal processes; storage of silt in lakes and wetlands.
Soil formation	Soil formation processes.	Weathering of rock and the accumulation of organic material.
Nutrient cycling	Storage, internal cycling, processing and acquisition of nutrients.	Nitrogen fixation; N, P, and other elemental or nutrient cycles.
Waste treatment	Recovery of mobile nutrients and removal or breakdown of excess or xenic nutrients and compounds.	Waste treatment, pollution control, detoxification.
Pollination	Movement of floral gametes.	Provisioning of pollinators for the reproduction of plant populations.
Biological control	Trophic–dynamic regulations of populations.	Keystone predator control of prey species, reduction of herbivory by top predators.
Refugia	Habitat for resident and transient populations.	Nurseries, habitat for migratory species, regional habitats for locally harvested species, or overwintering grounds.
Food production	That portion of gross primary production extractable as food.	Production of fish, game, crops, nuts, fruits by hunting, gathering, subsistence farming, or fishing.
Raw materials	That portion of gross primary production extractable as raw materials.	The production of lumber, fuel, or fodder.
Genetic resources	Sources of unique biological materials and products.	Medicine, products for materials science, genes for resistance to plant pathogens and crop pests, ornamental species (pets and horticultural varieties of plants).
Recreation	Providing opportunities for recreational activities.	Ecotourism, sport fishing, and other outdoor recreational activities.

Source: Data from R. Costanza, et al., "The Value of the World's Ecosystem Services and Natural Capital," *Nature* 387: 253–260.

The United States also used about 5.8 billion metric tons of materials in 2004 (Figure 10.7). About 634 million metric tons, or roughly 10 percent of the total, were imported, with crude oil accounting for half of that amount. This works out to over 50 kilograms per day per person, a mass roughly equal to a person's body weight every day or two. This totals 20 metric tons per person per year, or about 1,600 metric tons over an eighty-year life. This total does not include water because water use dwarfs the use of all other inputs.

What happens to all these materials and energy? Some of the materials are converted to the goods and services that we purchase or consume every day: newspapers, milk cartons, iPods, or food. However, most materials do not become part of finished goods and services. The majority are bulk inputs such as energy and construction materials, including commodities such as coal, oil, sand, stone, clay, steel, and grain. These materials operate in the background of what we typically see or think of in terms of economic activity by

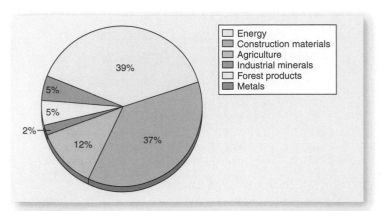

FIGURE 10.7 *U.S. Material Use* *Material use in the United States in 2004.* (Source: Data from U.S. Department of Interior.)

FIGURE 10.8 *Postconsumer Waste* *Postconsumer waste such as paper, bottles, and plastic is one of the largest sources of material wastes.*

providing the physical scaffolding for the entire economy: bridges, sidewalks, roads, buildings, and so forth.

The law of conservation of matter and the principle of entropy tell us that economic activity must generate wastes. Remember, no energy conversion process is 100 percent efficient. Similarly, materials embodied in goods and services wear out, decay, or otherwise succumb to the force of entropy. As such it is impossible to have an economy that generates no wastes. Nor would the effort to reduce emissions to zero be economically sensible. The economically sensible level of emissions is described in Chapter 11.

Physical and economic forces generate **wastes:** forms of energy and materials that no longer are useful to humans because they have been degraded in quality or because they are difficult or expensive to use again. Many wastes consist of unwanted by-products. **Material wastes** are forms of degraded matter that are released to the environment. They include wastes generated by mining operations, from the manufacture of goods and services, and during the combustion of energy, as well as solid wastes from households and commercial establishments. Material wastes come in three general forms—solid, liquid, and gaseous. Gaseous carbon dioxide emitted from fossil fuel combustion is by far the most abundant material released, followed by water vapor, methane, and carbon monoxide.

The production of wastes goes well beyond unwanted by-products. Consumers buy cars, MP3 players, and clothes. But all of these goods ultimately become wastes as they decay, decompose, break down, deteriorate, corrode, disintegrate, or otherwise wear out. Due to the forces of entropy, cars rust, your favorite shirt becomes threadbare, machines wear out, roads need to be repaired, bridges need to be repainted, and so on. Because matter cannot be destroyed, the ultimate output of the economy is waste—what we commonly call trash, junk, or garbage. **Postconsumer waste,** the next most abundant material waste, is the waste remaining after consumers use a product. Examples include bottles, food wrappers, newspaper, office paper, and many other items (Figure 10.8). **Dissipative wastes** are wastes that are not techologically

or economically feasible to collect and recycle. Examples include pigments, pesticides, herbicides, preservatives, antifreezes, propellants, and detergents.

Economic activity also generates **waste heat:** heat released to the environment from the combustion of energy. The second law of thermodynamics dictates the production of waste heat. Using energy to do work converts it from a high-quality to a low-quality state. Low-quality energy is waste heat. Most waste heat has few environmental consequences, such as the heat released by a lightbulb or computer. Power plants, on the other hand, produce significant quantities of waste heat as a by-product of electricity generation. At some plants this heat is released into adjacent rivers or lakes, where it may harm aquatic ecosystems.

Step 4: Waste Assimilation

The wastes released from the production and consumption of goods and services often are highly concentrated relative to their concentration in nature. Carbon monoxide, a highly toxic gas, makes up about 0.00001 percent of the atmosphere. In automobile exhaust it is found at about 0.1 percent, a ten thousandfold increase. **Waste assimilation** is the ability of the environment to absorb, detoxify, and disperse wastes in a way that makes them less harmful. Winds disperse carbon monoxide emissions, waters dilute pollutants, organisms detoxify harmful chemicals, and chemical transformations render many wastes less toxic to life.

One prominent example of this environmental service is sewage treatment (Figure 10.9). Most modern sewage treatment plants use biological processes to remove nitro-

FIGURE 10.9 *Environmental Services* *Modern sewage treatment plants use microorganisms to break down human waste.*

gen from wastewater that contains organic wastes such as feces and urine. Microorganisms are cultivated and added to the wastewater, where they concentrate nitrogen so that it can be removed from the wastewater before the water is returned to the environment. There typically are 4–40 million microorganisms per gallon of wastewater. In another example, pesticides applied to crops pose a potential health risk if they become incorporated into the food chain. Chemical and physical alterations by microorganisms transform many pesticides to less harmful forms over time.

The effectiveness of natural waste assimilation is determined by the physical, biological, and chemical nature of the wastes. **Biodegradable wastes** are broken down to less harmful forms by bacteria and other microorganisms. Human sewage released to a river, for example, can be broken down to less toxic forms by bacteria and protozoa. **Persistent wastes** are degraded very slowly and hence remain in the environment for long periods. Goods such as plastic containers, paper, and metal cans take hundreds of years to degrade. Plutonium and other radioactive wastes from nuclear power plants take thousands of years to decay to nontoxic forms. **Nondegradable wastes** cannot be assimilated and rendered harmless by the environment. For many years lead was intentionally added to gasoline to improve engine performance. Later we learned that lead and other metals such as mercury and aluminum are extremely toxic to living organisms and cannot be broken down by natural processes. (Can you explain why?)

ECONOMIC GROWTH

A predominant feature of modern society is **economic growth:** an increase in the amount of goods and services produced as measured by an increase in GNP, GDP, or GNI. Most industrial nations have enjoyed significant increases in GDP over the past two centuries (Figure 10.10). Economies in developing nations began to grow much more recently than industrial nations, but some now exhibit robust economic growth. China's GDP grew at an annual average rate

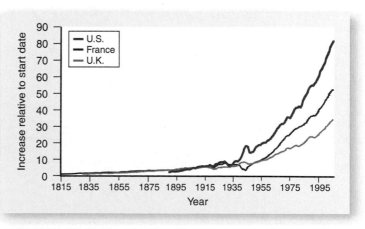

FIGURE 10.10 *Economic Growth* *Long-term growth of GDP in France, the United Kingdom, and the United States.*

of 8 percent from 2000 to 2005. This extremely rapid growth accounted for 15 percent of the total expansion of the world economy in that period.

Economic growth is the dominant economic and political goal in nearly every nation. Economies are "strong" when GDP is growing and "weak" when GDP is growing slowly or even declining. Why is growth so important? Increases in economic growth often lead to rising **affluence,** which is the average standard of living in a nation. Affluence typically is measured by the rate of economic activity per person, such as per capita GDP. Thus two things determine changes in affluence: the rate of population growth (or decline) and the rate of growth (or decline) in GDP.

In the United States population increased from 76 million in 1900 to 300 million in 2006 (Figure 10.11). But GDP increased even faster over this period, producing a steady increase in affluence. Per capita GDP increased sixfold in the twentieth century, reaching about $40,000 per person by 2004. The United States now has the largest economy in the world, producing about 20 percent of gross world output, and the most affluent lifestyle as measured by per capita GDP.

Comparing Levels of Affluence

A large GDP does not necessarily mean people in that society have a high standard of living. China has a larger GDP than Canada, but the average Canadian is more affluent than the average Chinese person. There are enormous variations in affluence among nations (Figure 10.12). Most low-income nations are located in Africa, Central America, Asia, or the Middle East. Most high-income nations are located in North America, western Europe, Australia, and parts of Asia (Japan, Singapore, Hong Kong).

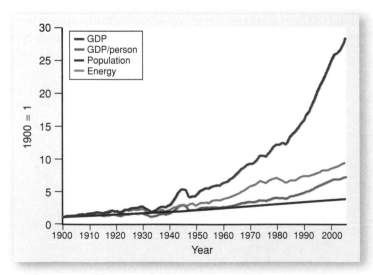

FIGURE 10.11 *Rising Affluence The growth in population, GDP, and per capita GDP in the United States.*

Most people in developed nations are far richer today than ever before. But it is important to remember that per capita GDP measures the national *average*. The level of affluence enjoyed by individuals varies greatly within rich and poor nations. The United States is very affluent as measured by its per capita GDP, but tens of millions of poor people do not have access to quality education, nutrition, or health care. Conversely, the per capita GDP in India is only $3,000, but people in the upper class have high incomes, own elegant homes, drive expensive automobiles, and take luxurious vacations.

The point is that national averages are useful indicators of general relationships, but they can mask large variations within nations. The standard of living enjoyed by a rich Indian is closer to that of a rich American than it is to that of a poor Indian. Similarly, a poor American suffers many of the same problems as a poor Indian. This is especially important to remember when we discuss the relationship between affluence and the environment in Chapter 11. See *Your Ecological Footprint: Comparing National Ecological Footprints.*

The Driving Forces Behind Economic Growth and Rising Affluence

The driving forces behind growth are complex, but two are particularly important: expanding the use of factors of production and natural resources, and changes in technology. The connection between factors of production and economic growth is obvious: more labor, capital, and natural resources increase economic production. As such, population growth is a critical component in economic growth. Changes in technology increase economic activity by increasing the amount of useful goods and services that can be produced from a given quantity of capital, labor, energy, and natural resources. As such, technological change is also responsible for the increase in affluence.

This leads to the question of what drives technological change. A more educated, better-trained, and healthier workforce is one factor, as is investment in research and development by firms and the government. But our ecological view of the economy leads us to another important factor. From a historical perspective technological change has taken the form of using increasing amounts of energy, more powerful energy converters, and more diverse types of materials to empower human labor in production.

Technological Change and Labor Productivity

Changes in the productivity of the workforce illustrate the connections among energy, materials, and technological change. **Labor productivity** measures the quantity of goods and services that a person can produce per unit of time, such as the number of hamburgers produced per day by a cook. The more productive people are, the higher their income tends to be. International differences in per capita GDP are due in large part to international differences in productivity. Workers in developed nations are more productive than workers in developing nations. Similarly, the historical increase in affluence is due mainly to increases in productivity. In the United States, for example, a worker in a manufacturing industry, such as textiles or steel, produces over three times more textiles and

steel than a worker in 1947. As a result, the average manufacturing wage today is more than $16.00 per hour compared to just $5.50 in 1947 when adjusted for inflation.

But why are workers in developed nations more productive than workers in developing nations, and why are current workers more productive than their historical counterparts? Workers in industrial nations are not stronger or inherently smarter than workers in the past or in developing nations. The differences in productivity are due largely to the quantities and types of energy and materials that are used to enhance their efforts in the workplace.

The rates at which workers can organize natural resources into useful goods and services depend on the rates at which they can do economic work. In a modern economy the rate at which a laborer can do economic work depends on the types and quantities of machines that enhance that laborer's effort, as well as the types and quantities of material used by the machines. In general, worker productivity is increased by enhancing human muscle power with machines that increase the amount of work each laborer can do, and by supplying those machines with natural resources that are easier to transform into goods and services. Developed nations are affluent because they supplement human muscle power with powerful machines and diverse, sophisticated materials. Developing nations are poor because they depend more on the muscle power of people and animals to manipulate a relatively small quantity of simple materials.

The productivity of a machine is determined mainly by its power—the rate at which it converts energy to work. Machines that have greater power are more productive than those with less power. For much of history, human labor powered very simple machines (Table 10.3). Examples include the use of a bow and arrow to hunt food or the use of a hoe to prepare a field for planting crops. But the human body has a limited capacity to do physical work. A healthy adult can generate only about 0.1 horsepower for sustained periods of work.

To increase labor productivity, humans developed technologies powered by domesticated animals. Humans breed and raise **draft animals** to help themselves do work (Figure 10.13). These are animals such as horses, oxen, and water buffalo that are used to pull heavy loads. Draft animals can convert greater quantities of food energy to useful work faster than humans, so they are more powerful than humans. Productivity rose when people learned how to make draft animals do tasks that are too strenuous or time-consuming for humans, such as plowing a field or pulling a cart. Another boost in productivity occurred when humans developed machines such as the paddle wheel and the propeller to harness water and wind energy. Harnessing inanimate energy sources made it possible to build more powerful machines, such as large grain mills and sailing ships.

Though these changes were important at one time, they pale in comparison to the increases in work that were made possible when humans learned to harness fossil fuels. Coal, oil, and natural gas contain more energy per unit mass than

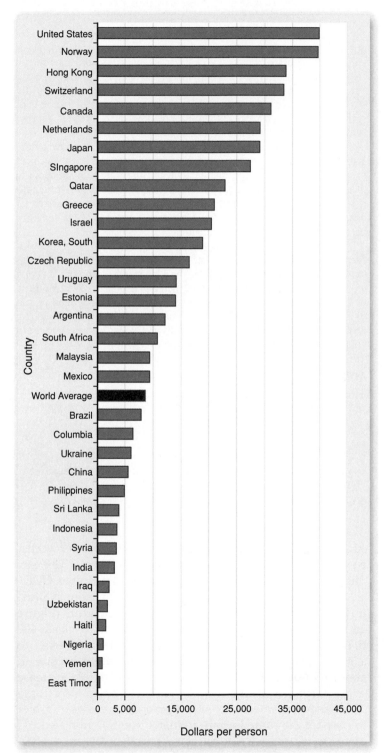

FIGURE 10.12 *International Variations in Affluence* A comparison of per capita GDP among nations in 2005.

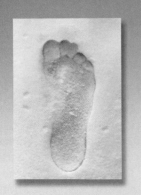

Your ECOLOGICAL *footprint*

Comparing National Ecological Footprints

Thus far we have focused on individual ecological footprints—the amount of natural resources you use or the amount of wastes you generate. This principle can be scaled up to calculate national ecological footprints and ultimately a global ecological footprint for humanity.

Go to http://www.ecologicalfootprint.org/ and download the spreadsheet "World Footprint Data (Excel)." This spreadsheet contains data for the ecological footprint of individual nations. The ecological footprint is a measure of how much biologically productive land and water area an individual, a city, a country, a region, or humanity uses to produce the resources it consumes and to absorb the waste it generates, using prevailing technology and resource management schemes. This land and water area can be anywhere in the world. This includes land used in other nations to supply food, fiber, or energy consumed domestically. (For example, the land in Costa Rica used to grow exported bananas shows up in the footprint for nations that import those bananas.)

The footprint for each nation is the sum of six categories of use of biologically productive areas. *Cropland* is land that is, or has the potential to be, cultivated for the production of food, animal feed, fiber, and oil. *Pasture land* is grassland and pasture area used to raise animals for meat, hides, wool, and milk. *Forest area* refers to natural or plantation forests that are harvested for timber, papermaking, and fuel. *Fisheries* are the area of ocean required to support fish populations that humans harvest. *Built space* includes infrastructure for housing, transportation, industrial production, and capturing hydroelectric power. *Energy land* is the biologically productive area needed to sequester enough CO_2 to avoid an increase in atmospheric CO_2 concentration. *Other land* is the land not included in any of the other categories.

The spreadsheet also contains data about the biological capacity (biocapacity) of nations. Biocapacity is the total usable biological production capacity in a given year of a biologically productive area—for example, within a country. Biologically productive area is land and sea area with significant photosynthetic activity and production of biomass. Marginal areas with patchy vegetation and nonproductive areas are not included. There are about 11.3 billion hectares of biologically productive area, corresponding to roughly a quarter of the planet's surface. These 11.3 billion hectares include 2.3 billion hectares of water (ocean shelves and inland water) and 9.0 billion hectares of land. The land area includes 1.5 billion hectares of cropland, 3.5 billion hectares of grazing land, 3.9 billion hectares of forest land, and 0.2 billion hectares of built-up land.

The remaining three-quarters of Earth's surface, including deserts, ice caps, and deep oceans, support comparatively low levels of bioproductivity that are too dispersed to be harvested. Bioproductivity (biological productivity) is equal to the biological production per hectare per year. Biological productivity is typically measured in terms of annual biomass accumulation. Both the biocapacity and ecological footprint data are expressed in units of hectares per person.

Use these data to calculate the ecological deficit or surplus for each nation. This is the difference between what a nation uses (the footprint) and what it has available (the biocapacity). Rank the nations in order of decreasing surplus and increasing deficit. What patterns emerge from this ranking in terms of geographic area, developed versus developing nations, population size, and so on? Calculate the global ecological deficit or surplus. What does that number suggest about the sustainability of our current way of life on the planet?

ADDITIONAL READING

WWF International. Living Planet Report 2004. *WWF Gland, Switzerland: UNEP World Conservation Monitoring Centre, Global Footprint Network, 2004.*

STUDENT LEARNING OUTCOME

- Students will be able to describe the components of a nation's ecological footprint.

---

renewable fuels such as fuel wood and animal feed. One kilogram of gasoline contains 11,228 kcals compared to only 3,822 kcals in one kilogram of animal feed. This means a kilogram of gasoline has the potential to do three times more economic work than animal feed. New machines and methods of production were developed to use fossil fuels. These included the internal combustion engine (used in automobiles), electric motors, the gas turbine (used in aircraft), and the steam turbine (used in electric power plants). The jet engine on a commercial aircraft can generate 100,000 times more horsepower than a draft animal (Table 10.3).

We can summarize these changes with the observation that the percentage of total work done in industrial economies by labor and draft animals has declined steadily since 1850. These animals have been replaced by more powerful energy converters—machines that use energy—that burn large amounts of fossil and nuclear fuels. The increase in the energy use per worker enabled GDP to increase faster than the number of laborers. This, in turn, increased labor productivity and per capita GDP.

The effect of a larger energy subsidy for labor is illustrated by the changes that took place in agriculture (Figure 10.14). Using only a hoe a single farmer may need 400 hours to till one hectare (2.5 acres). By hitching oxen to a plow a single farmer can prepare the same hectare in about 65 hours. This change represents a sixfold increase in the farmer's productivity. When attached to a tractor, a 50 horsepower internal combustion engine allows a farmer to prepare the same hectare in just 4 hours, a hundredfold increase in productivity. Similar improvements in productivity have occurred in every other sector of the economy when technologies capable of using fossil fuels have become widely available.

TABLE 10.3	Advances in the Power Output of Machines		
Energy Converter		**Date**	**Horsepower**
Animate			
Humans (moderate manual labor)			0.10
Draft animals (pulling a load)		3,000 B.C.	0.80
Wind and water power			
Water wheel		350 B.C.	3
		A.D. 1800	70
Windmill		1400	8
		1850	14
Steam engines			
Watt's steam engine		1800	40
Steam engine		1900	1,200
Piston engines			
Internal combustion engine (Ford's Model T)		1908	20
Internal combustion engine (passenger car)		1990	120
Internal combustion engine (diesel locomotive)		1990	4,700
Turbines			
Steam turbine (electric power plant)		1900	1,341
Steam turbine (electric power plant)		1980	1,341,000
Gas turbine (Boeing 747)		1970	80,460
Rocket engines			
Apollo 11 mission to the moon		1969	3,500,000

Sources: V. Smil, *General Energetics: Energy in the Biosphere and Civilization*, Wiley; E.F. Cook, *Man, Energy, Society*, Freeman.

FIGURE 10.13 *Draft Animals* *Animals such as the water buffalo in China improve the productivity of labor.*

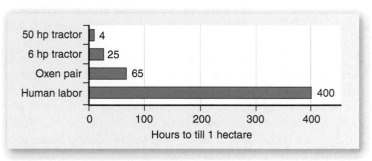

FIGURE 10.14 *Energy Subsidy of Labor* *The effect on labor productivity of increasing energy use per laborer in agriculture.* (Source: Data from D. Pimentel and M. Pimentel, *Food Energy and Society*, Colorado University Press.)

Affluence and Materials Use

The increase in labor productivity and affluence also is due to changes in the types of material used. Two features of this change are striking. The first is the sheer increase in the types of materials we now use. A century ago the United States used only about twenty elements; today the United States uses all ninety-two naturally occurring elements. In other words, every element in Earth's crust is harvested in some type of mining operation and converted into a raw material, which in turn is used to manufacture a good or service.

The second feature is the shift from renewable to nonrenewable materials. In 1900 the majority of materials were derived from renewable sources, such as forestry or agriculture. In the last forty years technologies based on metals, nonmetal minerals, and petroleum (**petrochemicals**) steadily replaced the use of many renewable resources. This substitution increased productivity in several ways. Materials made from nonrenewable resources can be fashioned into new products that match the task at hand. For example, new materials helped revolutionize the telecommunications industry. Telephone wires once were made exclusively from copper, a relatively scarce nonrenewable metal. Now they are increasingly made from fiber optic cables, some of which are made from silicon, one of Earth's most abundant materials. Fiber optic cables carry more information at less cost than copper wire.

Nonrenewable materials also increase productivity by freeing economic activity from limits imposed by the rate at which plants can convert sunlight into useful fibers. This is illustrated by changes in the types of fibers used to manufacture clothes, footwear, and bedding. In 1950 renewable natural fibers such as cotton, wool, and rayon accounted for 99 percent of all the fibers used in the world. But the explosion of the petrochemical industry in the last twenty-five years produced new synthetic fibers such as polyesters, acrylics, and nylon. Synthetic fibers often are cheaper to produce, have greater durability, and can be used to produce a greater range of products. Because of these advantages, and the rapid rates they can be produced from petroleum, abundant supplies of synthetic fibers now account for more than 50 percent of fiber use.

The Connection Among Economic Growth, Energy, and Materials

It should come as no surprise that there is a close connection between economic growth and the use of energy and materials. As the economies of individual nations grow over time, they tend to use more energy and materials. In the United States GDP has grown in tandem with the consumption of energy and materials (Figure 10.11).

The same connection appears in international comparisons of GDP and the use of energy and materials (Figure 10.15). In general, nations that have large economic systems also use great quantities of energy and materials. The world's biggest economies—the United States, China, Japan, and Germany—also use the most energy and materials. Small economies like Ghana and Suriname use less energy and materials.

You might be tempted to jump to the conclusion that there is an ironclad link between economic growth and the use of energy and materials. This is not the case. The United States and Japan have the two largest economies, but Japan uses considerably less energy than the United States. In fact, Japan uses about 6,500 Btu to produce each dollar of GNP, whereas the United States uses about 11,000 Btu. These differences stem from differences in the types of energy used, the types of goods and services produced, the general level of technological development, and a variety of social and cultural forces unique to each nation. One reason that Japan is more efficient in its use of energy is that it uses more high-quality energy such as electricity generated from nuclear energy. Energy prices are higher in Japan, which encourages the development of more energy-efficient technologies. We explore these forces in more detail in subsequent chapters.

THE ECONOMIC VALUE OF ENVIRONMENTAL GOODS AND SERVICES

Why Are Environmental Contributions Often Overlooked?

The four steps of economic production imply that the environment makes a significant contribution to economic well-being. Some of these contributions are measured easily. For example, GDP includes the contribution made by oil as measured by the prices and quantities that are bought and sold in the market. GDP also includes payments made by farmers to beekeepers who set up hives near the farmers' crops.

But other environmental goods and services are not included in GDP. For example, ozone protects you from the sun's ultraviolet rays, but this protection does not appear in GDP. Similarly, the water cycle provides clean water, but this purification process does not appear in GDP. Such omissions are termed **market failures.**

Market failures often are the result of the type of **property rights.** Property rights define ownership. In most developed nations **private property** is the dominant form of ownership. Private property gives the owner the right of control. Owners may change their property or exclude others from it as they see fit. Your car, home, and clothes are examples of private property. Market failures are not often associated with private property because owners manage their property with their long-term interests in mind.

The government owns **public property** at the local, state, or national level. Yellowstone National Park is an example of public property that is owned by the U.S. government. In Mexico and Saudi Arabia the national governments own crude oil resources. Market failures sometimes are associated with public property because the government may not manage public property in a way that is consistent with the long-term interests of

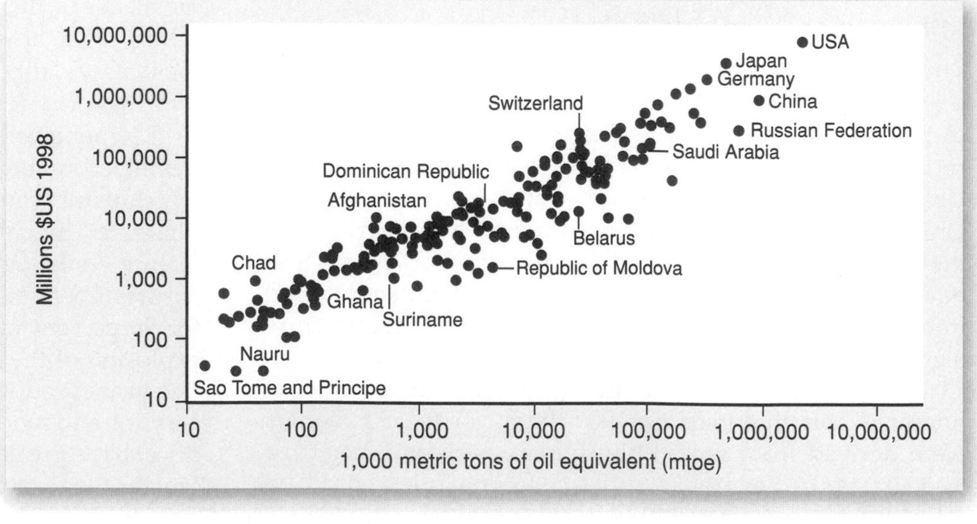

FIGURE 10.15 *International Comparison of Energy Use and GDP* Note that there is a strong correlation between energy use and the size of an economy. Data are for 2004.

the majority of people. Due to political corruption, for example, oil resources in Nigeria have been managed for the benefit of an elite ruling class while many people live in extreme poverty.

Common property refers to items that are not or cannot be owned. The lack of clearly defined ownership usually is associated with the nature of the object. In particular, if access cannot be restricted, an object cannot be owned. Ownership may be impossible for many aspects of the environment. We often cannot prevent people from dumping pollutants into the atmosphere, and it's difficult to establish property rights over fish that swim freely in the ocean.

Common property resources are especially vulnerable to degradation because economically rational decisions by individuals lead to environmentally destructive outcomes. This behavior is described in the "Tragedy of the Commons," a famous essay by ecologist Garrett Hardin. It is a parable about a village's pasture, or commons, which is open to any villager who wants to graze his or her cows. It is economically rational for each villager to graze as many cows as possible. But the sum of all the rational individual decisions leads to an irrational outcome: the complete destruction of the commons. The market cannot prevent this outcome because there is no incentive for an individual to limit the size of his or her herd. An individual who voluntarily limits the size of his or her herd simply makes room for others to increase the sizes of their herds.

Valuing Environmental Goods and Services

Because many environmental goods and services are public or common property, economists and ecologists have developed several methods to measure the economic value of environmental goods and services. **Direct market valuation** is the value that an environmental good and service makes to the production of an economic good and service that is traded in the market. For example, tropical wetlands increase the production of both wild fish and the productivity of fish farms. Both sets of fish are traded in the market. Using prices set by the market, the value of the wetland would be the value of the additional fish.

Calculating the many ways that environmental goods and services contribute to the production of economic goods and services requires more information than is currently available. To get around this roadblock, Amy Richmond, a geographer at the U.S. Military Academy at West Point, used net primary production to estimate the quantity of environmental goods and services that are provided by living organisms. Recall from Chapter 5 that net primary production represents the energy that is available to power the entire food chain, so any work done by biological organisms comes originally from net primary production. You might think that countries in the tropics with biomes such as tropical forests receive the greatest economic benefit from net primary production. In general, this is the case. Statistical analyses indicate that high rates of net primary production generate environmental services that increase GDP relative to nations with lower rates of net primary production.

However, an additional unit of net primary production generates the greatest value in developed nations, where superior technological abilities allow people to convert environmental goods and services to economic goods and services with greater effectiveness compared to developing nations (Figure 10.16). In addition, net primary production is more valuable in developed nations because the supply of net primary production is scarce relative to capital and labor. The reverse is true in developing nations, where local ecosystems generate large amounts of net primary production but economies have relatively little capital.

When no markets exist for environmental goods and services, several techniques are used to generate indirect market values. **Avoided costs** are based on services that are used to replace environmental goods or services. For

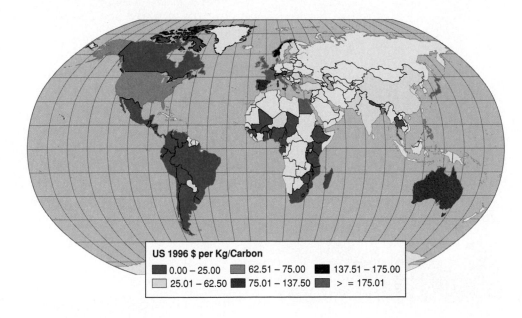

FIGURE 10.16 *The Economic Contribution of Net Primary Production* Note that net primary production generates the greatest value in developed nations, whose superior technological abilities allow them to convert environmental goods and services to economic goods and services with greater effectiveness compared to developing nations. *(Source: Data from A.K. Richmond, R.K. Kaufmann, and R.B. Myneni, "Valuing Environmental Services, a Shadow Price or Net Primary Production," Ecological Economics (in review).)*

US 1996 $ per Kg/Carbon

- 0.00 – 25.00
- 25.01 – 62.50
- 62.51 – 75.00
- 75.01 – 137.50
- 137.51 – 175.00
- > = 175.01

example, Brazilian free-tailed bats prey on several species of moths whose larvae are significant agricultural pests, including the cotton bollworm (Figure 10.17). Each moth eaten by a bat prevents many larvae from hatching and hence prevents damage to the cotton crop. In just an eight-county region in south central Texas, the value of the cotton crop protected by the bats may be as high as $1.7 million, compared to a $6 million per year annual cotton harvest (Figure 10.18). **Travel cost methods** value environmental goods and services based on how much people are willing to pay to travel to use such goods or services. For example, the marine ecosystems off the Florida Keys offer opportunities to catch tuna, sailfish, marlins, and groupers. The value of the marine ecosystems is much greater than the dollar price of the fish—it includes the cost of traveling to the Keys, the cost of renting boats, hotels supported by fishing tourists, and many other things. **Hedonic pricing** values ecosystem services based on the price that consumers are willing to pay for associated goods. For example, the price of clean air may be estimated based on the price difference between two similar houses—one located in an area with relatively clean air, another in an area with relatively contaminated air. **Contingent valuation** determines how much people are willing to pay for a good based on surveys in which people are asked how much they are willing to spend for an environmental good or service. For example, people may be asked how much they are willing to pay to clean up a local air pollution problem or pay for the protection offered by ozone.

Using all of these techniques, Robert Costanza, an ecologist at the University of Vermont, and a team of ecologists and economists estimated the value of the world's ecosystems. They included only the major services, such as the regulation of water and climate, soil formation, nutrient cycling, and habitat for biological diversity. Not surprisingly, the services are concentrated in the most biologically productive regions such as the coastal zone of the ocean, wetlands, and tropical rain forests (Figure 10.19). They estimated that the value of these services was about $33 trillion in the late 1990s. The gross world product—the sum of all gross national products—was about $18 trillion. These numbers suggest that the value of the services provided free by ecosystems is on the same order of magnitude as the value of all goods and services produced in the human economy.

Impacts of Environmental Degradation

Given the immense contribution that environmental goods and services make to economic well-being, it is important to understand how human activities reduce their contribution. Any reduction in the environment's contribution to economic well-being is termed **environmental degradation.** An extreme form of environmental degradation is **exhaustion,** which represents the complete loss of an environmental good or service. Examples include the extinction of a species, such as the passenger pigeon, or using up a natural resource, such as extracting all the anthracite coal in the United States.

FIGURE 10.17 *Natural Pest Control* Brazilian free-tailed bats emerging from a maternity colony in a cave in south central Texas.

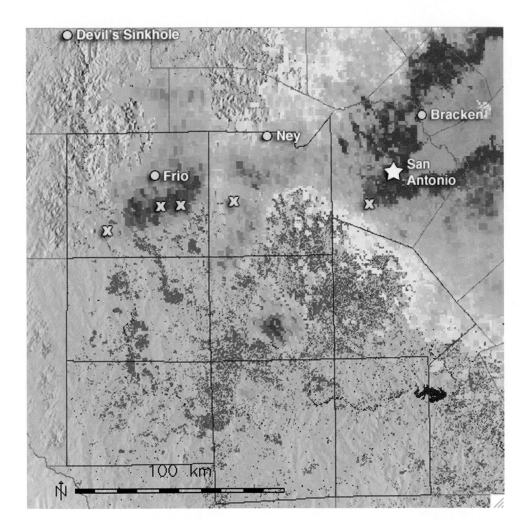

FIGURE 10.18 *Foraging by Bats* Radar *imagery of Brazilian free-tailed bats returning from nightly foraging over agricultural land (green) in south central Texas. Each pixel corresponds to approximately 1 km² of reflectivity from bats aloft. Darker blue colors indicate greater reflectivity and hence greater density of bats. Large areas of reflectivity are seen twice nightly—at the time of emergence and again when bats return to their roosts. The colonies include caves (circles) and concrete highway bridges (x's). The timing, directionality, and density of the reflectivity suggest that many bats forage over this area of agricultural production, consuming significant quantities of pest insects.* (Source: Data and map from J. Horn, Boston Unviersity.)

But exhaustion is only the most extreme form of environmental degradation. Well before a resource is exhausted, it is depleted. **Depletion** implies that the environmental good or service is still present, but its contribution to economic well-being is reduced. For example, U.S. resources of copper are depleted because the purity of ores extracted has declined over time. Similarly, the George's Bank fishery off the coast of Massachusetts is depleted because there are fewer fish to catch. But how does depletion reduce economic well-being? See *Case Study: Can Human Ingenuity Substitute for a Degraded Environment?*

We can measure the effect of depletion using the notion of **resource quality.** Resource quality can be measured by the amount of effort (capital, labor, and energy) that is required to extract a natural resource. Low-quality resources require more effort (more capital, labor, and energy) to extract and refine than high-quality resources (Figure 10.20). We can produce 99 percent pure copper metal from both 10 percent and 1 percent ore, but the 1 percent ore requires more effort to upgrade. A lot of effort is needed to crush very lean ores

and separate the valuable atoms of metal from the waste rock they are part of. The same principle applies to all natural resources. Fish in small, remote populations require more effort to locate and harvest. Growing food in soil with low concentrations of nutrients requires more effort than is needed with fertile soil.

The depletion of high-quality natural resources reduces the economy's ability to produce useful goods and services. As an economy uses up its high-quality deposits and moves on to lower-quality deposits, it must use more energy, capital, and labor to extract the same amount of natural resource. This leaves less capital, labor, and energy to produce other goods and services. The loss of these other goods and services represents **opportunity costs,** which are defined as what you give up to get the good or service you choose. These opportunity costs represent the economic losses associated with resource depletion.

We can illustrate these losses with a simple example. Suppose it takes 1,000 kcals of energy to convert copper at its average crustal abundance to wire, which is 99.9 percent

FIGURE 10.19 *The Global Value of Environmental Services* Note the high concentration of services in the highly productive waters of coastal environments. Source: R. Costanza, et al., "The Value of the World's Ecosystem Services and Natural Capital." Nature 387: 253–260.

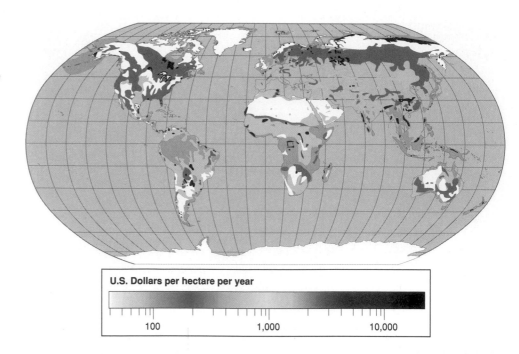

U.S. Dollars per hectare per year

100   1,000   10,000

copper. Suppose two copper mines are available—one that has ores that are 5 percent copper and one that has ores that are 10 percent copper. Suppose that 950 kcals are required to produce a gram of wire from ores that are 5 percent copper, whereas only 900 kcals are required to produce a gram of wire from ores that are 10 percent copper. If society shifts from the 10 percent ores to the 5 percent ores, each gram of wire requires an additional 50 kcals to produce. This increase implies that 50 fewer kcals are available to heat your home, run your car, or make peanut butter. These losses represent the opportunity costs of switching from 10 percent ores to 5 percent ores.

Given these opportunity costs, why would society switch from ores that are 10 percent copper to ores that are 5 percent copper? The answer is given by the best first principle. As defined in Chapter 1, the best first principle states that humans use the highest-quality sources of natural resources first. Given a choice, humans will grow crops on fertile (high-quality) soil before attempting to grow them on infertile (low-quality) soil. Humans prefer to use ores that are 10 percent copper rather than 5 percent copper, and deposits of oil that are 100 feet deep rather than 1,000 feet deep. Foresters cut trees in forests that are close to a sawmill before they cut trees that are a long distance from the mill. Fishers catch fish from large, highly concentrated schools in coastal waters before they harvest smaller, more random collections of fish in the open ocean.

Economies use lower-quality deposits only when forced to do so. For nonrenewable resources, an economy switches to lower-quality deposits after the higher-

quality deposits are depleted. Over the last century the average quality of copper ore extracted from U.S. mines declined from about 2 percent to about 0.5 percent (Figure 10.21). Economies also move to lower-quality deposits when demand exceeds the quantity that can be produced from high-quality deposits. This effect was most apparent during the Great Depression, when GDP declined by about 30 percent and about 25 percent of the workforce was unemployed. Because of the decline in economic activity, the demand for copper dropped, and the low-quality copper mines were closed. As a result,

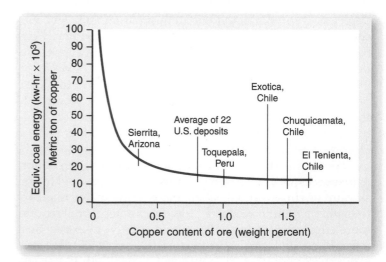

FIGURE 10.20 *Resource Quality* The relation between ore grade and the energy cost of metal extraction for copper. (Source: Data from N.J. Page and S.C. Creasey, "Ore Grade, Metal Production, and Energy," Journal Research U.S. Geological Survey, 3: 9–13.)

CASE STUDY

Can Human Ingenuity Substitute for a Degraded Environment?

A remarkable feature of human adaptation to the environment is our ability to adjust to the depletion of natural resources and environmental services. An important way in which we do this is by substituting human ingenuity, in the form of economic capital or social institutions, for reduced supplies of natural resources or environmental services. As crude oil becomes scarcer and motor gasoline prices rise, more efficient cars can be produced, which reduces our use of the scarce resource. This substitution lies behind the idea of weak sustainability that is discussed in Chapter 9.

This leads to an important question: To what extent can human capital and social institutions substitute for a degraded environment? Suppose that human ingenuity can replace any individual natural resource or environmental service. Fertilizers and genetically engineered crops can substitute for severely eroded soils; sea walls can protect low-lying coastal areas from rising sea levels caused by climate change; sewage treatment plants can substitute for the waste processing lost by drained wetlands; and so on. In short, does the concept of weak sustainability imply that environmental degradation will not constrain economic affluence?

The short answer is no. The reason is based on the physical interdependence among economic capital, social institutions, natural resources, and environmental services. Economic capital and social institutions require natural resources and environmental services. The construction, operation, and maintenance of tools, machines, and factories require energy, materials, and environmental services. Similarly,

the humans that direct machines require energy and materials (food and water). Thus the fuel-efficient car developed in response to oil scarcity in fact requires some oil, and other forms of energy, for its manufacture and maintenance. Similarly, the wastewater treatment plant built for Boston Harbor requires energy and natural resources to produce and operate.

This leads to an important observation: Substituting economic capital and social institutions for natural resources and environmental services requires some of the very resources and services whose scarcity is causing the search for a substitute. This fundamental aspect of our relationship with the environment stems from the basic physical laws governing the transformation of energy and matter described in this chapter and Chapter 2. All goods and services require the use of energy and materials, and all release some wastes.

In the short run, environmental degradation reduces economic output because that degradation increases the quantity of economic capital and social institutions used to produce a natural resource or an environmental service. This reduces the economic capital and social institutions that are available to produce other goods and services. Consistent with the notion of opportunity costs, this reduces economic output.

This loss of economic output grows over time. As total economic output shrinks, less is available to replace economic capital that is constantly wearing out. (Remember from Chapter 2 that the second law of thermodynamics dictates that the entropy of organized structures, such as economic capital, increases over time.) As

the stock of economic capital declines over time, economic production declines further. Alternatively, to avoid a reduction in economic capital, society can invest a greater fraction of the reduced output. But this reallocation reduces the quantity of goods and services that are available to consumers. And per capita consumption of these goods and services is the most direct measure of economic affluence.

This has important implications for evaluating the effects of large-scale environmental alteration such as climate change. We must evaluate the effects of climate change by answering these questions: (1) How will climate change affect the natural resources and environmental services we depend on? (2) How will changes in those resources and services affect the quantity of economic capital required to produce energy and raw materials? (3) How will such changes ripple through the economy over time?

ADDITIONAL READING

H.E. Daly and J. Farley. *Ecological Economics: Principles and Applications*. Washington, DC: Island Press, 2004.

R.K. Kaufmann. "The Economic Multiplier of Environmental Life Support: Can Capital Substitute for a Degraded Environment?" *Ecological Economics* 12 (1995): 67–79.

STUDENT LEARNING OUTCOME

- Students will be able to define the relationship between technological change and the environment.

the average quality of the ores mined in the United States increased from about 1.5 percent copper to about 2 percent during the 1930s.

The best first principle and opportunity costs associated with resource depletion are illustrated by the distribution and utilization of oil fields in the United States. Oil is found beneath the surface in single accumulations that geologists call fields. The size of an oil field is measured by the quantity of oil it contains. The average field discovered around the turn of the twentieth century contained 20–40 million barrels of oil, and the largest fields contained several billion barrels (Figure 10.22). But as the use of oil increased, the big high-quality deposits of oil were discovered and depleted. Today the average field discovered

contains less than a million barrels. The decline in quality of oil resources greatly increased the work required to find and extract oil. The cost to discover a barrel of oil in the early part of the century was less than $1.00 per barrel; today it is about $15.00 per barrel when adjusted for inflation (Figure 10.23). The $14 increase represents labor, capital, and energy that could have been used elsewhere in the economy to produce non-oil goods or services.

The best first principle also has implications for potentially renewable environmental services such as the dilution and detoxification of wastes. Environmental services can dilute and detoxify wastes at low levels of economic activity; but as economic activity expands, the production of wastes may exceed the environment's

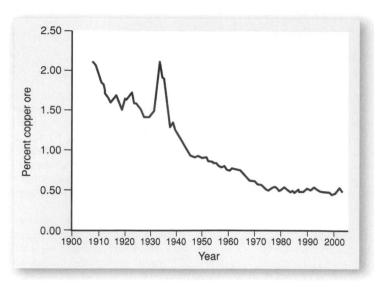

FIGURE 10.21 *Copper Depletion* The trend in ore grade for copper mined in the United States. *(Source: Data from U.S. Geological Survey.)*

ability to process them. Beyond this threshold, society may have to supplement these efforts by investing capital, labor, and energy. These expenditures are opportunity costs of exceeding the capabilities of environmental services.

Such costs are illustrated by changes in the Boston Harbor. For much of its existence the city of Boston and surrounding communities dumped sewage into Boston Harbor. These wastes had relatively little impact when the community was small—the amount of nitrogen in the sewage could be processed by the local flows and storages of the nitrogen cycle. But as the human population expanded, nitrogenous wastes were put into Boston Harbor faster than

the nitrogen cycle could convert them to less harmful forms. As a result, eutrophication lowered oxygen levels, which reduced fish and shellfish populations. To solve this problem, the Massachusetts Water Resources Authority built a $12 billion waste processing plant (Figure 10.24). As a result Boston area water and sewer bills are among the highest in the nation, which leaves local residents less money to spend on other goods and services.

ACCOUNTING FOR ENVIRONMENTAL DEGRADATION

Measuring the effects of environmental degradation is very useful, but calculating the opportunity costs associated with each form of environmental degradation is exceedingly difficult. To avoid these difficulties, economists use the concept of **natural resource** or **environmental accounting**. See *Policy in Action: Greening the GDP*. These techniques seek to measure the costs of environmental degradation so that they can be included in national measures of economic production, such as GDP. Returning to the opening story in this chapter, doing so would give the president a more accurate assessment of the economy.

Many of these techniques are based on the difference between income and wealth. You may use the terms income and wealth interchangeably, but there is an important difference. **Income** refers to the amount of money a person earns per unit time. For hourly workers, annual income is the wage rate (dollars per hour) times the number of hours worked per year. **Wealth** refers to the value of assets that one accumulates during a lifetime. Your wealth is the value of everything you own minus the value of all your debts.

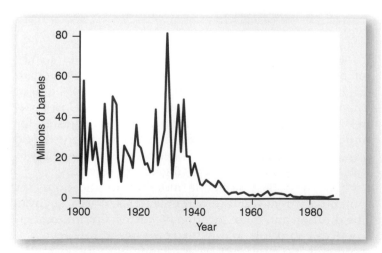

FIGURE 10.22 *Resource Quality for Crude Oil* The average size (barrels) of new oil fields discovered in the United States. *(Source: Data from U.S. Department of Energy.)*

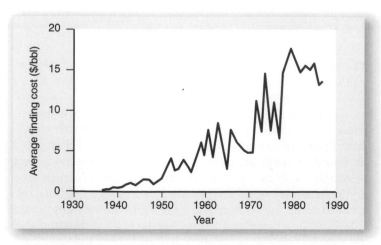

FIGURE 10.23 *The Cost of Crude Oil* The average cost ($/barrel) of new oil discoveries in the United States. *(Source: Modified from C.J. Cleveland, "Physical and Economic Aspects of Resource Quality: The Cost of Oil Supply in the Lower 48 United States, 1936–1988," Resources and Energy 13: 163–188.)*

FIGURE 10.24 *Sewage Treatment*
The Deer Island sewage treatment plant in Boston Harbor treats sewage generated by over 2 million people in 43 communities, plus 5,500 firms in the greater Boston area.

GDP measures annual rates of economic production or income. But GDP confuses income and wealth when it comes to the economic contribution of many environmental goods and services. Specifically, GDP overstates the income associated with extracting nonrenewable and renewable resources because part of this contribution comes from liquidating wealth. Liquidating means that a physical asset, such as a house or a car, is converted to its cash value. The full value of this cash payment should not be counted as income.

To illustrate, suppose you sell your home. You now have more money than you did before, but you should not treat all of it as income. In reality the sale reduces your future income. When you lived in your own house you did not pay rent. But after selling it you must pay rent, and these payments will reduce your future income. In short, you have sold a house that can be viewed as a stock of wealth, but doing so reduced the flow of benefits that contributed to your income—housing.

What does a house sale have to do with accounting for environmental degradation? Think of nonrenewable resources as finite stocks of wealth. As a finite stock, each barrel of oil or ton of metal extracted today leaves less for future extraction. When firms extract oil, they are liquidating the value of the stock. The value of resources extracted in any given year is included in GDP. Thus the liquidation of natural resources shows up as an increase in our income. This treatment contrasts with that accorded houses—GDP does not include the sale price of existing homes.

Natural resource accounting attempts to fix this flaw. Rather than count the full value of the resource that is extracted in a given year, modified versions of GDP, which are known as **green national accounts,** include only the portion that is sustainable. This sustainable portion corresponds to the income that the cash value of the extracted resource could generate. Suppose an economy liquidates $1 million of oil in a given year. Using the current system, GDP would include the full value of the oil, $1 million. But this quantity of oil cannot be produced on a sustainable basis. Eventually the oil will be exhausted. Now suppose that society puts the $1 million in the bank at 5 percent interest. Doing so would generate $50,000 per year on a sustainable basis. Only this $50,000 would be included in the revised version of GDP, regardless of whether the $1 million was put in the bank.

The same principle applies to the production of potentially renewable resources. For these resources, the full value of annual production could be included in GDP *so long as the resource is produced in a sustainable manner.* If production exceeds maximum sustainable yield (see Chapter 9), the resource will eventually be exhausted. Green versions of GDP include the value of potentially renewable resources that are produced sustainably; the value of production beyond maximum sustainable yield is not included in GDP. For example, suppose a forest can produce $1 million in timber annually, but firms cut timber worth $1.5 million. Green GDP includes only $1 million worth of timber sales.

Robert Repetto and his colleagues at the Word Resources Institute have applied natural resource and environmental accounting to calculate a revised version of GDP, which they called net domestic product, for Indonesia. Indonesia is a large developing nation that generates a significant portion of its GDP by selling agricultural products, timber, and oil.

POLICY IN ACTION

Greening the GDP

Most people who understand how GDP is calculated acknowledge that resource depletion and environmental degradation are not properly accounted for. So what is being done about this? Many nations are developing integrated environmental and economic satellite accounts (IEESA). This system includes the standard components of GDP (investment, production of goods and services), as well as a parallel account of natural assets (soils, timber, minerals, and the like) that tracks the annual rates at which they are degraded (soil erosion, overfishing, deforestation, and so on). IEESA lets a nation paint a much broader picture of its overall well-being.

The United States made a foray into environmental accounting at the start of the Clinton administration. In 1994 the Bureau of Economic Analysis released a full set of estimates for subsoil mineral assets—the value of mineral reserves. These accounts were prototype satellite accounts, designed to illustrate the method. But this preliminary attempt became embroiled in political controversy and faced opposition from the minerals industry. Congress directed the Bureau of Economic Analysis (BEA) to suspend further work in this area and to obtain an external review of environmental accounting. The National Research Council, part of the National Academy of Sciences, formed a panel to consider what the nation should do in the way of environmental accounting.

The National Research Council panel concluded that "extending the national income and product accounts to include assets and production activities associated with natural resources is an important goal" and provides useful information for decision making. Further, the panel determined that the rationale is "solidly grounded in mainstream economic analysis and that BEA's activities are consistent with the extensive domestic and international efforts to improve and extend the NIPA [National Income and Product Accounts]." Developing a set of comprehensive nonmarket economic accounts was determined to be a high priority for the nation. Despite these benefits, the Bureau of Economic Analysis was not allocated funds to continue its work. Instead congressional appropriations to the BEA expressly forbade further work on environmental accounting.

The reasons for this were political. Politicians from energy and mining states erroneously believed that a "green GDP" would hurt coal producers and power generators. Supporters of environmental accounting charged that mining firms, including oil and gas producers, were afraid that increased financial scrutiny would draw attention to the generous subsidies they receive from the federal government. In fact, environmental accounting does nothing more than treat natural assets used in the economy in the same manner as assets produced by society. "Greening" the GDP remains a critical ingredient for sound decision making regarding our use of all natural assets.

ADDITIONAL READING

National Academy of Sciences. *Nature's Numbers: Expanding the National Economics Accounts to Include the Environment.* Washington, DC: National Academy Press, June 1999.

STUDENT LEARNING OUTCOME

- Students will be able to discuss how GDP is modified to account for resource depletion and environmental degradation.

To calculate net domestic product, for Indonesia, three forms of environmental degradation are subtracted from GDP. The environmental degradation associated with soil erosion is based on the value of crop production lost due to the loss of soil productivity. The costs of deforestation are equal to the value of net reductions in the standing stock of timber (timber cut minus regrowth). The costs of oil depletion are equal to the value of the decline in the proved reserves of oil (discoveries minus extraction). Subtracting these costs produces an estimate of net domestic product that is generally lower than the traditional calculation of GDP (Figure 10.25).

Of course, trying to correct GDP for the effects of environmental degradation forces the analyst to make subjective judgments. Which interest rate should the analyst use to calculate sustainable income from revenues earned from nonrenewable resources? How does the analyst calculate the sustainable rate of producing a renewable resource? Clearly such decisions can cause analysts to overstate the effects of environmental degradation. On the other hand, making no corrections causes GDP to understate the effects of environmental degradation.

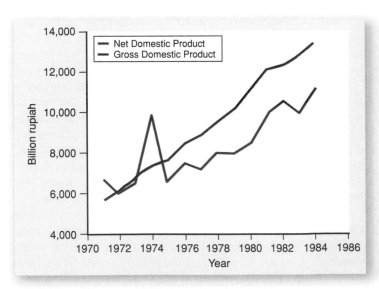

FIGURE 10.25 *Green National Accounts* Gross domestic product (GDP) and net domestic product (NDP) in Indonesia. Rupiah is the currency used in Indonesia. (Source: Modified from R. Repetto, Wasting Assets—Natural Resources in the National Income Accounts, World Resources Institute.)

SUMMARY OF KEY CONCEPTS

- The conventional view of the economy describes GNP as produced by two factors of production: capital and labor. This perspective is incomplete because it ignores the role of the environment in sustaining economic production.

- An ecological view of the economy describes the energy, materials, and environmental services that make the production of goods and services possible.

- Natural resources occur in differing grades or qualities. Lower-quality resources require more effort to upgrade to a useful state. Humans use the highest-quality sources of natural resources first.

- Higher-quality natural resources—those that are most concentrated—required greater environmental energy costs to create than their lower-quality counterparts.

- The opportunity costs of depletion and degradation tend to rise as we move from higher- to lower-quality natural resources.

- Rising affluence is associated with increasing labor productivity, which in turn is due to a greater energy subsidy of human labor.

- There is a close correlation between economic well-being, as measured by per capita GDP, and energy use per capita.

- Ecosystem services are a vital input to the economic process, but they tend be undervalued because they are not privately owned and thus are not traded in the market.

- Standard methods for calculating national production ignore the costs of resource depletion and environmental degradation. Accounting for such costs often reduces the apparent standard of living in a society, sometimes significantly.

REVIEW QUESTIONS

1. Draw a diagram that describes the circular flow model of the economy. Now modify it to include the role of the environment in sustaining economic production.
2. Explain the best first principle and its economic implications.
3. Explain the fundamental difference between natural resource inputs to the economy and ecosystem services.
4. Distinguish between postconsumer wastes and dissipative wastes.
5. Why are property rights an important determinant of how a natural resource or environmental service is used by society?
6. Should attempts be made to value all ecosystem services and include them in our systems of national accounts? What are the potential pitfalls in doing this?

KEY TERMS

affluence
avoided costs
biodegradable waste
capital
common property
consumption
contingent valuation
crustal abundance
depletion
direct market valuation
dissipative waste
draft animal
economic growth
ecosystem services
environmental accounting
environmental degradation
environmental energy cost
exhaustion

factors of production
firms
goods
green national accounts
gross domestic product (GDP)
gross national income (GNI)
gross national product (GNP)
hedonic pricing
income
labor
labor productivity
market failure
material waste
natural resource accounting
nondegradable wastes
opportunity costs
ore
persistent wastes

personal consumption expenditures
petrochemicals
postconsumer waste
private property
production
property rights
public property
resource quality
services
technology
travel cost method
utility
waste
waste assimilation
waste heat
wealth

11

THE DRIVING FORCES OF ENVIRONMENTAL CHANGE

Population, Affluence, and Technology

The new and the old *In China, ways of life that have existed for millenia, such as hand irrigation, coexist with industrialization. This chapter will explore the diverse ways that human activity impacts the environment.*

STUDENT LEARNING OUTCOMES

After reading this chapter, students will be able to

- Explain the terms that constitute the IPAT equation.

- Identify examples that illustrate how technology can reduce or increase environmental impact.

- Describe the net effect that population growth, affluence, and technological change have had on gasoline use in the United States.

- Cite the reasons why firms prefer market-based incentives in environmental policy to command and control approaches.

- Explain how personal and cultural beliefs affect environmental change.

INTRODUCTION

On New Year's Eve Tom and Mary Smith made a resolution to reduce their family's impact on the environment. That night they inventoried all the energy and materials they used and the wastes they produced. The totals were a bit surprising. Since they were married, their use of energy and materials, and hence their production of wastes, had increased steadily. They wondered what had caused those changes.

The first thing that struck them was the effect of changes in their income. Tom had received steady raises, and Mary's reentry into the workforce after the birth of their second child more than doubled their household income. As their income grew, they spent more money on clothes; they bought a second car and a new refrigerator; they ate in restaurants more often; and they took more expensive vacations. In addition, they spent more on water, electricity, natural gas for the furnace, and gasoline for their cars. They also noticed that they made more frequent trips to the recycling center and had to carry more bags of garbage to the curb every Monday morning.

The birth of their two children also increased their use of energy and materials and their production of wastes. Additional people in the household meant more food to purchase, more clothes to buy, and more hot water for showers, dishwashing, and laundry. They shuttled their kids to and from day care and later soccer practice, piano lessons, and overnights with their friends, and this increased gasoline consumption and generated more air pollution.

These changes surprised the Smiths because they prided themselves on being environmentally conscious. But reluctantly they admitted that they had chosen technologies that increased their environmental impact. Tom used to ride his bike to work, but several years ago he switched to driving a car. That change in transportation technology dramatically increased his use of fossil fuel energy and production of pollution. See *Your Ecological Footprint: Personal Transportation*. To reduce these impacts the Smiths chose a car that had a fuel-efficient engine and that used "ozone-friendly" refrigerants in the air conditioning system. The Smiths also tried to purchase products made from recycled materials.

The Smiths' environmental audit indicated that growing income, family size, and choice of technologies determined their use of resources and production of wastes. But their self-assessment revealed another factor: their attitudes, values, and beliefs. Tom and Mary admitted that they liked driving new air-conditioned vehicles, buying some of the latest electronic gadgets or fashions, eating out, and taking nice vacations. On the other hand, their environmental awareness revealed another set of values that emphasized environmental preservation. They realized that they needed to examine their attitudes about family size, choice of technology, and consumption of goods and services. They knew that the choices they made today would determine the quality of the environment for their children and grandchildren. ■

THE ROOT CAUSES OF ENVIRONMENTAL IMPACT

Families worldwide repeat the choices faced by the Smiths. We can generalize their experience and define environmental impact by the **IPAT equation:**

Impact (I) = Population (P) × Affluence (A) × Technology (T)

(11.1)

$$I = \text{People} \times \frac{\text{Economic activity}}{\text{People}} \times \frac{\text{Environmental impact}}{\text{Economic activity}}$$

According to this equation, environmental impact (I) is determined by population size (P); affluence (A), which determines the economic activity per person; and technology (T), which determines the amount of resources extracted or waste produced per unit of economic activity. Each of these factors is determined by decisions people make, which in turn are influenced by a larger system of rules, ideas, attitudes, and beliefs. These rules and beliefs are organized in social and political institutions. To understand the root causes of environmental impact, we describe how population, affluence, and technology affect the use of resources and the emission of wastes, and how social and political institutions influence people's choices about and attitudes toward family size, affluence, and technology.

The Smiths came to understand that population, affluence, and technology interact in complex ways, which sometimes make it difficult to identify the most important driving force of environmental change. In addition, a single action may cause dozens of environmental impacts. Thus to simplify our discussion, we focus on how population, affluence, technology, and social and political institutions determine the severity of environmental impacts associated with three important economic activities: food production, gasoline consumption, and solid waste disposal.

POPULATION

The chapter about carrying capacity discussed the reasons behind the rapid growth in the human population over the past two centuries. In the twentieth century the human population

Your ECOLOGICAL footprint

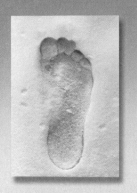

Personal Transportation

Transportation is an important aspect of modern life. People in developed nations spend at least one hour in travel each day, which consumes about one-sixth of the average household's income. The high degree of mobility enjoyed in developed nations is one of the defining characteristics of an affluent lifestyle.

Transportation also is a leading cause of resource depletion and environmental degradation. Globally transportation accounts for about a third of crude oil consumption and about 14 percent of emissions such as carbon dioxide. In the United States transportation is even more important—it accounts for about two-thirds of oil consumption, 32 percent of carbon dioxide emissions, 45 percent of nitrogen oxide emissions, and 77 percent of carbon monoxide emissions.

Why is transportation such a big environmental problem in the United States? One reason is that Americans use technologies that consume huge amounts of energy and release great quantities of pollutants. The single most important technology is the automobile. On a per capita basis, Americans own more cars and drive them more miles than all other nations. They also travel by air more frequently than inhabitants of most other nations. These choices are significant because cars and planes have large environmental impacts. Driving a single-occupant car to work uses 1,826 kcals of gasoline per passenger mile, whereas riding a bus uses less than 250 kcals of diesel fuel per passenger mile. Riding a bicycle is the most energy-efficient technology—it uses just 35 kcals per mile of relatively low-pollution carbohydrate energy (your morning breakfast) rather than fossil fuels such as gasoline and diesel fuel, which release harmful pollutants.

Why do Americans have a love affair with the automobile? Cultural values play an important role. Many Americans see their cars as a source of individual freedom and believe that a car says a lot about its owner. Advertising reinforces this belief with images of sleek, powerful vehicles streaking down deserted country roads, trees rustling in the wind, the sun majestically setting in the background.

These desires are supported by Americans' affluence. The rise in the affluence of the average American over the past several decades produced a surge in automobile use, and with it an increase in the use of gasoline and the release of air pollutants. In 1950 the average U.S. household owned one car—it now owns two. In 1950 each licensed driver drove an average of about 16,000 kilometers (10,000 miles) each year; now licensed drivers per household drive over 30,000 kilometers (19,000) miles each year.

Economic incentives also promote the use of cars. Due to market failures and government subsidies, car owners do not pay the full costs of operating a vehicle. External costs stemming from pollution, climate risks, a military presence in the Middle East, oil storage costs, accidents, and noise amount to $126 billion per year. Other costs of driving that are spread over the general population include road construction and repair, highway services, and parking, amount to $170 billion. The total cost not paid directly by drivers is nearly $300 billion, equivalent to more than 5 percent of the U.S. gross domestic product. If drivers had to pay these costs, they would probably drive less and choose alternative modes of transportation.

Calculating the Ecological Footprint of Your Transportation

1. Estimate the total number of miles you traveled last year by each transportation mode and enter them in column 2 of the accompanying worksheet. The easiest way to do this is to estimate the distance of your average commute from home to work or school and the number of trips per year. Also include trips you made to visit relatives, go on vacation, and so on.

2. In column 4 calculate the amount of energy consumed by your transportation choices. To do so, multiply column 2 (miles traveled) by column 3 (energy intensity). Each mode of transportation has a different energy intensity, expressed in kcal/passenger mile. This is the amount of energy required to transport you each mile.

3. In column 6 calculate the pollution generated by your transportation choices. To do so, multiply column 2 (miles traveled) by column 5 (pollution intensity). Each mode of transportation has a different pollution intensity, expressed in grams/passenger mile. This is the sum of six pollutants released by transporting you one mile in each mode. Those pollutants include carbon dioxide, sulfur dioxide, nitrogen oxides, total suspended particulates, carbon monoxide, and nonmethane hydrocarbons.

4. Calculate the annual total energy use and pollution released by your use of transportation (sum columns 2, 4, and 6).

Comparing Your Footprint to People in Other Nations

Reliance on various modes of transportation varies tremendously among nations. Table 1 shows the transportation modes used by the average Swede:

What would your footprint look like if you traveled as the average Swede does? The following exercise answers that question:

1. Multiply the Swedish fraction for each mode of transportation times the total number of miles you traveled last year (from the bottom of column 1 in the worksheet). This will give the number of miles traveled by each mode *if you had traveled like the average Swede.*

2. Recalculate the energy use and pollution released as you did in steps 2, 3, and 4.

TABLE 1	Mode of Transportation (fraction of total)				
Country	Bicycle	Walking	Motorcycle	Automobile	Public Transport
Sweden	0.1	0.39	0.02	0.36	0.13

3. Compare the results of the two exercises. Can you explain the differences in energy use and pollution release based on the differences in transportation technology used in the United States and in Sweden?

ADDITIONAL READING

Gordon, Deborah. *Steering a New Course: Transportation, Energy, and the Environment.* Washington, DC: Island Press, 1991.

Greene, David L., and Danilo J. Santini (Eds.). *Transportation and Global Climate Change.* Washington, DC: American Council for an Energy-Efficient Economy, 1993.

MacKenzie, James J., Roger C. Dower, and Donald D. T. Chen. *The Going Rate: What It Really Costs to Drive.* New York: World Resources Institute, 1992.

STUDENT LEARNING OUTCOME

- Students will be able to rank the major forms of transportation in terms of their energy requirements.

Worksheet for Calculating the Ecological Footprint of Transportation

(1) Transport Mode	(2) Miles Traveled	(3) Energy Intensity (kcal per mile traveled)	(4) Total Energy Used (kcal) (col 2 × 3)	(5) Pollution Intensity* (grams per mile traveled)	(6) Total Pollution Emission (grams) (cols 2 × 5)
Bicycle	_____	35	_____	0	_____
Walking	_____	76	_____	0	_____
Motorcycle	_____	629	_____	12.48	_____
Automobile					
Single occupancy	_____	1,826	_____	25.16	_____
Average occupancy	_____	968	_____	14.82	_____
Personal truck/van					
Single occupancy	_____	2,423	_____	33.65	_____
Average occupancy	_____	1,200	_____	17.71	_____
Rideshare					
3-person carpool	_____	623	_____	8.58	_____
9-person vanpool	_____	222	_____	3.75	_____
Bus (diesel)					
Transit	_____	741	_____	3.74	_____
Intercity	_____	237	_____	1.20	_____
Rail					
Intercity Amtrak (diesel)	_____	667	_____	2.41	_____
Transit (diesel)	_____	929	_____	7.73	_____
Commercial airline	_____	1,762	_____	2.26	_____
Annual totals:	_____		_____		_____

*Carbon dioxide, sulfur dioxide, nitrogen oxides, total suspended particulates, carbon monoxide, and nonmethane hydrocarbons.

Sources: Data from D. Gordon, *Steering a New Course: Transportation, Energy, and the Environment,* Island Press; data from S.C. Davis, *Energy Transportation Databook,* Oak Ridge National Laboratory.

increased from 1.7 billion to more than 6 billion people. At the same time enormous environmental change has occurred at local, regional, and global scales. Are the two phenomena connected? Is population growth the cause of environmental problems? This section explores this complex relationship.

The Contribution of Population Growth to Environmental Change

Population growth often is blamed for environmental problems because the two seem to go together. More people mean more mouths to feed, more roads and houses, more cars, more gasoline use, and so on. If we hold affluence and technology constant, the economic activity required to support an additional person means more resources are extracted and more wastes are generated.

Land Conversion for Food Production

Population growth is possible only if food production expands. For most of human history people increased food production by converting forests and grasslands to cropland and pastures. This process is well documented in the United States (Figure 11.1). The earliest settlers in New England quickly cleared forests to make way for crops and pasture. Around the time of the Civil War, pioneers were breaking the sod in the Midwest in what would become the corn belt (Figure 11.2). Between 1880 and 1910 settlers moved farther west, turning the prairie of the Great Plains into what is now the grain belt. The conversion of natural ecosystems to agricultural land slowed when the wave of expansion broke against the Rocky Mountains. By then much of all the forest and prairie ecosystems in the middle portion of the continent had been converted to croplands and pastures.

To varying degrees the same process occurred wherever human numbers increased. For each person added to the world's population since 1700, an average of 0.3 hectare (0.7 acre) of natural ecosystems was converted to agricultural land, predominantly cropland. As a result, about a third of the non-ice land surface of the planet has been converted from forest, prairie, or wetland to cropland or pasture. Land conversion is the principal means by which humans control

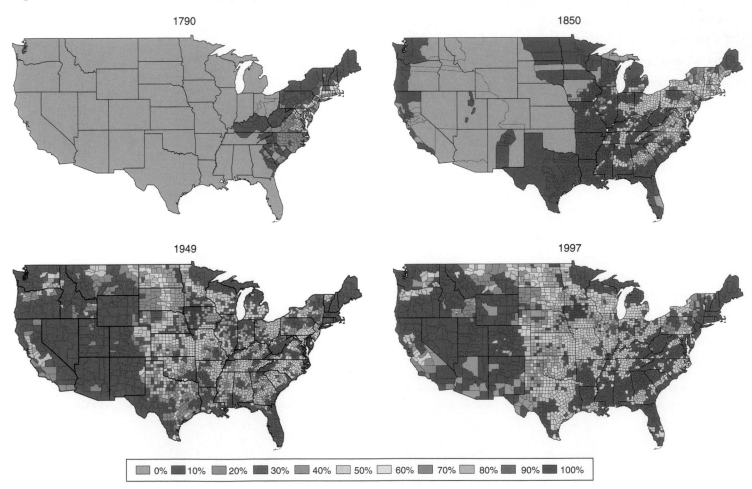

FIGURE 11.1 *Land Conversion for Agriculture* The conversion of native ecosystems to cropland in the United States from 1790 to 1997. *Colors represent the percentage of original ecosystem in each county that is converted to cropland. Note that the process works in both directions. For example, more than half of the forests in southern New England were eliminated by 1850, but by 1950 much of them had grown back.* (Source: Data from U.S. Geological Survey.)

(a)

(b)

(c)

(d)

FIGURE 11.2 *Land Conversion on the American Frontier* *A typical conversion of forest to cropland and pasture by one family in the American frontier in the nineteenth century. (a) The first six months; (b) the second year; (c) ten years later; (d) the work of a lifetime.*

about 40 percent of global net primary production. (See Chapter 6's Your Ecological Footprint.) This control alters or eliminates many of the important environmental services provided by ecosystems such as flood protection, habitat for biodiversity, and maintenance of soil fertility. Thus the expansion of food production systems arguably is one of the biggest environmental changes caused by humans, and it is directly tied to population growth.

Solid Waste

Solid wastes that are discarded by households, commercial establishments, and industry are called municipal solid wastes. There is a direct link between population size, population density, and the generation of solid wastes. For a given level of affluence and technology, more people living in an area

produce more waste. The 8 million people in New York City have a population density of 10,194 people per square kilometer (26,403 people per square mile). Each year households and firms in New York City generate more than 13.6 million metric tons (15 million tons) of solid waste and recyclables. *In a single day* the inhabitants of New York City produce enough waste to fill more than 6,000 garbage trucks, which lined up end to end would stretch 57 kilometers (35 miles).

All the major landfills in the city have been filled and closed, and there is no land available for new landfills. As a result New York City exports 67 percent of its waste to other states, some as far away as Virginia, at ever-escalating costs (Figure 11.3). In a report titled *No Room to Move: New York City's Impending Solid Waste Crisis*, the city warned that growing landfill shortages in those states would pose even more serious waste management issues in the coming years.

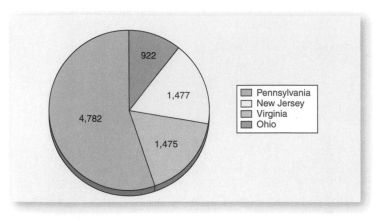

FIGURE 11.3 *Solid Waste Generation* *The amount of solid waste that New York City ships to other states each year. Units are thousands of tons.* (Source: Data from No Room to Move: New York City's Impending Solid Waste Crisis, New York City Comptroller's Office.)

This problem created a public stir when, in 1987, a barge named the *Mobro* left the New York area with 2,900 metric tons (3,200 tons) of trash (Figure 11.4). Because of shrinking landfill space, the plan was to ship the waste to North Carolina. However, officials there rejected the waste, and the *Mobro* began an unusual odyssey in search of a place to dump the waste. It made stops in Louisiana, Mexico, Belize, and Cuba, with no willing takers. After traveling 9,650 km (6,000 miles), the barge eventually was towed back to New York Harbor, and the trash was burned in Brooklyn.

Gasoline Consumption

The aging of the baby boom generation illustrates the effect of population size on gasoline consumption. Between 1962 and 1978, 78 million baby boomers reached the minimum driving age. The resulting surge in the number of registered drivers was much greater than the overall rate of population growth. In addition, the newly licensed baby boomers tended to drive more than their parents. More drivers helped double gasoline consumption from 1960 to the late 1970s (Figure 11.5). This accelerated the depletion of oil resources in the United States, increased the nation's dependence on oil imported from the Middle East, and increased the release of air pollutants caused by the combustion of gasoline.

Water Quality

Nearly four in ten people live within 100 kilometers (62 miles) of a coast. Rivers that run through coastal regions often receive significant inputs of pollutants. For example, human populations add nitrogen to rivers via industrial and automotive fossil fuel use, the application of fertilizers, and sewage discharge. Because nitrogen is a limiting factor for net primary production, significant increases in the riverine input of nitrogen to coastal waters can lead to problems such as algal blooms, hypoxia or anoxia, or the loss of native or commercially important species.

There is a close connection between the nitrate (NO_3^-) concentration in a river and the density of population living in that drainage area (Figure 11.6). More people packed

FIGURE 11.4 *The Garbage Barge* *The barge named the* Mobro, *famous for its ill-fated 6,000-mile journey to find a home for 3,200 tons of garbage.*

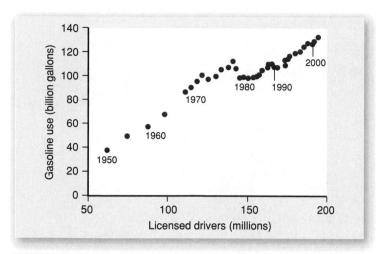

FIGURE 11.5 *Population and Gasoline Use* *The relationship between the number of licensed drivers and motor gasoline use in the United States.* *(Source: Data from U.S. Departments of Transportation and Energy.)*

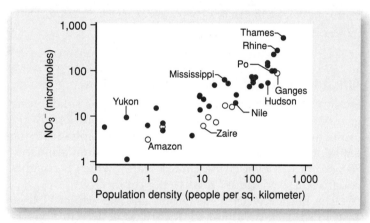

FIGURE 11.6 *Population and Water Quality* *The average annual nitrate concentration (NO_3^-) in rivers versus human population density in the watershed.* *(Source: Data from N.J. Cole, et al., "Nitrogen Loading of Rivers as a Human-Driven Process," in M.J. McDonnell and S.T.A. Pickett (Eds.), Humans as Components of Ecosystems, Springer-Verlag, pp 141–157.)*

into a drainage area release more nitrates into a relatively constant flow of water, thus harming water quality. Note in Figure 11.6 that the relation holds across temperate and tropical river systems and across rich and poor nations.

AFFLUENCE

Chapter 10 described the driving forces behind the rise in affluence and the differences in affluence among and within nations. Affluence is a critical determinant of environmental degradation because high rates of economic activity are associated with rapid rates of resource use and waste production. For this reason it is possible for one rich person to have a greater environmental impact than several poor people.

The Contribution of Affluence to Environmental Change

Affluence affects the environmental impact per person. In general, increasing affluence tends to exacerbate environmental impacts. We demonstrate this by returning to our examples of food production, motor gasoline use, and solid waste disposal.

Affluence and Food Production

On average you need to eat 2,500–3,000 kcals of food each day. But the types of food used to satisfy this requirement differ greatly among nations. In addition to local tastes, affluence influences the types of foods eaten. In general, rich people get a larger fraction of their diet from meat than poorer people, who tend to get most of their diet from grains. People in developed nations such as the United States and France consume 90 to 100 kilograms (198–220 lbs) of meat per year, whereas people in developing nations such as India and Egypt consume less than 20 kilograms (44 lbs).

Diets vary in their environmental impacts. Eating higher on the food chain causes more environmental degradation because meat requires more land, energy, and materials to produce (Figure 11.7). Because most of the energy is lost when transferred from one trophic level to the next (Chapter 5), more land is required to support a meat-based diet than a grain-based diet. In Europe nearly 60 percent of all grain grown is fed to livestock, in the United States that figure is 70 percent. Only 2 percent of the grain grown in India is fed to animals.

Compared to the production of grain, meat uses more fossil fuels and water. In developed nations each kilogram of corn requires 1,400 kilocalories to produce; it takes about 30,000 kcals of fossil fuel energy to produce a kilogram of pork. Other meats require nearly as much energy. Most of this energy comes from oil and natural gas that are used to power farm machinery and to manufacture fertilizers, pes-

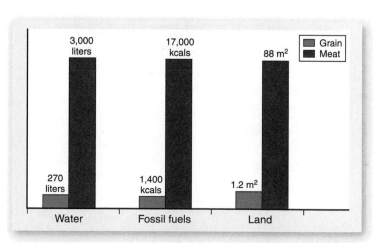

FIGURE 11.7 *Diet and the Environment* *The quantity of water, fossil fuels, and land required to produce one kilogram of grain and one kilogram of meat in the United States.*

ticides, and other agricultural chemicals used to grow the grains that are fed to livestock. All told, producing the meat eaten by an average American each year uses the equivalent of 190 liters (50 gallons) of gasoline. Meat production also uses large quantities of water. In the United States more than 3,000 liters (793 gallons) of water are used to produce a kilogram of beef. In the major beef-producing region in the United States—Colorado, Kansas, Nebraska, and the Texas Panhandle—pumping this water depletes underground supplies (see Chapter 18).

Affluence and Gasoline Consumption

The ability to own and drive a car is strongly correlated with income. In 2004 there were close to 500 cars registered in the United States per 1,000 people, compared to 6 in India. Most of the world's population is too poor to own cars, so they walk, ride bikes, or take buses, all of which have a smaller environmental impact than personal cars (Figure 11.8). People who own cars drive them roughly in proportion to their income. For example, the average person in Cameroon drives about 120 kilometers (75 miles) each year. But as incomes rise, people tend to drive more miles. The average person in the United States drives nearly seventy-five times farther in a year than people in Cameroon, thereby consuming much more gasoline (Figure 11.9).

Even within a nation, gasoline consumption is closely correlated with affluence. Higher incomes boost gasoline consumption in two ways. First, people drive more as they get richer. Economists estimate that a 10 percent increase in GNP per driver increases motor gasoline consumption by about 4 percent. Second, as people get richer they tend to purchase less fuel-efficient vehicles. One out of every two new vehicles now sold in the United States is a so-called utility vehicle: minivan, pickup truck, or four-wheel-drive vehicle. The gas mileage of these vehicles is much lower than the cars they replace.

FIGURE 11.9 *Affluence and Transportation* In low-income nations, bicycles, buses, and walking are common forms of transportation. Here a rickshaw driver rests on a street in Dhaka, Bangladesh.

Affluence and Solid Waste Generation

Studies show that the amount of municipal solid waste generated per person rises roughly in proportion to per capita income. Consistent with this link, per capita methane emissions from landfills in the United Kingdom are about 70 percent the level in the United States, which is roughly equal to these countries' relative per capita GNP.

The link between solid wastes and per capita GNP is easy to see (Figure 11.10). Affluent people live in a world filled with packaging, disposable goods, and products that rapidly wear out, break down, or become obsolete. These

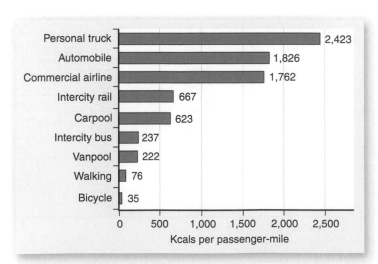

FIGURE 11.8 *Energy and Transportation* The energy intensity of various transportation modes. (Source: Data from U.S. Department of Transportation.)

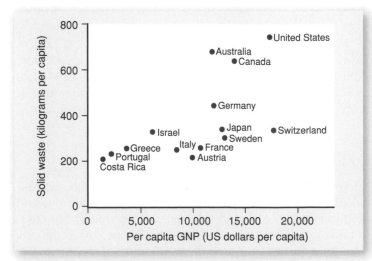

FIGURE 11.10 *Affluence and Waste* The relation between affluence (GDP/capita) and the quantity of solid waste produced per capita for various nations. (Source: Data from World Resources Institute.)

FIGURE 11.11 *Packaging Waste* *Packing is a major constituent of solid waste in developed nations.*

materials become the major components of solid waste: paper products (including cardboard) and yard wastes, followed by metals, plastics, glass, wood, and food wastes. Packaging consumes enormous volumes of materials and accounts for much of the paper, cardboard, plastic, metal, and wood discarded by consumers (Figure 11.11). Every conceivable consumer good—from toys to food and beverages to toiletries—comes wrapped in cardboard, paper, or plastic (or all three). Yet most packaging is purely cosmetic or is used to help advertise the product.

Poverty and Exposure to Environmental Health Risks

Affluence—or its opposite, poverty—is an important factor that determines a person's exposure to environmental health risks. Poor people are more likely to be exposed to pollution and degraded environments than

are affluent people, and poor people are more likely to have occupations that expose them to environmental hazards. Researchers at the University of California at Santa Cruz analyzed the relationship between the release of toxic chemicals from industrial facilities and income in Santa Clara County, California. This county is home to Silicon Valley, birthplace of the electronics industry. Santa Clara County has one of the highest average household incomes in the country ($75,000 in 2004), although there are a substantial number of low-income households. Research indicates that the most affluent neighborhoods are less likely to be located near a facility that produces or handles hazardous waste, and these residents are less likely to be exposed to toxic wastes (Figure 11.12). Instead the release of toxic wastes is concentrated in lower-income neighborhoods.

Concern for poor peoples' disproportionate exposure to environmental hazards and degradation has spawned scientific research and a social movement known as environmental justice. **Environmental justice** is the fair treatment and meaningful involvement of all people regardless of race, color, national origin, or income with respect to the development, implementation, and enforcement of environmental laws, regulations, and policies. Efforts to end environmental *in*justice are important because they are one of the first broad-based efforts by economically disadvantaged people to reshape one of the largest social movements of modern time: the environmental movement.

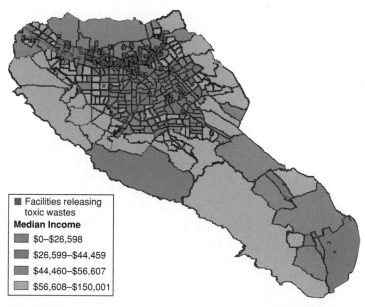

■ Facilities releasing
  toxic wastes
Median Income
■ $0–$26,598
■ $26,599–$44,459
■ $44,460–$56,607
☐ $56,608–$150,001

FIGURE 11.12 *Poverty and Environmental Risk* *The distribution of household income and facilities that release toxic materials in Santa Clara County, California. Black lines represent census tracts.* (Source: Data from M.R. Mausar and A. Szasz, "Environmental Inequality in Silicon Valley," www.mapcruzin.com/EI/index.hml.)

233

TECHNOLOGY

Population and affluence determine the level of economic activity—that is, the quantity of goods and services produced. All things being equal, higher levels of economic activity imply higher rates of resource extraction and waste generation. But all things are not equal. Technology determines the types and quantities of resources extracted and wastes generated by the production and consumption of goods and services.

In Chapter 10 we defined technology as a recipe—the combination of capital, labor, energy, materials, and information used to produce a good or service. For most goods and services, different combinations are possible. This flexibility is critical. Each combination extracts different quantities of resources and generates different quantities of waste. Steel made from an open-hearth furnace requires large amounts of coal and thereby damages the environment through the extraction and use of coal, whereas steel made from an electric arc furnace requires large amounts of electricity, which implies a different combination of environmental impacts. Similarly, producing and operating cars made from steel causes environmental impacts that differ considerably from those associated with cars made from aluminum and plastic.

The differences in environmental impacts imply that technology is a double-edged sword. Technology can exacerbate the environmental degradation that is associated with production and consumption by increasing the amount of resources extracted or wastes generated per unit of output. Or technology can lessen environmental impacts by reducing the amount of resources extracted or wastes generated per unit of output. We illustrate this potential by returning to the environmental impacts associated with food production, motor gasoline consumption, and solid waste disposal.

Technologies That Ease Environmental Problems: Fuel-Efficient Cars and Waste Recovery

More drivers driving more miles have increased gasoline consumption in the United States. But the greater gasoline consumption caused by the increase in population and affluence has been offset in part by changes in automotive technology. Rising gasoline prices in the 1970s and competition from Japanese manufacturers forced U.S. automakers to produce lighter cars with smaller engines. These changes in technology reduced the weight of new cars and the size of their engines, which helped increase average fuel efficiency to more than 20 miles per gallon by the 1990s (Figure 11.13). If energy efficiency had remained at 1960 levels (14.3 miles per gallon) the United States would have consumed 60 percent more motor gasoline than it did in 2004. This technological improvement slowed the rate at which we deplete oil resources and the rates at which our cars emit

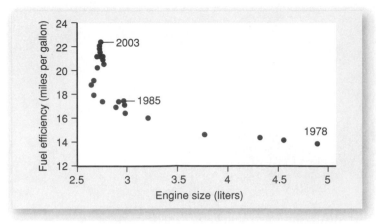

FIGURE 11.13 *Technology and Fuel Economy* The relationship between engine size and fuel economy for U.S. automobiles, 1978–2003. *(Source: Data from U.S. Department of Transportation.)*

harmful pollutants. But that improvement in fuel efficiency has triggered other changes that actually *increase* gasoline use. As a result, the reduction in gasoline consumption that results from more efficient cars is smaller than you might think. See *Case Study: Do More Efficient Automobiles Reduce Motor Gasoline Use?* Since the mid-1980s, however, another technological change has worked to reduce fuel economy: the rise in popularity of light trucks, minivans, and sport utility vehicles (SUVs). American consumers' preference for large, heavy vehicles hurt overall fuel economy because light trucks averaged just 17 miles per gallon compared to 22 miles per gallon for passenger cars in 2005.

New technologies also reduce the amount of solid waste. **Waste recovery** removes materials from the waste stream for recycling and composting (Figure 11.14).

FIGURE 11.14 *Waste Recovery* Composting is a form of waste recovery that reduces the need for landfill space. Here yard and food waste are being composted to produce valuable fertilizers and other forms of soil improvement.

CASE STUDY

Do More Efficient Automobiles Reduce Motor Gasoline Use?

This seems like a silly question. The fuel efficiency of the average U.S. automobile increased from 13.3 mpg in 1973 to about 22 mpg in 2005. All things being equal, this increase in energy efficiency should have reduced the use of motor gasoline in 2005 by nearly 50 percent relative to the level that would have prevailed had fuel efficiency remained at 13.3 mpg.

Of course all things are not equal. Because the economy is a complex system, changes in automotive technology initiate a series of feedback loops. The introduction of energy-efficient automobiles initiates a series of negative feedback loops that offset some of the environmental and economic benefits associated with the improved technology.

An increase in the energy efficiency of an automobile reduces the cost of driving. As the energy efficiency of automobiles increases from 13.3 to 22 mpg, the amount of gasoline needed to drive a mile declines significantly. At about $2.50 per gallon, the increased efficiency saves drivers a considerable amount of money; car owners perceive this savings as a decline in the price of gasoline.

The apparent decline in the price of motor gasoline can increase the demand for gasoline (and other forms of energy) in two ways. If a car owner continues to drive the same number of miles, the increase in energy efficiency increases the owner's disposable income. Suppose a car owner used to buy $100 of motor gasoline per month; but now that energy efficiency has increased, the owner spends only $50. This gives the owner $50 more each month to spend on other goods and services. The production and consumption of these goods and services requires gasoline (and other forms of energy). This indirect increase in the use of gasoline offsets some of the reduction in gasoline use by the car owner.

Alternatively, the car owner may decide to drive the car more often. That is, now that it costs less to drive the car, the owner may be willing to buy a house in the suburbs with a longer commute or take a weekend drive. In this case the owner will spend some of the $50 savings on additional gasoline. The increased purchase of gasoline will offset some of the savings associated with the increase in energy efficiency. This direct increase in gasoline use is called the "rebound effect."

Several economists have estimated the size of the rebound effect that is caused by efficiency gains in the U.S. automobile fleet. The effect is measured by the percentage increase in miles driven that is associated with a 1 percent increase in the energy efficiency of automobiles. Value range from 0.05 to 0.40, with most estimates between 0.1 and 0.2. This means that 10 to 20 percent of the motor gasoline saved due to increased energy efficiency is "lost" to increased driving.

Although the feedback effect does not eliminate the energy savings, it is large enough to be considered by policy makers when they choose between market-based incentives and command and control strategies for reducing motor gasoline use. If the feedback effects are small, command and control strategies such as government-mandated increases in CAFE (corporate average fuel efficiency) standards might be an effective policy for reducing motor gasoline use. On the other hand, if motor gasoline use is very sensitive to the price effect and the rebound effect is large, market-based incentives may be more effective at reducing motor gasoline use. (Can you explain why? Think about changes in price.) Choosing the correct strategy is critical to society's ability to reduce gasoline use in a cost-effective manner.

ADDITIONAL READING

Greene, D.L. "Vehicle Use and Fuel Economy: How Big Is the "Rebound" Effect?" *The Energy Journal* 13, no. 1 (1992): 117–143.

Khazzoom, D.J. "Energy Savings from More Efficient Appliances." *The Energy Journal* 8, no. 4 (1987): 85–89.

Saunders, H.D. "The Khazzoom-Brookes Postulate and Neoclassical Growth." *The Energy Journal* 13, no. 4 (1992): 131–148.

STUDENT LEARNING OUTCOME

- Students will be able to define the rebound effect in the context of improved fuel efficiency of motor vehicles.

Recovery improves environmental quality by reducing the need for landfill space and by reducing pollutants released by burning waste. Recovery also slows the depletion of nonrenewable resources. Producing a can from recycled aluminum uses 50 percent less fossil fuel than making the same can from bauxite (the ore used to make aluminum). Waste recovery in the United States has increased steadily from 7 percent of municipal solid waste in 1960 to about 28 percent in 2004. This increase is due in part to new cost-effective recovery technologies, consumer demand for recycled products, and environmental policies.

Technologies That Worsen Environmental Problems: Feedlots

We already have seen that the preference for meat that is associated with higher income increases the use of land, energy, and water. The size of these increases depends on the technology used to raise livestock. One option is **range-fed meat** production. Here livestock roam across rangeland ecosystems, foraging for grasses, shrubs, and other native plants. The manure produced by the animals is a valuable source of nutrients for plants. Meat produced in this way uses little fossil fuel. This method is sustainable if the number of animals and their feeding behaviors are managed properly.

Another technology is **feedlots** (Figure 11.15). Here animals are kept in an enclosed area for the last two to three months of their lives to accelerate growth and enhance fat content. Feedlots exacerbate the environmental impacts of meat production because they are extremely energy- and material-intensive. Cattle are fed specially prepared mixtures of corn, barley, wheat, molasses, and other ingredients that require lots of energy and water to manufacture. Trucks, front-end loaders, conveyors, and grinders burn more fossil fuels to prepare and deliver the

FIGURE 11.15 *Feedlots and Land Use* Feedlots house a large number of cattle in a relatively small space by using large amounts of energy, water, and nutrients obtained from much larger areas beyond the feedlot.

feed. As a result, a kilogram of beef produced in a feedlot uses three to five times more fossil fuel than that used in a range-fed operation.

Feedlots also release many harmful wastes. The fertilizers and agricultural chemicals that are used to grow feed grains can degrade water supplies and harm the health of agricultural workers and the consumers of the food itself. Feedlots produce concentrated volumes of animal wastes that can contaminate rivers and groundwater. Animal waste is a major ecological problem in the Netherlands, which produces much of the pork for western Europe. The millions of animals in feeding houses produce more manure than the environment can assimilate, which leads to what officials call a "manure surplus." The surplus includes excess nitrates and phosphates that become concentrated in the upper layers of the soil and groundwater, impairing many freshwater ecosystems. Manure nitrogen also escapes into the atmosphere, where it eventually causes acid precipitation. The damage to Dutch soils from acid precipitation that originates in animal production is greater than that caused by the fossil fuels burned by Dutch automobiles and factories. Similar problems are caused by feedlots throughout Europe and North America.

TYING IT ALL TOGETHER

Perhaps you noticed our use of the phrase "if you hold the effects of technology and affluence constant" in discussing the effect of population on the environment. We did this to simplify our discussion and to focus on the environmental effects of each driving force. In the real world factors change at the same time and interact with one another. For example, population growth and rising incomes both have adverse environmental impacts, but they often occur together. At the same time improved technologies, environmental policies, and heightened awareness and concern about the environment damp environmental impacts. There also are feedback loops among the driving forces. Our discussions revealed that family size is affected by income (affluence) and that technological development is affected by society's attitudes and beliefs. These interactions make it hard to attribute a given environmental change to a specific change in population, affluence, technology, institutions, or attitudes.

Despite these linkages, it is possible to disentangle the effects of population, affluence, and technology for some environmental issues. We demonstrate this by returning to our example of gasoline consumption by passenger cars. (*Note:* This does not include minivans, pickups, or sport

TABLE 11.1	The Effect of Population, Affluence, and Technology on the Amount of Motor Gasoline Consumed in the United States, 1960–2003			
	I =	**P ×**	**A ×**	**T**
	Gallons Consumed (billion)	Number of Cars (million)	Miles Driven per Car	Gallons Consumed per Mile
1960	40.81	61.7	9,446	0.07
2003	75.0	135.9	11,099	0.05

utility vehicles.) Between 1960 and 2003 gasoline consumption by cars in the United States increased by about 34 billion gallons. What caused this increase? To answer this we return to the IPAT equation:

$$\text{Impact} = \text{Population} \times \text{Affluence} \times \text{Technology} \qquad (11.2)$$

$$\text{Gallons consumed} = \text{Number of vehicles} \times$$

$$\frac{\text{Miles driven}}{\text{Vehicle}} \times \frac{\text{Gallons consumed}}{\text{Miles driven}}$$

The number of cars is determined primarily by population (although ownership rises with income), which affects ownership through the number of registered drivers. Affluence is represented by the number of miles each car is driven per year, which is correlated closely with household income. Technology is represented by the fuel efficiency of the average car, which is measured by the gallons consumed per mile driven (this is the inverse of the familiar miles per gallon (mpg)).

The rise in the number of cars and the number of miles driven per car clearly contributed to the increase in gasoline consumption (Table 11.1). But the improvement in fuel efficiency had the opposite effect. How can we sort out these effects? One way is to compare the actual change in gasoline consumption with the change that *would* have occurred due to the change in each of the factors (Figure 11.16). The

120.3 percent increase in the number of cars boosted consumption by about 49.1 billion gallons to 89.9 billion gallons. This total rises by another 15.7 billion gallons due to the 17.5 percent increase in the number of miles driven per car. But this 105.6 billion gallon total was lowered by about 30.6 billion gallons due to the 29 percent increase in average fuel efficiency. The net result of the three driving forces is the actual increase of 34.1 billion gallons. Thus we can conclude that the increase in the number of cars associated with the general increase in population was the principal driving force behind the increase in gasoline consumption. However, our analysis shows that the increase would have been much greater had it not been for changes in automotive technology that boosted fuel efficiency.

HOW SOCIETIES CHOOSE TECHNOLOGIES: POLITICAL–ECONOMIC INSTITUTIONS

Some societies drive SUVs while others drive fuel-efficient hybrid vehicles. Some use feedlots use to produce meat; others depend on rangelands. Some societies recycle their solid wastes; others dump them into landfills or incinerate them. How do people make these choices? To understand this we need to examine how social institutions influence the types of technologies used to produce and consume goods and services.

The Market

Most nations use the **market** to guide economic decisions. The market is an invisible network that links firms and households that wish to buy and sell goods, services, and factors of production. As described in Chapter 10, households spend their income in a way that maximizes their utility. A similar rule describes the behavior of firms. Firms try to maximize the profits they can earn from the production of goods and services. To maximize profits, firms choose the technology that minimizes the cost of production. The least expensive technology is chosen based on the quantities of capital, labor, energy, materials, and information associated

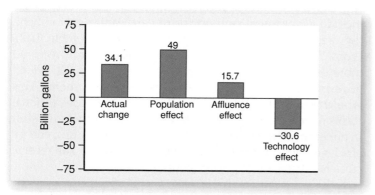

FIGURE 11.16 *The IPAT Equation for Gasoline Use* *The effects of population, affluence, and technology on gasoline consumption in the United States from 1960 to 2003.*

with a technology times the price for these inputs. For example, if energy is cheap relative to labor, firms tend to choose technologies that favor energy use relative to labor.

The interaction of utility maximization by consumers and profit maximization by firms determines the type, quantity, price, and allocation of goods and services. This combination is said to be **economically efficient** because it is impossible to change production and allocation in a way that makes some people better off without making others worse off. Put simply, an efficient allocation is the best (as measured by profits and utility) an economy can do with a given set of producers, consumers, and resources.

The incentive to minimize costs and produce what people want to buy tends to reduce the quantity of natural resources that are extracted and wastes generated per unit of economic activity. That is why market economies tend to use less energy and produce fewer wastes than **centrally planned economies.** In centrally planned economies such as North Korea and Cuba, state planners determine what goods get produced, what technologies are used to produce them, and their prices. Government bureaucrats have no incentive to minimize costs and have little information about the types of goods and services that people prefer (or they may ignore consumer preferences). Without these incentives or information, centrally planned economies often choose inefficient and more environmentally destructive technologies than market economies. For example, when Hungary was a centrally planned economy (it is now in the process of becoming a market economy), it used about twice as much energy to produce a unit of economic output than neighboring Austria, which is a market economy. Hungary also emitted about twice the amount of CO_2 per unit of economic output.

Choosing the "Right" Technology: Market Failures

Even though market economies tend to be more efficient than centrally planned economies, markets frequently do not work as effectively as we would like them to. The market's ability to generate an efficient outcome hinges on a critical assumption: The prices of economic goods and services include all costs associated with their production and consumption. This means that when a firm maximizes its profits by choosing the combination of capital, labor, energy, materials, and information that minimizes cost, it is also choosing the combination of capital, labor, energy, materials, and information that minimizes the cost to society to make the goods available. Put simply, decisions that work best for individual firms and households work best for society as a whole.

In many cases the prices of capital, labor, energy, materials, and information to individual firms and households reflect the effort made elsewhere in the economy to make them available. But in some cases prices do not reflect the total effort required to make goods or services available. "Wrong prices" represent a **market failure,** which occurs when a market does not allocate resources efficiently. Under these conditions, the production and consumption of goods and services could be changed in a way that makes some firms or households better off without making any other firms or households worse off.

Economists identify four general causes for market failures. **Market power** refers to cases in which individual households or firms can set prices, such as a monopoly, in which a single firm controls the output of a good. **Asymmetric information** refers to cases in which buyers and sellers have unequal information. For example, chemical firms may have information about toxicity that is not available to their workers who face occupational exposure. As a result, workers may not ask for pay that is commensurate with the hazards they face on the job. As described in the previous chapter, **public goods** refer to goods owned by the government. Under these conditions it is hard to price them. Finally, **externalities** occur when the market does not take into account the effect of an activity on others. For example, the price for electricity generated by plants in the Midwest portion of the United States does not include the environmental damages generated by acid deposition in the northeastern United States and Canada (Figure 19.6).

These causes for market failure are often unique to a good or service. Politicians create an additional cause for market failures via **subsidies.** Subsidies are money paid, usually by the government, to keep prices below what they would be in a competitive market, or to save businesses that would otherwise go bankrupt, or to make activities happen that otherwise would not take place. Subsidies often produce undesirable environmental outcomes that would not exist in the absence of the subsidies. For example, a federal U.S. law passed in 1976 made it possible for fishers to receive very low-interest loans to purchase fishing vessels. This enabled fishers to borrow money and buy bigger boats than would have been justified by the profits from fishing. The subsidy contributed to a huge increase in the size and capacity of the U.S. fishing fleet, which, in turn, led to the depletion of many commercial fish populations in the 1980s and 1990s.

The energy sector receives enormous subsidies. These include government support for research and development, tax credits, and transportation infrastructure (Table 11.2). From 1950 to 1990 fossil fuels and nuclear energy received more than 80 percent of the money given out by the federal government for energy research and development. This would tend to "tilt the playing field" in favor of conventional forms of energy. Since the 1990s there has been a relative shift in government support toward renewable energy and energy efficiency, although conventional forms of energy still receive close to 70 percent of government research support.

Market Failures and the Environment

Market failures associated with public goods, externalities, and subsidies are an important cause of environmental degradation. The prices for many factors of production, goods,

TABLE 11.2	Impact of Transportation Subsidies on the Energy Sector	
Transport Mode	**Issues**	**Energy Sector Impacts**
Water: inland	Waterway maintenance is often provided by governments; user fees may not recover costs.	Reduces delivered price of bulk oil and coal.
Water: coastal and international	Coastal ports, harbors, and shipping oversight are subsidized by federal and other government entities; user fees might not recover costs. Fuel consumed during shipment in international waters is generally tax free.	
Roads	Most roadways are municipally owned and operated; user fees (primarily from fuel taxes) are often insufficient to cover costs. Large trucks often pay proportionately less in taxes than the damage they cause roadways.	Primarily benefit refined petroleum products. Waste products from coal combustion or waste-to-energy plants may sometimes move by truck as well.
Rail	Many rail lines do not recover their full costs.	Largest beneficiary is coal, with some benefits to oil.
Pipelines	Rights of way, safety and security, and environmental cleanup contribute to reduced costs of pipeline ownership and operation. Depending on circumstances, government ownership may generate large subsidies to users and use the government monopoly to levy high taxes on users.	Primarily benefit oil and natural gas.
Electrical transmission grids	Rights of way, tax breaks for municipal ownership or capital investment, and government research and development can generate subsidies to electrical distribution.	Benefit all sources of centralized electricity in proportion to their share as a prime mover in generating stations; coal, nuclear, natural gas, hydroelectricity, and oil are the main beneficiaries.

Source: Data from Earth Track Inc., Cambridge, MA: http://earthtrack.net

and services do not have complete information about how the environment contributes to their production (step 1 of the economic process—see Chapter 10), how they affect the environment's ability to provide habitable living conditions (step 2), or how much work the environment does to dilute or detoxify them or their by-products (step 4). Without this information their prices are too low, which fools firms and households into using more natural resources and environmental services than they should. Under these conditions rational behavior by individuals leads to levels of environmental degradation that lower overall economic well-being.

The price of motor gasoline illustrates many causes for market failures and how they lead to environmental degradation. In 2005 the U.S. price for a gallon of gasoline varied between $2.00 and $4.00. That price included the costs incurred by oil companies to pump crude oil out of the ground, refine it into motor gasoline, ship it to local filling stations, and pay attendants. It also included taxes paid to support the government. But the price of gasoline did not include environmental impacts that are associated with every step of this process, such as the loss of barrier islands off the coast of Louisiana that would have lessened the impact of Hurricane Katrina in August 2005 or pipeline leaks that damage local ecosystems. Nor did the price include the environmental harm caused by burning the fuel, such as global climate change (Chapter 13) or the impairment of human health that is associated with air pollutants (Figure 11.17; see Chapter 19).

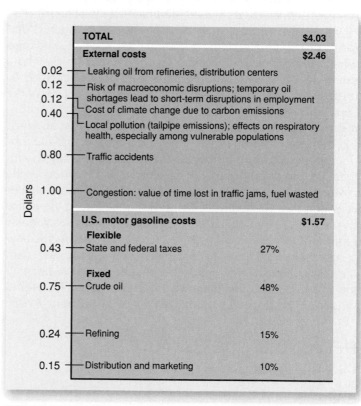

FIGURE 11.17 Oil and Externalities The external costs associated with the production and use of one gallon of gasoline in the United States in 2004. (Source: Data from National Geographic, "The End of Cheap Oil," June 2004, pp. 80–109.)

Because the price of motor gasoline does not include environmental impacts, the price of motor gasoline is too low. As a result, consumers purchase too much gasoline. What do we mean by too much? Economic efficiency requires that the benefits derived from the last good or service purchased equal the costs associated with its production. But if the market price does not include environmental costs, the market price is too low. When consumers equalize this "too low" market price to the benefits, the true costs are greater than these benefits. If the consumer knew these real costs, he or she would have bought less. Returning to the gasoline example, if a gallon cost two to three times what it does currently, higher prices would encourage people to drive less and to choose technologies that use less motor gasoline, such as public transportation. Higher prices also would create an incentive to develop alternative energy technologies.

If too much of a particular good or service is purchased, the level of environmental degradation that is associated with its production or consumption is too high. But this highlights an important point: Some environmental degradation is acceptable from an economic perspective. As described in Chapter 10, all goods and services ultimately come from the environment as natural resources or environmental services and return to the environment as wastes. In addition, some wastes are generated in every step of economic production and consumption. The only way to eliminate environmental degradation would be to stop economic activity. And stopping economic activity would eliminate all the benefits.

This leaves policy makers with a tricky question: What forms of environmental degradation should be allowed, and which should be prohibited? And of those allowed, how much degradation should be permitted? Partial answers to these questions are provided by the economic concept of the "optimal level of pollution." This concept states that environmental degradation should be allowed to the point at which the damages associated with the last unit of economic production or consumption equal the benefits associated with that last unit of production or consumption. Returning to the motor gasoline example, the use of motor gasoline should not be prohibited. Rather, consumers should be forced to pay for the environmental degradation caused by their use of motor gasoline. Once they realize the true cost, they will decide how to use motor gasoline so that the benefits associated with the last gallon burned equal the total costs of the last gallon burned, including the environmental costs. This would reduce environmental degradation relative to current levels.

DESIGNING SUSTAINABLE INSTITUTIONS

Attitudes and Beliefs

Slowing environmental degradation sounds easy—simply include the costs of environmental degradation in the price of economic goods and services and let the market determine the optimal level of pollution. However, for reasons described in Chapter 10, calculating the costs associated with environmental degradation is exceedingly difficult. One cause for this difficulty is scientific uncertainty. Put simply, we can't measure all the ways in which the environment contributes to economic activity.

But even if we could, there remains the problem of human attitudes and beliefs—the environment contributes to our well-being in ways that extend beyond our material needs. Many people draw aesthetic and spiritual fulfillment from unspoiled environments. Most U.S. citizens would not allow Yosemite or Yellowstone National Parks to be disturbed, regardless of the profit that might be made. And some people are happy knowing that the Arctic National Wildlife Refuge will not be disturbed by oil production, even if they never plan to visit this area (Figure 11.18).

This illustrates the importance of how cultural values and beliefs shape the relationship between society and the environment. Some people attribute environmental problems in the Western world to the separation of spirit and nature in the Judeo–Christian tradition. They claim that this attitude has led to an excessively human-centered or **anthropocentric** perspective on the natural environment. An anthropocentric society views itself as separate from the natural environment and places human wants and needs above those of nonhuman species. An anthropocentric society seeks to control nature for the purpose of satisfying human wants. Underground accumulations of crude oil and groundwater exist to be extracted and used to power our automobiles and to water our crops. Forests exist to be cut and converted into lumber and newspapers. Wetlands can be drained and converted to "useful" purposes such as suburbs and shopping malls. Rivers serve as waste treatment plants that detoxify and dilute pollutants from chemical plants and our bathrooms.

This attitude toward nature has helped shape technological change over the past several hundred years. Many technologies are designed to exert ever-increasing control over environmental storages and flows of energy and materials. We have developed more sophisticated methods of finding and extracting oil, building dams to control the flow of large rivers, and replacing forests with tree plantations. These technological changes push back environmental limits that constrain the size of the human population and its material standard of living. This seeming success has led some people to believe that there are no limits on the size and affluence of the human population.

Without apparent limits on what we can do, we have come to believe that there are no limits to what we can have. Many people believe that happiness is created by the accumulation of material goods and services. A society that values a high material standard of living will build and accumulate many machines, factories, and technologies that use fossil fuels to produce the goods and services desired by consumers. These desires help drive the exponential rise in

FIGURE 11.18 *The Arctic National Wildlife Refuge* *Many people who will never visit the Arctic National Wildlife Refuge in Alaska still prefer that it not be disturbed by oil production activity.*

the use of energy and materials and the release of wastes. Materialism is growing in many developing nations where people want a taste of the "good life." Indeed, the word "development" has come to mean one thing: increased consumption of material goods and services. The emergence of a middle class in India, China, and Indonesia has produced explosive growth in the number of televisions, cars, refrigerators, and other consumer goods.

There are alternatives to this anthropocentric perspective. A **biocentric** perspective holds that all living beings have a right to exist and that human decisions must respect these rights. Still others adopt an **ecocentric** view, which holds that entire ecosystems have a right to exist. Clearly either of these attitudes would generate much less environmental degradation than the anthropocentric point of view. Yet adapting these attitudes would require a sweeping change in how people view themselves and the other inhabitants of the planet.

Formulating Environmental Policy

Given the importance of attitudes and beliefs, it may be impossible to calculate the optimal level of pollution. Nonetheless, the optimal level of environmental degradation is a goal for environmental policy in much of the developed world. That is, natural scientists seek to quantify the impacts of economic activity on the environment. Economists use this information to calculate the economic value of environmental degradation. Decision makers incorporate

this information, along with social attitudes and beliefs, to choose environmental goals that move society toward a "best guess" for the optimal level of pollution while satisfying the needs and aspirations of various aspects of society. For example, Chapter 13 describes the Kyoto Protocol, an international treaty designed to slow global climate change. This treaty is based on scientific information from climate scientists regarding the effects of climate, estimates of how these changes in climate will affect economic well-being, and international negotiations regarding the willingness of nations to reduce their emissions. Regardless of what you think about this agreement, one thing is clear: The goals set by the Kyoto Protocol represent a political compromise of an uncertain estimate for the optimal level of emissions.

Once the political process generates a consensus, decision makers must design policy by which the government can intervene in the market to correct failures. Market intervention is unavoidable—by definition, the market cannot fix market failures. Without outside intervention, no firm or household has an incentive to reduce environmental degradation. Those that do will find themselves at a competitive disadvantage. For example, firms that install expensive pollution control equipment will have smaller profits than firms that do not install pollution control equipment. Over time, smaller profits would either force environmentally friendly firms out of business or would allow more polluting firms to buy out less polluting firms. Even dramatic changes in attitude by firms or consumers cannot fix market failures.

To fix market failures, the government intervenes using two general strategies. The **command and control** strategy uses legal means to specify an environmental standard and the methods for complying with that standard. "Command" refers to the laws or regulations that require all power plants, factories, machines, or other sources of pollution to meet an emission standard for a particular pollutant. "Control" refers to the technologies that are required to meet the standard. Much environmental legislation in the United States uses the command and control approach, including the Clean Air Act and the Clean Water Act, which are two of the nation's most important environmental laws. For example, the Clean Air Act reduced air pollutants by requiring manufacturers to install catalytic converters in automobiles. It also eliminated lead emissions by prohibiting refineries from adding lead to motor gasoline (Figure 11.19).

The other strategy for correcting market failures relies on **market-based incentives.** This approach uses economic incentives to steer economic decisions in a direction that reduces environmental degradation. Market-based incentives work by internalizing externalities, which means that the market price is changed (usually increased) to reflect the environmental degradation associated with a good or service. By including these externalities, firms and households equate the benefits of a good or service with its true cost, which moves environmental degradation toward the optimal level.

Market-based incentives are implemented using one of two policy instruments. **Pollution taxes** increase the price for goods and services based on the environmental degradation associated with them. Returning to the price of motor gasoline, a pollution tax would be set to the environmental degradation associated with that gasoline. Alternatively, **tradable permits** limit emissions to levels determined by the scientific, economic, and political processes just described. These quantities can be implemented by a **cap and trade system,** in which the government issues permits that allow the holder to emit a certain quantity of the pollutant. The cap is ensured by limiting the number of permits to the target chosen by society. "Trade" refers to the ability of firms or households to buy and sell the permits.

Market-Based Incentives versus Command and Control

At first glance there seems to be little difference between market-based incentives and command and control methods. But economic theory reveals an important difference. Market-based incentives can reduce emissions at lower cost; even efforts to reduce environmental degradation can be economically efficient. This efficiency is due to the genius of the market. The market can marshal more information than can any scientist, economist, or government bureaucrat, and the market can use this information to allocate emission reductions more efficiently than environmental policies that are based on command and control.

We can illustrate the cost savings of market-based incentives with a simple example. Suppose two firms produce copper. Total copper production is 4 metric tons; each firm produces 2 metric tons of copper, and each firm emits 2 metric tons of sulfur dioxide (1 metric ton of sulfur is emitted for each metric ton of copper produced). Firm A uses a relatively old technology, so reducing this firm's emissions costs $10 per metric ton; firm B uses a newer technology that lets it reduce emissions for $5 per metric ton. This difference in costs is vital—it is the key to success for market-based incentives.

FIGURE 11.19 *Environmental Laws* The increase in environmental regulations and laws from 1870 demonstrates the popularity of the command and control approach to environmental policy in the United States. *(Courtesy of the U.S. Environmental Protection Agency.)*

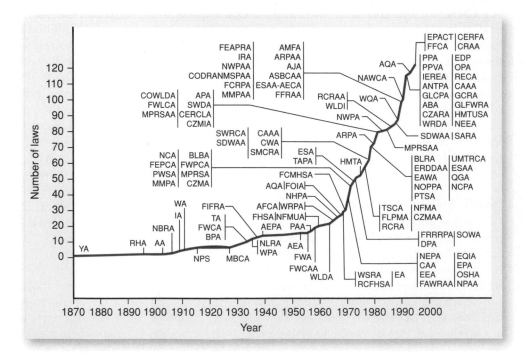

Suppose that the scientific, economic, and political process decides that emissions of sulfur dioxide should be reduced by 2 metric tons. Because the government cannot favor one firm over another, command and control legislation would require that all firms reduce their emissions by 1 metric ton. This policy will reduce emissions by 2 metric tons and will cost society $15—($10 paid by firm A and $5 paid by firm B).

Using a market-based incentive, suppose the government sets up a cap and trade system and issues a total of two permits for 1 metric ton of sulfur dioxide, giving one such permit to each firm. At first the result seems the same as the command and control policy: Because each firm has only one permit, firm A will spend $10 to reduce emissions by 1 metric ton and firm B will spend $5 to reduce emissions by 1 metric ton.

But the market would recognize an opportunity to reduce costs to firm A and increase profits for firm B. (Do you see this opportunity?) Firm A would be willing to buy firm B's permit for any price less than $10. Buying the permit for less than $10 would reduce firm A's costs because it costs firm A $10 to buy the equipment needed to reduce sulfur emissions by 1 metric ton. Similarly, firm B would be willing to sell firm A its permit for any price more than $5. A price greater than $5 would mean that firm B would have money left over after it paid the $5 to reduce its emissions by another metric ton.

To illustrate these benefits, suppose that firm B agrees to sell firm A its permit for $7.50. This sale would allow firm A to pay firm B $7.50 for a second permit (firm A still has the original permit it received from the government) and emit 2 metric tons of sulfur dioxide. Firm B has no permits, which means it would spend $10 to reduce its emissions by 2 metric tons. This reduction costs firm B only $2.50 because it received $7.50 from firm A. For society has a whole, the cap and trade system reduces emission by 2 metric tons, which is the same reduction made by the command and control strategy. But the total reduction would cost $10 as opposed to the $15 incurred by the command and control policy. (The cost represents the $10 of capital, labor, energy, and materials used by firm B to reduce emissions. The $7.50 paid by firm A for the permit is an intermediate transaction that does not represent factors of production.)

Similar savings are possible if the government imposes a tax of $7.50 on each metric ton of sulfur dioxide emitted. Firm B will reduce its emissions by 2 metric tons because it is less expensive to reduce emissions ($10) than it is to pay the tax ($15). On the other hand, firm A will pay the tax because it is less expensive to pay the tax ($15) than to reduce emissions ($20). Like the cap and trade policy, the tax has a total cost of $10. But what about the $15 tax paid by firm A? This payment does not represent a cost to society because it does not represent the use of capital or labor. Instead the $15 is what economists call a transfer. It could be used to increase government services or reduce other taxes (see Chapter 24).

There is considerable disagreement about whether environmental policy should use command and control techniques or market-based incentives, as is illustrated by the debate over gasoline consumption. See *Policy in Action: Reducing Gasoline Consumption: CAFE Standards or Higher Prices?* The command and control strategy is popular with politicians because it allows them to claim success for laws that reduce pollution. Many environmentalists prefer command and control strategies because they offer some degree of certainty about the quantity of pollution abated. Experience with the Clean Air Act demonstrates that this approach can work. On the other hand, the command and control approach tends to be expensive, and firms have no incentive to further reduce emissions once they satisfy the standards.

Many businesspeople prefer market-based incentives because they reduce the economic burden associated with environmental cleanup and give managers the flexibility to choose how they reduce emissions (Figure 11.20). Experience with the sulfur dioxide cap and trade system shows that market-based approaches can reduce emissions at a lower cost than command and control programs. As described in Chapter 19's Policy in Action, the sulfur dioxide cap and trade program has been successful, performing even better than expected. Furthermore, market-based incentives provide a continuous impetus for firms to reduce emissions—they will reduce emissions so long as they can make money by doing so. On the other hand, market-based incentives have considerable uncertainty. If a pollution tax is levied, politicians cannot know the quantity of emissions reduced. If a cap and trade program is implemented, politicians cannot be sure what price the market will set for the permits.

In practice effective environmental policy will use both approaches, applying each to the appropriate environmental problems. The choice may depend on the geography of pollutants. For pollutants emitted by firms that tend to cluster, a cap and trade policy may lead to dangerous levels of pollution in areas where most of the firms have bought permits. In such cases a command and control approach may be best. For pollutants that are emitted over a wide area by many firms and households, policies that use market-based incentives may be most effective.

Powernext Carbon

FIGURE 11.20 *Tradable Permits* The Powernext Carbon logo is the trademark of a company by the same name that trades permits for the right to emit carbon dioxide in the European Union.

POLICY IN ACTION

Reducing Motor Gasoline Consumption: CAFE Standards or Higher Prices?

Many of your parents remember the 1950s and 1960s as a time of cheap oil and big cars. Motor gasoline cost less than 30 cents per gallon, and cars had eight-cylinder engines, tailfins, and 30-gallon fuel tanks. These large tanks were needed because the average car could travel only about 13 miles on a gallon. Big cars and cheap energy were driving forces behind the increase in gasoline consumption. This depleted domestic resources of crude oil, increased U.S. dependence on oil imported from the Middle East, and caused severe air pollution problems in many cities.

The era of cheap oil ended in the fall of 1973 when the Organization of Petroleum Exporting Countries (OPEC) withheld oil supplies and increased world oil prices by about 250 percent. These price hikes caused anxiety about the security of U.S. oil supplies and the flow of money from U.S. consumers to foreign oil producers. At the same time the emerging environmental movement raised concerns about the environmental impacts of automobile emissions.

These issues generated a consensus for policy to reduce the consumption of motor gasoline. Several options were available, most of which could be classified as either command and control or market-based incentives. Market-based policies reduce motor gasoline use by raising taxes and thereby increasing prices to consumers. Command and control strategies reduce gasoline use by forcing automakers to make more efficient cars.

These options generated considerable debate among policy makers. Proponents of market-based policies argued that consumers would adjust their driving habits to higher prices. As prices rose, they would drive less. Higher prices also would increase consumer demand for cars that got more miles per gallon (mpg), which would prompt automakers to increase the energy efficiency of new cars. Proponents of market-based incentives asserted that imposing regulations would not be effective because carmakers would increase their energy efficiency just enough to meet government standards. Furthermore, should oil prices become cheap once again (which they eventually did), consumers would be "forced" to buy energy-efficient cars when they no longer were "needed" or wanted.

Proponents of command and control strategies said that the market needed help to reduce motor gasoline consumption. The cost of gasoline is small relative to the cost of insurance and maintenance, therefore buying motor gasoline is a small component of the total cost of operating a car. This means that even a large price increase would have a relatively small effect on driving habits. Furthermore, even after the OPEC price increase, the higher price of motor gasoline did not include the environmental effects of automobile emissions, so consumers still would not have the information needed to purchase the efficient amount.

Ultimately proponents of the command and control strategies won the day. Congress passed the Energy Policy and Conservation Act of 1975 (PL 94-163), which required firms that make more than 10,000 cars per year to increase the corporate average fuel efficiency (CAFE) of new cars to 18 mpg in 1978 and 27.5 mpg in 1985. If they failed to meet these goals, automakers would be fined $50 per car sold per 1.0 mpg shortfall—a so-called gas guzzler tax. Automakers sell millions of cars per year, so these fines could total tens of millions of dollars.

Because of the fines and the bad publicity associated with failure, most automakers satisfied the CAFE standards (or at least came close). These standards rose as scheduled by the 1975 law. As a result, the average fuel efficiency of all cars (new and old) increased from about 13.3 mpg in 1973 to 22 mpg in 2005.

Given these changes, the debate now focuses on how much of the increase in fuel efficiency is due to government standards versus the OPEC price increases. Some antiregulation analysts claim that the OPEC price increases, not the CAFE standards, caused the improved efficiency. But another analysis indicates that the CAFE standards were about twice as important as the oil price hikes. The price of motor gasoline has a very small effect—a 10 percent increase in motor gasoline prices increases the average efficiency by only about 2.1 percent. At this rate gasoline prices would have to rise from $3.00 to about $4.75 per gallon to increase fuel efficiency by about 10 percent.

Another important question is how much the CAFE standards cost consumers. Estimates vary over a wide range. One analyst estimates that the government CAFE regulations cost consumers $10–20 for each gallon of gasoline saved. Another analyst estimates that the standards increased the average price of a new car by about 1.4 percent, which is about $280 for a $20,000 new car. If CAFE standards increased the efficiency of the auto fleet by 10 percent, this implies that the CAFE standards cost about $2 per gallon saved (most cars use fewer than 6,000 gallons over their useful lives).

ADDITIONAL READING

Greene, David, L. "CAFE or Price? An Analysis of the Effects of Federal Fuel Economy Regulations and Gasoline Price on New Car MPG, 1978–1989." *The Energy Journal* 11, no. 3 (1990): 37–57.

Kleit, A. "The Effect of Annual Changes in Automobile Fuel Economy Standards." *Journal of Regulatory Economics* 2 (1990): 151–172.

Scott, Robert E. "The Effects of Protection on a Domestic Oligopoly: The Case of the U.S. Auto Market." *Journal of Policy Modeling* 16, no. 3 (1994): 299–325.

STUDENT LEARNING OUTCOME

• Students will be able to compare the relative strengths and weaknesses of fuel efficiency standards and higher prices as means to improve fuel efficiency.

SUMMARY OF KEY CONCEPTS

- Environmental change is caused by a combination of several forces that act together, most notably population growth, technology, affluence, institutions, and attitudes. Because these driving forces do not operate separately, we should be suspicious of any claim that one factor is the culprit.

- Population growth degrades the environment by increasing the demand for natural resources and waste assimilation services. Land conversion for agriculture is one of the most important environmental changes driven by population growth.

- Poor people tend to be exposed to pollution and degraded environments with greater frequency than the more affluent, and they are more likely to have occupations that expose them to workplace health hazards.

- Affluence tends to increase environmental change because of the increased use of energy and materials and release of wastes associated with higher incomes.

- The lack of well-defined ownership for common property resources causes them to be underpriced. These low prices increase consumption relative to the optimal level. Such excess consumption increases environmental degradation.

- Subsidies lower the prices of goods and services and often encourage economic activities that are not economically viable and are environmentally destructive.

- A command and control policy tends to work best for pollutants emitted by clustered firms, whereas market-based incentives may be most effective for pollutants that are emitted over a wide area by many firms or households.

REVIEW QUESTIONS

1. A nation with the world's largest GDP may not have the most affluent standard of living. Explain why.

2. Refer to the IPAT equation, and assume that total carbon dioxide emissions (in tons) for a nation is the "I" part of the equation. What variables would be on the right side of the equation? What would the units of those variables be?

3. Historically, what has been the principal effect of food production on the environment?

4. Why does eating higher on the food chain have greater effects on the environment?

5. What is the principal way in which market failures contribute to environmental change?

KEY TERMS

anthropocentric

asymmetric information

biocentric

cap and trade system

centrally planned economies

command and control

ecocentric

economic efficiency

environmental justice

externalities

feedlots

IPAT equation

market

market failure

market power

market-based incentives

pollution taxes

public goods

range-fed meat

subsidies

tradable permits

waste recovery

12

BIODIVERSITY
Species and So Much More

Biosphere 2 *The Biosphere in Oracle, Arizona, where scientists tried to assemble a system that would support humans for extended periods without any inputs of food, water, or atmospheric gases.*

STUDENT LEARNING OUTCOMES

After reading this chapter, students will be able to

- Describe the evolutionary mechanisms that generate new species and the factors that influence biodiversity.

- Explain how the loss of biodiversity reduces economic well-being in ways that are often not captured fully in the market.

- Describe how human actions reduce biodiversity by both purposeful action and inadvertent changes.

- Compare and contrast three mechanisms for preserving biodiversity.

In July 1991 eight "Biospherians" started a daring biodiversity experiment that called for the eight people to live in the airtight Biosphere 2 for two years (opening photograph). The Biosphere 2 had a footprint of 13,000 m^2 and enclosed 204,000 m^3. This area housed simplified versions of a desert, savanna, tropical forest, wetland, and a small ocean. The Biospherians hoped that the plants, animals, and other organisms in these simplified biomes would generate their air, food, and water.

Measured against this yardstick, the experiment was a spectacular failure. By January 1993 the oxygen content of the atmosphere dropped from 19 to 14 percent (Figure 12.1). This is close to the minimum needed by humans. The atmospheric concentration of carbon dioxide rose. Before the two-year anniversary, oxygen was added to prolong the experiment. After three years the atmospheric concentration of N_2O reached 79 parts per million (compared to 310 parts per billion in Earth's atmosphere). Such levels reduce the production of vitamin B_{12} and can interfere with or damage the human brain.

Nonhuman populations also changed. Vines such as morning glory were included to remove carbon dioxide from the atmosphere via photosynthesis. But the vines grew too quickly. Despite extensive weeding by the Biospherians the vines overran other plants. The water system became overloaded with nutrients, and this polluted the aquatic systems. All of the pollinators died. This meant that even if the atmosphere did not fail, many of the flowering plants would not produce food for the Biospherians. Nineteen of twenty-five vertebrate species went extinct, as did most of the insects other than cockroaches.

Why did the experiment fail to support human life? Some blamed the arrogance of the Biospherians. Had they asked ecologists for help, they could have been warned about many of the problems that surprised them. But this claim has been denied by ecologists who have visited the facility. They state that the experiment probably would have failed even if the experiment had been designed using state-of-the-art ecology. Nor did the experiment fail due to a lack of effort. Constructing Biosphere 2 cost about $200 million, and operating it cost several million dollars per year.

The experiment failed for reasons that reflect the technical and scientific shortcomings of both the Biospherians and the wider ecological community. The rise in atmospheric carbon dioxide was caused in part by microbial degradation of organic carbon in the soils (human food production requires soils that have a high organic content—see Chapter 15). Apparently the Biospherians forgot that respiration reduces oxygen levels. (Can you explain why? *Hint:* See Equation 5.3 in Chapter 5, page 89.)

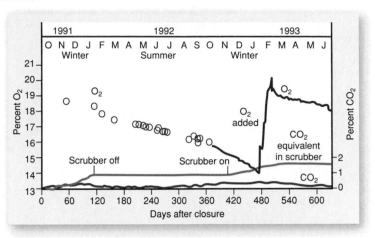

FIGURE 12.1 *Biosphere's Atmosphere After the experiment began, the concentration of oxygen in the biosphere declined while the concentration of carbon dioxide rose steadily. Because of these changes, additional oxygen was added after 480 days, and a scrubber was used to remove excess carbon dioxide.* (Source: Data from J.E. Cohen and D. Tilman, "Enhanced: Biosphere 2 and Biodiversity: The Lessons So Far." Science 274: 1150–1151.)

Oxygen levels were reduced further when some of this carbon dioxide combined with the calcium in the cement used to build Biosphere 2. This reaction locked up some of the original oxygen in the cement walls. Arbitrarily picking pieces of biomes did not create a small working version of Earth. No one knows what combinations of species are needed to reproduce the environmental services provided by planet Earth.

The failure teaches a valuable lesson. At this time it is not possible to build a simplified ecosystem that can supply and recycle the food, water, and gases that people need. This ignorance illustrates the danger of losing biodiversity. Currently humans are running the Biosphere experiment in reverse with the only working version of the Biosphere available—Earth. Humans are driving a wave of mass extinctions with little understanding of the implications for biogeochemical cycles, food webs, diversity, resistance, or resilience. The loss of some species may have a relatively small effect on the functioning of global ecosystems, including the environmental services that are critical to humans. But other species may be critical; therefore, their extinction could have important effects on both the functioning of the global ecosystem and human well-being. ■

DEFINING BIODIVERSITY

Biological diversity, or **biodiversity,** refers to the number and variety of living organisms. It is an umbrella term that encompasses all plants, animals, microorganisms, and ecosystems. Biodiversity often is measured at three levels. **Genetic diversity** refers to the genetic information in the DNA of plants, animals, and microorganisms. DNA contains many genes, which can

be combined in nearly infinite ways that lead to the amazing variety of shapes, sizes, colors, and behaviors of organisms. As described in Chapter 3, genetic diversity is the raw material for natural selection, which plays an important role in evolution.

At another level **ecosystem diversity** describes the variety of ecosystems on Earth such as coral reefs, forests, and wetlands. Ecosystem diversity also involves variation

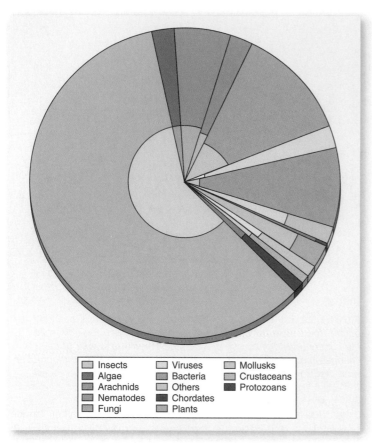

FIGURE 12.2 *Species Richness* Species richness in major groups of organisms. The inside portion of the pie represents the number of species known to exist; the outside portion represents the number of species thought to exist. *(Source: Data from A. Purvis and A. Hector, "Getting the Measure of Biodiversity." Nature 405: 212–219).*

Legend:
- Insects
- Algae
- Arachnids
- Nematodes
- Fungi
- Viruses
- Bacteria
- Others
- Chordates
- Plants
- Mollusks
- Crustaceans
- Protozoans

within ecosystems regarding the numbers and types of organisms, habitats, and ecological processes such as nutrient cycling. A grassland ecosystem with a variety of shrubs, trees, fungi, bacteria, insects, and mammals has greater diversity than a cornfield.

This aspect of biodiversity is measured by **species diversity,** which refers to the total number of living species. Biologists have identified about 1.7 million species, the majority of which are insects. New species are identified at the rate of about 300 per day. No one is quite sure just how many species exist; estimates range from 10 million to 100 million species.

Species diversity is measured in two ways. **Species richness** refers to the number of species present (Figure 12.2). An area with an especially large number of species is called a **biodiversity hot spot. Species evenness** refers to the distribution of individuals among species and measures whether individuals are distributed evenly among species or represent a few of the many species present. Evenness decreases as the concentration of individuals in a few species increases.

Differences between richness and evenness can be illustrated by the three hypothetical lizard communities in Figure 12.3. Communities A and B have three species, so they have the same species richness. But they differ in evenness. Evenness is greatest in community A, where each species makes up one-third of the total lizard population. All else being equal, you could say

Community C

1 Horned lizard

2 Collared lizards

1 Gila monster

4 Desert iguanas

Community A

3 Horned lizards

3 Collared lizards

3 Gila monsters

Community B

1 Horned lizard

2 Collared lizards

3 Gila monsters

FIGURE 12.3 *Richness versus Evenness* Communities A and B each have three species, which gives them the same species richness. But community A has greater evenness because it has three individuals of each species. Community C has the greatest richness because it has four species, but it has less evenness because two species (horned lizard and Gila monster) are represented by only one individual, while one species has four individuals (desert iguana).

that community A is more diverse than community B. But now look at community C, which has four lizard species, among which one species accounts for 50 percent of all lizards. Community C has greater richness than community A but has less evenness. So which community, A or C, has the greatest diversity? There is no clear answer. The lack of a single best measure for biodiversity poses difficulties for policy makers who want to preserve biodiversity. Should they preserve community A or community C? We return to this question later.

PATTERNS OF AND MECHANISMS FOR BIODIVERSITY

Tropical forests are legendary for their biodiversity. A 14 km^2 area of tropical forest in Costa Rica has about 1,500 plant species—more than all of Great Britain, which covers 243,500 km^2 (94,015 mi^2). At the global scale, biodiversity varies by latitude. In general, species richness is greatest near the equator and declines toward the poles, as seen in the distribution of bird species (Figure 12.4). There is some evidence that this poleward decline may occur faster in the Northern Hemisphere.

How can we explain this pattern? In any given area, species richness depends on the balance between the gain in new species and the loss of existing species. Areas gain species via **speciation,** which is the process by which evolution generates new species, and **immigration,** which occurs when new species spread into an area from elsewhere. Areas lose species via **extinction,** which occurs when a species fails to reproduce and no individuals remain.

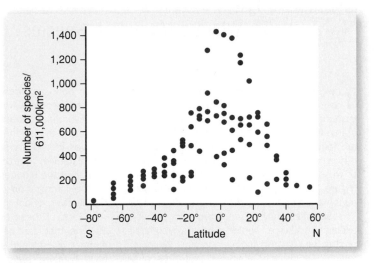

FIGURE 12.4   *Richness and Latitude* *The number of bird species declines as we move north (positive values) or south (negative values) of the equator.* (Source: Data from K.J. Gaston, "Global Patterns in Biodiversity." Nature 405: 220–227.)

How do these factors vary over space and determine geographical patterns of biodiversity? We can get an idea by understanding why species richness increases with island size. A classic study by Robert MacArthur and Edward O. Wilson found that the number of lizard species on Caribbean islands varies positively with the area of the island (Figure 12.5). Larger islands such as Cuba (100,860 km^2) have more species than smaller islands such as Saba (13 km^2). This positive relationship is not unique to Caribbean islands or lizards. Similar relationships are found with other animals, such as birds and plants, and in other environments, such as lakes and mountaintops.

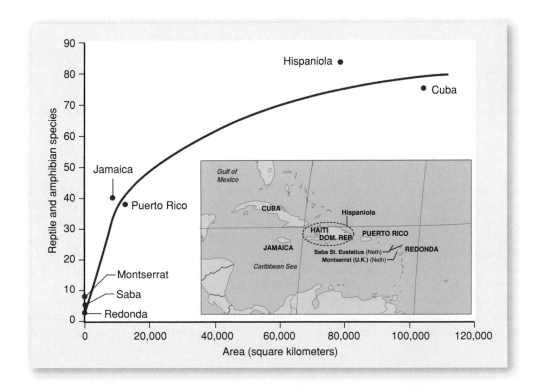

FIGURE 12.5   *Species–Area Relationship* *The number of reptile and amphibian species on West Indian Islands increases with island size, but at an ever-decreasing rate.* (Source: Graph modified from R.H. MacArthur and E.O. Wilson, The Theory of Island Biogeography. Princeton University Press. with data from P.J. Darlington, Zoogeography. Wiley.)

The coral reefs in the Indonesian and Philippine region are a wonderful place to dive. A diver is treated to a wide variety of reef fish species, many of which are endemic, which means they are found only there. Diversity quickly declines if one dives in areas farther north, south, east, or west of the Indonesian and Philippine region (Figure 1). Why is the Indonesian and Philippine region a hot spot for fish biodiversity?

Scientists have formulated two possible explanations. The center-of-origin hypothesis postulates that the Indonesian and Philippine region has a high speciation rate. Because some of these species can disperse over greater distances than others, biodiversity declines with distance from the IPR. Movements in the opposite direction are described by the center-of-accumulation hypothesis. According to this hypothesis, many species evolve in areas away from the Indonesian and Philippine region. Species then disperse from these other areas. The distances that these species can disperse overlap at the Indonesian and Philippine region, so the immigration from many areas creates the biodiversity hot spot.

Each of these hypotheses generates predictions that can be tested by analyzing the number of endemic species in the Indonesian and Philippine region and surrounding areas. The center-of-origin hypothesis argues that the IPR is a biodiversity hot spot because it has a high rate of speciation. If this is correct, the number of endemic species in the Indonesian and Philippine region should be greater than surrounding areas. Conversely, if the center-of-accumulation theory is correct, the areas surrounding the Indonesian and Philippine region hot spot should have the greatest number of endemic species.

A census of local populations indicates that the largest number of endemic reef fish inhabit the Indonesian and Philippine region (Figure 2). This region is home to about ninety endemic species—more than the number of endemic species that inhabit the areas that surround the Indonesian and Philippine region. Outside this region, the next highest area of endemism contains about forty-three species. These results

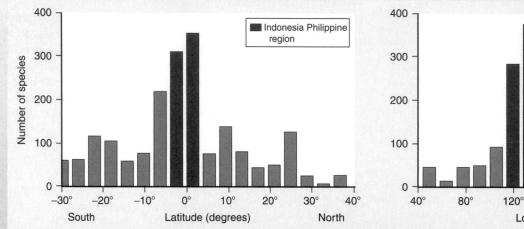
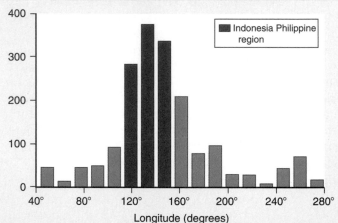

FIGURE 1 *Biodiversity Hot Spot* *The many species present in the Indonesian and Philippine region constitute a biodiversity hot spot. The number of species present declines east, west, north, or south of the Indonesian and Philippine region.* (Source: Data from C. Mora et al., "Patterns and Processes in Reef Fish Diversity. Nature 421: 933–936.)

Richness also is influenced by isolation. Isolated islands tend to have fewer species, which can be explained by immigration. Few new species arrive at islands that are far from the mainland or other islands because organisms or seeds must travel a long distance. The meaning of *long distance* varies among species. Most lizards are not good swimmers; they get to islands by hitching a ride on floating material. On the other hand, most birds can fly—some over great distances. Light seeds can be blown over long distances, whereas other seeds (such as coconuts) float. The ability of organisms to reach new environments is termed **dispersal,** which is defined as the distance a species can travel to find new environments. See *Case Study: Why So Many Fish Species?* Species that have relatively little ability to disperse often are found in only a single place. A species that is restricted to a certain geographic region and is thought to have originated there is said to be an **endemic** species.

Over much longer periods, richness is increased by speciation. Speciation occurs via two mechanisms. **Allopatric speciation** occurs when a population becomes geographi-

clearly show that local rates of speciation are responsible for the biodiversity hot spot in the Indonesian and Philippine region.

Understanding the mechanism that generated the biodiversity hot spot is important to efforts to preserve biodiversity. Although the center-of-origin hypothesis does not explain the hot spot, it implies that the Indonesian and Philippine region is an important source of fish species for the reefs throughout the Indian and Pacific oceans. To understand how this is possible, we offer a short explanation for the dispersal of fish. When they first hatch, the young of many fish species are so small that they float along with ocean currents in a pelagic larval stage. After a certain amount of time (the pelagic larval duration) the young fish become large enough to swim against currents. At this point they establish residency.

The Indonesian and Philippine region may be critical to species richness throughout the Pacific and Indian Oceans. Scientists have tested this hypothesis by examining the relationship between the ability of fish to disperse and local levels of species diversity. As expected, species richness in areas away from the Indonesian and Philippine Region declines with their distance from the Indonesian and Philippine region. Furthermore, there is a correlation between the mean pelagic larval duration of species in an area and the area's distance from the Indonesian and Philippine region. On average, fish species that live far from the Indonesian and Philippine region have a longer mean pelagic larval duration than fish species that live close to the Indonesian and Philippine region. This correlation implies that the number of species from the Indonesian and Philippine region whose pelagic larval duration is long enough to reach that area determines the species richness in an area.

These results show that the Indonesian and Philippine region is an important source of fish species for many areas of the Indian and Pacific oceans. Thus preserving biodiversity in the Indonesian and Philippine region is critical for both this hot spot and reefs throughout the Indian and Pacific oceans.

ADDITIONAL READING

Mora, C.P., M. Chittaro, P.F. Sale, J.P. Kritzer, and S.A. Ludson. "Patterns and Processes in Reef Fish Diversity." *Nature* 421 (2003): 933–936.

STUDENT LEARNING OUTCOME

- Students will be able to explain why biodiversity hot spots in some marine environments are critical to maintaining biodiversity well beyond those hot spots.

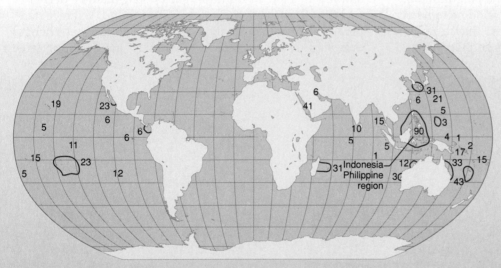

FIGURE 2 *Endemic Species* One reason for the Indonesian and Philippine region's biodiversity is the presence of endemic species. The large number of endemic species relative to other areas invalidates the center-of-accumulation explanation for the Indonesian and Philippine hot spot. (Source: Data from C. Mora et al., "Patterns and Processes in Reef Fish Diversity." Nature 421: 933–936.)

cally isolated from its parent population and then accumulates genetic or behavioral changes that differentiate it from the original population. For example, a few individuals may find their way to a new island, or a large population may be separated into smaller subpopulations by the formation of a new mountain, land bridge, or some other barrier. Both processes isolate the subpopulation, which allows it to accumulate differences relative to the original or parent population. The accumulation may be driven by random changes in the gene pool, which are known as **genetic** or **random drift**, or by natural selection, which may favor traits different than those favored in the original location. Either way, these differences may accumulate to the point that the subpopulation becomes a new species.

The Panama land bridge, which rose from the ocean about 3 million years ago, may have accelerated allopatric speciation of snapping shrimp. Before the land bridge rose, snapping shrimp in the western Caribbean and eastern Pacific were members of the same species. Although the physical appearance of the snapping shrimp off the eastern and western coasts of Panama is still nearly identical, separation by the land bridge allowed the populations to change both genetically and behaviorally (Figure 12.6). If male and female shrimp from opposite coasts are put in the same tank, they do not breed—they snap at each other. The inability to produce offspring makes them separate species.

Alternatively, new species may arise within the parent population, which is known as **sympatric speciation.** According to this explanation, individual traits isolate a subpopulation from the parent population, which allows the populations to evolve separately. Sympatric speciation may be powered by mutations that occur during cell division. In other cases different species can interbreed to create new species. For example, modern wheat plants that provide the flour for our bread may be the offspring of earlier wheat plants and a wild grass. In animals sympatric speciation may occur if genetic changes cause individuals to focus on resources that are not emphasized by the parent population. For example, Lake Victoria in Africa is home to nearly 200 species of cichlid fish. The lake is less than a million years old, which has led many to hypothesize that the rapid rate of speciation may have been driven by the exploitation of new food resources and habitats.

On the other hand, extinction reduces species richness. In general, the rate of extinction declines as the size of an island increases. Big islands generally have more diverse habitats than smaller islands, and this variety creates more niches. Conversely, extinction rates increase with the number of species present. Having more species increases the number of species that can go extinct. Furthermore, the presence of more species usually implies a smaller population size for each species, which increases the likelihood of a species going extinct. Finally, increasing species richness increases the potential for competition.

How can we use speciation, immigration, and extinction to explain the global distribution of species richness? One explanation applies to the species–area relationship to three stylized facts of Earth's physical geography. First, the amount of land in a latitudinal band generally decreases as latitude increases (Figure 12.7). Second, average temperature decreases nonlinearly with latitude—it changes only slightly between 20° north and 20° south. Third, large

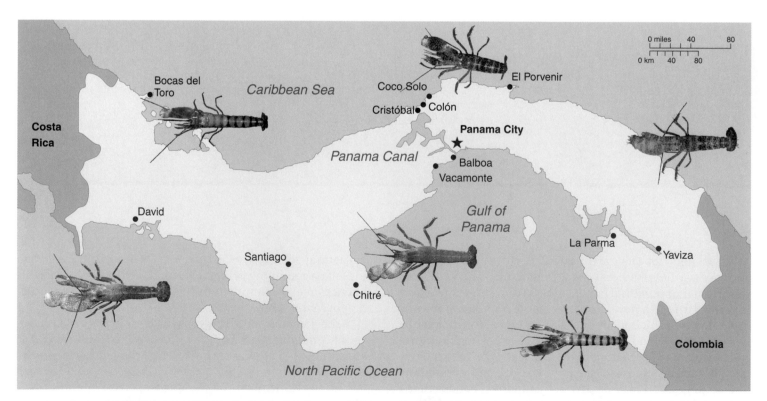

FIGURE 12.6   *Allopatric Speciation* *The rise of the Panama land bridge separated shrimp populations. Over time the populations have become increasingly different so that individuals from the Atlantic and Pacific populations will not breed with each other. As such, they are now separate species.* (Source: Redrawn from of The Smithsonian Institution/Carl Hansen and Nancy Knowlton http://www.pbs.org/wgbh/evolution/library/05/2/1_052_03.html.)

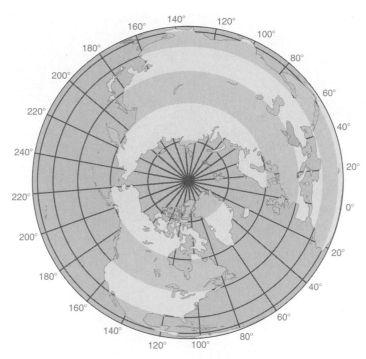

FIGURE 12.7 *Continuous Landmass* When viewed from the Arctic, land in the Northern Hemisphere appears as bands. The largest bands occur at lower latitudes, and these large areas can support many species, as implied by the species–area relationship shown in Figure 12.5.

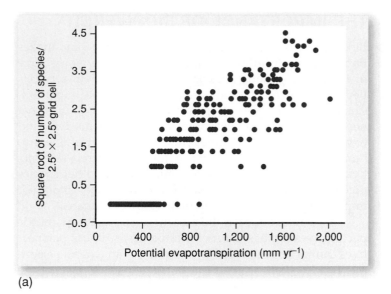

(a)

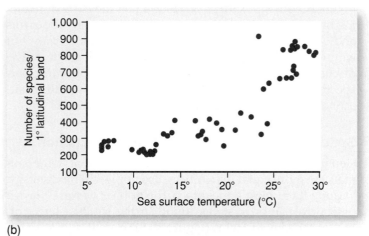

(b)

FIGURE 12.8 *Richness and Energy* (a) In North America the number of beetle species is positively related to potential evapotranspiration, which is a proxy for net primary production. (b) In the eastern Pacific the richness of marine snail species is positively related to sea surface temperature, which positively affects net primary productivity. *(Source: Data from K.J. Gaston, "Global Patterns in Biodiversity." Nature 405: 220–227.)*

land areas between 20° north and 20° south are in contact. Together these features create a huge continuous land area with a similar climate.

So why are big, continuous land areas with similar climates important to global biodiversity? These features reduce extinction rates by supporting larger populations and by giving individual populations more areas where they can survive following a disturbance. These large areas also increase speciation, especially allopatric speciation. For example, genetic analyses indicate speciation is greatest on large islands. Finally, geographic barriers are more likely to divide wide-ranging species into smaller populations.

Another explanation for the latitudinal pattern of species richness is based on the importance of energy to biological organisms. As discussed in Chapter 5, organisms need a continuous supply of energy. All other things being equal, increasing the supply of energy should increase the number of individuals and species that can be supported in an area. Consequently, high rates of net primary production by terrestrial biomes near the equator that decline with latitude (see Chapter 7) imply a similar pattern in species richness. This hypothesis is consistent with the positive relationships between potential evapotranspiration (on land) and the richness of beetle species (Figure 12.8a) and between sea surface temperature and the species richness of snails (Figure 12.8b).

The species–energy relationship is a hotly debated topic in the study of biodiversity. Empirical analyses identify some significant contradictions. First, the positive relationship between available energy and the number of individuals may not apply to plants. Some analyses indicate that the number of adult plants decreases as biomass increases. For animals, food supplies decline at higher trophic positions, and this may reduce the effect of net primary production on biodiversity. For example, birds in the Hubbard Brook Experimental Forest in New Hampshire (see Chapter 8) consume only about 0.17 percent of net annual production. This small fraction makes it unlikely that the diversity of bird species could be linked to overall productivity. Because of these contradictions, some believe that the link between available energy and diversity is merely a correlation (see Chapter 4's Case Study on page 68).

THE IMPORTANCE OF BIODIVERSITY

Biodiversity is important for ecosystem function and makes important contributions to human well-being as a source of insurance, genetic knowledge, and ecosystem services. Beyond these roles, biodiversity may itself be important. Consistent with this notion, a **biocentric** view argues that nonhuman species have value in and of themselves and have the right to exist independent of their usefulness to humans. An **anthropocentric** or human-centered perspective on the relationship between humans and other species holds that other species exist for human use, and their importance is determined by their value to humans.

Differences between the biocentric and anthropocentric perspectives are based on ethical beliefs, so these perspectives cannot be reconciled by science. As an extreme example, in 1999 there was ethical debate about destroying the last known sample of the smallpox virus: Could the extinction of the smallpox virus be justified on the basis of reducing human suffering? We have personal opinions about this and other ethical issues about the value of biodiversity, but we prefer to keep our opinions separate from the text. The discussion that follows emphasizes an anthropocentric perspective based on our efforts to integrate the scientific and economic underpinnings of environmental science.

Species Interactions and Ecosystem Function

George Orwell's statement in *Animal Farm* that some animals are more equal than others also applies to biodiversity. Despite its modest appearance, the gopher tortoise plays a critical role in ecosystems found in the southeastern portion of the United States (Figure 12.9). Here it digs burrows that can reach 2 meters below the surface. At these depths the microclimate remains relatively constant year-round. Constancy provides the tortoise a refuge from inhospitable surface conditions during certain parts of the year. These burrows also shelter many other species, including the indigo snake and the burrowing owl. It is probably no coincidence that the populations of indigo snakes, burrowing owls, and many other occupants of tortoise burrows have declined with the population of gopher tortoises.

This effect has prompted scientists to categorize species according to their role in ecosystem function. **Keystone species** are species whose presence and numbers control the integrity of a community or ecosystem and allow that system to persist within its nature range of environmental conditions. Based on this definition, scientists consider the gopher tortoise a keystone species.

The effect of a keystone species on ecosystem function is illustrated by the largemouth bass. In an experiment, scientists added phosphorus and nitrogen fertilizers to two lakes. Adding fertilizer to the lake that retained its top-level predators, such as largemouth bass, increased net primary production, which allowed the lake to remove carbon from the atmosphere.

FIGURE 12.9 *Keystone Species—Gopher Tortoise The gopher tortoise is a keystone species in many ecosystems of the southeastern United States. It digs burrows that provide habitat for a variety of reptile, amphibian, mammal, and bird species.*

Scientists added the same amount of fertilizer to another lake from which they had removed all top-level predators, including the largemouth bass. The bass feed on smaller fish such as fathead minnows. Fathead minnows and other small fish eat zooplankton, which feed on algae. Without the bass, the population of smaller fish exploded, which reduced the population of zooplankton, which allowed the fertilizer to increase the population of algae. The algal bloom caused the lake to become eutrophic, which caused it to emit carbon dioxide. Thus largemouth bass have an effect, albeit small, on global climate change. (Atmospheric carbon dioxide enhances global warming: See Chapter 13.)

Of course not all species are critical to ecosystem function. The loss of noncritical species would have a relatively small effect on an ecosystem. Such species are known as **passenger** or **redundant species.** Although no two species occupy the same niche (see Chapter 7), sometimes the degree of overlap is sufficient to minimize the effect of their loss. The concept of passenger species is related to the notion of functional biodiversity described in Chapter 8. Remember that scrapers are critical to the health of coral reefs. But having six species of scrapers instead of seven may have relatively little effect on the reef's ability to recover from a disturbance. The loss of a species within this functional group may have less effect than the loss of a keystone species.

Biodiversity as Insurance

The importance of insurance is formalized in several well-known clichés, such as "Don't put all your eggs in one basket" or "Set something aside for a rainy day." To understand the role of biodiversity as insurance, it is necessary to understand **insurance,** which spreads the potential effects of risk. When you buy insurance, you sign a contract that entitles you to compensation should some event happen. To maintain insurance you make payments that are called an insurance premium. You pay that premium without knowing whether the event that triggers compensation will occur. With regard to biodiversity, society may be willing to buy insurance against the ongoing loss of biodiversity by paying now to preserve biodiversity that may be of use in the future.

Society bought such insurance by supporting the International Rice Research Institute. For many years the International Rice Research Institute used funds from international organizations such as the Food and Agriculture Organization of the United Nations and the World Bank to preserve seeds from rice varieties that were considered uneconomic. This label applies to varieties that produce lower yields than varieties currently grown by farmers. There was no immediate economic reason to preserve the seeds of uneconomic varieties. Their potential value lay in the future, when some unknown aspect of their genetic diversity could be of economic value.

This value was proven by some varieties' resistance to the grassy stunt virus, which first appeared in the 1970s. At that time the virus threatened much of the Asian rice harvest. In some years the virus eliminated nearly a quarter of the potential harvest.

To develop a resistant strain of rice, the International Rice Research Institute launched a breeding program by growing rice from the uneconomic seeds it had preserved. Scientists hoped that one of the varieties had a genetic component that would allow it to resist the grassy stunt virus. After several years they identified a resistant variety. Originally this variety was found in only one location, which was now submerged by the construction of a hydroelectric dam. In short, the variety would have been extinct if not for the International Rice Research Institute. The expense associated with the institute's effort to collect and preserve many uneconomic varieties can be thought of as insurance for biodiversity.

Apart from this ability to grow a variety of rice that is resistant to the grassy stunt virus, why is biodiversity insurance valuable? It allows farmers to practice agricultural techniques that are highly productive but would be too risky without insurance. As described in Chapter 16, modern farmers plant a few highly productive varieties of corn, wheat, or rice over large areas. Although this generates large amounts of money for farmers and food for people in most years, it is also risky. By creating one huge, genetically identical food source, farmers speed the evolution of new pests and diseases. These new pests or diseases could cause major crop failures—a risk that is mitigated by preserving the genetic diversity of crops.

Genetic Knowledge

The genetic diversity of Earth's species contains information to make a wide variety of chemical compounds, some of which can be used as human medicines. Of the 150 most prescribed drugs in the United States, about 57 percent contain one or more compounds derived from or patterned after living organisms. Such compounds also serve as models for human-made molecules that appear in nearly a quarter of all prescription drugs sold in the United States. Examples include the anticancer drug taxol, which is derived from the Pacific yew tree, and the heart drug digitalis, which is derived from the foxglove plant.

You may be tempted to ask why plants and other organisms make chemical compounds that are good for what ails humans. These benefits are an unintended consequence of **diffuse chemical coevolution,** in which natural selection favors individuals that accumulate compounds effective against a wide variety of enemies. Such selective pressures tend to be especially strong in sessile organisms such as plants, which use toxic or distasteful chemicals to discourage potential predators (see Chapter 5). Therefore, the search for medicine from living organisms often starts with a screening process that looks for chemical compounds that are toxic to entire organisms or specific cell types.

Understanding evolutionary history may help drug companies find potential medicines. Instead of screening compounds from species chosen at random, drug companies often screen organisms that are genetically related to species that already provide commercially important drugs. Alternatively, drug companies may screen compounds from species that live in ecosystems where chemical defenses could be useful. Examples include tropical rain forests, where high levels of biodiversity imply a large number of potential predators, or hydrothermal vents, hot springs, and the like, where temperature, pH, or salinity is extreme. Organisms that live in extreme environments, known as **extremophiles,** require special adaptations. Some organisms that live near hydrothermal vents, known as *Archaeans,* can withstand temperatures greater than 100°C, which implies that their enzymes are very stable. Such enzymes have proven critical to the biotechnology industry.

Finally, companies may screen organisms based on **ethnobotany,** which describes how different groups of people, including indigenous cultures, use plants and animals. Through the process of trial and error, and perhaps direct observation of animals, people have identified plants and animals that have medicinal uses. This knowledge has been passed from one generation to the next and now is a valuable source of information for drug companies.

For example, legend says that the ancient people of India noticed that mongooses eat rauvolfia root before attacking cobras. Local people found that this plant provides an effective antivenom. Based on this ethnobotanical description, recent experiments found that the roots contain a compound that lowers blood pressure, and this compound is an active ingredient in the commercial drug reserpine.

Environmental Services

Biodiversity provides environmental services such as stabilizing climate, pollination, and pest control. These roles imply that biodiversity can be viewed as a capital asset. If the asset is managed properly, it will provide a sustainable stream of economically important services.

Living organisms have a stabilizing effect on the climate via a negative feedback loop. There is some evidence that autotrophs increased their uptake of carbon dioxide as the sun grew stronger (over millions of years). As will be described in the next chapter, carbon dioxide absorbs heat and therefore warms the planet. Reducing the atmospheric concentration of carbon dioxide thereby offsets some of the warming effect of increased solar radiation. On a regional scale, trees have an important effect on temperature and precipitation. About 50 percent of the rainfall in the Amazon rain forest is water that was transpired by forest trees. Without the forest, rainfall would be greatly reduced. (For more information about the climate effects of forests, see Chapter 17.)

Biodiversity is critical to soil formation, and soils provide essential services, such as water purification. (Chapter 15 will describe in detail the role that soil plays in plant growth.) As water percolates through the soil, impurities are removed. Some estimates indicate that large cities use up to 15 percent of Earth's land area for water supply and purification. Soils also store water and nutrients for crops and provide an anchor for plants. Replacing these services would be expensive. Trays and stands that are used in hydroponic agriculture cost about $55,000 per hectare; good agricultural land in Iowa costs less than $10,000 per hectare.

Scientists estimate that over 100,000 different species of bats, insects, and birds act as pollinators. **Pollinators** are animals that place pollen on the stigma of a plant. By doing so, pollinators are critical to sexual reproduction by flowering plants. About 70 percent of the crop species grown by farmers depend on animal pollinators. Pollination services are crucial to agriculture because many vegetables and fruits grow from fertilized flowers. Farmers pay beekeepers $40 to $70 to set up a colony near their crops; for each hectare of crop, two to four colonies are needed. These purchased services are essential because the population of native pollinators is declining.

WHY IS BIODIVERSITY DECLINING?

The Rate of Extinction

Extinction occurs when there are no more living representatives of a given species. Extinction can be either global—the species is gone forever—or local, in which case the species could migrate back into the area from other populations. Extinction has a negative connotation, but it is an integral part of evolution. Species have a finite lifetime—95 to 99 percent of the species that have ever existed are now extinct. Is the current rate of extinction faster than normal?

The average rate of extinction over long periods can be estimated from the fossil record, which shows that the average life spans of species range from 1 million to 10 million years. Using such information, scientists estimate that the average rate of extinction is about one species every four years. This continuous, low-level rate of extinction is known as **background extinction.**

The fossil record also identifies five periods, known as **mass extinctions,** when the extinction rate was much greater than the background rate (Figure 12.10). At these times huge numbers of both land and marine species became extinct. During some events more than three-quarters of the living species disappeared. One of the more famous mass extinctions wiped out dinosaurs and many other species about 65 million years ago.

Although mass extinctions show up clearly in the fossil record, the causes are debated. Some scientists hypothesize that major changes in climate and increased volcanic activity are responsible. Other scientists hypothesize that cata-

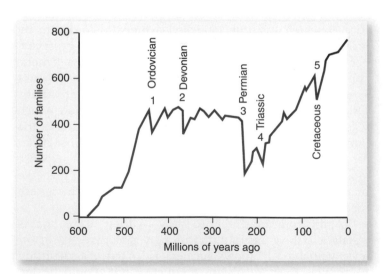

FIGURE 12.10   *Mass Extinctions Over the last 600 million years there have been five periods when large numbers of species have gone extinct. These mass extinctions are indicated by abrupt drops in the line that represents the number of families. The last dip, about 65 million years ago, represents the demise of dinosaurs.*

strophic events such as a collision with a meteorite could have triggered mass extinctions. Even a small meteorite can kick enough dust into the atmosphere to reduce sunlight, lower net primary production, and lower temperature. There is considerable evidence for the impact of a meteorite. The geological record shows a jump in the concentration of iridium (an element that is rare on Earth but common in meteorites) at the same time as the mass extinction about 65 million years ago (Figure 12.11).

Current estimates show that the number of species is declining. Simply put, the current global rate of extinction is greater than the rate of speciation. Since 1600 more than 1,000 extinctions have been documented—and many more probably have not been noticed. During the twentieth century more than twenty species of mammals have become extinct. Evolution would require about 20,000 years to replace them. Although the exact number of extinctions is in dispute, most scientists agree that the current rate of disappearance is greater than the rate of loss during the mass extinctions. This unprecedented rate of extinction is associated with human activities that alter habitats, introduce alien species, change biogeochemical cycles, and overharvest plants and animals for their commercial value.

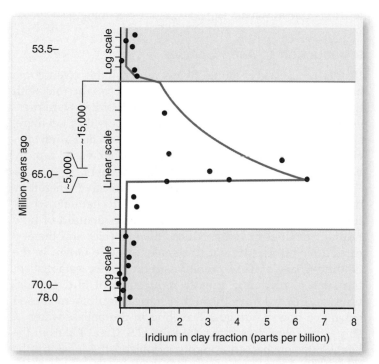

FIGURE 12.11 *Meteorite Impact* *The concentration of iridium increases sharply in a layer of clay residues from limestone that was laid down about 65 million years ago. This increase in concentration suggests a meteorite impact because iridium is rare on Earth but is present in greater concentrations on meteorites.* (Source: Data from W. Alvarez et al., "Iridium Anomaly Approximately Synchronous with Terminal Eocene Extinctions." Science 216: 886–888.)

Habitat Alteration

The single greatest threat to biodiversity is the large-scale alteration of natural ecosystems by humans. As described in Chapter 8's Your Ecological Footprint on pages 148–149, such conversions are known as changes in land use or land cover. To review, land cover describes the type of biome present, whereas land use describes the purpose to which land is put. Changes in land use or land cover are known as **habitat conversion,** which is the change of land quality—for example, through land transformation or intensification of land use. Common habitat conversions include deforestation and reforestation; urbanization; desertification and conversion to agriculture, such as wetland drainage; irrigation; and degradation due to overgrazing.

Habitat conversion destroys or degrades natural ecosystems. In the last three centuries habitat conversion has eliminated nearly 20 percent of Earth's forests, altered 8 percent of Earth's grasslands, and expanded croplands by nearly 500 percent. During the twentieth century the fastest rates of habitat conversion occurred in developing nations in the tropics. Because biodiversity in the tropics is much greater than in temperate regions, extensive habitat conversion in developing nations eliminates more biodiversity than do agricultural and industrial conversions in developed nations.

Habitat conversion reduces biodiversity by eliminating habitats in a given biome. The Philippines lost about 3.5 percent of its forests in 1999. Based on the species–area relationship discussed earlier in this chapter, the loss of forest habitat reduced the number of species that can be supported in that area—an effect amplified by the spatial pattern of the habitat loss. This pattern, described as **fragmentation,** is the breakup of a continuous habitat, ecosystem, or land use type into smaller areas.

Fragmentation's effect on biodiversity is determined in part by the size of the remaining habitat. Generally, breaking habitat into many smaller pieces has a greater effect on biodiversity than retaining one large area. The species–area relationship implies that breaking a 10,000 km^2 forest into ten 500 km^2 forests eliminates more biodiversity than retaining a single 5,000 km^2 forest (Figure 12.12). Breaking up the forest into smaller fragments also generates longer **edges,** which create transition zones between different land covers. In some cases these edges can improve species diversity by creating areas that support species that cannot exist in either of the adjacent land covers.

But in many other cases fragmentation can disrupt ecosystem processes such as nutrient cycling and predator–prey relationships. These disruptions often are caused by **edge effects,** which expose forest species to unfavorable conditions such as stronger winds, temperature changes, increased incidence of fire, and increased predation and competition from exotic and pest species. Edges increase the damage and death rates of trees living in the Amazon forest—even those located up to 500 m from an edge. Edge effects may be as important as the outright loss of habitat. Only 39 percent of the loss of

FIGURE 12.12 *Fragmentation and the Edge Effect* A forest that is 100 km on a side has an area of 10,000 km² and an edge of 400 km. If half of the forest is cut as one continuous area, the remaining forest has an area of 5,000 km² and an edge of 300 km. But if half the forest is cut into fifty fragments each with an area of 100 km², the remaining forest has an area of 5,000 km² as before but now has a total edge of 2,000 km (10 × 4 × 50). If we assume that species are spread evenly across the original forest, the number of species present declines with size of the individual forest patches according to the simple species–area relationship. Furthermore, the edge effect increases by nearly a factor of 10.

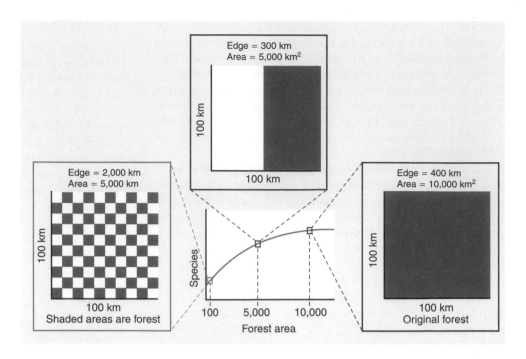

habitat in the Amazon forest from 1978 to 1988 was caused by outright forest clearing; the rest occurred through fragmentation and edge effects (Figure 12.13).

Scientists use the species–area relationship and information about the rate and spatial pattern of habitat conversion to estimate the rate of extinction associated with habitat destruction. One detailed estimate by a pioneer in the study of biodiversity, Edward O. Wilson of Harvard University,

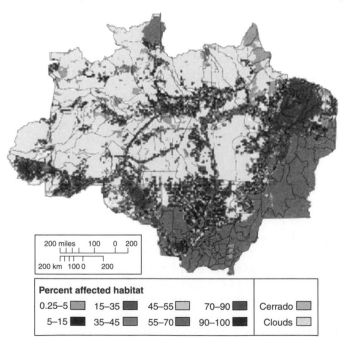

FIGURE 12.13 *Amazonian Fragmentation* Locations where Amazonian biodiversity was negatively affected by deforestation, fragmentation, and edge effects in 1988. (Source: Data from D. Skole and C.J. Tucker, "Tropical Deforestation and Habitat Fragmentation in the Amazon—Satellite Data from 1978 to 1988." Science 260: 1905–1910.)

indicates that habitat conversion caused the extinction of 27,000 species per year during the 1980s. This number has been criticized as too high. Nonetheless, high rates of extinction implied by the species–area relationship are consistent with the rate at which populations are declining.

Introduction of Alien Species

People rarely travel alone. Plants and animals often come along, invited or not. The Polynesians brought rats with them to Easter Island. No terrestrial vertebrates were present before humans arrived. Regardless of how they get there, species that are new to an environment are called **alien species.** The introduction of alien species has increased rapidly with the boom in international trade and travel.

In some cases alien species become **invasive** when they displace indigenous species or spread into habitats where they were not previously common. Stories abound of how human activities have introduced alien species and thereby caused havoc. The brown tree snake came to Guam in the landing gears of U.S. military aircraft. Once established, the snakes' predation greatly reduced local bird populations, and their forays through electrical equipment caused extensive power outages. Traveling in the ballast water of ships, the zebra mussel reached the U.S. Great Lakes in the late 1980s and had thoroughly colonized the lakes by the mid-1990s. There it grows rapidly, clogging the water intake pipes of industry and drinking systems.

Together alien species cause huge economic damages. In the United States the federal government estimated that seventy-nine alien species caused $97 billion in damage between 1901 and 1991. A more recent estimate by Cornell scientists put the cost at $138 billion per year. For example, efforts to remove zebra mussels cost about $5 billion per year. Another $500 million is spent in Texas to control fire ants.

The disruptive effect of alien species can be explained using evolutionary theory. Most species evolve with other species. In their original habitat, which is known as their **native range,** a variety of predatory, parasitic, and disease organisms often keep the populations in check. But many of these limiting factors are left behind when species are introduced to a new environment, which is known as their **naturalized range.** This allows the population to grow rapidly. The lack of predators, parasites, or disease allows the population of alien species to grow and increase its use of energy and nutrients, which leaves less for native species, causing native populations to decline.

This explanation for the disruptive effect of alien species is described by the **enemy release hypothesis,** which states that the population of an alien species can grow rapidly (escape) if the number of pathogens it leaves behind in its native range exceeds the new pathogens it accumulates in its naturalized rage. This hypothesis was examined by determining the factors that allow alien plant species to become invasive. As you might expect, the number of pathogens in the naturalized range was less than the number in the native range. On average, plants living in their naturalized range had 84 percent fewer fungi and 24 percent fewer virus species than individuals living in their native range. So the plants did indeed escape. Furthermore, the degree to which a species became invasive increased with the reduction in the number of pathogens (Figure 12.14). Together these results show that the impacts of alien species depend in part on the degree to which they can escape their enemies with which they have evolved in their native range.

Changes in Biogeochemical Cycles

Previously we noted that greater supplies of energy may increase the number of animal species in an area but reduce the number of plant species. According to this notion, initial increases in net primary production increase biodiversity; but as net primary production exceeds some threshold, further gains in net primary production reduce plant biodiversity (Figure 12.15a). Biodiversity declines at high rates of net primary production because only a few tall, fast-growing species can successfully compete for light. These species cannot dominate in less productive environments in which some limiting nutrient slows their growth. As described in Chapter 6, this nutrient is thought to be either nitrogen or phosphorus.

Humans have disrupted both the nitrogen and phosphorus cycles in ways that increase their availability to plants, which has increased the amount of biomass per square meter. This general increase in biomass has eliminated habitat for plants that are **endangered,** a term that refers to any species that is in danger of extinction throughout all or a significant portion of its range. Endangered plant species are most abundant where biomass per m^2 is relatively low (Figure 12.15b). This loss has been amplified in western Europe, where high rates of fossil fuel use have greatly increased nitrogen deposition. The increased availability of nitrogen has driven endangered plant species from sites that were previously limited by nitrogen. The shift in nitrogen-limited ecosystems to higher production levels has caused species loss in U.K. grasslands.

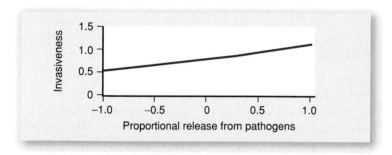

FIGURE 12.14   *Enemy Release   The degree to which plant species become invasive in their naturalized range depends on the degree to which they escape pathogens in their native range. Plants that leave many their pathogens behind are more likely to become invasive.* (Source: Data of C.E. Mitchell and A.G. Power, "Release of Invasive Plants from Fungal and Viral Pathogens." Nature 421: 625–627.)

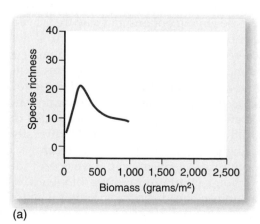

(a)

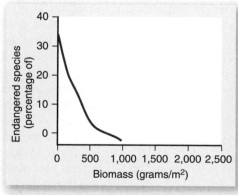

(b)

FIGURE 12.15   *Species Richness and Biomass   (a) Species richness in plants tends to be greatest at intermediate levels of biomass. It declines at high levels of biomass because only a few tall, fast-growing species can successfully compete for light. (b) The largest percentage of endangered species live in areas with relatively small biomass. They may be forced into extinction if human activity increases nutrient supplies in these areas, which would increase biomass and reduce the availability of light to these endangered species.* (Source: Data from M.J. Wassen et al., "Endangered Plants Persist Under Phosphorus Limitation. Nature 437 (7058): 547–550.)

The availability of phosphorus has also increased greatly across Europe. The increasing availability of phosphorus has reduced the biodiversity of herbaceous plants (plants that do not have a permanent woody stem) in wet and moist environments more than has the increased availability of nitrogen. The increased availability of phosphorus is important because there are far more endangered pants in low-phosphorus ecosystems. Furthermore, very low levels of phosphorus are correlated with very high numbers of endangered plant species. This implies that great numbers of endangered plants could be lost if human additions of phosphorus reach these ecosystems and increase their net primary production.

These effects suggest that efforts to preserve biodiversity should focus on phosphorus. Conservation managers have tried to preserve plant biodiversity by reducing nutrients, but these efforts have often failed because they have not reestablished phosphorus as the limiting nutrient. For example, some European efforts have focused on reducing nitrate. Because this left large amounts of phosphorus applied by farmers, efforts to reestablish endangered species in what were previously agricultural fields have failed. To be successful, environmental restoration managers must reestablish the balance of nutrients that existed in the original ecosystem. Based on this need, researchers are experimenting to determine the sources of the excess phosphorus.

Hunting and Harvesting

Humans hunt and harvest many plant and animal species. The rate of hunting or harvesting often exceeds the maximum sustainable yield, and populations then decline. If this decline continues long enough, species may become extinct. But overhunting is not just a modern achievement. Human migrations to previously uninhabited continents have historically coincided with rapid increases in extinction. The arrival of aboriginal human populations in Australia about 30,000 years ago contributed to the extinction of many large mammals, including marsupial lions and kangaroos that reached 2.5 meters (8 feet) tall. Similarly, the arrival of people in Madagascar around A.D. 500 initiated a wave of extinction.

Hunting now extends well beyond the need for food. People try to reduce the populations of species that compete with humans for crops or game in a process known as **predator control.** In an attempt to help ranchers, who argued that wolves, mountain lions, and eagles killed their livestock, U.S. federal and state governments used to pay hunters to kill top-level predators. But the effectiveness of predator control is sometimes uncertain. In Britain badgers are a source of the bacterial agent responsible for bovine tuberculosis, a disease that infects cows. To reduce such infections, the British government allowed farmers to kill badgers. However, bovine tuberculosis increased in areas where farmers reduced the badger population.

Plants and animals are also important commodities for the international wildlife market. Collectors in developing nations sell plants and animals to buyers, who generally live in developed nations. There the plants and animals are valued for their fur, skins, teeth, horns, shells, beauty, or rarity. A rhinoceros horn can sell for thousands of dollars in the Middle East and the Far East, where the horns are fashioned into handles for daggers or used in medicines. The world's largest parrot, the bright blue hyacinth macaw, sells for up to $8,000. Less glamorous species sell more cheaply but are harvested in larger numbers. A typical year's trade includes about 1.5 million birds, 150,000 furs, and 53,000 live wild orchids.

About is a key word here because trade in wildlife is poorly documented. The best data come from reporting requirements that are associated with the Convention on International Trade in Endangered Species of Wild Fauna and Flora. These data indicate that the trade is worth about $4–$5 billion per year. Another $5–$8 billion of plants and animals are traded illegally. This activity is spurred by the high prices paid by buyers in developed nations and low incomes in developing nations. Illegal killing or collecting of plants and animals, which is known as **poaching,** threatens to undermine efforts to protect biodiversity. See *Policy in Action: Preserving Biodiversity in the Face of Corruption.*

PRESERVING BIODIVERSITY

Current efforts to preserve biodiversity are motivated by the following question: If society is willing to spend money to preserve biodiversity, how can this money be directed to save the greatest amount of biodiversity? From this anthropocentric perspective, policy to preserve biodiversity is like many other environmental challenges. The optimal solution is for society to allocate its preservation efforts so that the benefits associated with the last species saved equal the effort spent saving it.

Finding this balance is difficult. In an ideal world policy makers would examine their options for preserving biodiversity. Each option would have a cost, and each would preserve a certain amount of biodiversity. Policy makers would be able to choose among options in a way that saved the greatest amount of biodiversity. This approach is not feasible because there is no single measure of biodiversity. Should policy makers seek to maximize species richness, species evenness, or some combination of the two? Furthermore, economists cannot measure the economic benefits generated by plants and animals that are not traded in markets. As a result, the problem of preserving biodiversity cannot be solved using a simple cost–benefit approach.

Short of this optimum, scientists have developed four criteria to choose among efforts to preserve biodiversity. The first criterion examines how distinct a species is. Higher priority is given to unique species. This ranking is based on their potential contribution to the growing collection of genetic knowledge: Genetically unique species may contain information that is not available in other species. The second criterion examines the usefulness of a species. This usefulness can be either commercial value (such as that offered by wild salmon) or the degree to which humans like the species (consider the bald eagle or the giant panda). The third criterion is the degree to which human actions can reduce the probability that a species will go extinct. That is, what are the odds that a species will go extinct if society does nothing? Or if some effort is expended, will that effort lower the probability that the species will go extinct? Clearly it does not make sense to try to preserve a species that is unlikely to become extinct or that will become extinct even if society tries to save it. Finally, society must evaluate the cost of preserving a species. Conservation efforts for some species may be small compared with the effort required to save others. Following these four criteria, it may be possible to compare conservation efforts in a way that preserves the greatest biodiversity for the least cost. Toward this end, policy makers have developed three general approaches to preserving biodiversity: legal protections, ecosystem protections, and market-based mechanisms.

Legal Protections

Society has sought to preserve biodiversity through a series of national and international laws. In the United States the most visible of these laws is the Endangered Species Act of 1973. The Endangered Species Act authorizes the U.S. Fish and Wildlife Service and the National Marine Fisheries Service to protect endangered and threatened species. A **threatened species** is any species that is likely to become an endangered species in the foreseeable future throughout all or a significant portion of its range. Endangered species include the whooping crane and the American crocodile. Threatened species include the northern spotted owl.

The Endangered Species Act protects threatened or endangered species via several mechanisms. The act prohibits hunting or commercial harvesting, including species that are not found in the United States. The Endangered Species Act also can mandate protection of habitat that is deemed critical to a species' survival. That protection may include blocking public or private construction that would damage habitat. The act states that any threat to a protected species supersedes all economic activities regardless of the activities' value.

Protecting a species regardless of its value implies that the process by which a species is listed as threatened or endangered is critical. Although the law dictates that biological criteria alone be used to assess the threat to a species, a close examination of the real-world political process shows that economic arguments about the costs and benefits of preserving a species often occur before it is listed as threatened or endangered.

Once listed, endangered and threatened species compete for federal and state funds aimed at preservation. Economists have compared these expenditures to the four criteria just listed. During the relatively short period for which data are available, there is little evidence that policy makers use any of the four criteria to allocate funds among species. Instead research indicates that funds are allocated idiosyncratically. For example, large sums have been spent to preserve popular species like the bald eagle, but much less money has been spent on less appealing reptiles and amphibians.

The seemingly arbitrary nature of the allocation process does not mean that the Endangered Species Act is a failure. The act has saved several species from possible extinction. One of its more visible successes is the American bald eagle. The use of DDT (Chapter 5), hunting, and destruction of the eagles' habitat reduced the number of nesting pairs in the continental United States to fewer than 500. In 1967 bald eagles south of the 40°N parallel were listed under the Endangered Species Preservation Act (the forerunner of the Endangered Species Act). By 1993 the number of nesting pairs had rebounded to more than 4,000, which prompted the U.S. Fish and Wildlife Service to upgrade the bald eagle's status from endangered to threatened in 1995. In 2004 there were 7,678 breeding pairs, and some groups have called for the bald eagle to be delisted.

Internationally, species are protected by the 1973 Convention on International Trade in Endangered Species. At the start of 2006 more than 160 parties had entered the convention, which monitors and regulates international trade of endangered plants and animals and the products made from them. The convention has effectively reduced trade in ivory, pelts, and other products that threaten certain species. Enforcement varies among countries, and illegal trade and corruption affect the success of preservation efforts.

Preserving Species and Habitat

Maintaining a species away from its natural habitat is called **ex situ** conservation. Examples of ex situ conservation techniques include breeding animals in zoos or aquariums and storing genetically diverse plants in seed banks such as the International Rice Research Institute.

Efforts to preserve species in functioning ecosystems are called **in situ** conservation. Preserving habitat is a growing focus of international cooperation. The Convention on Wetlands of International Importance, signed in Ravisar, Iran, in 1971, provides an international framework for conserving wetlands. As of 2004 more than 540 listed wetlands covered some 32 million hectares. The Convention

POLICY IN ACTION

Preserving Biodiversity in the Face of Corruption

In this chapter and many that follow, environmental policy is described in terms of what needs to be done. This simplification implies that decision makers develop a policy, it is implemented, and the policy either succeeds or fails. But this description overlooks an important determinant of success—**governance,** which is defined as the act of governing and exercising authority. Governance determines the degree to which the laws associated with environmental policy are enforced. As such, governance plays a critical role in the success or failure of environmental policy.

An important determinant of governance is **corruption,** which is defined as the unlawful use of public office for private gain. By outlawing certain actions, environmental policy often creates a black market, which generates a high price for an environmental good or service. For example, banning the sale of ivory raises its price, and prohibiting logging raises the price of wood. High prices generate enough money for lawbreakers to pay government officials to ignore illegal activities. For example, poachers or illegal loggers may pay park rangers to ignore activities that are forbidden by environmental policy. So corruption can reduce the effectiveness of even the best-conceived environmental policy.

We wish that our discussion of corruption could include the names of the guilty parties. But no one is willing to admit illegal actions. Instead political scientists try to estimate the level of corruption by surveying local officials and using the information to develop numerical indexes that measure the degree of corruption. Indexes range between 1 and 10. A score of 1 indicates a high degree of corruption, while a score of 10 indicates a low degree of corruption.

A simple analysis shows a negative correlation between the corruption index and nations' species richness for birds and mammals (Figure 1). This negative correlation indicates that species richness tends to be greatest in the most corrupt nations. Similarly, biodiversity hot spots tend to be located in nations that are plagued by corruption.

The link between biodiversity and corruption is troubling because there are several reasons to believe that the government officials who enforce laws to preserve biodiversity are vulnerable to corruption. First, many of these government officials are poorly paid, which makes them vulnerable to bribes. Second, many conservation projects run for relatively short periods and are funded by organizations outside the host nations. The brief employment opportunities and the lack of close connections with the source of funds provide a rationale for corruption. Finally, it is difficult to measure the success of most conservation projects. Without such criteria, it is hard for funding organizations to determine whether their money is being spent effectively.

Corruption's influence on biodiversity has been tested by examining the link between corruption and the success of efforts to protect the remaining populations of African elephants and black rhinoceroses, which are protected by the Convention on International Trade in Endangered Species. To evaluate the convention's success, scientists measured

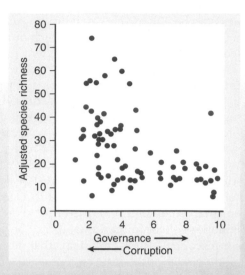

FIGURE 1 *Biodiversity and Corruption*
Species richness for birds and mammals is greatest in nations with low levels of governance, which have a high level of corruption. This corruption could interfere with efforts to preserve these species. (Source: Data from R.J. Smith et al., "Governance and the Loss of Biodiversity." Nature 426: 67–70.)

the percentages of change in the population of African elephants and black rhinoceroses between 1987 and 1994 in individual African nations and compared these changes to government efforts to save the species, such as national spending on protected areas (per square kilometer), per capita GDP, a measure of the human development index (a measure of the quality of life), and the index of governmental corruption.

Concerning the Protection of the World Cultural and Natural Heritage was signed in Paris in 1972. This convention designates protected areas as World Heritage Sites. They can be a unique ecological resource such as the Everglades National Park in Florida or a unique cultural resource such as the archaeological remains in Petra, Jordan. There are now almost 100 World Heritage Sites throughout the world.

Just as with species, preserving habitat requires policy makers to choose among competing sites. To date these choices have been based in part on goals for both the total area covered and representation of the various biomes. In 1993 the World Conservation Union set a goal of preserving

10 percent of the land surface that is covered by each of the fourteen major biomes. In 2003 the same group announced that protected areas covered 11.5 percent of Earth's land surface in nine of fourteen biomes.

Nongovernmental organizations (NGOs) play an important role in preserving habitat. The World Bank defines nongovernmental organizations as private organizations that pursue activities to relieve suffering, promote the interests of the poor, protect the environment, provide basic social services, or undertake community development. In this role many nongovernmental organizations raise money and make agreements with national governments to pre-

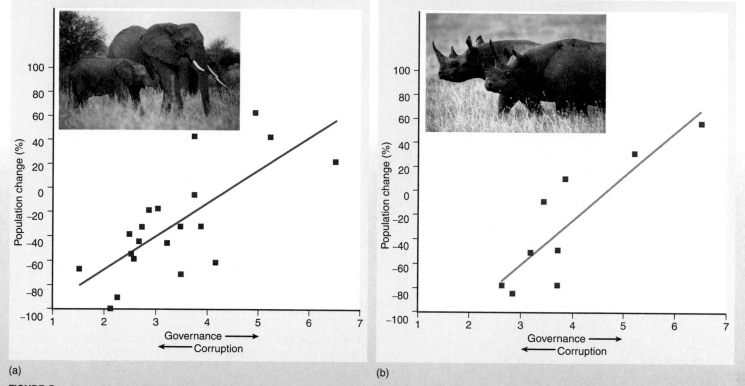

FIGURE 2 *Corruption and Species Preservation* *Efforts to preserve the (a) African elephant and (b) black rhinoceros are more successful, as indicated by higher rates of population growth, in nations with high levels of governance, which have low levels of corruption.* (Source: Data from R.J. Smith et al., "Governance and the Loss of Biodiversity." Nature 426: 67–70.)

The results show that spending on protected areas, per capita GDP, and the human development index are not related to changes in the population size of African elephants and black rhinoceroses. Rather, changes in these populations are positively related to governance scores. Populations of African elephants and black rhinoceroses shrank in nations where corruption was high, and the populations grew in nations with relatively little corruption (Figure 2). This result demonstrates the role of corruption. Shrinking populations in corrupt nations imply that poachers can convince government officials to ignore their illegal activities. On the other hand, the observation that populations grew in nations with less corruption implies that the Convention on International Trade in Endangered Species can be successful when local officials enforce the law.

ADDITIONAL READING

Smith, R.J, R.D.J. Moiré, M.J. Wallop, A. Balmford, and N. Leader-Williams. "Governance and the Loss of Biodiversity." Nature 426 (2003): 67–70.

STUDENT LEARNING OUTCOME

- Students will be able to explain why and how corruption affects efforts to preserve biodiversity.

serve land. For example, the Nature Conservancy's mission is to preserve the plants, animals, and natural communities that represent the diversity of life on Earth by protecting the lands and waters they need to survive. To do so the organization joins with governments, private corporations and individuals, nonprofit organizations, and indigenous people to purchase land they deem critical to biodiversity. The land is then managed with respect to land and marine conservation, freshwater supplies, global climate change, fire, and invasive species. As of 2005 the Nature Conservancy had protected more than 117 million acres of land and 5,000 miles of rivers and had established over 100 marine conservation projects in twenty-one countries and twenty-two U.S. states.

Despite efforts by governments, nongovernmental organizations, and private individuals, choosing among areas to protect is difficult because biodiversity varies greatly by biome. The range of many species does not coincide with existing protected areas. Species whose range falls outside protected species are termed **gap species.** Most of these gap species live in tropical areas (Figure 12.16). Based on these results, scientists have recommended that the global network of protected areas be expanded based on geographic patterns of biodiversity.

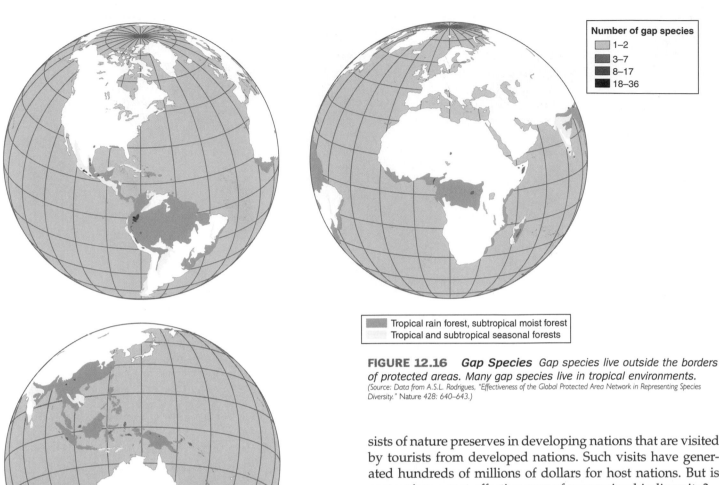

Tropical rain forest, subtropical moist forest
Tropical and subtropical seasonal forests

Number of gap species
- 1–2
- 3–7
- 8–17
- 18–36

FIGURE 12.16 *Gap Species Gap species live outside the borders of protected areas. Many gap species live in tropical environments.* (Source: Data from A.S.L. Rodrigues, "Effectiveness of the Global Protected Area Network in Representing Species Diversity." Nature 428: 640–643.)

Market-Based Mechanisms

As with many environmental challenges, some of the most effective efforts to preserve biodiversity use market forces. From an anthropocentric perspective biodiversity is valuable because it provides economically valuable goods and services. If policy can make consumers pay for these goods and services, these payments would serve as an incentive to preserve biodiversity. See *Your Ecological Footprint: Biodiversity-Friendly Coffee.*

One of the most successful market-based mechanisms to preserve biodiversity is **ecotourism,** which the United Nations describes as tourism contributing to the conservation of natural environments that is planned, developed, and operated with local communities in a way that contributes to the well-being of local communities. Much ecotourism con-

sists of nature preserves in developing nations that are visited by tourists from developed nations. Such visits have generated hundreds of millions of dollars for host nations. But is ecotourism a cost-effective way of preserving biodiversity?

To answer this question, ecologists and economists measured the costs and benefits of the Mabira Forest Reserve in southern Uganda. The benefits to local inhabitants are measured by the revenue generated by the entrance fee, which is currently about $5. But this fee probably is too low. A questionnaire filled out by foreign tourists indicated that an entrance fee of about $47 would maximize the funds earned by local inhabitants. Visitors are willing to spend $47 because the reserve has so many bird species (143); the same questionnaire indicates that visitation rates are positively related to the number of birds an ecotourist is likely to see. Given these preferences, the current fee understates the value of biodiversity at the Mabira Reserve.

What are the costs of the Mabira Reserve? Aside from operating costs, preserving the forest means that the local people cannot use the land for agriculture. This cost can be measured by the loss in rent, which is the difference between the value of the agricultural products grown on a hectare of land and the cost of growing those crops and transporting them to market. Economic analysis shows that much of the land in the reserve would generate relatively little rent, in part because the lack of roads makes it expensive to transport goods to the market (Figure 12.17). As we will see in Chapter 17, access to the market has an important effect on the rates at which people cut down forests.

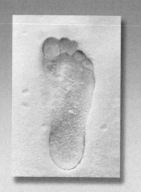

Biodiversity-Friendly Coffee

In many coffee shops it is possible to spend a few extra cents for a cup of "biodiversity-friendly" or "bird-friendly" coffee. Already about 10 percent of the U.S. coffee market consists of biodiversity-friendly coffee that is grown in a way to protect migratory birds. Does the type of coffee you drink matter?

An understanding of the potential benefits of biodiversity-friendly coffee begins with an explanation of how coffee is cultivated. Coffee beans are grown on small trees. Nearly all coffee comes from two tree species, *Coffee arabica* and *Coffee robustus*. Both species evolved in Africa and, in the wild, grow on the forest floor. As such, coffee is shade adapted. (What does this imply about its light compensation and light saturation points?) Consistent with this adaptation, agricultural studies indicate that coffee trees grown in shade produce up to 50 percent more beans than trees grown in full sunlight. In addition, coffee trees grown in the shade have larger, tastier beans.

Nonetheless, coffee can be grown under a variety of conditions (Figure 1). At one extreme coffee trees can be grown in rows in full sunshine. At the other extreme coffee trees can be grown in full shade under a nearly unaltered forest canopy. Between these extremes are various cultivation techniques, which vary according to the density and types of shade trees. Traditional polyculture (growing more than one species) simplifies the forest canopy but retains many native tree species. Commercial polyculture uses commercially valuable species, such as citrus or banana plants, for shade. Alternatively, shade can be provided by a very simplified canopy of nitrogen-fixing tree species.

Over the last decade the fraction of coffee that is grown in full sun has increased dramatically. In Mexico, Colombia, the Caribbean, and Central America about 40 percent of the 2.8 million hectares of shade coffee has been converted to full-sun production. This conversion has been stimulated by higher coffee prices, which make it economically viable to increase the application of agricultural inputs such as fertilizers. With these additional inputs, yields of full-sun coffee can exceed those for shade coffee.

The change to full-sun coffee exacerbates the loss of biodiversity associated with coffee production. In general, full-sun coffee reduces the structural complexity of the ecosystem, reduces leaf litter, and increases the incidence of disease. Consistent with these differences, conversion to full-sun cultivation reduces the diversity of invertebrates, mammals, and local and migratory bird species. Among these species are pollinators, which are critical to yield. Without pollinators the fruit does not set, and yields are reduced. In Indonesia, fruit set rises from 60 percent to 90 percent as the number of bee species increases from three to twenty.

The degree to which biodiversity can be preserved is not an all-or-nothing proposition. Much of a forest's original biodiversity can be preserved depending on the type of shade trees used and how those trees are managed. Trees that have a complex structure and lots of flowers, such as the ice cream bean tree, support higher levels of biodiversity than simpler species such as the gliricidia.

Given these advantages, how much extra does biodiversity-friendly coffee cost? To ensure that the "biodiversity-friendly" label is meaningful, several U.S. conservation organizations have set up the Conservation Principles for Coffee Production. Economic analyses indicate that all growers can satisfy these principles if they charge a small fee. This fee varies from 3 percent (for expensive coffees) to 25 percent (for inexpensive coffees). Given the small fraction of the price of a cup of coffee that goes to the grower, the biodiversity fee adds little to what you pay. So a small investment in biodiversity-friendly coffee could preserve biodiversity in much of the world's coffee-growing regions, where most of the world's biodiversity is located.

ADDITIONAL READING

Donald, P.F. "Biodiversity Impacts of Some Agricultural Commodity Production Systems." *Conservation Biology* 18, no. 1 (2004): 17–37.

Gobbi, Jose A. "Is Biodiversity-Friendly Coffee Financially Viable? An Analysis of Five Different Coffee Production Systems in Western El Salvador." *Ecological Economics* 33 (2000): 267–281.

STUDENT LEARNING OUTCOME

- Students will be able to explain how the method used to cultivate coffee affects biodiversity and why high coffee prices may either reduce or preserve biodiversity.

FIGURE 1 *Coffee Production and Biodiversity* Coffee production systems vary according to the amount of shade they provide to coffee plants. Coffee grown under a full canopy (rustic) preserves more bird species than coffee grown in full sun (unshaded monoculture). (Source: Redrawn from J.A. Gobbi, "Is Biodiversity-Friendly Coffee Financially Viable? An Analysis of Five Different Coffee Production Systems in Western El Salvador," Ecological Economics 33: 267–281.)

Rustic | Traditional polyculture | Commercial polyculture | Technical shade | Unshaded monoculture

Traditional — Modern

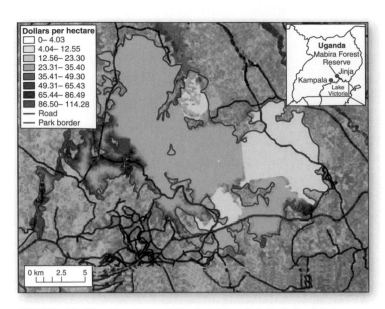

FIGURE 12.17 *Roads and Ecotourism Relatively few roads go through the Mabira Reserve in Uganda. As a result, transportation costs to and from local lands are high, which reduces their economic attractiveness for agriculture. The rent that farmers could earn from converting the land to agriculture is low—mostly less than $4 per hectare. A well-functioning market would encourage the continued use of the reserve's land for ecotourism.* (Source: Data from R. Naidoo and W.L. Adamowicz, "Economic Benefits of Biodiversity Exceed Costs of Conservation at an African Rain Forest Reserve." Proceedings of the National Academy of Sciences of the United States of America 102 (46): 16712–16716.)

We can use these costs and benefits to determine whether local inhabitants would be better off if they converted the Mabira Reserve to agricultural land. As the best areas of the preserve (for agricultural use) are converted to agriculture, income generated by agriculture rises. But the income generated from the preserve declines because the reduced forest area cuts the biodiversity of birds (via the species–area relationship described previously). And the reduction in bird biodiversity reduces the number of visits by ecotourists. As more of the preserve is converted to agricultural land, the increase in income from agriculture will equal the loss in income from ecotourism. The equality corresponds to the size of the Mabira Reserve that would be supported by the market; these results show

that the market should retain much of the preserve. At the current entrance fee of $5, the market would support a preserve that retains about 114 of the 143 bird species. If the current entrance fee of $5 is raised to $47, which is what should be charged, the market will support a reserve that retains 131 of the bird species. Clearly the Mabira Reserve is an efficient market-based mechanism for preserving biodiversity.

Biodiversity also is a valuable source of genetic information, but uncertainty about who owns that information has caused the market to undervalue the information. If a pharmaceutical company that finds a drug owns its genetic information, then governments have no incentive to preserve natural ecosystems. But if local governments own the genetic information and charge a high price for its use, pharmaceutical companies have less incentive to analyze species for medicinal compounds. As a compromise, the value of genetic diversity has been brought into the market through a series of agreements between pharmaceutical companies and nations with high biodiversity. For example, Merck, Inc., one of the world's largest pharmaceutical companies, recently paid $1.35 million to Costa Rica for the rights to the genetic information of local species. This payment allows Merck to comb local forests, looking for species with characteristics that can be used to produce marketable drugs. Should Merck be successful, Costa Rica would get a royalty. For its part, Costa Rica must use the $1.35 million to preserve habitat that is critical to biodiversity.

Market-based mechanisms may also provide an effective method for preserving species that have a history of being hunted or harvested commercially. Prohibiting the hunting or trade of an endangered species tends to drive up the price for a species, sometimes leading to the formation of a **black market,** which is a market where goods or services are sold illegally. Here the high price provides an incentive for poaching (illegal hunting or fishing). At the same time the lack of a legal market means there is no economic incentive to preserve the endangered species' habitat. If a well-regulated market can be established, prices for the protected species can be reduced, and landowners will be willing to set aside habitat for endangered species.

SUMMARY OF KEY CONCEPTS

- Biodiversity consists of three components: genetic diversity, ecosystem diversity, and species diversity. Diversity can be measured by the number of species present or the distribution of individuals among species. In general, species richness is greatest near the equator and declines toward the poles. This pattern is explained by latitudinal changes in the amount of land and in net primary production.

- The number of species present depends on the rates of extinction, immigration, and speciation. Speciation occurs when subpopulations

are separated from their parent populations by geographic barriers, by genetic barriers, or by changing their resource base.

- The importance of biodiversity can be evaluated from a biocentric view, in which each species has a right to exist, or an anthropocentric view, in which a species' value is determined by its usefulness, directly or indirectly, to human beings. These uses include an insurance policy for economic activity, the provision of environmental services, and genetic resources.

- The fossil record identifies five periods of mass extinctions during which the extinction rate was much greater than the background rate. Most scientists agree that the current extinction rate is greater than the rate of loss during the mass extinctions. Current extinctions are caused by habitat conversion and fragmentation, introduction of alien species, changes in biogeochemical cycles, and hunting and harvesting.

- The economics of preserving biodiversity can be envisioned as preserving the greatest amount of biodiversity per dollar. Economists have four criteria to choose among efforts to preserve biodiversity. Higher priority is given to unique species, species that are useful to humans, species that have a high likelihood of being saved, and species that cost relatively little to save.

- Policy to save biodiversity is implemented through a series of national and international laws that protect species, market-based mechanisms that try to have users pay for the benefits of biodiversity, and efforts to preserve natural ecosystems.

REVIEW QUESTIONS

1. Did the Biosphere experiment fail because it housed too little or too much biodiversity? Explain.

2. Explain the differences between barriers that lead to sympatric versus allopatric speciation.

3. How could you test the species–area relationship versus the species–energy relationship with regard to the global pattern of diversity?

4. How would you calculate the costs and benefits to rice producers provided by the International Rice Research Institute with regard to the grassy stunt virus?

5. How can the preservation of biodiversity save society money?

KEY TERMS

alien species	endemic	mass extinctions
allopatric speciation	enemy release hypothesis	native range
anthropocentric	ethnobotany	naturalized range
background extinction	ex situ	nongovernmental organizations
biocentric	extinction	passenger species
biodiversity	extremophiles	poaching
biodiversity hot spot	fragmentation	pollinators
black market	gap species	predator control
corruption	genetic diversity	random drift
diffuse chemical coevolution	governance	redundant species
dispersal	habitat conversion	speciation
ecosystem diversity	immigration	species diversity
ecotourism	in situ	species evenness
edge effects	insurance	species richness
edges	invasive	sympatric speciation
endangered species	keystone species	threatened species

13

GLOBAL CLIMATE CHANGE
A Warming Planet

Norse Greenland *Hvalsey Church is one of the best-preserved buildings left from about 450 years of Norse settlements in Greenland.*

STUDENT LEARNING OUTCOMES

After reading this chapter, students will be able to

- Identify the factors affecting the amount of energy that enters and leaves Earth's atmosphere and how this energy flow affects global temperature.

- Explain the effects of human activity on climate based on changes in storages and flows in global biogeochemical cycles.

- Explain how detection and attribution are critical to climate change policy.

- Describe the positive and negative effects of climate change on environmental and economic systems.

- Describe the factors that determine economically efficient climate change policy and what this policy implies about the Kyoto Protocol.

Popular culture portrays early inhabitants of Scandinavia, the Norse, as Vikings, who raided large areas of Europe starting in about A.D. 800. Eastern Europe and Russia were raided by people who lived in what is today Sweden. People from what is now Norway headed west, raiding France, England, and Ireland.

The Norse were also farmers who set up colonies in areas that are now part of Iceland, Greenland, and Newfoundland on the North American continent. In Greenland Norse settlements persisted for about 450 years, from A.D. 984 to sometime in the early 1400s (opening photograph). We say sometime in the early 1400s because the Greenland Norse disappeared suddenly. Archeologists have found remains of people who died in their houses and were simply left there. Why did the Norse inhabitants of what is now Greenland disappear?

Changes in climate are partially responsible for the early success and ultimate failure of the settlements. Settlement started at a time known as the medieval warm period, when Greenland's weather was perhaps a little warmer than now. Under these conditions about 5,000 Norse were able to transplant much of Norwegian agriculture to Greenland. Specifically the Norse brought cows, pigs, sheeps, and goats, which they used as sources of meat, dairy products, and wool. There were no grains for bread or beer.

Even at its warmest Greenland was too cold for wheat or barley. Instead the Greenland Norse supplemented their diet with the meat of local animals, such as caribou and seals. These animals were also eaten by the Inuit, a tribe of North American people who arrived in Greenland in about 1200, when the medieval warm period melted enough ice for the Inuit to follow their prey through the waters that separated Greenland from the islands north of what is today Canada.

Even in these good years, life in Greenland for the Norse inhabitants was difficult. In late summer and fall farmers had to store enough hay to ensure their animals an adequate food supply through the long Greenland winters. But the arrival of the Inuit and the little ice age, which was an extended period of cooling, probably made life too difficult for the Norse settlers on Greenland. Colder weather shortened the summer and stretched the winter. Both of these changes made it increasingly difficult for Norse farmers to store enough hay, which made them more reliant on local wildlife. However, the Inuit were better hunters than the Greenland Norse, and competition may have prevented the Norse from gathering enough food. The lack of food is confirmed by the bones of those who lived toward the end of the Norse settlements on Greenland. By the mid-1400s no Norse settlers remained in Greenland. ■

CLIMATE AND CLIMATE CHANGE

The Difference between Climate and Weather

There is a subtle but important difference between climate and weather. **Weather** describes atmospheric conditions at a particular place and point in time and how they change from day to day. Weather is described by a series of measurements such as temperature, precipitation, wind speed, cloud cover, and humidity. These measurements and their rate of change depend on where you live. If you live in Costa Rica near the equator, you hardly need a daily temperature forecast because changes are relatively small throughout the year. If you live far from the equator, weather changes significantly. In New England temperatures can vary by as much as 50°C from July to January. Dramatic changes are possible from one day to the next. Indeed, New Englanders say, "If you don't like the weather, just wait a couple of hours."

Regardless of location, the annual pattern of weather repeats itself year after year. This repeating pattern of weather, known as **climate**, represents the average weather conditions over a long period. For example, weather includes the temperature on November 3, 1957. Climate describes the average temperature on November 3 between 1851 and 2006.

For us climate seems to be more or less constant. Some years are colder and wetter than average; but over time these years are balanced by years that are warmer and drier than average. As a result, average temperature and precipitation seem to remain constant over the long run.

But what if weather does not go back to the average? What if winters become colder and snowier or summers become hotter and drier year after year? Under these conditions, scientists would say that climate has changed. **Climate change** is a shift in the long-term average weather. This shift may generate warmer or cooler temperatures, more or less precipitation, higher or lower humidity, or stronger or gentler winds.

A Changing Climate

Climate change is the rule rather than the exception. Climate may appear stable over your lifetime, but this stability disappears when scientists compare temperature and precipitation over longer periods. During some periods, Earth was warmer than it is today. During other periods, Earth was cooler; some of the cool periods are called "ice ages." The most recent ice age lasted from 30,000 years before today until 12,000 years ago (Figure 13.1a). During this period huge sheets of ice covered much of Europe and North America. In North America ice covered what is now Chicago. Only during the last 10,000 years have the ice sheets shrunk significantly. The retreat of the ice sheets corresponds to the beginning of agriculture and the evolution of complex human societies.

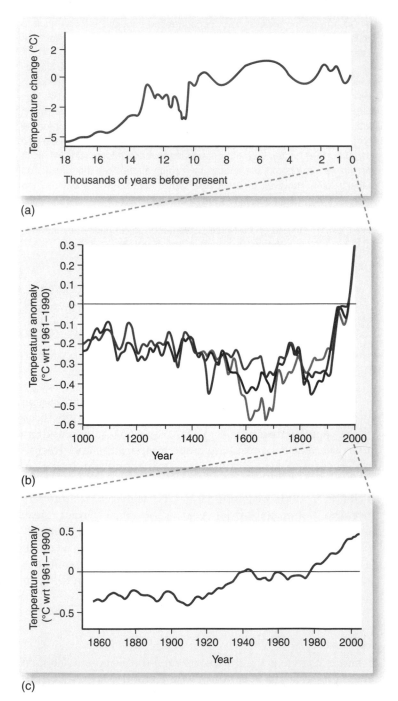

(a)

(b)

(c)

FIGURE 13.1 *Temperature (a) Surface temperature has risen by about 4–5°C over the last 18,000 years before the present, which is commonly termed 1950. (b) The change in temperature over the last 1,000 years, as compiled by three different groups of scientists. (c) The increase in average surface temperature since 1860, as indicated by direct measurements.*

During the past 1,000 years, climate has warmed and cooled several times (Figure 13.1b). As the Greenland Vikings found out, temperatures were relatively warm during the medieval warm period. Conversely, temperatures were relatively cold during the little ice age of 1300–1860. These changes are recorded in paintings: An analysis by art historians shows that the skies painted during the little ice age differ from those painted in the twentieth century.

THE HEAT BALANCE OF PLANET EARTH: THE CAUSE OF A CHANGING CLIMATE

Climatic warming and cooling are caused by changes in the planet's **heat balance**—the difference between the amounts of energy that enter and leave the atmosphere. If more energy enters the atmosphere than leaves it, heat accumulates and the climate warms. On the other hand, if more energy escapes than enters the atmosphere, heat is lost and the climate cools.

How Much Energy Reaches Earth's Surface?

As described in Chapter 4, electromagnetic radiation from the sun is the main source of energy to Earth's surface. Recall that the solar constant is the amount of energy that reaches the outer layers of the atmosphere. Over Earth's lifetime the solar constant has varied. These variations continued through the last 150 years and are one cause for recent changes in Earth's climate. The little ice age was caused in part by a slowdown in solar activity.

Once solar energy reaches the outer atmosphere, the amount of energy that reaches Earth's surface is determined by the composition of the atmosphere. Clouds reflect sunlight; therefore, increases in cloud cover reduce the amount of solar energy that reaches the surface. This quantity also is reduced by particles and **aerosols,** which are tiny solid particles or liquid droplets that remain suspended in the atmosphere for a long time. These particles and aerosols **reflect** (turn back to space) or **scatter** light in different directions before it reaches Earth's surface. One important natural source of aerosols is sulfur emitted by volcanoes. Sulfate aerosols affect the heat balance in two ways. These aerosols reflect and scatter solar energy before it reaches Earth's surface. They also increase the formation of clouds, which further reflect and scatter solar energy. Volcanoes periodically eject large amounts of sulfur, ash, and other particles into the atmosphere, where they can remain for several years. During this period they cool the climate significantly. The average temperature of the planet cooled by about 1°C in 1991 and 1992 following the eruption of Mt. Pinatubo in the Philippines.

The sun's fusion reaction burns at a high temperature—remember we stated that the photosphere radiates energy to Earth at 5,800 K. Now we will explain why this temperature is important. Temperature determines the wavelength and energy content of the radiation. The sun's high temperature

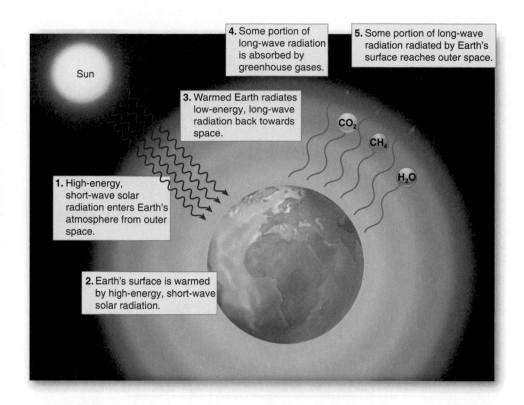

1. High-energy, short-wave solar radiation enters Earth's atmosphere from outer space.

2. Earth's surface is warmed by high-energy, short-wave solar radiation.

3. Warmed Earth radiates low-energy, long-wave radiation back towards space.

4. Some portion of long-wave radiation is absorbed by greenhouse gases.

5. Some portion of long-wave radiation radiated by Earth's surface reaches outer space.

FIGURE 13.2 *Earth's Heat Balance* *The balance is determined by the amount of short-wave solar energy that enters the atmosphere and long-wave radiation that leaves the atmosphere. High-energy, short-wave solar energy (red lines) passes through Earth's atmosphere and warms Earth's surface. That energy is reradiated back toward space as low-energy, long-wave radiation (blue lines). A significant fraction of that long-wave radiation is absorbed by greenhouse gases, such as water vapor (H_2O), carbon dioxide (CO_2), and methane (CH_4), before it reaches space. The amount of energy absorbed is measured by radiative forcing.*

implies that the radiation leaving the sun contains a lot of energy and has a short wavelength (Figure 13.2). This short wavelength is vital—the gases in Earth's atmosphere do not absorb significant amounts of radiation with short wavelengths (the atmosphere does absorb a large fraction of the energy with very short wavelengths, as described in Chapter 14). This allows short-wave radiation to reach the ground if it is not scattered or reflected.

How Much Energy Escapes Back to Space? The Greenhouse Effect

The land and ocean surface absorb incoming high-energy, short-wave energy. The surface warms as it absorbs energy faster than it radiates energy back to the atmosphere. Because Earth's surface is not as warm as the sun (about 6°C–43°F), the energy radiated to the atmosphere contains less energy and has a longer wavelength than the incoming radiation from the sun. The change from short-wave radiation to long-wave radiation is significant because some atmospheric gases, which are known as **greenhouse gases,** absorb much of this long-wave radiation. Greenhouse gases include water vapor (H_2O), carbon dioxide (CO_2), methane (CH_4), chlorofluorocarbons (CFCs), and nitrous oxide (N_2O). Together with particles and aerosols, these gases absorb and reflect radiation and therefore affect the energy balance of the atmosphere.

The atmosphere's ability to absorb energy with longer wavelengths and convert it to heat is known as the **greenhouse effect.** The greenhouse effect makes the atmosphere warmer than it would be if the *outgoing* long-wave radiation passed through the atmosphere in the same way as *incoming* short-

wave radiation. The greenhouse effect is a natural phenomenon that warms the lower atmosphere by about 35°C. Without the greenhouse effect, Earth's average temperature would be a chilly –15°C. Natural greenhouse warming is essential to life. If the average temperature of the planet was below the freezing point of water, life probably would not have evolved.

The strength of the greenhouse effect is measured by **radiative forcing.** Radiative forcing, measured in units of watts per square meter, can be thought of as the total amount of energy (watts) that is absorbed by the gases that lie above a square meter of Earth's surface, from ground level to the top of the atmosphere (Figure 13.2).

The types and quantities of gases and particles in the atmosphere determine its radiative forcing. The ability to absorb long-wave energy often is measured relative to the quantity that is absorbed by a molecule of carbon dioxide (Table 13.1). Chlorofluorocarbons (there are many types—see

TABLE 13.1	**Characteristics of Greenhouse Gases**		
Gas	Radiative Forcing Relative to Carbon Dioxide	Atmospheric Lifetime	Concentration (Units of Measure)
Carbon dioxide	1	120 years	Parts per million
Methane	11	10.5 years	Parts per billion
Nitrous oxide	270	12.2 years	Parts per billion
CFC-11	3,400	55 years	Parts per trillion
CFC-12	7,100	116 years	Parts per trillion

Chapter 14) are the most effective absorbers of long-wave radiation: Some types of chlorofluorocarbon molecules can absorb thousands of times more energy than a molecule of carbon dioxide.

A gas's ability to absorb energy and its concentration determine which gases play the most important role in the greenhouse effect. Apart from water vapor, carbon dioxide is the most abundant greenhouse gas. At the beginning of 2005 there were about 375 molecules of carbon dioxide for every million molecules in the atmosphere (parts per million, written as ppm). There were about 1,800 parts per billion (ppb) of methane and 318 ppb of nitrous oxide. The concentration of chlorofluorocarbons is so small that it is measured in parts per trillion. A part per trillion is about equal to a single drop in a very large swimming pool. Multiplying a gas's concentration by its ability to trap heat determines its contribution to the greenhouse effect. A single molecule of CFC can absorb thousands of times more energy than a molecule of carbon dioxide, but there are about a million molecules of carbon dioxide for every one molecule of CFC. The net result is that carbon dioxide absorbs more heat than CFCs and all other greenhouse gases, except for water vapor (Figure 13.3).

Concentrations depend in part on the time that the average molecule spends in the atmosphere, which is called **residence time.** The residence time for most greenhouse gases is decades or longer (Table 13.1). As a result, the concentration of greenhouse gases is relatively well mixed throughout the entire atmosphere. This means that the concentration of carbon dioxide is about the same everywhere within a given layer of the atmosphere. On the other hand, the residence time for a sulfate aerosol is a couple of years if injected high

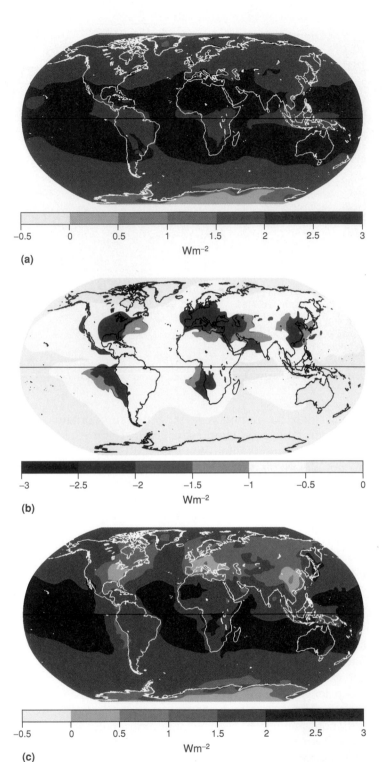

(a)

(b)

(c)

FIGURE 13.4   *Hemispheric Radiative Forcing* (a) Radiative forcing associated with greenhouse gases is similar in the Northern and Southern Hemispheres. (b) The radiative forcing associated with sulfate aerosols is concentrated near their point of origin. Most economic activity occurs in the Northern Hemisphere, which is where they have their greatest effect. (c) Due to the spatial differences in the radiative forcing of sulfate aerosols, radiative forcing in the atmosphere over the Southern Hemisphere is greater than radiative forcing of the atmosphere over the Northern Hemisphere. *(Source: Data from J.T. Kiehl and B.P. Briegleb, "The Relative Roles of Sulfate Aerosols and Greenhouse Gases in Climate Forcing." Science 260: 311–314.)*

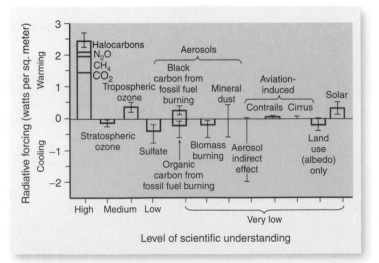

FIGURE 13.3   *Radiative Forcing* Radiative forcing is determined by the amount of incoming solar energy, atmospheric particles and aerosols, and greenhouse gases. Greenhouse gases increase radiative forcing while particles, aerosols, and Earth's albedo reduce radiative forcing. Scientists' understanding of these contributions varies from high to very low. *(Source: Data from V. Ramaswamy et al., "Radiative Forcing of Climate Change," Climate Change 2001, The Scientific Basis.)*

into the atmosphere, but only a week to ten days if emitted near ground level. This short residence time prevents complete mixing, so sulfate aerosols have their greatest effect near their point of origin. Sulfate aerosols are emitted mainly in the Northern Hemisphere (see the next section), so the radiative forcing of the atmosphere above the Southern Hemisphere is greater than the radiative forcing of the atmosphere over the Northern Hemisphere (Figure 13.4c). As discussed in *Case Study: Hemispheric Patterns in Temperature Change*, this difference is used in efforts to determine whether human actions are responsible for the recent increase in global temperature.

RADIATIVE FORCING AND HUMAN ACTIVITY

Concentrations of Greenhouse Gases

Atmospheric concentrations of greenhouse gases are increasing steadily (Figure 13.5a–c). Remember from Chapter 6 that measurements at Mauna Loa show that the atmospheric

concentration of carbon dioxide has increased from about 316 parts per million in 1958 to about 375 parts per million in 2005. This increase started well before the 1950s. Scientists track the atmospheric concentration of carbon dioxide prior to 1958 by analyzing ice cores, which are tubes of ice withdrawn from glaciers in Greenland and the Antarctic. One recent core from Antarctica is about 3,000 meters long; ice at its bottom formed nearly 740,000 years ago (but its contents have not been fully analyzed). The ice contains tiny bubbles of air that were trapped when the ice was formed. By analyzing air in the bubbles from different layers of the core, scientists can reconstruct the atmospheric concentration of carbon dioxide while the core was formed. During the last half million years the concentration of carbon dioxide has varied between 200 and 400 ppm (Figure 13.6). This means that the 2005 value of 375 ppm is near the largest value over the last 420,000 years.

The other greenhouse gases show a similar pattern of increase. The concentration of methane has risen significantly over the last 100 years (Figure 13.5) and also is near

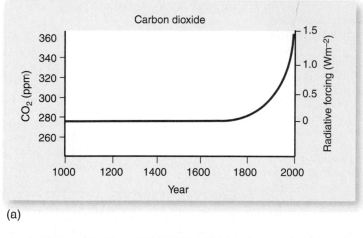

(a)

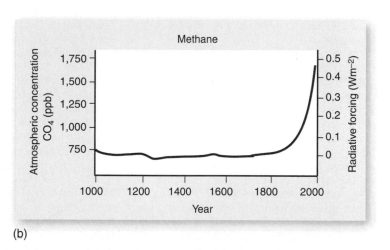

(b)

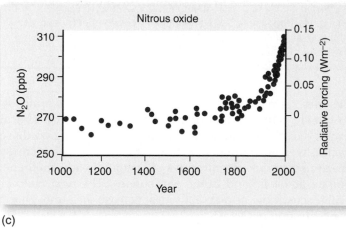

(c)

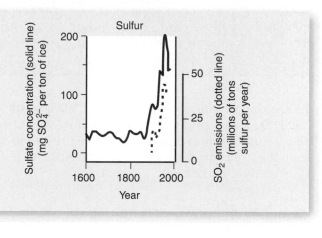

(d)

FIGURE 13.5 *Greenhouse Gases and Sulfate Aerosols* The atmospheric concentration of (a) carbon dioxide (b) methane (c) nitrous oxide (d) and sulfate aerosols. The concentration of greenhouse gases has risen steadily over time, while air pollution laws have recently reduced the atmospheric concentration of sulfur. Notice too that the absolute levels of their concentrations differ greatly. (Source: Data from I. Prentice et al., "The Carbon Cycle and Atmospheric Carbon Dioxide," Climate Change 2001: The Scientific Basis.)

CASE STUDY

Hemispheric Patterns in Temperature Change

The Northern and Southern Hemispheres differ in many ways. There are great differences in the ratio of land to water and the amount of economic activity. These differences, combined with differences in the residence times of greenhouse gases and sulfate aerosols, cause the atmosphere over the Northern and Southern Hemispheres to warm at different rates in ways that indicate humans are partially responsible for the increase in temperature since 1860.

The Northern and Southern Hemispheres may warm at different rates due to differences in the ratio of land to water. Remember from Chapter 4 that the specific heat of water is about ten times greater than land. This implies that land warms faster than water. Most of the world's landmass is located in the Northern Hemisphere (the land–water ratio is higher); so the Northern Hemisphere should warm faster than the Southern Hemisphere, all other things being equal.

But of course all other things are not equal. Most of the world's population lives in the Northern Hemisphere. Furthermore, nearly all of the world's developed nations are in the Northern Hemisphere. As a result, most of the world's economic activity and emissions of greenhouse gases and sulfur occur in the Northern Hemisphere. This too would seem to imply that the Northern Hemisphere should warm faster than the Southern Hemisphere.

But this conclusion ignores important differences in residence time between greenhouse gases and sulfate aerosols. The residence times of most greenhouse gases are measured in decades or centuries (see Table 13.1). Long residence times allow these gases to mix across the equator. Due to this mixing, the atmospheric concentrations of various greenhouse gases are about the same in the Northern and Southern Hemispheres (see Figure 13.4(a)).

On the other hand, the atmospheric concentrations of sulfate aerosols that are associated with human activity differ significantly between hemispheres. This difference is caused by the relatively short residence time of human sulfur emissions. A residence time of about ten days to two weeks does not allow the sulfate aerosols to mix evenly throughout the atmosphere. Instead the atmospheric concentration of sulfate aerosols varies greatly in space. The highest concentrations occur in the American Midwest, central and eastern Europe, and India and China, all of which burn lots of coal.

In these areas the cooling effect of sulfur emissions exceeds the warming effect of greenhouse gases (see Figure 13.4(b)). As a result, the net effect of human activity cools some areas of the Northern Hemisphere. On average the radiative forcing of the atmosphere over the Northern Hemisphere is less than the radiative forcing of the atmosphere over the Southern Hemisphere (see Figure 13.4(c)). We have unwittingly set up an experiment that may allow us to identify the effect of human activity on temperature. If human-induced increases in radiative forcing are responsible for the global increase in temperature, such temperature increases should appear in the Southern Hemisphere before they appear in the Northern Hemisphere.

The hypothesis that differences in radiative forcing cause the Northern and Southern Hemispheres to warm at different rates is confirmed by statistical analyses of the historical record and model simulations. Statistical analyses of hemispheric temperature data signify that temperature increases appear in the Southern Hemisphere before they appear in the Northern Hemisphere. Furthermore, the strength of this pattern becomes stronger as radiative forcing increases. Other statistical analyses show that the south-to-north pattern of temperature changes is generated by the historical pattern of greenhouse gases and sulfur emissions.

The link between the pattern of hemispheric temperature increases and the radiative forcing of greenhouse gases and human sulfur emissions is confirmed by climate models. Climate models show no pattern in hemispheric temperature changes when radiative forcing is held constant—the so-called reference scenario. When used with the historical pattern of radiative forcing, the climate models simulate the same south-to-north pattern that appears in the historical temperature data. As with the historical data, the strength of the pattern is positively correlated with radiative forcing. Together these results provide additional evidence that human activity is partially responsible for the change in temperature over the last 150 years.

ADDITIONAL READING

Kaufmann, R.K., and D.I. Stern. "Evidence for Human Influence on Climate from Hemispheric Temperature Relations." *Nature* 388 (1997): 39–44.

Kiehl, J.T., and B.P. Briegleb. "The Relative Roles of Sulfate Aerosols and Greenhouse Gases in Climate Forcing." *Science* 260 (1993): 311–314.

STUDENT LEARNING OUTCOME

• Students will be able to explain how residence times of greenhouse gases and sulfur emissions cause the climates in the Northern and Southern Hemispheres to change at different rates.

its maximum value over the last 420,000 years (Figure 13.6). Data that proxy the concentration of N_2O and sulfate aerosols are available from 1860. They too indicate that the concentration of these molecules is increasing (Figure 13.5c and d).

Changing Climate by Disrupting Global Biogeochemical Cycles

The atmospheric concentration of nearly all greenhouse gases and sulfate aerosols is at or near the maximum value observed over the last half million years. Is the simultaneous increase a coincidence? No. The increase is due mostly to human activities that disrupt the global biogeochemical cycles of carbon, nitrogen, and sulfur. By focusing on these cycles, we can understand how human activities increase the atmospheric concentrations of carbon dioxide, methane, sulfate aerosols, and other particulates. The next chapter will discuss how human activities increase the atmospheric concentration of CFCs.

Humans alter the global carbon cycle by accelerating some existing flows and by creating entirely new flows. On net, these changes increase the amount of carbon stored in the atmosphere as carbon dioxide and methane.

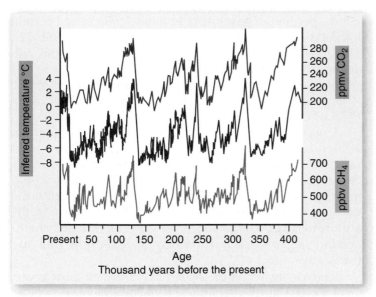

FIGURE 13.6 *Carbon Dioxide, Methane, and Temperature*
The atmospheric concentration of carbon dioxide (blue) and methane (green) and temperature (red) over the previous 420,000 years as indicated by the Vostok ice core. Notice the positive correlation among these three variables. Temperature rises when the atmospheric concentration of carbon dioxide and methane rises. (Source: Data from J.R. Petit et al., "Climate and Atmospheric History of the Past 420,000 Years from the Vostok Ice Core, Antarctica." Nature 399: 429–436.)

Great quantities of carbon are stored in Earth's crust as fossil fuels, which include coal, oil, natural gas, oil shale, and tar sands. These fuels are the carbon remains of plants and microscopic organisms that accumulated in the crust over millions of years (Chapter 20). This accumulation has been partially reversed by the Industrial Revolution. By burning fossil fuels, society creates new flows that return carbon to the atmosphere in the form of carbon dioxide and methane.

Most of the carbon in fossil fuels returns to the atmosphere as carbon dioxide, a by-product of combustion (see Equation 2.2 on page 25 of Chapter 2). The amount of carbon dioxide emitted varies among fossil fuels (Table 13.2). Of the fuels currently in use, natural gas emits the least carbon dioxide per unit of energy, and coal emits the most. Alternative sources of fossil fuels, such as oil shale and liquefied coal, emit even more carbon dioxide per energy unit. The increased emissions associated with coal and alternative sources of fossil fuels are important because the world may become increasingly dependent on these fuels as we deplete the supply of oil and natural gas (Chapter 20).

TABLE 13.2	Carbon Emission Rates for Fossil Fuels
Fuel	**Emission Rate (grams per thousand Btus)**
Coal	26.0
Oil	21.4
Natural gas	14.5

Some of the carbon stored in fossil fuels is returned to the atmosphere as methane. Methane is an important component of natural gas, so the production, transport, and consumption of natural gas allows methane to leak into the atmosphere. Methane also escapes from coal mines. Coal miners have long feared methane, which is known as coal gas because it can suffocate miners and cause explosions. As human use of coal has increased, so too has the amount of methane that escapes from coal mines.

Fossil fuels, especially coal, contain significant amounts of sulfur, so burning coal emits sulfur. Over the last 150 years there has been a significant increase in the amount of sulfur emitted by human activity (Figure 13.5d). Sulfur emissions reflect sunlight and enhance cloud formation (much like volcanic sulfates) and have a cooling effect. This cooling effect is enhanced by nitrate, ammonium, black carbon, and organic carbon aerosols that are also produced by the combustion of fossil fuels and biomass. The totality of these effects has been amplified by building taller smokestacks: Because they are ejected higher into the atmosphere, the aerosols and particles remain in the atmosphere longer, which increases their cooling effect.

Humans also have reduced the amount of carbon stored by terrestrial biota. The general process by which humans change the amount of carbon stored in terrestrial biota is known as **land use change.** This practice replaces natural ecosystems with others that meet human needs and wants. The replacement ecosystems often store less carbon than their natural predecessors. For example, the vegetation on 1 square kilometer of tropical rain forest may store 12.3 million metric tons of carbon, whereas 1 square kilometer of cropland may store only 0.2 million metric tons of carbon.

Where does the carbon in the natural vegetation go? Most of it goes directly into the atmosphere as carbon dioxide. In many places farmers create new cropland by burning the forest. Forest fires in the Brazilian Amazon cause great amounts of carbon to flow from the forest into the atmosphere. Wood that does not burn immediately decays over time; and as it decays, it too releases carbon dioxide. Similarly, harvesting timber for use as a raw material also causes carbon to flow into the atmosphere. Wooden building materials and paper products that are dumped in landfills eventually decay, which allows more carbon to reach the atmosphere.

Agriculture alters the carbon cycle in ways that also increase the atmospheric concentration of methane. Methane is produced by anaerobic decomposition of organic material. The anaerobic microorganisms that live in rice paddies produce methane as a by-product. Livestock have an extra stomach, called a rumen, which contains bacteria that break down plant material anaerobically. Each time a cow burps, it emits a small amount of methane to the atmosphere. Before you laugh about the role of cow burps in global climate change, consider this—in 2004 there were 1.3 billion cattle globally. Agriculture accounts for about half of the methane released by human activity.

Human activity also has created a new storage of carbon—landfills—and this storage generates a new flow of methane to the atmosphere. Much of the organic waste from kitchens, yards, and restaurants is dumped in landfills and covered with soil, which cuts it off from oxygen. Anaerobic bacteria break down the organic wastes and produce methane. Some landfills produce so much methane that holes are drilled to capture the methane as fuel.

Will Emissions Grow?

The combination of a growing population and rising incomes foreshadows increasing rates of energy use, agriculture, and other activities that emit greenhouse gases, sulfate aerosols, and other particles. The rate at which society uses fossil fuels will have the greatest effect on the atmospheric concentration of carbon dioxide because this storage holds the greatest amount of carbon. As shown in Figure 6.6 on page 105, there are about 10,000 petagrams of carbon stored in fossil fuels. If all of this carbon is burned, atmospheric concentrations could increase fifteenfold. On the other hand, there are only about 1,800 petagrams of carbon in biota and soils. Even if we chopped down every tree (we do not advise this!), this would only double the atmospheric concentration of carbon dioxide.

How quickly will carbon emissions grow in the next century? The IPAT equation from Chapter 11 (Equation 11.1 on page 179) implies that emissions (I) will depend on population growth (P), economic development (A), and changes in energy efficiency (T). Human population is growing rapidly, and as it does so too will emissions. The United Nations expects the world's population to grow from 6.3 billion people in 2002 to about 9 billion people by 2050 (see Figure 9.9 on page 142). This implies that emissions of carbon dioxide will increase by about 50 percent if affluence and technology remain the same.

But affluence will not remain the same. People also will become richer (or at least try to), and rising incomes may increase emissions even faster than population growth. People use more energy and emit more carbon as they become richer. The average U.S. citizen emitted 5.5 metric tons of carbon dioxide in 2001—about eight times the 0.65 metric tons emitted by the average Chinese citizen, or about twenty-two times the 0.25 metric tons emitted by the average citizen of India. These differences imply that carbon dioxide emissions could increase significantly as income levels in China, India, and other developing nations rise toward those of the United States and other developed nations. To calculate your own carbon dioxide emissions, see *Your Ecological Footprint: How Much Carbon Dioxide Do You Emit?*

Nor will technology remain the same. Increases in energy efficiency may slow emissions. Engineers have developed a host of technologies that use less energy to provide the same service. Some cars can travel 63.8 kilo-

meters per liter of gasoline (150 miles/gallon), and certain lightbulbs use 90 percent less electricity (see Chapter 22). The impact of these technologies on carbon emissions will depend on industry's willingness to produce them and people's willingness to buy them. See *Policy in Action: Which Comes First—The Supply or Demand for Energy-Efficient Capital?* on page 236.

Uncertainty about future changes in population, affluence, technology, and energy prices leads to a wide range of forecasts for carbon emissions. Emissions may double or triple over the next century, or they could increase by a factor of seven (Figure 13.7). Despite the uncertainty, one conclusion is clear: The increase in the atmospheric concentration of carbon dioxide and methane that began in the past century will continue well into the next century.

Conversely, sulfur emissions may increase or decrease. Over the last thirty years the shift away from coal to other fuels and the development of technologies that remove sulfur from coal and its combustion by-products have reduced the quantity of sulfur emitted (Figure 13.5d). This decline could continue if the use of pollution abatement technologies spreads to developing nations. On the other hand, sulfur emissions could grow if nations increase their use of coal as the world depletes its inexpensive sources of oil and natural gas.

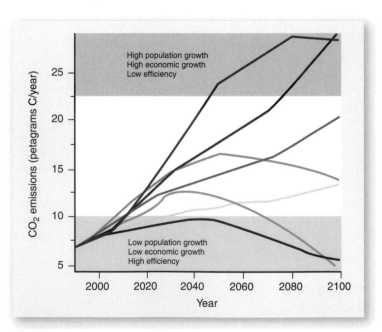

FIGURE 13.7 *Carbon Dioxide Futures To forecast rates of carbon emissions, scientists make different assumptions about future rates of population growth, affluence, and technological change. Because there is a great deal of uncertainty about each of these, rates of carbon emissions differ greatly among scenarios.* (Source: Data from I. Prentice et al., "The Carbon Cycle and Atmospheric Carbon Dioxide." In Climate Change 2001: The Scientific Basis.)

DETECTING CLIMATE CHANGE AND ATTRIBUTING IT TO HUMAN ACTIVITY

A basic understanding indicates that the increasing concentration of greenhouse gases should cause the atmosphere to absorb more outgoing long-wave radiation. This change is confirmed by satellites that measure the amount of energy that leaves Earth's atmosphere. Between 1970 and 1997 the amount of long-wave radiation leaving the atmosphere decreased (Figure 13.8). This reduction occurred mainly at wavelengths that are absorbed by carbon dioxide and other greenhouse gases.

Is this increased absorption responsible for some or all of the 0.7°C rise in global temperature over the last century? To investigate this possibility, scientists seek to answer two questions: (1) Is climate changing? (2) If climate is changing, are the changes caused by human activity?

Detecting Climate Change

Detection seeks to determine whether climate is changing. Detection consists of three steps. First scientists choose an indicator that represents some aspect of Earth's climate. Next they measure its mean and natural variability. Finally scientists determine whether recent changes are greater than expected based on natural variability. A sustained change that is greater than natural variability is interpreted as evidence for a change in climate.

Several indicators are used to detect changes in global climate. In February 2005 scientists announced that 2004 was the fourth warmest year since 1860, the first year for

which scientists have reasonably reliable direct measurements of Earth's average temperature (Figure 13.1c). Nine of the warmest years since 1860 have occurred since 1990. This string of warm years is unlikely to be the result of random chance. Scientists calculate that the odds are between 30:1 and 100:1 that the recent warming is due to natural variability. Furthermore, the recent increase is large relative to the natural variability during the preceding 1,000 years (Figure 13.1b). This too suggests that the long-term average temperature is increasing. This interpretation is consistent with the global increase in sea levels (Figure 13.9a). As the world's oceans warm, the water becomes less dense, which causes it to expand. This expansion is partially responsible for the rise in global sea level, which is enhanced by water from melting glaciers. This melting has also reduced snow cover in the Northern Hemisphere (Figure 13.9b).

Other indicators measure the response of biological systems to changes in climate. At the level of entire biomes, scientists analyze remotely sensed images of temperate and boreal forests taken between 1982 and 1999—the entire period for which data are available. Satellite measurements of vegetation NDVI (see Chapter 7's Case Study on page 107)

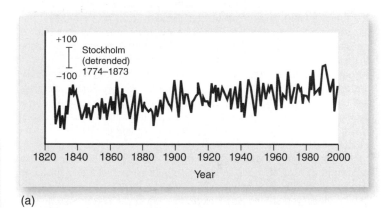

(a)

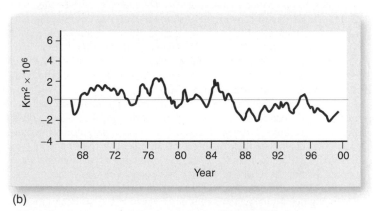

(b)

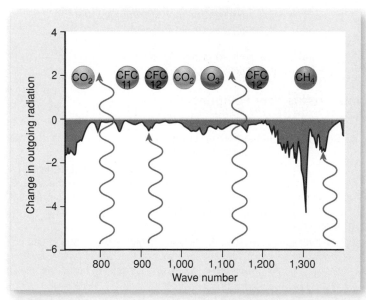

FIGURE 13.8 *Reductions in Outgoing Radiation* Scientists used satellites to measure the change in outgoing long-wave radiation between 1970 and 1997. Outgoing radiation has declined where the line dips below zero. These dips correspond to wavelengths that are absorbed by greenhouse gases. *(Source: Data from J.E. Harries et al., "Increases in Greenhouse Forcing Inferred from the Outgoing Long-Wave Radiation Spectra of the Earth in 1970 and 1997." Nature 410: 355–357.)*

FIGURE 13.9 *Detecting Climate Change* (a) The steady rise in sea level in Stockholm suggests that warming temperatures are melting glaciers and are expanding seawater. *(From J.A. Church and J.M. Gregory, et al., "Changes in Sea Level." In Climate Change 2001: The Scientific Basis, 2001).* (b) *The steady decline in snow cover in the Northern Hemisphere suggests that the planet is warming.* *(Source: (b) Data from C.K. Folland and T.R. Karl, "Observed Climate Variability and Change," Climate Change 2001: The Scientific Basis.)*

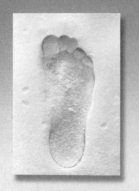

Your ECOLOGICAL *footprint*

How Much Carbon Dioxide Do You Emit?

In 2002 the average U.S. citizen emitted about 20 metric tons of carbon dioxide to the atmosphere. At first glance this average may appear much greater than your emissions. Other than running your car, it seems like days or weeks can go by between buying and using energy. But a careful look at how you spend money shows that you emit carbon dioxide almost constantly, albeit at a relatively low rate. These slow but steady emissions add up to the average of 20 metric tons per person per year.

To identify how your actions lead to emissions of carbon dioxide, you need to classify your purchases into two categories: direct purchases of energy and purchases of nonenergy goods and services. Calculating the quantity of carbon dioxide emitted by your direct purchases of energy might seem relatively straightforward. Simply multiply your purchases of motor gasoline, home heating oil, and natural gas by the carbon emission factors that are listed in Table 1. But this approach is fraught with difficulties.

One problem is how to account for the carbon emitted in the process of generating the electricity you use. In some regions of the United States, such as the Northwest, much electricity is generated using hydropower. This source emits little or no carbon dioxide. So carbon emissions per kilowatt of electricity (kwh) use would be low. High rates of carbon dioxide emissions per kilowatt hour prevail in the Midwest, where much electricity is generated from coal, which emits lots of carbon per kilowatt.

To account for these differences, look at the quantity of carbon dioxide emitted per unit of electricity generated (Table 2). These averages are calculated by dividing the total quantity of electricity generated in a region, which includes the quantity generated from fossil fuel, hydroelectric, and nuclear power stations, by the total quantity of carbon dioxide emitted by the fossil fuels that were burned to generate electricity in that region. Multiply these averages by the quantity of electricity you use (shown in your electric bill) to calculate your carbon dioxide emissions associated with electricity use.

In theory you should also include the quantity of carbon dioxide that is emitted to make these fuels available. For example, oil refineries use oil and natural gas to convert crude oil (oil that comes from the ground) into motor gasoline (a form of oil your car's internal combustion engine can use). Similarly, energy is used to mine and transport coal. Luckily for our calculations, the amount of carbon dioxide that is emitted by these activities is small relative to the amount emitted when we burn the coal, oil, or natural gas, so we will ignore them.

But you cannot ignore these indirect emissions when you examine purchases of nonenergy goods and services. Because such goods do not emit carbon dioxide directly, emissions associated with their production constitute their main effect. To calculate these effects, we use input–output tables to trace the use of energy as raw materials are extracted from the environment and fashioned into useful goods and services. (See Chapter 3's Your Ecological Footprint on page 49.) By multiplying these energy flows by their emission rates, you can tabulate the quantity of carbon dioxide emitted as you spend each dollar. The amounts of carbon dioxide emitted per dollar spent on various goods and services are given in Table 3. For example, 0.197 kilograms of carbon dioxide are emitted for every dollar you spend on electrical equipment, which is a type of durable manufactured product (Table 4).

Calculations

Direct Uses

___ Carbon dioxide emitted (gasoline) =
___ gallons/week × 2.4 kg/gal × 52 weeks/year

___ Carbon dioxide emitted (heating oil) = ___ gallons/month × 2.8 kg/gal × ___ months/heating season

___ Carbon dioxide emitted (natural gas) = ___ ft^3/month × 0.015 kg/ft^3 × 12 months/year

___ Carbon dioxide emitted (electricity) = ___ kwh/month × ___ kg/kwh × 12 months/year

Indirect Uses

___ Carbon dioxide emitted (biological resources) =
___ $/month × 0.22 kg/$ × 12 months/year

___ Carbon dioxide emitted (mineral resources) =
___ $/month × 0.76 kg/$ × 12 months/year

___ Carbon dioxide emitted (durable manufacturing and construction) = ___ $/month × 0.20 kg/$ × 12 months/year

___ Carbon dioxide emitted (transportation services) =
___ $/month × 0.60 kg/$ × 12 months/year

___ Carbon dioxide emitted (other services) =
___ $/month × 0.14 kg/$ × 12 months/year

Interpreting Your Footprint

1. Are your emissions greater or less than the U.S. national average?
2. By what percentage would your emissions decline if you used a car that traveled twice as far per gallon?
3. Table 1 in Chapter 8's Your Ecological Footprint (page 148) lists the net primary production of various biomes. Calculate the area of biome that would be needed to take up your carbon emissions. (To calculate this, divide your carbon emissions by net primary production.)
4. The average citizen in Mexico emitted about 4 metric tons of carbon in 2003. How could you rearrange your direct uses of energy and your purchases of nonenergy goods and services to reduce your emissions to a level similar to the average citizen of Mexico?

Student Learning Outcome

- Students will be able to identify how their activities contribute to emissions of carbon dioxide and how they could reduce their emissions should they wish to do so.

TABLE 1   Carbon Dioxide Emissions from Commonly Used Fuels

Fuel	Physical Units Consumed	Carbon Dioxide Emitted
Gasoline	Gallon	2.42 kg
Heating oil	Gallon	2.77 kg
Natural gas	Cubic foot	0.015 kg

TABLE 2   Carbon Dioxide Emissions per Unit of Electricity by U.S. Region for 2004

Regions (States)	Kilograms of Carbon Dioxide per Kilowatt-Hour
New England (Connecticut, Maine, Massachusetts, New Hampshire, Rhode Island, Vermont)	0.79
Middle Atlantic (New Jersey, New York, Pennsylvania)	0.41
East North Central (Illinois, Indiana, Michigan, Ohio, Wisconsin)	0.81
West North Central (Iowa, Kansas, Minnesota, Missouri, Nebraska, North Dakota, South Dakota)	0.80
South Atlantic (Delaware, District of Columbia, Florida, Georgia, Maryland, North Carolina, South Carolina, Virginia, West Virginia)	0.59
East South Central (Alabama, Kentucky, Mississippi, Tennessee)	0.66
West South Central (Arkansas, Louisiana, Oklahoma, Texas)	0.70
Mountain (Arizona, Colorado, Idaho, Montana, Nevada, New Mexico, Utah, Wyoming)	0.73
Pacific Contiguous (California, Oregon, Washington)	0.05
Pacific Noncontiguous (Alaska, Hawaii)	0.63

Source: Data for electricity generation and quantity of fossil fuels used from *Electric Power Monthly*, U.S. Department of Energy. Data for energy and carbon dioxide emissions from *Electronic Power Annual*, U.S. Department of Energy.

TABLE 3   Carbon Dioxide Emitted by Producing a Dollar's Worth of Goods or Services

Category	Kilograms of Carbon Dioxide per Dollar
Biological resources	0.223
Mineral resources	0.726
Durable manufacturers and construction	0.197
Nondurable manufacturers	0.297
Transportation services	0.595
Other services	0.139

Source: Data modified from R.W. England and S.D. Casper, "Fossil Fuel Use and Sustainable Development: Evidence from U.S. Input–Output Data, 1972–1985." *Advances in the Economics of Energy and Resources* 9: 21–44. Data were adjusted for the effects of inflation and gains in energy efficiency.

TABLE 4   Examples of Goods and Services Produced by Sectors

Category	Example 1	Example 2	Example 3
Biological resources	Agricultural products	Forestry products	Fishery products
Mineral resources	Metals	Stone and clay	Chemicals and fertilizers
Durable manufacturers and construction	Machinery	Electrical equipment	Construction
Nondurable manufacturers	Textiles and clothes	Rubber and leather	Plastics
Transportation services	Railroads	Motor freight	Air and water transport
Other services	Radio and TV broadcast	Wholesale and retail trade	Medical and hotel services

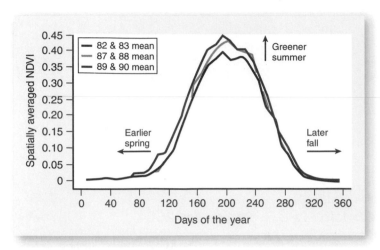

FIGURE 13.10   Longer Growing Seasons, Greener Summers *The bell-shaped curve for NDVI show how Northern Hemisphere vegetation greens up in spring, reaches a peak in summer, and browns in the fall. Over time the bell-shaped curves rise sooner, which indicates an earlier spring green-up, and drop back toward zero later, which indicates a later fall. Together these changes show a longer growing season. In addition, the height of the bell-shaped curves increase over time, so Northern Hemisphere summers are becoming greener.* (Source: Data from R.B. Myneni et al., "Increased Plant Growth in the Northern High Latitudes from 1981 to 1991." Nature 386: 698–701.)

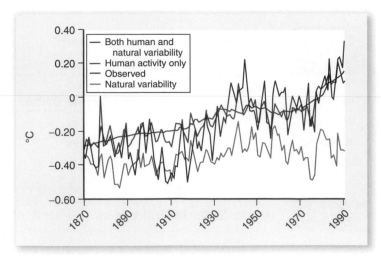

FIGURE 13.11   Attribution *Statistical analyses identify changes in global temperature (black line) that are associated with natural variability (green line), human activity (red line), and the combination of natural variability and human activity (blue line). In this analysis natural variability includes changes in solar activity, volcanic activity, and El Niño events. Human activities include changes in greenhouse gases and sulfur emissions.* (Source: Data from R.K. Kaufmann et al., "Emissions, Concentrations, and Temperature: A Time Series Analysis." Climatic Change, 77: 249–278.)

show that the length of the growing season in the Northern Hemisphere has increased by about a week in the spring and fall and that the maximum amount of biomass per hectare present in midsummer also has increased (Figure 13.10).

Behavioral changes by individual species also may signal a change in climate. As Earth warms, many Northern Hemisphere species have moved their ranges north at a rate of about 6.1 km per decade. Similarly, the times at which plants leaf out or migratory species start or finish their travels or begin their breeding seasons have been starting earlier by about 2.3 days per decade. Statistical analyses indicate that these changes in reproductive range and life cycle by so many species in the direction predicted by climate change are highly unlikely. This too suggests that climate is changing.

Attributing Climate Change to Human Activity

Even if scientists can detect climate change, the cause of the change still is in question. **Attribution** is the process of establishing a cause-and-effect relationship between human activity and the observed change in climate. The consensus of scientists involved in the Intergovernmental Panel on Climate Change is that the balance of evidence suggests that there is a discernible human influence on climate.

Efforts to attribute climate change to human activities use two methodologies: statistical analyses of historical data and computer models of the climate system. Statistical analyses of historical data demonstrate a strong link between radiative forcing and temperature since 1860. During this period temperature changes in the Northern and Southern Hemispheres have been driven by changes in the radiative

forcing of greenhouse gases, human sulfur emissions, and solar activity. This shows that temperature is associated with both human activity (greenhouse gases and sulfur emissions) and natural variability (solar activity and sulfur emitted by volcanoes). Of these factors, human activity accounts for most of the increase in temperature over the last century (Figure 13.11).

These empirical results are confirmed by computer models, which simulate climate using the physical principles described in Chapter 4. The most complex of these models are ocean atmosphere general circulation models; they simulate the patterns of atmospheric and oceanic circulation based on the flows of energy and water. To track these flows, the model lays a three-dimensional grid across the land and ocean (Figure 13.12). At each grid box the model tracks the flow of energy horizontally from one box to the next, as well as vertically from the top of the atmosphere to the surface and on to the bottom of the ocean. Differences in the rate at which energy heats the atmosphere, land surface, and ocean create gradients that cause heat and water to flow in patterns we recognize as atmospheric and oceanic circulation (both surface and deep water).

The role of human activity is demonstrated via scenario analysis. In one scenario the climate model is run with only natural changes in radiative forcing, such as solar activity and volcanic sulfates. This scenario represents a world in which human activity has no effect on climate. Without human intervention, temperature shows a great deal of variability but no tendency to increase (Figure 13.13a). This pattern represents a base case that is used to evaluate a scenario that represents the effect of human activity. In this scenario

Variables simulated

Ocean grid
Current vectors
Temperature
Salinity

Surface grid
Ground temperature
Water
Energy
CO_2 flux

Atmosphere grid
Wind vectors
Humidity
Clouds
Temperature
Chemical species

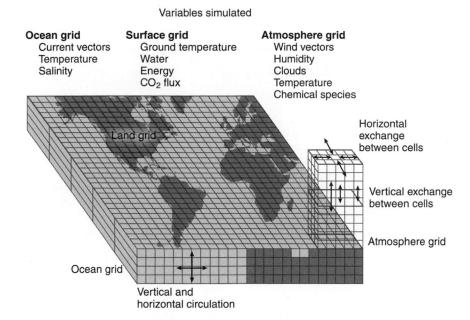

FIGURE 13.12 *Climate Models Climate models divide Earth's surface into a series of boxes. These boxes are stacked on top of one another to divide the atmosphere into vertical layers. Boxes extend below the surface under the ocean surface. The computer models track the amount of energy and water in each box; differences between boxes create a gradient that causes energy and/or water to flow between boxes. These flows simulate climate variables such as temperature and precipitation.*

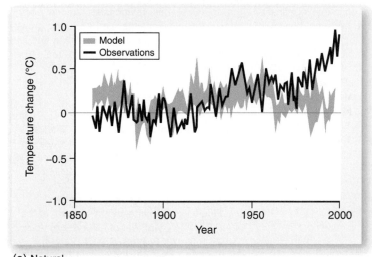

(a) Natural

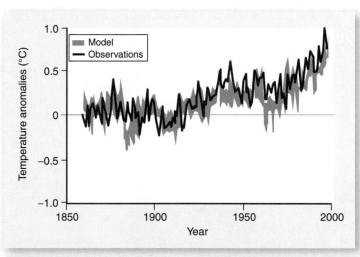

(c) All forcings

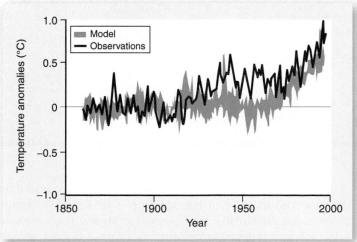

(b) Anthropogenic

FIGURE 13.13 *Attribution (a) Climate models can separate the effects of natural variability from human activities by simulating natural changes in forcing, such as changes in solar activity. This scenario is similar to the green line in Figure 13.11; like the green line, this scenario shows no general increase in temperature. (b) The model is rerun by changing greenhouse gases and sulfur as caused by human activity. This scenario is similar to the red line in Figure 13.11; like the red line, this scenario indicates that human activity accounts for most of the observed increase in temperature. (c) Finally the model simulates both natural and human changes in radiative forcing. This scenario is similar to the blue line in Figure 13.11; like the blue line, this scenario indicates that the model can account for most of the observed change in temperature.* (Source: Data from J.F.B. Mitchell and D.J. Karoly, "Detection of Climate Change and Attribution of Causes," Climate Change 2001: The Scientific Basis.)

POLICY IN ACTION

Which Comes First: The Supply or Demand for Energy-Efficient Capital?

The forecast for human emissions of carbon dioxide and other greenhouse gases depends in part on the energy efficiency of capital equipment. Will people drive hybrid vehicles that travel nearly 26 kilometers per liter (60 miles per gallon)? Or will they drive SUVs that travel only 7 km/l (17 mpg)? The answer depends in part on the degree to which firms develop energy-efficient technologies and the degree to which people are willing to buy them. Your willingness will depend on energy prices—higher energy prices increase the attractiveness of energy-efficient technologies. But is the reverse also true—is the production of energy-efficient technologies driven by energy prices? The answer is critical to policy makers, who hope to reduce emissions by spurring the innovation and adoption of energy-efficient technologies.

Despite its central importance to economic systems, there is much uncertainty about the rate and nature of technical change. Many economists assume that technical change is exogenous—determined by conditions outside the economic system. According to this perspective, technical know-how increases year after year and can be forecast by assuming that it will grow at some predetermined rate—say 3 percent per year. If this view of technical progress is correct, policy makers can do relatively little to encourage the development of energy-efficient capital equipment.

Recently this view has been challenged by some economists, who argue that technical change is endogenous—determined by conditions inside the economic system. Greater investment can increase research and development. A bit more controversial is the hypothesis that the relative prices of capital, labor, and energy can move technical change in a specific direction. Higher energy prices can stimulate technical innovations that reduce energy use. This type of innovation is known as price-induced technical change.

Price-induced technical change sets up an interesting question for policy makers concerned with global climate change. Which comes first: energy-efficient technologies or the high energy prices that would encourage consumers to switch from energy-inefficient to energy-efficient capital? If technical innovation is relatively unaffected by energy prices, then policy makers must raise energy prices slowly. If energy prices rise before energy-efficient capital is available, energy costs for firms and households will rise because energy-efficient capital is not yet available. Higher energy costs will reduce the amount of money available to buy other goods and services, which will slow the economy. Given these effects, one group of economists opposes significant increases in energy prices because they will harm economic growth.

Economists who argue for the importance of price-induced technical change ask how energy-saving technologies will become available if energy prices rise slowly or remain level. Developing new technologies is expensive. Firms must hire scientists, engineers, and economists to develop new technologies. These investments are financially uncertain because there is a long lag between the money spent developing energy-efficient technologies and the money earned from selling the new technology. As a result, firms are reluctant to invest in energy-saving technologies unless there is a clear demand for that technology. This demand will materialize only if energy prices are high enough that firms and households will save money by purchasing energy-efficient technology. If technical change is endogenous, higher energy prices are the only way to spur the development and commercialization of energy-saving technologies.

But even if this second group is correct, other economists worry that raising energy prices could slow economic growth in the long run. They assert that money spent to develop energy-saving technologies reduces funds spent to develop technologies that speed the growth of the economy. This reallocation of investment could generate a future in which economic growth slows.

Currently economists are trying to sort out how important these various effects are in the real world. An analysis of patent data shows that higher energy prices stimulate energy-saving innovation. Consistent with this result, analysts find that a $1 increase in energy prices generates new energy-efficient technology that is not abandoned when energy prices decline. On the other hand, computer models and statistical analyses of the U.S. economy verify that high energy prices slow long-term economic growth. Until this relationship among energy prices, technical change, and economic growth is clarified, policy makers must walk a fine line between increasing energy prices too slowly and too quickly.

ADDITIONAL READING

Goulder, L.H., and S.H. Schneider. "Induced Technical Change and the Attractiveness of CO_2 Abatement Policies." *Resource and Energy Economics* 21 (1999): 211–253.

Kaufmann, R.K. "The Mechanisms for Autonomous Increases in Energy Efficiency: A Cointegration Analysis of the U.S. Energy/GDP Ratio." *The Energy Journal* 25, no.1 (2004): 63–86.

Popp, D. "Induced Innovation and Energy Prices." *American Economic Review* 92, no. 1 (2002): 160–180.

STUDENT LEARNING OUTCOME

• Students will be able to describe the factors that influence the development of energy-efficient technologies and why their availability does not depend solely on engineering capabilities.

the atmospheric concentrations of greenhouse gases, sulfate aerosols, and other particles increase as caused by human activity. This scenario generates an increase in temperature that looks more like the observed increase in temperature (Figure 13.13b). A third scenario, which includes the effects of both human activities and natural factors, looks most like the observed increase in temperature. This similarity leads to the same conclusion as statistical analyses: Both human activities and natural forces are responsible for the recent increase in global temperature.

Why Are Many Skeptical?

Statistical analyses and climate models have convinced the overwhelming majority of scientists that humans are partially responsible for the increase in global temperature over the previous century. The Intergovernmental Panel on Climate Change (see Chapter 3's Policy in Action on page 53) states,

> The warming over the part 100 years is very unlikely to be due to internal variability alone . . . and is unlikely to be entirely natural in origin. . . . Detection and attribution studies consistently find evidence for an anthropogenic signal in the climate record of the last 25 to 50 years. . . . Most of the observed warming over the last 50 years is likely to have been due to the increase in greenhouse gas concentrations.

Despite such pronouncements, many nonscientists remain skeptical. Some of this doubt may be caused by reluctance to slow emissions—but the issue of what should be done about climate change is a separate question. Another source of doubt stems from an incomplete understanding of the science that has been popularized by the media. This misunderstanding is best illustrated in *The State of Fear* by Michael Crichton, in which the main character is a climate scientist who fights the scientific establishment's (supposedly) mistaken notion that humans are partially responsible for climate change. Although *The State of Fear* is a work of fiction, the claims of its main character often appear in arguments about attribution.

One of the more popular arguments against the role of human activity is mouthed by Crichton's main character: "If rising carbon dioxide is the cause of rising temperatures, why didn't it cause temperatures to rise from 1940 to 1970?" You should be able to answer this question. Remember that humans emit two types of gases: greenhouse gases, which tend to warm the atmosphere, and sulfur and other particles, which tend to cool the atmosphere. So the fictional character's comment is partially correct—carbon dioxide concentrations did rise from 1940 to 1970. But so too did the emission of sulfur. These emissions rose faster than carbon dioxide and other greenhouse gases, so radiative forcing of the atmosphere decreased between 1940 and 1970 (Figure 13.14), which stopped temperatures from rising (they actually cooled slightly—see Figure 13.1c). Thus the full answer to the question reinforces the conclusion that human activity is partially responsible for changes in global temperature—both warming and cooling.

Another criticism focuses on the supposed mismatch between the rise in surface temperature as measured by satellites and thermometers on the ground. Crichton's main character states, "Trust me, the satellite data have been reanalyzed dozens of times, . . . they show much less warming than expected by theory." This statement was once correct, but this contradiction no longer exists.

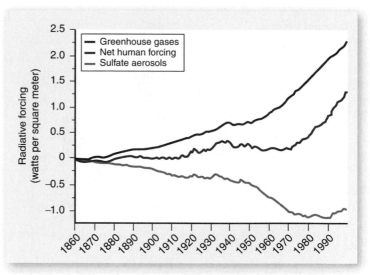

FIGURE 13.14   *Erratic Changes in Radiative Forcing* *The radiative forcing associated with greenhouse gases, human sulfur emissions, and the net effect of all human activities. Notice that radiative forcing does not increase between the 1940s and 1970s because the increase in greenhouse gas emissions is offset by increased emissions of sulfur particles. After the 1970s legislation in the United States and Europe aimed at reducing acid deposition reduced sulfur emissions. As a result, sulfur emissions declined, emissions of greenhouse gases increased, and net forcing rose.*

Scientists now know that the satellite data, which they use to measure surface temperatures, include temperature changes in the next layer of the atmosphere, which is known as the stratosphere (layers of the atmosphere will be discussed in Chapter 14 in relation to ozone depletion). Climate models predict that the stratosphere will cool as the surface layer warms because less long-wave radiation reaches the stratosphere. Consistent with these models, the stratosphere is cooling. Because satellites measure temperature both at the surface and in the stratosphere, falling temperatures in the stratosphere offset warming at the surface. Therefore satellite data show little change in temperature. If temperature changes in the stratosphere are removed, increases in surface temperature that are measured by satellites are similar to those measured by thermometers on the ground (Figure 13.15).

When evaluating arguments that dispute the scientific consensus that human activity plays a role in increasing global temperature, remember that the consensus is not based on any piece of evidence that constitutes a "smoking gun." Rather, the scientific consensus is generated by the totality of evidence. It is highly unlikely that all of this evidence is wrong. Few political decisions have a level of scientific certainty equal to that about the effect of human activity on climate.

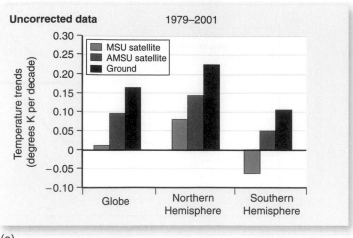

(a)

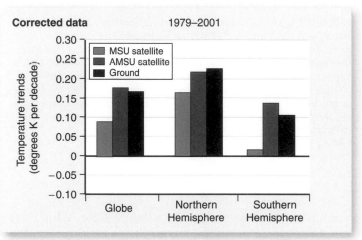

(b)

FIGURE 13.15   *Reconciling Satellite and Ground-Based Measures of Temperature* *(a) Skeptics of climate change argue that decadal changes in surface temperature measured at the ground (black bars) are much greater than those measured by uncorrected data from the MSU satellite (blue bars) and its successor the AMSU satellite (red bars). Subsequent analyses show that cooling in the upper atmosphere contaminated the satellite measures. (b) When this cooling is eliminated from the satellite data, satellite measures for ground-level warming agree with measurements taken at ground level.* (Source: Data from Q. Fu et al., "Contribution of Stratospheric Cooling to Satellite-Inferred Tropospheric Temperature Trends." Nature 429: 55–58.)

HOW WILL HUMAN ACTIVITY AFFECT CLIMATE?

If human emissions of greenhouse gases, sulfate aerosols, and particles continue, how will climate change? Will the planet warm or cool, and by how much? How will the change affect other aspects of climate, such as precipitation and the frequency and severity of hurricanes, tornadoes, and other storms?

The Past as the Future

It may be possible to project the effect of human activity on climate by examining the historical relationship between climate and the atmospheric concentration of greenhouse gases. There is a strong correlation between Earth's temperature and the atmospheric concentration of greenhouse gases over the last 420,000 years. Warm periods have correlated with high concentrations of carbon dioxide and methane, whereas cool periods have correlated with low concentrations (Figure 13.6).

But as we discussed in the Chapter 4 Case Study (page 68), correlation is not the same as causation. Did the increased concentration of carbon dioxide and methane cause Earth's temperature to rise, or did a rise in Earth's temperature increase the concentration of these gases? Convincing arguments can be made either way. We know that an increased concentration of greenhouse gases absorbs more heat and therefore can increase Earth's temperature.

But it is also possible that rising temperatures increased the atmospheric concentration of greenhouse gases. Permafrost stores lots of methane that could be released to the atmosphere as temperature rises. Higher temperatures also speed the decay of organic material, which increases the flow of carbon from soils to the atmosphere. Finally, there is a negative relationship between temperature and the solubility of carbon dioxide in water, so rising temperatures could increase the flow of carbon dioxide from the ocean to the atmosphere (Figure 13.16). Thus scientists are

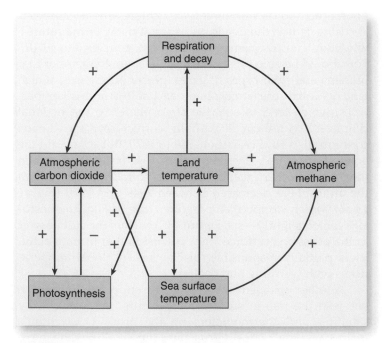

FIGURE 13.16   *Feedback Loops, Temperature, Carbon Dioxide, and Methane* *Feedback loops make it difficult to determine whether increases in the atmospheric concentration of carbon dioxide or methane raise temperature, or increases in temperature raise the atmospheric concentrations of carbon dioxide and methane.*

not sure whether changes in greenhouse gas concentrations cause temperature to change, or whether changes in global temperature alter the atmospheric concentration of greenhouse gases. Without this information, analysis of historical climate cannot generate conclusive information about how human activity will affect future climate.

Computers as Crystal Balls

Suppose scientists could build a replica of Earth; add greenhouse gases, sulfate aerosols, and particulates to its atmosphere; and study how its climate changes. Obviously scientists cannot run this experiment; but they use a similar approach in scenario analyses to analyze changes in climate that are simulated by computer models. Scenario analysis is needed because there is a great deal of uncertainty about how emissions will change in the future and how these emissions will affect concentrations. (Remember from Chapter 6 that an unknown carbon sink prevents scientists from balancing the flow of carbon dioxide into and out of the atmosphere. Because of this mystery, scientists cannot translate emissions of carbon dioxide into atmospheric concentrations of carbon dioxide.) Future concentrations are simulated using different scenarios for emissions (Figure 13.7). These scenarios allow us to evaluate how climate will change if population grows more slowly than anticipated, if people become richer than anticipated, or if society develops and adapts energy-efficient technologies.

Uncertainty about human actions is not the only source of uncertainty about the forecasts generated by climate models. To check the models' ability to simulate the relationship between radiative forcing and climate, scientists investigate how temperature may increase if the troposphere's concentration of carbon dioxide doubles relative to the start of the Industrial Revolution. This increase, termed **temperature sensitivity**, is the long-term change in temperature given a doubling in atmospheric concentration. Comparisons indicate that this doubling will increase global temperature 3.5°C plus or minus 0.9°C (3.5°C ± 0.9°C). The plus or minus 0.9°C is one way to measure the uncertainty about the models' ability to simulate climate. This range of values for temperature sensitivity is consistent with results generated by simpler models and by statistical analyses of the historical record. On the other hand, some more recent simulations specify that doubling the atmospheric concentration of carbon dioxide could increase global temperature by much greater amounts.

Why Does Temperature Rise?

The increase in temperature that is represented by temperature sensitivity is caused by the direct effects of doubling the atmospheric carbon dioxide concentration and a variety of feedback effects. In one model, doubling the atmospheric carbon dioxide concentration increases temperature by

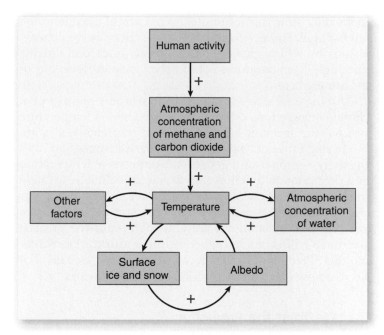

FIGURE 13.17 *Feedback Loops, Human Activity, and Temperature Change* Human activities that increase the atmospheric concentration of carbon dioxide and methane trigger feedback loops that raise temperature by an amount greater than the increase associated with the initial increase in radiative forcing. Warmer temperatures increase the atmosphere's ability to hold water vapor, which is a potent greenhouse gas, and melt ice and snow, which reduces albedo.

4.2°C. Of this increase, the direct effects of doubling atmospheric carbon dioxide increase temperature by 1.2°C. This temperature increase triggers a series of positive feedback loops that indirectly increase temperature by another 3.0°C (Figure 13.17). One of these feedback loops includes water vapor. Higher temperatures increase the atmosphere's ability to hold water vapor. Water vapor is a potent greenhouse gas; therefore, increasing its concentration raises temperature. This feedback loop increases global temperature by another 1.6°C. (Notice that this temperature increase is larger than the direct effect of doubling the atmospheric concentration of CO_2.) Another 0.8°C rise is due to a positive feedback loop that includes the melting of snow and ice. As temperature rises, ice and snow melt, which reduces Earth's albedo. Lower albedo increases the amount of solar energy absorbed by Earth's surface, which increases the amount of outgoing long-wave radiation that can be captured by greenhouse gases in the atmosphere. The remaining 0.8°C is caused by other small changes in the climate system.

The temperature effects of these positive feedback loops could be slowed by a series of negative feedback loops. Increasing water vapor in the atmosphere could increase cloud cover, which would tend to cool the planet. The scale of this effect is highly uncertain because it is difficult to model the formation of clouds, yet the quantity and types of clouds formed will have a significant effect on the size of the temperature increase. In addition, several feedback loops may slow the accumulation of carbon dioxide in the atmosphere. Figure 6.8 on page 108 described several mechanisms by which greater concentrations of carbon dioxide and higher temperatures could speed carbon flows out of the atmosphere.

An increase in average temperature is only one aspect of climate change forecast by climate models. As temperature warms, the location and strength of gradients will shift, modifying flows of heat and water. These changes may alter the daily, seasonal, and geographic patterns of temperature and precipitation that make up the current climate. Relative to today's climate, climate models forecast that the upper reaches of Earth's atmosphere will cool, surface temperature will increase most significantly at night, and the greatest warming will occur in higher latitudes during the winter months. Precipitation is forecast to increase globally but decrease over the middle portions of the continents.

Can We Trust the Predictions?

Ocean atmosphere general circulation models are among the most complex tools used by environmental scientists because they test the limits of our knowledge of atmospheric science, mathematics, and computer science. Even though computers do not make mathematical errors, they still can be wrong. Mistakes occur because scientists may simulate the wrong processes, use the wrong form of an equation, or omit an important process.

Scientists know that all climate models contain errors because there is a great deal of uncertainty about how the climate system really works. Because the mathematics of cloud formation is so complex, some climate models do not include equations to represent how a changing climate will affect cloud formation. These and other shortcomings create a great deal of uncertainty about the forecast for future changes in climate.

To help policy makers understand the effects of climate change, scientists evaluate this uncertainty by ranking their forecasts. Scientists are most confident that the forecasts for near surface warming and cooling in the stratosphere (the next layer up) are correct. Scientists are sure of other changes, such as extreme precipitation events (storms) increasing more than average precipitation. Scientists are less sure about other changes, such as whether storms will become more severe.

THE IMPACTS OF GLOBAL CLIMATE CHANGE

The 3.5°C (± 0.9°C) rise in average temperature that is associated with a doubling of the atmosphere's radiative forcing seems trivial. It would be difficult to feel this difference on any given day. But the significance of a 3.5°C rise in average temperature can be appreciated through the following

fact: Earth's average temperature was only 3–5°C (5–9°F) cooler than today during the last ice age, when much of North America was covered by ice and boreal forests covered what is now North Carolina. Clearly small changes in average temperature can dramatically alter climate and the geographic distribution of biomes.

Can Biomes Move Faster Than Climate Changes?

Responses to previous variations in climate may help us understand how biomes will change in response to human-induced changes in climate. As the climate warmed between 5,000 and 10,000 years ago, biomes changed their geographic position in North America (Figure 13.18). Boreal forests shifted north out of the United States into Canada, and temperate forests expanded their range from the Gulf of Mexico northward along the Atlantic seaboard.

For biomes to follow changes in climate, individual species must be able to disperse. When temperature or precipitation changes, a species must move to a new region with favorable conditions. If it cannot, it will become trapped in an unfavorable climate and go extinct. During the warming that began 10,000 years ago, temperatures along the northern U.S. border became too warm for spruce trees. As they moved north to follow the cooler climate, spruce trees were replaced with hardwood species that thrive in warmer climates.

Biomes require a long time to shift their range. Trees in particular change their range very slowly. As the climate warmed about 10,000 years ago, tree species in North America shifted their range north between 100 and 400 meters per year. Of course individual trees cannot move—dispersal occurs via the success and failure of seedlings. For a tree species to move north, seedlings that are scattered in a northerly direction must succeed in areas where they were previously unsuccessful. Similarly, seedlings that are scattered in a southerly direction must fail where they were previously successful. Trees mature slowly, so such movements may require a long time. It took about 5,000 years for spruce trees to move the southern portion of their range from southern Michigan to northern Michigan—a distance of only 500 kilometers.

The slow movement is significant because the rate of climate change forecast by climate models is about ten times faster than the rates of change that occurred after the last ice age. Complicating matters, other factors may block the path of biome relocation. Cities, highways, and agricultural fields block the transport of seeds by animals and wind. Equally important, the soil in a region with a favorable climate may not be able to support the new biome. Each species requires specific types of soil, and it may take a long time for soil to change in a way that can support new species. In North America the sugar maple and yellow birch should be able to migrate from areas around the Great Lakes to areas just south of Hudson Bay. But this migration may be prevented

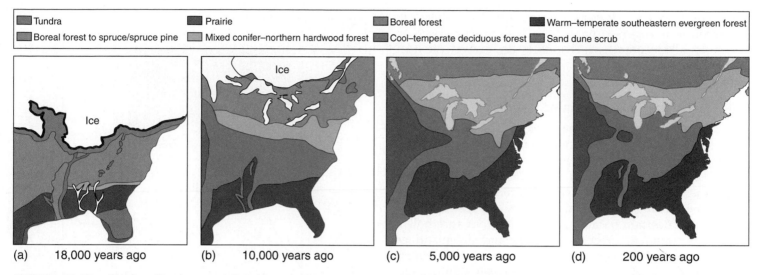

| Tundra | Prairie | Boreal forest | Warm–temperate southeastern evergreen forest |
| Boreal forest to spruce/spruce pine | Mixed conifer–northern hardwood forest | Cool–temperate deciduous forest | Sand dune scrub |

(a) 18,000 years ago   (b) 10,000 years ago   (c) 5,000 years ago   (d) 200 years ago

FIGURE 13.18 *Moving Biomes* As global climate changed over the last 18,000 years, so did the distribution of terrestrial biomes in eastern North America. (a) 18,000 years ago boreal forests reached as far south as Louisiana. (b) 10,000 years ago the boreal forest and ice sheet moved north into Canada. (c) 5,000 years ago the ice sheet was confined to northernmost Canada, and the distribution of biomes was similar to its position 200 years ago, which is shown in panel (d). (Source: Data from D.M. Gates, Climate Change and Its Biological Consequences.)

by soil conditions. The sandy soils common just south of Hudson Bay cannot store enough water for sugar maples and yellow birch trees (see Chapter 15).

Rapid climate change and slow ecosystem migration imply that forest biomes may become trapped in areas where climate conditions cannot support them. This could cause ecosystems to collapse and be replaced with much simpler systems. The biodiversity supported by the forests would be lost or reduced, and this would reduce the environmental goods and services provided by the forest. For example, the loss of forest area could reduce timber supplies, increase the frequency and severity of floods, and increase soil erosion and the subsequent clogging of waterways (see Chapter 17).

Will Food Supplies Decrease or Increase?

Some of the changes forecast by climate models could increase food supply, whereas others could reduce food supply. Currently it is not possible to calculate the net effect of the positive and negative changes, but it is highly unlikely that the positive effects will be balanced exactly by the negative effects. Instead food production will increase in some regions and decrease in other regions.

Increasing atmospheric carbon dioxide can raise crop production directly. Remember that carbon dioxide is the raw material for photosynthesis; therefore an increase in the atmospheric concentration of carbon dioxide may act as a fertilizer and increase the production of wheat, rice, and other crops. Doubling the atmospheric concentration of carbon dioxide may increase the yield of spring wheat by 8–10 percent. Such increases are uncertain because it is difficult to extrapolate results of experiments in growth chambers to results in the field, where other factors may limit yields.

Higher concentrations of atmospheric carbon dioxide also may allow farmers to increase production in arid areas. Plants lose water through their stomata when they take in carbon dioxide. If plants can take in more carbon dioxide (due to the increased concentration) each time they open their stomata, the stomata will be open for less time. This would reduce water loss and allow plants to produce more biomass per unit of water, which is known as **water use efficiency.** This would increase yields in areas where water is the limiting factor.

A general warming also will let farmers grow crops in areas where the growing season currently is too short. Corn requires a relatively long growing season (in some areas more than 100 days). This requirement restricts European production to the central and southern portions of the continent. A temperature increase of 1°C could expand the area where corn is grown by nearly 1.3 million square kilometers.

Alternatively, changes in climate could reduce food production. Many of the world's best agricultural areas are located in the middle of continents. But these areas are forecast to dry as the result of climate changes. This drying will include less precipitation and more evaporation, both of which will reduce water supplies to plants.

Climate change also could reduce food supplies by providing more suitable habitats for pests and diseases that attack crops. Mild winters and longer growing seasons may allow insect pests such as the soybean cyst nematode and the corn gray leaf blight to expand their range and speed their reproduction. Even the boost in yields associated with higher concentrations of carbon dioxide could increase agricultural yields lost to pests. Experiments show that plants grown in atmospheres with elevated concentrations of carbon dioxide have a lower nitrogen content (a higher ratio of carbon to nitrogen—Chapter 6). This induces insects to eat more of the plant, which offsets the carbon dioxide-induced increase in yield.

Will We Drown under a Rising Sea?

Warmer temperatures will raise sea level. The midrange scenario for climate change reported by the Intergovernmental Panel on Climate Change forecasts that sea level will rise by about 0.384 meter by 2100. Of this, 0.288 meter will be due to the thermal expansion of water and 0.106 meter will be due to the melting of mountain glaciers and ice caps. This estimate is uncertain because scientists are not sure how a warmer climate will affect the Antarctic and Greenland ice sheets. Because this ice is not currently in the ocean, runoff from the melting sheets could raise sea levels significantly. Complete melting of these sheets would raise sea levels by about 70 meters. Fortunately such tremendous rises in sea level are not likely.

Rising sea levels could affect society directly by destroying infrastructure such as homes, factories, bridges, and roads. Historically societies have located their farms and cities near the ocean. Currently about 20 percent of the world's population lives within 30 km of the ocean, and about 40 percent lives within 100 km. Economists estimate that a 0.5 meter rise in sea level could destroy about $20–150 billion worth of human infrastructure in the United States. To avoid such damages, several scientists have suggested building dikes around heavily populated areas that are close to sea level. Such an effort would require an enormous amount of time and money. Alternatively, entire towns may be relocated.

Rising sea levels also would affect society indirectly by reducing the area and productivity of estuaries and coastal wetlands. Estuaries are among the most productive ecosystems (as discussed in Chapter 7). Estuaries process organic and toxic wastes, serve as nurseries for many commercial species of fish, and reduce the strength of and damage done by coastal storms. These environmental services could be reduced if estuaries and wetlands are drowned by rising seas. As with forests and farmlands, estuaries may not be able to migrate inland as sea level rises.

Will Large Number of Species Go Extinct?

The first species known to have become extinct due to recent climate change come from a family of toads that are mislabeled harlequin frogs. These colorful toads, including the famous golden toad, live at relatively high latitudes in Central American cloud forests, such as Monteverde in Costa Rica. Between 1975 and 2000 field researchers noticed that entire species would suddenly disappear. Naturalists would see the Jambato toad during 64 percent of their visits to its habitats throughout its 10,234 km² range prior to 1988;

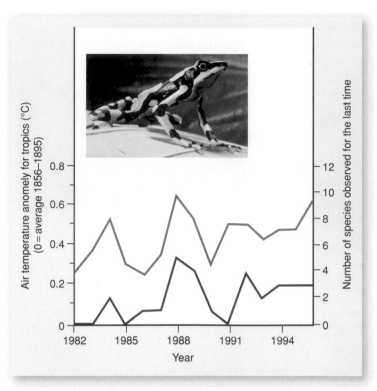

FIGURE 13.19 *Warming and Amphibian Extinctions* Warm years, indicated by higher air temperatures (green line), are followed by increased extinctions of amphibian species (red line). (Source: Data from Pounds et al., "Widespread Amphibian Extinctions from Epidemic Disease Driven by Global Warming," Nature 439: 161–167.)

but since 1988 it has not been seen anywhere. Statistical analyses of 104 species that have disappeared indicate that they went extinct following an especially warm year (Figure 13.19). These years enhance the growth of a fungus that causes a fatal disease for many amphibian species.

Over time, scientists estimate that global climate change could cause 9–52 percent of the world's known terrestrial species to go extinct. How can anyone make such a startling estimate? Projections for the extinction rate are generated by combining forecasts for future climates, the species–area relationship, and the notion of a **climate envelope.** The climate envelope represents the conditions under which populations of a species can persist in the face of competitors and natural enemies (this definition is similar to the concept of reproductive range in Chapter 7). Next scientists use climate models to project how climate change will cause the areas that satisfy these climate envelopes to increase or decrease over the next century. This change in area is then combined with the species–area relationship (Chapter 12) to project the number of species that will become extinct due to a reduction in the area that is consistent with a given climate envelope.

Of course this number depends on the scenario for climate change, the scale at which the species–area relationship is applied, and the degree to which species can disperse. For the worst-case scenario, which assumes rapid climate change and that species cannot disperse, the extinction rate varies between 38–52 percent. For the best-case scenario, which assumes minimal climate change and that a species can find its climate envelope wherever it exists, the extinction rate is 9–13 percent.

These ranges are not meant to be precise forecasts; rather they reflect a range of what could happen given the current understanding of the factors that determine biodiversity. Estimates could be too high or low if the models have left out important determinants. One such determinant may be the degree to which species can change their climatic envelopes. The model assumes that this is relatively fixed, but a recent analysis indicates that for some coral species, this envelope may be more flexible than previously thought. Corals that are stressed by high water temperatures lose algae, which provide much of their energy (Chapter 7). Without this energy the corals die in a process known as bleaching. Many coral species can host a variety of algal species, some of which are more resistant to high temperatures. Analyses indicate that natural selection tends to increase the abundance of heat-resistant algal species following bleaching events that were caused by an El Niño warming. Their increased abundance implies that models that assume the climate envelope for a species is relatively fixed would tend to overstate the extinction rate for coral species and the reef species that depend on the corals for food and habitat.

CLIMATE CHANGE POLICY

What Should Be Done?

Human activity is changing climate. Because climate takes a long time to adjust completely to changes in radiative forcing, climate will continue to change even if we stop future increases in radiative forcing. As a result, human activity has already committed the world to future changes in climate. Policy can only slow future changes.

The question of what should be done about global climate change illustrates the strengths and weaknesses of environmental policy making that are discussed in Chapter 11. Most important, it highlights the idea that zero emissions is not the optimal policy in the near term. To eliminate all carbon dioxide emissions society would have to eliminate the use of coal, oil, and natural gas and stop converting forests to agriculture, residential, and commercial uses. But such measures would impose huge costs on society. As described in Chapter 10, fossil fuels support all aspects of human existence, from biological survival to material comfort. Halting deforestation will be nearly as difficult—increasing amounts of land will be required to feed and house the world's growing population.

If it is not economically sensible to eliminate all activities that emit greenhouse gases, which activities should be stopped or reduced? In theory, society should internalize the cost of climate change so that consumers forgo activities that generate benefits that are smaller than the damages they cause by changing the climate. Internalizing these costs would lead to a world in which the benefits of consuming the last unit of fossil fuel and clearing the last unit of forest equal the damages caused by the last unit of greenhouse gas emitted. Alternatively, the cost of reducing the last unit of carbon dioxide emitted would equal the damage caused by the last unit of carbon dioxide emitted. Under these conditions consumers would emit the so-called optimal level of pollution.

To calculate this optimal level of pollution, scientists are working to estimate the cost of reducing emissions and the cost of climate-related environmental damage. A host of technologies can reduce emissions of carbon dioxide (the most important greenhouse gas). The cost of reducing emissions by one metric ton varies greatly among nations and studies. The cost of climate-related environmental damage also is uncertain. Because of this uncertainty, scientists do not try to add up damages associated with individual changes in climate. Rather they formulate damage functions, which show the percentage reduction in GDP that is associated with a given temperature change (Figure 13.20). Even at this aggregated level considerable uncertainty remains about how damage changes with increasing temperature. Specifically, is the relationship linear or exponential? Or do damages increase rapidly beyond some threshold?

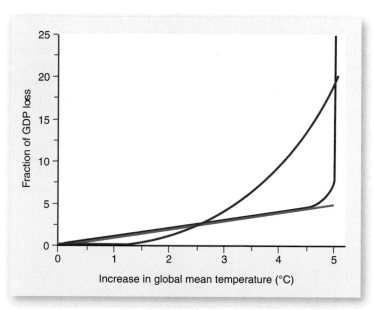

FIGURE 13.20   *Damage Functions* *Damage functions indicate the fraction of GDP lost as global temperature rises. Considerable uncertainty exists over whether the loss in GDP rises linearly with temperature (green line), rises exponentially with temperature (blue curve), or increases sharply after temperature exceeds some threshold (red curve).*
(Source: Data from S. Fankhauser et al., "Vulnerability to Climate Change and Reasons for Concern: A Synthesis," eds. J.J. McCarthey et al., in Climate Change 2001: Impacts, Adaptation, and Vulnerability.)

The long time horizon associated with climate change also makes it difficult to calculate the optimal level of emissions. Activities that emit greenhouse gases benefit people alive today, but damages generated by those emissions will be felt most strongly by future generations. How should society weigh certain current benefits against uncertain costs in the future?

But even if scientists and economists could eliminate all uncertainty and calculate the optimal level of emissions, issues of equity would prevent a simple solution. Suppose scientists and economists could agree that the optimal level of carbon dioxide emissions is 4 petagrams, which represents a 40 percent reduction in emissions relative to 2001. How should this reduction be allocated internationally? If the 4 petagrams of emissions are allocated equally among all people, people in developed nations would have to reduce their emissions more than people in developing nations because per capita emissions in developed nations are much greater than in developing nations. People in developed nations would argue that a policy of equal emissions is not fair. Instead developed nations might propose that everyone's emissions be reduced by 40 percent. In this case per capita emissions in developing nations would remain well below those in developed nations. This would limit the ability of people in developing nations to expand their material standard of living—and developing nations would complain that the policy of equal percentage reductions is not fair.

Scientists in the Intergovernmental Panel on Climate Change point out that the scientific and political difficulties make it pointless for policy makers to argue about what might be the best course for the next 100 years. Rather policy makers need to consider the best course for the next decade given some long-term objective. That long-term objective is stated in Article 2 of the United Nations Framework Convention on Climate Change, agreed to in Rio de Janeiro in 1992:

> . . . stabilization of greenhouse gases in the atmosphere at a level that would prevent dangerous anthropogenic interference with the climate system. Such a level would be achieved within a time frame sufficient to allow ecosystems to adapt naturally to climate change, to ensure that food production is not threatened, and to enable economic development to proceed in a sustainable manner.

No Silver Bullet

How can society achieve this objective, given the conflicting demands of economic development and environmental protection? There is no single answer. Rather policy must be flexible, take advantage of all options, and not be constrained by national borders. Decision makers should continually establish flexible short-term policy objectives. Scientific under-standing of climate change always is improving, so policy will be effective only if it can change along with science.

To achieve these short-term goals, decision makers need to implement cost-effective policies. These measures will be drawn from four categories: (1) reductions in emissions of greenhouse gases; (2) research and development for new supply and conservation technologies that lower the cost of reducing greenhouse gas emissions; (3) research to reduce critical areas of uncertainty; and (4) investment in actions that help human and natural systems adapt to climate change. Each category contains many options, and option effectiveness varies greatly within a category and among categories. As a result, effective policy will include a portfolio of options.

The benefits of a portfolio of options can be illustrated as follows. Carbon dioxide is the largest contributor to the increase in radiative forcing and therefore is the focus of much research on reducing emissions of greenhouse gases. But recent research shows that reducing methane emissions may be a less expensive way of slowing the increase in radiative forcing. Other options have both environmental and economic benefits. New conservation technologies may reduce emissions and reduce energy bills.

To make full use of these options, policy must be coordinated internationally. The cost effectiveness of each option varies among nations. Nations with few cost-effective options should be allowed to take advantage of cost-effective options in other nations. For example, much capital equipment in developing nations is very energy inefficient. Replacing it with new energy-efficient capital is a cost-effective means for reducing emissions, but many developing nations cannot afford to replace energy-inefficient capital. On the other hand, developed nations can afford to purchase the most energy-efficient capital; but much of their capital equipment already is relatively energy efficient, so replacing their existing capital would reduce emissions only slightly.

To overcome these obstacles, policy makers must devise ways that developed nations can invest in energy-efficient equipment in developing nations and allow the developed nations to count these reductions in emissions toward their emission reduction target. The United Nations has developed one such scheme, known as the **clean development mechanism.** The clean development mechanism allows a nation to earn credit for reducing emissions in another nation. One such example is an agreement between Chile's largest pork producer and power companies in Japan and Canada. The power companies purchased the right to emit more carbon dioxide when they bought the pig farmers equipment to capture methane and burn it as a renewable source of energy.

The Kyoto Protocol: A First Step?

The first step toward an active policy to slow climate change is embodied in the Kyoto Protocol, which went into effect in February 2005. Under this treaty, industrialized nations are obligated to reduce their carbon dioxide emissions below 1990 levels by between 2008 and 2012. Developing nations have a voluntary commitment to slow the growth of their emissions. If the participating nations comply, carbon dioxide emissions will be reduced relative to the levels that would have prevailed in the second decade of the twenty-first century. None of the participants argue that this reduction represents the economically efficient level of emissions. Rather, they assert that it is a first step on a road that will eventually require more reductions if we are to avoid the damages associated with large changes in climate.

To reduce the costs of compliance, the methods used to reduce emissions can vary among nations. Techniques include efficiency standards, taxes on carbon dioxide emissions, and trading schemes among nations. The European Union has set up an international market where firms can buy and sell the right to emit carbon dioxide. This effort is modeled on the successful market for trading emissions of sulfur dioxide, which was set up by legislative efforts to reduce acid deposition in the United States (see Chapter 19's Policy in Action on page 392). Already the number of trades to comply with the Kyoto Protocol has grown (Figure 13.21).

The United States and Australia are not part of the Kyoto Protocol. President George W. Bush withdrew from the treaty shortly after he took office in 2001, asserting that reducing carbon dioxide emissions would hurt the U.S. economy. He is probably correct about that. But will the economic damage caused by climate change exceed the possible loss in economic activity due to the treaty? We cannot calculate the impacts of the Kyoto Protocol on the U.S. economy or the economic damages it might cause or prevent, but we know one thing for sure—the damage done by climate change will be greater than zero.

President Bush also argues that the Kyoto Protocol does not force developing nations such as China and India, whose emissions are increasing greatly, to reduce their emissions. This too is correct. But should developing nations be forced to reduce their emissions? Carbon dioxide has a long residence time in the atmosphere, so most of the increase in the atmospheric concentration of carbon dioxide since the Industrial Revolution is due to economic activity by devel-

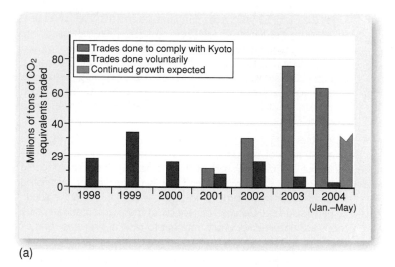

(a)

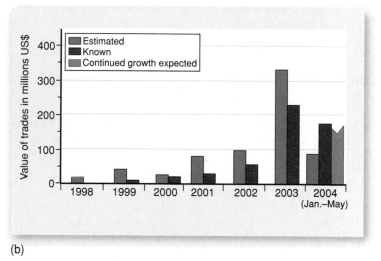

(b)

FIGURE 13.21   *A Market for Carbon Dioxide Emissions* The European Union has established a market in which carbon dioxide emissions can be bought and sold. Notice that (a) the quantity of carbon dioxide traded and (b) the value of these trades have increased significantly and are expected to continue. *(Source: Data from M. Hopkin, "The Carbon Game." Nature 432: 268–270.)*

oped nations. Should developing nations be forced to help clean up the mess made largely by industrial nations? This question has no objective answer—the answer must come from ethical beliefs that indicate what is fair.

SUMMARY OF KEY CONCEPTS

- Climate is determined by the planet's heat balance, which is determined by the amounts of incoming short-wave solar radiation and retained outgoing long-wave radiation. The former is determined by solar activity, clouds, aerosols, and other particles in the atmosphere, while the latter is determined by greenhouse gases. These greenhouse gases occur naturally; without them the planet would be inhabitable.

- Human activities increase the concentrations of greenhouse gases, sulfate aerosols, and other particles. Concentrations of greenhouse gases are near the highest values observed over the last half million years due to changes in global biogeochemical cycles that increase the flows of greenhouse gases into the atmosphere and reduce their flows out of the atmosphere. The rates at which these flows will change depend on population growth, affluence, technology, and energy prices.

- Over the past 150 years both temperature and sea level have risen. In addition, the length of the growing season has increased, and Northern Hemisphere species have migrated north and moved up in altitude. Together these changes show that Earth's climate is changing. Statistical analyses of historical data and computer models that simulate Earth's climate indicate that some climate change is due to human activities that increase the concentrations of greenhouse gases, sulfate aerosols, and other particles.

- Computer simulation models predict that doubling the atmospheric concentration of carbon dioxide will increase global temperature by about 3.5°C. Only part of this increase is due directly to the increase in radiative forcing. The direct effect is enhanced by positive feedback loops that include atmospheric water vapor and the albedo of Earth's surface. Because the effects of these and other feedback loops are uncertain, scientists attempt to rank the certainty of their predictions about climate change.

- Climate change will have both positive and negative effects on people. If temperatures rise faster than biomes can migrate, species may get trapped in inhospitable climates and become extinct.

- Ecosystems may collapse and be replaced with simpler, less productive systems, which would reduce the availability of natural resources and environmental services. Rising ocean temperatures and melting ice will raise sea levels, which could force societies to invest huge sums to protect existing infrastructure or help millions of people to move.

- Climate change could enhance food production because the greater concentration of carbon dioxide could increase net primary production and water use efficiency. Warmer temperatures could lengthen the growing season, which might expand agricultural acreage. On the other hand, agricultural production may decrease due to a general drying in midcontinental areas and increasing losses due to pests and disease.

- Policy makers need to evaluate the costs of climate change so that consumers can balance the costs and benefits of activities that emit greenhouse gases. Many uncertainties make this difficult, so policy makers need to develop a suite of options for reducing emissions of greenhouse gases. The Kyoto Protocol is not perfect but may be a first step in this direction.

REVIEW QUESTONS

1. Suppose that economic activity emits quantities of carbon dioxide and methane such that each gas will raise radiative forcing by 1 watt per square meter. Do these changes have the same effect on climate?

2. Explain how human disruptions of biogeochemical cycles affect radiative forcing of the atmosphere.

3. Discuss the difference between detection and attribution. What is the relevance of each for policy decisions about what should be done about climate change?

4. Explain the role of feedback loops in climate change. How do these feedback loops affect radiative forcing and temperature change?

5. Define the optimal level of greenhouse gas emissions. Describe why defining this level and restricting emissions are so difficult.

6. Discuss the benefits and drawbacks of the Kyoto Protocol.

KEY TERMS

aerosols

attribution

clean development mechanism

climate

climate change

climate envelope

detection

greenhouse effect

greenhouse gases

heat balance

land use change

radiative forcing

reflect

residence time

scatter

temperature sensitivity

water use efficiency

weather

A REDUCTION IN ATMOSPHERIC OZONE

Let the Sunshine In

14

STUDENT LEARNING OUTCOMES

After reading this chapter, students will be able to

- Explain why temperature rises or falls with altitude in different layers of the atmosphere.
- Compare and contrast three hypotheses for the depletion of stratospheric ozone and how the scientific method was used to choose among them.
- Compare and contrast the depletion of stratospheric ozone above the northern and southern poles.
- Describe how a reduction in stratospheric ozone affects organisms that live in both marine and terrestrial ecosystems.
- Describe the role of scientific information in the process that culminated in agreements to end the production of CFCs.

Monitoring Earth's atmosphere from space *Satellites measured sharp declines in ozone concentrations, but were initially ignored because scientists thought the calculations were flawed.*

In the spring of 1986 scientists announced that the atmosphere above the south pole had 40 percent less ozone than expected. The announcement shocked much of the public. But for many scientists the ozone reduction was another piece of evidence in a scientific controversy that started more than a decade earlier. In the spring of 1974 F. Sherwood Rowland and Mario Molina hypothesized that a group of chemicals known as chlorofluorocarbons (CFCs) could destroy the protective ozone in Earth's atmosphere. This hypothesis started a controversy that illustrates all the ingredients of modern environmental challenges. Because the chemical reactions by which CFCs could destroy ozone were complicated and uncertain, both environmentalists and the chemical industry tried to portray themselves as protecting the public interest. Environmentalists argued that life could be wiped out if Earth lost much of its ozone. Industry responded that CFCs were critical to modern life and that companies producing CFCs employed thousands of people and generated many economic benefits.

The controversy also illustrates how the scientific method can reduce uncertainty about a potential environmental threat. Originally the hypothesis that CFCs could reduce the atmospheric concentration of ozone was based on a simple model. Scientists on both sides of the controversy agreed that the model could not be used as proof that CFCs created a problem, but they agreed that the scientific method could help test this and alternative hypotheses for lessened ozone concentrations. Scientists who argued that CFCs were the cause tried to find evidence that supported the model's conclusions (opening photograph). At the same time scientists who did not support the theory tried to find evidence that contradicted the model.

The ozone controversy also illustrates the need for cooperation among consumers, industry, and government at the individual, local, national, and international levels. Some consumer groups wanted to ban the production or use of CFCs at the local and national levels. Others asserted that only international agreements could solve the problem. Some industry groups said that CFCs should not be banned until suitable replacements were developed, whereas others claimed that without a specific timetable for a ban, the chemical industry would not have an economic incentive to develop alternatives. Negotiation, compromise, and consensus among these groups were critical to the solution.

Finally, and perhaps most important, the controversy illustrates the potential for a happy ending. Rigorous science eliminated much uncertainty about the causes for the lower ozone concentrations. Sherwood Rowland and Mario Molina (along with Paul Crutzen) won the Nobel Prize for chemistry in 1995. Patient negotiation culminated in a series of international agreements to eliminate the production of CFCs. The chemical industry developed alternatives that are nearly as effective as and not much more expensive than CFCs. Although the environmental threat posed by the reduction in ozone concentrations has not disappeared, the threat is much less severe than it appeared in the spring of 1986. ■

THE ATMOSPHERE

The atmosphere is composed of the gases that surround Earth. These gases provide many environmental services. As described in Chapter 13, greenhouse gases absorb longwave radiation and thereby warm the planet in a way that makes life possible. Other gases protect Earth's surface from high-energy solar radiation that can kill or damage living organisms.

Components of the Atmosphere

Gases that make up the atmosphere are classified as either main gases or trace gases. Nitrogen and oxygen are the two main gases. **Molecular nitrogen,** N_2, makes up about 78 percent of the atmosphere while **molecular oxygen,** O_2, makes up another 21 percent. The concentration of these gases is relatively constant over time. Compared to the 20 percent increase in the atmospheric concentration of carbon dioxide since 1959, the atmospheric concentration of oxygen has varied by less than 0.03 percent over the last fifty years. This stability is necessary for life. High oxygen concentrations could cause the smallest spark to set off a worldwide fire. Conversely, low oxygen concentrations would make it progressively harder to breathe. Remember, low oxygen levels posed a problem for the Biosphere experiment described in Chapter 12.

The remaining 1 percent of the atmosphere includes more than twenty **trace gases.** Although these gases are a small part of the atmosphere, they often play a vital role. The greenhouse gases described in Chapter 13, such as carbon dioxide, methane, and nitrous oxide, are trace gases that are measured in parts per million or parts per billion; yet their role in the greenhouse gas effect is essential for life. Similarly, ozone is a crucial trace gas.

Layers of the Atmosphere

The concentrations of gases just described are based on measurements taken at or near sea level. But Earth's atmosphere rises more than 100 kilometers above the surface, and the concentrations of gases change with altitude. Although there are no actual boundaries, the temperature of the atmosphere changes with altitude, and these differences tend to keep gases in their respective layers (Figure 14.1).

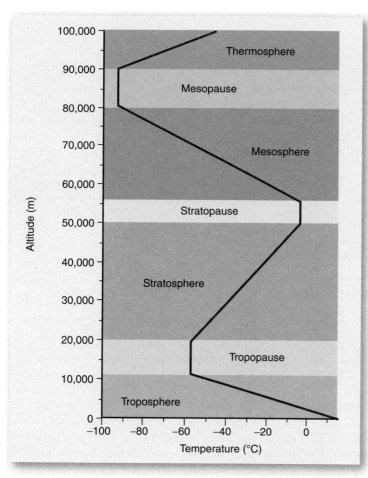

FIGURE 14.1 *Layers of the Atmosphere* *Earth's atmosphere is separated into several layers that are defined by the change in temperature with altitude. In some layers, such as the troposphere, temperature falls with altitude. In other layers, such as the stratosphere, temperature increases with altitude.*

The lowest layer of the atmosphere is called the **troposphere.** The troposphere extends from the surface to an altitude of 10–20 kilometers. In this layer temperature declines with altitude at a rate of about 6.5°C per km. The change in temperature with altitude is known as the **lapse rate.** After about 15 km in tropical regions or about 8 km in polar regions, air temperature reaches about –50°C. Beyond this point, temperature starts to rise again. This change in the lapse rate, so that temperature no longer declines with altitude, defines the **tropopause,** which marks the end of the troposphere.

The tropopause is essential to life because it keeps water vapor in the troposphere, where it can fall back to the surface as rain or snow (see Chapter 18). Without the tropopause, water vapor would rise until solar energy split water molecules into oxygen and hydrogen, which would escape into space. This would make Earth lose some water with each turn of the water cycle, and eventually the planet would run out of water.

Beyond the tropopause, the atmosphere warms with altitude. The layer of rising temperatures is known as the **stratosphere,** which extends from about 20 to 50 km above the

surface. Within these 30 km temperature rises by about 50°C. The atmosphere warms because solar energy is converted to heat by the chemical reactions that create and destroy ozone. Above the stratosphere temperature starts to decline again. This zone of cooling, known as the **mesosphere,** extends from about 50 to 100 km above the surface. The end of this layer is marked by temperatures of about –100°C.

The atmosphere begins to warm again beyond 100 km in the layer known as the **thermosphere.** In this outermost layer solar energy splits molecular oxygen and nitrogen into single atoms. The solar energy that drives these reactions is converted to heat, and this heat generates very high temperatures (about 1,200°C).

STRATOSPHERIC OZONE

Ozone is a special form of oxygen. The most common form of oxygen is molecular oxygen. In the stratosphere however, solar energy powers a reaction that converts molecular oxygen to **ozone,** which is a molecule that consists of three atoms of oxygen O_3. About 90 percent of Earth's ozone is found in the stratosphere. Smaller amounts of ozone are found at ground level in the troposphere, where it is an air pollutant caused by photochemical smog (see Chapter 19).

The Formation of Ozone

Solar energy powers the formation ozone in a series of chemical reactions (Figure 14.2). In the first step molecular oxygen absorbs solar energy, which splits it into two individual

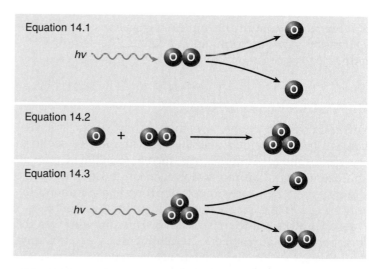

FIGURE 14.2 *Ozone Formation* *The formation of ozone starts when solar energy splits molecular oxygen into two single atoms of oxygen (Equation 14.1). In the next step one of these atoms combines with molecular oxygen to form ozone (Equation 14.2). The cycle is completed when ozone absorbs solar radiation and is split into an oxygen atom and molecular oxygen (Equation 14.3).*

oxygen atoms. The process by which solar energy splits a molecule is known as **photodissociation.** The photodissociation of molecular oxygen is given by Equation 14.1:

$$O_2 + hv \longrightarrow O + O \qquad (14.1)$$

The term hv stands for ultraviolet radiation, a type of solar energy that has a short wavelength and great energy (Chapter 2).

In the next step ozone is formed by combining the single oxygen atoms generated in Equation 14.1 with molecular oxygen. This reaction is given by Equation 14.2:

$$O + O_2 \longrightarrow O_3 \qquad (14.2)$$

When ultraviolet radiation is absorbed by ozone, the ozone molecules are split back into the parts that formed them—single atoms of oxygen and molecular oxygen. (As described in more detail later in this chapter, ozone provides an environmental service by absorbing ultraviolet radiation during this reaction and thus shields Earth from the sun's ultraviolet rays.) This reaction is given by Equation 14.3:

$$O_3 + hv \longrightarrow O + O_2 \qquad (14.3)$$

The reactions described by Equations 14.1–14.3 set up a cycle that forms and destroys ozone. The reactions described by Equations 14.2 and 14.3 occur more rapidly than the reaction described by Equation 14.1. This cycle implies that the rate at which ozone is formed and destroyed depends on the amount of single-atom oxygen (O) available. Any process that reduces the supply of single oxygen atoms in the stratosphere can slow the formation of ozone. This is how CFCs reduce the concentration of ozone in the stratosphere.

The Distribution of Ozone

The chemistry of ozone formation is part of the information we need to understand the concentration of ozone in the stratosphere. Although the media sometimes refer to an "ozone layer," ozone is not a major constituent of the stratosphere. The concentration of ozone in a column of the atmosphere is measured in Dobson units (Figure 14.3). A **Dobson unit** is named after G.M.B. Dobson, who was the first to measure atmospheric concentrations of ozone during the 1920s and 1930s. To envision a Dobson unit, suppose you could separate all of the molecules in the stratosphere into individual layers, such as a molecular oxygen layer, a molecular nitrogen layer, and an ozone layer. If all the molecules in these layers were compressed at standard temperature and pressure (standard pressure is one atmosphere—760 mm mercury or 101.3 kPa—and standard temperature is the freez-

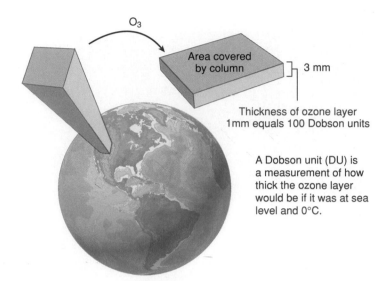

FIGURE 14.3 *Dobson Unit The concentration of ozone in the stratosphere is measured in Dobson units. A Dobson unit measures the thickness of ozone molecules at standard pressure and temperature if these ozone molecules could be separated from all other gases. On average, all the ozone molecules in the stratosphere would form a layer about 3 millimeters thick.*

ing point of water, 0°C), one hundred Dobson units would be a "pile" of molecules that is 1 millimeter high. There are about 300 Dobson units (DU) of ozone, which is equivalent to about 3 millimeters. Put another way, there are only about 3 millimeters of ozone protecting life from the sun's ultraviolet radiation. Of course these ozone molecules are not in a single layer that is 3 millimeters thick. Instead most of these ozone molecules are spread throughout the stratosphere.

Another important characteristic of ozone is its geographic distribution. Most ozone is formed near the equator, which is to be expected. As described in Chapter 4, the sun's rays are most direct over the equator, so that is where the most energy is available to create the single atoms of oxygen (Equation 14.1) that form ozone. Ozone formed at the equator is distributed throughout the stratosphere by global air circulation. This circulation generates the greatest concentration of ozone at midlatitudes, especially in the Northern Hemisphere.

A REDUCTION IN STRATOSPHERIC OZONE

Like many environmental crises, the media term "ozone hole" is misleading because there is no ozone "hole." Rather, the problem is a reduction below its average of about 300 Dobson units. That is, the "pile" of ozone molecules is less than 3 millimeters thick.

There are two indicators that the stratospheric concentration of ozone is declining. One is a reduction in the amount of ozone in the entire atmosphere (Figure 14.4). Instruments aboard satellites have been measuring the amount of ozone in the stratosphere since 1979, and the information they collect

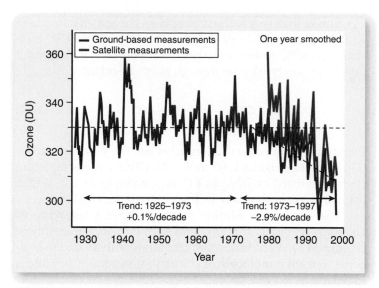

FIGURE 14.4 *A General Decline in Ozone* Satellite measurements over Switzerland indicate that the concentration of ozone in the stratosphere has declined steadily since 1979, at about 0.41 percent per year. *(Source: Data from NASA, Studying Earth's Environment from Space, June 2006. http://www.ccpo.odu.edu/SEES/index.html)*

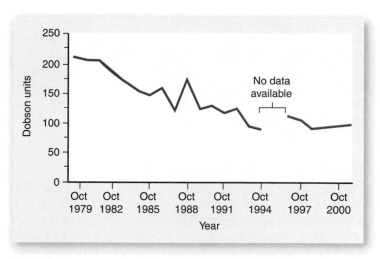

FIGURE 14.5 *An Antarctic October Decline* The concentration of ozone in the stratosphere during October has declined from about 210 Dobson units during the 1980s to less than 100 Dobson units during the first decade of the twenty-first century. *(Data from U.S. Environmental Protection Agency.)*

shows that the amount of ozone is declining steadily (despite annual fluctuations). Since 1979 the amount of ozone in the atmosphere has declined about 0.41 percent per year.

A second indicator points to a significant ozone reduction over the Antarctic every October. The reduction over the south pole is a relatively recent phenomenon that became more severe during the 1980s and early 1990s (Figure 14.5). Measurements of the ozone layer over the Antarctic show a minimum value of about 220 Dobson units until the mid-1960s. Since then October measurements over the Antarctic have frequently dropped below 100 Dobson units. Equally important, the area covered by this reduced concentration has grown from a few million square kilometers in the early 1980s to over 26 million km² in the early 2000s, which is roughly the size of North America (Figure 14.6).

The extent of ozone reductions over the Antarctic surprised everyone, including scientists who thought that human activities could lower concentrations. Reductions over the Antarctic were first reported in 1985 by a British scientist, Joseph Farman, who measured ozone concentrations over the south pole each year since the late 1950s. As concentrations declined during the 1970s and 1980s, Farman thought something was wrong with his equipment. To avoid embarrassment, he checked and rechecked his equipment and measured several large drops before going public with his results.

A group of NASA (National Aeronautics and Space Administration) scientists could not believe Farman's results. They too were using instruments to monitor the ozone layer, and their satellite measurements did not show such a reduction. But when they rechecked their data, they discovered that their satellite *was* sensing a big reduction in ozone. Measuring ozone from satellites requires precise measurements and complicated calculations. The slightest error can generate mislead-

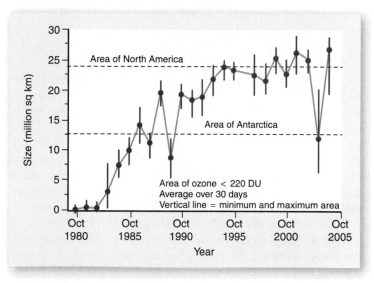

FIGURE 14.6 *An Expanded Area of Decline* The portion of the stratosphere where the October concentration of ozone drops below 220 Dobson units has increased from nearly zero in the early 1980s, to an area the size of Antarctica in the late 1980s, to an area the size of North America in the mid 1990s. *(Source: Data from U.S. Environmental Protection Agency.)*

ing results. To avoid confusion, scientists had programmed their computers to ignore any results that indicated ozone concentrations less than 180 Dobson units. Because NASA scientists had assumed that ozone concentrations could not drop below a certain level, they overlooked this vital information.

WHY IS STRATOSPHERIC OZONE DECLINING?

As evidence mounted to confirm ozone depletion, scientists began to debate its cause. Identifying the cause was more than intellectual curiosity. The answer would dictate the need for and type of policy. If human activity was responsible, potentially expensive policy would be needed. If the reductions were part of a natural process, such as changes in solar activity or atmospheric circulation, no action would be required.

With such high stakes, society had to agree on the process to identify the cause of the reduction. Many means were possible. The process by which society came to understand the causes for the reduction in stratospheric ozone illustrates how the scientific method can help solve environmental challenges.

The Halogen Depletion Hypothesis

Concern about ozone concentrations was sparked by a 1974 paper in the journal *Nature* in which Rowland and Molina hypothesized that concentrations could be reduced by a group of molecules known as chlorofluorocarbons (CFCs). CFCs are a group of chemicals in the halogen family, so this hypothesis became known as the halogen depletion hypothesis.

Chlorofluorocarbons are a relatively simple family of molecules (Table 14.1). Many CFCs consist of a single atom of carbon (C) that is bonded with one or more molecules of chlorine. For example, CFC-11 is one atom of carbon, one atom of fluorine, and three atoms of chlorine ($CFCl_3$).

Chemists designed CFCs to be chemically inert, meaning that they do not break down rapidly and do not react with other chemicals. These properties make CFCs ideal for use as refrigerants (in refrigerators and air conditioners), as propellants in aerosol spray cans, for cleaning electronic equipment, and as fire extinguishers.

From an environmental perspective, inert chemicals are desirable because they do not break down to forms that are toxic or dangerous. For example, the pesticide DDT becomes more dangerous when it breaks down into dieldrin. Because they are nonreactive, CFCs do not interfere with biological processes.

The nonreactive nature of CFCs attracted the attention of scientists. They wondered what happens to CFCs after they are released from spray cans or leaking refrigerators. Extensive investigations indicated that CFCs in the troposphere do not break down, nor do they react with other chemicals. Eventually

CFCs make their way to the stratosphere. It is here that Rowland and Molina identified a process that could destroy the CFC molecule. As a CFC molecule reaches the troposphere, the powerful rays of the sun break it up through photodissociation:

$$CFCl_3 + hv \longrightarrow Cl + CFCl_2 \qquad (14.4)$$

Eventually photodissociation frees all the chlorine atoms (Cl) from the original CFC molecule. Other ozone-depleting chemicals listed in Table 14.1 release bromine atoms, which act like chlorine atoms with respect to ozone depletion.

The release of chlorine atoms triggers a sequence of chemical reactions that depletes ozone in the stratosphere (Figure 14.7). A chlorine atom combines with an ozone molecule to form a molecule of chlorine monoxide (ClO) and a molecule of oxygen (O_2):

$$Cl + O_3 \longrightarrow ClO + O_2 \qquad (14.5)$$

In the next step chlorine monoxide (ClO) can react with a single oxygen atom to reform the chlorine atom (Cl) and molecular oxygen:

$$O + ClO \longrightarrow Cl + O_2 \qquad (14.6)$$

TABLE 14.1	Chemicals with the Potential to Deplete Ozone	
Name	Formula	Ozone Depletion Potential
CFC-11	$CFCl_3$	1.0
CFC-12	CF_2Cl_2	0.82
CFC-113	CF_3Cl_3	0.9
Halon-1301	CF_3Br	12
Halon-1211	CF_2ClBr	5.1
Methyl chloroform	CH_3CCl_3	0.12
HCFC-22	CF_2HCl	0.04

Source: Data from Staehelin, J. et al., Ozone trends: a review, Review of Geophysics 39: 231–290.

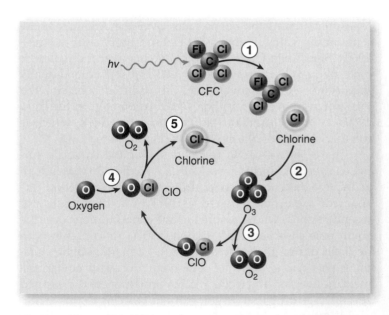

FIGURE 14.7 *Ozone Depletion Cycle* The cycle starts when a CFC molecule reaches the stratosphere. Here solar energy separates a chlorine atom from the CFC molecule (Equation 14.4). In the second step the chlorine atom combines with ozone. In the third step the ozone molecule splits into molecular oxygen and chlorine monoxide (Equation 14.5). In the fourth step chlorine monoxide combines with an atom of oxygen, which liberates the chlorine atom and molecular oxygen. For each turn of the cycle, one oxygen atom and one ozone molecule are converted to two molecules of molecular oxygen (Equation 14.7)

Because the chlorine atom is regenerated by Equation 14.6, Equations 14.5 and 14.6 constitute a cycle that destroys ozone. In that cycle a chlorine atom converts an ozone molecule to molecular oxygen; then the chlorine atom is freed from the chlorine monoxide, which enables it to destroy another molecule of ozone. The reaction also converts a single oxygen atom to molecular oxygen. This too reduces ozone concentrations because it reduces the number of single oxygen atoms that are available to form new ozone molecules. Every turn of the cycle reduces the number of ozone molecules as follows:

$$O + O_3 \longrightarrow O_2 + O_2 \qquad (14.7)$$

This cycle demonstrates that the number of ozone molecules destroyed by CFCs depends on the quantities of chlorine atoms that they contain and the layer of the atmosphere where these atoms are released. The ability of a chemical to destroy ozone is measured by its ozone depletion potential (Table 14.1). **Ozone depletion potential** measures the number of ozone molecules destroyed by a molecule relative to the number destroyed by a molecule of CFC-11 (just as the abilities of greenhouse gases to absorb long-wave radiation are measured relative to carbon dioxide).

Once the chemistry of the ozone depletion cycle was established, scientists turned to another important question: How fast does the cycle turn? If the cycle turns quickly, many ozone molecules could be converted to molecular oxygen. On the other hand, if the reactions proceed slowly, fewer ozone molecules would be converted.

A group of scientists put together a relatively simple model that was based on the rate at which reactions occur. Their model predicted that the stratospheric concentrations of ozone would decline by 5 to 7 percent by 1995 if the production of CFCs continued to grow. Although significant, the 5 to 7 percent loss is much smaller than the large reductions that were observed over the south pole. The model's inability to forecast the actual loss of ozone led some scientists to question the halogen depletion hypothesis. In response, scientists developed two alternative hypotheses to account for the ozone reduction over the south pole.

The Odd Nitrogen Hypothesis

The odd nitrogen hypothesis theorizes that changes in solar activity cause the reduction in stratospheric ozone over the south pole (Figure 14.8). According to this hypothesis, the reduction is a natural phenomenon that poses little threat to humans and requires no policy response.

The odd nitrogen hypothesis is based on natural cycles in solar activity. One of these cycles is related to the appearance of sunspots and solar flares. During these periods the sun emits more energy. Such increases in activity occur on a regular eleven-year cycle. The cyclic increase in solar activity increases the amount of photodissociation that occurs in the

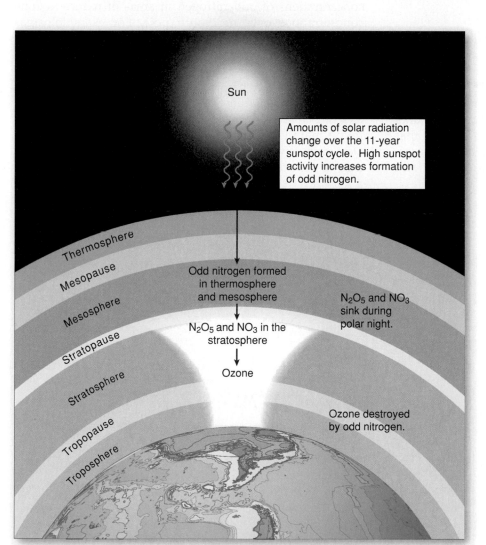

FIGURE 14.8   *Odd Nitrogen Hypothesis*
According to this hypothesis, natural increases in solar activity increase the energy available to generate odd forms of nitrogen (N_2O_5 and NO_3) in the mesosphere. The odd forms of nitrogen move down into the stratosphere and are trapped above the Antarctic by the polar vortex. Here they destroy ozone molecules in much the same way that chlorine atoms destroy ozone molecules. (Source: Redrawn from L.B. Callis and M. Natarajan, "The Antarctic Ozone Minimum—Relationship to Odd Nitrogen, Odd Chlorine, the Final Warming, and the 11-year Solar Cycle." Journal of Geophysical Research—Atmospheres 91 (d10): 771–796.)

upper layers of Earth's atmosphere. This photodissociation increases the amount of "odd nitrogen," a term that refers to forms of nitrogen other than molecular nitrogen. In particular, more intense solar energy rearranges molecular nitrogen to form molecules such as N_2O_5 and NO_3.

Much of this photodissociation occurs in the upper levels of the atmosphere: the thermosphere and the mesosphere. According to the odd nitrogen hypothesis, the N_2O_5 and NO_3 are transported down to the stratosphere, where they are trapped above the south pole by the polar vortex. The **polar vortex** is a wind that blows in a circular pattern around the south pole during the winter (see Figure 4.8 on page 61). This circular pattern isolates the air over the south pole from the rest of the atmosphere.

As winter draws to a close, the first appearance of the sun changes the N_2O_5 and NO_3 to NO and NO_2. The polar vortex traps these forms of nitrogen in the air above the south pole. This conversion and confinement are critical because they concentrate forms of nitrogen that destroy ozone molecules in much the same way that chlorine atoms destroy ozone.

The Dynamic Uplift Hypothesis

The dynamic uplift hypothesis postulates that the pattern of atmospheric circulation changed after 1979 and that these changes are responsible for the reduction in ozone concentrations over the south pole (Figure 14.9). The hypothesis starts with a simple observation. There are significant quantities of ozone in the stratosphere but only minor amounts of ozone in the troposphere. Some scientists suggest that upward movement (hence the name "dynamic uplift") carries air with little ozone from the troposphere to the strato-

sphere. This upward movement dilutes the concentration of ozone in the stratosphere and makes it appear as if stratospheric concentration of ozone is declining. Air displaced from the stratosphere sinks downward around the edges of the polar vortex into the troposphere.

Which Hypothesis Is Correct?

Because of the environmental and economic threats posed by a reduction in stratospheric ozone, the U.S. government and the chemical industry sponsored research expeditions to the south pole. There scientists performed experiments designed to test the competing hypotheses. Each hypothesis predicted specific events; scientists hoped that observing (or not observing) these events would enable them to validate one hypothesis and falsify the other two. Consistent with these hopes, the expeditions were a scientific success.

All three hypotheses were tested by flying a plane from the area outside the reduced ozone concentrations into the area of the reduced ozone concentrations, and then back out. During these flights scientists did not find increased concentrations of odd nitrogen in areas of reduced ozone concentrations over the south pole. For example, areas where ozone concentrations were low also have low concentrations of NO_2. Without any evidence of odd nitrogen, scientists were able to falsify the odd nitrogen hypothesis.

To test the dynamic uplift hypothesis, the plane's instruments measured the concentration of nitrous oxide (N_2O). Scientists focused on nitrous oxide because it is found mainly in the troposphere, not in the stratosphere. Data from several flights indicated that the air inside the area of reduced ozone concentrations had very low con-

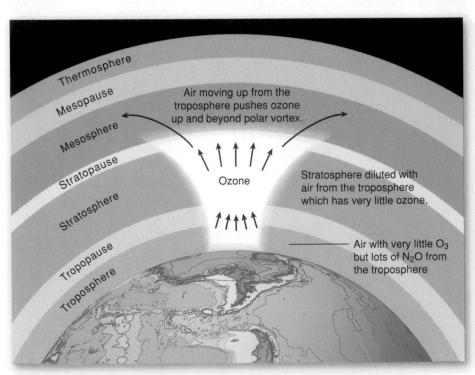

FIGURE 14.9 *Dynamic Uplift Hypothesis*
According to this hypothesis, ozone-poor air rises from the troposphere into the stratosphere. As it rises, it pushes out ozone-rich air. On net, this reduces the concentration of ozone in the stratosphere.

centrations of nitrous oxide. These low concentrations implied that the ozone-poor air could not have originated in the troposphere. Further analysis showed that the air over the south pole was sinking rather than rising, which contradicted the movements predicted by the dynamic uplift hypothesis. Finally, the stratospheric concentration of nitrous oxide was the same before and after the annual reduction in ozone. This indicated that there was no change in the pattern of air flows. Together these results led to agreement among scientists that the dynamic uplift theory could not account for the reduction in stratospheric ozone concentrations above the south pole.

The halogen depletion hypothesis was relatively simple to test. If the halogen depletion hypothesis was correct, areas where stratospheric ozone concentrations were lowest would have high concentrations of chlorine atoms (Cl) or chlorine monoxide (ClO). If either of these were absent from the air associated with reduced ozone concentrations, the halogen depletion theory would be falsified.

During the flights in and out of the area of reduced ozone concentrations, scientists measured the concentration of chlorine monoxide. The relationship between the concentration of ozone and chlorine monoxide was striking (Figure 14.10). In areas where ozone concentrations were reduced significantly, from about 68°S to the pole (90°S), low levels of ozone were associated with high levels of chlorine monoxide. Low levels of chlorine monoxide were associated with normal levels of ozone in areas north of 68°S. Clearly these results (and others) are consistent with the hypothesis that chlorine (and bromine atoms) from CFCs and other chemicals with the potential to deplete ozone are responsible for the declining concentration of stratospheric ozone over the south pole.

The Complete Explanation for the Reduction in Stratospheric Ozone

Experiments also showed that the south pole's ozone reduction during early spring is caused by a sequence of events that involves human-made chemicals and seasonal fluctuations in winds, temperature, and solar radiation. The polar vortex that forms over the south pole during the winter traps air at very low temperatures (−80°C to −70°C). Cold temperatures spur the formation of **polar stratospheric clouds,** which consist of very small droplets of water and nitric acid. These clouds can cover the entire Antarctic; their bright colors lead to the nickname "mother of pearl" clouds. The formation of polar stratospheric clouds is crucial to the reduction in ozone because they accelerate reactions that liberate chlorine from molecules that normally remove chlorine atoms from the atmosphere. Freeing these chlorine atoms allows them to react with and destroy ozone.

These reactions also explain the seasonal nature of the reduction. As the polar vortex forms during winter, polar stratospheric clouds liberate chlorine atoms (originally from CFCs) into the atmosphere. The concentration of chlorine atoms rises because the polar vortex traps them over the Antarctic. The concentration of ozone remains near normal levels during the winter because there is no solar energy to drive the cycle of chemical reactions (remember that there is little or no sunlight during the polar winter). As spring comes, the increase in solar energy powers the reactions that destroy ozone. Because the polar vortex remains intact until late spring, the concentration of ozone in the air trapped by the vortex becomes progressively lower as the cycle of chemical reactions reoccurs.

Eventually rising temperatures break up the polar vortex, which allows air with very low concentrations of ozone to mix with the rest of the atmosphere in the Southern Hemisphere. The initial release of low-ozone air causes stratospheric ozone concentrations to decline in regions near the south pole. For example, ozone concentrations over parts of Australia sometimes are reduced by about 20 percent. After several months the ozone-poor air is mixed throughout the Southern Hemisphere. During this time the reactions described in Equations 14.1–14.3 return ozone to its normal concentration. Currently the reduction in ozone concentrations over the south pole does not carry over from one year to the next.

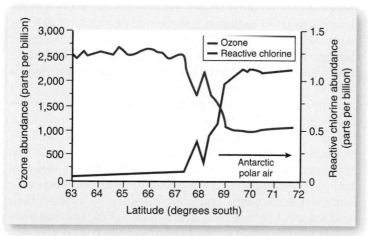

FIGURE 14.10   *Validating the Halogen Depletion Hypothesis*
North of 68°S, the stratosphere has normal levels of ozone and low levels of chlorine monoxide. South of 68°S, low concentrations of ozone in the stratosphere are associated with high levels of chlorine. This negative correlation is consistent with the notion that chlorine atoms from CFC molecules are responsible for the reduction in stratospheric ozone. (Source: Data from J.G. Anderson et al., "Ozone Destruction by Chlorine Radicals within the Antarctic Vortex: The Spatial and Temporal Evolution of C10-O3 Anticorrelation Based on in situ ER-2 Data," Journal of Geophysical Research 94 (11): 465–475)

What about a Reduction over the North Pole?

The reduction in ozone over the south pole is severe, but the impact on humanity is lessened by the fact that relatively few people live in the extreme Southern Hemisphere. Luckily for people living in the Northern Hemisphere, the reduction in ozone over the north pole has been small relative to the south pole. The largest reductions in the Northern Hemisphere are between 13 and 20 percent, compared to 40 to 60 percent over the Southern Hemisphere.

The relatively small decline over the north pole is caused by several differences relative to the south pole. Warm temperatures over the north pole inhibit the formation of polar stratospheric clouds. With fewer polar stratospheric clouds, the reactions that lead to the destruction of the ozone layer occur less rapidly. The pattern of mountains, land, and oceans weakens the polar vortex over the north pole. As a result, the polar vortex in the Northern Hemisphere tends to break up earlier in the spring, which means that the air over the Arctic is mixed before significant ozone reduction can occur.

THE EFFECTS OF LESS STRATOSPHERIC OZONE

Ozone absorbs solar energy that is dangerous to living organisms (Figure 14.11). Solar radiation that reaches the outer layer of the atmosphere can be grouped into categories according to its strength or intensity. Solar radiation with the shortest wavelength, ultraviolet radiation A (UV-A), is the most dangerous because it has the highest energy. Radiation in this category is absorbed strongly by the atmosphere so that little reaches the surface regardless of the atmospheric concentration of ozone. Another category of radiation with longer wavelengths, which is absorbed and used by plants in photosynthesis, is known

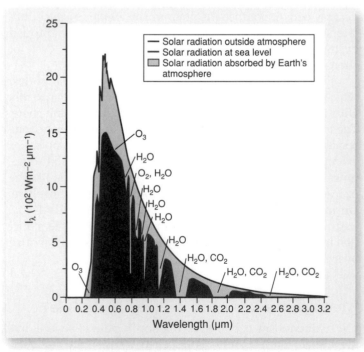

FIGURE 14.11   *Ozone Absorption*   *The wavelengths of solar energy that reach Earth's surface (dark blue area) are different than the solar energy that reaches the outer layer of the atmosphere (red line) because gases in the atmosphere, such as ozone and water, absorb solar energy (light blue area). Ozone absorbs solar energy with short wavelengths, including ultraviolet radiation B (UV-B).*

as photosynthetically active radiation. A significant fraction of photosynthetically active radiation reaches Earth's surface—a condition that enables life to exist.

Solar radiation with wavelengths between these two categories is ultraviolet radiation B (UV-B). The amount of UV-B that reaches Earth's surface depends in part on the concentration of ozone in the stratosphere. As the concentration of ozone has declined, the amount of biologically significant UV-B radiation that reaches Earth's surface has increased. Since 1980 the amount of UV-B that reaches the surface has increased 4–7 percent in the middle latitudes of the Northern Hemisphere, increased 6 percent at the middle latitudes of the Southern Hemisphere, and increased 130 and 22 percent in the Antarctic and Arctic springtimes, respectively.

These increases merit attention because they pose a threat to living organisms. UV-B's relatively short wavelength allows it to penetrate deep into living tissue, where it can cause considerable damage by breaking apart molecules. Exposure to high levels of UV-B can kill. Lesser amounts can cause a variety of damages. For example, increased levels of UV-B radiation impair the function of enzymes that are part of the photosynthetic process in several tree species. Another problem caused by UV-B radiation is **mutations,** which are breaks and rearrangements of DNA molecules. Mutations are especially dangerous for living organisms because DNA contains the information that is needed to make the components of the body, to turn cell growth on and off, and to reproduce.

DNA's importance has favored the evolution of mechanisms that protect individuals against UV-B radiation and that repair the damage done by UV-B radiation. To protect themselves, organisms use some of their energy and materials to synthesize compounds that absorb UV-B radiation. Many animals have some form of hair, fur, shell, or skin pigmentation that protects their inner tissues and organs against UV-B radiation. When UV-B radiation gets past this first line of defense, many organisms have mechanisms that identify and repair damage. For example, most living organisms have DNA repair mechanisms. When an error is found, the incorrect sequence is "cut out" and the correct sequence is "put back in."

The evolution of these protection and repair mechanisms is driven in part by the amount of UV-B that an individual normally is exposed to. Before the springtime ozone declines, exposure to UV-B declined with latitude. This implies that organisms at the poles tend to have the fewest defenses against UV-B radiation. For example, increased exposure stimulates both plants and phytoplankton to increase their production of compounds that absorb UV-B. The effectiveness of this responses varies. In plants, protective compounds reduce the effects of UV-B radiation on the photosynthetic machinery and DNA. The strategy is less effective in phytoplankton, partly because of the short distance between the outer boundary of the cell and its photosynthetic machinery compared to the larger distance between the outer layers of the leaf and the inner layers where photosynthesis occurs.

Impact on Marine Ecosystems along Antarctica

The vulnerability of phytoplankton in general and the timing and location of the greatest increases in ground-level UV-B radiation make marine ecosystems along Antarctica especially vulnerable. Blooms occur at the edge of the Antarctic ice pack, where melting ice forms a shallow layer of relatively freshwater. Because freshwater is less dense than salt water, it floats above the salt water. This area, known as the **marginal ice zone,** is ideal for phytoplankton because there are plenty of light and nutrients. But this habitat makes phytoplankton vulnerable to UV-B. Because the marginal ice zone is shallow, the phytoplankton cannot escape the UV-B radiation by moving to deeper water.

The increase in UV-B may decrease net primary production, alter food webs, and slow biogeochemical cycles. Laboratory experiments demonstrate that greater UV-B radiation could reduce net primary production by about 25 percent if phytoplankton have no defense mechanisms. But most phytoplankton do have some mechanisms to cope with higher levels of UV-B. Studies of phytoplankton in the ocean show that the UV-B increase may reduce net primary production by up to 10 percent.

As described in Chapter 5, reductions in net primary production are amplified in higher trophic positions. This can be illustrated by the possible effect of lower net primary production on penguins. Penguin parents take turns caring for their young. While one returns to sea to forage for food (penguins eat fish), the other remains with the chicks on land. A reduction in net primary production lengthens the time that the parent spends in the water catching the fish that are needed by both the parent and the chicks. Because the parent with the chicks can wait only so long for the other parent to return, the greater time needed to capture enough fish increases the probability that the chicks will be left alone. Left alone, penguin chicks are much more likely to die. Thus a reduction in net primary production may reduce the penguin population.

Higher UV-B levels also may change the composition of phytoplankton communities. Such changes are important because not all species of phytoplankton are edible by zooplankton. That is, zooplankton can eat only certain types of phytoplankton. If the composition of phytoplankton changes in favor of less edible species, the amount of energy that flows through the food chain could be reduced by an amount that is greater than the reduction in net primary production. This could change the entire food chain.

Changes in the Antarctic marine ecosystem could ripple across the entire planet. Ecologists estimate that net primary production in Antarctic seas accounts for 10–20 percent of the total generated by the world's oceans. This production supports marine and terrestrial ecosystems well beyond the south pole. Many birds (Figure 5.7 on page 85) and marine mammals migrate to Antarctica and rear their young there to take advantage of the abundant spring food supplies. Should net primary production diminish, the energy return on

investment for these migrations would be reduced, leaving less surplus energy to produce and raise offspring.

The increase in UV-B also could alter biogeochemical cycles. Bacteria, an important component of the nitrogen and sulfur cycles (Chapter 6), generally are vulnerable to UV-B radiation; so reductions in stratospheric ozone could reduce microbial activity such as decomposition. This would lengthen the time that nutrients remain tied up in dead organic matter, slowing several biogeochemical cycles.

The Impact on Terrestrial Organisms (Including People)

Terrestrial ecosystems on Antarctica are very sparse, so the reduction in stratospheric ozone should have a relatively small effect on plants and animals. On the other hand, the springtime breakup of the polar vortex and the subsequent spread of ozone-poor air over the Southern Hemisphere could have a significant effect on terrestrial ecosystems, including people.

Similar to its effect on marine ecosystems, a reduction in stratospheric ozone could reduce net primary production in terrestrial ecosystems. Leaves of conifer trees generally are less vulnerable to UV-B than the leaves of deciduous trees and herbaceous plants. More UV-B radiation slows leaf elongation rates, reduces leaf area, and reduces aboveground biomass. Nonetheless, these reductions are expected to be small relative to aquatic ecosystems because protective mechanisms by plants are more effective.

Photosynthesizers are not the only vulnerable organisms. Many animals lay eggs and pass through juvenile stages that are vulnerable to UV-B radiation. Some herpetologists (scientists who study amphibians and reptiles) think that some of the worldwide reduction in frog and toad populations may be caused in part by a reduction in stratospheric ozone (Figure 14.12). A study in Oregon indicated that elevated UV-B levels

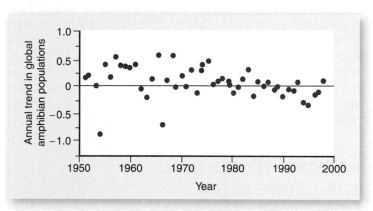

FIGURE 14.12   *Global Decline in Amphibian Populations*
The annual change in global amphibian populations. Before 1990 the values generally are positive, which indicates that populations were increasing. After 1990 negative values indicate that populations are shrinking. (Source: Data from R.A. Alford et al., "Ecology: Global Amphibian Population Declines," Nature 412: 499–500.)

reduced the fraction of eggs that hatched. Species that had the lowest success also had the least ability to repair mutations.

Nor does this effect have to be so direct. Effects of elevated levels of UV-B radiation may be exacerbated by climate change. Embryos of some amphibian species are vulnerable to fatal fungal infections following exposure to UV-B radiation. Exposure is determined by both ozone concentrations and water depth (water absorbs UV-B radiation, so eggs laid in deeper waters are more protected). Water levels in Pacific Northwest ponds are relatively low during El Niño events (the change in atmospheric circulation steers storms away from the Pacific Northwest toward southern California), and this increases the mortality rates for amphibian embryos. If climate change increases the frequency of El Niño events or warms and dries the Pacific Northwest, these changes would enhance the exposure to UV-B radiation that is associated with a reduction in stratospheric ozone.

Finally, UV-B radiation can have direct health effects on terrestrial animals, including people. Research shows that higher UV-B levels damage internal organs, suppress the immune system, and increase the formation of cataracts and lead to blindness. The most visible concern for human health focuses on skin cancer. The U.S. Environmental Protection Agency estimates that every 1 percent reduction in ozone concentrations increases basal and squamous cell skin cancers (the least dangerous kinds) by 2–3 percent and increases malignant melanoma skin cancers (the most dangerous kind) by 1–2 percent. Trends in the rates of skin cancer seem to justify this concern. Melanoma is the fastest-increasing cancer in the U.S. population of Caucasians.

But is this increase caused by the reduction in stratospheric ozone? Changing fashions also could be responsible. During the last several decades exposure to UV-B radiation has increased as people pursue the perfect tan. Tanned skin is viewed as a status symbol that implies a person can travel to warmer climates during the winter and has sufficient leisure time to rest in the sun. This attitude contrasts to earlier times, when pale skin was the ideal. Pale skin used to indicate that you didn't have to work outdoors in the sun (such as on a farm). Because of this change in fashion, scientists can't say if the rise in skin cancer is due to ozone depletion or people deliberately spending more time in the sun. Undoubtedly both are key. See *Your Ecological Footprint: Managing Your Fun in the Sun on an Ozone-Depleted Planet.*

POLICIES TO RESTORE THE OZONE LAYER

The political response to the reduction in stratospheric ozone is a rare environmental success story. Although ozone concentrations will continue to decline for many years (see *Case Study: The Link between Climate Change and the Reduction in Stratospheric Ozone*), the global community has taken actions that will cause the springtime reduction over the Antarctic to disappear some-

time in the second half of the twenty-first century. This success is based on institutional arrangements that highlighted scientific research, fostered international policy, and encouraged cooperation between government and industry.

Policy Deadlock

CFCs were invented in 1928 by the chemist Thomas Midgely Jr. at Dupont, a major multinational chemical company. CFCs became a critical part of refrigerators because they replaced ammonia and SO_2, which made home refrigerators safer and more affordable. This success led to other uses, such as in air conditioners, spray cans, fire extinguishers, cleaners for electrical equipment, and foams. By the time that Rowland and Molina published their hypothesis in 1974, producing CFCs generated $8 billion in revenues and employed 200,000 people.

The coupling of these facts set up a heated debate. If the halogen depletion hypothesis was correct, CFCs posed a severe threat and had to be banned. But if the halogen depletion hypothesis was false, banning CFCs would impair the global economy and the modern way of life. To avoid such a ban, the chemical industry claimed that CFCs were innocent until proven guilty. Should they be proven guilty, the industry promised to stop their production. In 1975 Dupont promised to halt production of CFCs if "credible scientific evidence" showed that CFC harmed the ozone layer. On the other hand, scientists argued that it might be too dangerous to wait until CFCs were proven guilty. By that time chlorine and bromine would accumulate in the stratosphere and cause long-term reductions in ozone.

The economic and environmental threats created a deadlock. No nation was willing to eliminate CFC production. Instead policy makers banned the least important uses of CFCs, a process made easier by consumer advocacy groups. Activists urged consumers to stop buying spray cans, arguing that they were expensive and damaged the environment. In general, consumers and industry responded. After increasing steadily for several decades, the production of CFCs for use as aerosol propellants declined dramatically in the mid-1970s. Part of this success was based on a new type of valve that allowed manufacturers to use water and butane gas instead of CFCs (see "substitution" in Figure 9.19 on page 153). In May 1977 the U.S. government formalized this reduction by announcing a ban on nonessential CFC uses that would take effect in 1978. But the federal government delayed action on "essential" uses of CFCs—uses for which no substitutes were readily available.

The lack of scientific certainty about the cause of the stratospheric ozone reduction and the lack of available substitutes led to a period of deadlock in international efforts to phase out CFCs. Without international agreements, national bans could not prevent a severe reduction in ozone concentrations. The rise in living standards in developing nations, combined with high use in industrial nations, increased demand for CFCs for uses other than aerosols. As a result, the aerosol ban had little long-term effect on the total global production of CFCs (Figure 14.13).

Breaking the Deadlock

Ongoing negotiations about CFCs took on an added sense of urgency starting in 1986, when NASA and the World Meteorological Organization published a series of "blue books" showing that CFC production would grow and that this increase would reduce stratospheric ozone. Although these findings were not original, policy makers were impressed by publications authored by an international panel of distinguished scientists. The panel included nearly all scientists most knowledgeable about the topic—so no scientists outside the panel had sufficient credibility to challenge the findings. Equally important, the international makeup of the panel prevented individual nations from arguing that the science was driven by a single nation's political agenda. Finally, the findings satisfied the condition of the promise made by industry to stop producing CFCs when science could establish the link between CFCs and ozone depletion.

Industry lived up to its promise, and this shift allowed national governments to act. Action culminated in the 1987 Montreal Protocol. Among its provisions, the Montreal Protocol called for a 50 percent reduction in the production of CFCs by 1997–1998. Interestingly, the 50 percent target was not chosen based on arguments about economic efficiency. The negotiators of the Montreal Protocol and the subsequent

revisions rarely considered the economic costs associated with a reduction in stratospheric ozone. Furthermore, lowering CFC production would do little to slow the increase in the stratospheric concentration of chlorine or bromine (Figure 14.14). At best it was an important first step that would postpone the buildup by about a decade.

Instead the 50 percent reduction was a political compromise that turned out to be ideal. The targets seemed achievable because industry had nearly a decade to develop alternatives. The reduction was great enough to convince industry that it had to develop alternatives—business as usual would not allow them to meet the targets. In addition, the protocol was flexible. Goals could be changed as scientific knowledge accumulated. Industry realized that further knowledge probably would boost calls to eliminate CFCs, so it anticipated further tightening of the requirements.

Institutional Determinants of Success

The Montreal Protocol initiated and strengthened a series of institutions that were needed to verify compliance and modify the agreement. The protocol required updated scientific knowledge about the effects of CFCs on ozone and industry's ability to reduce its use of and develop alternatives to CFCs. The need for current information spawned scientific and technical advisory committees, which took on a life of their own. Meetings were held frequently, often under the

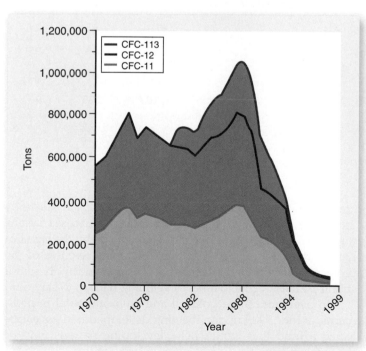

FIGURE 14.13 **CFC Production** Despite a U.S. ban on the use of CFCs as an aerosol propellant in the 1970s, total production continued to rise because other uses increased rapidly. After the Montreal Protocol and its amendments, production declined sharply.

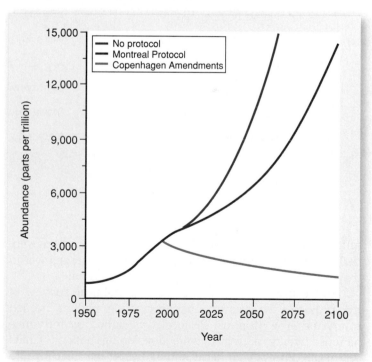

FIGURE 14.14 **Future Levels of Chlorine/Bromine** Although hailed as a success, the original Montreal Protocol would have barely slowed future concentrations of chlorine and bromine. Future increases have been avoided by subsequent revisions, such as the Copenhagen Amendments in 1992, which will reduce concentrations of chlorine and bromine.

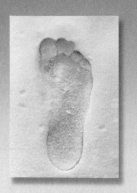

Your ECOLOGICAL *footprint*

Managing Your Fun in the Sun on an Ozone-Depleted Planet

After reading this chapter you may be alarmed by reductions in stratospheric ozone and increased skin cancer rates. In response you may want to manage your time in the sun. Before going out, you can check the local weather forecast for the UV index. But what does this number mean? How can you use it to reduce the negative health effects that are associated with too much sun exposure?

The UV index measures the intensity of ultraviolet radiation at Earth's surface that can affect human health. The UV index varies between 1 and 11+. Values represent the maximum exposure during a thirty-minute period. For sunny days, a single number is reported. But users should remember that the UV index varies over the day. Exposure is greatest a couple of hours before and after noon. For the first and last hours of daylight, UV exposure can be low. For example, the UV index may be extreme (11+) at noon and low (1–2) at 9:00 A.M. and 5:00 P.M. A range of values is reported for cloudy days.

Values are grouped to represent the degree of exposure. Values below 2 indicate that the sun's rays pose relatively little danger. At the other extreme, the value 11+ signifies extreme exposure. The UV index simply represents the degree of exposure. The index designers specifically caution against using the index as burn times. They worry that you will interpret index values as the time that you can spend safely in the sun. They warn that extending exposure time is unacceptable—damage is determined by cumulative exposure and skin type. We list burn times to help identify your skin type (see Table 1).

What damages are you trying to avoid? UV radiation affects your skin, eyes, and immune system. The immune system protects you against invading microorganisms (infections) and incipient tumors. Both of these capabilities are reduced by prolonged exposure. Prolonged exposure also can reduce the effectiveness of vaccinations. Many vaccinations inject a weakened form of a potentially invasive microorganism. These agents allow the immune system to recognize the invader and produce antibodies—responses that are reduced by exposure to UV radiation.

TABLE 1 Skin Types and Sun Exposure

Skin Type	Skin's Reaction to One Hour of Summer Midday Sun	Characteristics	Time to Redden (Minutes)		
			UV 6	UV 8	UV 10
I	Always burn; no tan	Pale skin, blue/green/hazel eyes, blond/red hair	28	21	17
II	Usually burn; minimal tan	Fair skin, blue/green/hazel eyes, blond/brown hair	33	25	20
III	Mild burn; moderate tan	Average Caucasian skin	44	33	27
IV	No burn; good tan	Light brown skin	58	44	35
V	Never burn; usually tan	Brown skin	89	67	53
VI	Never burn; always tan	Black skin	167	125	99

official banner of the United Nations. These committees were dedicated to the implementation of the protocol, and it was in their interest to make sure the protocol did not fail.

To ensure the environmental and political success of the protocol, the number of participating nations needed to expand. The Montreal Protocol was signed by forty-nine nations, twenty of which were developing nations. China and India did not sign because they feared the treaty would stifle efforts to increase their standard of living. The absence of China and India was a major impediment. (Does this issue sound familiar? The Kyoto Protocol does not require China and India to reduce their emissions of greenhouse gases; see *Policy in Action: Why Was the Solution to the Reduction in*

Stratospheric Ozone Simple Relative to Global Climate Change?) Increasing production and consumption by these nations threatened to overwhelm the reductions made by the signatories, thereby reducing the environmental effectiveness of the protocol. Furthermore, the absence of China and India sent a strong political signal to other developing nations to stay out.

To overcome these political and environmental impediments, at the 1990 London meeting developed nations established a fund that would help developing nations pay for new ozone-friendly technologies. These efforts convinced China and India to join the protocol. Their cooperation enhanced the global standing of the protocol so that the number of participating nations rose to sixty-six. The London

Elevated exposure to UV radiation also can damage your eyes. Your eyes reduce exposure to bright light by squinting and narrowing your pupils. These reactions are less effective when UV radiation is elevated. Allowing more UV radiation to enter your eyes can cause sunburnlike effects such as photoconjunctivitis. Although painful, these conditions are reversible.

Irreversible damage comes in the form of cataracts. Increased levels of UV radiation accelerate the rate at which proteins in your eye unravel. Once they unravel, the proteins tangle and accumulate pigments that cloud the lens. Eventually these clouds block light and cause blindness. Cataracts are responsible for blindness in 12–15 million people; about 20 percent of these cases are due to elevated exposure to UV radiation.

Many people extend their exposure to get a suntan. A tan is an increase in the production of protective pigments that impart "a healthy glow." Many try to get a tan without burning. Your ability to do so depends on your skin type (Table 1). Very light-skinned people, known as melano-compromised (skin types I and II), always burn and rarely tan in the sun. Slightly darker skin, known as melano-competent (skin types III and IV), occasionally burns and usually tans in the sun. Naturally brown or black skins, known as melano-protected (skin types V and VI), never burn and usually tan. You should use Table 1 to identify your skin type.

Contrary to popular opinion, a tan does not protect you or your skin from exposure. A dark tan on white skin is equivalent to a skin protection factor of about 4. "Skin protection factor" is a term that appears on suntan lotions and clothing and refers to the reduction in exposure. Specifically, skin protection factor refers to the increased time to reddening associated with that product. This definition is vague but can be clarified with a simple example. Suppose you have skin type II; it takes you about twenty minutes to redden when the UV index is 10. Applying a suntan lotion with a skin protection factor of 4 would increase the time until you redden fourfold to eighty minutes. Clearly a tan would not help much if you were spending the day at the beach.

Long-term exposure can increase skin's thickness, reduce its elasticity, increase wrinkling, and generate freckles. The greatest health concerns focus on UV exposure and skin cancer. Your vulnerability to skin cancer depends on your skin type. People with skin types I–IV suffer over 90 percent of nonmelanoma skin cancers. Nonmelanoma skin cancers are rarely fatal, but removing them can be painful and leave you permanently disfigured. Even more dangerous are malignant melanoma skin cancers, which can be fatal. The incidence of these cancers has increased by about 4 percent annually since the 1970s. These cancers are more frequent among skin types I–II.

Consult your local newspaper for today's UV index. If it does not appear in your local newspaper, you can get it from the NOAA Web site listed below. Use the information in Table 1 to calculate how long it would take you to redden. Calculate the skin protection factor that you would need to prevent reddening if you were exposed to the sun all day. This is just an exercise—you should not try to maximize your time in the sun. Even if you do not redden on a single day, the risk of skin cancer is determined by cumulative exposure.

ADDITIONAL READING

World Health Organization. *Global Solar UV Index: A Practical Guide.* Geneva, Switzerland, 2002.

NOAA Web site for current UV indexes: www.cpc.ncep.noaa.gov/products/ stratosphere/uv_index/uv_current_map.shtml.

STUDENT LEARNING OUTCOME

- Students will be able to explain their vulnerability to damage from the sun's rays and how they can evaluate the potential for damage on any given day.

meeting also expanded the number of chemicals regulated and accelerated the phase-out of CFCs—production would be banned by 2000. Expanding the chemicals regulated and accelerating the timetable for stopping their production continued at subsequent meetings in Copenhagen (1992), Vienna (1995), Montreal (1997), and Beijing (1999). As a result the concentration of some CFCs in the troposphere is declining (Figure 14.15).

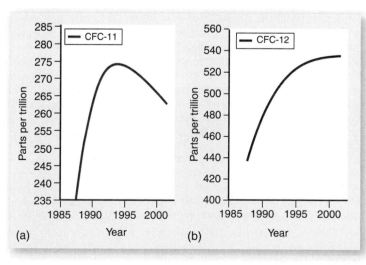

FIGURE 14.15 *Declining Concentrations* (a) After increasing steadily, the atmospheric concentration of CFC-11 started to decline in 1995. (b) On the other hand, the atmospheric concentration of CFC-12 appears to be stabilizing.

CASE STUDY

The Link between Climate Change and the Reduction of Stratospheric Ozone

Climate change and the reduction of stratospheric ozone often are presented as separate environmental challenges. In this book we discuss them in separate chapters. But contrary to this separation, the atmosphere is a dynamic system that includes many feedback loops. These loops imply that human-induced changes in climate and ozone cannot be separated. Anthropogenic climate change will postpone the date at which the reduction in stratospheric ozone is reversed; and the reduction in ozone may damp the ongoing increase in global temperature (Figure 1).

The stratospheric concentrations of chlorine and bromine probably peaked in 1997, but stratospheric ozone is expected to decline for many years. This ongoing reduction is due in part to global climate change. As described in Chapter 13, climate models forecast (with a high degree of certainty) that climate change will cool the stratosphere. A cooler stratosphere enhances the formation of polar stratospheric clouds. These clouds speed the chemical reactions that destroy ozone—so a cooler stratosphere will increase the number of ozone molecules that each atom of chlorine or bromine can destroy.

The formation of polar stratospheric clouds will be enhanced by another aspect of climate change: the greater concentration of water vapor in the stratosphere. Most water vapor in the stratosphere comes from the chemical destruction of methane. As described in Chapter 13, methane is an important greenhouse gas whose concentration has nearly doubled. This doubling increases the amount of methane that can be converted to water vapor. And more water vapor will increase the formation of polar stratospheric clouds, which will accelerate the destruction of ozone.

Climate change also may alter the patterns of atmospheric circulation in a way that enhances the destruction of ozone. Models indicate that climate change will increase the stability and persistence of the polar vortex. If the polar vortex continues later into the spring, the loss of ozone over the poles would accumulate longer. Allowing ozone-poor air to accumulate longer would lower the minimum ozone value and elongate the life span of the springtime reduction.

The link between climate change and the reduction in stratospheric ozone is a two-way street, so lower ozone concentrations will affect climate change. As indicated by Table 13.1 on page 271, CFCs are potent greenhouse gases.

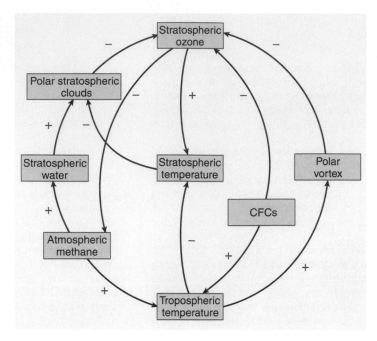

FIGURE 1 *Ozone Reductions and Climate Change Changes in radiative forcing will lead to changes in tropospheric temperature that will enhance the reduction in stratospheric ozone. The reduction in stratospheric ozone will affect tropospheric temperature by reducing radiative forcing and changing the residence time of methane.*

But the ability of CFCs and other ozone-destroying chemicals to absorb long-wave radiation may overstate their effect on climate. CFCs and other halogens destroy ozone. Because ozone also is a potent greenhouse gas, its loss offsets some of the increase in radiative forcing associated with CFCs and other ozone-destroying chemicals. After accounting for the loss of ozone, CFCs have a positive effect on radiative forcing while other ozone-destroying chemicals, such as halons and methyl bromide, have a negative effect on radiative forcing. That is, the reduction in radiative forcing that is associated with the loss of ozone is greater than these chemicals' direct contribution to radiative forcing.

The reduction in stratospheric ozone also may slow warming by shortening the residence times of greenhouse gases. Less ozone increases the amount of ultraviolet radiation that passes through the atmosphere. This radiation can increase the concentration of the hydroxyl radical (OH). Hydroxyl radicals are a small but important component of the atmosphere because they destroy many gases, including methane. An increase in the concentration of hydroxyl radicals could reduce the residence time of methane and thereby shorten its effect on radiative forcing.

The feedback effects of less stratospheric ozone on climate change are small relative to human emissions of greenhouse gases. Conversely, the effects of climate change on the mechanisms responsible for the reduction of stratospheric ozone will determine the date at which the areal extent and degree of reduction start to decline. For the next couple of decades the feedback effects of climate change that enhance the reduction will dominate the policy-driven decline in the atmospheric concentration of chlorine and bromine atoms. So despite the success of the Montreal Protocol and its subsequent amendments, the concentration of ozone in the stratosphere probably will decline for several more decades.

ADDITIONAL READING

Staehelin, J., N.R.P. Harris, C. Appenzeller, and J. Eberhard. "Ozone Trends: A Review." *Reviews of Geophysics* 39, no. 2 (2001): 231–290.

STUDENT LEARNING OUTCOME

• Students will be able to explain how global climate change affects the depletion of stratospheric ozone and how the depletion of stratospheric ozone affects global climate change.

Why Was the Solution to the Reduction in Stratospheric Ozone Simple Relative to Global Climate Change?

In hindsight, the policy solution to the reduction in stratospheric ozone was relatively easy. In little more than a decade the international community agreed to a treaty and strengthened it so that the production of chemicals that destroy ozone was essentially eliminated. Conversely, little progress has been made to slow the emissions of gases that cause global climate change. Why was the solution to ozone reduction easy relative to global climate change? The answers are found in the economic roles of the pollutants, the role of economic theory in policy formulation, the degree of scientific uncertainty, and the stance of the U.S. government.

There are important differences in the economic roles of CFCs and the fossil fuels that emit greenhouse gases. Despite the economic significance of the CFC industry, their production was a relatively small component of total economic activity in both the U.S. and foreign economies. Once produced, CFCs were used by only a few sectors. Together these characteristics lowered the economic costs of banning the production of CFCs and eliminating their use.

The limited role of CFCs contrasts sharply with the overarching importance of fossil fuels that is described in Chapters 10 and 11. In the United States and many other nations, the production of fossil fuels accounts for a significant fraction of GDP. More important, fossil fuels are used by nearly every sector of the economy. Under these conditions, efforts to reduce or eliminate the use of fossil fuels can be expensive.

The high costs of some methods for reducing greenhouse gas emissions may have altered the criteria used to develop climate change policy relative to the criteria used to develop ozone policy. The most important change concerns the role of economic efficiency. The targets set by the Montreal Protocol were not determined by the economic costs imposed by reduced ozone concentrations. Nor did negotiators evaluate the economic costs of potential alternatives. Instead political compromise was used to develop the targets, whose feasibility was unclear.

Conversely, arbitrary targets for the emission of greenhouse gases, such as those set by the Kyoto Protocol, are derided by economists. Debate about climate change policy is guided by concerns about economic efficiency. Decision makers have tried to develop policy that equalizes the marginal damages associated with climate change with the marginal costs of reducing emissions, but both aspects are surrounded by considerable uncertainty. Uncertainty has been used to paralyze international efforts that would set targets for a climate change treaty comparable to the Montreal Protocol.

Another difference concerns the perceived degree of scientific certainty. Based on statements by some politicians, you might think that there is a great deal of uncertainty about the effect of greenhouse gases on global surface temperature. But the scientific community has no such doubt. The certainty expressed by the IPCC regarding the effect of greenhouse gases on temperature is comparable to the certainty expressed in the blue books published by the NASA and WMO assessments regarding the effect of CFCs on stratospheric ozone. Nonetheless, the public remains skeptical.

One reason for such skepticism may be the nature of the scientific evidence. The effect of CFCs on ozone was demonstrated by simple experiments that indicated an unambiguous negative relationship between chlorine monoxide and ozone depletion (see Figure 14.10). No similarly plain evidence is available to demonstrate the effect of greenhouse gases on temperature. Instead scientific evidence for the effect of greenhouse gases on surface temperature is provided by simulation models and statistical analyses.

Together these factors have led many U.S. politicians to argue against international efforts to reduce greenhouse gas emissions. This opposition is critical because the United States is the world's largest emitter. International agreements that do not include the United States will not work. The obfuscatory role of the United States in climate change policy stands in stark contrast to the role the United States played in negotiations concerning efforts to ban CFCs. In that case U.S. negotiators argued strongly for international cooperation, and U.S. pressure on foreign governments and industry was critical to the final agreement.

Given these differences from the conditions that led to the Montreal Protocol, many policy makers are not optimistic about a similar agreement concerning climate change. To achieve such an agreement, the scientific community must find a way to convince the public that the dangers of climate change are real and that the costs of doing nothing exceed the costs of setting the policy incorrectly—and must convince the United States that it is in its own self-interest to act.

STUDENT LEARNING OUTCOME

- Students will be able to describe the scientific, economic, and political factors that led to the success of policy aimed at banning CFCs and how these factors affect the outlook for climate change policy.

The Role of Technology

Given the subsequent tightening of the original Montreal Protocol, it is tempting to say that the treaty's success was driven by technical breakthroughs. No doubt industry developed alternatives that facilitated the replacement of CFCs; the newer chemicals had little or no chlorine or bromine. For example, one group of chemicals, the hydrofluorocarbons, uses fluorine in place of chlorine. The chemical industry also designed chlorine-based chemicals that break up in the troposphere rather than the stratosphere.

But these changes happened *after* the protocol was signed. At the time, the signatories could not be sure that alternatives were possible. Rather, the technical changes that allowed negotiators to speed the phase-out of CFCs were facilitated by the treaty. The timetable for a ban on CFCs provided an incentive for firms to change production processes in a way that reduced or eliminated CFCs. Redesign turned out to be more effective than just looking for chemicals to replace CFCs.

Redesign and replacement were facilitated by the Technology and Economics Assessment Panel. This panel was set up by the protocol to gather and disseminate information about ways to reduce the use of CFCs and the development of alternatives. As such, it played a pivotal role in technical efforts to reduce and eliminate CFCs and other chemicals that destroyed ozone. Most important, the panel assembled experts from affected industries and exempted them from antitrust rules (rules that forbid firms from working together like a monopoly). Cooperation allowed industry to advance the goals of the treaty by advising the committee about what controls were feasible, by establishing collaborative efforts to solve technical problems with alternatives, and by spreading information about efforts to replace ozone-destroying chemicals. Experts participated because what they learned helped their firms. This encouraged industry to send their best people, make good-faith efforts to contribute, and collaborate with policy makers seeking to regulate them. Without anyone working against the treaty, and with a large bureaucracy working for it, the treaty has become a blueprint for a successful response to an environmental challenge.

SUMMARY OF KEY CONCEPTS

- The atmosphere consists of two main gases, molecular nitrogen and oxygen, and several trace gases, such as ozone. Their concentration varies among layers of the atmosphere, which are defined by positive or negative lapse rates. The tropopause is an especially important boundary that preserves the planet's total supply of water. The stratosphere contains most of our atmosphere's ozone, which is formed by the photodissociation of molecular oxygen.

- Over the last twenty years the amount of ozone in the stratosphere has declined. There has been a slow decline in global ozone. In addition, there have been sharp declines over Antarctica in the Southern Hemisphere spring. The reduction starts with the photodissociation of CFC molecules, which liberates chlorine and bromine atoms in the stratosphere. There they react with ozone to convert ozone molecules and oxygen atoms to molecular oxygen. These reactions are accelerated by polar stratospheric clouds, and the reduction in ozone is concentrated by the polar vortex.

- The reduction in stratospheric ozone increases the amount of UV-B radiation that reaches Earth's surface. This radiation can disrupt photosynthesis and mutate genetic material. Many terrestrial plants reduce these effects by generating protective chemicals or increasing leaf thickness. Terrestrial animals that lay eggs without shells may suffer the greatest effects.

- In aquatic environments, protective chemicals are less effective due to the small size of phytoplankton. Increased UV-B may reduce net primary production, especially around the Antarctic. This reduction is amplified at higher trophic positions. Furthermore, these changes may be felt globally because many animals migrate to the Antarctic to eat some of the springtime burst in net primary production.

- Scientific certainty about the effect of CFCs on atmospheric ozone led to an international agreement to slow their production—the Montreal Protocol. The protocol has been strengthened, and the number of participating nations has increased. As such, the reduction in stratospheric ozone is the first global environmental problem for which international cooperation has been mostly successful.

REVIEW QUESTIONS

1. Explain how the stability of the CFC molecule reduces stratospheric ozone.

2. Explain how the scientific method was able to test competing hypotheses for the thinning of the ozone layer.

3. Trace the changes in the food chain through which reductions in net primary production in Antarctic aquatic biomes affect the reproductive success of penguins.

4. Explain the role of scientific understanding and technology in efforts to craft policy that reduced the production of CFCs.

5. How did developed nations entice developing nations to participate in efforts to ban the production of CFCs? Could a similar approach be used to entice developing nations to participate in efforts to lower emissions of greenhouse gases?

KEY TERMS

Dobson unit	mutations	stratosphere
lapse rate	ozone	thermosphere
marginal ice zone	ozone depletion potential	trace gases
mesosphere	photodissociation	tropopause
molecular nitrogen	polar stratospheric clouds	troposphere
molecular oxygen	polar vortex	

SOIL

A Potentially Sustainable Resource

15

STUDENT LEARNING OUTCOMES

After reading this chapter, students will be able to

- Describe the physical and biological processes that form soil and how differences among processes generate different soil types.
- Describe the factors that determine the quantity of water and nutrients a soil can hold.
- Compare and contrast the factors affecting the amount of soil eroded by water and wind.
- Explain the factors that influence the soil conservation measures that are adopted by farmers and why farmers tend to underinvest in soil conservation measures.

Dirt Storms *During the Dust Bowl, the wind picked up huge amounts of soil, which blackened the sky. The soil was deposited on both nearby farms and faraway cities.*

THE DUST BOWL

Stockbrokers jumping out of windows—that is one vision of the Great Depression. The Great Depression was an economic crisis that persisted through much of the 1930s. After a stock market collapse in October 1929, by 1934 the U.S. economy shrank by nearly 30 percent. By that time unemployment was nearly 25 percent, compared to the current rate of about 5 percent.

In addition to being a financial crisis, the Great Depression was exacerbated by soil erosion—the Dust Bowl. The physical loss of soil reduced wheat yields 15 to 25 percent in 1933. These losses were aggravated by a reduction in the area of wheat harvested from 23.5 million hectares (58 million acres) in 1929 to 16.6 million hectares (41 million acres) in 1934. As a result, total production of wheat dropped nearly 45 percent. Clearly agriculture's contribution to economic activity declined dramatically.

The losses associated with the Dust Bowl were driven by a decade-long period of very high rates of wind-driven soil erosion in the U.S. southern Great Plains. The erosion was caused by extensive cultivation, which reduced barriers to the wind, and an extended drought. Below-average rainfall started in the summer of 1931. By 1934–1936 the Great Plains experienced record dryness.

The drought made it easier for the wind to carry away soil. People described dirt storms so thick that they darkened the sky (opening photograph). Some of the soil (in some cases several meters deep) was dropped on nearby farms, killing crops and livestock. Huge storms such as that of May 1934 blew soil from Texas, Oklahoma, and Kansas to New York City and Washington, D.C. The worst dust storm, called the "black blizzard," occurred on April 14, 1935, which became known as Black Sunday.

Because some of these storms affected much of the nation, some people thought the world was coming to an end. For some, it nearly did. The U.S. government reported that 21 percent of all rural families in the Great Plains received federal emergency relief. Despite these efforts, many farmers were forced off their land. Some voluntarily forfeited their farms to banks. In 1933–1934 nearly 10 percent of U.S. farms changed possession, half of them involuntarily.

With nothing to eat and nowhere to go, hundreds of thousands packed their possessions in old cars and drove west, looking for work and new land. Much like the victims of Hurricane Katrina, they became refugees. Their plight was described by John Steinbeck in his 1939 novel *The Grapes of Wrath.* ■

LAND USE, SOIL, AND BIOLOGICAL ACTIVITY

Most of Earth's fertile soils already are used for agriculture. Although vast tracts of lands are unused, relatively little of this area is suitable for agriculture (Figure 15.1). Of the lands that remain, climatic and soil conditions limit their use for agriculture. Crops need time to be planted, grow, mature, and be harvested. To complete this cycle, many crops require 100–120 days. This requirement precludes agriculture on about 25 percent of Earth's land surface.

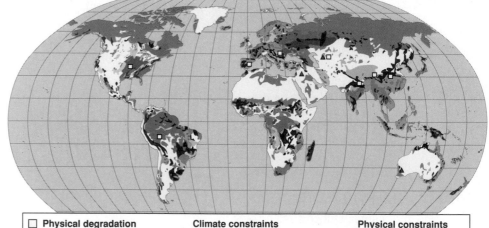

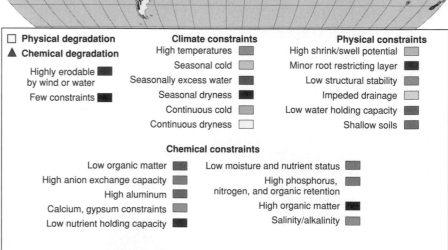

FIGURE 15.1 *Constraints on Soil Use* *Some soils are highly desirable because there are few constraints on their use. Most of the world's soils are less versatile because their use is constrained by climatic factors, physical characteristics, or their chemical makeup. In addition, some soils are highly vulnerable to erosion. Soils in many areas are already considered highly degraded.*
(Sources: Redrawn and adapted from major Land Resource Constraints Map by P. Reich and H. Eswaran of USDA/NRCS Soil Survey Division, World Soil Resources, Washington, D.C., from WSR Soil Climate Map and FAO Soil Map of the World; GLASOD data provided by K.Sebastain, IFPRI; data on compaction in Europe from SOVEUR/ISRIC.)

Crops also need lots of water. Crops transpire about 1,000 kilograms of water for each kilogram of dry material harvested by farmers. Conversely, crops cannot grow where the soil is constantly saturated. Too little or too much water precludes agriculture on a little more than 20 percent of Earth's surface.

Other areas cannot support agriculture because the soils are simply too poor; these soils account for about 5 percent of Earth's surface. Nor can crops be grown on steep slopes—it is difficult for farmers to work these areas. Mountainous areas exclude agriculture on another 18 percent of Earth's surface.

Because of these requirements, agriculture is feasible on a relatively small portion of Earth's surface, about 20 percent. Here climatic and soil conditions influence agricultural output. High levels of agricultural productivity are possible on only about 3 percent of Earth's surface. This area expands to between 6 and 13 percent as the level of productivity declines. Because of the small area where climate and soil conditions support high levels of agricultural production, it is essential to understand how soil is formed and how it contributes to agriculture.

SOIL FORMATION

The relatively small area that can support high levels of agricultural production is determined in part by the geological and biological factors that create soil. Soil is created from below by the breakup of underlying rock material and from above by the biological actions of production and decay.

When it was first formed, Earth's surface was barren—there was no soil. Soil formation starts in the unconsolidated material that lies above bedrock, which is known as the **regolith.** This regolith serves as the **parent material**, which is the mineral material from which a soil forms. The breakup of solid rock is termed **weathering**. Weathering includes both physical changes in the solid material, which are known as **disintegration**, and chemical changes, which are known as **decomposition**. Water often combines with carbon dioxide to form a weak acid called carbonic acid. This acid converts some minerals such as calcium and potassium to carbonates, which dissolve in water and thereby open spaces within rock. This form of decomposition is known as **carbonation and solution.** In other cases, water can disrupt the minerals in the parent material in a process known as **hydrolysis. Oxidation** occurs when oxygen combines with compounds in the parent material. Oxidation often weakens a material, which makes it more vulnerable to weathering.

Disintegration occurs with and without plants and animals. The freeze–thaw cycle breaks the regolith into smaller pieces. Liquid water seeps into cracks and freezes, which causes the ice to expand (remember that water is densest at 4°C). This expansion increases the size of the cracks and can cause portions of the rock to split apart. This process is painfully apparent to anyone who drives a car in areas with a long cold winter: The freeze–thaw cycle opens up large potholes in the road surface.

As materials on the surface move, they reduce the pressure on the materials below them. Rocks are slightly elastic (although this is not apparent to anyone who has fallen on a rock!), and this pressure reduction allows the newly exposed portion of rock to expand upward. This expansion can cause cracks that eventually allow large slabs of rock to break off.

Once they are able to gain a foothold, plants and animals accelerate disintegration. Anyone who has tripped on a sidewalk that has been lifted by plant roots knows that roots are so powerful that they can break rocks and pavement apart. Burrowing animals often move material, and these actions add to disintegration. Finally, fungi and lichens often produce acids that contribute to decomposition.

Soil Horizons

Soil is defined as the upper portion of the regolith that has been changed both chemically and biologically. These processes are associated with six general layers known as **horizons** (Figure 15.2). Horizons are layers that are approximately parallel to the surface and have distinct characteristics that are related to the process of soil formation.

The top horizon is known as the **O horizon.** The O horizon consists primarily of organic material, which serves as a precursor for soil formation. Within the O horizon there often are different layers. Soils in many deciduous forests

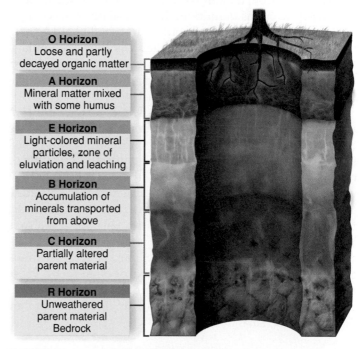

O Horizon
Loose and partly decayed organic matter

A Horizon
Mineral matter mixed with some humus

E Horizon
Light-colored mineral particles, zone of eluviation and leaching

B Horizon
Accumulation of minerals transported from above

C Horizon
Partially altered parent material

R Horizon
Unweathered parent material
Bedrock

FIGURE 15.2 *Soil Horizons Most soils are made up of six horizons. The O horizon consists largely of organic matter. As it decays and mixes with mineral material, it becomes the A horizon. The E horizon consists of materials that lose minerals as water percolates through. These minerals are deposited in the B horizon. The C horizon consists largely of the parent material. All of these layers sit atop bedrock, which is the R horizon.*

have three layers: a surface layer that includes leaves and twigs; a middle layer that includes partially decomposed organic material; and a bottom layer that consists of **humus,** which is partially decomposed plant or animal matter. The organic portion of soil does not include the roots of living plants or the organisms that live in the soil.

The next layer is termed the **A horizon.** The A horizon consists of organic matter mixed with mineral material. That most of the material in the A horizon is mineral, as opposed to organic, distinguishes the A horizon from the O horizon. The A horizon commonly is known as **topsoil.**

Below the A horizon is the **eluvial** or **E horizon.** The term *eluvial* comes from the Latin word that means "to wash out." Consistent with this meaning, the eluvial horizon is defined as the layer from which minerals are leached as water percolates through the soil. As gravity pulls water down through the eluvial horizon, the water dissolves some of the nonresistant nutrients and minerals, which include iron, aluminum oxides, and certain clays. Leaching causes their concentration to decrease, which can give the eluvial layer a lighter washed-out look. Conversely, some minerals do not dissolve, and concentrations of these resistant minerals increase in the eluvial horizon. Minerals that resist dissolution include quartz.

Once dissolved, minerals move with the water down to the next horizon, which is known as the **illuvial** or **B horizon.** This horizon also is known as **subsoil.** The term *illuvial* comes from a Latin word that means "to wash into." Consistent with this definition, the illuvial horizon accumulates the minerals that wash out from the eluvial horizon, such as iron and aluminum oxides. As they accumulate, these minerals often give the illuvial layer a striking color. For example, the iron oxides that accumulate in the illuvial layer give some Georgia soils a reddish color.

Farther down, the **C horizon** includes the regolith. That is, the C horizon consists of the parent material from which the soil formed. This horizon shows little or no sign of soil formation. As such, the C horizon usually supports little or no biological activity. Furthermore, there is relatively little weathering. Some materials, such as calcium or magnesium, accumulate in the C horizon; but generally this layer has a relatively minor role in soil formation. This horizon also is known as the substratum.

Finally, the **R horizon** consists of hard bedrock. Bedrock can consist of basalt, granite, quartzite, or other rocks. These materials usually are continuous and they cannot be penetrated with hand tools. Cracks appear in this layer, but they are usually too small for roots.

Soil Formation

As you dig a hole, the depths at which you encounter the six horizons and the depth at which you hit bedrock are determined by five factors: climate, living organisms, parent material, topography, and time. **Climate** has an important effect on the rate of soil formation. Decomposition rates are determined mostly by temperature and precipitation. Warm temperatures speed many chemical reactions. The rates of many reactions that form soil double for every 10°C increase in temperature.

In the United States the effect of precipitation is illustrated by two general soil types, pedalfers and pedocals (Figure 15.3). Pedalfers are located primarily in the eastern United States, where moderate rates of precipitation and organic decay leach minerals that contain iron from the topsoil (A horizon) and allow it to precipitate in the B horizon. In contrast, calcium carbonate (limestone) accumulates in the B horizon of pedocals from water that is drawn up through the soil due to high rates of evaporation and low rates of precipitation. These conditions are found mostly in the western United States.

Climate also affects soil formation indirectly by influencing the types of organisms that live in an area. As described in Chapter 7, temperature and precipitation are among the most important factors that influence the distribution of terrestrial biomes. The plants, animals, and fungi that live in tropical rain forests are very different from those that live in deserts. The species present determine the nature of the organic material that makes up the O horizon and how that organic material decays. These two aspects of biological activity determine many soil characteristics, such as its organic matter, structure, and nutrient content.

The effect of living organisms is illustrated by differences in the soil below grasslands and forests. Soils associated with grasslands generally have a higher organic content than soils associated with forests. This is partially due to the materials that come from trees as opposed to grasses. Leaves of coniferous trees are relatively resistant to decay and so generate soils with relatively thin A layers. Grassland soils tend to have a greater nutrient content and have a thicker A horizon because they are inhabited by a greater population of nonsymbiotic nitrogen-fixing bacteria.

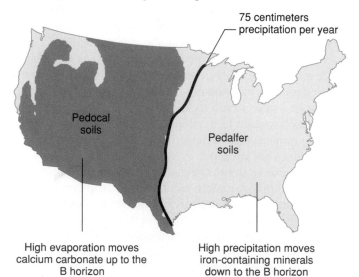

75 centimeters precipitation per year

Pedocal soils

Pedalfer soils

High evaporation moves calcium carbonate up to the B horizon

High precipitation moves iron-containing minerals down to the B horizon

FIGURE 15.3 *Soil Types and Precipitation* *Pedalfers are found mainly in the eastern United States, where annual precipitation is 75 cm (30 in) or more. Below this threshold, pedocals are the predominant soil types in the western United States.*

Soil formation also is influenced by parent material. Parent material that consists of softer rocks tends to disintegrate faster than parent material that consists of harder rocks. Similarly, parent material that consists of soluble minerals tends to decompose faster than parent material that consists of insoluble minerals.

Topography, which describes the surface features of an area, also exerts considerable influence on the rate of soil formation and the depth to bedrock. Soils on steep slopes tend be shallower than soils in flatter areas. Steep slopes encourage erosion, which slows the accumulation of surface layers. Topography also affects the rate of soil formation by influencing moisture content, which can speed or slow soil formation. In relatively rainy areas, small depressions may accumulate water. This water may saturate the soil, which can retard plant growth and slow soil formation. In arid areas that same small depression may accumulate enough water to support a tree or two, which accelerates the formation of soil relative to drier areas that are mostly grass.

Finally, time plays an important role in soil formation and depth to bedrock. All things being equal, time enhances the formation and depth of soil. This effect can be seen most clearly by comparing soils in areas that were recently glaciated (covered by ice) to soil from areas that remained ice-free. Relative to ice-free areas, areas that were glaciated tend to have thin, poorly developed soils.

SOIL TYPE

Geographic variations in climate, living organisms, parent material, topography, and time of formation alter the thickness of soil horizons. The vertical arrangement of the horizons is known as the **soil profile.** The soil profile is used to classify soils into twelve general types that are known as orders. These orders are distributed around the globe, much like terrestrial biomes

(Figure 15.4). Orders vary from entisols, which have little or no profile development, to mollisols, which have an especially well-developed O horizon. We illustrate the effects of climate, living organisms, parent material, topography, and time of formation by describing the soil profile for four soil types.

Mollisols generally are formed beneath grasslands, where the soil receives lots of organic matter annually because many grasses die at the end of each growing season. In addition, the large fraction of net primary production that is consumed by grassland herbivores relative to other biomes (see Chapter 7) is added to the surface layer as fecal material. Together these additions generate a deep A horizon (Figure 15.5a). Surface layers of some mollisols in the midwestern United States can be 2 meters or more. Such deep soils are possible because the low levels of precipitation (grasslands do not receive enough precipitation to support trees) and cold winters tend to slow the rates at which organic materials decay.

Relative to mollisols, **spodosols** tend to have a thinner O horizon and a more thoroughly leached eluvial horizon (Figure 15.5b). Spodosols often are formed under coniferous forests. Coniferous trees tend to contribute less organic matter than grasses, which limits the thickness of the O horizon. The higher rates of precipitation, which support the coniferous trees, tend to increase leaching from the eluvial layer. Higher rates of leaching are enhanced by the acidic nature of both the parent material and the organic matter from coniferous trees. Higher acidity increases the amount of minerals that are dissolved as water percolates through the soil profile (see page 320 for more explanation). These differences, especially in the O horizon, tend to reduce the fertility of spodosols relative to mollisols.

Oxisols often have a thin O layer and small amounts of nutrients (Figure 15.5c). Oxisols usually are considered to have a limited potential for use in agriculture. This does not imply that agriculture is impossible. Rather, oxisols must be managed carefully. What is surprising about oxisols is their

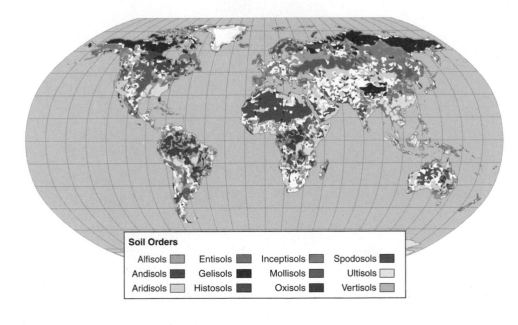

FIGURE 15.4 *Global Soils Soils are classified into twelve categories that are known as orders. The location of these orders parallels the global distribution of terrestrial biomes in Figure 7.9 on pages 108–109 because soil is formed by the interplay of climate and biological activity.* (Source: Data from Natural Resources Conservation Service, USDA.)

Soil Orders

Alfisols	Entisols	Inceptisols	Spodosols
Andisols	Gelisols	Mollisols	Ultisols
Aridisols	Histosols	Oxisols	Vertisols

FIGURE 15.5 *Soil Profiles* The thickness of soil horizons differs among (a) mollisols, (b) spodosols, (c) oxisols, and (d) aridisols. Mollisols tend to have the thickest A horizons, which thin progressively in spodosols, oxisols, and aridisols.

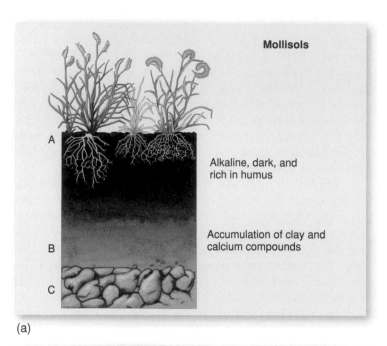

Mollisols

A — Alkaline, dark, and rich in humus

B — Accumulation of clay and calcium compounds

C

(a)

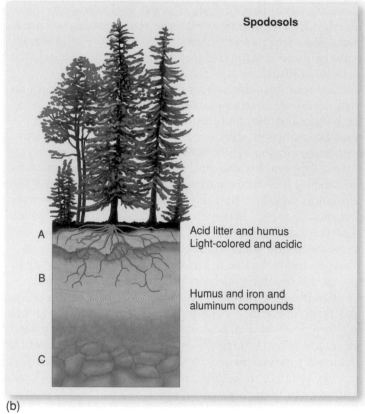

Spodosols

A — Acid litter and humus
Light-colored and acidic

B — Humus and iron and aluminum compounds

C

(b)

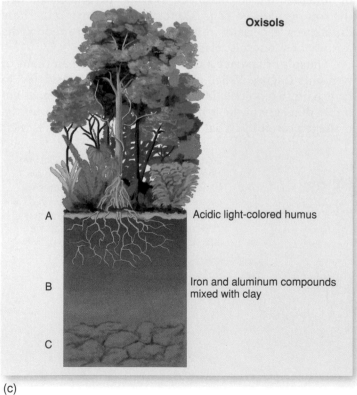

Oxisols

A — Acidic light-colored humus

B — Iron and aluminum compounds mixed with clay

C

(c)

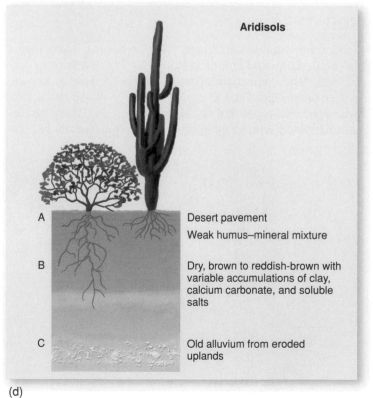

Aridisols

A — Desert pavement
Weak humus–mineral mixture

B — Dry, brown to reddish-brown with variable accumulations of clay, calcium carbonate, and soluble salts

C — Old alluvium from eroded uplands

(d)

location—tropical rain forests. The juxtaposition of a lush canopy and a thin O horizon seems contradictory. Tropical rain forests generally have very high rates of net primary production and hefty amounts of biomass per unit area; therefore oxisols receive lots of organic material each year. But these additions do not accumulate. Warm temperatures and plentiful rain cause organic matter to decay quickly. Furthermore, abundant rainfall leaches many minerals from the upper layers, the old age of which exacerbates the leaching. Because glaciers never reached the tropics, these soils have been weathered over a long period.

Dry climates favor the formation of soils known as **aridisols**. These soils are characterized by eluvial horizons that undergo relatively little leaching (Figure 15.5d) because of scant precipitation. Normally aridisols are thought to have little potential for agricultural production. This perception is correct; but if aridisols are irrigated they can support high levels of agricultural production. For example, aridisols are present in the southwest corner of Arizona, where water from the Colorado River irrigates fields that grow large quantities of lettuce.

SOIL FUNCTION

The phrase "form follows function" is credited to Louis Sullivan, who was an architect. But he could just as easily have been a soil scientist: Differences in soil profile that allow scientists to assign soil to orders are more than cosmetic. Soil orders influence patterns of land use. A soil's contribution to plant growth in general and agricultural production in particular is determined mostly by the soil's ability to provide water and nutrients. Both of these abilities are determined in part by the physical characteristics that are used to classify soils.

Storing Water

Plants obtain their water from the soil, and soils obtain water from precipitation. This obvious truth seems to imply that soil is simply an intermediary for precipitation. But in many locations soil's role in water storage may be as important as total precipitation. Even in areas where annual rates of precipitation are sufficient to support high levels of agricultural production, rainfall (or snowfall) often is relatively infrequent and episodic. Just after a rain, water supplies may be ample. But in the days or weeks between rains, there would be no water for plants without soil. Soil smoothes the peaks and valleys in a plant's water supply by storing water between precipitation events. Remember from Policy in Action in Chapter 7 (page 102) that agriculture is possible in eastern Africa because its soil stores enough water from a single rainy season for a single crop.

The quantity of water soil can deliver to a plant is determined by the amount of water the soil can store and how tightly this water is held by the soil. These abilities are to some degree mutually exclusive. Some soils can store a lot of water, but they hold onto it so tightly that much of the water is not available to plants. At the other extreme are soils that hold their water loosely but don't have much capacity to store water. These soils don't provide much water either. The most desirable soils are those that combine the best of both worlds.

Soil's ability to store and relinquish water is based on a simple but not so obvious fact—soil is not solid. If you examine soil closely, you will see that soil consists of particles and spaces between particles, which are termed **pores** (Figure 15.6). Soil stores water in the pores among its particles. Similarly, oil is stored in the spaces among the sedimentary rock particles that make up an oil field. Indeed, the forces that govern a soil's ability to provide water also influence how much oil we get from an oil field (see Chapter 20).

Because of the importance of both water and oil, we describe how particle size determines the pore space in soils (and oil fields) and the tenacity with which they hold their

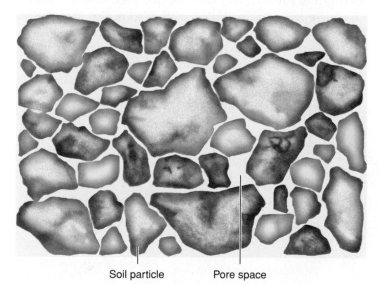

FIGURE 15.6 *Soil Pores* *A microscopic view of soil indicates that it consists of tiny particles that are separated by spaces known as pores. These pores hold water and air that are vital for plants.*

Soil particle   Pore space

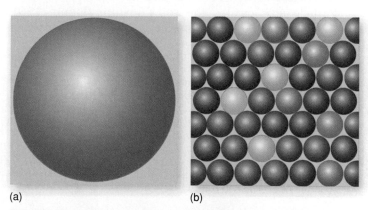

(a)   (b)

FIGURE 15.7 *Pore Space and Particle Size* *(a) Soils that consist of large particles have a few very large pores that are known as macropores. (b) Soils that consist of many small particles have many smaller pores known as micropores. On net, these many smaller pores represent more space than the few large pores.*

water (and oil). We illustrate these effects with boxes filled with marbles of various sizes (Figure 15.7). The box on the left side of Figure 15.7 contains one very large marble. This marble takes up most of the space in the box, so there are few pores and relatively little pore space. The few pores present are relatively large and are known as **macropores.** On the right side of Figure 15.7 is a box that is filled with many small marbles. Between these marbles are many small spaces, known as **micropores.** Micropores create much more total pore space, which means that soils and oil fields that consist of large particles store less water or oil than soils and oil fields that consist of small particles.

Particle size also influences the tenacity with which the soil or oil field holds its water or oil. Macropores in the box with the single marble allow water or oil to flow quickly. In soils this flow is known as **gravitational water,** which is water that moves into, through, or out of the soil by gravity within a day or two (Figure 15.8). Micropores in the box with the many small marbles create two additional types of water. **Capillary water** is the water that fills a soil's micropores. Here water is held with a moderate force—between 0.3 and 31 bars of suction. (*Soil tension* is the formal term for suction, which is measured in bars. A large value, in absolute terms, indicates that the water is held tightly.) This suction implies that capillary water flows less rapidly than gravitational water. Water that is held very tightly (more than 31 bars) is known as **hygroscopic water,** which is water that forms a thin film around individual soil particles. Because it is held so tightly, hygroscopic water flows slowly at best. These three types of water imply that soils and oil fields that consist of small particles hold their water or oil more strongly than soils or oil fields that consist of large particles.

To understand these effects in soil, scientists classify soil particles among sands, silts, and clays. **Sands** include the largest soil particles—they can be seen with the unaided eye. Particles that can be seen only under a microscope are termed **silt.** The smallest particles, **clays,** include particles that can be seen only with an electron microscope.

Particle size creates the soil's texture (Figure 15.9). Consistent with the particles they are made of, soils that

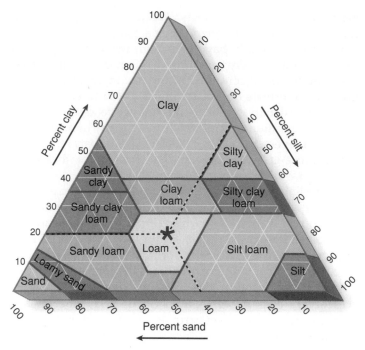

FIGURE 15.9 *Soil Texture Soils consist of particles of three general sizes: sand (largest), silt, and clay (smallest). To determine the makeup of a given soil, first find the percentage of clay along the left side of the triangle. Next find the percentage of sand on the triangle's base by reading the lines parallel to the triangle side labeled "percent silt." Finally, determine the percentage of silt directly from the triangle by finding the percentage silt along the right side of the triangle by reading the lines parallel to the "percent clay" side of the triangle. For example, the asterisk represents a loam soil that is 20 percent clay, 40 percent sand, and 40 percent silt.*

consist mostly of large particles are termed sandy soils, whereas soils that consist mostly of very small particles are termed clays. You can feel these differences. Sandy soil feels gritty, silt soils feel smooth, and clay soils feel sticky. Of course more sophisticated methods are also available to classify soil texture.

To illustrate how texture affects soil's ability to store water, we follow the flow of water after a rainstorm. Following a heavy rain, water may fill all pore spaces. **Total saturation**

FIGURE 15.8 *Three Types of Soil Water Gravitational water is the water that fills macropores. This water drains quickly. Capillary water fills the micropores and drains slowly. This makes it available to plants between precipitation events. Hygroscopic water is water that forms a thin film around individual particles. These particles hold it tightly in place, so much of this water is not available to plants.*

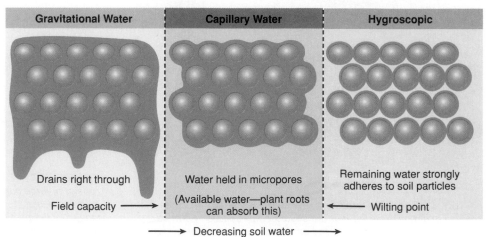

represents the maximum water capacity of a soil. Because of its rapid flow, gravitational water drains from the soil within a day or two. As such, gravitational water is of relatively little use to plants. Nonetheless, some drainage is necessary. As this water drains, air fills the larger pore spaces. Air is critical to plant growth: Without it plants can "drown" in a soil whose pore spaces are filled completely by water.

Two or three days after the rain, the moisture that remains is measured by a soil's **field capacity.** Field capacity includes both capillary and hygroscopic water. Of these, plant roots have enough strength to pull capillary water from micropores. Capillary water is therefore an important source of water for plants between rainfall events.

As roots remove capillary water, micropores dry out. This depletes the plant's water supply. Eventually only hygroscopic water remains. Because this water is held tightly, roots cannot pull it from the soil. Without additional water supplies, a plant loses its ability to support itself, which is termed wilting. When plants wilt, the soil is said to have reached its **wilting point.** At its wilting point, the soil has not run out of water—hygroscopic water remains, but roots cannot pull it from the soil. The same idea applies to oil fields. Significant amounts of oil remain in a field after it has run dry. We cannot get it because too much effort is required to pull it from the field (Chapter 20).

The three types of soil water and the total amount of pore space imply that the greatest amount of water is available to plants that live in loam soils. **Loam soil** is a roughly equal mixture of clay, sand, and silt (Figure 15.9), and these particles create an ideal mixture of macro and micropores. You can understand this ideal by looking at how the wilting points and storage capacities of soils change with particle size. On the left side of Figure 15.10, sandy soils have a very low wilting point; therefore, a large portion of the water in a sandy soil is available to plants. But the total amount of water in sandy soils is low due to the small amount of pore space. A large fraction of a small number is still a small number, so only a small amount of water is available to plants that live in sandy soils. A similar result applies to clay soils. The small particles in clay soils create a lot of pore space, which means the total amount of water in these soils is large. But the small soil particles hold the water tightly in place. A small fraction of a large number is a small number; therefore, a small amount of water is available to plants that live in clay soils.

Soils with intermediate particle sizes balance storage space and the strength with which the water is held. Notice that field capacity rises rapidly throughout the left side of Figure 15.10 before leveling off at intermediate and small particle sizes, whereas wilting point rises steadily as particle size decreases. The distance between field capacity and wilting point represents the total amount of water that is available to plants. The most water is available to plants from loam soils, in which field capacity has pulled far ahead of wilting point.

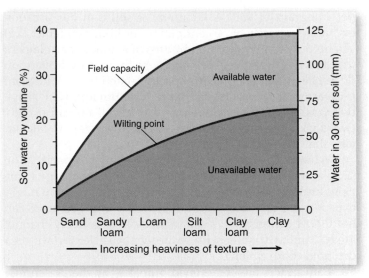

FIGURE 15.10 *Soil Texture and Soil Water The total amount of water in a soil is given by the top line, which is labeled field capacity. This total includes both water that is available to plants and water that is unavailable to plants. The amount of water that is unavailable is given by the dark blue area at the bottom of the graph. This quantity is determined by the wilting point—the plants' ability to pull hygroscopic water from soil particles. Water held less tightly than the wilting point is available to the plant and is given by the light blue area above the wilting point but below field capacity. Notice that the total amount of available water is greatest in loam soils, which consist of a mixture of particle sizes.*

Storing Nutrients

As discussed in Chapter 6, nutrients are critical to plant growth. The deficiency of a single nutrient, the so-called limiting nutrient, can slow growth. Other than a few insectivorous species such as the Venus flytrap, plants obtain their nutrients from the soil. Therefore, the amount of nutrients in the soil and the soil's ability to hold them are crucial aspects of soil function.

How do soils hold nutrients in place? We can get a hint from the nutrients, such as potassium (K^+) and ammonium (NH_4^+). Notice that these nutrients are positively charged ions, which are known as **cations.** Soil's ability to hold a cation in place is based on the laws of attraction—a negatively charged component of the soil holds a positively charged cation in place. Without this attraction, nutrients would be lost as water percolated through the soil and carried the nutrients below the rooting depth or flowed across the soil surface and washed the nutrients into streams and rivers.

Soil's ability to hold positively charged nutrients is known as **cation exchange capacity.** Cation exchange capacity is the total amount of exchangeable cations that a soil can absorb. You can think of this in terms of a parking lot. Cation exchange capacity represents the number of parking spaces that are available to positively charged ions. The number of parking spaces often is measured in centimoles of charge per kilogram of soil. A kilogram of soil that has a cation exchange capacity of 50 centimoles per kilogram (cmol/kg) can hold 50 centimoles (hundredths of a mole) of an element that has a single positive charge, such as a hydrogen ion (H^+). The number of parking spaces needed by each nutrient is determined in part by its charge. Calcium ions (Ca^{2+}) have two positive charges; therefore, a calcium ion can occupy two parking spaces. This reduces the number of calcium ions that the soil can hold.

What determines a soil's cation exchange capacity? The simple answer is the prevalence of negatively charged particles and the surface area of the soil particles. As you might expect, a soil that has a lot of negatively charged areas can hold more cations than a soil with fewer negatively charged areas. Similarly, soils with extensive particle surface area can hold more cations.

Both factors are determined primarily by the presence of inorganic clays. Clay particles have a large surface area. A gram of clay soil has a combined surface area of 10 to 1,000 square meters, which is thousands of times greater than the surface area of a gram of sandy soil. Because of these differences, sandy soils usually have the lowest cation exchange capacity (less than 5 cmol/kg), whereas clay and loam soils often have the greatest cation exchange capacity (above 30 cmol/kg).

A soil's cation exchange capacity also is determined in part by its pH. pH measures the abundance of positive hydrogen ions (H^+) relative to negative hydroxyl ions (OH^-). A solution is said to be an acid if its pH is less than 7. A change in pH from 7 to 6 represents a tenfold increase in hydrogen ions. As the abundance of hydrogen ions increases relative to hydroxyl ions, a soil becomes increasingly acidic. The greater numbers of positively charged hydrogen ions take up parking spaces that would otherwise be available to cations. Because of these effects, a soil's cation exchange capacity often is positively related to its pH. To grow crops in acid soils, farmers often add lime. For example, a major impediment to soybean production in the Amazon was overcome when farmers found a local inexpensive source of lime.

SOIL EROSION

Even under the most favorable conditions, soil forms very slowly. Centuries or even millennia are required to generate a centimeter of soil. That same centimeter can be lost in years or decades. The rate at which soil is lost is termed **soil erosion.** Soil erosion occurs in three steps. In the first step soil particles are detached from other particles. Detachment makes the individual particles easier to transport, which is the second step. Transport is the step that moves the particles from their original location to a new location, where they are deposited. Deposition is the third step.

Detachment and transport are powered by water or wind. To understand water-based erosion, envision big raindrops falling on tiny soil particles, creating very small craters in the soil. The particles of soil from these craters

are loosened by the drops and detached (step 1) from their original locations, which means they are free to be transported (step 2) by running water.

Both natural factors and human activities determine the rate of water-driven soil erosion, which is summarized by the **universal soil loss equation.** This equation states that the rate of erosion is determined by (1) rainfall, (2) soil erodibility, (3) slope length, (4) land cover, and (5) erosion control practices. The power of raindrops implies that erosion increases with total precipitation. But the distribution of water among precipitation events also is important. Areas that receive their annual precipitation in a few intense storms tend to have higher rates of erosion than locations where precipitation is distributed evenly among less intense events. The timing of precipitation also is important. Temperate latitudes receive much of their rainfall in the early spring and late fall, when soil is most vulnerable to erosion because it is not frozen and there is little plant cover.

Erosion also is influenced by the soil's **erodibility factor,** which represents how easily soil particles can be detached and transported. Erosion is lower on soils with stable structures that can resist the effects of raindrops. Soil erosion also is influenced by the ease with which water can percolate through the soil. If water percolates through the soil easily, water is less likely to pool on the soil surface and run horizontally. Horizontal water movements can detach and transport soil particles.

Erosion also is determined by topography. Steep slopes speed the rate at which water flows across the surface. Simple physics dictates that rapidly moving water, which has a lot of momentum, can transport more soil particles than slowly moving water. This effect is reinforced by slope length. A long slope allows surface water to gather momentum and wash soil away.

Plant cover (or its absence) has an important effect on the rate of erosion. See *Your Ecological Footprint: How Much Soil Do You Erode?* Erosion tends to be lowest on land covered by trees or grasses. Tree leaves intercept raindrops before they hit the soil. This slows the rate at which raindrops hit the soil. With less momentum, raindrops have less power to detach soil particles. Similarly, grasses tend to slow raindrops and impede the flow of water across the surface.

Crops provide less protection against soil erosion than natural vegetation. Crops such as wheat are only slightly less protective than the grasslands they replace because surface water flows are slowed by their tight planting. But crops that are planted in rows, such as corn and soybeans, offer less protection. Rows between plants expose soil to direct hits by raindrops and provide channels for water to flow across the soil surface.

The quantity and speed of the water that flows across the soil surface influence the amount of soil transported. If rainfall exceeds the rate at which water percolates into the soil, a film of water will build up on the surface. As this film moves downhill, it will scrape and transport the topsoil in

(a)

FIGURE 15.11 *Water-Driven Erosion* (a) Soil particles detached and transported by water that flows evenly across the surface is known as sheet erosion. (b) Rill erosion occurs where soil is detached from small channels (rills) filled with running water. (c) Gully erosion occurs where soil particles are detached from deeper channels (gullies) filled with flowing water.

(b)

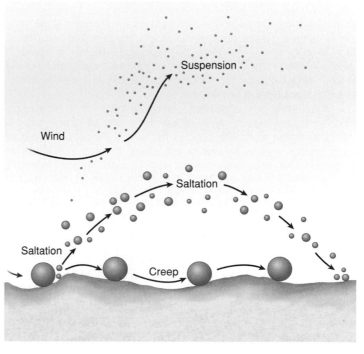

(c)

a fairly even layer. This loss is termed **sheet erosion** (Figure 15.11a). If the sheet erosion becomes concentrated in small channels, which are known as rills, the result is **rill erosion** (Figure 15.11b). Rill erosion sometimes is viewed as relatively harmless because the rills can be smoothed by plowing. But this cannot fix the fact that soil is lost. If the rills are concentrated into deeper channels, the erosion is known as **gully erosion** (Figure 15.11c). Gully erosion is the most visible form of erosion, but in many areas more soil is lost via sheet and rill erosion.

Wind detaches and transports soil particles in three ways. Most of the soil that is transported by the wind moves via **saltation,** which bounces particles along the ground in a series of short hops (Figure 15.12). As the soil particles

FIGURE 15.12 *Wind-Driven Erosion* Suspension occurs when small particles of detached soil are carried by the wind. The wind is not strong enough to lift larger particles, which are moved either by saltation, in which the wind bounces medium-sized particles along the surface, or creep, in which the wind pushes larger particles along the surface. (Source: Redrawn from USDA Agricultural Research Service in cooperation with Kansas State University Wind Erosion Research Unit.)

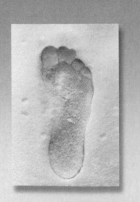

Your ECOLOGICAL footprint

How Much Soil Do You Erode?

For many of you this seems like a silly question. On most days you hardly have any contact with the soil. Perhaps when you rake leaves or mow the lawn you increase the rate of soil erosion. More daring readers may increase soil erosion by riding a bike or driving a car off-road.

But like most of the other Your Ecological Footprint calculations, here we provide information that allows you to quantify the effects of your everyday activities on soil erosion. Some of these activities are fairly obvious. Growing your food increases the rate of soil erosion. But so do other activities, such as harvesting trees for use as paper or wood products. The effects are caused by a simple fact—harvesting environmental goods that are generated via net primary production usually changes land cover in a way that increases the rate of soil erosion.

We would like to provide a table here that shows the amount of soil eroded during each of these activities. Unfortunately we can't. One reason is that the soil loss equations described in this chapter require too much information. There is simply no way to know the slope, surface roughness, soil erodibility, and so on for each parcel of land that is used to grow your food and provide your wood and paper products. Even if we could know these factors, the equations are too imprecise. As was discussed in this chapter's case study, scientists who have detailed information about each of the independent variables in the soil loss equations produce estimates for soil erosion that don't match the quantities deposited in waterways.

To get around these difficulties, we use the approach developed by the International Geosphere–Biosphere Program, which is an international group of scientists whose mission is to deliver scientific knowledge to help human societies develop in harmony with Earth's environment. They calculate the relative rate of erosion with the following equation:

Relative erosion rate = $Slope^{1.5} \times$ (Precipitation/1,000) $\times$ Biome factor

in which slope is the slope of a parcel of land, precipitation is the mean monthly rainfall for the peak rainfall month (measured in millimeters), and biome factor is a parameter that represents the rate of erosion in various biomes. This equation does not calculate the quantity of soil eroded. Rather, it calculates the increase or decrease in erosion relative to erosion in a tropical rain forest.

When using this equation to calculate your effect on soil erosion, we do not expect you to include the effect of slope—there is little information about the slope of the land that you use in distant environments. Rather, we include the effect of slope to illustrate its importance. As slope steepens, its effect on soil erosion increases exponentially (it is raised to the 1.5 power). In contrast, the equation represents the effect of precipitation as linear: Doubling precipitation doubles erosion.

The biome factor represents the effect of plant cover on soil erosion. Notice that the values increase as you read down Table 1. This pattern should make intuitive sense. Because of their closed canopy and many layers of large leaves, tropical rain forests provide extensive protection for the underlying soil. Further down the table are biomes such as open savannas, where the lack of trees offers less protection, so soil erosion is greater than in a tropical rain forest. Even less protective are croplands and bare soils. Without plant cover during some or all of the year, soil is vulnerable to wind and rain.

TABLE 1 Relative Rates of Erosion

Land Cover	Maximum Monthly Precipitation (mm)	Biome Factor
Evergreen broadleaf forest	320	1.0
Evergreen needleleaf forest	42	1.5
Deciduous needleleaf forest	100	2.0
Closed shrubland	98	4.0
Open shrubland	58	5.0
Woody savanna	320	6.0
Savannas	85	8.0
Croplands, natural mix	100	12.0
Grasslands	100	12.5
Croplands	100	21
Urban and built up	100	21
Barren or sparse vegetation	10	21

Sources: Data for maximum rates of precipitation from M.C. Molles, Jr., Ecology: Concepts and Applications, McGraw-Hill; data from biome factor from the International Geosphere Biosphere Programme.

bounce, they may hit larger particles (up to 1 mm in diameter), which can cause the larger particles to **creep** along the surface. The most spectacular type of soil movement is **suspension,** in which soil particles may be lifted high into the air and carried long distances. This is the process that generated the dirt storms associated with the Dust Bowl.

The rate at which these three processes erode soil is represented by an equation that is similar to the universal soil loss equation. The **wind erosion equation** states that the rate of wind-driven soil erosion is determined by (1) soil erodibility, (2) climate, (3) roughness, and (4) vegetative cover. The interpretation of soil erodibility is similar

Because the biome factor represents relative rates of soil erosion, we need to make assumptions about how your use of net primary production changed land cover. For crops this transformation is relatively simple—the original biome was converted to croplands. If you think the pastureland used to support the livestock that provided your meat was overgrazed, perhaps the grassland was converted to the land cover "barren or sparse vegetation." For timber and paper products, we have to make some assumption about what happened to the forest. If the forest was allowed to regrow, the forces of succession changed the community from barren or sparse vegetation, to grasslands, to open shrublands, to closed shrublands, and back to forest. In other cases timber is harvested and the forest is converted to cropland or pastureland.

To convert the relative rate of erosion to a quantity, you need an estimate for soil erosion on a flat hectare of tropical rain forest. Estimates for this rate of erosion indicate that about 0.2 kilograms of soil are lost per square meter per year. If you assume that all of your products of net primary production come from flat land, we can use this estimate to approximate the quantity of soil that is eroded due to the environmental goods you use.

Calculation and Example

The formula used to calculate the amount of soil eroded is based on the amount of net primary production you use, the biomes from which these environmental goods originate, and their rates of net primary production:

Soil eroded = [(__#1__ Good × __#2__ Grams carbon/good)/__#3__ Grams/m^2]

*(__#4__ Converted biome factor/__#5__ Original biome factor) × 0.2 kg/m^2/year

Notice that this calculation does not include the effects of slope because we could not determine the slope of the land from which you obtained your goods. Nor does it include the effect of precipitation—we assume that the change in ecosystem type does not change local precipitation. (For more about this assumption see Chapter 17.)

This equation seems complex, but it will seem simpler after we step through how the equation is used to calculate the quantity of soil eroded due to the trees harvested to provide the paper you use. Suppose that you use 10 kilograms of paper per year: Put a 10 in blank #1. According to Table 1 in Chapter 6's Your Ecological Footprint (page 106), each kilogram of paper requires 990.83 grams of carbon, so put 990.83 in blank #2. The product of these two numbers indicates that your use of paper represents net primary production of 9,908.3 grams of carbon. The area needed to generate this net primary production can be calculated by dividing 9,908.3 by the rate of net primary production in temperate forests, which can be obtained from Table 1 in Chapter 8's Your Ecological Footprint (page 148); put 702 grams of carbon per square meter in blank #3. This quotient indicates that you need 14.1 square meters of temperate forest to grow the trees that are harvested to make your paper.

Suppose that the trees used to produce your paper were grown in an evergreen needleleaf forest: Put a 1.5 in blank #5. After the trees were removed, the land becomes an open shrubland: Put a 5.0 in blank #4. This quotient (5/1.5) shows that converting the evergreen needleleaf forest to an open shrubland increased soil erosion by a factor of 3.33.

Now we can put all these numbers together to calculate the total quantity of soil eroded that is associated with your use of paper. The total area used to grow these trees is 14.1 m^2. The rate of erosion on these factors increases by a factor of 3.33 relative to a square meter of tropical rain forest, which loses 0.2 kg/m^2/year of soil. This product indicates that your use of paper increases soil erosion by 9.4 kilograms (14.1 m^2 × 3.33 × 0.2 kg/m^2/year). Now repeat these calculations with the other products of net primary production that you use (listed in Your Ecological Footprint in Chapters 6, 7, and 8) and sum them.

Interpreting Your Footprint

These calculations can be used to evaluate the importance of land degradation. Repeat your calculations by assuming that the original ecosystem is converted to barren land, which has an erosion factor of 21 relative to tropical rain forests. The difference between this total and your original represents the increase in soil erosion if the natural ecosystem is degraded after harvest.

We can also use these calculations to evaluate what happens when farmers are forced to grow food on land with ever-increasing slope. Repeat all of your calculations with a slope of 3° instead of 0°. By how much does soil erosion increase? How about when the slope rises from 3° to 6°? As you can see, the relationship is nonlinear: Moving to steeper slopes increases erosion at increasing rates.

STUDENT LEARNING OUTCOME

- Students will be able to explain the factors that determine how their activities affect the rate of soil erosion.

to that for the universal equation. Soil particles that stick together tend to be more resistant to wind erosion. (What types of soil tend to be "sticky"? Think back to the section about soil texture.)

Climate variables that affect wind erosion include precipitation. Wet soil is stickier than dry soil and therefore more resistant to wind; but high wind speeds, both average and maximum, tend nevertheless to increase soil erosion. Soil erosion also is increased by **turbulent** flow, which moves across the surface but also has an up-and-down component. These movements can pick up soil particles and move them over long distances.

The degree to which wind speed and turbulence affect soil is determined by surface roughness, field width, and vegetative cover. **Surface roughness** refers to irregularities (bumps) in the soil surface. Irregularities slow soil erosion by slowing wind speed. The degree to which winds are slowed depends in part on the scale of the irregularity. At a small scale, irregularities can be caused by the soil structure. Larger bumps include living plants and **crop residues** (stalks, leaves, and the like). At an even larger scale, hedges and trees can be thought of as giant bumps that slow wind speeds for a considerable distance downwind. As you might expect, increasing roughness or plant cover reduces soil erosion.

The universal soil loss equation and its counterpart for wind erosion must be used carefully. These equations predict the amount of soil moved. However, such movements, which include the transport of particles within a field and between fields, cannot be translated directly into quantities of soil that are lost. It is difficult to measure exactly how much soil is lost due to erosion. See *Case Study—Where Has All the Soil Gone?*

Impacts of Soil Erosion

Soil erosion reduces the economic well-being of farmers and nonfarmers alike. Society as a whole suffers **social costs** when soil is transported beyond the farm and causes damage where it is deposited. Social costs include eutrophication, siltation of streams (and its negative effect on fisheries and shipping), and siltation of dam reservoirs (and its damage to hydroelectric facilities). Social costs of soil erosion are discussed in Chapter 17 in relation to deforestation.

Private costs, which are incurred by the individual who causes the erosion, occur on the farm, and farmers suffer their effects. Soil erosion reduces crop production by reducing the soil's ability to store water and nutrients. Reducing the thickness of soil's upper layers can reduce rooting depth, which cuts the supply of water between rainfall events. Less water shortens the time during which plants can take in carbon dioxide, which is essential for photosynthesis. Lower rates of net primary production reduce the energy plants can allocate toward their parts that people eat, such as ears of corn or heads of wheat. These reductions are amplified by loss of nutrients. Erosion removes both nutrients and the soil's cation exchange capacity.

These private costs squeeze farmers in two ways. They reduce farm revenue by lowering yield and boost costs by increasing the farmers' purchases of water and fertilizer. Both mechanisms are used to measure the private costs of soil erosion. Yield reductions caused by erosion vary greatly by site and crop. Losing about 35 kg of soil per hectare (about 3 cm) per year in the northern Great Plains of the United States reduces revenue by about $50 per hectare over a twenty-year period and about $200 per hectare over sixty-eight years (Figure 15.13). These losses depend on the discount rate. If farmers don't value future income highly (a

5 percent discount rate), the losses after twenty years would be only $45 per hectare. If farmers value future harvest more dearly (a 1 percent discount rate), the twenty-year loss rises to $79 per hectare.

The sharp reduction in revenue is caused by the accelerating effects of soil erosion. The first centimeter of soil lost reduces wheat yields by about 10 kg per hectare—from 1,719 to 1,709 kg per hectare—whereas losing the nineteenth centimeter reduces yields by about 31 kg per hectare—from 1,362 to 1,331 kg per hectare (Figure 15.14).

Alternatively, the private costs of soil erosion can be measured by the value of the nutrients that are lost with the eroded soil. This cost can be calculated by measuring the physical quantities of nitrogen, phosphorus, potassium, and other nutrients that are lost with the eroded soil and multiplying them by the cost of fertilizers that would be needed to replace them. Again, these costs vary greatly by

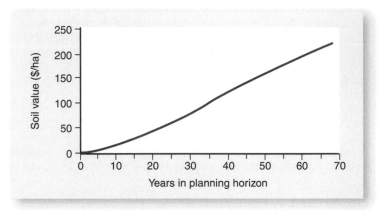

FIGURE 15.13 *Dollar Costs of Soil Erosion* In the northern U.S. Great Plains, losing about 3 cm of soil per year per hectare reduces farmers' income by about $60 over twenty years. These losses increase over time. The rate of this increase depends on the farmer's discount rate. High discount rates imply that the farmer does not value future revenues highly, so losses add up slowly over time. *(Source: Data from J.R. Williams and D.L. Tanaka, "Economic Evaluation of Topsoil Loss in Spring Wheat Production in the Northern Great Plains, USA." Soil and Tillage Research, 37: 95–112.)*

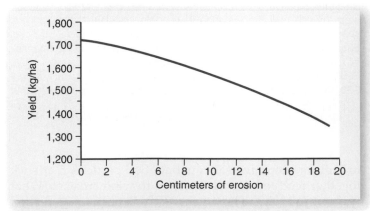

FIGURE 15.14 *Lower Yields* Yields decrease over time due to an annual loss of about 3 cm of soil per year. *(Source: Data from J.R. Williams and D.L. Tanaka, "Economic Evaluation of Topsoil Loss in Spring Wheat Production in the Northern Great Plains, USA." Soil and Tillage Research, 37: 95–112.)*

CASE STUDY

Where Has All the Soil Gone?

The U.S. government has spent billions of dollars to reduce soil erosion. But are such expenditures warranted? Scientists disagree about how much soil is being lost from U.S. agricultural fields, the degree to which these losses are sustainable, and the impact of these losses on agricultural output. Solving these uncertainties is critical to understanding how much money should be spent to slow soil erosion.

Estimates for the rate of soil erosion vary greatly. Average annual rates of soil erosion on U.S. croplands are thought to be 2.0–6.8 billion metric tons per year (2.2–7.5 billion tons/year). There is even some uncertainty about whether erosion is increasing or decreasing. The U.S. Department of Agriculture suggests that soil erosion has slowed since the 1970s and 1980s. Others assert that soil erosion is now more severe than it was during the Dust Bowl.

Before debating these numbers, it is important to describe how soil erosion is measured and what erosion measures. National estimates for soil erosion do not come from direct measurements. Rather, these estimates are generated using computer models and revised versions of the universal soil loss equation and the wind erosion equation. These equations use site-specific information, such as precipitation, wind speed, slope, and crops planted, to calculate erosion. Site-specific rates are summed to generate an estimate for the U.S. total.

To interpret these estimates, it is important to remember what the soil loss equations calculate. These equations calculate the quantity of soil detached and transported, but they ignore deposition. As such, transport from one area of an agricultural field to another or from one agricultural field to another is classified as soil erosion. Interpreting the quantity of soil moved as the quantity of soil lost may overstate the magnitude of the problem.

Controversy about soil erosion focuses on the fraction of eroded soil that is lost from agriculture. Soil that is lost from agriculture due to sheet, rill, or gully erosion ends up in streams as sediments. This implies that the fate of eroded soil can be proxied by the sediment delivery ratio, which is the fraction of eroded soil that ends up in sediments. The sediment delivery ratio can vary greatly. Detailed studies of Coon Creek in Wisconsin show that streams receive only about 8 percent of the eroded soil that is simulated by the universal soil loss equation. This ratio was well over 100 percent in the 1930s, when nearly everyone agrees that soil erosion was more severe than today.

The universal soil loss equation indicates that extended areas of the United States lose more than 20 metric tons per hectare (9 tons/acre—see Figure 15.18a). But the amount found in sediments implies a loss of 0.45–1.8 metric tons (0.5–2.0 tons) per hectare. For the entire United States, 2.4–3.6 billion metric tons (2.7–4.0 billion tons) of soil are eroded, but only about 0.45 billion metric tons (.5 billion tons) appear as sediment. This would imply that only a small fraction of eroded soil is lost from agriculture. If this is correct, soil erosion is a much smaller problem than is indicated by calculated rates of erosion.

Allowing for the possibility that much of the eroded soil is not lost, soil movements may still be unsustainable if soil is lost faster than it is generated. Studies indicate that the U.S. average rate of soil erosion has declined from about 17 metric tons per hectare per year (7.6 tons/acre/year) to less than 13 metric tons per hectare per year (5.4 tons/acre/year). Even these lower rates of erosion may be worrisome because they are about ten times greater than the average rate of soil formation. But averages can be misleading. Some studies show that most agricultural activity occurs on soils that can tolerate a loss of about 11 metric tons per hectare per year (4.9 tons/acre/year). If this is correct, about a third of U.S. agricultural land is being eroded faster than the sustainable rate (assuming all of the soil eroded is lost).

Differences in the rates and severity of erosion generate uncertainty about the economic impacts of erosion. High rates of soil erosion imply a loss of nearly $30 billion per year. Lower rates of soil erosion would reduce yields by only 2–4 percent, even if those lower rates continued for a hundred years. If these lower estimates for the rate of soil erosion are correct, the U.S. government is spending too much money trying to control soil erosion. But if the higher rates are correct, the U.S. government needs to do more to slow erosion. Currently policy makers are not sure what to do because scientists cannot narrow the range of uncertainty about rates of soil erosion and soil loss.

STUDENT LEARNING OUTCOME

- Students will be able to explain the source of uncertainty about rates of soil erosion and how this affects efforts to implement policy to slow soil erosion.

site. During the 1980s and 1990s farmers in India lost about 6.6 billion metric tons of soil annually. This soil contained about 5.4 million metric tons of fertilizers, which were worth about 2.2 billion rupees ($1–2 billion). During the same period in Costa Rica the value of the nutrients lost via soil erosion represented about 17 percent of the value of the crops produced.

Conserving Soil

The private and social costs of soil erosion can be either moderated or exacerbated by farming techniques. Farming techniques that are designed to slow soil erosion are termed **soil conservation techniques.** One soil conservation technique increases the crop residues left on the soil after harvest. Crop residues protect soil from water erosion by reducing the power of raindrops and protect soil from wind by increasing surface roughness. Despite these benefits, farmers often harvest crop residues because they can be sold as animal feed.

Decisions about the direction of crop rows also have an important effect on rates of soil erosion. Economically, farmers prefer to plant crops in long straight rows because this reduces tractor work and thus costs of labor and fuel. Such rows often do not follow topography—that is, they run uphill and downhill. As such, the long (and sometimes steep) slope accelerates surface flow. Farmers can reduce this cause of soil erosion via **contour plowing** (Figure 15.15),

FIGURE 15.15 *Contour Plowing* Using the contour plowing technique, the farmer plows across the slope such that there is little change in elevation along a row.

in which the farmers plant crops in rows that cut across slopes. Rather than moving up and down hills, rows wrap around rises so that there is no slope along each row. This greatly slows surface flow (and hence erosion). In spite of these benefits, farmers are reluctant to adopt contour plowing because it increases plowing time, raises costs, and reduces the area that can be planted.

The next step in contour plowing is contour terracing (which differs from terraces that are cut into steep slopes, which are important conservation techniques in developing nations where there is plenty of cheap labor to build and maintain terraces on very steep slopes). **Terraces** consist of ridges and channels that are constructed across the slope to prevent rainfall runoff from accumulating and causing serious erosion. Surface runoff that accumulates behind a terrace drains slowly, which causes water to build up temporarily behind the inlet. The water sits long enough for sediment to settle out, but not long enough to damage the crop. Indeed, under some conditions, the gently sloped

ridges of the terrace are cultivated as a part of the field, which maintains total crop production.

As in fashion, old is new. Strip planting and shelterbelts put an end to the Dust Bowl and prevented its return, but they became less popular as the size of farm machinery increased. Economies of scale (cost savings) are possible when farm machines are used over large areas. Although big farm machines still dominate, strip planting and shelterbelts are making a comeback. Following this procedure, farmers divide their land into strips and plant crops on alternate strips (Figure 15.16). Vegetation covers the intervening strips and protects the cultivated strips from wind. This protection is enhanced by **shelterbelts,** which can be a row of trees that line every second strip. These trees also provide wildlife habitat.

Regardless of a row's shape or direction, plowing enhances soil erosion. Turning soil over breaks soil structures that resist water and wind and creates channels for surface flow. To reduce these effects, farmers have developed

FIGURE 15.16 *Strip Planting and Shelterbelts* Using the strip planting technique, farmers divide their land into strips and plant crops on alternate strips.

FIGURE 15.17 *No-Till Agriculture* Using no-till agriculture, the farmer plants the seed without turning over the soil. To prevent the seeds of unwanted plants from reaching the soil or seedlings obtaining sunlight, the farmer will sometimes cover the ground with straw or other plant material.

USING SOIL SUSTAINABLY

The potential to lose soil faster than it is formed implies that soil is a *potentially* renewable resource. In many areas farmers allow soils to erode faster than they are created and therefore "mine" soil much as copper is mined (Figure 15.18). In other areas farmers manage their land so that soil is generated and lost at about the same rate. In these areas soil is used as a renewable resource.

Farmers make decisions about how much erosion to allow based on the costs of slowing erosion and the benefits they earn by slowing erosion. Farmers can reduce the costs associated with soil erosion by investing in soil conservation techniques. Conservation techniques are not free: They cost farmers money now, but they also increase current and future benefits, which include higher agricultural production and lower costs. Farmers choose conservation techniques based on the balance between the current costs of soil conservation practices and the current and future benefits of lower soil erosion rates. Given this trade-off, the economically optimal rate of soil erosion is the level at which the marginal costs of soil conservation practices equal the marginal benefits of soil conservation. As with other environmental challenges, the optimal rate of soil erosion is not zero.

Why Don't Farmers Use Optimal Soil Conservation?

Farmers probably understand how erosion affects their revenues better than policy makers; therefore, farmers should adapt soil conservation techniques that offset the private costs

practices that reduce plowing. Practices vary from low-till, which reduces the number of times a tractor passes through a field, to no-till, in which the new crop is planted without plowing and without removing crop residues (Figure 15.17). For example, conventional tillage causes the loss of 17.9 metric tons (19.7 tons) of soil from each hectare planted in a conventional wheat and fallow system. Minimum tillage reduces erosion to 8.7 metric tons (9.6 tons). That rate is reduced to 1 metric ton (1.1 tons) by no-till farming. Low-till and no-till practices are possible due to the development of powerful herbicides, which remove weeds that compete with the crops for sunlight, nutrients, and water.

FIGURE 15.18 *Erosion-Prone Areas* Regional differences in climate, slope, and soil type make some areas vulnerable to (a) water-driven erosion or (b) wind-driven erosion. *(Source: Data from Natural Resources Conservation Service, USDA.)*

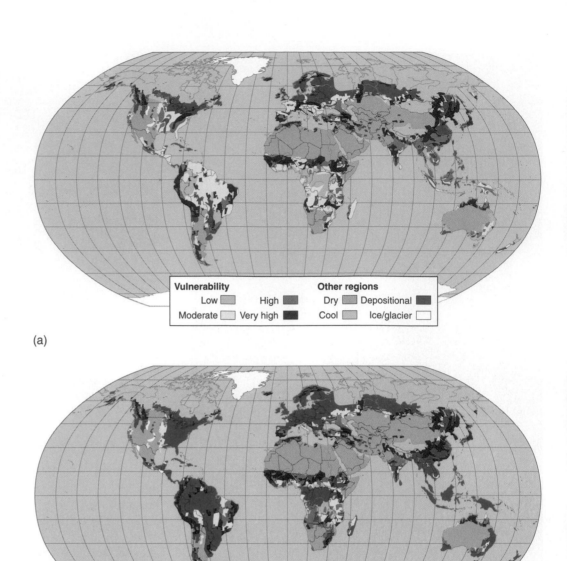

(a)

(b)

of soil erosion. (There is no incentive for farmers to offset the social costs of soil erosion; see Chapter 17.) Yet studies show that farmers do not invest in soil conservation techniques at an optimal rate. Why do farmers underinvest in soil conservation techniques? Such underinvestment leads to rates of soil erosion that reduce the economic well-being of farmers, which are known as **economically excessive rates of soil erosion.**

Population growth is an important cause for economically excessive rates of soil erosion. Thinking back to Malthus, population tends to grow exponentially while arable land grows linearly (at best). So population growth pushes farmers onto land of decreasing quality. Land quality is determined in part by the soil's vulnerability to soil erosion. In many developing nations, population growth increases soil erosion by bringing ever more vulnerable soils under cultivation.

The effect of population growth on soil erosion depends on the technologies that farmers choose. Sometimes population growth can actually reduce erosion. As population grows, land becomes more valuable. High land values encourage farmers to increase profits by using more fertilizers, pesticides, tractors, and so on. (See Chapter 16 for an explanation of this process.) These investments spur the development of social institutions that enforce private property rights. Ownership reduces farmers' fears that their land can be confiscated, which gives them confidence to invest in soil conservation techniques.

In very poor nations farmers do not have enough money to buy tractors or significant quantities of fertilizers and pesticides. Under these conditions population growth simply means that farmers add more workers. Workers often are hired in institutional arrangements that tend to discourage soil conservation practices. In many cases workers do not own the land—they are tenant farmers. As tenants they have no economic incentive to invest in conservation practices (the landowner could terminate the lease and capture the benefits of any conservation measures implemented by the tenants). The seasonal nature of the added demand for labor also discourages soil conservation measures: High demand for labor during spring planting may preclude soil conservation practices, which often are best installed in the spring.

Agricultural subsidies that reduce the price of fertilizer and water encourage their use and accelerate soil erosion. For example, the price of nitrogen fertilizers in several Asian nations is 30–60 percent below their cost of production. Low prices allow farmers to hide the cost of soil erosion—they simply replace the lost soil and nutrients with artificially cheap fertilizers. Low fertilizer prices also blind farmers to the benefits of cost-effective soil conservation measures. In 1987 farmers in Java used 1,000 kg of subsidized fertilizer to produce a crop that was 50 percent smaller than what they could have grown had they applied green manure and reduced erosion.

Policies that seek to help farmers or consumers also contribute to soil erosion. In developed nations governments often help farmers by establishing a price floor: If the market price for a crop such as corn drops too far, the government will pay the farmer the difference between the floor price and the market price. These payments increase farmer profits, which they use to expand production onto marginal land. Marginal lands often are more vulnerable to soil erosion because their soils have a higher erodibility factor and often have a steeper slope. Similar effects are generated by limiting imports, which raise domestic prices. In Indonesia limiting food imports has encouraged farmers to expand vegetable production by planting on steep slopes.

Policies that lower food prices are equally troublesome. Many developing nations subsidize food prices in the hope that lower prices will alleviate poverty and malnutrition. These policies help city dwellers but hurt farmers by reducing their profits. Lower profits mean that farmers have less money to invest in soil conservation practices that would otherwise make economic sense.

Recall from page 278 how the discount rate affects the present value of the crops lost to soil erosion. This effect implies that the rate at which farmers value future income also affects their willingness to invest in soil conservation practices. Economists find that the discount rate used by individual farmers, who have a relatively short lifetime, is greater than the discount rate used by society, which has a much longer lifetime. If farmers don't value future harvests as highly as society, the initial costs of soil conservation practices will discourage investment. Under these conditions soil erosion will exceed the rate that society would deem optimal.

Finally, poverty and lack of credit may exacerbate soil erosion. Poor farmers may not be able to afford soil conservation practices, even if they are economically justified,

because they do not have access to credit. A survey of farmers in Java indicated that 87 percent of the poor farmers in an upland area could not build terraces to control erosion because they were too poor and did not have sufficient access to credit.

Implementing Soil Erosion Policy—Limits to Market-Based Mechanisms?

In theory, policy makers can slow economically excessive rates of soil erosion by forcing farmers to bear the full cost of their actions. In some cases doing so involves seemingly simple fixes like removing subsidies. But such a solution often is difficult to implement. Once institutionalized, subsidies create a group of like-minded individuals who will resist efforts to curtail the programs. Consider the efforts to eliminate agricultural subsidies in the United States. Rather than eliminate subsidies, the government has tweaked existing programs and added new subsidies. As a result, government efforts to reduce soil erosion have become a confusing hodgepodge of laws and subsidies that may do more harm than good. See *Policy in Action: Contradictions in U.S. Soil Erosion Policy.*

Alternatively, policy makers may strive to reduce soil erosion in an economically efficient manner using market-based mechanisms such as taxes or tradable permits. To date these policy instruments have not been effective—perhaps because soil erosion differs from the other environmental challenges in which market-based mechanisms have been successful. Think back to the description of taxes and tradable permits for carbon dioxide emissions (Chapter 13). At their heart these instruments were relatively simple because the relationship between burning fossil fuels and the emission of carbon dioxide is easily measured (Table 13.2). Comparable measurements for soil erosion are not possible. Rather, soil erosion is approximated from improved versions of the universal soil loss equation and the wind erosion equation. But even these equations are approximations whose estimates are uncertain.

Approximations create uncertainty for those who must pay taxes or buy tradable permits. To illustrate, suppose you are a farmer who uses the erosion equations to calculate how much tax you must pay for the soil erosion on your fields. For each variable in the equation, a range of values is possible. Given this range, you are likely to choose a value that reduces your taxes. In short, the system is too vulnerable to abuse to establish an effective market.

The ability to manipulate the system creates the **second best problem,** which says that when the theoretical conditions that underlie a policy are not satisfied, implementing the policy may make things worse. Because of this danger, it may be better to choose a second best solution that is less damaging. Although market-based mechanisms for reducing excessive soil erosion may be best in theory, in reality the difficulties associated with their implementation may imply that command and control policies may be better.

POLICY IN ACTION

Contradictions in U.S. Soil Erosion Policy

The U.S. government spends more than $1 billion annually to reduce soil erosion. One important component of this effort is the Conservation Reserve Program. Farmers in this program stop planting on highly erodible cropland for ten years. In return, farmers receive an annual payment. But the effectiveness of the Conservation Reserve Program is uncertain. It is not clear whether the Conservation Reserve Program actually targets highly erodible soil and whether its effectiveness is diminished by contradictions within the legislation and with other government programs.

Studies show that most agricultural losses associated with soil erosion occur on a relatively small fraction of cropland. The concentrated nature of the problem implies that to be effective, the Conservation Reserve Program must target cropland that is vulnerable to erosion. But as with many government programs, the Conservation Reserve Program can be abused. Some evaluators argue that the process by which land is brought into the program does not target vulnerable soils. Rather, the lands enrolled seem to maximize the number of participating acres and the money paid to farmers.

Concerns about the lands enrolled in the Conservation Reserve Program are highlighted by geographic differences in participation rates. The highest rates of participation occur in the Great Plains and the Mississippi River Delta. But the most severe problems associated with soil erosion occur in the Southwest. Quantitative analysis of the Conservation Reserve Program indicates that the presence of soils with a high erodibility factor increases the probability that the cropland will be enrolled. But this probability is not related to the cropland's tolerance to soil erosion. These results imply that the money spent on the program may be directed to areas that are vulnerable to erosion, but not necessarily the parcels that are least able to resist its effect.

But even if the Conservation Reserve Program were reformed to enroll only the least tolerant soils, the program might initiate a negative feedback loop that would offset some of the reduction in soil erosion (Figure 1). If the program idles large areas of cropland, this will reduce the area planted and ultimately the crops supplied. Lower supplies will increase prices. And higher prices will encourage farmers to increase production on remaining croplands and expand production into new areas. Together such changes would increase soil erosion and thereby offset some of

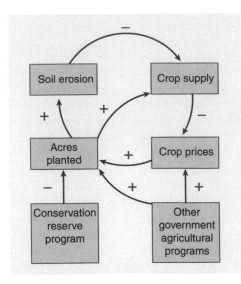

FIGURE 1 *More Soil Erosion Policy* Feedback loops *diminish the ability of the Conservation Reserve Program to slow soil erosion. For example, enrolling land in the reserve reduces crop supply, which drives up prices. The increase in price induces farmers to bring new land under cultivation. This new land often is more vulnerable to soil erosion. This effect is amplified by other government programs that bring new land under cultivation.*

the reductions generated by the Conservation Reserve Program. A detailed study finds that twenty new hectares are brought into production for every hundred hectares that are idled.

Even more important are other government agricultural programs that tend to offset the reductions in soil erosion generated by the Conservation Reserve Program. The U.S. federal government pays farmers directly and indirectly through a variety of programs that are designed to encourage production. By doing so the government hopes to keep food prices low for U.S. consumers and to preserve family farms.

But a quantitative analysis shows that government payments tend to offset the benefits of the Conservation Reserve Program. A 1 percentage point increase in the total hectares that are enrolled in the program reduces erosion by about 0.63 metric tons/hectare (0.28 tons/acre). Conversely, a 1 percentage point increase in the proportion of farm revenues associated with direct payments from farm programs tends to increase cropland, which increases soil erosion by 0.31 metric tons/hectare (0.14 tons/acre).

We can evaluate the size of these effects by comparing the reduction in soil erosion associated with the Conservation Reserve Program

to the increase in soil erosion associated with the general increase in direct government payments to farmers. Between 1982 and 1992 participation in the Conservation Reserve Program grew from 0 to 3.6 percent. This rise implies an average reduction in soil erosion of 2.3 metric tons/hectare (1.02 tons/acre). Between 1982 and 1992 the fraction of gross farm income from government payments increased from 3.5 percent to 7.6 percent. This rise implies an increase in soil erosion of about 1.3 metric tons per hectare (0.59 tons/acre). This increase represents a little more than half of the reduction in soil erosion associated with the Conservation Reserve Program.

Together these results indicate that programs to reduce soil erosion are less effective than often thought. The lands idled as part of the Conservation Reserve Program are not necessarily those most in need of protection. Of the lands that are idled, reductions in erosion are offset by bringing new land into production and increases in erosion associated with other government programs. Congress has modified these programs since the analysis of their effects was published, but these changes are too recent to generate the many observations needed to evaluate their overall effectiveness.

ADDITIONAL READING

Goodwin, B.K., and V.H. Smith. "An Ex Post Evaluation of the Conservation Reserve, Federal Crop Insurance, and Other Government Programs: Program Participation and Soil Erosion." *Journal of Agricultural and Resource Economics* 28, no. 2 (2003): 201–216.

Roberts, M.S., and S. Bucholtz. "Slippage in the Conservation Reserve Program or Spurious Correlation? A Comment." *American Journal of Agricultural Economics* 87 (2005): 244–250.

Wu, J. "Slippage Effects of the Conservation Reserve Program." *American Journal of Agricultural Economics* 82 (2000): 979–992.

Wu, J.J. "Slippage Effects of the Conservation Reserve Program: Reply." *American Journal of Agricultural Economics* 87 (2005): 251–254.

STUDENT LEARNING OUTCOME

- Students will be able to explain how efforts to slow soil erosion may have exacerbated soil erosion.

Command and control lay at the heart of perhaps the most important example of successful soil erosion policy—alleviating the Dust Bowl. At the time erosion was controlled primarily by strip cropping: alternating strips of crops and fallow land (land planted in native vegetation) and windbreaks of trees or brush. These techniques were not used efficiently because average farm sizes were too small to capture the benefits. Eroding soil represented a negative externality—farmers did not have to pay for crops that were destroyed downwind by the soil that eroded from their land. Because of this externality, the benefits to individual farmers of erosion control mechanisms were less than their costs. The poor cost–benefit ratio discouraged farmers from implementing the optimal level of erosion control devices even when demonstration projects highlighted the benefits to the region as a whole.

To overcome this externality, the government strongly encouraged farmers to implement control mechanisms. The Soil Conservation Service set up soil conservation districts. These districts, large relative to individual farms, were used to determine the optimal level of erosion control. This optimal level was then recommended to individual farmers through a series of subsidies and penalties. Their subsequent collective actions reduced erosion rates. The success of this program can be judged by the effects of subsequent droughts. During the droughts of the 1950s and the 1970s, which were similar to those of the 1930s, soil erosion levels remained well below those that prevailed during the Dust Bowl.

SUMMARY OF KEY CONCEPTS

- Land use is closely tied to soil. Soil is formed by weathering and chemical changes that are driven by climate and biological organisms. These processes generate different layers, which are used to classify soils and influence a soil's ability to support biological activity and agricultural production.

- Soil supports plant growth by storing water between precipitation events. The amount of water available to plants depends on the amount of water a soil can store and the tenacity with which the soil holds that water. Both traits are determined by the size of the particles that make up the soil. Soils with the greatest capacity to store water include particles of many different sizes.

- Many plant nutrients are positively charged ions; therefore, a soil's ability to store them is determined by the presence of negatively charged areas. These areas increase with the surface area of the soil particles. The ability of acid soils to store nutrients is diminished by the presence of positively charged hydrogen ions, which occupy negatively charged areas that otherwise would hold nutrients.

- The rate of soil erosion is determined by the rates at which soil particles are detached and transported. The rates of detachment and transport are influenced by the degree to which soil is exposed to wind and water. This exposure is determined by local climate, soil type, vegetative cover, and slope. Erosion rates also are influenced by agricultural practices.

- Soil erosion inflicts damages on both society and the farmer. Even though farmers understand these effects, spending on soil conservation practices often is less than the benefits that these practices would generate. For these cases, rates of soil erosion are too high from an economic perspective.

- Ideally policy to slow soil erosion would internalize the externalities that cause farmers to spend too little on soil conservation measures. But measurements of soil erosion are fraught with uncertainty, so policies based on market-based incentives may be ineffective. Instead historical experience with the Dust Bowl indicates that policies based on command and control mechanisms may be more effective.

REVIEW QUESTIONS

1. Explain why soil types often are found in connection with specific biomes.

2. Explain how particle size determines a soil's wilting point and field capacity and how these factors affect the amount of water available to a plant.

3. Explain the similarities and differences in the factors that determine the rate of water-driven soil erosion and wind-based soil erosion. Discuss how these factors influence the effectiveness of various soil conservation measures.

4. Explain how social institutions affect the willingness of farmers to adopt soil conservation measures.

KEY TERMS

A horizon	capillary water	clay
aridisols	carbonation and solution	climate
B horizon	cation exchange capacity	contour plowing
C horizon	cations	creep

crop residues

decomposition

disintegration

E horizon

economically excessive rates of soil erosion

eluvial

erodibility factor

field capacity

gravitational water

gully erosion

horizons

humus

hydrolysis

hygroscopic water

illuvial

loam soil

macropores

micropores

mollisols

O horizon

oxidation

oxisols

parent material

pores

private costs

R horizon

regolith

rill erosion

rooting depth

saltation

sand

second best problem

sheet erosion

shelterbelts

silt

social costs

soil

soil conservation techniques

soil erosion

soil profile

spodosols

subsoil

surface roughness

suspension

terraces

topsoil

total saturation turbulent

universal soil loss equation

weathering

wind erosion equation

wilting point

CHAPTER

16

AGRICULTURE
The Ecology of Growing Food

STUDENT LEARNING OUTCOMES

After reading this chapter, students will be able to

- Compare and contrast the advantages and disadvantages of agriculture relative to hunting and gathering.
- Explain how economic and ecological conditions affect agriculture practices and the technology used.
- Discuss whether Green Revolution agriculture contributes to or detracts from sustainability.
- Discuss whether agriculture will be able to feed a population that is projected to grow in numbers and affluence over the next century.
- Assess the contradiction that India used to import food to alleviate famines but now exports food, but its number of malnourished individuals has increased.

Abu Hureyra *Excavation of the ancient city of Abu Hureyra. This is what remains of a five room house. The inhabitants of this town changed from hunting and gathering to agriculture about 11,500 years ago.*

The land of milk and honey may be a euphemism for an abundance of oaks and pistachios that supported many people in areas that are now Israel, Lebanon, and Syria. Among the most understood of these people are those who lived in Abu Hureyra in what is now Syria (opening photograph), who left enough artifacts to follow their change from hunter and gatherers to agriculturalists.

During the ice age people who lived in the Middle East were hunters and gatherers. They gathered seeds from April to June and fruits from September to November. They hunted animals such as gazelles, deer, wild boar, and aurochs (wild oxen—aurochs became extinct in 1627). Of these foods plants were less important because the local climate was dry.

About 15,000 years ago Earth's climate warmed, which brought more rainfall to the Middle East. This rainfall supported an extensive area of oak and pistachio trees. People took advantage of this abundance, including those who lived at Abu Hureyra, which started as a small village about 13,500 years ago. After about 500 years the abundance of acorns, pistachio nuts, and desert gazelles supported relatively large numbers of people who lived in permanent villages.

Earth's climate changed again about 13,000 years ago: The melting of the ice sheet in North America allowed a pulse of fresh water to drain into the North Atlantic, where it slowed thermohaline circulation. This change caused a cooler period known as the younger dryas (see Figure 13.1(a) on page 270). This period is associated with a thousand years of cooler and drier climate in much of the Middle East.

This drought hit the hunter–gatherers of the Middle East hard. Dryer conditions replaced forests with grasslands, and the people of Abu Hureyra started to eat the seeds of wild grasses. These seeds required more work than the acorns and pistachios, so people worked harder for less food. Over time the local population dropped. About 12,000 years ago Abu Hureyra was abandoned as people dispersed to smaller settlements.

Many hunter–gatherers realize that plants come from seeds. This knowledge and the hard times imposed by drier conditions generated the next step: Instead of gathering wild grasses, they started to grow wild grasses from seed. The cultivation of wild grasses marked the start of agriculture in the Middle East. Agriculture spread quickly among the people of the region. When the climate warmed again and people returned to Abu Hureyra about 11,500 years ago, the new, larger settlement was based on agriculture. ■

A BRIEF HISTORY OF FOOD PRODUCTION

Hunting and Gathering versus Agriculture

For most of their existence humans obtained food by hunting and gathering. **Hunting** refers to the capture of animals for use as food, and **gathering** refers to the collection of plant material for use as food. Hunting and gathering occur in relatively unmodified ecosystems. That is, hunters and gatherers obtain their food from the edible energy flows of natural ecosystems.

The fraction of total energy flows in natural ecosystems that can be eaten by people is relatively small. Humans cannot digest cellulose, which is the main component in plant cell walls. Therefore, humans eat only a small fraction of the plant species on Earth; and of these species, humans consume only small portions of the plants. The term **vegetable** is generally defined as the edible seeds, roots, stems, leaves, bulbs, tubers, or nonsweet fruits of herbaceous plants. The most commonly eaten portions include nuts, berries, and seeds. Other edible portions include roots such as carrots, tubers such as potatoes, and some types of leaves such as lettuce. Meat from animals is more digestible than vegetable matter, but the amount of food available from herbivores and carnivores generally is smaller than that available from plants due to losses associated with trophic inefficiencies (Chapter 5).

The low rate at which natural ecosystems generate edible energy influences many aspects of hunter–gatherer societies. Most notably, the number of people per unit area, or **population density,** of hunter–gatherer societies is low (Figure 16.1). On average, population density is less than 1 person per square kilometer (fewer than 3 people per square mile). Variations in population density among hunter–gatherer

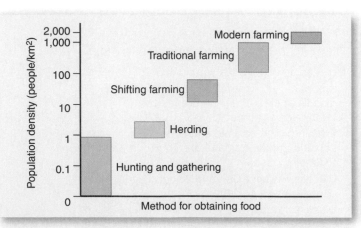

FIGURE 16.1 *Food Strategy and Population Density* The density of hunters and gatherers is small relative to agriculturalists. The y-axis is in log 10, showing that traditional farming methods support 100–1,000 people per km² compared to hunting and gathering, which support at most 1 person per km².

societies are determined in part by net primary production. In productive tropical rain forests the population density of hunter–gatherers can exceed 1 person/km². Population densities often are less than 0.1 people/km² (0.3 people/mi²) in less productive ecosystems, such as boreal forests.

Small population densities do not mean that hunter–gatherers live on the edge of starvation. Many hunter–gatherers are well fed because they are very efficient in getting food from the environment. Many hunter–gatherers spend only four to six hours per day obtaining food. As a result, many hunter–gatherer societies have considerable spare time. (How does this compare to the amount of spare time you have?) Some anthropologists call hunter–gatherers the original leisure society.

The relatively small fraction of the day that is spent obtaining food implies that hunting and gathering have a relatively high energy return on investment (Chapter 5, page 77). Measurements indicate that the energy return on investment for hunting and gathering ranges between 10:1 and 20:1. This ratio implies that on average, a hunter–gatherer obtains between 10 and 20 kcal of food for every 1 kcal used to obtain food. Some hunting techniques, such as the capture of large marine mammals, can have an energy return on investment of well over 100:1.

Agriculturists obtain food by changing natural ecosystems in a way that increases the amount of edible energy generated. To do so, agriculturists replace species that produce relatively little edible energy, such as trees, with species that generate more edible energy such as **cereals**, which are cultivated members of the grass family whose seeds (**grains**) are eaten by people or domesticated animals.

Agriculturists also change the characteristics of the species they use for food through **domestication.** Domestication modifies a species relative to its wild ancestors by **selective breeding,** in which only individuals with traits desired by agriculturists are allowed to reproduce. As such, people act in place of natural selection. For example, the wild ancestor of wheat produces its seeds at the top of a stalk that "shatters" spontaneously, thereby dropping its seeds to the ground. This makes it difficult to harvest. The domesticated version of wheat is bred from a form that does not drop its seeds and is therefore easier to harvest.

Agriculture influences many aspects of agricultural societies. Replacing inedible species with edible species increases the amount of food that is available per unit area; therefore, the population densities of agricultural societies are much larger than those of hunter–gatherer societies. Even traditional farming methods can support 100–1,000 people/km² (260–2,600 people/mi²) (Figure 16.1), which is two to three orders of magnitude greater than hunter–gatherers.

Relatively high population densities do not imply that agriculture is easy. The stereotype of a hardworking farmer is accurate. For reasons described in this chapter, farming requires a lot of time and effort. Therefore, the energy return on investment for agriculture can be relatively low, and often it is smaller than the energy return on investment for hunting and gathering. The transition from hunting and gathering to agriculture often is associated with smaller body size, poorer nutritional status, and increased disease. So why did hunter–gatherers switch to agriculture?

Why Agriculture?

Agriculture is a relatively recent innovation. Although humans have been around for nearly 4 million years, people started to practice agriculture between 5,000 and 12,000 years ago. Agriculture developed independently in at least nine areas (Figure 16.2). The earliest sites included southeast Asia (such as Papua New Guinea), where agriculture arose around the production of root crops, and the Fertile Crescent (Iraq, Syria, and Israel), where agriculture arose around wheat or barley. Later sites included sub-Saharan Africa and South America, where agriculture arose around the production of root crops.

Why did agriculture originate in these geographically disparate areas? Some scientists assert that these areas were home to species that could be domesticated. Relatively few

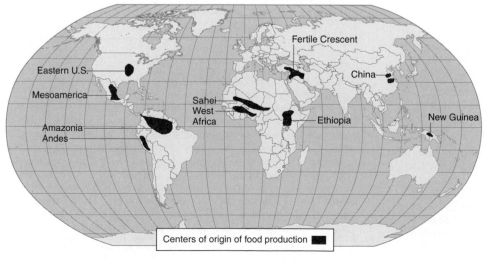

FIGURE 16.2 *The Origins of Agriculture*
Agriculture started independently in at least nine sites over the previous 12,000 years. (Source: Redrawn from J. Diamond, "Evolution: Consequences and Future of Plant and Animal Domestication," Nature 418: 700–707.)

Fertile Crescent

Eastern U.S.

China

Mesoamerica

Sahel West Africa

Ethiopia

New Guinea

Amazonia Andes

Centers of origin of food production ■

species can be domesticated. For example, the wild ancestors of almond and oak trees produce nuts with a bitter taste. Bitterness is controlled by a single gene in almond trees, so selective breeding was able to eliminate the bitter taste. On the other hand, the bitter taste in acorns is controlled by many genes, so oak trees cannot be domesticated (the people of Abu Hureyra worked hard to process acorns into a form that can be eaten). Similarly, zebras cannot be domesticated, whereas wild horses can be and were domesticated.

Given the opportunity, anthropologists and demographers suggest three reasons why a given society will adopt agriculture. One theory, the **technical change hypothesis,** holds that agriculture arose with increasing human technical capabilities. This hypothesis seems consistent with the observation that agriculture requires specialized tools and technologies. But there are some important contradictions. Some agriculturists use relatively simple tools, such as digging sticks, whereas some hunter–gatherer societies use sophisticated tools like poison arrows or complex fishing equipment.

Another explanation, the **coevolutionary hypothesis,** holds that agriculture coevolved with humans. According to this view, there is a positive feedback loop that includes the human population and the plants and animals it eats (Figure 16.3). The positive feedback loop works as follows. In the process of providing wood for fires and building materials, humans remove trees. Removing trees creates sunlit areas. Sunlit areas provide ideal habitat for edible plants. In addition, the organic wastes and human wastes generated by hunter–gatherers contain seeds of plants that they eat as well as nutrients that increase soil fertility. The combination of sunlight, seeds, and nutrients enhances the growth of edible plants near human settlements. The prox-

imity and abundance of edible plants increases food supply and lets the local population grow. Population growth reinforces this process by speeding the rate at which trees are removed and the soil is fertilized.

A third explanation, the **resource depletion hypothesis,** holds that agriculture is a response to population growth and the best first principle. We can understand the role of resource depletion by examining the relationship between population size and the energy return on investment of hunting and gathering versus agriculture (Figure 16.4). Making the tools needed for hunting and gathering requires relatively little effort. And the costs of hunting and gathering also are relatively small so long as the human population is small. As such, hunting and gathering is the easiest way for small populations to obtain food. But as a hunter–gatherer population grows, it depletes the local supply of edible plants and animals. This forces them to look longer and walk farther to obtain food. Such efforts increase the cost and lower the energy return on investment for hunting and gathering.

The energy return on investment for small-scale agriculture is low because of the high costs of establishing an agricultural system (the reasons are explained in the next section). But once such a system is established, relatively little effort is required to expand food production. As a result, the costs of obtaining food via agriculture rise slowly as population increases. Because the costs of agriculture grow less rapidly than those of hunting and gathering, the cost of obtaining food via agriculture for large populations is lower than hunting and gathering. Agriculture may have

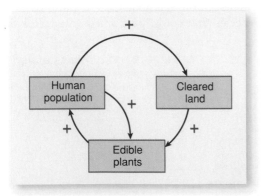

FIGURE 16.3   *The Coevolutionary Hypothesis* *The coevolutionary hypothesis postulates that agriculture originated with a positive feedback loop that includes human populations, edible plants, and cleared land. According to this hypothesis, people clear land, which opens space for edible plant species, whose seeds are contained in human fecal material. The local availability of plants enhances population growth, which opens more land and accelerates the process.*

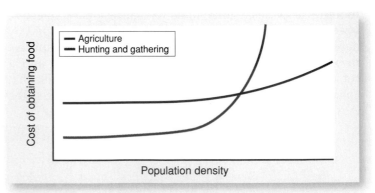

FIGURE 16.4   *The Costs of Hunting and Gathering versus Agriculture* *Hunters and gatherers invest relatively little in producing their tools, which implies that the cost of obtaining food is low initially. But as population density increases and local plants and animals have been collected, hunters and gatherers must walk long distances to obtain food, which raises the energy cost of obtaining food. Setting up an agricultural system requires a lot of initial effort. But increasing output from an agricultural field requires relatively little extra effort; so greater population density increases the cost of obtaining food relatively slowly. At some intermediate population size the cost of obtaining food from hunting and gathering exceeds the cost of agriculture, and people convert from hunting and gathering to agriculture.* (Source: Data from C.A. Hastorf, "Changing Resources Use in Subsistence Agricultural Groups in the Prehistoric Mimbres River Valley, New Mexico," eds. T.K. Earle and A.L. Christenson, in Modeling Change in Prehistoric Subsistence Economies, Academic Press.)

developed in hunter–gatherer societies whose population grew beyond the carrying capacities of their local ecosystems. Consistent with this hypothesis, there are historical examples in which an agricultural society returned to hunting and gathering when their population size shrank.

Regardless of the cause, the area occupied by agriculturists spread from each of the nine points of origination (Figure 16.2). In general, agriculture spread in an east–west direction, as would be dictated by the latitudinal bands in temperature, precipitation, and solar radiation that are described in Chapter 4. Like all species, domesticated plants and animals thrive under a relatively narrow range of environmental conditions (Chapter 7). These conditions remain relatively constant moving east and west, compared to the greater changes in temperature, precipitation, and solar energy that occur in movement north and south.

Despite these difficulties, agriculture also moved north and south; hunters and gatherers now are found only where it is too cold for agriculture (such as northern Canada) or too dry for agriculture (such as central Australia), or where

hunter–gatherers have been isolated from agriculturists, such as the interior portions of the Amazon rain forest (Figure 16.5). Nearly everywhere that agriculturists have encountered hunter–gatherers, hunting and gathering practices have been displaced.

There are two ways in which agriculturists can displace hunters and gatherers: the idea (and the domesticated plants and animals) of agriculture can spread from group to group, or the agriculturists themselves can expand their range. The historical record shows that agriculture spread largely through the movement of people. Specifically, agriculturists took over land used previously by hunter–gatherers either by spreading disease or by military conquest. Due to the coevolution of agriculturists and their diseases, hunter–gatherers often are more susceptible to the diseases that afflict agriculturists, such as chickenpox or smallpox. As a result, hunter–gatherer populations often suffer epidemics when first contacted by agriculturists. And in the case of violent conflict, hunter–gatherers usually retreat because their small numbers are no match for the higher population densities of agriculturists.

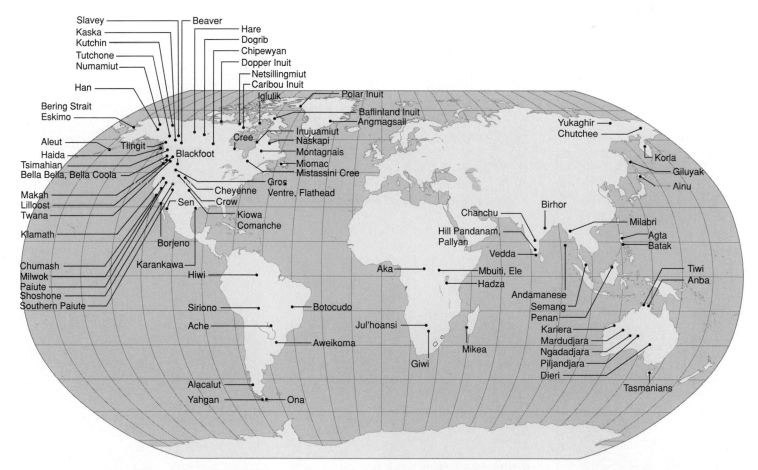

FIGURE 16.5   *Remaining Hunters and Gatherers*   *The remaining populations of hunters and gatherers tend to live in areas too cold or too dry for agriculture. Or they live in remote areas where they have not yet come into contact with agriculturalists.*

THE ECOLOGY AND ECONOMICS OF AGRICULTURE

The process of modifying a natural ecosystem so that it produces more edible energy is governed by ecological and economic principles. Ecological principles describe seven steps by which people manage and convert natu-

ral systems into agricultural ecosystems: clearing land, sowing seeds, fertilizing soil, irrigation, suppressing succession, pest control, and harvest (Figure 16.6). The mix of land, labor, capital, and materials that people use to create and operate various types of agricultural ecosystems is determined by a combination of ecological and economic conditions.

(Step 1)

(Step 2)

(Step 3)

(Step 4)

(Step 5)

(Step 6)

(Step 7)

FIGURE 16.6   *The Ecological Steps of Agriculture* In the first step, natural ecosystems are removed to make room for agroecosystems, which are set up in the second step. In the third and fourth steps, farmers add nutrients and water. In the fifth and sixth steps, farmers suppress succession and competitors, which increases the amount of food that can be harvested, which is the seventh step.

The Ecology of Agriculture

Agroecosystems are simplified ecosystems that are set up and operated to produce food for people. Their setup and operation are determined by the ecological principles described in Chapters 5–8, such as disturbance, succession, biogeochemical cycling, and energy allocation strategies. The first step requires agriculturists to remove the natural ecosystem. If left intact, the original plants and animals would compete with domesticated species for light, nutrients, and water. This competition would reduce the production of edible components. Chopping down trees and removing native grasses require considerable effort and constitute the large start-up costs of agriculture that are described in Figure 16.4. From an ecological perspective, eliminating the natural ecosystem and replacing it with an agroecosystem can be viewed as a disturbance that replaces a late successional community with an early successional community of the farmer's choosing.

This choosing is the second step in establishing the agroecological community. To establish their agricultural community, farmers must plant the seeds of the desired species. Sowing seeds requires that soil be moved, seeds dropped into place, and the soil replaced. In the simplest agricultural systems seeds are sown with sticks. The development of the plow, which opens space for seeds as it is pulled through the soil, was an important technical advance. Pulling a plow, whether by farmers, animals, or tractors, requires considerable effort.

For the seedlings to mature and generate sufficient edible energy, the third step requires that nutrients be added to the soil. This step is dictated by a change in the biogeochemical cycles of agroecosystems relative to the original ecosystem. As plants or animals are removed from the agroecosystem to be eaten, so too are their nutrients. This breaks the local cycling of nutrients that characterizes many ecosystems. Every harvest reduces nitrogen, phosphorus, and other nutrients stored by the soil. Eventually the declining supply of nutrients reduces the amount of food that can be grown.

To avoid these reductions, agriculturists add nutrients to the soil. Originally this consisted of returning organic wastes to the soil. Recycling became less feasible when farmers started to sell much of their food to consumers who lived in cities. To replace the nutrients that are removed with the previous year's harvest (and to replace nutrients lost via soil erosion), industrial societies add fertilizers. The production and application of these nutrients requires much energy.

In many agricultural areas the amount of precipitation or the soils' ability to store water is so low that water often is a limiting factor. Plants need a lot of water to produce dry material. Cereal crops use 300 to 1,000 kg of water to produce each kilogram of dry matter. This implies that wheat plants transpire hundreds of kilograms of water to produce the flour used to make your pasta. Without this water, agricultural **yields** (the amount of edible food grown per land area) are much lower. For example, a 2 cm reduction in rainfall while corn plants fills their ears with seeds can reduce corn production by about 10 percent.

To mitigate the effects of insufficient precipitation, many farmers bring water to their plants. This process, termed **irrigation,** can be considered a fourth step of agricultural production. The simplest irrigation techniques involve a bucket. But such manual methods generally are infeasible because plants require a lot of water and water is heavy (1 liter of water weighs 1 kilogram; 1 gallon weighs about 8.3 lb). Instead irrigation usually involves the construction of canals and wells and the use of animals or motorized pumps. As such, irrigation uses the most energy in some agricultural systems.

Because replacing the original ecosystem can be viewed as a disturbance, the fifth step of agriculture includes efforts to suppress ecological succession. Should an early successional community become established on an agricultural field, these early succession plants would compete with crops for light and nutrients. In some cases the crops would be outcompeted because early successional species often are characterized by rapid growth rates.

To suppress the process of ecological succession, humans originally removed unwanted plants by hand—that is, weeding. As the area planted increased, hand weeding was replaced by **herbicides,** which are chemicals designed to kill unwanted plants. Herbicides often are applied several times each growing season via tractors or airplanes; therefore, their use requires considerable effort.

Many of the plant parts that are edible by people also are an attractive source of food for other species, such as insects, birds, or small mammals. These competitors are termed **pests.** As a result, farmers are forced to compete, in an ecological sense, for the net primary production of their fields. If farmers are to reap (harvest) what they sow, the sixth step of agriculture often includes efforts to reduce competition from pests by either denying them access to the crop plants or killing them.

The most famous icon of reducing competition is the scarecrow. As the name sounds, it was designed to look like a person and therefore keep crows away from the harvest. As technical capabilities increased, industrial societies developed **pesticides,** which are chemicals designed to kill insect pests or plant diseases. Like herbicides, applying pesticides requires considerable effort.

The seventh step of agriculture is identical to that of gathering—harvesting the edible portions of the crop. Agricultural harvests are much simpler than gathering because the food plants are concentrated in a single area. Originally harvesting techniques were identical to those used by gatherers. Technical innovations have produced giant machines that pick a wide variety of crops, such as wheat, corn, or rice.

The Economics of Agriculture

Nearly all forms of agriculture include some or all of the steps just described, but there is a deal of variation in how each step is accomplished. Some variation is associated with the type of crops grown. Clearly the methods used to grow corn differ from those used to grow rice. Other variations are associated with the availability of land, labor, machinery, energy, and materials. Consistent with the economic principle of minimizing costs, farmers rely on factors that are locally abundant and inexpensive and reduce the use of factors that are locally scarce and expensive. Geographic differences in the types of crops grown and the abundance of land, labor, machinery, energy, and materials have led to agricultural techniques that are adapted to local environments and economies (Figure 16.7).

In areas where land is abundant and population density is relatively low, such as tropical rain forests, slash and burn is a common form of agriculture. As implied by its name, **slash and burn** agriculture creates the agroecosystem by burning a patch of forest. This burning removes the trees and transfers nutrients from the vegetation to the soil (most nutrients in tropical biomes are stored in the vegetation, not the soil—Chapter 15). Crops are then planted and harvested for several seasons. Eventually yields decline due to the loss of nutrients caused in part by high rates of precipitation and decay. When yields drop below some threshold, farmers abandon the land, burn a new patch of forest, and start the process anew.

Slash and burn agriculture has a bad reputation because of the highly visible means by which the original ecosystem is removed and the short time during which that area is farmed. Nonetheless, slash and burn agriculture is a sustainable form of agriculture that has been practiced for thousands of years. Recently the time between burnings has decreased due to increasing population densities. Without enough time for succession to regenerate the forest, slash and burn agriculture now causes severe damage and often is not sustainable.

In other areas where land is ample but rates of net primary production are low, agriculture often takes the form of nomadic herding. Following this practice, agriculturists herd their **livestock,** which are domestic animals raised for food and fiber, over large areas. In this process the livestock harvest small amounts of net primary production over wide areas. Equally important, the livestock convert inedible plants to edible proteins. As such, nomadic herding is a form of agriculture that is well adapted to dry and cold areas. For example, nomadic herders live in areas of sub-Saharan Africa where precipitation is insufficient for crops (see Chapter 7's Policy in Action on page 126) and northern Scandinavia, where summers are too short for most crops (Figure 16.8).

A very different mix of inputs is used to grow rice in Asia. Here population densities are among the highest on Earth. This implies that land is scarce and labor is abundant. Under these conditions, agriculture is labor-intensive. Hefty amounts of labor are used to create rice paddies, which are human-made wetlands. Water flows in and out of these paddies often are carefully managed. Individual rice plants are planted by hand. Similarly, the rice is harvested by hand.

In temperate climates corn and wheat farmers supplement their efforts with domesticated animals. Using domesticated animals is possible only where land is plentiful because additional land is needed to feed the animals. This requirement implies that work animals have a high oppor-

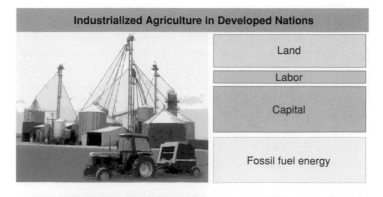

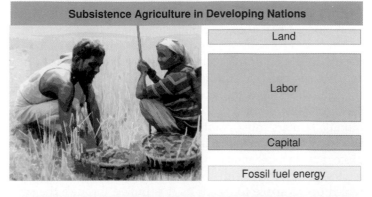

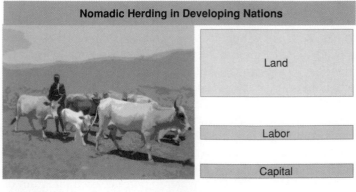

FIGURE 16.7 *Types of Agriculture Agricultural systems can be classified by their use of land, labor, and capital. Land-intensive agricultural systems include slash and burn and herding livestock. Labor-intensive agricultural systems include rice production in Asia. Green Revolution agriculture uses large quantities of capital, such as tractors, and energy that are used to drive them.*

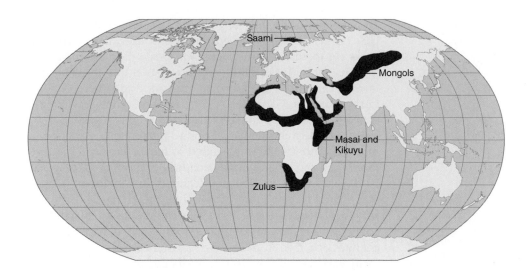

FIGURE 16.8 *Herders* Herders persist in areas that are too cold for agriculture, such as the Lapps in Northern Scandinavia, or in areas that are too dry for agriculture, such as sub-Saharan Africa or Central Asia.

tunity cost—land used to feed animals cannot be used to feed people. Why then do some forms of agriculture rely on work animals? There are two general reasons. Work animals often are used where human labor is scarce. For example, work animals were common in colonial America, where there was plenty of land but labor was in short supply.

Equally important, animals work faster and longer than people. Speed and endurance are critical in temperate regions where the growing season is short. Because frost kills many crop plants and most crop plants require three to four months to produce edible portions (mature), farmers value speed in temperate climates. Seeds must be sown quickly in spring so that crops have sufficient time to mature. In the fall the edible portions must be harvested quickly before the first frost ruins the crop (freezing ruins most fruits and vegetables by bursting individual cells and giving them a "mealy" taste). By reducing the time required to sow and harvest, domesticated animals increase the amount of land a farmer can plant and harvest in a single season.

Green Revolution Agriculture

Green Revolution agriculture evolved in the United States during the second half of the twentieth century. At that time and location the prices of land and labor were high relative to the prices of machinery, energy, and materials. Relative prices favored the substitution of machinery, energy, and materials for labor and land. Such substitutions were made possible by breeding new **cultivars,** which is a shortened term for cultivated varieties that refers to plants people have bred for a specific trait or characteristic. Cultivars bred to produce more edible food per unit area are called **high-yield varieties** and were first produced by Norman Borlaug at the International Maize and Wheat Improvement Center in Mexico. For these efforts Borlaug won the Nobel Peace Price in 1970.

To generate these high yields, plant breeders change the physiology of high-yield varieties. Three breeding strategies are possible: (1) increasing the efficiency of photosynthesis;

(2) increasing net primary production; and (3) changing the plant's allocation of energy toward parts humans can eat. As described in Chapter 4, photosynthesis is a relatively inefficient process. Less than 1.0 percent of the solar energy absorbed by a leaf is converted to chemical energy in a glucose molecule. If this efficiency could be increased, crops would have more energy, and some of this energy could be used to increase the production of edible portions. To date breeders have not been able to increase the photosynthetic efficiency of crops significantly. Increasing this efficiency probably will remain difficult. A simple genetic change that increases photosynthetic efficiency probably would be strongly favored by natural selection; therefore, any such simple change likely would already be part of the genetic makeup of existing crops.

Instead breeders increase yields by increasing net primary production and by changing the energy allocation strategy of plants. To increase corn yields, breeders have developed cultivars that produce more net primary production. For rice and wheat, breeders have developed cultivars that allocate more energy toward edible plant parts. This allocation is summarized by the **harvest index,** which is the ratio of grain to total crop biomass. For example, the harvest index for rice was increased by making the plant shorter and making its stalks stronger (Figure 16.9).

The productivity of high-yield varieties can be seen in the increasing yields over time. Prior to the Green Revolution, breeders increased crop yields, albeit slowly (Figure 16.10). After development of the high-yield varieties, yields increased dramatically. For example, U.S. wheat yields averaged about 1 ton per hectare during the first half of the twentieth century. Since the 1990s yields have averaged between 2 and 2.5 tons/hectare. Similar increases have been observed for other crops such as corn, rice, and soybeans.

Of course farmers always wanted plants that produced more food. So why was the Green Revolution so successful? Genetic changes associated with high-yield varieties are responsible for about one-quarter to one-third of the increase in yield. But the greatest portion of the increase

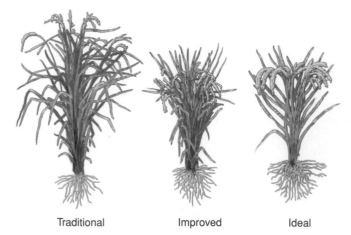

Traditional   Improved   Ideal

FIGURE 16.9 *High-Yield Varieties of Rice* Traditional varieties tend to have many thin stalks, each of which can support relatively few seeds. The ideal variety reduces the number of stalks and makes each stalk thicker, which allows it to support many seeds. As such, the ideal variety allocates less energy to stalks and more to seeds that humans eat. *(Source: Redrawn from a figure in Huang et al., "Enhancing the Crops to Feed the Poor." Nature 418 (2002): 678–684.)*

has come from increased use of energy and material inputs. Specifically, farmers now add lots of fertilizer and have expanded irrigation. These inputs allow high-yield varieties to increase their net primary production and allocate more energy to parts that people eat. Increased use of nutrients and water is responsible for two-thirds to three-quarters of the increase in yields generated by high-yield varieties.

Based on this relationship, the Green Revolution changed the mix of land, labor, machinery, energy, and materials (Figure 16.11). To ensure adequate supplies of water, the area of cropland that is irrigated in the United States increased.

To alleviate limits imposed by the availability of nutrients, U.S. farmers have increased their use of fertilizers. To suppress competition from weeds and pests, farmers increased their use of herbicides and pesticides. To apply these inputs, farmers increased their use of tractors and other types of farm machinery. This machinery is powered by inanimate sources of energy, so the use of oil, natural gas, and electricity increased. Because the use of domesticated animals has been reduced (and less land is needed to feed them) and because more food can be grown on each unit of land, the amount of land used by U.S. farmers has declined. In sum these changes make each farmer more productive; therefore, labor inputs have declined. Changes in the mix of land, labor, machinery, energy, and water are not unique to the United States: Similar changes occurred in India when farmers there adopted high-yield varieties (Figure 16.12). Reducing the use of land and labor while increasing the use of energy, materials, and water is known as **intensification.**

This same intensification philosophy has been extended to livestock. Before the Green Revolution livestock grazed mainly on **rangelands** (relatively unaltered natural ecosystems where the natural vegetation is predominantly native grasses, glasslike plants, forbs, or shrubs) or in **pasture** (land used for grazing or cultivated to produce animal feed such as hay or

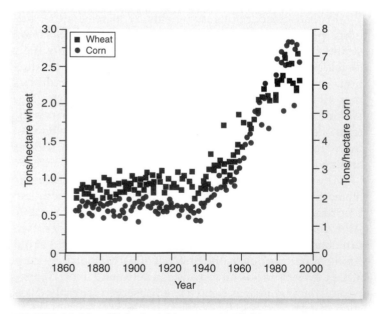

FIGURE 16.10 *Historical Changes in Yield* Yields of wheat (squares) and corn (circles) in the United States changed little between 1860 and 1940. After 1940 the Green Revolution increased their yield severalfold. *(Source: Redrawn from http://images.google.com/imgres?imgurl = http://www-formal.stanford.edu/jmc/progress/picts1-20/Figure s-16.gif&imgrefurl = http://www-formal.stanford.edu/jmc/nature/node26.html&h = 293& w = 337&sz = 46&tbnid = BVyr5BjwgQ8J:&tbnh = 99&tbnw = 114&start = 6&prev = /images%3Fq%3DUS%2Bwheat%2Byields%26hl%3Den%26lr%3D%26ie%3DUTF-8.)*

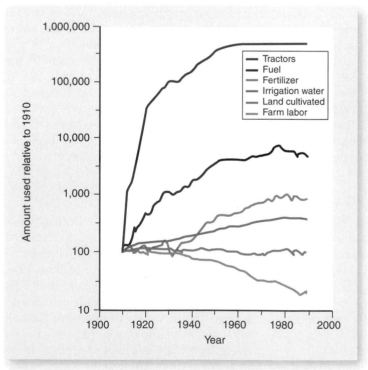

FIGURE 16.11 *Green Revolution Inputs to U.S. Agriculture* The Green Revolution increased U.S. use of tractors, fuel, fertilizer, and irrigation between 1901 and 1990. During the same period the amount of cultivated land remained roughly constant, and the amount of farm labor dropped by about 70 percent. *(Source: Redrawn from C.J. Cleveland, "Reallocating Work between Human and Natural Capital in Agriculture: Examples from India and the United States. In A. Jansson, M. Hammer, C. Folke, and R. Costanza (Eds.), Investing in Natural Capital. Island Press, Washington, DC: 1994.)*

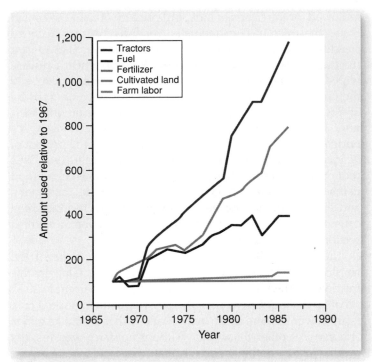

FIGURE 16.12   *Green Revolution Inputs to Indian Agriculture* *The Green Revolution increased the use of tractors, fuel, and fertilizer in India between 1967 and 1990. During the same period the amount of cultivated land and labor inputs remained roughly constant.* *(Source: Data from C.J. Cleveland, "Reallocating Work Between Human and Natural Capital in Agriculture: Examples from India and the United States" eds. A. Jansson et al., in Investing in Natural Capital, Island Press.)*

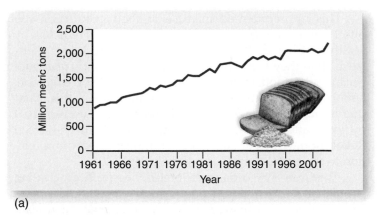

(a)

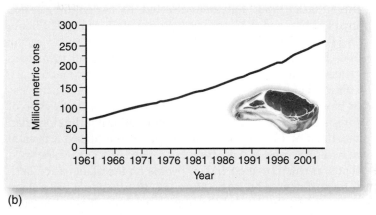

(b)

FIGURE 16.13   *Green Revolution Increases* *Global increase in the production of (a) grains and (b) meat between 1961 and 2002.* *(Source: Data from Food and Agricultural Organization.)*

alfalfa). The Green Revolution took these efforts a step further with so-called landless operations known as **feedlots,** which are confined yards where livestock eat prepared or manufactured feed. Hogs and chickens may spend their entire lives in feedlots; cattle often spend a few weeks or months in feedlots, where they gain additional weight before being slaughtered. Most feedlots produce less than 10 percent of their own feed and are known as **industrial livestock production systems.**

Industrial livestock production systems have much in common with Green Revolution agriculture. They both depend heavily on outside supplies of feed, energy, and other inputs. They require large investments in transportation and capital infrastructure and are highly productive when measured in terms of output per unit of land or labor, but are relatively unproductive when measured in terms of output per unit of energy. As such, the costs and benefits of industrial livestock systems are similar to those for Green Revolution agriculture.

THE BENEFITS AND COSTS OF THE GREEN REVOLUTION

Increases in crop yields or meat production have prompted farmers in many nations to intensify their agricultural practices. Changes in the use of land, labor, machinery, energy, and materials generate important benefits, such as slowing agricultural land conversion, increasing food supply, reducing the price of food, and enlarging the nonagricultural workforce. But changing inputs also generate important costs, such as undermining small farmers, increasing energy use, increasing agricultural pollution, and spreading pest resistance.

Benefits of the Green Revolution

The greatest benefit of the Green Revolution probably has been the rapid increase in food production (Figure 16.13). Increases in yield powered big increases in food production. Between 1961 and 2003 global production of cereals increased by about 150 percent. Between 1961 and 2002 global production of **coarse grains,** which are cereals used to feed livestock (like sorghum), doubled. This, along with industrial livestock production systems, allowed the global production of meat to triple. These increases are critical: During the same period global population doubled. The Green Revolution has helped increase the food supply for a rapidly growing population. Such increases are most dramatic in developing nations. For example, India switched from importing its grains in the early 1960s to being a net exporter of food. Without the Green Revolution per capita food intake in developing nations would be about 14 percent lower, and the fraction of malnourished children would be about 7 percent higher.

This unprecedented increase in food production has offset (or delayed) the population collapse predicted by Malthus (Chapter 9). *On average*, people are better nourished than they were fifty years ago. We emphasize *average* because the food supply has not increased for everyone. Unfortunately many people are still hungry. But this is not due to an overall lack of food. Rather, it is caused in part by social and economic changes associated with the Green Revolution, which are described in the next section in relation to the costs of the Green Revolution.

The Green Revolution ameliorated the effects of increasing population and therefore food production on natural ecosystems. Higher yields allowed agricultural production to increase much faster than the area devoted to agriculture, which increased by about 11 percent between 1961 and 2002 (Figure 16.14). See *Your Ecological Footprint: The Land Requirements of Your Diet* on pages 346–347. This relatively small increase is crucial because the conversion of natural ecosystems to agricultural land is one of the most important causes for the loss in biodiversity described in Chapter 12 and the destruction of forests described in Chapter 17. If agricultural technology had remained at 1961 levels, an additional 910 million hectares of agricultural land would have been needed to grow the food that was eaten by the 6 billion people on the planet as of 2000. (The area devoted to agriculture in 2001 was about 559 million hectares.) In the United States higher yields allowed farmers to abandon about 80 million hectares, much of which succession has returned to natural ecosystems. As such, the Green Revolution has been among the most successful, albeit accidental, policies for slowing the conversion of natural ecosystems.

The Green Revolution also powered a significant decline in the price of food. The reduction in the cost of producing food and the increase in supply combined to reduce the real (adjusted for the effects of inflation) price of food. In the United States the price of corn (adjusted for inflation) has declined steadily since 1950 (Figure 16.15). Furthermore, year-to-year fluctuations were reduced—except for 1970, when a disease wiped out a significant fraction of the U.S. corn crop. Globally the Green Revolution has reduced the real price of crops by 35–66 percent. Together with higher incomes, buying food now requires a smaller fraction of the *average* household's budget. In 2002 the average U.S. household spent 13 percent of its income on food, down from about 26 percent in 1960. This reduction implies that households can spend more on nonfood items such as housing, clothing, cars, electronic equipment, and entertainment. The decline in food prices has benefited consumers and large farmers—but as we describe in the next section, the same decline has hurt small farmers.

Equally significant, the Green Revolution reduced the fraction of the labor force that works on the farm. Globally this fraction declined from 65 percent in 1960 to 42 percent in 2000. As of 2003 about 2 percent of the U.S. population worked on a farm. Fewer farmers mean more workers that can make goods and services other than food. Without nonfarm workers, the production of nonfood goods and services would decline. And many people consider nonfood goods and services to be the basis of a high material standard of living.

The Costs of the Green Revolution

Like all technological changes, the Green Revolution also imposes considerable costs on some segments of society and the environment. Small farmers in both developed and developing nations have been hurt by the Green Revolution. From an economic perspective, the Green Revolution must be viewed as a package. To participate, farmers must buy the more expensive

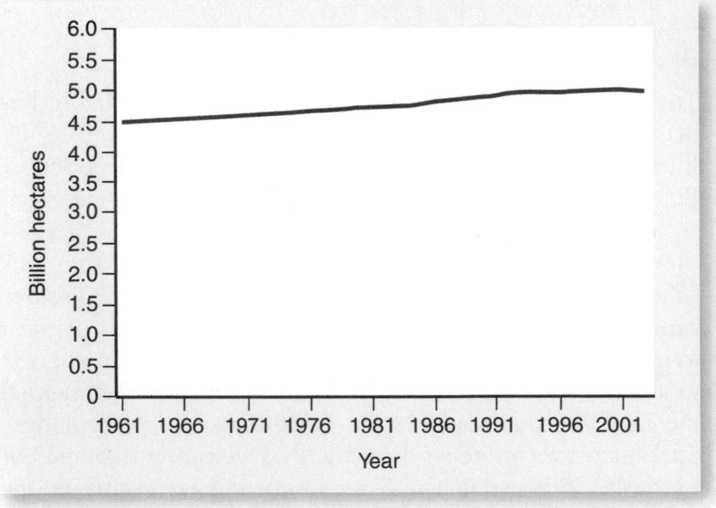

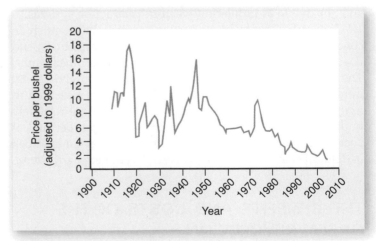

FIGURE 16.14 *Agricultural Area* *Agricultural area increased by about 11 percent between 1961 and 2003. This is much smaller than the large increases in grain and meat production that are shown in Figure 16.13.* *(Source: Data from Food and Agriculture Organization.)*

FIGURE 16.15 *Declining Corn Prices* *Before 1950 corn prices (adjusted for inflation) were high and fluctuated from year to year. After 1950 corn prices declined and the year-to-year fluctuations were reduced, except for 1970, when a disease destroyed about 15 percent of the U.S. crop.*

seeds of high-yield varieties; sow, manage, and harvest crops on large tracts of lands using mechanized equipment; and use this equipment to apply more fertilizers, pesticides, herbicides, and water. This coupling was emphasized by Norman Borlaug himself: "Once you've put together the jigsaw of production, you've got to further link it to economic policy that permits the little farmer to apply the technology."

But this is easier said than done. Large up-front cost for seeds, machinery, and inputs make it difficult for many small farmers to participate. Without access to Green Revolution technologies, farmers suffer from lower crop prices without being able to enjoy **economies of scale,** in which the unit costs of production decline as the quantity produced increases. Lower unit costs mean that large farmers make more profit than small farmers. This gives large farmers an economic incentive to buy out smaller farmers—an effect amplified by greater production and steadily dropping food prices.

Green Revolution agriculture has contributed to a class of landless farmers. Without land, they cannot grow their own food. Because they are poor, they cannot afford to buy food produced by the larger farms, even at the lower prices. This has created some disturbing contradictions. Specifically, the world grows more than enough food to feed everyone, but there are still hungry people who cannot afford to purchase food. For example, India became self-sufficient in grain production during the late 1970s, but the number of hungry people increased between 1980 and 2002.

Green Revolution agriculture also has negative environmental effects that are caused by the increased use of energy, materials, and machinery. As shown in Figure 16.11, the direct and indirect use of energy by U.S. agriculture increased by more than an order of magnitude. Since the start of the Green Revolution, energy use has increased faster than production. As a result, the amount of energy used to produce edible crops and livestock has increased.

This rapid increase in energy use has reduced the energy return on investment for our food supply. For most of human history the energy return on investment for agriculture has been greater than 1:1. By definition, if effort to grow food exceeds the amount of edible energy produced, people would eventually starve. But the Green Revolution eliminated that barrier. Now agricultural systems in developed nations produce fewer edible kilocalories than the kilocalories of coal, oil, natural gas, and electricity that they use to grow food. The energy return on investment for the U.S. agricultural system is about 1:10. That is, it takes about 10 kilocalories of inanimate energy to grow 1 kilocalorie of edible energy.

There is nothing inherently wrong with an agricultural energy return on investment that is less than 1 so long as energy supplies are abundant, energy prices are inexpensive, and energy use has little effect on the environment. But as described in Chapters 13 and 19, energy use emits carbon dioxide and methane, which are partially responsible for global climate change, and nitrogen and sulfur oxides, which cause photochemical smog and acid deposition. The financial

sustainability of Green Revolution agriculture is also threatened by the finite supply of oil and natural gas (Chapter 20). As supply declines, the price of energy will increase, and this could raise food prices and reduce supply.

In addition, Green Revolution agriculture uses huge quantities of materials—mostly nitrogen and phosphorus fertilizers. These nutrients are responsible for many environmental challenges. See *Case Study: Agricultural Pollution via the Nitrogen Cycle* on page 348. Among the most important of these challenges is eutrophication. Crops use only about 30–50 percent of the nitrogen fertilizers and about 45 percent of the phosphorus fertilizers that are applied to agricultural land. The other 50–70 percent can wash out of the soil and into local waterways. Similarly, pigs and poultry excrete about 70 percent of the nitrogen and phosphorus in their diet. In the United States much of this fertilizer and nitrogenous waste ends up in the Mississippi River and ultimately the Gulf of Mexico, where it creates the large anoxic areas that are described in Chapter 6 on page 120.

Nitrogen runoff may be even more damaging in tropical coastal areas. Here denitrification makes the water especially deficient in nitrogen; so nitrogen from agricultural activities may support large blooms of phytoplankton. This effect is illustrated by the link between wheat production in the Yaqui Valley of Mexico and phytoplankton biomass in adjacent waters in the Gulf of California (Figure 16.16).

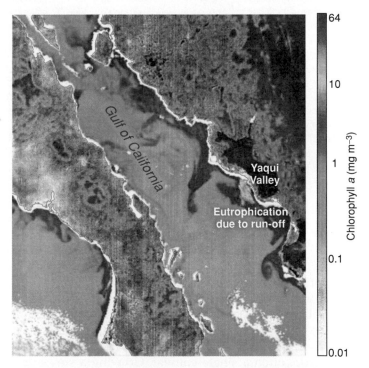

FIGURE 16.16 *Eutrophication in the Gulf of California* Some of the nitrogen fertilizer that is used to support highly productive agriculture in the Yaqui Valley of Mexico, which is indicated by dark green area, washes into the adjacent waters of the Gulf of California. Here it generates blooms of phytoplankton, which are indicated by areas of red and yellow just off the coast. *(Source: Redrawn from J.M. Beman et al., "Agricultural Runoff Fuels Large Phytoplankton Blooms in Vulnerable Areas of the Ocean," Nature 434: 211–214.)*

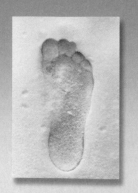

Your ECOLOGICAL *footprint*

The Land Requirements of Your Diet

How much land is used to grow the food you eat? At first glance this question seems simple to answer. It is relatively easy to obtain information about the yields of most crops. Indeed, this chapter contains information about yields for several important crops, such as corn and wheat (see Figure 16.10). Information about yields (bushels per acre) could be combined with information about your diet to calculate the amount of land used.

But compiling the relevant information about your diet is more difficult than it seems. You know what you eat—but try answering the following question: How much corn, wheat, or soybeans did you eat today? You might answer none, but this answer probably is wrong. Most of us don't eat much corn, wheat, or soybeans directly. Instead we eat these crops indirectly as bread, pasta, and other foods. The quantity of wheat cannot be deduced from the nutritional information on a box of spaghetti.

Similarly, most of us eat some corn on the cob in late summer and early fall. But this seasonal consumption does not come close to the total amount of corn grown in the United States. Most of the corn grown in the United States and other developed nations is fed to cows, pigs, and chickens. You cannot estimate this quantity of corn without knowing the diet fed to livestock and the physiological efficiency with which livestock convert various foods to meat.

The highly processed nature of the foods consumed in developed nations makes it difficult to calculate the land used to grow food. To estimate this quantity, information about two stages of food production is needed. In the first step, food is grown using land. The importance of land has long been recognized, and careful records are kept for agricultural yields.

In the second step, raw foods are processed into forms we eat. Corn, wheat, eggs, sugar, and many other ingredients are used to make breakfast cereals. Similar "recipes" are required for nearly all other foods. So the second set of information includes the recipes that are used to produce processed foods.

Food processing is a highly interconnected, technically sophisticated sector of the economy. While the number of people who work on farms to grow food has declined over the last century, the number of people who work to process food has increased. Hence input–output tables sometimes are used to trace the agricultural products that are used to produce the foods we consume (see Chapter 3's Your Ecological Footprint on page 49).

Using these tables, estimates have been compiled for the amount of land per year that is needed to produce a kilogram of food in five groups (Table 1). Though the land use requirements are calculated with data from the Netherlands, they should be applicable to other developed nations. For these nations, the conversion factors in Table 1 may understate land use because agricultural yields in the Netherlands are among the highest in the world. For example, wheat yields averaged 8.9 metric tons per hectare in the Netherlands during 1996; the European Union average was about 3 metric tons per hectare, while U.S. yields were about 2.5 metric tons per hectare.

Calculating Your Footprint

You can use the data in Table 1 to calculate the amount of land that is required per year to grow the food you eat. For Chapter 5's Your Ecological Footprint (pages 80–81) you compiled information about the food you eat in a day. Many of those foods are listed in Table 1. Calculate the area of agricultural land that is used to support one day's worth of food. For example, suppose you had a quarter-pound cheeseburger with fries for lunch. The land requirements associated with some of the more visible ingredients would be as follows:

Beef patty (.25 lb) $2.29 \text{ m}^2 \text{ year/kg} = 0.11 \text{ kg} \times 20.9 \text{ m}^2 \text{ year/kg}$

Cheese (1 oz) $0.306 \text{ m}^2 \text{ year/kg} = 0.03 \text{ kg} \times 10.2 \text{ m}^2 \text{ year/kg}$

Bun (flour, .25 cup) $0.02 \text{ m}^2 \text{ year/kg} = 0.01 \text{ kg} \times 1.6 \text{ m}^2 \text{ year/kg}$

French fries (.25 lb) $0.02 \text{ m}^2 \text{ year/kg} = 0.11 \text{ kg} \times 0.2 \text{ m}^2 \text{ year/kg}$

Much of the nitrogen that is applied to the fields is washed into waterways by irrigation (see *Policy in Action: Reducing Agricultural Pollution and Increasing Farmer Profit*), where it eventually reaches the Gulf of California. There the timing and size of phytoplankton blooms coincide with irrigation.

Most Green Revolution croplands are planted with a single crop, often a single cultivar. Growing a single crop over a large area is known as **monoculture.** Although this approach reduces costs by increasing the efficiency of farm machinery and material inputs, monoculture increases the crop's vulnerability to pests and disease. That is, a mono-culture is the ideal feeding or breeding ground for a pest or disease that specializes in that crop. Under these conditions huge economic losses are possible. For example, the corn leaf blight of 1970 ruined 15–25 percent of the U.S. corn crop and caused about $1 billion in losses.

To prevent such losses Green Revolution agriculture depends on the development and use of pesticides, herbicides, and **fungicides,** which are chemicals designed to kill fungal diseases. Over time the application of these chemicals has increased. Without these chemicals, insects and fungi would reduce the world food crop by about 70

TABLE 1 — Land Requirements for Selected Food Types

Food Item	Land (m² year kg⁻¹)	Food Item	Land (m² year kg⁻¹)
Beverages		**Milk products and eggs**	
Beer	0.5	Whole milk	1.2
Wine	1.5	Semiskim milk	0.9
Coffee	15.8	Cheese	10.2
Tea	35.2	Eggs	3.5
		Cereals, sugar, potatoes, vegetables, and fruits	
Fats			
Fats for frying	21.5	Flour	1.6
Margarine	21.5	Sugar	1.2
Lowfat spread	10.3	Potatoes	0.2
Meat		Vegetables	0.3
Beef	20.9	Fruits	0.5
Pork	8.9		
Minced meat	16		
Sausage	12.1		

Source: Data from P.W. Gerbens-Leenes et al., "A Methods to Determine Land Requirements Relating to Food Consumption Patterns," *Agriculture, Ecosystems, and Enviroments* 90: 47–58.

Fats (frying, 4 tbsp) $1.20 \text{ m}^2 \text{ year/kg} = 0.0.056 \times 56 \text{ kg} \times 21.5 \text{ m}^2 \text{ year/kg}$

Total = 3.84 m² year

Now try similar calculations for all the foods you eat in a single day. Add their total land requirements to get the amount of land that must be used for a year to grow a day's worth of food. Multiply this total by 365: This is about the amount of land that is used per year to grow your food.

Interpreting Your Footprint

As described in Chapter 5, a vegetarian diet can increase food supplies if the land used to support livestock can also be used to grow crops. To evaluate this claim, suppose you replace the burger and fries lunch with ramen noodles. Suppose the noodles are made with .22 kg of wheat flour: Your lunch would require about 0.35 m² year (0.22 kg × 1.6 m² year/kg), which is about a tenth of the value of the quarter-pound cheeseburger. (Notice that this is consistent with the trophic efficiencies discussed in Chapter 5.)

The average U.S. citizen eats about 43 kg of beef per year. This is much greater than the world average of 9.7 kg. How much more land does U.S. beef consumption represent relative to the world average? The United States is not, however, the largest consumer of beef. That honor goes to Argentina, where the average person eats 62.3 kg per year. How much more land does the average Argentinean use for beef consumption relative to the United States?

ADDITIONAL READING

Gebhardt, S.E., and R.G. Thomas. *Nutritive Value of Foods.* U.S. Department of Agriculture, Agricultural Research Service, Home and Garden Bulletin Number 72, http://www.primeindia.com/manav/food9.html.@Data on energy content of foods.

Gerbens-Leenes, P.W., S. Nonhebel, and W.P.M.F. Ivens. "A Method to Determine Land Requirements Relating to Food Consumption Patterns." *Agriculture, Ecosystems, and Environment* 90 (2002): 47–58.

STUDENT LEARNING OUTCOME

- Students will be able to explain how their dietary choices affect the amount of land required to grow their food.

percent. Agricultural chemicals reduce that loss to about 42 percent.

Chemicals may reduce losses in the short term, but they initiate an evolutionary positive feedback loop that causes farmers to increase their use of pesticides and pressures firms to develop new pesticides. The positive feedback loop works as follows. The application of a pesticide kills most individuals in a pest population. But individuals that are less susceptible to the pesticide are more likely to survive and reproduce. If resistance is based on an individual's genetic makeup, repeated application will increase the fre-

quency of genes that confer resistance in subsequent generations. The increasing number of resistant individuals forces the farmer to apply more pesticide, which accelerates the selection process. Eventually the genetic basis for resistance is present in most of the pest population. At this point the pesticide is ineffective, and firms must develop a new pesticide. Because of this feedback loop, the number of insects and mites, weeds, and plant diseases that are resistant has increased since the 1950s (Figure 16.17).

Furthermore, most pesticides and herbicides are not specific. Pesticides kill many species, including predators

CASE STUDY

Agricultural Pollution via the Nitrogen Cycle

To the casual observer, agricultural fields seem an unlikely source of pollution. There is no smokestack emitting air pollutants, nor is there a pipe spilling pollutants into a local waterway. Nonetheless, each square meter of agricultural land emits air pollutants and is the source of several water pollutants. As such, agriculture is a nonpoint source of pollution. That is, agricultural fields emit pollution, but there is no single location where the pollution originates and potentially can be controlled.

Chapters 6 and 15 describe one of the most significant sources of pollution from agricultural fields: runoff from nitrogen fertilizers that causes eutrophication. But this is only one of many environmental challenges associated with agriculturally driven changes in the nitrogen cycle. Many of these other challenges are associated with forms of nitrogen that are derived from fertilizers.

Urea is an important component of many nitrogen fertilizers. Once applied, urea is quickly converted to ammonium (NH_4^+). When the soil is moist, ammonium is converted to nitrates (NO_3^-). If the soil lacks sufficient oxygen, denitrification predominates, and the nitrates are converted to nitrous oxide (N_2O). Nitrous oxide is a greenhouse gas that is the fourth largest source of radiative forcing (see Figure 13.3 on

page 272). The atmospheric concentration of nitrous oxide has increased significantly over the last century (see Figure 13.5(c) on page 273). Much of this increase is associated with the increasing use of nitrogen fertilizers.

Nitrous oxide also has a negative effect on stratospheric concentrations of ozone. Nitrous oxide is relatively resistant to change in the troposphere. Like CFCs, it too rises slowly to the stratosphere, where short-wave radiation splits it into nitric oxide (NO) (much as short-wave solar radiation separates the chlorine atoms from CFCs). Once formed, the nitric oxide serves as a catalyst that speeds the destruction of stratospheric ozone.

Agriculture also plays a role in the formation of photochemical smog. If the soil dries out after a farmer applies fertilizer, nitrification can generate significant quantities of nitric oxide. As will be described in Chapter 19, nitric oxide plays a role in the formation of photochemical smog. This form of air pollution is hazardous to human health, reduces plant growth, and damages materials.

Nitrogen fertilizers also serve as precursors to important water pollutants, some of which are dangerous to human health. The most notable threat may be methehemoglobinemia, which is also known as blue baby syn-

drome. This syndrome is associated with the ingestion of nitrate ions (NO_3^-). When nitrates are ingested, bacteria in the gut convert them to nitrites (NO_2^-). These ions reduce hemoglobin molecules' ability to transport oxygen. Infants are especially vulnerable to such changes. The Environmental Protection Agency estimates that tens of thousands of infants drink water with nitrate concentrations above the U.S. health standard. Such violations are relatively widespread because removing nitrates from drinking water is expensive.

ADDITIONAL READING

Socolow, R.H. "Nitrogen Management and the Future of Food: Lessons from the Management of Energy and Carbon." *Proceedings of the National Academy of Science* 96 (1999): 6001–6008.

STUDENT LEARNING OUTCOME

- Students will be able to explain how agricultural activities change the nitrogen cycle in a way that threatens human health and the provision of environmental goods and services.

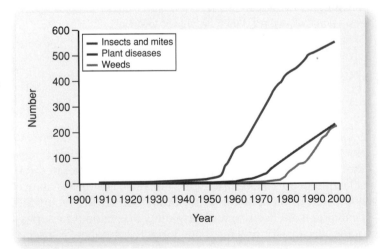

FIGURE 16.17 *Resistance* Historical increase in the number of insect species, plant diseases, and weed species that are resistant to pesticides, herbicides, and fungicides. *(Source: Data from Worldwatch Institute.)*

that eat and in some cases control the population of pests. Due to the loss of predators, pesticides sometimes increase the pest population. For example, in the 1890s California citrus farmers imported the Videlia beetle from Australia to control the cottony cushion scale. In 1947 farmers sought to supplement that control with the newly developed pesticide DDT. But DDT killed the Videlia beetle faster than it killed the cottony cushion scale, and within a year losses associated with the cottony cushion scale increased. To avoid further losses, farmers restricted their use of DDT.

The Green Revolution also increases the use of water for irrigation. In many areas farmers use water faster than the supply is replenished. As result, water bodies have shrunk and wells have been depleted. These costs are described in Chapter 18.

Reducing Agricultural Pollution and Increasing Farmer Profit

If successful, reducing the material and energy inputs to agriculture will ameliorate environmental challenges and increase farmer profits. Such changes are called win–win solutions because everyone benefits. Win–win solutions sound too good to be true, but ecological and economic studies of wheat production in the Yaqui Valley of Mexico show that precision agriculture may benefit both farmers and the environment.

Following current practices, farmers in the valley apply 250 kg (551 lb) of nitrogen fertilizer per hectare. Of this total, 187 kg (412 lb) are applied to dry soils one month before planting. The soil is then irrigated, and the remaining 63 kg (139 lb) are applied about six weeks after planting.

This pattern of fertilization and irrigation causes big losses of nitrogen. Between the initial fertilization and irrigation, there is little change in the concentration of inorganic nitrogen in the soil. After the soil is irrigated, urea is converted to ammonia and then to nitrate (NO_3). In the anaerobic wet soil conditions, much of the nitrogen is converted to nitrous oxide (N_2O), which enters the atmosphere and therefore is lost from the soil (and not available to the plants). As the soil dries, the emission of nitric oxide (NO) increases. Once the seeds are sown, nitrogen emissions drop because plowing aerates the soil (and slows denitrification), and dwindling supplies of ammonia slow the production of nitric oxide.

Processes that form nitric oxide and nitrous oxide suggest that losses can be reduced by lowering the quantity of fertilizer applied and changing the timing of fertilization, planting, and irrigation. One alternative has farmers apply the same amount of nitrogen, 250 kg/ha, but reverses the timing so that 63 kg are applied before planting and 187 kg/ha after planting. This pattern reduces the loss of nitrogen 50 percent relative to current practices. An even greater reduction in the amount lost, about 90 percent relative to current practices, is achieved by reducing the total amount of fertilizer applied by 70 kg/ha to 180 kg/ha (160 lb/acre) and applying a third of this quantity at planting and the remaining two-thirds six weeks

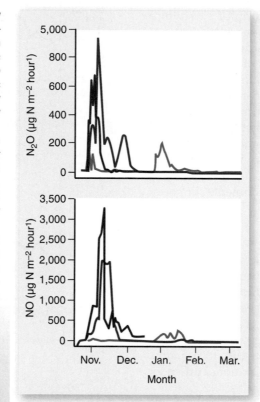

FIGURE 1 *Reducing Nitrogen Losses, Increasing Profit The emissions of nitrous oxide (N_2O) and nitric oxide (NO) from the soil surface in the farmers' practice over the 1994/1995 (in blue) and 1995/1996 (in red) wheat cycles, and for the best alternative (in green) in the 1995/1996 wheat cycle. The best alternative increased farmers' profits because yields did not decline despite efforts to reduce nitrogen losses by reducing fertilizer use. (Source: Data from P.A. Matson et al., "Integration of Environmental, Agronomic, and Economic Aspects of Fertilizer Management." Science 280: 112–115.)*

after planting (Figure 1). Concurrent with the reduced loss of nitrogen to the atmosphere, the alternative practices increase the fraction of nitrogen that is taken up by the wheat plants and increase the fraction that remains in the soil for future plantings.

Alternative practices also increase the profitability of wheat production. Despite the nearly 30 percent reduction in the quantity of nitrogen applied by one of the alternative practices, neither alternative reduces wheat yields. Similarly, neither alternative reduces the protein content of the wheat. (How could nitrogen fertilizer affect protein content?) But because alternative practices increase soil nitrogen and reduce applications, they reduce the costs of producing wheat by $0.55–0.76 per hectare ($0.22–0.31/ acre). Although these numbers seem small, they represent 12 to 17 percent of the aftertax profit generated by growing wheat in the Yacqui Valley of Mexico. As such, the alternative fertilization practices increase farmer profits.

Despite this seeming win–win solution, there may be some drawbacks to alternative practices, especially the one that reduces the total amount of nitrogen applied. Reducing nitrogen application increases the risk to farmers. Wheat plants need nitrogen at certain stages of their development. Without enough, lack of nitrogen could reduce yields regardless of what farmers do later. If less fertilizer is applied before planting, rains after planting could delay the second application of fertilizer relative to the time that the plants need the nitrogen, which could reduce yields.

Despite this increased risk, the results show that changing agricultural practices can reduce the environmental impacts of agriculture and increase farmer profits. Precision agriculture may be able to generate win–win solutions.

ADDITIONAL READING

Matson, P.A., R. Naylor, and I. Ortiz-Monasterio. "Integration of Environmental, Agronomic, and Economic Aspects of Fertilizer Management." *Science* 280 (1998): 112–115.

STUDENT LEARNING OUTCOME

- Students will be able to explain how changing fertilizer practices can generate a win–win solution that increases farmers' profits and enhances environmental quality.

THE FUTURE OF AGRICULTURE

The benefits and costs of the Green Revolution create a conundrum for modern societies. On one hand, society depends on Green Revolution agriculture for food. It is highly unlikely that a more traditional form of agriculture could feed 6 billion people without a major restructuring of the farm and nonfarm populations and a dramatic increase in agricultural land (and a resultant loss of natural ecosystems). On the other hand, the costs of Green Revolution agriculture cause concerns about its economic and ecological sustainability. Projections for a growing population and rising incomes imply that the demand for food will grow significantly over the next fifty years. Many wonder whether Green Revolution agriculture can be changed in a way that will enable it to grow more food, reduce its energy and material inputs, and reduce its environmental impacts.

Increasing Food Production

Despite the great increase in food production over the last fifty years, there are some signs the gains may be slowing. The increase in per capita cereal production is slowing (Figure 16.18). Furthermore, global changes hide important regional differences. Gains have been greatest in developed nations, while production has stagnated or declined in developing nations. Although such results foreshadow problems for those living in developing nations, this poor performance offers a strategy for future gains.

The Green Revolution increased yields in two ways: by increasing the theoretical yield that a high-yield variety could produce, which is determined on experimental farms, and by increasing the use of inputs so that real farmers could come close to this maximum. Evidence suggests that the ability to breed cultivars with higher yields is diminishing. The harvest index of rice cultivars, which is between 50 and 55 percent, appears to be near a physiological maximum. Plants seem to have reached a minimum level for leaf area and stem biomass that is needed to capture sunlight, store the products of photosynthesis, and support the grain that is produced. Small gains may still be possible for wheat. Increases in harvest index are relatively unimportant for increasing corn yields. Instead farmers have found ways to increase the plants' net primary production. But this strategy also has limits.

Alternatively, farmers can increase food production by changing their management techniques so their yield comes closer to the cultivar's theoretical maximum. The degree to which farmers' yields approach this theoretical maximum is measured by **crop yield potential.** Although difficult to measure, approximations indicate that the yields obtained by farmers in many large developing nations are well below those obtained on experimental farms. For example, rice yield potential for high-yield varieties in Indonesia and Thailand are well below 70 percent. In Indonesia crop yield potential is estimated to be about 60 percent—just 5 metric tons per hect-

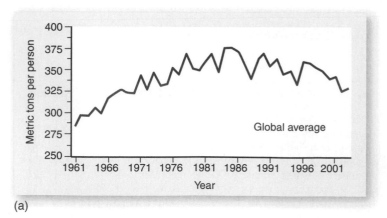

(a)

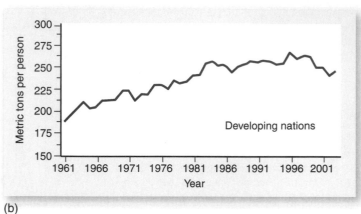

(b)

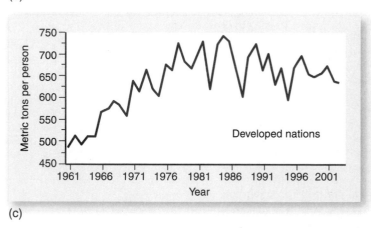

(c)

FIGURE 16.18 *Per Capita Cereal Production* *Per capita production of cereals for (a) the world as a whole, (b) developing nations, and (c) developed nations.* (Source: Data from Food and Agricultural Organization.)

are versus 8 metric tons per hectare on experimental farms. These gaps imply that farmers in developing nations may be able to increase yields significantly by increasing their use of irrigation and managing water more effectively. Similar gains are possible in the irrigated rice–wheat systems of Pakistan and rain-fed cereal production in Brazil and Argentina.

On the other hand, rice yields from high-yield varieties in the Republic of Korea, Japan, and parts of the United States are above 70 percent of their yield potential. As such, further gains in yield by farmers will be difficult. Under these conditions a different strategy is needed to boost production. One such strategy is to change the genetic makeup of crop species, creating so-called **transgenic cultivars.** To grow these cultivars, scientists move genes from other species into crop species. This transfer gives these crops new capabilities that they could never have obtained following traditional breeding techniques. Cultivars with genes from other species, known as **genetically modified** crops (the popular media use the acronym GMO for "genetically modified organisms"), were developed in the 1990s. Starting in the late 1990s they were planted by farmers in developed nations; farmers in some developing nations started to plant them the following decade. In some cases genetically modified crops are designed to increase their nutritional value. For example, golden rice was given a gene that allows it to produce vitamin A, which may help prevent blindness. But in other cases the crops are modified to reduce material inputs.

Can Farmers Reduce Material Inputs?

In many genetically modified crops the new genes are designed to increase the plants' resistance to insect pests or to herbicides. For example, Roundup Ready© soybeans have genes that allow the soybean plants to resist the herbicide Roundup. Elevated resistance allows farmers to kill weeds by spraying Roundup with little or no effect on the soybean plants. Similarly, there are Bt strains of cotton and maize that allow the plants to produce their own pesticide. This pesticide, produced by a gene that was obtained from the bacterium *Bacillus thuringiensis,* allows farmers to reduce the application of pesticides. To slow the process by which pests become resistant to the Bt poison, the sellers of Bt strains limit the fraction of cropland that a farmer may plant with Bt varieties. Areas planted with non-Bt strains allow nonresistant pests to survive and reproduce, which slows the rate at which Bt resistance spreads through the pest population.

Genetically modified crops have reduced the use of pesticides and herbicides (fewer applications), which lowers the cost of producing crops. For example, Bt cotton reduces the number of times that farmers spray pesticides for the Asian boll worm from twenty to six times per year. In China a kilogram of cotton that is grown from Bt strains cost farmers 28 percent less than a kilogram of cotton grown from non-Bt strains. Similar reductions have been recorded in both developed and developing nations.

Despite these benefits, many people are concerned about the use of genetically modified crops. New genes may cause the plant to produce new proteins, which may cause an allergic reaction in some consumers. Because of such concerns, consumers in some countries, especially in Europe, are reluctant to eat genetically modified foods and have asked their governments to force manufacturers to label foods produced from genetically modified crops. But the feasibility of such labeling has been undermined by the accidental mixing of foods grown from genetically modified and non-genetically modified crops.

The potential for such mixing also creates health concerns. Antibiotic-resistant marker genes were used to create genetically modified crops, and they could be transferred to bacteria that inhabit the human digestive system. This could create antibiotic-resistant strains. To date there are no documented cases of this, and many scientists argue that such a transfer is highly unlikely. To reduce this likelihood further, the practice of using antibiotic-resistant marker genes now is widely discouraged.

Similar worries exist about the transfer of genes to other plant species, thereby creating superweeds. It is possible that pollen from genetically modified crops could reach local plants and create an invasive hybrid. This outcome depends on whether the new gene gives the hybrid an advantage. Even if an individual becomes resistant to a specific agricultural chemical, this would not be an advantage unless the population is suppressed by that chemical. To date there is no evidence of gene transfer from crops to wild species. Nonetheless, farmers are working to reduce this possibility by not planting genetically modified crops where wild relatives are present or by establishing buffer zones between genetically modified crops and conventional varieties.

The unpredictability of evolution creates another danger. Giving genetically modified plants new capabilities may generate surprises when they are introduced into ecosystems where species have evolved together. Laboratory research generated just such a surprise—eating pollen from Bt corn kills the caterpillars of monarch butterflies. But these laboratory deaths may not occur in the field. Caterpillars eat milkweed, which is not often found near cornfields. Furthermore, caterpillars do much of their feeding before corn plants release their pollen, and when the pollen is released, the amount the caterpillars consume is unlikely to be toxic.

To increase yields and reduce the environmental impacts of using fertilizers and irrigation, some farmers (mostly in developed nations) are starting to practice **precision agriculture.** Precision agriculture seeks to provide the inputs that are needed for crop growth (water and nutrients) and crop protection without deficiency or excess at each point in time during the growing season. Put simply, give the crop exactly what it needs when it needs it so that its yield can approach the theoretical maximum. By providing "just enough," farmers hope to reduce the environmental impacts of using excess amounts of fertilizer and water.

Evidence suggests that precision farming could be most effective in developing nations, especially those located in the tropics. Scientists now understand that the fertility of tropical soils varies greatly over short distances. For example, rice yields were found to vary from 2,400 to 6,000 kg/ha in forty-two fields that surrounded a single Filipino village, and these variations were associated with differences in the soil nitrogen content. Variations in soil nitrogen imply that the optimal rate of applying nitrogen differs among farms. If farmers know this quantity, they can increase output on croplands with low soil nitrogen without increasing nitrogen runoff from croplands that already have sufficient quantities.

To be successful, precision agriculture depends on an "information revolution." That is, farmers need real-time information on the availability of water and nutrients in their soil and the status of their crops. Technologies can now measure the nutrient and water content of soils and relay that information to farmers almost instantly. Similarly, the resolution of satellites has been honed to 1 meter. This allows farmers to receive information about the status of their crops in locations they cannot see from the ground. Through better education and training, farmers will be able to diagnose problem crops and solve them through the precise (in both quantity and location) application of water, fertilizers, and pesticides. Of course, this implies that the skill level of farmers needs to be upgraded. Similarly, agricultural extension agents, the local experts who dispense advice to farmers, must tailor their recommendations to specific farms rather than just making general suggestions.

Efforts to reduce the use of water and fertilizer are complemented by efforts to reduce the use of pesticides, herbicides, and fungicides via **integrated pest management.** This is defined as the coordinated use of pest and environmental information along with available pest control methods to prevent unacceptable levels of pest damage by the most economical means, with the least possible hazard to people, property, and the environment. Integrated pest management grew from the realization that the widespread use of pesticides, herbicides, and fungicides was failing, both because they were becoming less effective and because they had significant environmental impacts beyond the farm. These impacts were first publicized by Rachel Carson, who described how the disappearance of songbirds was related to the use of pesticides in her landmark book *Silent Spring*.

Integrated pest management uses pests' ecology against them. This approach is illustrated by ongoing efforts to control the cabbage root fly and the carrot fly, which are two important root-eating pests in northern Europe. During the last twenty years the amount of insecticide used to control these pests has declined significantly, but the techniques that powered these reductions no longer are economically viable; therefore, farmers have turned to biological controls. Some farmers inoculate their fields with a fungus that kills the fly larvae. Other farmers release beetles that eat the flies; farmers in Belgium release beetles that parasitize the flies. In the Netherlands farmers release sterile flies to disrupt the mating of the wild flies. Yet other farmers plant clover among the cabbage, which discourages the flies. Together options like these constitute integrated pest management.

SUMMARY OF KEY CONCEPTS

- Hunter–gatherers obtain food from unaltered ecosystems, whereas agriculturists change ecosystems to increase food availability. Agriculturists replace local ecosystems with domesticated plants or animals. From an ecological perspective, agriculture represents a disturbance that changes biogeochemical and hydrological cycles and forces farmers to compete with other species for the edible energy produced by their crops and livestock.

- The types and technologies of agriculture are implemented using combinations of land, capital, labor, and domesticated animals that depend on their local availability. Green Revolution agriculture arose during the second half of the twentieth century, when energy and materials were inexpensive relative to labor. Breeders developed high-yield varieties that allocate more energy toward parts that people can eat. To grow these varieties farmers must fertilize and irrigate their crops. Labor is replaced with inanimate forms of energy, which are used to produce fertilizers, power farm machinery, and pump water.

- Green Revolution agriculture has increased food supplies faster than population has grown, thereby increasing average food supply with relatively little increase in agricultural land. Substituting energy and machinery for labor reduced the number of farmers, which increased the number of workers that could produce nonfood

goods and services. Consumers can purchase more of these goods and services because increased food supply reduced prices, which increased disposable income.

- Green Revolution agriculture requires a large investment, displays increasing returns to scale, and lowers food prices—results that together drive many small farmers from their land and create large numbers of landless poor. Large energy requirements threaten the sustainability of agriculture as inexpensive supplies of oil and natural gas are depleted. High rates of fertilizer use create several forms of pollution.

- Gains associated with high-yield varieties are slowing, which forces farmers to pursue other strategies. One strategy seeks to improve farm management so that yields approach the maximum values obtained on experimental farms. Genetically modified crops are designed to increase nutritional value and reduce farmers' use of inputs.

- Precision farming seeks to reduce the use of material inputs by applying the amounts needed when they are needed. This effort is complemented by integrated pest management, in which pesticide use is replaced by emphasizing biological controls that are based on ecological relationships between a pest, its food sources, and potential predators.

REVIEW QUESTIONS

1. If hunter–gatherers are so energy-efficient, why are their populations so small?

2. Explain why Green Revolution agriculture and industrial feedlot operations enjoy economies of scale.

3. Many areas of the Brazilian Amazon are being cut and replaced with soybean production for export to the world market. What combination of land, labor, capital, and domestic animals probably is used to grow the soybeans?

4. Discuss how Green Revolution agriculture has affected the lives of small farmers in the U.S. Midwest and large farmers in the Brazilian Amazon.

5. Some say that the Green Revolution invalidated Malthus's predictions; others argue that it has only postponed the inevitable. Give two arguments for both positions.

KEY TERMS

agriculturists

agroecosystems

cereals

coarse grains

coevolutionary hypothesis

crop yield potential

cultivars

domestication

economies of scale

feedlots

fungicides

gathering

genetically modified

grains

harvest index

herbicides

high-yield varieties

hunting

industrial livestock production systems

integrated pest management

intensification

irrigation

livestock

monoculture

pasture

pesticides

pests

population density

precision agriculture

rangelands

resource depletion hypothesis

selective breeding

slash and burn

technical change hypothesis

transgenic cultivars

vegetable

yields

CHAPTER

17

FORESTS

So Much More than Wood

STUDENT LEARNING OUTCOMES

After reading this chapter, students will be able to

- Describe the economic factors that drive deforestation.
- Compare and contrast the economic costs and benefits of deforestation.
- Describe why the market often generates rates of deforestation in which the costs of deforestation exceed the benefits.
- Explain why efforts to slow the rate of deforestation probably cannot depend solely on raising the price of wood.

Deforestation in Seventeenth-Century England
As early as the 1500s much of England's trees were cut. This map shows the area that remained of the Rockingham Forest in the seventeenth century. Some of the forest has been converted to pasture (see the Kirby Pasture) while other areas, designed as coppices (see the Lord's Coppice), are managed by cutting to provide wood. Some of the same forces drive deforestation today.

DEFORESTATION IN SEVENTEENTH-CENTURY ENGLAND

In many places the area covered by forests has declined rapidly. Such a reduction in forest area is known as **deforestation.** Deforestation is an important environmental challenge in many nations. But this is not the first time that deforestation has threatened economic well-being: Deforestation was also a problem in sixteenth-century England (opening photograph). Retelling this story illustrates its causes and consequences.

As it does today, deforestation in that period eliminated large areas of forest. In one area around Derbyshire, England, the number of mature oak trees declined from 59,412 in 1560 to 2,716 in 1587, and the number of small oaks declined from 32,820 to 3,032. In a little more than one generation about 93 percent of the trees were cut down.

Why did the forest area decline so quickly? The wood was used for a variety of economic purposes. Industry used wood as a source of energy to produce a variety of materials such as glass, copper, and iron. People used wood to heat their homes and cook their food. Wood also was used as a raw material to produce paper and build homes.

Deforestation had an inflationary impact. Growing demand for wood caused wide areas of the forest to be cut, which reduced supply. A growing imbalance between supply and demand caused wood prices to increase significantly. Around Derbyshire the price of fuel wood to homeowners increased 280 percent, the price to industry increased 400 percent, and the price of wood planks to boat builders increased 300 percent.

Rising prices had important social consequences. As wood became increasingly expensive, consumers were forced to choose between "heating and eating." In response, many directed their anger against those they blamed for the loss of the forest. In some areas local inhabitants prevented new industries from opening. In other areas inhabitants forced existing industries to close. In some places the construction of new housing was forbidden unless the new homes were made of brick or stone. In response, government established regulations regarding which trees industry and consumers could harvest.

The story of deforestation in sixteenth-century England could come from today's headlines. In developing nations great areas of the forest are being cut for both domestic use and export. ■

HOW QUICKLY ARE FORESTS BEING CUT (AND REGROWING)?

Over the last 8,000 years much of the world's forests have been lost (Figure 17.1). These losses vary across continents (Figure 17.2). The largest fractions of original forest cover have been eliminated in Asia and Africa, especially in tropical areas. Today much land remains devoid of forests. In Russia, Europe, and North and Central America, a greater fraction of the original forest remains. Furthermore, succes-sion in temperate forests has reestablished some areas of the disturbed forest. South America still has a significant portion of its original forest area.

But how reliable are these estimates? Many estimates of forest cover are generated by local governments, industry, or conservation groups. Each group has an incentive to distort estimates. Industry tends to understate deforestation to avoid government regulation and attacks by conservation groups; conservation groups tend to overstate deforestation to prompt government action and raise money. Even

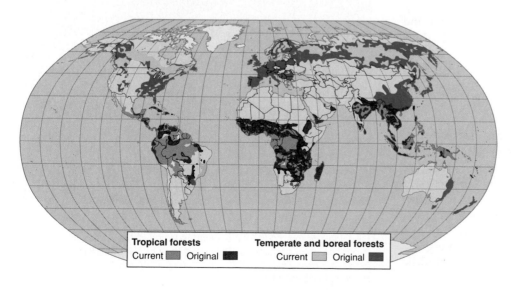

FIGURE 17.1 *Deforestation* *The change in the area covered by forests over the last 8,000 years. Areas originally covered by tropical or temperate forests are indicated by dark areas of brown and green. The remaining areas of these forests are indicated by pale green and brown.* (Source: Data from United Nations Environment Programme.)

Tropical forests
Current ■ Original ■
Temperate and boreal forests
Current ■ Original ■

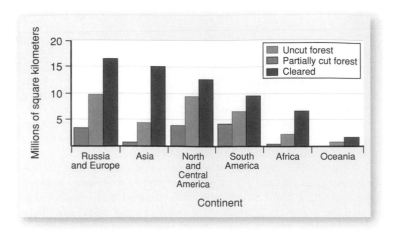

FIGURE 17.2 *Geographic Variation in Deforestation Status of forests by continent.* (*Source: Data from World Resources Institute, World Resources: A Guide to the Global Environment.*)

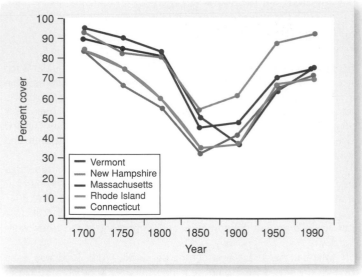

FIGURE 17.3 *Deforestation in Colonial New England* *Estimates for the fraction of forest remaining in Vermont (dark blue line), New Hampshire (orange line), Massachusetts (red line), Rhode Island (turquoise line), and Connecticut (green line) between 1700 and 1990.* (*Source: Data from D.R. Foster. "Land-Use History and Four Hundred Years of Vegetation Change in New England," eds. B.L. Turner et al., in Global Land Use Change: A Perspective from the Columbian Encounter, Consejo Superior de Investigaciones Científicas, Spain.*)

without deliberate manipulation, estimates of deforestation often are unreliable. Tropical rain forests cover vast areas, but only a small portion can be visited by those charged with estimating rates of deforestation. Often scientists have traveled along roads to estimate deforestation, only to realize later that deforestation is greatest along roads.

Over the last two decades scientists have been able to estimate deforestation more accurately by analyzing remotely sensed images (see Chapter 7's Case Study on page 131). Rates of deforestation estimated from remotely sensed images generally are smaller than the rates estimated using ground-based methods of measurement. For example, annual rates of deforestation in the Brazilian Amazon for the 1980s estimated from remotely sensed images are about 20,000 km² per year, compared to 50,000–60,000 km² estimated using ground-based methods.

Because satellite images are available for the entire planet, scientists can calculate global rates of deforestation. Forest area in developed nations generally is stable or increasing, while forest area in developing nations generally is decreasing. These differences may be associated with the period during which forest area is measured. Two hundred years ago an analysis of U.S. forests would have measured rapid rates of deforestation. About half of the forests in the U.S. states of Vermont, New Hampshire, Connecticut, Rhode Island, and Massachusetts had been cut by the mid-1800s (Figure 17.3).

Since then much of the New England forest has regrown. Hikers often come across stone walls in the middle of a forest (Figure 17.4). This odd juxtaposition shows that the land was previously used for agriculture; the wall marked the property line of adjoining farms. In temperate forests succession has been able to regenerate forests in areas disturbed by farmers. But temperate forests may be more resilient than their tropical counterparts—there is no guarantee that future hikers will find stone walls in the middle of tropical rain forests.

FIGURE 17.4 *Stone Walls* *Stone walls often are found in the middle of forests in New England, such as in Harvard Forest in Petersham, Massachusetts. They used to mark the boundary between farms. Since then, the farms have been abandoned and the forest has regrown.*

CAUSES FOR DEFORESTATION

The oddity of a stone wall in the middle of a forest highlights the forces that cause deforestation in developing nations and cause **afforestation,** which is defined as forest regrowth after a disturbance, in developed nations. Afforestation in developed nations generally is not purposeful. Rather, forests regrow because changes in affluence, population, and technology reduce the profitability of activities that drive deforestation. But these changes are the exception, not the rule. Many of the economic and technical conditions that caused developed nations to cut their forests earlier in their history now prevail in developing nations. There the conversion of forest to agricultural land, the timber trade, uncertainty about property rights, mineral and energy production, and investments in transportation infrastructure are responsible for rapid rates of deforestation during the last fifty years.

Forests to Agriculture

Changes in land use from forest to agriculture are one of the most important causes of deforestation. As described in Chapter 16, subsistence farmers use relatively large areas to feed their families because yields tend to be relatively small and relatively constant over time as technology remains stable. Under these conditions, population growth tends to elicit a proportional increase in agricultural land. Much of this new agricultural land is created by clearing forests.

The link between population growth and the increased demand for agricultural land is illustrated by changes in U.S. forest area. Before European settlers arrived in North America, much of the East Coast was covered by temperate deciduous and evergreen forests (Figure 17.5). As the population of settlers increased, so did the demand for food and the area needed to grow it. Much of this agricultural land was created by clearing forest. For example, the town of Petersham, Massachusetts, cleared about 77 percent of its forest by 1830. By 1910 much of the forest along the East Coast had been cleared (Figure 17.5).

Technological changes that reversed this trend started in the middle of the nineteenth century. At that time, westward migration converted large tracts of fertile grasslands in the American Midwest to agricultural use. Grassland soils (mollisols, Chapter 15) are more suitable to grain agriculture and therefore generate larger yields at lower costs. Over time, the opening of the Erie Canal and the construction of railroads reduced the cost of moving this grain east. Inexpensive grain made it cheaper for New Englanders to buy midwestern grain than to grow it themselves. Consequently, New England farmers abandoned their land. By 1870 half of Petersham's farms were abandoned. After several decades, ecological succession (Chapter 8) replaced agroecosystems with communities similar to those that originally occupied the land. As a result, the area covered by forest on the U.S. East Coast now is greater than it was in 1910 (Figure 17.5).

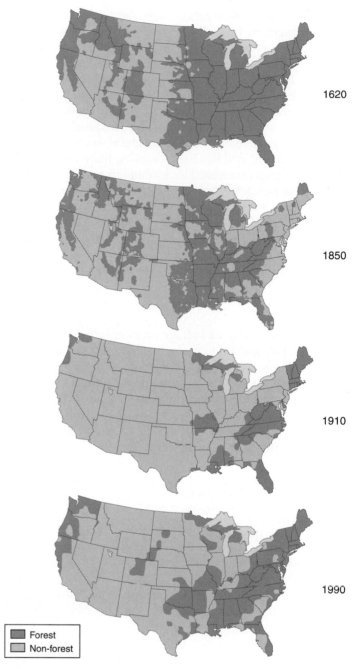

FIGURE 17.5   *Changes in U.S. Forested Area*   *Area of forest in the conterminous United States in 1620 is given in green. By 1850 the areas covered by forest had shrunk considerably, especially on the East Coast. By 1910 forested area was reduced further: Relatively little forest area remained on the East Coast. Much of this area had been afforested by 1990.* (Source: Data from S.B. Roy et al., "Impact of Historical Land Cover Change on the July Climate of the United States," Journal of Geophysical Research-Atmosphere 108 (D24).)

The same forces that drove the conversion of forests to agricultural land in developed nations hundreds of years ago are now partially responsible for the rapid rates of deforestation in developing nations. There the population of subsistence farmers is growing rapidly. Many of these farmers are poor, so they cannot increase their crop yields by purchasing tractors, fuel, fertilizers, and other inputs needed by Green Revolution high-yield varieties. Nor can they supplement their diets by purchasing imported food. Under these conditions, the increasing demand for food has increased the demand for agricultural land. Increasing demand for agricultural land in Nigeria, Madagascar, and Rwanda has reduced forest area by 90 percent.

Affluence also is responsible for the conversion of forests to agricultural land. Increasing incomes in developed nations stimulate the demand for exotic foods such as pistachio nuts, pineapples, and bananas. Growing these food items for export occupies much land in developing nations that otherwise could be used by subsistence farmers. Without access to this land, subsistence farmers are forced to create new agricultural land by clearing existing forests. As described in the *Case Study: Deforestation and Oil Prices,* the production of soybeans for export is partially responsible for the rapid rates of deforestation in the Brazilian Amazon since the 1980s.

Timber

Timber harvests also clear huge areas of forest. The production of wood and its manufacture account for about $400 billion, which is roughly 2 percent of the world's GDP. Much of this economic activity is concentrated in North America, Scandinavia, and tropical Asia, such as Indonesia and Malaysia. In developing nations, timber can account for more than 2 percent of GDP. To do so, the timber industry removes millions of hectares of forests.

The area covered by the trees that are harvested often is only a small fraction of the forested area that is affected by the timber sector. To get equipment into the forest and to remove trees, the road network often is expanded into new areas. In the United States forests managed by the federal government contain more than 600,000 kilometers of roads. Building these roads sometimes clears more of the forest than will be harvested. For one project in Indonesia, building 500 km of new roads cleared 40,000 more hectares than the area that was harvested.

The construction of roads and the clearing of land by the timber sector initiate a series of positive feedback loops that accelerate deforestation (Figure 17.6). New roads make it easier for farmers to enter the forest and clear land for agriculture, as well as facilitating the transport of agricultural goods for domestic consumption and export. The indirect effects of building new roads may be most important in Central and South America. Scientists estimate that each kilometer of road into the forest may cause 400–2,000 hectares of forest to be cleared. In Brazilian state of Para, the construction of new roads helped increase the rate of deforestation from 0.16 percent in 1972 to 17.3 percent in 1985.

Harvesting timber reduces forest cover, which initiates a positive feedback loop that enhances deforestation. Logging opens gaps in the forest. This allows more sunlight to reach the ground, which dries the soil. As the soil dries, the ground becomes more vulnerable to fires. In addition, edge effects kill big trees near forest gaps at a rate three

FIGURE 17.6 *Deforestation Feedback Loops* Deforestation is part of a larger system that includes many feedback loops. These feedback loops include road paving, transportation costs, fire, local climate, and the profits earned from various land uses. (Source: Redrawn from figure by D. Nepstad et al., "Road Paving, Fire Regime Feedbacks, and the Future of Amazon Forests," Forest Ecology and Management, 154: 395–407.)

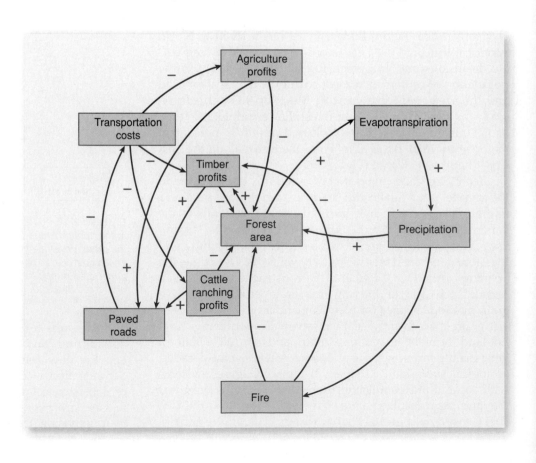

CASE STUDY

Deforestation and Oil Prices

As described in this chapter, clearing forests to make room for agriculture is one of the most important causes of deforestation. Demand for agricultural land is motivated by a variety of causes. Large increases in oil prices during the 1970s were one of the less obvious but more important reasons why great areas of the Brazilian Amazon were cleared for agriculture during the 1980s.

The 1973–1974 and 1979–1981 oil price increases sent shock waves through the global financial system. One of the most important was a change in the net flow of money among nations. In general, oil and other natural resources are extracted in developing nations and sold to developed nations. Developed nations use oil and other natural resources to manufacture goods and services, some of which they sell back to developing nations. Prior to the oil price shocks, prices for oil and other natural resources generally were low relative to the prices for the goods and services that they were used to manufacture. Under these conditions, capital generally flowed from developing nations to developed nations (Figure 1).

The oil price shocks reversed this flow. After the price rise, developed nations consumed roughly the same quantity of oil, but the cost of that oil increased significantly. As a result, paying for imported oil meant that much more money flowed from developed nations to developing nations. At the same time, the ability of developing nations to purchase more goods and services, which is known as absorptive capacity, was limited. Put simply, developing nations could not spend their oil revenues as quickly as the money came in.

The money that oil-exporting nations could not spend was put into the international banking system. This cash accumulated, which caused a dilemma for bank managers. Banks have to pay interest on their deposits, so the banks' interest payments also increased. To make these payments, the banks needed to lend the money to borrowers who would repay the bank at an interest rate greater than the interest rate that the bank paid to its depositors (this is the basis for the banking system).

But it was difficult to find borrowers for the billions of dollars that suddenly made their way into the banking system. The oil price shocks created an economic slowdown (a recession) in developed nations, so they did not want to borrow money. Developing nations needed huge amounts of money to fund their economic aspirations, but they had little collateral. Desperate to lend the huge sums that were accumulating

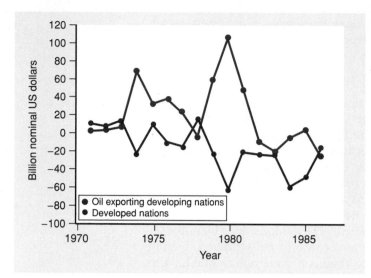

FIGURE 1 *Current Accounts and Oil Prices* Current accounts for oil-exporting developing nations (blue line) increased dramatically in 1973–1974 and 1979–1981 when the price of oil increased dramatically. These increases were matched by declines in the current accounts for developed nations, whose bills for imported oil caused money to flow out of their economies (red line). (Source: Data from R.K. Kaufmann, Higher Oil Prices: Can OPEC Raise Prices by Cutting Production? PhD Dissertation, University of Pennsylvania.)

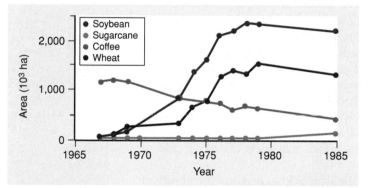

FIGURE 2 *Acres Planted* Acres planted of major crops in the Brazilian state of Paraná. Notice that the area planted in soybeans and wheat increased significantly between 1970 and 1980 while the area planted in coffee declined. (Source: Data from D.L. Skole, et al., "Physical and Human Dimensions of Deforestation in Amazonia." Bioscience 44, (5): 314–322.)

in their banks, managers lent billions of dollars to developing nations during the 1980s.

One developing nation that borrowed heavily during the 1980s was Brazil, which used some of these funds to modernize its agricultural sector. To do so, and to earn money that could be used to repay the loans, agricultural modernization programs focused on export crops such as soybeans and wheat. Efforts to expand production of soybeans were particularly successful. Planted acreage increased sixfold during the 1970s, and yield (output per acre) increased fivefold (Figure 2).

Much of the increase in soybean acreage occurred in the state of Paraná. This state has little rain forest, so not much forest was cleared to expand soybean production. Instead an increase in soybean production changed the pattern of agriculture in Paraná, which had focused on small-scale production of labor-intensive crops such as coffee. Growing coffee requires more labor than soybeans, so converting agricultural land from coffee to soybeans displaced many workers. Workers were forced to emigrate because soybean produc-

tion pushed land prices beyond levels that workers could afford. Some of these workers went to the Amazonian frontier. There they cleared land for crops, livestock, and small-scale production. This migration into the area was facilitated by the construction of the Trans-Amazon highway.

The final cause for much of the deforestation in the Brazilian Amazon was the conversion of forest to agricultural land. But this conversion was motivated by a long chain of events that was triggered by rapid increases in oil prices and was transmitted by the international financial system.

ADDITIONAL READING

Skole, D.L., W.H. Chomentowski, W.A. Sulas, and A.D. Nobre. "Physical and Human Dimensions of Deforestation in Amazonia." *Bioscience* 44, no. 5 (1994): 314–322.

STUDENT LEARNING OUTCOME

• Students will be able to explain how the world banking system can accelerate or slow deforestation.

times greater than the death rate of similar trees in the forest interior. The deaths of mature trees at the forest edge create fuel for fires and expose more area to soil drying. Together fires and tree mortality open more gaps in the forest, which create additional opportunities for fire.

Property Rights and Fire

Establishing property rights also is a significant cause of deforestation. Much of Western law regarding property rights evolved from the notion of improvements to land. Early in the British settlement of New England, owners had to demonstrate that they made their land productive. "Productive use" often required that forests be cleared and converted to another purpose such as agriculture.

Although the definition of property rights has changed, the notion that ownership must be demonstrated by improvement remains an important cause of deforestation. During the 1980s and 1990s the Brazilian government offered free land in Amazonia to settlers. A settler could claim three times the amount of land that he or she cleared. To maximize their claims, settlers burned extensive areas of the forest. These fires were partially responsible for the rapid rates of deforestation in the Amazon during the 1980s (Figure 17.7).

These fires, as well as accidental ones, initiate a series of positive feedback loops that also increase deforestation. Fires destroy trees, reducing evapotranspiration and adding aerosol particles to the atmosphere, which increase the ability of clouds to hold water. Both factors reduce rainfall, which leads to drought, which increases the likelihood of fires that deforest land.

This positive feedback loop is enhanced by the ability of fires to jump from one property to the next. Fire often is used to manage land, such as clearing forests to create pasture or agricultural fields or to destroy pests. Once set, fires can sometimes burn beyond their intended boundaries. This risk discourages **intensive land uses,** which are systems that use a lot of capital per unit area. For example, landowners are reluctant to set up tree plantations because a neighbor's fire could destroy their investment. Instead the potential for accidental fires encourages **extensive land uses,** which use greater areas of land per unit capital, such as cattle ranches. Extensive land uses tend to accelerate the rate of deforestation.

Mineral and Energy Production

Mining and energy firms also clear quite a bit of forested land. The construction of a hydroelectric facility often creates a reservoir behind the dam, and this reservoir can flood extensive areas of forest. For example, the reservoir associated with the Tucurui Dam destroyed 2,400 km² of the Amazon forest.

Gold mining can have an especially severe effect on forests in Central and South America. Recovering gold reduces forest area directly by clearing small amounts of land, but this

FIGURE 17.7 *Amazonian Fires* Fires, both accidental and those set to establish property rights, were a major cause of deforestation in the Brazilian Amazon during the 1980s.

disturbance is amplified by pollutants associated with mining. Gold often is separated from its parent material through the use of mercury. Much of this mercury escapes from the mining area via runoff. As mercury makes its way through the forest, its toxicity kills plants and animals over great areas. Some organizations estimate that gold mining destroys about 80,000 hectares of rain forest in Colombia annually.

Although mining and energy production are directly responsible for much deforestation, there is some evidence that mineral production may reduce total deforestation. A negative correlation exists between a nation's annual percentage reduction in forest cover and the value of its mineral exports. That is, nations that generate a large fraction of their GDP from mineral exports have lower rates of deforestation than nations that generate a small fraction of their GDP from mineral exports. The cause of this negative relationship is uncertain. It may be associated with exchange rates—greater mineral exports may increase the value of the local currency and therefore discourage international trade in products such as timber or agricultural goods.

Roads and Other Transportation Infrastructure

Much of the time people cut down forests to make money. The profitability of deforestation depends heavily on **transportation costs,** which is the price paid for moving capital, labor, or other inputs into the forest or moving agricultural goods, timber, or minerals out of the forest. Even the most fertile agricultural land or productive timber area has little economic value if the costs of moving crops or timber to the market increase their price relative to other sources of supply. For example, many areas of the Brazilian Amazon are being converted to soybean fields. This conversion would not be possible if high transportation costs made Brazilian soybeans more expensive than soybeans grown in the United States or Argentina.

Transportation costs are determined primarily by the **transportation infrastructure,** which includes roads, railroad lines, and ports. Although transportation networks do not shorten the distance between forests and markets, they can reduce transportation costs, which makes it seem as though the forest is closer. For example, paving 753 kilometers of the 1,736-kilometer highway between Santarem and Cuiba in Brazil shortened the trip by several hours and reduced the cost of moving a metric ton of soybeans by several dollars. Similar reductions are made possible by railroads and pipelines. Consistent with the effect of lower transportation costs, deforestation along paved roads is greater than along unpaved roads (Figure 17.8).

The effect of roads is illustrated by changes in forest area that accompanied the construction of Brazil's main highway, BR-364 (Figure 17.9). The 1975 satellite image shows that the road has relatively little effect on forest cover. Over time the land along the main road becomes deforested, and feeder roads spread outward from the main highway.

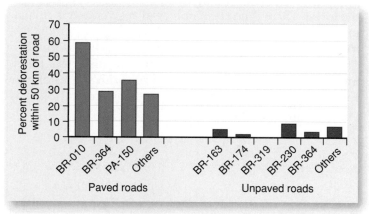

FIGURE 17.8   *Deforestation and Paving* In the Brazilian Amazon, the percentage of deforestation along paved roads is much greater than the percentage of deforestation along unpaved roads. (Source: Redrawn from figure from D. Nepstad et al., "Road Paving, Fire Regime Feedbacks, and the Future of Amazon Forests," Forest Ecology and Management, 154: 395–407.)

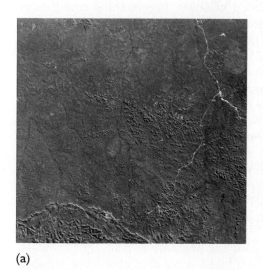

(a)

(b)

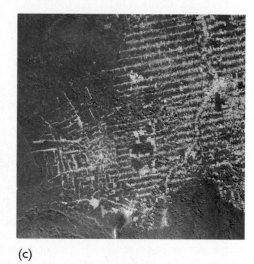

(c)

FIGURE 17.9   *Roads and Deforestation* Deforestation along the Cuiabá-Pôrto Velho highway in the province of Rhondônia (Brazil) in (a) 1975, (b) 1986, and (c) 2000. Forested lands in green, deforested land in white, forest regrowth in light green.

Deforestation follows the feeder roads. Eventually deforestation spreads outward from the feeder roads. This creates a fishbone pattern of deforestation.

Roads usually are built by the government; therefore, government decisions affect deforestation rates. Currently the Brazilian government has an extensive plan to build roads, called "Avanca Brazil" (Figure 17.10), which includes a paved highway that will cross the Andes. Direct access to the Pacific Ocean will reduce the cost of shipping agricultural goods and timber to Asia, which will increase exports. But increasing exports of agricultural goods and timber also will increase deforestation. And that is something that the government hopes to manage carefully.

To manage the effect of roads on deforestation, decision makers must determine whether construction is driven by the economic potential of cutting the forest, or whether the construction of a road determines the economic potential of the forest. There is no clear, universal answer to this circular problem. In some cases roads are built to access productive forest areas. But in other areas the construction of roads is driven by politics and personal gain. The history of building U.S. railroads and highways is rife with stories of how routes were altered to connect towns that were controlled by powerful individuals.

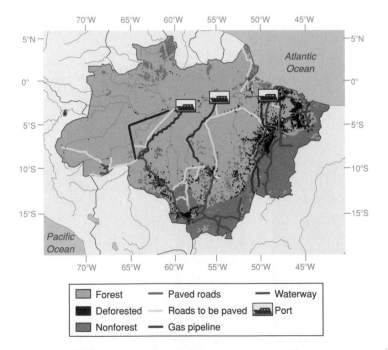

FIGURE 17.10 *Avanca Brazil Avanca Brazil is a plan for extensive investments in roads, gas pipelines, waterways, and ports. These investments will reduce transportation costs and the cost of local inputs, which will reduce the price of goods for export to North America, Europe, and Asia. Increasing these exports will accelerate deforestation.* (Source: Redrawn from G. Carvalho et al., "Sensitive Development Could Protect Amazonia Instead of Destroying it," Nature 409 (6817): 131.)

THE CONTRIBUTION OF FORESTS TO HUMAN WELL-BEING

Unlike many environmental challenges, deforestation is a purposeful act. People cut forests because they believe that the benefits outweigh the losses. Consistent with this belief, many of the factors that cause deforestation contribute directly to economic well-being by generating a valuable natural resource—wood. But forests also enhance economic well-being indirectly by providing environmental services, such as regulating the climate and holding soil in place. These services often do not have a price; thus these externalities often are overlooked when people compare the benefits and costs of deforestation.

Direct Contributions

Forests contribute directly to economic well-being by providing wood, which can be used in two ways: as a raw material for the production of goods and services or as a source of energy. These uses are approximately equal—about half of the wood cut in the 1990s was used as a raw material and about half was used as a source of energy. These fractions are global averages—developed nations use most of their wood as a raw material, whereas developing nations use most of their wood as a source of energy.

For most of human existence wood was the most important nonfood resource. Many of the earliest hunting tools such as spears and agricultural tools such as digging sticks were made of wood. As technical capabilities increased, so did the use of wood. Where trees were available, people fashioned wood into furniture, houses, and ships.

The economic importance of wood for shipbuilding reached a peak in the seventeenth and eighteenth centuries. During this period, the economic might of many European nations was built on trade. Profitability depended mostly on the size and speed of ships. Bigger ships could carry more cargo and generate greater profits per trip, while fast ships could make more trips per year. The size and speed of ships depended in part on the size of trees available. Because there was no way to join individual pieces of wood strongly enough to support sails or the hull, ship size was determined by the largest trees that could be used for the masts and keel. Because of their importance, the British Navy controlled the tallest trees in the forests of colonial North America.

Since the eighteenth century, economic and military might no longer depend heavily on the availability of trees. Advances in metalworking and the use of petrochemicals have eliminated many uses of wood. The handle of a screwdriver now is plastic—it was once wood. Similarly, the dashboard in your car probably is made of plastic—it too was once wood. Only luxury cars now have wooden dashboards.

Despite these changes, wood remains an important raw material. Wood is used by the construction industry for structural support, walls, and roofs, or as forms for pouring concrete. Large quantities of forest products are used to produce paper, especially in developed nations. The average North American consumer uses over 150 kilograms of paper and paperboard per year, compared to about 10 kilograms for the average Asian. To calculate how much paper you use, see *Your Ecological Footprint: How Much Wood Do You Use for Paper?* In both developed and developing nations, paper consumption is growing rapidly. Since 1913 global demand for paper has increased more than twentyfold.

Ironically, some of this increase is associated with technologies that once promised to create a paperless office. Word processors allow endless revisions (we revised this chapter more than twenty times before arriving at the version you are reading now), and copy machines let consumers make many copies (each time we revised the chapter, we printed out copies to read, revise, and send to others for review). Even the Internet has increased paper demand. How many times have you printed information that you downloaded from the Web?

Developing nations use most of their wood as a source of energy for heating, cooking, and lighting. Wood is the main energy source for many African nations, where wood fuels supplied 60–80 percent of total primary energy consumption in 1999. During the same period wood supplied 1–2 percent of total primary energy consumption in North America and much of western Europe. The greater fraction of wood used as fuel also represents a larger amount of energy. The average person in Ghana used 0.114 million metric tons (0.125 million tons) of oil equivalent of energy from fuel wood in 1999. This is about 61 percent greater than the amount of energy the average U.S. citizen obtained from wood fuels in 1999.

Wood is burned both directly as fuel wood and indirectly as charcoal. Freshly cut wood is difficult to burn because it has a relatively high water content. As freshly cut wood burns, much of the energy is used to evaporate the water in the wood. As a result, the flame from unseasoned fuel wood cooks food slowly and may not be hot enough for some industrial purposes, such as making glass. Furthermore, the relatively high water content makes fuel wood heavy and expensive to transport. To reduce the water content, wood is split and left to dry. This process reduces its water content 10–60 percent and therefore doubles its energy content. Alternatively, wood can be converted to charcoal by heating it above 300°C in the absence of air. Heating reduces the water content of charcoal to about 1–10 percent, which increases its energy content by a factor of three to four relative to freshly cut wood. These advantages are offset by the large amounts of wood that are needed to produce charcoal—about 6 m³ of wood are required to produce a little less than a metric ton of charcoal.

Indirect Contributions

Forests contribute indirectly to economic well-being by generating a variety of environmental services. One important service is habitat for biodiversity. As described in Chapter 12, most of the world's terrestrial species are found in forests, especially tropical rain forests. As described by the species–area curve in Figure 12.5 on page 249, deforestation reduces biodiversity by reducing the area covered by forests.

Forests also play an essential direct role in climate by influencing the rate at which Earth's surface absorbs solar energy and the rate at which water evaporates from the surface, which influence temperature and precipitation. This combination tends to cool temperatures. If you live on the U.S. East Coast you may have noticed that the springtime warmup suddenly slows when deciduous trees leaf out (Figure 17.11). This is no coincidence. The greening of deciduous forests in the eastern U.S. cools springtime temperatures by about 3.5°C (6.3°F). This cooling is associated with increased transpiration. Solar energy that would have warmed the surface is used instead to evaporate water. As explained in Chapter 4, evaporation requires a lot of energy and therefore has a significant cooling effect. This cooling effect operates at an even larger scale and has slowed global warming. As temperatures have warmed, forests have become greener, which increases the surface area for water to evaporate from leaves.

In other forest biomes, removing trees and replacing them with grasslands can either warm or cool temperatures. Scientists find that replacing portions of the Amazon forest

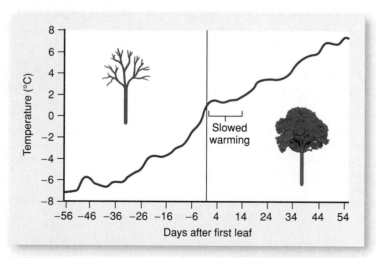

FIGURE 17.11   *Leaf Out and Spring Temperature* *Spring temperatures increase steadily until the leaves of deciduous trees emerge. At this point increased evapotranspiration offsets the warming so that the warming slows for about two weeks. Once the leaves have emerged fully, temperature resumes its rise toward the summer peak. (Source: Data from M.D. Schwartz and T.O. Karl, "Spring Phenology: Nature's Experiment to Detect the effect of "Green-up" on Surface Maximum Temperatures." Monthly Weather Review 118: 883–890.)*

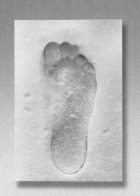

Your ECOLOGICAL *footprint*

How Much Wood Do You Use for Paper?

Much of the wood that is used in developed nations is converted to paper. In 2002 an average U.S. citizen used about 314 kg of paper. How much wood is required to supply this paper? Information given here will allow you to calculate the amount of wood used to produce the paper you consume.

It might seem that calculating this quantity should be simple. Some fact sheets report that approximately thirty-four trees are needed to make one metric ton of paper. This simple conversion is misleading because the amount of wood required depends on the type of tree used, the fraction of postconsumer fiber used, and the type of paper produced.

Much of the paper produced in the United States is made from wood fiber. The amount of fiber in a tree's wood varies by species. Hardwood species, such as northern red oak or the eastern cottonwood, tend to have more fiber than softwoods, such as jack pines or Douglas firs. These differences imply that more softwood is needed to produce a given quantity of paper.

Fibers from newly cut trees come from two sources: round wood and residues. Round wood includes whole trees. Normally the paper industry obtains its fiber from small trees because large trees are more valuable as lumber. Residues are the materials that are left over from the normal operation of the timber industry, such as wood chips and scraps. Typically residues are burned or wasted if they are not used to make paper. But not all wood fibers have to come from newly cut trees. A third source of wood fibers is recovered paper, which includes old paper that is made available to the industry through recycling.

These three sources affect the amount of wood required to produce paper. The amount of new wood required to produce paper can be calculated from its content of virgin materials, which are fibers that come from trees. On the other hand, recovered paper does not require any newly cut wood. As indicated by the calculations that follow, increasing the fraction of fiber that comes from recovered paper reduces the amount of new wood required to produce paper.

Finally, the amount of wood needed to produce paper differs according to the type of paper. Papers vary in many aspects, including their lignin content. Lignin is a natural component of wood that binds the fibers. The presence of lignin enhances the printing properties of paper but reduces its strength and makes it more likely to discolor. Including lignin allows a large fraction of wood to be converted to paper. High-lignin papers include newsprint and office paper, such as copier paper, envelopes, and the like. Stronger papers, such as corrugated paper, are produced by lowering the lignin content. Removing lignin reduces the conversion efficiency of wood and increases the amount of wood needed to produce corrugated paper relative to newsprint.

Calculating Your Footprint

Each day you read newspapers, print material from the Web, and use paper for many other purposes. Here are three equations to let you calculate the quantity of wood required to produce three types of paper:

Newsprint   __ kg wood/kg paper = 2.09 − 0.0209 × % recycled

Office paper   __ kg wood/kg paper = 3.47 − 0.0347 × % recycled

Unbleached paperboard   __ kg wood/kg paper = 3.04 − 0.034 × % recycled

In these equations % recycled is the percentage of fiber that comes from recovered paper.

To help you convert kilograms of wood to volume of wood, Table 1 lists the density of some softwoods and hardwoods that are used to produce paper. To illustrate, suppose you read the Sunday *New York Times*, which weighs about 1 kilogram. Assume that the newsprint is produced with 50 percent recovered fiber. The amount of wood required to produce the paper is 1.04 kg (2.09 − 0.0209 × 50). If this amount of wood were cut from a northern red oak, about 0.23 cubic meters (1.04 kg/4.52 kg/m³) of wood would be required.

Compile information about the amount of paper you use. Newsprint includes paper used to produce your newspaper; office paper includes the paper you use in printers, envelopes, letterhead, and printed forms. Paperboard is used in folded cartons that hold the things you buy in a store. In case you have trouble compiling your use, the average person in the United States in 2003 used about 35.4 kg of newsprint, 94.5 kg of office paper, and 148.2 kg of paperboard.

Interpreting Your Footprint

Suppose the paper you used was produced completely from virgin materials. In other words, 0 percent of its fiber came from recovered paper. By how much could you reduce your use of wood if 40 percent of the fiber used to produce the paper you consumed came from recovered paper?

The global economy produced about 16 kg of office paper, about 6 kg of newsprint, and about 32 kg of paperboard for every person on Earth in 2003. If we assume that all of this paper was produced from virgin material (it was not), by how much would you have to switch your paper usage to recycled paper to lower your use of wood so that it matched the global average?

ADDITIONAL READING

Bierman, C.J. *Handbook of Pulping and Papermaking*. San Diego: Academic Press, 1996.
The "Paper Calculator" at www.ofee.gov/recycled/descript.htm will allow you to calculate many of the material and energy flows required to produce paper.

STUDENT LEARNING OUTCOME

- Students will be able to explain how using recycled paper slows the rate of deforestation.

TABLE 1	Wood Density		
Softwoods	**Density kg/m³**	**Hardwoods**	**Density kg/m³**
Jack pine	3.2	Eastern cottonwood	2.97
Douglas fir	3.61	Northern red oak	4.52

with pasture warms local temperatures by up to 3°C. This warming is due to increased amounts of long-wave radiation coming from the pasture (remember from Chapter 13 that long-wave radiation is absorbed by greenhouse gases and warms the atmosphere). In the central portion of the United States, replacing forests with agroecosystems has cooled summer temperatures by about 2°C. This cooling is associated with an increase in albedo and transpiration.

Forests affect climate indirectly by storing great quantities of carbon. A hectare of tropical rain forest may store 300–600 metric tons (330–660 tons) of carbon in both plants and soil. Similarly, a hectare of boreal forests may store 300–400 metric tons (330–440 tons) of carbon in plants and soil. Together the world's forests store 1,140–1,250 billion metric tons of carbon (1,254–1,375 billion tons). If all of this carbon were released to the atmosphere, the atmospheric concentration of CO_2 would approximately double. As discussed in Chapter 13, climate models indicate that doubling the preindustrial concentration of CO_2 would increase Earth's average surface temperature by about 3.5°C (6.3°F). Based on these effects, some scientists estimate that the value of carbon stored by forests is between $10 and $20 per metric ton ($11–22 per ton).

The ability of trees to store carbon suggests that society could slow climate change with an extensive effort to plant trees. But such a policy may not be an effective means for removing carbon that was put into the atmosphere by burning fossil fuels. After planting, a growing tree removes carbon that was put into the atmosphere when the tree that previously grew in that spot was removed. That is, the new tree removes carbon that was put into the atmosphere by deforestation. It does not remove carbon that was emitted by burning fossil fuels. The source of carbon removed by a growing tree is not trivial. Fossil fuels store more carbon (and hence have the potential to release more carbon) than do all of the world's forests (see Chapter 6).

This is not to say that planting trees is futile. In addition to storing carbon, trees perform many other environmental services. Among the most important is soil retention. Roots hold soil in place, and trees transpire lots of water. If this water remained in the soil, rainstorms would be more likely to saturate the soil, which would allow water to move horizontally across the surface and erode soil. In addition, the trees protect the soil from wind.

As described in Chapter 15, the damages done by eroded soil beyond an owner's property line are termed the social costs of soil erosion. Social costs can exceed private costs. Some eroded soil washes into nearby rivers and streams, where the soil is suspended in the running water and increases its **turbidity** (which lessens light penetration). Turbidity reduces photosynthesis and ultimately net primary production. Soil particles also interfere with the gills of fish, reduce the rates at which fish eggs hatch, and reduce the survival rates of young fish. The U.S. Forest Service estimates that logging operations destroy significant areas of fish habitat in the western portion of the United States.

Eroded soil particles also reduce the productivity of economic infrastructure. Where stream flow slows, soil particles accumulate on the bottom, which is termed **siltation.** Siltation kills small benthic organisms, reduces the storage capacity of reservoirs, and shortens the life spans of hydroelectric dams. Siltation also interferes with water transportation by making waterways shallower, which reduces the size of ships that can navigate the waterways.

The siltation of harbors due to deforestation is a recurring theme throughout history. The city of Ephesus was a bustling harbor on the Aegean Sea in the seventh century B.C. (Figure 17.12). Regional development spurred the rate of deforestation. As a result, the sediment load carried by the Cayster River increased. By the third century B.C. the port of Cayster had to be moved because silt made the harbor too shallow. In colonial New England, soil erosion caused by rapid rates of deforestation along the Connecticut River silted the harbor in New Haven. Local inhabitants had to extend the wharf to maintain their ability to service large ships. And to maintain water traffic in 1998 the U.S. Army Corps of Engineers removed about 153 billion cubic meters (200 billion cubic yards) of dredged material from federal channels at a cost of about $650 million.

Deforestation and soil erosion may threaten the ability of the Panama Canal to move ships between the Atlantic

Seventh century B.C.

Third century B.C.

Second century A.D.

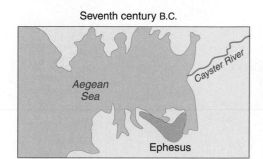
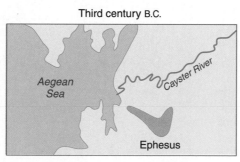
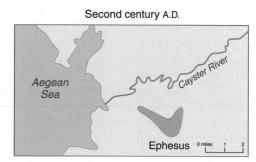

FIGURE 17.12 *Erosion and Siltation In the seventh century B.C. Ephesus was a port city on the Aegean Sea. Over time, eroded soil from the Cayster River built new land around the river's mouth. This new land gradually cut off Ephesus from the sea, so that by the third century B.C. Ephesus no longer was a port.* (Source: Redrawn from Perlin, A Forest Journey, p. 80.)

FIGURE 17.13 *Deforestation and the Panama Canal* *Moving a ship between the Caribbean Sea and the Gulf of Panama requires large amounts of freshwater, which come from Gatún Lake. But the lake's ability to provide this water is threatened by deforestation, which increases the amount of soil that is eroded into the lake and ultimately reduces its ability to store water.* (Source: Redrawn from The New York Times.)

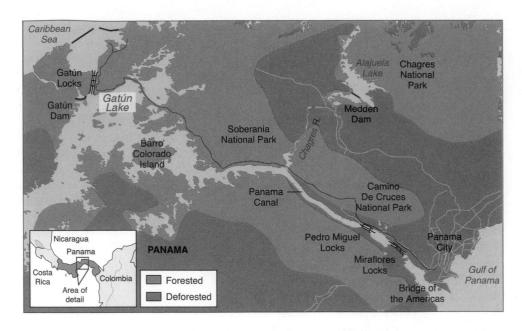

and Pacific Ocean. Moving a ship between the Caribbean Sea and the Gulf of Panama requires nearly 100 million liters (264,000 gallons) of freshwater (Figure 17.13). This water comes from Gatún Lake. But the amount of water the lake can hold is being reduced by soil that has eroded from land cleared in the surrounding forests.

Forests also affect the flow of local streams and rivers by regulating the rate at which soils absorb and release water. Soils covered by trees generally store more water than bare soils, releasing that water steadily throughout the year. This slow but steady contribution supplements ongoing precipitation and allows streams to flow year-round. Without trees, streams tend to go through periods of flood and drought. The effect of trees on stream flow is illustrated by increases in the number and frequency of floods in colonial New England. When European settlers first arrived, floods were relatively rare. Historical records indicate that there was only one severe flood in New England between 1635 and 1720. But the next eighty years witnessed six severe floods. At the same time the frequency with which streams dried up in the summer increased. This loss of stream flow damaged the colonial economy because water power was an important source of energy.

The effect of deforestation on a forest's ability to modulate the hydrological cycle was illustrated more recently by the effects of Hurricane Mitch in 1998. Some of the heaviest damage occurred in Honduras, where 91,125 hectares (225,000 acres) of forest are cleared every year. The combination of a treeless landscape and heavy rains increased the frequency and severity of floods and mudslides, which were the greatest cause of the loss of lives and property (Figure 17.14). Forestry engineers found that the storm caused less damage along rivers that were protected by trees.

Finally, forests contribute to human well-being even when the forests are unused. This contribution is known as **existence value.** Forests have existence (intrinsic) value because many people say it is important to them that tropical rain forests or old growth forests continue to exist, even though they may never visit them.

FIGURE 17.14 *Hurricane Mitch* *The damages inflicted by Hurricane Mitch were exacerbated by deforestation. Without trees to hold soil in place, the storm caused large mudslides, which killed many people and caused considerable damage.*

ARE RATES OF DEFORESTATION TOO HIGH?

Comparing Direct and Indirect Contributions to Economic Well-Being

Clearing forests generates both benefits and costs to society. Cutting trees raises economic well-being by supplying wood that can be used as raw material or as a source of energy (direct benefits). Conversely, cutting trees lowers economic well-being by reducing environmental services generated by an intact forest (indirect benefits).

To maximize economic well-being, society should weigh the benefits associated with wood as a natural resource against the losses of environmental services. Ideally society would harvest wood so that the gains and losses are balanced. In many nations, however, forests are being cut too rapidly so that the reduction in well-being caused by the loss of environmental services outweighs the benefits associated with an increased supply of wood and energy. Why do some societies reduce their economic well-being by cutting their forests too quickly?

One reason is associated with an important difference between the benefits and costs associated with cutting forests. The benefits obtained from cutting forests generally are concrete and obvious. Trees and agricultural land are valuable. As such, they can be sold in the market and generate income for those who cut forests. Activities that increase income are encouraged in a market economy.

On the other hand, the loss of a forest's environmental services may be less immediately noticeable. For example, forest owners cannot charge the U.S. Army Corps of Engineers for the soil that their forests hold in place. This reduces the economic incentives to preserve the forest. In other words, environmental services provided by forests are externalities. Users do not pay for the environmental services that are provided by forests; therefore, the value of those services is not captured by the market or is priced too low.

Difficulties in comparing the direct economic benefits of deforestation and the economic costs (due to the loss of indirect contributions) can generate conflicts about whether a forest should be cut or preserved. One such conflict occurred in California over the Headwaters Forest—a virgin redwood forest that was owned by Pacific Lumber Company. Its fate became a national story when Pacific Lumber was purchased by a larger company, MAXXAM, in a hostile takeover, and MAXAAM announced that Pacific Lumber Company would accelerate its harvest from remaining tracts of virgin redwood forest, including the Headwaters Forest. This announcement touched a nerve because much of California's original redwood forests already had been cut. Only about 4.4 percent of the original forest remained, and most of that 4.4 percent was protected in parks. The Headwaters Forest represented a significant portion of the virgin redwood forest in private hands.

To save the Headwaters Forest, environmentalists took several actions, including having people live in trees, sabotag-

ing equipment, and "spiking" trees (inserting metal spikes into trees that could injure workers when they cut them). Eventually the indirect contributions of the Headwaters Forest won out when the government purchased the 3,034 hectares (7,500 acres) Headwaters Forest and 1,012 hectares (2,500 acres) of old-growth woodlands for $480 million. This total represented the value of timber in the forest.

But was buying the Headwaters Forest the most effective way to maximize the indirect contributions of redwood forests? In California cutover areas (areas that have been previously logged) of redwood forest sell for about $243 per hectare ($600 per acre). At this price the government could have purchased 256,166 hectares (633,000 acres) and allowed them to regrow. In several hundred years these 256,166 hectares could have represented nearly a third of the land originally covered by redwood forest. Instead society spent $480 million to save 3,034 hectres of the virgin Headwaters Forest, which represents about 1.5 percent of the original forest. Was this the best way to spend money and preserve forest? The answer depends in part on the degree to which succession can return cutover areas to their original state. If the original ecosystem is highly resilient, the money may have been better spent on the 256,166 hectares of cutover land. But if succession regenerates only a portion of the original plant and animal community, preserving the virgin Headwaters Forest may have been the most effective use of funds. Beyond economic considerations, many people consider the ancient redwood forests (some of which contain trees that are thousands of years old) irreplaceable. But this is an ethical position that cannot be evaluated using scientific methodology.

Policies

The policy environment is filled with government subsidies that encourage rapid rates of deforestation. Subsidies artificially inflate the financial returns of activities. By doing so, government subsidies can make poor investments profitable. Subsidies take many forms, including tax reductions, low-interest loans, or government services that are priced below cost. For example, governments often build roads into the forest at little or no cost to timber companies. If timber companies paid the full costs of road construction, lower profits would slow deforestation.

Agricultural subsidies also encourage farmers to clear forests. These subsidies often constitute a significant fraction of farm income in developing nations. In Mexico, Brazil, and South Korea government subsidies account for about half of agriculture's contribution to GDP. These large sums can have visible effects on deforestation. In some developed nations governments tried to stimulate agricultural production by taxing undeveloped forest lands more than developed lands. This tax structure had the unintended effect of encouraging farmers to clear forests (without increasing agricultural production) to reduce their tax bills.

In nations where the government owns much of the forest, rules that control timber harvests encourage rapid deforestation. When a firm wishes to harvest trees from a government forest, it must obtain a **timber concession,** which is a contract that defines the rules for harvesting trees. To prevent firms from controlling large areas of the forest for long periods, timber concessions often run for relatively short periods. Under these conditions firms have little incentive to consider long-term costs and benefits. Instead they tend to cut as many trees as they can during the concession period.

Social Structure

About 350 million of the world's poorest people depend entirely on forests for their food, fuel, and materials that they use for shelter and clothing. Another 60 million indigenous people live in forests in ways much like their ancestors. These people often have no property rights and have little or no political power. Under these conditions, the benefits that they derive from the forest carry little weight in economic and political decisions regarding the fate of the forest. As such, decisions to cut forests often ignore environmental services that support just under 10 percent of the world's population.

The lack of economic and political power of indigenous peoples relative to those who govern encourages corruption, which also speeds deforestation. In many developing nations, a small group of rich elites controls political and economic power. This power creates opportunities for government officials to negotiate "sweetheart deals" in which family members and friends obtain timber concessions at costs well below market value (often in return for kickbacks). These deals rarely come to light because the poor do not have the political power to make the terms of concessions public.

Differences in the political power of rich and poor sometimes mirror differences in the political power of men and women. In many societies women are responsible for collecting fuel wood, but men control decisions about the forest. Under these conditions women who bear the costs that result from the use of the forest have little control over forest management, and this lack of connection can lead to rapid deforestation.

HOW CAN DEFORESTATION BE SLOWED?

Getting Prices Correct

According to economic theory, the most effective way to slow deforestation is to address its root cause—the market's inability to correctly price the environmental services provided by an intact forest. If the market could price these services correctly and charge consumers for their use, problems of deforestation would be eased. People would cut the forest at a rate that balanced the benefits against the loss in environmental services. But as described here, forest policy that focuses only on getting prices correct probably will not succeed.

It is difficult to fix the market so that it can price the environmental services provided by an intact forest. First, it is extremely difficult to value environmental services. Scientists estimate that tropical forests generate $630–4,500 of environmental services

per hectare ($243–1,822 per acre). The exact value depends on site-specific characteristics such as biomass and proximity to economic markets. Nor are scientists sure about the totality of ways in which forests contribute to human well-being.

But even if scientists could identify all these contributions and calculate the corresponding prices, it is challenging to identify all consumers and charge them for the environmental services provided by the forest. For example, forests control floods, but who benefits from flood control? People that live downstream from a forest benefit from its ability to control floods—but how far downstream and how far from the banks of the river do these benefits extend?

Even if scientists could identify the people living downstream who enjoy the benefits of flood control, how can society make users pay for environmental services such as flood control? Many of a forest's environmental services are common property: Forest owners cannot deny flood control to those who do not pay. Without the ability to withhold flood control services, there is little incentive for those who benefit from flood protection to pay the forest owners.

Finally, there is an intergenerational conflict about the rate of deforestation. The current generation enjoys the benefits of timber or fuel wood. Future generations will not enjoy these benefits but will suffer the loss of indirect benefits. Because they are not yet born, future generations do not get to choose the optimal rate of deforestation. Nor does anyone know exactly what they would prefer. For example, our ancestors deforested much of the United States. Some of these efforts have increased our economic standard of living. Would you have preferred our ancestors to cut less forest if it meant a lower standard of living? If so, how were our ancestors to know your preference?

Even if society cannot create a so-called efficient market for forest products and environmental services, deforestation can be slowed by eliminating subsidies that encourage deforestation. Removing subsidies already has slowed deforestation in many parts of the world. In 1995 the Russian government reduced subsidies to the transportation system, which forced the timber industry to pay the full cost of transporting logs. Higher transportation costs shrank the area over which it was economically profitable to harvest trees, which slowed deforestation.

Sustainable Forestry Practices

Sustainable forestry practices also would slow the rate of deforestation, but developing and implementing sustainable forestry practices is problematic. One difficulty concerns the definition of *sustainable.* Sustainable forestry practices have many definitions. For those in the timber industry the term implies a **sustainable yield of timber,** which means harvesting timber no faster than the rate at which trees produce new supplies. But timber is only one of the benefits generated by forests. For example, tree plantations can generate a sustainable supply of timber. Yet many people argue that replacing forests with tree plantations is not a sustainable forestry practice because tree plantations generate fewer environmental services than the forests they replace. For example, some tree plantations reduce an area's water supply. Tree plantations support fewer species, which also reduces the forest's environmental services.

Such contradictions imply that a more comprehensive definition of sustainable forestry practices is needed. In response, the notion of sustainable forest management has been expanded into the concept of **ecosystem management.** Ecosystem management seeks to balance human needs for wood as a raw material and an energy source with the forest's provision of environmental services. To do so, ecosystem management seeks to ensure a sustainable supply of timber while conserving soil and water, sustaining the resistance and resilience of the forest, supporting the food security of indigenous people, conserving biological diversity, and so on.

These are lofty ideals, but like getting prices correct, ecosystem management is challenging to implement. For example, there are many ways to harvest trees. To choose among techniques, forest managers need to quantify the ecological characteristics of the site, such as soil type, climatic variability, and vegetation type. In addition, managers need to identify the many ways in which people use the forest directly and indirectly. As a result, there is no agreement on what constitutes ecosystem management. Practices vilified by many environmentalists may not be inherently unsustainable.

For example, many environmentalists are especially critical of **clear-cutting**—a practice in which all commercially valuable trees are harvested at the same time (Figure 17.15). In many cases clear-cutting generates severe environmental damage such as soil erosion, the loss of timber production for many decades until trees regrow, the loss of habitat, and the possibility that unwanted plant species will become established. But clear-cutting also has some advantages, which include reducing the cost of harvesting trees and speeding the regrowth of tree species whose seedlings cannot grow in the shade, such as red pines. Because the effects of clear-cutting are site-specific, the World Commission on Forests and Sustainable Development does not universally condemn or condone clear-cutting.

Alternatively, several techniques are available for cutting individual trees or smaller groups of trees, which are known as **selective harvesting.** For example, **high-grading** harvests only the trees that give the highest immediate economic return. Individual trees are chosen by species or diameter. This practice often is used because it generates cash for the owner in the short term; but these benefits sometimes are offset by long-term losses. Harvesting trees with the highest value may imply that the remaining trees are diseased, are weakened by the harvest process and the loss of their more mature neighbors, or are members of less valuable species. These are not the individuals from which the landowner wants to regrow the forest. Leaving the weakest and smallest individuals also reduces the forest's resistance—the trees that remain are more vulnerable to pests, disease, and storms.

Between these two extremes, the **shelterwood method** removes trees in a series of two or three partial cuts. This

FIGURE 17.15   *Clear-Cutting* *In some areas, all of the commercially valuable trees are cut simultaneously. This creates a large open area. In some areas, such as the one pictured here, the forest is able to regrow.*

method is designed to simulate a natural disturbance that leaves a stand of trees that are nearly the same age. These trees provide shelter that speed the regrowth of shade-tolerant trees, such as red oak or American beech. In addition, these trees provide habitat for wildlife. Despite these advantages, harvesting via the shelterwood method is difficult because it requires expert management. For example, removing large trees can damage the remaining trees and make them more vulnerable to high winds.

This lack of agreement about the best harvest method or the meaning of ecosystem management does not imply that nothing can be done. Just as the getting prices correct approach starts with policy that removes subsidies, sustainable forestry practices start by encouraging harvest methods that soften environmental impacts. In tropical dry forests such practices include cutting the tops of trees to produce more digestible fodder. Similarly, harvesting fuel wood from cut stumps may generate more usable biomass than the large logs that are generated by traditional timber-oriented management practices.

Expanding the definition of sustainable forestry practices to include the forests' environmental services and changing the techniques used to harvest trees often are costly, which reduces the likelihood that firms will adopt sustainable forest practices. See *Policy in Action: Sustainable Forestry Practices: What Price and Who Pays?* How can forest owners and managers be encouraged to adopt sustainable forestry techniques?

Incentives for implementing environmentally benign timber practices may come from consumers. In many developed nations consumers want timber products that are produced using sustainable forestry practices. In response, several organizations such as the Forest Stewardship Council now certify forestry operations that satisfy some basic criteria for sustainability. Their products carry labels certifying the environmental integrity of the production process. These labels increase demand relative to products with no labels. Because consumers in developed countries account for about 60 percent of the timber market, the threat of a reduction in market share is a powerful incentive for producers to adopt sustainable forestry practices.

POLICY IN ACTION

Sustainable Forestry Practices: What Price and Who Pays?

Scientists have identified several sustainable forestry practices. Rarely are these practices implemented because they are less attractive financially than traditional options. The increase in costs associated with sustainable forestry practices and who will pay for their implementation can be understood by following a landowner's decision-making process in the forests of Costa Rica.

Forest owners view their property as a means for supporting themselves and their families now and in the future. The ways in which land can be used to generate this support differ in the timing of the income stream (now or in the future) and the degree of environmental degradation. One option would be to "mine" the forest. Following this option, the owner cuts and sells all of the economically valuable trees. This creates a short-term infusion of income, but this income disappears when all of the trees are cut. Consequently, this income will not reappear until the trees regrow, which requires many decades. Alternatively, landowners may stagger the harvest. By leaving some trees in place, this strategy reduces immediate income but increases future earnings. If the rate of harvest is sufficiently slow, this practice may be sustainable.

Alternatively, the owner can sell the valuable trees and replace the entire forest with a tree plantation. This option generates income in the short term but also requires a large initial investment to establish the plantation. Once established, the plantation can generate a steady income stream by producing harvestable trees more rapidly than the forest. Because tree plantations generate a lot of income per unit of land, they are an intensive land use option. However, even though trees are replanted, such a monocultural plantation can inflict considerable environmental damage and reduce environmental services.

As another option, the owner may convert the forest to a cattle ranch. To do so, the economically valuable trees are cut and sold. The rest of the forest is burned; the ground is seeded; and the resultant pasture is used to

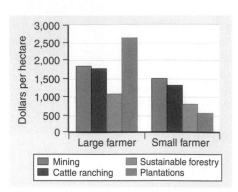

FIGURE 1 *Return on Investment* The profitability of different land use options that are available to large farmers and small farmers. In general, the returns to small farmers are lower than those to large farmers because small farmers do not have access to credit or must pay higher interest rates when they borrow money. These differences affect the land use that generates the highest returns to large farmers (plantations) and small farmers (mining the forest). (Source: Data from from N. Kishor and L. Constantino, "Sustainable Forestry: Can It Compete." Finance and Development 34(4): 36–39.)

graze cattle. Cattle ranches generate immediate income (selling the trees) and a steady income (selling cattle) and are less costly (than the tree plantation). Because cattle ranches generate relatively little income per unit of land, they are an extensive land use option. By removing the forest, a cattle ranch has the most severe environmental impact of the options just listed.

Which option will the forest owner choose? The answer depends in part on the landowner's credit rating. Small landowners often have little or no access to credit and so must choose an option that has relatively little cost and generates an immediate income stream. Given this restriction, "mining" the forest or converting it to a cattle ranch generates the greatest revenue stream to small landowners (Figure 1).

Conversely, credit at or near market rates is available to large landowners. They can use these borrowed funds to set up a more expensive land use option (such as tree plantations) and use the funds to support their families until

the option generates a steady income stream. Access to credit increases the net present value of economically viable land use options relative to those available to small landowners, as well as eliminating the biases that push small landowners toward more extensive, environmentally destructive options.

Regardless of access to credit, sustainable forestry practices generate the income stream with the lowest net present value to both large and small landowners. To choose the sustainable forestry option, large landowners in Costa Rica would require a payment. But where would such payments come from? Environmental services of local forests generate relatively few benefits to Costa Ricans: about $200 per hectare ($81 per acre). On the other hand, environmental services generated by Costa Rican forests have a high value to foreigners: $1,000–2,000 per hectare ($405–810 per acre). Most of this value is associated with carbon storage and the existence value of tropical rain forests. If Costa Rica could charge foreigners, these payments could be used by local farmers to adopt sustainable forestry practices. Consistent with this possibility, policy makers have focused on ways to collect these funds. Options include tradable permits by which firms in developed nations would pay Costa Rica to preserve areas of the forest that store amounts of carbon proportional to the quantities that the firms emit (see Chapter 13). This option is one reason why many smaller nations signed the Kyoto Protocol.

ADDITIONAL READING

Kishor, N., and L. Constantino. "Sustainable Forestry: Can It Compete?" *Finance and Development* 31, no. 4 (1994): 36–39.

STUDENT LEARNING OUTCOME

- Students will be able to explain why developing new technologies does not automatically lead to lower rates of deforestation.

Debt for Nature Swaps

Debt for nature swaps also are an effective mechanism by which both developing and developed nations can slow deforestation and benefit economically. Swaps are based on the fact that many developing nations owe large sums to banks in developed nations. Some developing nations owe so much money that banks have little hope the loans will ever be repaid; therefore, the banks are willing to sell the debt for a few cents on the dol-

lar. For example, it is possible to buy a million dollars worth of Peru's debt from a bank for several thousand dollars.

Taking advantage of these bargains, several environmental nongovernmental organizations have bought portions of developing nations' debt. They too have little hope of receiving payment. Rather, the nongovernmental organizations use the debt as a bargaining chip. The final agreement, which is termed a **debt for nature swap,** has the following form. The developing nation uses some of the money that it would have used to repay its debt to designate a portion of its forests as a national park where timber harvesting, agriculture, and mining are not allowed. Indigenous people are allowed to live there and are encouraged to maintain their lifestyle. In this way nongovernmental organizations achieve their goal of slowing deforestation. In return for the setting up of the park, the nongovernmental organization forgives the loan. That is, the developing nation no longer has to pay back the portion of the loan that the nongovernmental organization bought from the bank. This reduces the developing nation's debt burden. Debt relief increases the funds that are available to improve the material well-being of the country's inhabitants.

Debt for nature swaps were pioneered by the World Wildlife Federation in the late 1980s. In one prominent swap, finalized in July 2002, the U.S. government canceled $5.5 million of Peru's debt. (The use of government funds was authorized by the Tropical Forest Conservation Act of 1998.) Paying this debt would have cost Peru about $14 million over time. In return, Peru designated 11.1 million hectares (27.5 million acres) in the western Amazon as a conservation area and agreed to set up a $10 million trust fund to support Peruvian conservation groups. This effort was supported by a $1.1 million contribution from three conservation groups: Conservation International, the Nature Conservancy, and the World Wildlife Fund.

SUMMARY OF KEY CONCEPTS

- Rates of deforestation vary over time and among nations. Developing nations generally cut their forests faster than developed nations. Local rates of deforestation are associated with the conversion of forests to agricultural land, timber harvests, fires (both accidental and deliberately set), and energy and mineral production. These causes of deforestation are enhanced by actions that reduce transportation costs to and from the forest.

- Forests supply an important natural resource: wood. Wood is used primarily as a material in developed nations. In developing nations wood is used mostly as a source of energy. To increase its energy content, wood often is converted to charcoal. Forests also generate many environmental services, such as habitat, climate control, carbon storage, and preserving soil.

- It is difficult to compare the direct and indirect contributions that forests make to economic well-being. Externalities tend to increase deforestation because they cause the market to undervalue environmental services. Conversely, the direct benefits associated with cutting forests tend to be priced by the market. In addition, decisions to cut forests often ignore the environmental services that support indigenous and poor people because they lack economic and political power.

- Rates of deforestation could be slowed if the market could price environmental services correctly and charge consumers for their use. The market does not provide a mechanism by which owners can charge those who use services that are generated by the forest. Similarly, sustainable forestry practices would slow the rate of deforestation, but developing and implementing sustainable forestry practices is difficult because there is no single criterion for defining sustainability, nor is there a single harmless harvesting technique.

REVIEW QUESTIONS

1. Explain how investments in roads and other transportation infrastructure affect the rates of deforestation associated with agriculture, timber, property rights, and mineral and energy production.

2. Compare and contrast who benefits (landowners versus the public) from the direct and indirect contributions of forests. How do these differences contribute to rates of deforestation that exceed the optimal level?

3. Explain why market-based mechanisms may not be able to generate the optimal rate of deforestation. Is this another case of the second best solution, in which command and control mechanisms may be most effective at slowing deforestation?

4. Explain the economic basis for a debt for nature swap and how this process works.

5. Why do externalities tend to increase the rate of deforestation?

KEY TERMS

afforestation

clear-cutting

debt for nature swap

deforestation

ecosystem management

existence value

extensive land uses

high-grading

intensive land uses

selective harvesting

shelterwood method

siltation

sustainable yield of timber

timber concession

transportation costs

transportation infrastructure

turbidity

CHAPTER

18

WATER RESOURCES

The Legend of Ubar *This depression in the sand contains remnants of the city of Ubar, which collapsed into the desert after it depleted its underground water supplies, which helped hold the city up.*

STUDENT LEARNING OUTCOMES

After reading this chapter, students will be able to

- Explain how linkages between surface water and groundwater constrain human efforts to increase water supply.

- Explain the effects of overdrafts on water supply in particular and the environment in general.

- Explain how the hydrologic cycle contributes to the potential for conflict over water.

- Describe why the ability to buy and sell permits to emit water pollutants could dramatically reduce the effectiveness of the Clean Water Act.

- Explain how the supply and cleanliness of water supplies affect economic development.

THE LEGEND OF UBAR

Was it real? According to legend, Ubar was a city–state in the Arabian Desert, where it was known as the "Atlantis of the Sands." It was a stopover for merchants and their camels in the frankincense trade. (Frankincense is an aromatic resin that was highly prized from Rome to India for use in religious rituals and cosmetic and medicinal uses.) The city was important because it was the only source of water for hundreds of kilometers. To protect its liquid wealth, the city built fortified walls and watchtowers.

The water and caravans made the locals wealthy. The Holy Quran describes Ubar as "the many-columned city whose like has not been built in the whole land." Like other wealthy cities of the time, it supposedly became a hotbed of wickedness. Residents ignored commands to stop their evil ways, and Ubar was destroyed in a single night. If this legend is true, the city of Ubar disappeared sometime between A.D. 300 and 500.

The myth of Ubar fascinated the filmmaker Nicholas Clapp, whose enthusiasm led him to contact NASA and ask that the space shuttle help him look for Ubar. Radar images from the shuttle highlighted an ancient network of roads, many of which came together around the current city of Shisur. Further investigations headed by the archaeologist Juris Zarins

located ruins buried deep below the surface (opening photograph). Excavation revealed a city complete with 9.1 meter (30 foot) towers, thick walls, storerooms, frankincense burners, and pottery shards, all about 2,000 years old.

In addition to confirming Ubar's existence, the excavations revealed the cause for its demise. Much of Ubar's water came from a spring-fed well. As the inhabitants used the water to support travelers and irrigate crops, water levels dropped. Unknown to local inhabitants, this underground water supported a limestone shelf on which the city of Ubar was built. Without the water below it, the limestone shelf became unstable. A small earthquake may have cracked the shelf, and the city of Ubar and its inhabitants fell into the underlying void. The void was quickly covered with sand, and the city became a legend.

The true story of Ubar illustrates many important lessons of water use. Water is a precious commodity that can make people wealthy. To generate this wealth, people will use water as quickly as possible, even if this threatens their long-term survival. And in some cases entire civilizations can collapse when they lose their water supply. ■

THE HYDROLOGIC CYCLE

The **hydrologic cycle** describes the flow of water from the ocean through the air to the land and back to the ocean. These flows obey the same basic rules described in Chapter 6. As with carbon and nitrogen, Earth has a relatively constant supply of water. But unlike biogeochemical cycles, the

important flows in the hydrologic cycle involve changes in the location and physical state of water.

We start our trip through the hydrologic cycle with the largest storage—the ocean. Oceans hold about 1.35 billion km³ (0.32 billion mi³) of water (Figure 18.1). Ocean water, known as **salt water**, contains more than 35,000 milligrams

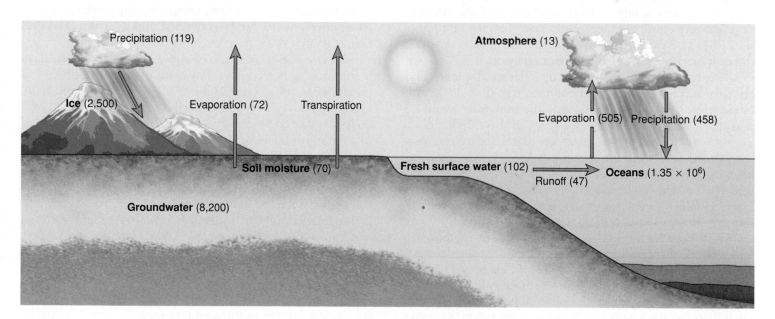

FIGURE 18.1 *The Hydrologic Cycle* The major storages and flows in the hydrologic cycle. Storages and flows are measured in thousand km³. Storages and flows do not balance because there is considerable uncertainty regarding their size; not all storages and flows are shown.

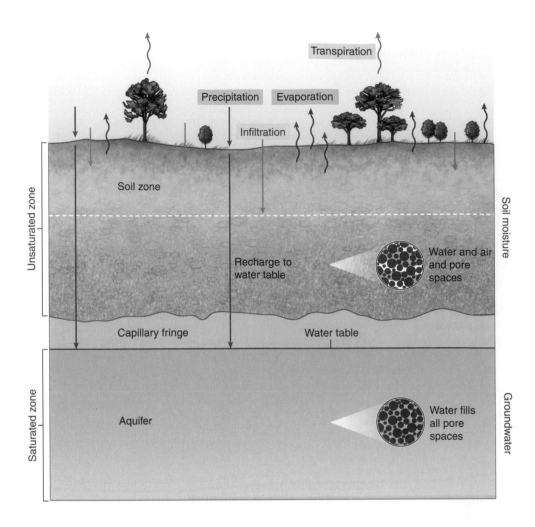

Transpiration

Precipitation   Evaporation

Infiltration

Soil zone

Unsaturated zone

Soil moisture

Recharge to water table

Water and air and pore spaces

Capillary fringe

Water table

Saturated zone

Aquifer

Water fills all pore spaces

Groundwater

FIGURE 18.2 *Infiltration Water that falls on soil can either move horizontally or downward. Downward movement is possible through the unsaturated layer where air remains within the pores. Here the water can be taken up by plant roots or move farther down toward the water table, which is the upper portion of the saturated zone.*

per liter of dissolved solids, most often salt. High concentrations of salts, such as sodium, make salt water of little use to humans and most terrestrial plants and animals. You cannot drink salt water safely—to rid your body of the sodium, your kidneys eliminate more freshwater than the amount of salt water you drink. Similarly, high sodium levels make it difficult for plants to pull water from the soil.

The type of water needed by terrestrial plants and animals is **freshwater,** which contains less than 1,000 milligrams per liter of dissolved solids and makes up about 2.5 percent of Earth's total water storage. Of these 36.9 million km^3 (8.8 million mi^3), two-thirds are stored as ice and permanent snow cover in the Arctic and Antarctic regions. Less than 1 percent of the world's freshwater, or about 0.007 percent of all water, can be used by people. This includes water found in lakes, rivers, reservoirs, and underground sources that are shallow enough to be used economically.

Freshwater is produced from salt water via **evaporation,** in which water changes from a liquid to a gas. As water warms, molecules at the surface become sufficiently energized to break free of the attractive force that binds them. Only the water molecules become a gas—salts and other materials remain. This separation creates freshwater. In total, evaporation generates 500–600 million km^3 (120–144 million mi^3) of freshwater each year. Of this about 90 percent evaporates from oceans, seas,

lakes, and rivers. The remaining 10 percent is transpired by plants or evaporates from the land's surface.

Once in the atmosphere, water vapor cools and some **condenses** (changes into a liquid), usually on tiny particles of dust. Alternatively, water can be converted directly into a solid (ice, hail, or snow). Water droplets collect and form clouds.

On average, a molecule of water spends about ten days in the atmosphere. Water flows from the atmosphere via **precipitation** as a liquid (rain) or solid (ice, snow, and hail). Precipitation moves water to Earth's surface. Water that falls on land is moved by gravity in two directions: vertically down into the soil or laterally across the surface. Vertical movements are known as **infiltration** (Figure 18.2). In the upper layers this water is known as soil moisture (see Chapter 15). On average, a water molecule spends about one year as soil moisture. Some soil moisture returns to the atmosphere via transpiration.

A fraction of soil moisture continues to move down until it reaches an impermeable layer. Here it accumulates, filling all spaces among soil particles and creating a zone of saturated soil known as an **aquifer** (Figure 18.2). Aquifers contain **groundwater,** and the top of the aquifer is known as the **water table.** The area from which an aquifer receives its water is known as the **recharge area.**

Groundwater flows relatively slowly in the direction dictated by gravity. Following these movements, the aver-

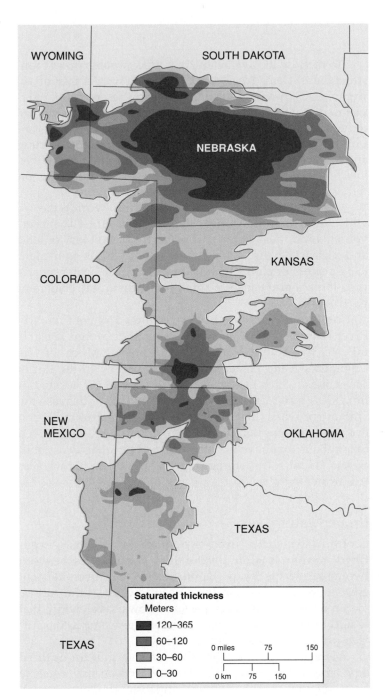

FIGURE 18.3 *Ogallala Aquifer* *Located in the central portion of the United States, the Ogallala Aquifer contains water from glaciers that melted thousands of years ago. These glaciers were so thick that their meltwater created a saturated zone that can be hundreds of meters thick.*

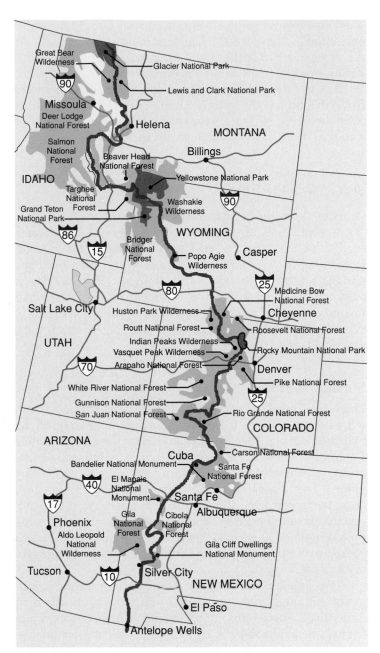

FIGURE 18.4 *Continental Divide* *The Continental Divide is a series of mountain ridges that determines whether water flows west toward the Pacific Ocean or east toward the Atlantic Ocean. As such, it divides the United States into two very large drainage basins.*

age water molecule spends about 1,400 years as groundwater. For example, the Ogallala Aquifer contains water that melted from glaciers 10,000–25,000 years ago, infiltrating the soils below the Great Plains (Figure 18.3). Sooner or later groundwater resurfaces at points known as **groundwater discharge.** You may know some of these points as springs.

Water that flows horizontally across the land is known as **runoff.** Gravity moves runoff toward the nearest **surface water,** such as a river, lake, or ocean. The average water molecule spends about sixteen days in a river or seven-

teen years in a lake. The area from which surface waters derive surface runoff and groundwater flows are known as a **drainage basin** or **watershed.** Their boundaries are defined by elevation. In the contiguous United States, the **Continental Divide** is a series of ridges through the Rocky Mountains that divides the country into two drainage basins (Figure 18.4). Raindrops or snowmelt that fall west of the divide flow toward the Pacific Ocean, whereas water east of the divide flows toward the Gulf of Mexico or Atlantic Ocean.

These flows complete water's journey back to the ocean and explain why ocean water contains so much salt. As water washes across the land surface and through the soil, it dissolves salts. Some of these salts are carried from streams to lakes to rivers and on to the ocean. As water evaporates from the ocean, the salts remain. Oceans become saltier over time because water carries salt to the ocean each time it moves through the hydrologic cycle.

WATER SUPPLY

Our trip through the hydrologic cycle indicates that water flows seamlessly among the atmosphere, groundwater, surface waters, and ocean. Understanding these linkages is essential for evaluating how water withdrawn from one storage affects other flows and storages. See *Case Study: Climate Change and the U.S. Water Supply.* As we discuss later in this chapter, a flawed understanding of the hydrologic cycle led to laws governing water use that contradict the mass balance of the hydrologic cycle.

Surface Water

The main source of freshwater for humans is surface runoff, whose most important feature is its annual renewal in the hydrologic cycle. About 72 million km^3 of the approximately 119 million km^3 (28.5 million mi^3) of precipitation that falls on the continents each year evaporates or is transpired by plants. This leaves about 47 million km^3 (12 million mi^3) of water to run from the land to the ocean. As it runs off, people capture a portion of it. Globally surface runoff adds up to about 7,300 m^3 (258,000 ft^3) of water per person each year. But as with many other natural resources, averages are misleading.

The availability of freshwater varies among continents. Although more than half of Earth's runoff occurs in Asia, this is shared among billions of people, so per capita freshwater availability is the lowest of all the continents. Conversely, Australia is relatively dry; but its small population implies that plenty of runoff is available to each person. Another 19 percent of runoff occurs in remote regions such as the upper reaches of the Amazon River, where it is unavailable for human use.

Even continental averages are misleading: Most of the world's runoff flows through a limited number of rivers. The Amazon carries 16 percent of global runoff, and the Congo–Zaire River basin carries one-third of African river flow. Arid and semi-arid zones, which constitute 40 percent of landmass, enjoy only 2 percent of global runoff.

Seasonal variations in runoff also affect water availability. Nearly half the world's runoff occurs following snowmelt or heavy rains. Such bursts of water flow quickly to the ocean unused. This loss is amplified by urban areas, which accelerate the rate at which precipitation reaches surface waters.

To increase the availability of surface waters, people construct dams. Dams block surface water flows. As they do so, water backs up behind the dams, creating **reservoirs.** A reservoir is a natural or artificial pond or lake that is used for storage or regulation. Water can be released from the reservoir at controlled intervals. These intervals are chosen to prevent flooding and to increase water supply when flows would normally be low. In effect, dams spread a short-term peak in surface runoff throughout the year, which increases usable supply.

Over 800,000 dams have been built worldwide, of which more than 45,000 are categorized as large. The reservoirs behind these dams cover nearly 500,000 km^2 of land and store 6,000 km^3 (1,439 mi^3) of water. This is equivalent to 14 percent of global freshwater runoff. The water's mass is so large that it causes a small but measurable change in Earth's orbit.

Perhaps no river has been so dammed as the Colorado River (Figure 18.5). This river drains 632,000 km^2 (244,000 mi^2) in the southwestern United States. Much of the Colorado's water comes from melting snow in the Rocky Mountains, which generates a rapid flow in the spring but much less water during the rest of the year. To increase the amount of water that can be delivered year-round to an ever-growing population, the U.S. government built seven major dams that have a combined storage capacity of about 80 km^3 (19 mi^3). This storage represents a three- to four-year freshwater supply.

Groundwater

Groundwater comes from two types of aquifers. A **confined aquifer** is groundwater that accumulates between two impermeable layers (Figure 18.6(a)). Such confinement creates pressure that can force groundwater to the surface via an **artesian well.** Because such flows save energy that would otherwise be needed to move the water to the surface, artesian wells are highly desirable.

More common are wells that are dug into **unconfined aquifers,** which are aquifers that sit atop an impermeable layer (Figure 18.6(b)). Without intrinsic aquifer pressure, water must be brought to the surface. Water is relatively heavy—each liter weighs a kilogram (2.2 lb). So the effort required to pump groundwater limited its use for much of human history. During the later decades of the nineteenth century, farmers in the United States used windmills to bring groundwater to the surface. But these pumps, being relatively weak, could move less than 150 liters per minute (40 gal/min) where the water table was within 20–25 meters (66–82 ft) of the surface.

This changed in the 1950s. The U.S. Rural Electrification Administration brought electricity to the farms of the high plains, and the development of new turbine pumps allowed farmers to pump 4,700 liters per minute (1,241 gal/min) from groundwater up to 1 kilometer (0.62 mi) below the surface. This fueled a rapid increase in the use of ground-

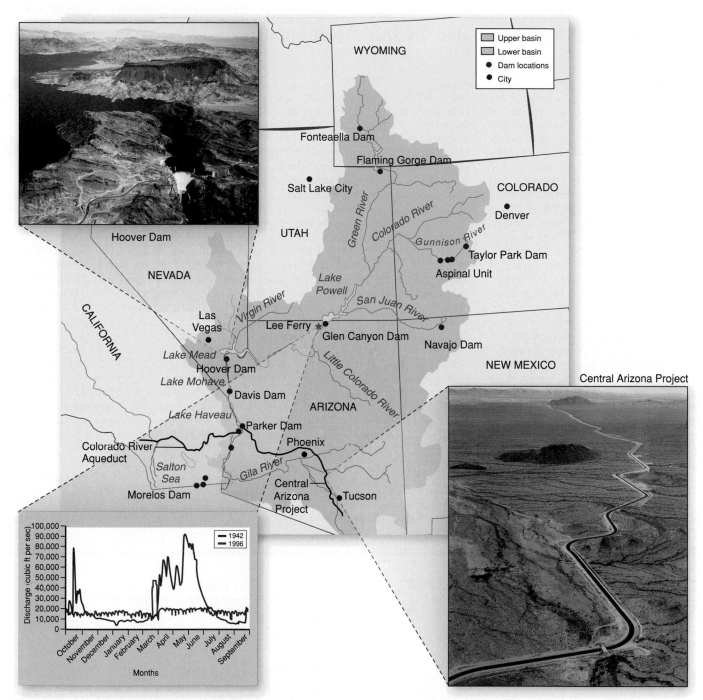

FIGURE 18.5 *The Colorado River* The Colorado River drains the Southwestern United States and supplies much of that region's surface water. *To increase usable supply, the river has several dams, such as Hoover Dam. These dams damp fluctuations in supply by storing snowmelt and making it available during the summer. Water is moved to users via pipelines and aqueducts, such as the Colorado River Aqueduct or the Central Arizona Project.* (Source: Data from U.S. Geological Survey.)

water. In the United States the use of groundwater increased from about 132 billion liters (32 billion gal) per day in 1950 to about 320 billion liters (85 billion gal) per day in 2000, the most recent year for which data are available. The Green Revolution (Chapter 16) wrought similar changes in India, where the number of groundwater pumps increased from 87,000 in 1950 to more than 13 million in the 1990s.

As pumps push water to the surface, groundwater moves through the soil to take its place. But groundwater flows relatively slowly. As a result, the water table tends to drop around the well in what is known as the **cone of depression.** This cone of depression lowers the water table and increases the distance that groundwater must be pumped to the surface (Figure 18.6(b)).

CASE STUDY

Climate Change and the U.S. Water Supply

As described in Chapter 13, human-driven increases in radiative forcing will change many aspects of climate, including precipitation. Such changes will affect water supply, but how? Precipitation is correlated with water supply because it is the amount of water that flows from the atmosphere to Earth's surface; but this flow is only one determinant of supply. By now you should realize that the effects of disturbing a highly complex, nonlinear system rarely are predictable. Climate change will affect human water supplies by altering temperature, precipitation, vegetation, and the atmospheric concentration of carbon dioxide.

Of the many adjustments associated with global climate change, scientists are most certain that surface temperature will increase. But the effect of higher temperature on water supply is uncertain. Higher temperatures mean that the atmosphere can hold more water, which will generate more precipitation. On the other hand, higher temperatures increase evaporation, which reduces the fraction of precipitation that is available for humans to use.

Evaporation also will be affected by climate-driven changes in terrestrial vegetation. As described in Chapter 13, warmer temperatures and increased precipitation have lengthened the growing season and increased summer greenness of forests in the Northern Hemisphere (Figure 13.10 on page 280). Such changes increase the quantity of water transpired by plants. If these climate-driven changes in terrestrial vegetation continue, greater transpiration will reduce usable water supply. On the other hand, an increase in temperature without a corresponding increase in precipitation may slow plant activity, which would reduce transpiration and increase water supply.

Water supply also will be affected by increasing concentrations of the gas that is mostly responsible for climate change—carbon dioxide. Increasing concentrations of carbon dioxide will increase water use efficiency by plants (see page 287 in Chapter 13). Increased efficiency means that plants will transpire less water. This would boost the quantity of soil water that can either discharge into surface waters or recharge aquifers. Both changes would increase water supply.

Tracking these many possible effects takes a computer. Allison Thomson and her colleagues have linked scenarios for climate change generated by general circulation models with a model of the hydrologic cycle. For each of more than 2,000 watersheds in the coterminous United States, the hydrologic model uses climate data to track precipitation as it hits the surface, flows along the surface, infiltrates the soil, evaporates, or is transpired. The net effect of climate change on these flows is evaluated by tracking changes in the annual supply of freshwater and seasonal changes in its availability.

To evaluate the effects of uncertainty about future rates of climate change, Thomson uses two scenarios for temperature increases: a 1°C (1.8°F) change in global mean temperature and a 2.5°C (4.5°F) rise in global mean temperature. Each of these scenarios is simulated with the current concentration of atmospheric carbon dioxide, 365 parts per million, and double the pre-industrial concentration, 560 parts per million, to assess the effect of changes in water use efficiency. To evaluate uncertainty about how global changes affect regional temperature and precipitation, each of the four scenarios is simulated by three general circulation models. This makes a total of twelve simulations.

Differences among these twelve scenarios are so great that the models cannot generate a consensus about the effect of climate change on U.S. water supply. As shown in Figure 1, one climate model predicts a general reduction in water supply, with the most severe losses in the lower Mississippi basin and the Pacific Northwest. On the other hand, two models forecast a general increase in water supply, with the greatest increases in the lower Mississippi and a larger area of the Pacific Northwest. Nor can the models generate a consensus about the effect of changes in atmospheric carbon dioxide and vegetation.

The scenarios do generate one common result: Water supply will become more vari-

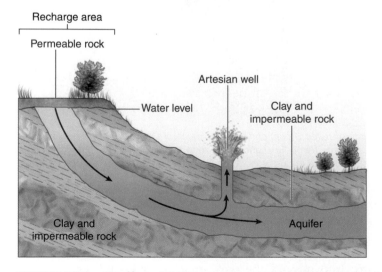
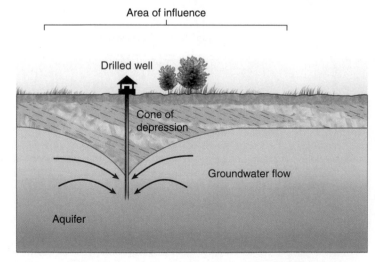

FIGURE 18.6 *Aquifers* **(a)** *The recharge area for confined aquifers is at relatively high elevation. As the water infiltrates between two impermeable layers, the water at the top pushes down on the water at the bottom. This pressure allows water to flow to the surface on its own via an artesian well.* **(b)** *Water in an unconfined aquifer is not under pressure, so energy must be used to bring it to the surface. As it is brought to the surface, the water table around the well drops, which creates a cone of depression.*

able if climate change leads to drying and will become less variable if climate change leads to an increase in water supply. This result is consistent with observations of the present climate.

The lack of a consensus about the impact of climate change on water supply does not mean the effect is unimportant. Rather, current uncertainty means that we can't know this effect yet. This uncertainty, coupled with the potential for a significant effect, may require decision makers to err on the side of caution.

ADDITIONAL READING

Thomson, A.M., R.A. Brown, N.J. Rosenberg, R. Srinivasan, and R.C. Izaurralde. "Climate Change Impacts for the Coterminous U.S.: An Integrated Assessment. Part 4: Water Resources." *Climatic Change* 69 (2005): 67–88.

STUDENT LEARNING OUTCOME

- Students will be able to explain why changes in precipitation do not translate directly into changes in human water supplies.

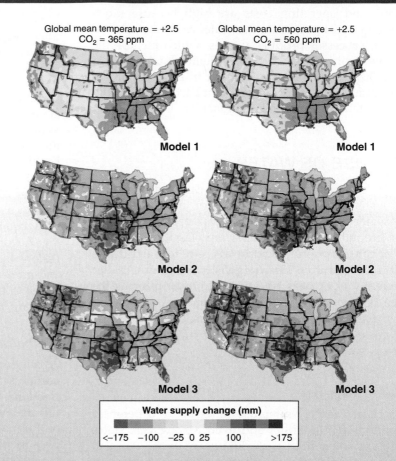

Global mean temperature = +2.5
CO_2 = 365 ppm

Global mean temperature = +2.5
CO_2 = 560 ppm

Model 1   Model 1

Model 2   Model 2

Model 3   Model 3

Water supply change (mm)

<-175 -100 -25 0 25 100 >175

FIGURE 1   *Climate Change and Water Supply* *The effect of climate change on water supply as projected by three general circulation models. The results vary among models, so it is not possible to determine whether climate change will increase or decrease water supply. (Source: Data from A.M. Thomson et al., "Climate Change Impacts for the Coterminous U.S.: An Integrated Assessment, Part 4: Water Resources," Climatic Change 69: 67–88.)*

Water Diversion

Water diversion is the movement of water from surface water or groundwater over some distance to its point of use. To reach users, water is transported by canal, aqueduct, pipeline, barge, or truck. Water diversion attempts to correct imbalances between where and when people want water and where, when, and how much water is available locally.

One of the largest water diversion systems uses canals, aqueducts, and pipelines to transport water from the Colorado River to cities and farmers in six western states. In southern California, Los Angeles and the agricultural fields of the Imperial Valley depend on the 400 kilometer Colorado River Aqueduct to supply more than 60 percent of their water (Figure 18.5). Similarly, Tucson and Phoenix in Arizona rely on the 560 km (348 mi), $3.6 billion Central Arizona Project to supply nearly one-third of their water.

Desalinization

Desalinization produces freshwater by removing the salt from salt water and brackish waters (brackish water has a salt content between freshwater and salt water). It is used in some arid and semiarid areas that have access to seawater—and increasingly in wetter regions such as Tampa, Florida, where freshwater resources have been depleted or where population and economic growth have outstripped available water supplies.

There are two basic desalination technologies. Much of the world's desalted water is produced via **thermal desalting,** in which seawater is boiled or evaporated and the steam or evaporate is drawn off as pure water. **Membrane separation** physically separates salt from water by pushing seawater through thin filters that do not allow the minerals to pass.

Desalination is much more expensive than tapping natural freshwater. A liter of water from a desalination plant can cost five to ten times more than a liter from traditional sources. Building a desalination facility requires a great investment, and operating costs are high because much energy is required to separate the water from the salt. All desalination processes produce wastes, which include the salts removed from the seawater and chemicals used to treat the water. Disposing of this wastewater in an environmentally appropriate manner is an important part of the feasibility and operation of a desalinization facility.

HUMAN USE OF WATER

Human systems for supplying water are called **waterworks.** Modern waterworks provide water for **offstream** uses, which is water that is withdrawn or diverted from surface water or groundwater. These uses are classified as withdrawals or consumption. **Withdrawals** refer to water that is removed from its source. **Discharge** is water returned after use, frequently at or near its source. **Consumption** is the difference between the quantity of water withdrawn and the quantity discharged. Consumption includes withdrawals that are evaporated, transpired, or incorporated into products or crops.

Offstream Water Uses

A glance at per capita water use finds many differences among nations (Figure 18.7) and even among U.S. states (Figure 18.8). Why should such differences exist? Don't all people use the same amount of water to drink, bathe, and prepare food? See *Your Ecological Footprint: How Much Water Do You Use?* These domestic uses of water constitute a small fraction of withdrawals. Developing nations use most of their water for agriculture or municipal uses. Developed nations use most of their water for purposes other than agriculture or municipal uses, which include industry and thermal power generation.

Municipal Water Use

This category of water use includes that by households, businesses, and government. How much water does a person need? Physical survival requires that you replace the water lost through routine metabolic processes. Individual needs vary, but people consume about three liters per day in temperate climates and up to six liters per day in hot, dry climates. Water is also used to dispose of human wastes. About 20 liters (5.3 gal) per person per day generates measurable public health benefits. Water also is needed for basic hygiene (washing, showering, and bathing) and food preparation. This adds another 25 liters (6.6 gal) per person per day. Meeting basic human needs—drinking, sanitation, bathing, and cooking—thus requires about 50 liters (13.2 gal) of water per person per day. Beyond these basic needs, municipal uses in developed nations include water to fill swimming pools, nurture lawns and gardens, wash cars, and so on. Public agencies also use water to meet community needs such as firefighting, street washing, municipal parks, and swimming pools.

Agriculture

Agriculture is the largest user of water. Globally, agriculture accounts for about two-thirds of human water use; this fraction is higher in developing nations. Such copious use is dictated by plant physiology: Remember that 300 to 1,000 kg of water are needed to produce each kilogram of dry matter. Because of this relatively low efficiency, most water used for agriculture is consumed.

Much of the water that is used in agriculture is applied via irrigation, which is the controlled application of water to arable land to supply requirements of crops not satisfied by precipitation. Before the Green Revolution the increase in irrigated acreage was driven by the need to increase arable land. Since then the increase has been driven primarily by the requirements of the high-yield varieties described

FIGURE 18.7   *Per Capita Water Withdrawals* *The amount of water withdrawn varies greatly among nations.* (Source: Redrawn from "Atlas of a Thirsty Planet," Nature 422, as provided by P. Gleick of the Pacific Institute for Studies in Development, Environment and Security.)

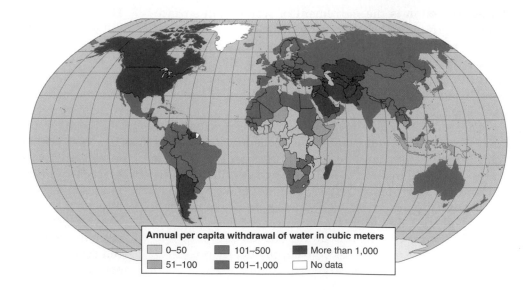

Annual per capita withdrawal of water in cubic meters

0–50	101–500	More than 1,000
51–100	501–1,000	No data

in Chapter 16. Globally, irrigated land has increased from 139 million hectares or ha (343 million acres) in 1961 to 279 million ha (689 million acres) in 2002. These 279 million ha account for 20 percent of the world's farmland but produce about 40 percent of the world's food supply.

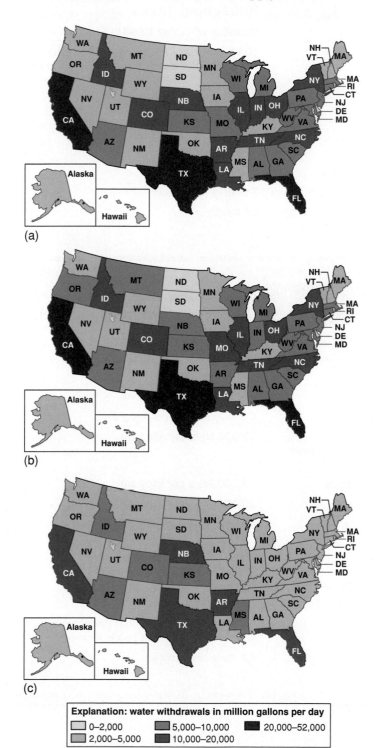

FIGURE 18.8   *U.S. Water Withdrawal* **(a)** *Water withdrawals by U.S. states in 2000.* **(b)** *Withdrawals from surface waters in 2000.* **(c)** *Withdrawals from groundwater in 2000.* *(Source: Data from U.S. Geological Survey.)*

Energy

Significant quantities of water are used to generate electricity. In the United States thermoelectric power is the largest single use of water; generating electricity used about 50 percent more water than irrigation in 2000. An electricity generating plant that is fired by coal, oil, or natural gas requires about 140 liters (37 gal) of water to produce 1 kilowatt-hour of electricity. That same kilowatt-hour generated by a nuclear power plant requires 205 liters (54 gal) of water. Some of that water is converted to steam that drives the generator producing the electricity. (See Figure 2.14 on page 35.) Most of the water, however, is used for condenser cooling. For a coal-fired plant, the weight of the water that is used for cooling is many times the weight of the coal used for fuel, so it makes economic sense to locate the plant near water rather than the coal mine. The lack of water makes it difficult to build large fossil fuel power plants in northern Africa.

Industry

Industrial uses claim extensive quantities of water to produce the goods and services you use every day—from clothes to plastics to automobiles. Water is used as a raw material, a coolant, a solvent, a transport agent, and as a source of energy. Water-intensive industries include paper and allied products, chemicals, primary metals, and petroleum refining. Producing a metric ton of steel requires up to 280 metric tons (308 tons) of water, while an automobile coming off the assembly line requires about 250,000 liters (66,000 gal). Unlike agriculture, water used for most industrial tasks normally is not consumed: Much of this water is returned to the hydrologic cycle, albeit in an altered state (see the section about industrial pollution).

Instream Uses

Some water use occurs without the water being diverted or withdrawn from surface water or groundwaters. Such uses are known as **instream** uses. Examples of instream uses include environmental services that rely on water, hydroelectric power generation, navigation, and recreation.

Environmental Services of Water

As water moves through the hydrologic cycle, it plays an important role in many environmental services. Chapters 4 and 13 describe how water and energy flows are important drivers of weather and climate. Water vapor in the atmosphere is the most important greenhouse gas as measured by its contribution to total radiative forcing. Plants rely on water's ability to dissolve many nutrients. Wetlands recycle nutrients, protect cities against storm surges, and provide habitat for biodiversity.

Water Transport

Barges, freighters, and other modes of water transport are a vital cog in the world's transportation system. Water transport often is the least expensive way to move bulky raw materials

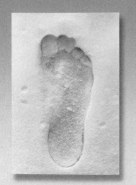

Your ECOLOGICAL footprint

How Much Water Do You Use?

The question in the title seems easy to answer. Simply look at your water bill—it will list how much water your household consumed over the last month (or quarter), the price of water, and how much you owe. Each of these numbers contains important but incomplete information about your water footprint.

Let's start with the price of water. How much does your local waterworks charge? How does this price compare to the $4.26 per thousand gallons (page 393 of this chapter) that is needed to fund the expansion of existing waterworks? By how much would your bill increase (or decrease) if the price were $4.26 per thousand gallons? If your local waterworks charged $4.26 per thousand gallons, would you change the amount of water you consume?

Calculating Your Footprint

Next look at the quantity of water you consume. The bill records the quantity of water used, but it doesn't reveal how you used it. The water used to flush the toilet, wash your hands, take a shower, and other daily activities is given in Table 1. To determine how much water you use, multiply these uses by the frequency of these activities. To illustrate, suppose each day you take a ten-minute shower (19 liters/minute × 10 minutes = 190 liters), flush the toilet four times (4 flushes × 15 liters/flush = 60 liters), run the faucet for five minutes (5 minutes × 4 liters/minute = 20 liters), wash dishes by hand for ten minutes (10 minutes × 11 liters/minute = 110 liters), and do an average of one-seventh of a load of laundry (1/7 × 208 liters/load = 30 liters); you also have one dripping faucet (86 liters/day). Together these activities use about 500 liters per day.

Now calculate the water used by the agricultural and industrial sectors to produce the goods and services you buy. Nearly all the goods and services you purchase require water to be produced (Table 2). These goods and services include those examined in Your Ecological Footprint pages in other chapters, such as the food energy you consume (Chapter 5, pages 80–81), paper (Chapter 17, page 364), and fossil fuels and electricity (Chapter 20, pages 423–425).

Use the values in Table 2 to calculate the quantity of water required to grow your food and produce your paper using the items listed in Chapters 5 and 17. To illustrate, suppose you ate one egg (454 liters) and drank a quarter liter of milk (62 liters) for breakfast. For lunch you ate a hamburger (4,900 liters) and french fries (23 liters). For dinner you ate a quarter kilogram of meat (0.25 × 41,500 = 10,125 liters) and a quarter kilogram of rice (0.25 × 5,000). If you did (this would not be a balanced diet), you would consume 16,814 liters. In addition, suppose you consumed 0.75 kg of paper per day, which would require 17,550 liters (0.75 kg × 23,400 liters/kg = 17,550 liters).

Because we cannot include information about the water used to produce every type of good and service, Table 3 lists the amount of water used to produce a dollar's worth of output by various sectors of the U.S. economy. These quantities, known as water intensities, are calculated using input–output tables as described in Chapter 3's Your Ecological Footprint (page 49). You can use these intensities, along with information about how you spend your money, to calculate the quantity of water you use indirectly. For example, if the price of your cell phone was about $100, about 79 liters of water were used to produce it ($100 × 0.79 liter/$ = 79 liters).

Now calculate your total use of water. For simplicity, include only your household uses and the water used to grow your food and produce your

TABLE 1	Water Intensities of Common Household Tasks
Task	**Quantity of Water Used**
Shower	19 liters per minute (standard showerhead) 7.5 liters per minute (low-flow showerhead)
Toilet	15 liters per flush (standard) 6 liters per flush (low-volume toilet)
Faucet	4 liters per minute
Washing dishes	11 liters per minute (by hand) 57 liters per load
Laundry	208 liters per load
Watering the lawn	7.6 liters per minute
Dripping faucet (60 drips per minute)	86 liters per day

Source: Data on coefficients from www.tampagov.net/dept_water/conservation_education.

TABLE 2	Water Intensities of Common Goods and Services
Good or Service	**Quantity of Water Used**
Chicken	1,512 liters per serving
Rice	5,000 liters per kilogram
Meat	41,500 liters per kilogram
Aluminum	1.3 million liters per kilogram
Heavy manufactured goods	47,600 liters per kilogram
Textiles	13,100 liters per kilogram
Paper	23,400 liters per kilogram
Egg	454 liters per egg
Bread	567 liters per loaf
Milk	246 liters per liter
One car	245,700 liters
Hamburger	4,900 liters per serving
Electricity	140 liters per kilowatt-hour (fossil fuels) 205 liters per kilowatt-hour (nuclear power)
French fries	23 liters per serving

paper. Continuing with the previous example, daily water use would be 34,864 liters (500 liters + 16,814 liters + 17,550 liters = 34,864 liters).

Interpreting Your Footprint

Compare the quantity of water you use with the amount of water used by the average person in the United States in 2000. According to the U.S. Geological Survey, the United States withdrew 408 billion gallons per day for 285.3 million people, which works out to about 5,413 liters per person

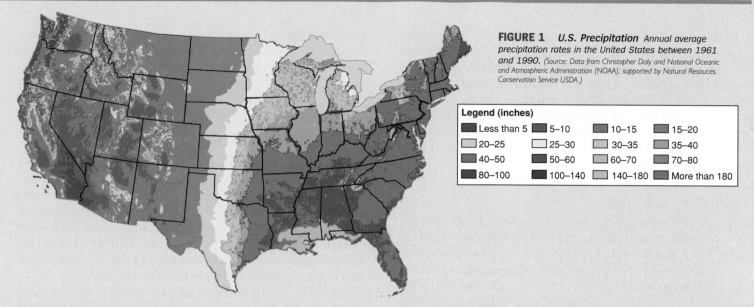

FIGURE 1 *U.S. Precipitation Annual average precipitation rates in the United States between 1961 and 1990.* (Source: Data from Christopher Daly and National Oceanic and Atmospheric Administration (NOAA), supported by Natural Resouces Conservation Service USDA.)

Legend (inches)

Less than 5	5–10	10–15	15–20
20–25	25–30	30–35	35–40
40–50	50–60	60–70	70–80
80–100	100–140	140–180	More than 180

per day (3.78 liters/gal × 408 × 10⁹ gal/day/285 × 10⁶ people = 5,413 liters/person/day). This average is much smaller than the value in our example and is probably smaller than the value you calculated.

Have we made a mistake? No! Can you explain the difference? The answer has to do with the way that U.S. water use is calculated. The U.S. Geological Survey calculates water use based on withdrawals. But withdrawals refer to water that is removed from its source and does not include much of the water used to grow food and trees. According to the Food and Agriculture Organization, only about 5 percent of U.S. agricultural area is irrigated. Water used to irrigate this land is considered a withdrawal, but the U.S. Geological Survey's estimate of water use does not include the precipitation that falls on the remaining 95 percent of agricultural area. Nor is irrigation used to grow trees—so relatively little of the water used to produce paper is included in withdrawals.

So let's calculate the land area that "collects" this water. Continuing with our previous example, your annual use would be about 12.7 million liters per year (34,864 liters/day × 365 days/year = 12,725,360 liters/year). Now convert your annual water use to acre-feet (1 acre-foot = 1,233,481.84

liters). For our example, 12.7 million liters represents 10.3 acre-feet (12,725,360 liters/1,233,482 liters/acre-foot = 10.3 acre-feet). Locate your home or college in Figure 1 and match the color to the average rainfall in the legend. Convert total inches to feet—for example, if you attend Boston University, the 45 inches in the Boston area represents 45/12 = 3.75 feet. Next divide the acre-feet of water used by local precipitation. This number represents the acres of land area needed to collect the water you consume. Continuing with the previous example, you would need 2.75 acres in the Boston area to capture 12.7 million liters of water.

How could you reduce the area of your water footprint? One way to reduce your use of water would be to reduce your consumption of beef. How would your water footprint change if you replaced beef with chicken or stopped eating meat altogether?

STUDENT LEARNING OUTCOME

- Students will be able to explain why their water footprint is larger than per capita water withdrawals.

TABLE 3 Water Intensity of Selected Goods

Goods	Water Intensity (liters per dollar)	Goods	Water Intensity (liters per dollar)
Meatpacking plants	2.09	Glass and glass products	4.57
Natural, processed, and imitation cheese	2.62	Steel	149.54
Ice cream and frozen desserts	1.22	Aluminum	33.86
Canned fruits, vegetables, preserves, jams	8.10	Construction machinery	3.27
Cereal breakfast foods	2.88	Motors and generators	2.73
Bread, cake, and related products	0.28	Household cooking equipment	2.56
Malt beverages	10.78	Lighting fixtures and equipment	0.29
Pasta	0.30	Household audio and video equipment	1.10
Fabrics	2.88	Prerecorded records and tapes	1.19
Paper products	100.32	Telephone equipment	0.79
Chemicals	101.11	Batteries	2.18
Drugs	6.51	Bicycles and motorcycles	3.59
Petroleum products	14.86	Motor vehicles	0.60

Source: Data on water intensities from Professor H. Scott Matthews, Carnegie Mellon University.

such as grain, pulp, crude oil, lumber, and minerals. In the United States more than a billion tons of freight are moved by water annually, which accounts for about 20 percent of total freight. Grains and oil are moved by barge down the Mississippi River to New Orleans and Baton Rouge, Louisiana, where they are shipped all over the world. Minerals, grains, and steel are shipped through the Great Lakes and out the St. Lawrence River to Europe and beyond. And West Coast ports play a significant role in U.S. international trade.

Hydroelectric Energy

Hydroelectric energy is produced by the force of falling water. As water builds up behind a dam, it accumulates potential energy. This potential energy is transformed into mechanical energy when water rushes through a tunnel called a penstock and strikes the rotary blades of turbines (Figure 18.9). The turbines' rotation spins electromagnets, which generate current. About 20 percent of the world's electricity is generated by hydropower. In the United States falling water generates about 12 percent of the nation's electricity.

Recreation

Rivers, lakes, and reservoirs provide recreation opportunities. Swimming, boating, and fishing are among the most popular leisure activities. As such, they generate big economic benefits as measured by willingness to spend (see Chapter 10). In 2001 nearly 50 million anglers in the United States spent $24 billion on tackle, equipment, food, lodging, and other goods and services related to fishing. These expenditures provided jobs for 1.3 million people and generated $2.1 billion in federal income tax revenue. More than 80 million people living in

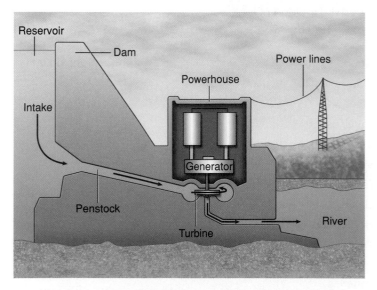

FIGURE 18.9 *Hydroelectric Dam Gravity causes water in the reservoir to flow through the intake and down the penstock. As it moves through the penstock, it causes the turbine to spin, which generates electricity. The water then flows into the river below the dam.*

the United States participate in recreational boating. Together they own more than 16 million boats and spend in excess of $19 billion each year on boats and boating accessories.

THREATS TO SUSTAINABLE SUPPLIES OF CLEAN WATER

As population and affluence grow, so does water use. In some places people use nearly 100 percent of surface waters and pump groundwater faster than it is replenished. Many uses of water involve some discharge. These discharges often are degraded relative to the water that was withdrawn. Such **water pollution** (the purposeful or accidental addition of materials that contaminate water) threatens the health of humans and their environment and reduces the usefulness of water.

A nation's vulnerability to water scarcity can be measured by **absolute water scarcity,** which is the ratio of annual water availability to population. As a general rule, water-stressed countries have less than 1,700 m³ (59,989 ft³) per person per year. In 2000 thirty-six nations fell below this threshold (Figure 18.10). Below this threshold water becomes a severe constraint on food production, economic development, and the protection of natural systems.

Diverting Surface Waters

Surface waters are the easiest place to obtain water. This ease and growing demand have increased withdrawals. Some rivers are now dry for part of the year because of excessive withdrawals. But in most cases the impacts are not as obvious because much of the water withdrawn is discharged back into the same body. This discharge rarely is as pure as the water that is diverted, which reduces its suitability for users downstream.

The effect of such withdrawals can be seen in the Colorado River, which has been extensively dammed to increase usable water supplies. Formal agreements among states regarding withdrawals from the Colorado River started during Herbert Hoover's presidency (1929–1933). States along the river agreed that the upper and lower basins would each get 7.5 million acre-feet (9,251 billion liters) of water. Later the basins allocated water among the states. For example, states in the lower basin allocated their 7.5 million acre-feet among Arizona (2.8 million acre-feet—3,453 billion liters), California (4.4 million acre-feet—5,427 billion liters) and Nevada (0.3 million acre-feet—370 billion liters). These totals were then divided among users in each state.

At first adhering to these agreements was relatively easy because no one envisioned that these allotments would ever be used. But that changed as population and affluence increased. In the 1930s fewer than 2 million people lived in Los Angeles, and its allocation of 550,000 acre-feet (678 billion liters) seemed sufficient. But as the city grew, demand collided with its allocation, and the city became notorious for trying to secure additional supplies.

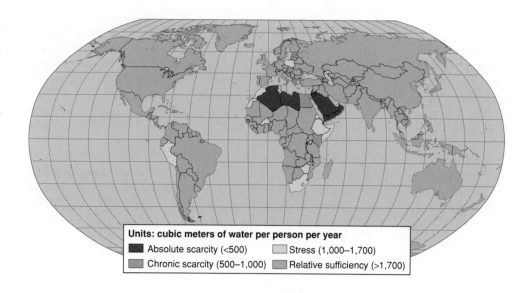

FIGURE 18.10 *Per Capita Water Supply* Less than 1,700 m³ of water is available per person in thirty-six nations. These nations are considered to suffer from water stress. This implies that water is a severe constraint on food production, economic development, and the protection of natural systems. *(Source: Redrawn from "Atlas of a Thirsty Planet," Nature 422, as provided by P. Gleick of the Pacific Institute for Studies in Development, Environment and Security.)*

Units: cubic meters of water per person per year

Absolute scarcity (<500)
Chronic scarcity (500–1,000)
Stress (1,000–1,700)
Relative sufficiency (>1,700)

As California grew, so did its use of water. Eventually California's use exceeded its Colorado River water allotment. These excesses were relatively unimportant until the mid-1990s, when Arizona completed its Central Arizona Project and the city of Las Vegas grew explosively. Now both Arizona and Nevada wanted their full share of Colorado River water. This meant that California had to reduce its use. To help settle the dispute, Secretary of the Interior Bruce Babbitt became involved. Eventually California returned to its allotment by shifting water among users and reducing agricultural water use.

Now users in the lower basin withdraw all of the water allocated in the 1944 Colorado River Compact. A significant fraction of this water is consumed, so the amount of Colorado River water that flows into Mexico barely meets the 1.5 million acre-feet (1,850 billion liters) that was promised in a 1944 treaty between the United States and Mexico. In Mexico even more water is withdrawn and consumed; in some years the Colorado River dries up before it reaches the Gulf of California.

This undermines an important habitat for migrating birds in North America. Such an impact is demonstrated by the correlation between water withdrawals from the Truckee River in Nevada and the local population of white pelicans—increasing withdrawals reduces the pelican population.

Such effects are not unique to the United States. Perhaps the most notorious example of surface water diversion is the Aral Sea in the former Soviet Union (Figure 18.11). Before 1960 the Aral Sea was the world's fourth largest lake. Since then farmers irrigated increasing areas with water diverted from the rivers that flowed into the sea. By the 1980s inflows were only 13 percent of their pre-irrigation levels. As a result, the volume of freshwater in the Aral Sea has declined by 75 percent. All its native fish species went extinct, as did the 60,000 jobs associated with fishing. However, recently better management and the construction of dams have allowed water levels to rise, and a fish hatchery has allowed fishers to catch 1,000 tons of fish, compared to 20,000 tons during the sea's best years.

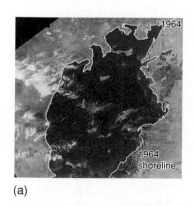
(a)

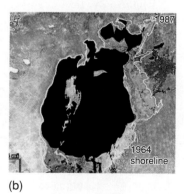
(b)

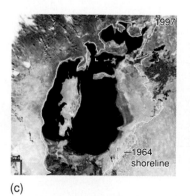
(c)

(d)

FIGURE 18.11   *A Changing Aral Sea* The surface area of the Aral Sea shrank as the amount of land irrigated with its waters increased.

Mining Groundwater

In places where demand grows beyond the level that can be supplied by surface waters, people increase supply by pumping groundwater. Groundwater has the potential to be used sustainably. As shown in Figure 18.2, groundwater is recharged by water that infiltrates from the surface. If the recharge rate exceeds the rate at which people pump groundwater, the aquifer is a sustainable source of water. Should the rate of pumping exceed the rate of recharge (termed an **overdraft**), groundwater is being "mined" like any other nonrenewable resource, such as copper or oil. Figure 18.12 displays areas where the Ogallala Aquifer has been subjected to overdraft.

Well before the last drop is pumped, the best first principle dictates that overdrafts have adverse economic and environmental effects. The most immediate effect is a drop in the water table. In some places in the United States the water table has dropped dozens of meters.

As the water table drops, so may water quality. In general, groundwater is highly desirable because it is relatively free of contaminants—pollutants are filtered as water percolates through the soil. But water from the lower layers of the aquifer has higher levels of potentially dangerous elements such as arsenic or radon because higher temperatures increase the quantity that can be dissolved. (Why is the temperature warmer in lower Earth layers? See Chapter 4.)

FIGURE 18.12   *Overdrafts in the Ogallala Aquifer* *The amount of water pumped from the Ogallala Aquifer has been greater than the recharge rate. Between 1980 and 1994 this has caused the water table to drop, in some areas by more than 40 feet (13 m).*

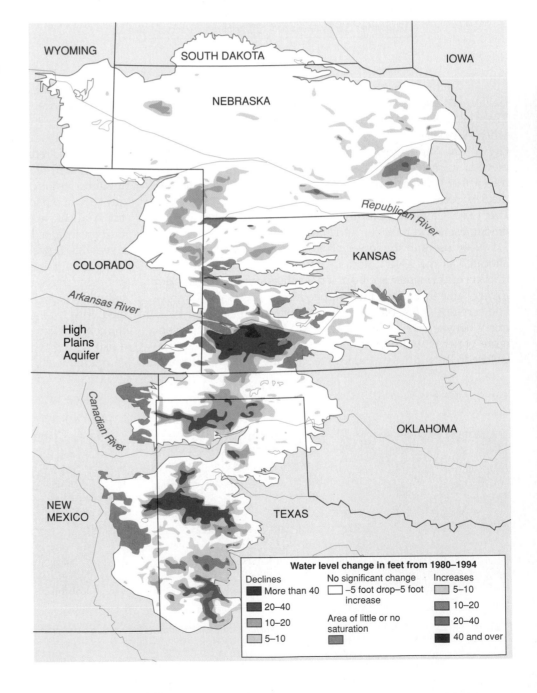

Water level change in feet from 1980–1994

Declines
More than 40
20–40
10–20
5–10

No significant change
−5 foot drop–5 foot increase

Area of little or no saturation

Increases
5–10
10–20
20–40
40 and over

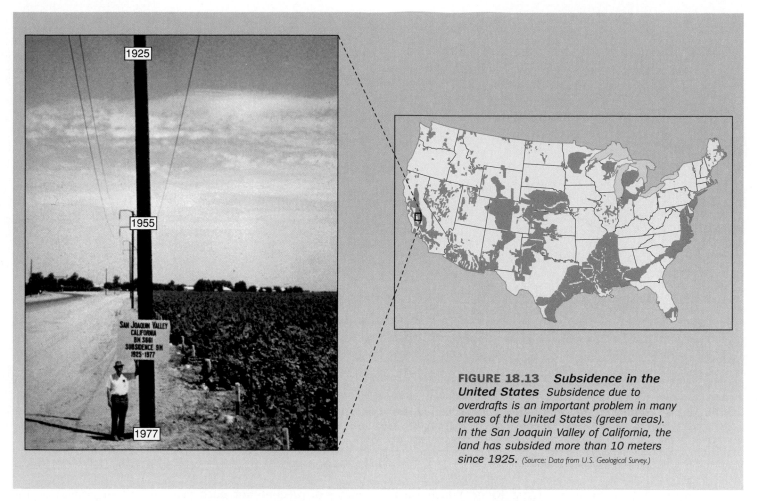

FIGURE 18.13 *Subsidence in the United States* Subsidence due to overdrafts is an important problem in many areas of the United States (green areas). In the San Joaquin Valley of California, the land has subsided more than 10 meters since 1925. *(Source: Data from U.S. Geological Survey.)*

A dropping water table also affects the economic viability of agriculture. As the water table drops, the amounts of energy and money used to bring water to the surface increase, which cuts the profitability of agriculture. Turbine pumps have allowed farmers to increase their income by growing corn, which needs more water than wheat. But the increasing cost of pumping groundwater from depleted aquifers reduces the profits of growing corn, which has prompted some farmers to go back to wheat. In other places the increasing cost of irrigation water has caused farmers to stop planting crops.

As was the case for Ubar, groundwater helps support the surface. As the water table drops, pore spaces among soil particles dry out. With nothing but air among them, the weight of the overlying material compresses the soil particles. As a result, the surface drops, which is known as **subsidence.** In some locations the surface has dropped nearly 10 meters (32.8 ft) such as the San Joaquin Valley of California, where farmers have pumped significant amounts of water from the Central Valley Aquifer, which can be up to 1,000 meters (3,281 feet) thick (Figure 18.13).

Subsidence causes many problems. The dropping land surface can change the flow of surface water. In coastal areas subsidence may allow tides to move salt water into areas that were above the high-tide mark. It also can damage public infrastructure, such as roads, bridges, and sewers, and

public and private buildings. Together these effects impose economic damages in every state. The U.S. Geological Survey reports that subsidence imposes more than $250 million of damages annually in Houston and Galveston, Texas; New Orleans, Louisiana; Santa Clara County, California; and the San Joaquin Valley, California.

Near the coast, overdrafts may allow salt water to flow into aquifer pore spaces that were previously occupied by freshwater, which is known as **saltwater intrusion** (Figure 18.14). Saltwater intrusion threatens water supplies in coastal areas such as Long Island. Much of this island is densely populated, has relatively little surface water, and is bounded by salt water on three sides. As the number of people living on Long Island grew, so did their use of groundwater. This allowed significant quantities of salt water to enter the aquifer in parts of southwestern Nassau County, the Great Neck peninsula, and the Port Washington peninsula.

Overdrafts can also affect surface water bodies, such as streams, rivers, and lakes. For example, rivers are supplied by water that flows across the surface toward the stream. Rivers will gain additional water if the water table is above the surface of the stream (Figure 18.15). But if overdrafts drop the water table below the stream surface, surface water will infiltrate into the aquifer, thereby lowering the river's water level. In extreme cases a river can dry up.

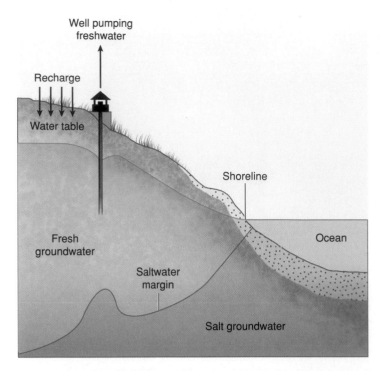

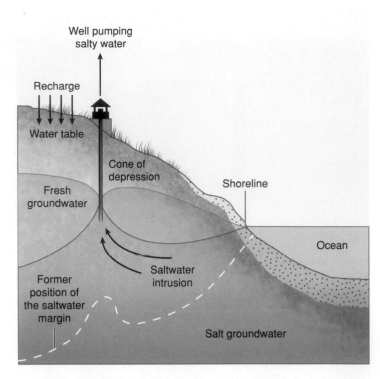

FIGURE 18.14   *Saltwater Intrusion*   **(a)** *As water is pumped from groundwater faster than it is recharged, the water table around the well drops, and the dividing line between freshwater and salt water in the aquifer rises toward the well. (b) The dividing line may eventually reach the well, at which point the water reaching the surface contains too much salt to drink or use for household, industrial, or agricultural purposes. This has occurred in several areas along the East Coast, such as Great Neck, Long Island, New York state, USA.*

One such case is the Santa Cruz River, which runs along the west side of Tucson, Arizona (Figure 18.15). Although Tucson receives only about 28 centimeters (11 in) of precipitation per year, the Santa Cruz River used to flow year-round because groundwater discharged into its bed. Now the river flows only occasionally: during spring snowmelt and summer rains, and when the city's water treatment plant releases effluent. Most of the time it is an empty wash because the growing population and economic base pumped increasing amounts of groundwater. In the last fifty years the water table has dropped about 60 meters (197 ft); so when water is present, it flows from the channel of the Santa Cruz River into the ground.

The loss of surface water eliminates important habitat (Figure 18.15). The Santa Cruz River was home to a variety of fish and invertebrates. The banks provided habitat for small mammals and waterfowl. The banks were lined by cottonwood and willow trees; further inland the water supported mesquite bosquets, which are near-river forests. Together the various habitats supported populations of large mammals, including mountain lions. Much of this wildlife has disappeared with the water.

But the Santa Cruz is not the only river that sometimes dries up due to groundwater withdrawals. This problem also occurs in the relatively moist Northeast. Just north of Boston, the Ipswich River runs about 56 km (35 mi) to the Atlantic Ocean. Although this area receives more than a

meter of precipitation each year, suburban growth on the "north shore" increased groundwater withdrawals so much that the river went dry in 1995, 1997, and 1999.

Domestic and Municipal Sewage

Domestic and municipal water uses generate **sewage,** which is waste and wastewater produced by residential and commercial users that is discharged into sewers. Sewage usually includes human wastes (feces and urine) and waste materials that result from household activities, such as soaps, paints, and oils. Storms add to these wastes by washing oils and other materials from roadways into drainage systems. Together these sources provide organic material, microorganisms, and a variety of materials. Here we discuss organic materials and microorganisms; material wastes are discussed in the section about industrial water pollutants.

Microorganisms pose a significant threat to human health. Such organisms include viruses, bacteria, protozoa, and parasites, which can spread waterborne diseases via fecal–oral transmission. Through this mechanism people become sick after drinking water contaminated with animal or human **pathogens,** which are microorganisms that cause disease. Pathogens cause many of the diseases that have the greatest impact on human health, such as diarrhea, schistosomiasis, intestinal worms, and river blindness.

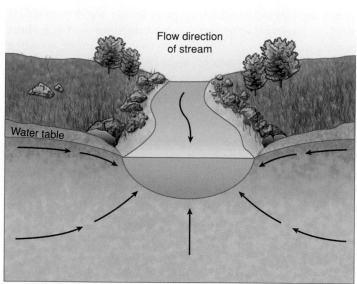

(a) Gaining stream

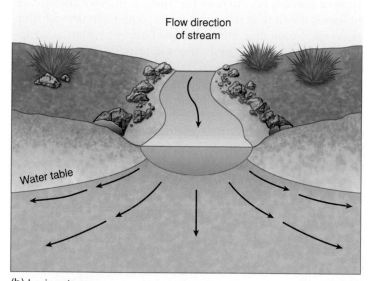

(b) Losing stream

FIGURE 18.15 *Rivers Run Dry* *The direction of discharge depends on the height of the water table relative to the surface of the stream.* **(a)** *If the water table is above the surface of a stream, groundwater will flow into the stream's channel. This process supported the Santa Cruz River, which flowed past the desert city of Tucson.* **(b)** *If overdrafts lower the water table below the surface of a stream, surface water will flow into the surrounding aquifer, causing the water level to drop. Groundwater pumping around the city of Tucson causes water to flow into the ground from the Santa Cruz River, which has dried up the river. Notice that the same rock appears in the foreground of both photos.*

Many of these diseases can be avoided by treating sewage, which eliminates up to 99.99 percent of the microorganisms. This protection is amplified by treating drinking water with chlorine, which kill microorganisms, and filtration, which removes particulate matter. The effectiveness of such treatments is evaluated by the **fecal coliform count**, which measures the number of coliform bacteria per 100

milliliters (0.002 gallons). Coliform bacteria live in the guts of most mammals. Although they usually do not cause disease, they are used as a proxy for the presence of other microorganisms that can cause disease. In general, drinking water should have a count of about zero. For swimming and other "full-body contact" activities, the fecal coliform count should be 200 or less.

Unfortunately such water treatment is not available in many developing nations (Figure 18.16). The World Health Organization estimates that 250 million cases of waterborne diseases occur each year, killing 5–10 million people. Children are particularly susceptible. Diarrheal diseases leave millions of children underweight, mentally and physically handicapped, and vulnerable to other diseases. In addition to the human tragedy, these health problems undermine social and economic development. Given these effects, now you should understand how the availability of clean drinking water can start the demographic transition (see Chapter 9).

Organic wastes are an important component of sewage. Discharging too much organic material into surface waters can depress oxygen levels via eutrophication. But rather than supplying nutrients for algae, which eventually die and are metabolized by decomposers, the organic material is decomposed directly. As microorganisms metabolize organic wastes, they use oxygen faster than it can reenter the water from the atmosphere. If the water becomes completely anoxic, decomposition proceeds anaerobically (see Equation 5.2 on page 88), which releases noxious gases.

The severity of these problems is determined by the amount of organic material in the wastewater and the water body that receives the wastes. The amount of organic material in wastewater can be measured in several ways. **Chemical oxygen demand** measures the amount of oxygen required to oxidize the organic material in a sample. Definitions of **biochemical oxygen demand** vary, but this measures the amount of oxygen required for aerobic organisms to decompose organic material in wastewater over a five- to twenty-day period; the usual measure is five days.

Differences between the definitions for chemical and biological oxygen demand help determine the effect of organic materials on surface waters. Some organic materials cannot be oxidized biologically, so their presence does not depress oxygen levels in the water. The rate of decomposition that is measured by biological oxygen demand determines the degree to which oxygen levels are depressed. If much of the organic material can be decomposed in a short period, this waste will depress the oxygen content. If only a small portion can be decomposed in the five- to twenty-day period, oxygen levels may be relatively unaffected.

The reduction in oxygen levels is measured by the **dissolved oxygen deficit,** which is the difference between the amount of oxygen in water when it is fully saturated (remember, warm water holds less oxygen than cold water) and the amount of oxygen actually present. As organic material is added to surface water, the dissolved oxygen deficit increases, which generates the oxygen sag curve (Figure 18.17). The location of this dip identifies where eutrophication has its greatest effect.

The rate at which the dissolved oxygen deficit is eliminated depends on the quantity of organic material added, the rate at which oxygen reenters the water, and the degree to which the organic waste can be diluted. Oxygen reenters water based on the size of the dissolved oxygen deficit: Large deficits speed the flow of oxygen from the atmosphere to the water. This rate of reaeration also is affected by flow conditions, such as turbulence: Surface turbulence increases the flow of oxygen into the water. Finally, diluting the waste by mixing it into a large body of water or a rapidly flowing stream speeds the rate at which the dissolved oxygen deficit is eliminated.

The size of the initial dip in the oxygen sag curve and the rate at which the deficit is eliminated determine the ability of a waterway to process organic wastes. A big initial dip or a slow rate of return may undermine a waterway's ability to support fish and other aerobic organisms. Because the shape of the oxygen sag curve is different for every waterway, the amounts of organic waste that can be processed vary among waterways. These unique limits are recognized

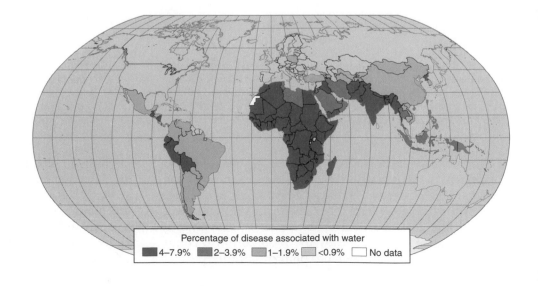

FIGURE 18.16   *Waterborne Disease*   *The percentage of total disease that is associated with waterborne disease in 2000. Notice that the incidence of disease is greatest in developing nations, both in areas of plentiful rainfall and in relatively dry areas.*   (Source: Redrawn from "Atlas of a Thirsty Planet," Nature 422, as provided by P. Gleick of the Pacific Institute for Studies in Development, Environment and Security.)

Percentage of disease associated with water
■ 4–7.9%   ■ 2–3.9%   ■ 1–1.9%   □ <0.9%   □ No data

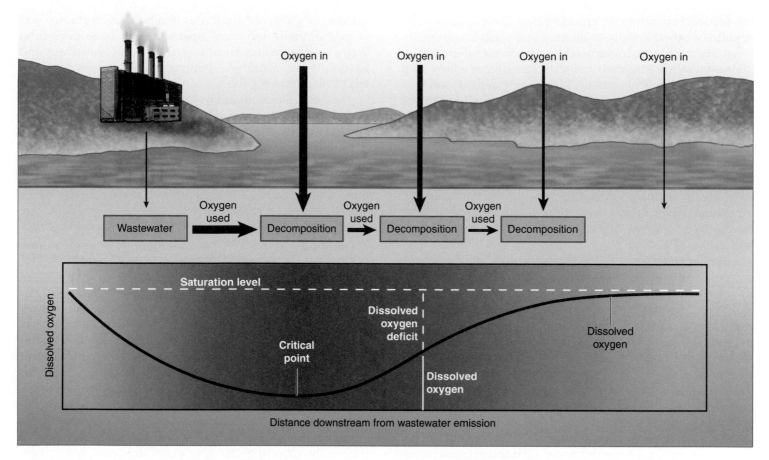

FIGURE 18.17 *Oxygen Sag Curve As organic waste is put into a river and carried downstream, oxygen is used to decompose it. Decomposition uses oxygen faster than oxygen can flow into the river, and the dissolved oxygen content of the river drops. The drop in oxygen increases the rate at which oxygen flows into the river. Nonetheless, decomposition continues to use oxygen faster than it flows into the river until the critical point, which represents the low point on the oxygen sag curve. Downstream from the critical point, oxygen flows into the river faster than it is used for decomposition, and the oxygen content of the river rises. Eventually the oxygen returns to saturation level, which is determined in part by water temperature.*

by the U.S. Clean Water Act, which is the most important law governing the pollution of surface waters, as described later in this chapter.

Industrial Water Pollutants

Many firms use great quantities of water directly in production processes or to clean and remove residues. The resulting wastewater contains a variety of pollutants, such as sediments, heavy metals, and **xenobiotics,** which are organic compounds that are synthesized by humans and therefore are relatively resistant to organic decay.

Of these, heavy metals may pose the greatest threat to human health. Heavy metals are linked to several illnesses, including cancer, damage to the nervous system, and birth defects. Metals of greatest concern include lead, mercury, cadmium, and arsenic. Although industrial wastewater usually emits these metals at very low levels, biological magnification (see Chapter 5) increases their concentra-

tion in the food chain to the point at which they may harm people. For example, Minamata disease is caused by people eating seafood contaminated with mercury from industrial discharges. The cumulative effect of long-term exposure at what seem like low concentrations is still uncertain.

Agricultural Water Pollutants

There are three general types of agricultural water pollutants: nutrients, organic wastes, and agricultural chemicals such as pesticides and herbicides. These wastes are classified as **nonpoint pollutants** because they are not discharged or emitted from a specific point, such as a pipe or smokestack. The first two categories of agricultural water pollutants are discussed in previous chapters. Eutrophication, which is associated with the use of fertilizers, is discussed in Chapters 6 and 15. Organic wastes, which are generated by large feedlots (Chapter 16), are similar to those associated with municipal wastes.

Agricultural chemicals in wastewater can leach through the soil and contaminate groundwater. Their impact there depends on their toxicity to humans and other living organisms. The ability to contaminate groundwater depends on their persistence and their mobility. Persistence—how long agricultural chemicals remain in the environment—often is measured by half-life, which is the time required for half of a pollutant's original mass to be transformed. Unlike the half-lives of elements (Chapter 2), the half-lives of agricultural chemicals are determined by how quickly they can be broken down by physical processes, such as photodissociation, or organic processes, such as decay. Half-lives vary greatly among agricultural chemicals. For example, malathion, which is an insecticide, has a half-life of one day. At the other extreme chlordane, also an insecticide, has a half-life of 3,500 days.

The mobility of an agricultural chemical depends on the environment where it is applied and its chemical composition. Groundwater contamination is usually more severe in areas with high rates of precipitation and irrigation because water moves chemicals through the soil. The composition of the soil is pertinent too. For example, soil organic carbon can soak up or attract agricultural chemicals, a process known as **sorption.** Many agricultural chemicals are organic, so the soil's tendency to hold these chemicals in place can be measured by its organic carbon sorption coefficient. A high coefficient indicates that the chemicals are held strongly and are less likely to reach the groundwater. On the other hand, they are more likely to contaminate surface water as erosion moves the soil and the agricultural chemicals.

WATER AND CONFLICT

Unsustainable and polluting uses of water create the potential for conflict. This potential is exacerbated by a mismatch between political boundaries and watersheds. Watersheds define the quantity of water available and who will be affected by polluting activities. But when watersheds are divided by political boundaries, the borders create two sets of people: those who live upstream and have unfettered access to clean water and those who live downstream and have only the water and pollutants left to them by those who live upstream.

A quick glance at a map indicates that political boundaries rarely follow watersheds (Figure 18.18). Continuing our story of the Colorado River, its watershed includes both the United States and Mexico. This is not an isolated example. About 40 percent of the world's population lives in nations that share watersheds with other nations. Together these shared watersheds account for about 60 percent of the world's supply of surface water.

Sharing a vital resource like water can lead to conflict. Water lies at the heart of many conflicts in the Middle East, the Indian subcontinent, and southeast Asia. The struggle for water is an integral part of these conflicts, but rarely is water the cause of the conflict. Disagreements about water did not start the conflict between Israel and its neighbors or between India and Pakistan. Because of water's importance to warring parties, cooperation over water supply often continues during overt conflicts. Israel and Jordan held secret negotiations over water rights while they were legally at war. The Indus River Commission continues to function despite two wars and ongoing hostility between India and Pakistan.

Fortunately a detailed study of international incidents demonstrates that conflicts over water often are solved peacefully. Over the last fifty years delegates have negotiated more than 157 international water treaties. These include treaties between Mexico and the United States regarding water in the Colorado River and the Rio Grande and treaties between Iran and Iraq regarding the Tigris River. Ironically, the most violent conflicts over water often occur among groups within nations. For example, California farmers have destroyed pipelines that were to bring water to Los Angeles, and disputes between farmers in the Indian states of Karnataka and Tamil Nadu led to riots and death.

FIGURE 18.18 *Potential for Conflict* *Many of the world's major water sheds contain more than one nation. Under these conditions, upstream users control the quantity and quality of the water that is available to downstream users.*

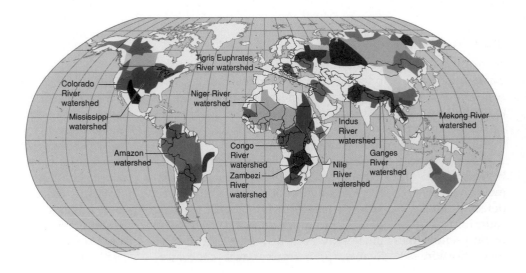

ENSURING ACCESS TO A SUSTAINABLE SUPPLY OF CLEAN WATER

Given this potential for violence, creative solutions to the depletion and pollution of water are critical. Three strategies have emerged. One seeks to create an economic market that allocates water among users based on price. Another seeks to increase the efficiency of water use technologies. And a third seeks to ensure the cleanliness of water supplies by treating wastewater.

A Market for Water?

Many of the problems associated with the depletion of surface water and groundwater can be traced back to how ownership of water is determined and prices are assigned. U.S. laws that govern water rights vary among states. On the East Coast British colonists brought the English system of water rights. Known as **riparian water rights,** these laws allow a landowner to use a share of the water that flows naturally past his or her property. Riparian rights do not entitle a landowner to divert water for storage in a reservoir for use in the dry season or to use water on land outside the watershed. This prohibition was ideal for the New England economy, where factories were powered by flowing water.

In the U.S. West water is allocated based on the **prior appropriation doctrine,** which dictates that no one owns the water in a stream and that all people, corporations, and municipalities have the right to use water for beneficial purposes. Allocation among users is based on the principle of "first-in-time, first-in-right." According to this system, the first person to use water (called a "senior appropriator") acquires the right (called a "priority") to its future use; later users (called "junior appropriators") also have rights to the water, but their water use rights are the first to be curtailed during shortages. The prior appropriation doctrine arose in California, where miners diverted water considerable distances in their search for gold. These diversions required large investments that might not have been made if other users could later divert water upstream. First-in-time, first-in-right guaranteed the first miners access to all the water needed, regardless of who came along later.

Rules that regulate the use of groundwater also vary among states. In most western states the prior appropriation doctrine also applies to groundwater. The **reasonable use doctrine,** which allows landowners to pump water for any beneficial use and does not recognize priority among users, is the law in some western states and many eastern states. The **rule of absolute ownership** allows landowners to pump as much water as they want. In California and Vermont the **correlative rights doctrine** forces landowners to share water. Many of these rules can be traced back to an explanation of groundwater flows that was written by a lawyer, Clesson S. Kinney. In 1894 Kinney wrote that underground water was separate from surface water flows

and was nearly infinite in supply. Based on this false notion, most states allow landowners to pump water as they see fit with little regard for others who pump groundwater or those who hold rights to surface water. Clearly these laws can lead to conflict among users, as we have seen for the Santa Cruz and Ipswich rivers.

The notion of first-in-time, first-in-right and various government regulations led to price differences among water users. Often senior water rights users, such as mining and agricultural firms, pay little for water relative to newer users such as households, municipalities, or other industries. For example, California farmers pay about $15 per acre-foot for Colorado River water while city residents pay up to $300 per acre-foot.

Different prices cause water to be used inefficiently. Efficiency is possible only if there is one price for water. A single price ensures that water is allocated to users that generate the greatest benefits to society. Returning to the prices just described, the willingness of industry to pay higher prices implies that they may be able to generate more value per liter than farmers. Under these conditions, total economic output could be increased if farmers used water more efficiently and sold some of their water to municipal and industrial users. Such sales helped California reduce its water use to the 4.4 million acre-feet limit described earlier in this chapter.

Furthermore, efficient allocation is possible only if prices reflect the costs of supply. This requirement often is not satisfied. Even when there is one price for water, prices often understate the cost of supply. One important cause is government subsidies. The U.S. government paid for the dams on the Colorado River, but these costs are not fully included in the price of water from that river. In other places the price of water does not include its scarcity value, which reflects the limits on supply. Finally, the ability of groundwater to flow across property lines of surface landowners gives individual landowners an incentive to pump groundwater as quickly as possible—if they don't, their neighbors' pumping will draw groundwater toward their side of the aquifer.

So the price of water generally is too low, which means that consumers use "too much" and producers supply "too little." The U.S. government estimates that the average household price of $2.30 per thousand gallons ($0.61 per thousand liters) will not generate enough revenue to fund the infrastructure needed to supply water to a growing population. To meet this growing demand, an extra $12 billion a year is needed, which would require an average price of $4.26 per thousand gallons ($1.13 per thousand liters).

Given governments' reluctance to increase taxes, many municipalities have privatized their drinking water systems. **Privatization** involves selling a state-owned business, such as the local waterworks, to private investors. In the early 1990s private companies provided water to about 51 million people in twelve nations. A decade later they supplied water to about 600 million people. At this growth rate 70 percent of the people living in North America and Europe will get their water from private firms within a decade.

Private firms price water so that supply equals demand. This price usually is greater than current prices. Higher prices increase supply by generating the funds to expand waterworks. Higher prices also slow demand growth by reducing uses that generate little value. Higher prices increase the efficiency of water use but may also make water too expensive for poor people. In Cochabamba, Bolivia, privatizing local waterworks raised prices to the point where a significant fraction of the population could not afford to purchase water, which caused social unrest. See *Policy in Action: Privatizing Water in Cochabamba, Bolivia.*

Such problems have led policy makers to consider whether access to clean water is a right or clean water is a commodity. If access to water is a right, a market may not be an effective way to allocate water use. Markets have a hard time allocating physical necessities because people will pay any price for survival. But beyond these quantities, water may be like other goods or services: A competitive market may allocate water effectively among industrial and agricultural sectors.

Increasing Efficiency

In some places water purchases are a significant fraction of total costs. Under these conditions raising water prices could induce users to improve efficiency. For farmers, new sprinkler designs, such as low-energy precision application, can boost efficiency from 60 to 95 percent. **Drip irrigation** is the slow, localized application of water just above the soil surface. Drip irrigation increases efficiency by preventing evaporation of irrigation spray before it reaches the ground and by reducing the wet soil surface area, which lessens loss via evaporation and infiltration beyond the root zone (Figure 18.19). Nutrients can be applied through the drip systems, which diminishes the use of fertilizer and improves the quality of returned water. In Israel, where it was invented, drip irrigation halved the water requirements of many crops.

For many households the water bill is a small component of costs, so higher prices may do little to reduce municipal use. In these cases command and control policies may be the most effective way to increase efficiency. Toilets use the most water in households, and requiring homes to install water-efficient toilets can reduce household use significantly. Before the 1990s most U.S. toilets used about 6 gallons (23 liters) per flush. In 1992 legislation required residential toilets sold after 1994 to use about 1.6 gallons (6 liters) per flush.

Despite these savings, there is no consensus for national standards to govern water use. Instead federal agencies support programs that encourage conservation. For example, the Environmental Protection Agency created WAVE (Water Alliances for Voluntary Efficiency), which is a nonregulatory water efficiency partnership whose mission

FIGURE 18.19 *Drip Irrigation This method increases the efficiency of water use in dry areas by reducing the surface area from which water can evaporate. It also increases the efficiency of fertilizer use by concentrating the fertilizer around the plant's roots.*

POLICY IN ACTION

Privatizing Water in Cochabomba, Bolivia

The World Bank views privatization as the most reliable way to ensure clean water. Putting its money where its mouth is, many of the bank's loans for water supply stipulate that public waterworks be privatized. Selling public waterworks to private firms has improved access to a clean, reliable supply of water for some users. For others, the loss of public control has reduced supply or reliability. And then there is Cochabamba, where civil unrest over privatization of the local waterworks resulted in death.

Cochabomba, Bolivia, lies at the eastern edge of the Amazon forest. Although the city has existed for more than 400 years, the population increased fourfold since the mid-1970s. Many of these people are poor and live in rudimentary housing that rings the town. The combination of rapid growth and poverty means that many people do not have access to basic services, such as electricity, sanitation, or water. The local waterworks is especially decrepit. Many poor neighborhoods are not connected to the local water system. As a result, government subsidies to the local waterworks do not help the poor—they help middle- and upper-class citizens. To supplement government efforts, international organizations funded the drilling of wells that were run by local cooperatives, which supplied inhabitants with relatively inexpensive water.

This patchwork supply generated a consensus that the local waterworks had to be improved. But such improvements would not be easy or cheap. After many years of neglect, repairing and expanding the local waterworks would cost more than the money that was available from either the city or national government.

To get around these financial constraints, the government auctioned the waterworks in Cochabomba to a private firm. The only bidder was a consortium (a group of companies),

Aquas del Tunari, which paid $2.5 billion for the waterworks and forty years of exclusive rights to sell water to people living in and around Cochabamba. For these services, the firm was guaranteed a 15 percent return on its investments. Because the agreement gave Aquas del Tunari the exclusive right to all water in the district, including groundwater, Aquas del Tunari could charge fees to local cooperatives that pumped water from local aquifers.

News of this agreement frightened the local residents, who realized that water prices would soon rise. Aquas del Tunari stated that the average price rise would not be more than 35 percent, but the government admitted that for some people water bills would triple. Just how high they would climb became clear in January 2000, when the average water bill represented 20 to 30 percent of a working person's monthly pay.

Such increases cut into people's meager standard of living. By February people started to protest in the main plaza. Over time the number of protesters grew, as did their diversity. Eventually most local residents opposed the privatization. Leaders of the protest movement sponsored an unofficial referendum, in which 96 percent of the 50,000 voters preferred to cancel the contract with Aquas del Tunari.

The government ignored this vote, and the size of the protest grew. On April 8 the government declared a state of siege, which allowed mass arrests. That same day a large protest led to the death of Victor Hugo Daza, a 17-year-old student who was shot in the face by an army sharpshooter. (The sharpshooter was later acquitted by a military court after no civilian judges were willing to hear the case: They feared for their lives if the sharpshooter was convicted.)

The killing of Victor Hugo Daza undermined all support for Aquas del Tunari. The mayor of Cochabamba switched his support to the protesters. Executives of Aquas del Tunari were informed that the police could no longer guarantee their safety. When they left Cochabamba, the government told the company that it lost its contract because it had abandoned its concession. Even the World Bank claimed it had nothing to do with the privatization.

If this were a movie, the story would end here with a victory by the "good guys." A return to local control might be desirable, but the water supply problem is not solved. Although the water utility was reorganized and corruption was reduced, it does not have enough money to expand the waterworks for the growing population of Cochabamba. To make matters worse, foreign investments in Bolivia have declined sharply—there were no bids at a recent auction to privatize the telephone company in La Paz, and the corporations behind Aquas del Tunari are suing the Bolivian government for $25 million for breaking the contract. The outcome highlights the fact that access to clean water may be a human right, but it cannot be provided for free.

ADDITIONAL READING

Finnegan, W. "Leasing the Rain." *The New Yorker* (April 8, 2002), p. 43.

STUDENT LEARNING OUTCOME

- Students will be able to compare and contrast the costs and benefits of efforts to increase water supply by privatizing water utilities.

---

is to encourage firms to reduce water consumption while increasing efficiency, profitability, and competitiveness. To join, firms agree to survey water-using equipment and, where profitable, install water-efficient upgrades within a prearranged time frame. Members also agree to design all new facilities with water-efficient equipment.

Controlling Water Pollution

Nonconsumptive uses of water often return water in a polluted form. Many of these pollutants can be removed by existing technologies, but firms have little economic

incentive to do so. Thus water pollution is another example of an externality that causes environmental degradation. As with many other externalities, water pollution can be controlled using either market mechanisms or command and control. To date, much of the water pollution policy in the United States and abroad has relied on command and control.

In the United States water pollution is regulated by the Clean Water Act, which was enacted originally in 1948 as the Federal Water Pollution Control Act. Today water pollution is mostly regulated by a new set of laws contained in a 1972 set of amendments. Their objective is the "restoration and maintenance of the chemical, physical, and biological integrity of the nation's water." To reach these objectives, the amendments established two goals: Improve water quality so that surface waters are "fishable" and "swimmable" by mid-1983 and by 1985, and ultimately achieve no discharge of pollutants. Although it sounds vague, **fishable** means that fish and shellfish can thrive in a water body and can be eaten safely by people. **Swimmable** means that recreation in and on the water will not threaten people's health. To reach these goals, the act consists of two parts: regulatory requirements that limit pollutants and provisions authorizing the U.S. government to fund construction of municipal treatment plants that are needed to meet the regulatory requirements.

The regulatory requirements have two components: water quality and a technology standard. Water quality standards establish overall water quality. Quality is defined by how the water is to be used, such as recreation, drinking supply, or industrial uses, and the maximum levels of individual pollutants that are consistent with such uses. These standards are established by individual states. To meet these standards, the act is said to be "technology-forcing" because it requires those who discharge wastewater to use legally mandated technologies. By 1977 firms and municipalities had to install the **best practicable control technology** to control the discharge of conventional pollutants, such as municipal wastes and suspended solids. By 1989 the requirement for the **best available technology** forced emitters to use technologies that can remove toxic materials from wastewater.

To ensure compliance, the Clean Water Act starts with the assumption that all discharges into the nation's waters are forbidden. To obtain permission to discharge pollutants, dischargers must obtain a permit from the Environmental Protection Agency. The permit specifies the type of technology that must be used and limits on the quantity of pollutants that can be discharged. These permits are valid for five years, after which they must be renewed.

States monitor whether firms have the needed permits and comply with their requirements (Figure 18.20). Violators can be fined up to $25,000 per day. These fines increase if the violations are due to negligence or are willful. If they are the latter, violators can be sentenced to prison.

In addition to these "sticks," the Clean Water Act also provides "carrots." The federal government helps municipalities pay for the required technologies. Depending on the sophistication of the technology, federal grants pay up to 75 percent of its cost. These grants do not have to be repaid by local governments, which pay the differences between total costs and those covered by the federal grant.

Most of these funds are used to build plants that treat municipal wastes. Treatment occurs in three general steps. **Primary treatment** removes large solids by mechanical techniques, such as screens and settling tanks. **Secondary treat-**

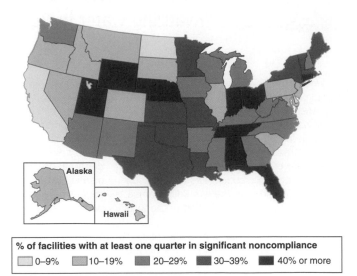

% of facilities with at least one quarter in significant noncompliance
0–9%   10–19%   20–29%   30–39%   40% or more

FIGURE 18.20   *Compliance with the Clean Water Act* *The percentage of facilities that violate the Clean Water Act for one or more quarters varies among states. Compliance is greatest in states such as California or North Dakota and is lowest in states such as Florida or Connecticut.* (Source: Data from J. Baumann and R.C Puchalsky, "Poisoning Our Water: How the Government Permits Pollution.")

ment reduces the number of pathogens and accelerates the decomposition of organic wastes by enhancing the actions of aerobic and anaerobic bacteria. In the final step, **tertiary treatment** separates undecomposed organics from the wastewater, which is discharged back to the environment.

Since 1972 Congress has spent more than $70 billion to help build wastewater treatment plants. As of 1996 another $140 billion was needed to meet the standards mandated by the Clean Water Act. Such spending caused policy makers to wonder whether the first $70 billion was well spent.

The preliminary answer seems to be yes. This answer was supplied by the Environmental Protection Agency, which performed a cost–benefit analysis. During the 1990s the Clean Water Act generated about $13 billion a year in spending on water pollution abatement (the costs). The benefits of cleaner water along the 907,000 km (603,000 miles) of U.S. rivers and streams totaled about $11 billion.

This balance is preliminary because calculating both the costs and benefits forced analysts to make a lot of assumptions. To estimate costs, analysts had to identify pollution control expenditures that would have been made without the Clean Water Act. This turns out to be a significant fraction of total expenditures. Of the $45 billion spent annually on water pollution abatement during the 1990s, analysts estimate that about $32 billion would have been spent even if the Clean Water Act were not implemented. Conversely, benefits do not include the reductions in other pollutants, such as toxic materials and nutrients, that were generated by the Clean Water Act.

SUMMARY OF KEY CONCEPTS

- The hydrologic cycle indicates that a relatively constant quantity of water flows seamlessly among a series of storages. Evaporation creates freshwater, which is the type of water needed by people. People increase their access to surface waters by building dams and constructing water diversions. In places where surface water supply is insufficient, people pump groundwater. Connections among storages and flows imply that increasing water use from one storage or flow reduces other storages and flows.

- Per capita water use varies greatly among nations due to differences in affluence. Affluent nations generally use lots of water in the industrial and energy sectors, whereas developing nations use most water in the household and agricultural sectors. Growing populations and affluence induce people to remove increasing quantities of surface waters, which leaves less water for natural ecosystems, and pump groundwater faster than it is recharged.

- Municipal pollutants contain microorganisms, which can spread disease, and organic wastes, which can cause eutrophication. Industrial wastewaters contain heavy metals that threaten human health. Agricultural water pollutants contain nutrients, organic material, and agricultural chemicals, which can contaminate surface water and groundwater.

- Water is often used inefficiently because prices vary among users and rarely include all costs. To eliminate these difficulties, policy makers have tried to create a market for water by privatizing waterworks. The efforts have had mixed results because poor people can sometimes be priced out of the market. When higher prices are not effective, governments use command and control mechanisms to increase the efficiency of water use.

- U.S. efforts to reduce water pollution are centered in the Clean Water Act, which establishes water quality based on the water's intended use. Dischargers must obtain permits that specify the quantities of pollutants emitted and the types of abatement technologies used. The federal government helps localities build wastewater treatment plants. To date these expenditures seem roughly equal to the benefits of cleaner water.

REVIEW QUESTIONS

1. Explain how changes in surface water use affect groundwater supplies and how changes in groundwater use affect surface water flows.

2. Explain the differences and similarities in nutrient runoff from agricultural fields and organic wastes from municipal wastewaters.

3. Explain the negative economic and environmental effects of overdrafts.

4. Explain how changes in the speed of rivers affect the size and location of the trough in the oxygen sag curve.

5. Explain why water is currently used inefficiently in many parts of the world.

6. Why would it be difficult to allow firms to buy and sell permits that are required by the Clean Water Act?

KEY TERMS

absolute water scarcity
aquifer
artesian well
best available technology
best practicable control technology
biochemical oxygen demand
chemical oxygen demand
condenses
cone of depression
confined aquifer
consumption
Continental Divide
correlative rights doctrine
desalinization
discharge
dissolved oxygen deficit
drainage basin
drip irrigation
evaporation
fecal coliform count

fishable
freshwater
groundwater
groundwater discharge
hydrologic cycle
infiltration
instream
membrane separation
nonpoint pollutants
offstream
overdraft
pathogens
precipitation
primary treatment
prior appropriation doctrine
privatization
reasonable use doctrine
recharge area
reservoir
riparian water rights

rule of absolute ownership
runoff
salt water
saltwater intrusion
secondary treatment
sewage
sorption
subsidence
surface water
swimmable
tertiary treatment
thermal desalting
unconfined aquifer
water diversion
water pollution
water table
watershed
waterworks
withdrawals
xenobiotics

19

AIR POLLUTION

Costs and Benefits of Clean Air

STUDENT LEARNING OUTCOMES

After reading this chapter, students will be able to

- Describe the factors that determine the atmospheric concentration of air pollutants.
- Compare and contrast the advantages and disadvantages of reducing air pollution via command and control versus market-based mechanisms.
- Explain why zero emissions may not be economically efficient and may not be necessary to protect sensitive groups from the effects of air pollution.
- Explain the benefits and costs of reducing air pollution and describe their relative size.

Air Pollution in Guangzhou, China *There are few sunny days in Guangzhou, China, because air pollution scatters and reflects much of the sun's light.*

Despite the hour, 5:30 a.m., I am very excited. I hurry to catch the 7:30 express train from Hong Kong to Guangzhou, China, which is a large city near the mouth of the Pearl River. I am going to Guangzhou as part of an effort to validate our team's analysis of satellite images. Prior to my trip we analyzed nine images to determine how quickly agricultural land was being converted to urban uses (the answer is very fast–urban areas increased by 300 percent in nine years!). Now we are going to determine whether what we "saw" in the images matches what is happening on the ground. In short, are the areas in and around Guangzhou that we classify as agriculture in the satellite image really being used for agriculture?

As the train pulled out of the station, I was forced to draw my shade. It was a sunny day. Not being a "morning person," I wanted to go back to sleep. I did not wake until the train reached Guangzhou. When I raised the shade, I noticed that the day had turned cloudy. I could no longer see the sun.

My research associate was waiting for me. As we loaded luggage into the car, I remarked that the day had turned cloudy. She smiled ironically. She told me every day is like this, but it is not really cloudy. Rather, the air above Guangzhou is so polluted that it blots out the sun. Over the next week I would see the effects of air pollution. One day I went to a meeting with Chinese scientists on the sixth floor of an office building. From the window I could not see the ground. People walking in the street wore surgical masks. While in the city, I never saw the sun. Only when we visited sites upwind from Guangzhou would the sun "reappear." My impressions of air pollution in Guangzhou are consistent with high levels of air pollution in cities in developing nations (Figure 19.1). Concentrations of particulates, which block out the sun, sulfur dioxide, which causes respiratory problems, and nitrogen oxides, which lead to ground-level ozone, are greatest in cities located in developing nations. ∎

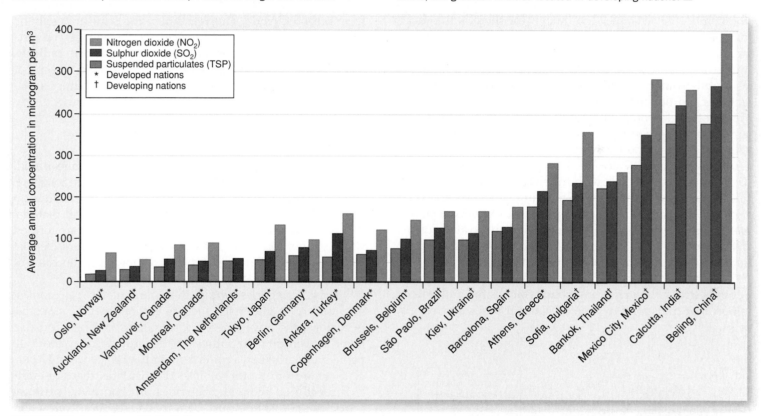

FIGURE 19.1 *Urban Air Pollution* *Between 1990 and 1995, the average concentration of three important air pollutants (particulates, sulfur dioxide, and nitrogen oxides) in the atmosphere of cities in developing nations was greater than their concentration in the atmosphere of cities in developed nations.* (Source: Data from World Resources Institute.)

POLLUTANTS

An **air pollutant** is any gas or particle in the atmosphere that has the potential to harm life or property. These gases and particles come in many forms and can be classified as either primary or secondary pollutants. **Primary air pollutants** enter the atmosphere in the form in which they harm life or property. For example, carbon monoxide (CO) comes directly out of automobile tailpipes. The same tailpipes also emit volatile organic compounds. These compounds are not directly hazardous, but under circumstances described later in this chapter, they help generate photochemical smog containing ground-level ozone

(O$_3$), which is hazardous to humans and ecosystems. Ozone, as well as other gases and particles that form when primary pollutants interact with sunlight and other gases in the atmosphere, are termed **secondary air pollutants.**

Within each category, air pollutants come in a variety of forms. In this section we focus on five primary pollutants that are regulated by the U.S. Environmental Protection Agency (EPA): carbon monoxide, hydrocarbons, particulate matter, sulfur dioxide, and nitrogen oxides, as well as the secondary pollutant ozone.

Carbon Monoxide

Carbon monoxide (CO) is a colorless, odorless gas that is formed when fossil fuels are not burned completely. Incomplete combustion occurs when there is not enough oxygen, flame temperature is too low, air passes through the combustion chamber too quickly, or there is too much turbulence in the combustion chamber. These characteristics are important because they define the major source of carbon monoxide—motor vehicles. Motor vehicles, which are known as **mobile sources,** emit lots of carbon monoxide because car engines are not managed for maximum energy efficiency. (How often do you have your car tuned up?) Automobiles emit about 60 percent of U.S. carbon monoxide (Figure 19.2). Compare that with the constant monitoring of large factories and power plants, which are known as **stationary sources.** For example, power plants burn about 30 percent of fossil fuels consumed in the United States but generate only 6 percent of total carbon monoxide emissions. Once formed, carbon monoxide can last one to two months in the atmosphere. This relatively long life span allows atmospheric circulation to transport carbon monoxide around the globe.

Carbon monoxide is hazardous to human health because it binds strongly with hemoglobin, which is the molecule in blood that transports oxygen. Because it binds so strongly, fewer hemoglobin molecules are available to transport oxygen. Instead oxygen is transported via plasma, which is much less effective than hemoglobin; therefore, high concentrations of carbon monoxide reduce the blood's ability to transport oxygen. Reduced oxygenation exacerbates the pain of angina pectoris and causes difficulties for those with cardiovascular disease. High concentrations of carbon monoxide also reduce the amount of oxygen that reaches the brain, which can interfere with your ability to concentrate, cause dizziness, and even kill.

The government recommends that people not breathe air with concentrations of carbon monoxide in excess of 35 parts per billion (ppb) for more than one hour. This level occurs in a room filled with cigarette smokers. Cigarette smoke contains more than 400 ppb of carbon monoxide.

Carbon monoxide may also contribute to global climate change by reacting with the hydroxyl molecule (OH) to form carbon dioxide. This reaction reduces the atmospheric concentration of the hydroxyl molecule, which converts methane to other forms. By reducing hydroxyl radicals, carbon monoxide may increase the atmospheric lifetime of methane, which would increase the warming that is associated with each molecule emitted by human activity.

Particulate Matter

Particulate matter is the general term used for a mixture of solid particles and liquid droplets that are found in the atmosphere. A mixture of solid particles and liquid droplets is known as an aerosol. Aerosols can consist of **dust,** which is solid material that is generated by crushing or grinding. **Fumes** are formed when vapors condense. **Mist** (or fog) refers to aerosols that consist of liquids. Particles that form during the combustion of fossil fuels are termed **smoke** or **soot.**

These categories imply that particulates can consist of a variety of materials, such as carbon, metals, asbestos, pesticides, and heavy metals. Particulates' composition also is influenced by their source. In the United States about 40 percent of particulates come from industrial process; 17 percent come from motor vehicles. Particulates that are emitted directly by economic activities are known as **primary particles** and include dust from roads and soot from wood combustion. In many cases particles emitted by human activity interact with natural components of the atmosphere to form **secondary particles,** such as nitrates, which are formed from nitrogen oxides that are emitted by fossil fuel combustion.

Particle size can vary greatly. Some are large enough to be seen with the naked eye, whereas others can be seen only with an electron microscope. Size differences are used to classify particulates (Figure 19.3). Fine particles are smaller than 2.5 microns in diameter (a micron is one millionth of a meter or about four ten-thousandths of an inch) and are termed PM$_{2.5}$. Particles that are smaller than 0.1 microns are known as **ultrafine particles.** Particles that are bigger than

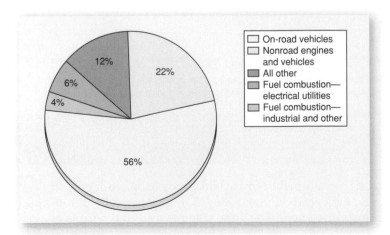

FIGURE 19.2 *Carbon Monoxide Emitters In the United States motor vehicles are responsible for nearly 60 percent of carbon monoxide emissions. (Source: Data from U.S. Environmental Protection Agency.)*

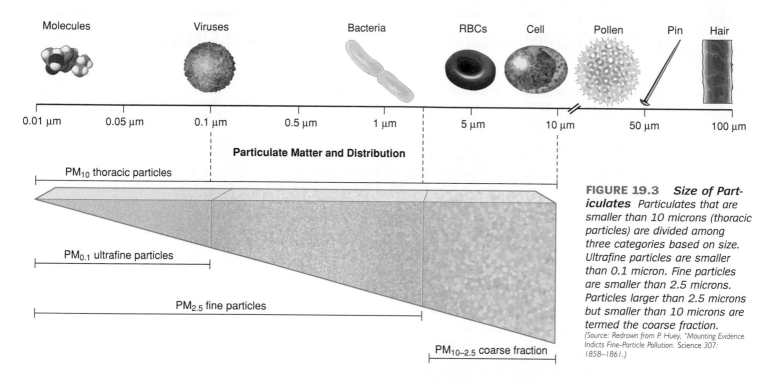

Particulate Matter and Distribution

PM$_{10}$ thoracic particles

PM$_{0.1}$ ultrafine particles

PM$_{2.5}$ fine particles

PM$_{10-2.5}$ coarse fraction

FIGURE 19.3 *Size of Particulates Particulates that are smaller than 10 microns (thoracic particles) are divided among three categories based on size. Ultrafine particles are smaller than 0.1 micron. Fine particles are smaller than 2.5 microns. Particles larger than 2.5 microns but smaller than 10 microns are termed the coarse fraction.* (Source: Redrawn from P. Huey, "Mounting Evidence Indicts Fine-Particle Pollution. Science 307: 1858–1861.)

2.5 microns but smaller than 10 microns are known as the **coarse fraction.** All particles that are smaller than 10 microns are known as **thoracic particles** and are labeled PM$_{10}$.

Size influences a particle's residence time in the atmosphere. Large particles, greater than 10 microns, tend to sink out of the atmosphere relatively quickly. Ultrafine particles have the potential to remain in the atmosphere for long periods because their aerodynamic resistance to sinking generally exceeds the force of gravity. Nonetheless, the residence time of ultrafine particles is relatively short because they collide with other particles and stick to form larger particles.

Size also influences particulates' ability to enter the respiratory system and how long they can remain there. Particles larger than 10 microns can be ejected after they are trapped by hairs that line the nose or by mucus that lines the trachea and bronchial tubes. These defenses are ineffective for particles smaller than 10 microns, which can reach the alveoli (air sacs in the lungs where the body exchanges oxygen and carbon dioxide). There they may be breathed back out if they are smaller than 2 microns. If they are slightly larger (2–4 microns), they may sink to the surface of the alveoli, where they can remain for prolonged periods.

Particulates can cause a variety of respiratory ailments, such as bronchitis, pneumonia, emphysema, and asthma. Some studies suggest that chronic exposure to fine particles is associated with rates of lung cancer that are similar to those associated with secondhand smoke. Statistical studies show that each 10 microgram per cubic meter increase in thoracic particles increases death rates by 0.21 percent, and there is a 4 percent increase for a similar increase in fine particles. Particulates also are implicated in genetic mutations. Babies born to mothers in New York City who are exposed to high levels of fine particles (and hydrocarbons, as we discuss in a bit) have higher rates of genetic mutations. Similarly, the offspring of male mice who are exposed to high levels of fine particulates have higher rates of genetic mutations. Finally, particulates are strongly correlated with deaths caused by cardiopulmonary and cardiovascular disease.

Scientists are not sure whether the health impacts of particles are determined by their size, composition, or both. Fine particles that are derived from average materials of Earth's crust seem to have few health effects, whereas fine particles that are derived from metals such as zinc or copper can inflame lungs and damage heart tissue. Other scientists argue that particle size is critical. Ultrafine particles penetrate deep into the lungs and have a large surface area.

Particulates (and other pollutants) also alter local climate. This effect was discovered by finding weekly patterns in climate variables such as temperature and precipitation. See *Case Study: A Link between Local Pollution and Global Climate Change.* There is no geophysical basis for climate variables to vary by the day of the week, but the concentrations of pollutants do vary this way—concentrations are greatest on weekdays and smallest on weekends. If pollution affects climate, climate variables may show a **weekend effect** in which climate observations for Saturday or Sunday differ from those observed for Monday through Friday. For example, just beyond the U.S. East Coast, less precipitation falls on Monday when pollution levels are relatively low, whereas the highest amount of precipitation falls on Friday and Saturday following a full workweek of polluting activities (Figure 19.4(a)). In that same area, the greatest wind speeds of hurricanes, and their less intense relatives tropical cyclones, occur on Thursdays and Fridays (Figure 19.4(b)).

CASE STUDY

A Link between Local Pollution and Global Climate Change

The Case Study in Chapter 14 described how the depletion of stratospheric ozone affects climate change. A similar linkage exists for air pollution: As described on page 401 of this chapter, air pollution affects local climate. Aerosols may change local temperature in a way that alters the daily temperature cycle.

To find this effect, scientists compiled more than forty years of daily information about maximum and minimum temperatures from 660 of the most reliable weather stations in the United States, as well as thousands of other stations from around the world. For each station and each day, the daily temperature range is the difference between the daily high temperature, which usually occurs during the day, and the daily low temperature, which often occurs during the night.

Daily values were compiled for each day of the week, and the fifty-two values for each year were averaged. This procedure generated values for the average daily temperature range on Mondays, the average daily temperature range on Tuesdays, and so on for the other days of the week. This process was repeated

for each of the forty-plus years, giving scientists information about the average daily temperature ranges for each day of the week between 1940 and 1990.

Analyses of individual stations demonstrate that human activities affect the daily temperature range. For example, in Carlsbad, New Mexico, the daily temperature range on Sunday and Monday is larger than values during the workweek (Figure 1). Furthermore, this effect has grown over time, especially since the mid-1970s, so that the average daily temperature range on Sundays and Mondays is statistically different from the average daily temperature range on other days. This statistically significant difference is another example of a weekend effect (see page 401 of this chapter).

The weekend effect need not be positive. In other locations, such as Katsuura, Japan, the average daily temperature range during the workweek is greater than the value during the weekend. Although the direction of the weekend effect is reversed, the interpretation is the same—the difference between the daily temperature range during the workweek and

the weekend shows that human activities affect local climate.

To get a broader view of this weekend effect, the authors of this study plotted changes at all stations in the United States and around the world (Figure 2). In the United States a weekend effect appears in more than 35 percent of the 660 stations. This percentage is much greater than would be expected based on random chance alone—indicating that human activity is changing the daily temperature range in many areas of the United States.

The weekend effect seems to vary geographically. In much of the West and the East, the daily temperature range on the weekends is greater than the range during the workweek. On the other hand, at stations in the Midwest daily temperature ranges on the weekends are smaller than the ranges during the weekdays.

Weekend effects are less clear outside of the United States. Taken as a whole, the global pattern of weekend effects is not statistically different from what would be generated by random chance. The lack of evidence is not

FIGURE 1 *Weekend Effect The daily temperature range (the daily high temperature minus the daily low temperature) varies by the day of the week.* **(a)** *In Carlsbad, New Mexico, the daily temperature range on Sunday and Monday is greater than the daily temperature range during other days.* **(b)** *In other locations, such as Katsuura, Japan, the average daily temperature range during the workweek is greater than average daily temperature range during the weekend.* (Source: Data from P.M. Forster and S. Solomon, "Observations of a 'Weekend' Effect in Diurnal Temperature Range." Proceedings National Academy of Science 100(20): 11225-11230.)

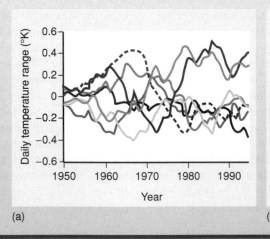

(a)

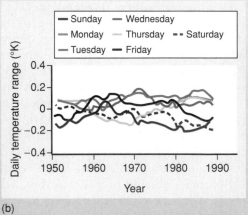

(b)

Sulfur Dioxide

Natural sources of sulfur include volcanic eruptions and sea spray. These natural sources are supplemented by emissions from burning fossil fuels or processing ores that contain sulfur (Figure 19.5). The sulfur content of fossil fuels varies by fuel. Sulfur may constitute nearly 7 percent of coal. Less sulfur is present in oil; the greatest amounts are in residual fuel oil, which may be 3 percent sulfur. Natural gas contains little or no sulfur. Consistent with these differences, coal-fired electric power plants are responsible for most of the sulfur emitted in the United States.

Sulfur is emitted as sulfur dioxide (SO_2). High concentrations of sulfur dioxide can constrict airways, change respiratory and pulse rates, and cause a variety of respiratory diseases (such as bronchitis). Sulfur oxides may be responsible for 50,000 U.S. deaths per year (about 2 percent of all deaths) when sulfur oxides are present with other air pollutants, such as particulates.

Sulfur dioxide also harms plants. High levels of sulfur dioxide change the way that soybeans allocate energy between roots and shoots, which reduces soybean yields. Sulfur dioxide also can affect plants indirectly. Insects prefer

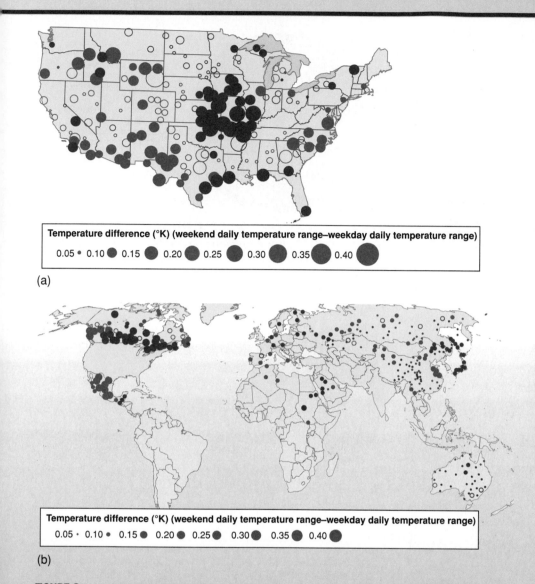

Temperature difference (°K) (weekend daily temperature range–weekday daily temperature range)

0.05 • 0.10 ● 0.15 ● 0.20 ● 0.25 ● 0.30 ● 0.35 ● 0.40 ●

(a)

Temperature difference (°K) (weekend daily temperature range–weekday daily temperature range)

0.05 · 0.10 • 0.15 ● 0.20 ● 0.25 ● 0.30 ● 0.35 ● 0.40 ●

(b)

FIGURE 2 *Size of the Weekend Effect in Daily Temperature Range.* **(a)** *In the United States, a weekend effect appears in more than 35 percent of the 660 stations, which is much greater than would be expected based on random chance alone. In much of the West and the East, solid green circles show that the daily temperature range on the weekends is greater than the range during the workweek. Stations in the Midwest are blue, which indicates that the daily temperature range on the weekends is smaller than the range during the weekdays. Open circles are locations where the weekend effect is not statistically significant.* **(b)** *Weekend effects are less clear outside the United States. Taken as a whole, the global pattern of weekend effects is not statistically different from what would be generated by random chance.* (Source: Data from P.M. Forster and S. Solomon, "Observations of a 'Weekend' Effect in Diurnal Temperature Range." Proceedings National Academy of Science 100(20): 11225-11230.)

too surprising. Many of the stations shown in Figure 2(b) are relatively uninhabited, so local human activity would have little effect on local climate. In some densely populated nations, such as China, Japan, and Mexico, weekend effects are statistically significant. On the other hand, the lack of weekend effects in densely populated areas of western Europe is puzzling.

Scientists, unsure how air pollution affects the daily temperature range, have focused on aerosols because their effects can vary by location. Aerosols can both increase and decrease cloud cover, and such changes could explain both positive and negative weekend effects. Alternatively, aerosols could change wider patterns of heating and cooling, which would alter weekly atmospheric circulation.

Regardless of the cause, weekend effects provide additional evidence that human activity is altering climate, which is an issue that lies at the heart of attribution. In addition, areas that do not have a weekend effect may be good places to look for a warming that is associated with greenhouse gases (in many areas aerosols offset much of the increase in radiative forcing that is associated with greenhouse gases—see Figure 13.4(c) on page 272). Finally, efforts to explain the mechanisms responsible for the weekend effect may improve climate models.

ADDITIONAL READING

Forster, P.M., and S. Solomon. "Observations of a 'Weekend' Effect in Diurnal Temperature Range." *Proceedings National Academy of Science* 100, no. 20 (2003): 11225–11230.

STUDENT LEARNING OUTCOME

• Students will be able to describe why a weekend effect shows that human activity affects local climate.

---

to feed on soybean plants that have been exposed to and damaged by sulfur dioxide.

Even more important is the damage done by secondary pollutants that originate from sulfur dioxide. Once in the atmosphere, sulfur dioxide can be quickly converted to sulfur trioxide (SO_3). Over several days, sulfur trioxide reacts with moisture

$$SO_3 + H_2O \longrightarrow H_2SO_4 \qquad (19.1)$$

to form sulfuric acid (H_2SO_4).

Sulfate particles (SO_4) are removed from the atmosphere by precipitation, a process known as **wet deposition** or **acid deposition**. Deposition can occur far from the source, and this distance is determined in part by smokestack height. Many of the sulfate particles that fall in the U.S. Northeast are emitted from tall smokestacks, which vent coal-fired electricity-generating plants in the Midwestern portion of the United States (Figure 19.6). Before the construction of these tall stacks, which can be 150 meters (492 ft) high, sulfate particles fell much closer to their source.

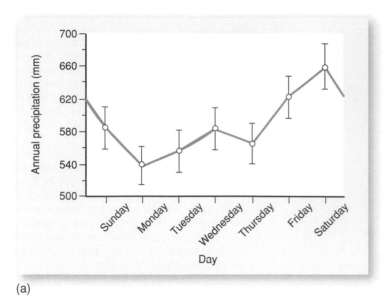

(a)

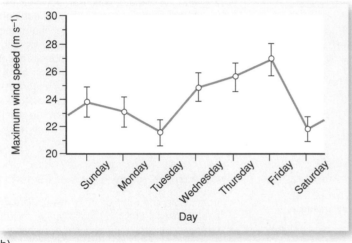

(b)

FIGURE 19.4 *Weekend Effect* Air pollution changes local climate, precipitation, and wind speed depending on the day of the week. **(a)** After a week of polluting economic activities, annual precipitation on Friday and Saturday between 1970 and 1996 is greater than on the other days of the week. **(b)** This effect extends to storms, Maximum wind speeds for hurricanes and tropical cyclones along the East Coast of the United States between 1970 and 1996 were greatest toward the end of the workweek. *(Source: Data from R.S. Cerveny and R.C. Balling, Jr., "Weekly Cycles of Air Pollutants, Precipitation, and Tropical Cyclones in the Coastal NW Atlantic Region," Nature 394: 561–563.)*

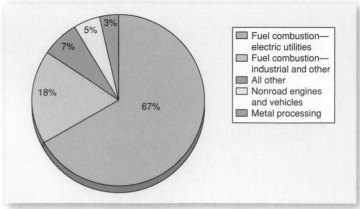

FIGURE 19.5 *Sulfur Emissions* About 67 percent of the sulfur emitted in the United States comes from electric utilities. *(Source: Data from U.S. Environmental Protection Agency.)*

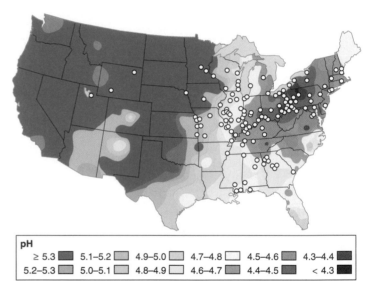

FIGURE 19.6 *Acid Deposition* Average pH of U.S. precipitation in 2003. Notice that the largest sulfur emitters, which are indicated by circles, are located in the Midwest USA. From here winds carry sulfur to the Northeast, where precipitation has the highest acid content. *(Sources: Data from National Atmospheric Deposition Program; data from National Oceanic and Atmospheric Administration (NOAA).)*

Sulfate particles lower the pH of precipitation (for an explanation of pH, see page 320 in Chapter 15). Due to carbon dioxide, the pH of precipitation from unpolluted air is somewhat acidic, with pH levels between 5.0 and 5.6. Sulfuric acid can reduce precipitation pH significantly to 4.2 or lower (remember that a one-unit decrease in pH, such as from 5 to 4, is a tenfold increase in acidity). For example, the pH of precipitation in the states of New York and Pennsylvania, which are downwind from large coal-fired electric power plants in the Midwest, can be below 4.5 (Figure 19.6). Acid deposition can damage

buildings and works of art, especially those made from marble, which is a metamorphosed (see Chapter 4) form of limestone. In Florence, Italy, Michelangelo's statue *David* had to be moved inside to protect it from the effects of acid deposition.

Acid deposition generally has a negative effect on ecosystems, although the impact varies due in part to geographical differences in acid neutralizing capacity. A high acid neutralizing capacity means that a soil or waterway can receive large amounts of acid with little or no change in pH. Acid neutralizing capacity often is associated with calcium carbonate, which is an important component of limestone. Following Equation 19.2, calcium carbonate ($CaCO_3$) supplies bicarbonate ions (HCO_3^-):

$$CaCO_3 + CO_2 \longrightarrow Ca^{2+} + 2HCO_3^- \qquad (19.2)$$

The bicarbonate ion (HCO_3^-) combines with the hydrogen ions from the acid deposition (low pH is associated with excess hydrogen (H^+) ions) as follows:

$$HCO_3^- + H^+ \longrightarrow CO_2 + H_2O \qquad (19.3)$$

In areas with abundant calcium carbonate, Equations 19.2 and 19.3 indicate that acid precipitation can add many hydrogen ions with little effect on pH. In other areas, such as New England or upstate New York, where soils are derived from granite, there is little calcium carbonate, so acid precipitation quickly lowers pH.

Even in areas with a high acid neutralizing capacity, the relationship between acid deposition and the pH of local waterways can be highly nonlinear. At first acid deposition may have little effect on pH because hydrogen ions are neutralized by carbonate ions. But as the supply of carbonate ions is reduced, further additions may quickly lower pH.

Lowering pH damages aquatic ecosystems by reducing the reproductive success of aquatic organisms. For example, low pH diminishes the size of salamander embryos and larvae, which cuts their reproductive success. Lower pH also affects ecosystems indirectly by reducing the nutrient holding capacity of soils (as discussed in Chapter 15, page 320). In western Pennsylvania, lower soil pH leads to the loss of cations and increases the mortality rates of sugar maples and red spruce. These effects can spill into aquatic ecosystems. A reduction in soil pH diminishes the soil's ability to hold aluminum ions, which leach into local waterways and kill freshwater fish.

Nitrogen Oxides

Nitrogen oxides is a catch-all term for a group of highly reactive gases that contain various proportions of nitrogen and oxygen, such as NO, NO_2, NO_3, N_2O, N_2O_3, N_2O_4, and N_2O_5. Of these nitric oxide (NO) and nitrogen dioxide (NO_2) are known as NO_x. These oxides are generated during the combustion of fossil fuels via two processes (Figure 19.7). Burning fossil fuels creates temperatures high enough (beyond 1,000 K—542°F) to oxidize atmospheric nitrogen (N_2) and oxygen (O_2), which creates **thermal NO_x**. The rate at which thermal NO_x is formed depends on an important trade-off with the formation of carbon monoxide. Efforts to reduce the emissions of carbon monoxide by motor vehicles include raising combustion temperatures and increasing air mixtures. Both changes

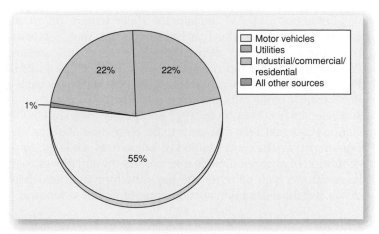

FIGURE 19.7 *Nitrogen Oxides* Motor vehicles accounted for slightly more than half of NO_x emissions in 2003. The rest was mostly divided between electric utilities and the industrial/residential sectors. *(Source: Data from U.S. Environmental Protection Agency.)*

increase the formation of thermal NO_x. **Fuel NO_x** is produced when the nitrogen in fossil fuels is oxidized. Nitrogen content varies by fuel; coal has the most and natural gas has the least.

Nearly all thermal and fuel NO_x is emitted as nitric oxide (NO), which is odorless and colorless and has no health effects. Once emitted, however, nitric oxide combines with oxygen to form nitrogen dioxide (NO_2). This molecule has a reddish-brown color, which gives city air a "dirty" appearance (Figure 19.8(a)); but it too has relatively little direct effect on human health. Nitrogen oxide is an important precursor to many secondary pollutants. Nitrogen oxides combine with water to form nitric acid (HNO_3), which reinforces the effects of sulfur emissions on acid deposition.

Nitrogen oxide also leads to the formation of ozone (O_3), which is an important secondary air pollutant. First, high temperatures associated with the combustion of fossil fuels split atmospheric oxygen and nitrogen and allow them to form nitric oxide (a similar set of equations can generate ozone from fuel NO_x):

$$N_2 + O_2 \longrightarrow 2NO \qquad (19.4)$$

Further contact with oxygen generates nitrogen dioxide:

$$2NO + O_2 \longrightarrow 2NO_2 \qquad (19.5)$$

During daylight hours, solar energy separates an oxygen radical from NO_2 that reacts with oxygen to form ozone:

$$NO_2 + h\nu \longrightarrow NO + O \qquad (19.6)$$

$$O + O_2 \longrightarrow O_3 \qquad (19.7)$$

Equations 19.4–19.7 explain the daily pattern of ozone concentrations in cities such as Los Angeles. As motorists start the morning commute, their car engines create nitric oxides (Equation 19.4). As the concentration of nitric oxide builds, Equation 19.5 becomes important. At this point the concentration of nitric oxide falls, and nitrogen dioxide builds. As the sun rises, its rays become stronger and power the photodissociation (Equation 19.6) that leads to the formation of ozone via Equation 19.7. The concentration of ozone rises while the concentration of nitrogen dioxide declines. As the sun moves past its zenith (its high point for the day), the formation of ozone slows and the destruction of ozone predominates as follows:

$$O_3 + NO \longrightarrow NO_2 + O_2 \qquad (19.8)$$

As a result, the concentration of ozone declines toward the end of the day.

Because car exhaust plays such an important role in ozone formation, high concentrations of ground-level ozone occur in urban areas. Ozone levels also are high in rural and suburban areas that are "downwind" from major urban areas because ozone and the chemicals from it (along with other particulates, sulfur oxides, and carbon monoxide) can be carried hundreds of kilometers. In the eastern United States this mix is carried from Washington, D.C., to Baltimore, Philadelphia, New York City, and Boston by summer winds that tend to blow from the southwest. Transport is also possible at the continental scale: Studies show that outflows of ozone from Asia increase concentrations in the western United States, and outflows from the U.S. East Coast increase concentrations

FIGURE 19.8 *Nitrogen Oxides and Ground-Level Ozone* **(a)** *Nitrogen oxides, which are emitted mainly near cities, contribute to the formation of ozone.* **(b)** *This ozone is then distributed globally by winds, as indicated by a model of atmospheric transport.* (Sources: (a) Data from the University of Toronto; (b) data from H. Akimoto, "Global Air Quality and Pollution," Science 302: 1716–1719.)

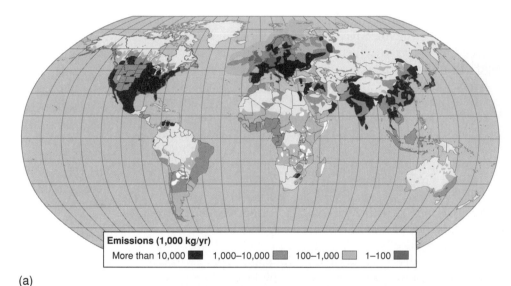

Emissions (1,000 kg/yr)

More than 10,000 ■   1,000–10,000 ■   100–1,000 ■   1–100 ■

(a)

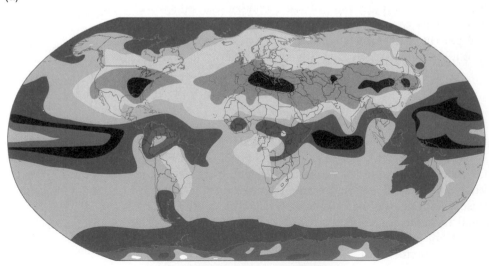

Concentrations (parts per billion volume)

0   5   10 15 20 25 30 40 50 60 70

(b)

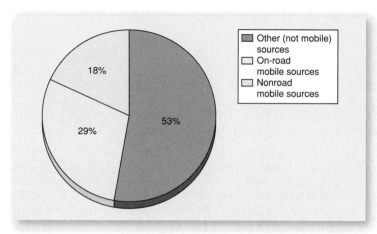

FIGURE 19.9 *Hydrocarbons Mobile sources, such as automobiles, trains, and planes, were responsible for nearly half of hydrocarbon emissions. The rest originated from nonmobile sources in 1999.* (Source: Data from U.S. Environmental Protection Agency.)

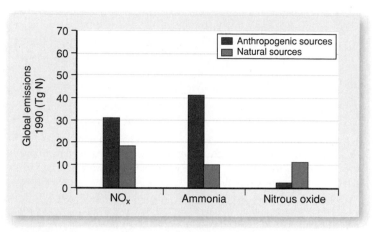

FIGURE 19.10 *Natural versus Human Emissions For several hydrocarbons, natural sources are comparable to human emissions. For nitrous oxide, natural sources are greater than human emissions.* (Source: Data from J.G.J. Olivier et al., "Global Air Emission Inventories for Anthropogenic Sources of NO$_x$, NH$_3$, and N$_2$O in 1990," Environmental Pollution 102: 135–148.)

in western Europe. Together these effects generate a band of elevated ozone that stretches across the Northern Hemisphere (Figure 19.8(b)), even though emissions of nitrogen oxides are concentrated near cities (Figure 19.8(a)).

Laboratory experiments can be used to estimate the rates at which the reactions given by Equations 19.4–19.8 occur. But if scientists use these rates to simulate ozone levels in cities, their projections are well below observations. This underprediction can be explained by including hydrocarbons in the pollution mix.

Hydrocarbons

Hydrocarbons are molecules that consist of carbon and hydrogen and result from incomplete fuel combustion or the evaporation of fuel (Figure 19.9). Some of these hydrocarbons are human emissions of volatile organic compounds, which are compounds that have a high vapor pressure and low water solubility (in other words, they evaporate at room temperature and do not dissolve in water). Human emissions add to natural emissions, which are associated with biological processes. In some cases natural emissions of volatile organic compounds are equal to or greater than human emissions (Figure 19.10).

Some volatile organic compounds are **carcinogens** (chemicals that cause cancer). Carcinogens include benzene and formaldehyde. Although dangerous on their own, volatile organic compounds have attracted the attention of environmental regulators because they accelerate the formation of ozone. Ozone is a key component of photochemical smog, which is responsible for the poor air quality over cities such as Los Angeles.

The chemistry of photochemical smog is complex, but one molecule is critical. Again that molecule is the hydroxyl radical (OH). Hydrocarbons combine with the hydroxyl radical and initiate a series of reactions that convert nitrous oxide to nitrogen dioxide. This "extra" nitrogen dioxide

serves as the raw material for ozone according to Equation 19.6. Furthermore, these reactions reform the hydroxyl radical much as the chlorine atom is reformed by the reactions that destroy ozone in the stratosphere (see Chapter 14), and this speeds the formation of ozone according to Equation 19.6. The reactions by which the hydroxyl radical forms nitrogen dioxide and is reformed also generate other components of photochemical smog, such as aldehydes. The simplest aldehyde is formaldehyde.

Ozone, aldehydes, and other components of photochemical smog are hazardous to human health, reduce plant growth, and damage materials. Eye irritation often is caused by formaldehyde and other aldehydes. As the concentrations of these and other irritating pollutants build, they can cause coughing and chest pain and trigger asthma attacks or bronchitis. Increased concentrations or prolonged exposure increase susceptibility to respiratory infections, decrease lung function, and diminish physical performance. Recent studies suggest an association between ozone concentrations and mortality rates. These effects lie behind the warnings not to engage in strenuous physical activity during air quality alerts. See *Your Ecological Footprint: How Much Air Pollution Do You Emit?*

You may want to remember that stratospheric ozone is "good" while ground-level ozone is "bad." In addition to health effects, ground-level ozone damages plants, as can be seen in flecks on the tops of leaves. Reductions in yield associated with elevated levels of ground-level ozone cost U.S. farmers about $500 million each year.

CONCENTRATIONS

The severity of damages caused by a given air pollutant is determined principally by its concentration and the duration of exposure. Exposure is determined by factors that often are unique to individuals, such as the amount of time

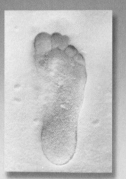

Your ECOLOGICAL *footprint*

How Much Air Pollution Do You Emit?

When we think of air pollution, many of us envision dark smoke belching from a tall stack. Because few of us see smoke coming from our own chimneys, we have a hard time identifying how we contribute to air pollution. Yet nearly all of our activities emit air pollutants, either directly or indirectly. Here we supply information that allows you to determine your emissions of sulfur dioxide and nitrogen oxides and how your emissions, along with your neighbors', affect local air quality.

If you read the section about the five major air pollutants carefully, you probably noticed that most are associated with energy consumption, especially fossil fuels. This implies that the types of fossil fuels you use and the rate at which you use them determine your emissions. Consequently, calculating your emissions of air pollutants starts with the data you used to calculate your emissions of greenhouse gases—the amount of coal, oil, natural gas, and electricity you use.

Calculating Your Footprint

Use this information in conjunction with the emission factors in Table 1 to calculate the amount of sulfur dioxide and nitrogen oxides you emit. Using these conversion factors to calculate your emissions is relatively straightforward. For example, burning a thousand cubic feet of natural gas in your hot water heater generates about 30 grams of sulfur dioxide (1,000 $ft^3 \times .03$ g/ft^3 = 30 grams). Notice that some of the emission factors are represented as a range instead of the single values for carbon dioxide. The range is dictated by technology. For example, the thermal production of nitrogen oxides depends on the temperature at which the fuel is burned. Because of this effect, burning a pound of bituminous coal in a fluidized bed emits about 0.7 grams of nitrogen oxides. If that same pound is burned in a cyclone boiler, it emits about 7.5 grams of nitrogen oxides.

As with carbon dioxide, you also emit air pollutants when you turn on a light switch or run your refrigerator. The amount of pollution you emit is determined by how the electricity is generated, which varies geographically in the United States. The Pacific Northwest generates much of its electricity from hydropower; therefore, consumers in the Pacific Northwest emit smaller amounts of sulfur dioxide and nitrogen oxides than consumers who use electricity that is generated in the Midwestern portions of the United States, where much electricity is generated using coal. To account for these differences, Tables 2 and 3 show the quantities of sulfur dioxide and nitrogen oxides that are emitted to generate a kilowatt-hour of electricity in various regions of the United States.

Use Tables 1–3 to approximate your emissions of sulfur dioxide and nitrogen dioxide. This number understates your effect because it does not include the pollutants associated with producing the goods and services you consume. You can approximate these numbers based on the total use of energy you will calculate for the next chapter's Your Ecological Footprint. Nor does it include emissions associated with transportation, either directly or indirectly. These emissions depend on the type and age of the vehicle and so cannot be calculated based on the use of energy alone.

Interpreting Your Footprint

The impact of your emissions depends on where you live. If you live in a sparsely populated area that has few natural precursors of air pollution and is upwind from major emitters, your emissions have a relatively small effect on local air quality. On the other hand, if you live in a densely populated area that has many natural precursors of air pollution and is downwind from major emitters, your emissions may be enough to push air quality into the unhealthful category.

The U.S. Environmental Protection Agency represents these geographic variations in air quality with an index for the five major air pollutants. Each day the EPA generates a map that represents air quality with an index that varies between 0 and 500, with 0 being the cleanest air (http://airnow.gov). Values are grouped in six categories that represent the health threat to various groups. For example, values between 0 and 50 fall within the "good" category, which means that air quality is considered satisfactory, and air pollution poses little or no risk to any group. Values between 101 and 150 fall into the category of "unhealthful for sensitive groups," which means that members of sensitive groups such as older people or people who suffer from heart or lung disease are at greater risk, but there is no elevated risk to the general public. At the other extreme, values between 301 and 500 belong in the "hazardous" category. Values in this category trigger health warnings of emergency conditions that are likely to affect the entire population. Under these conditions, the Environmental Protection Agency suggests restrictions on daily activities. For example, when ground-level ozone concentrations cause the air quality index to exceed 150, the air is classified as unhealthful, and this category carries a warning that people in sensitive groups, such as children and adults with respiratory disease, avoid outdoor exertion, while everyone else should limit outdoor exertion.

STUDENT LEARNING OUTCOME

- Students will be able to explain why it is more difficult to calculate their emissions of sulfur dioxide and nitrogen oxides compared to the calculation of their carbon dioxide emissions.

TABLE 1   Emissions per Unit of Fuel

Fuel	Units	Sulfur Dioxide (grams)	Nitrogen Oxides (grams)
Coal	Pounds	0.7–8.6	1.1–7.5
Heating oil	Gallons	32.2	5.4
Natural gas	Cubic feet	0.03	0.03

Source: Data from Electric Power Annual, Energy Information Administration, U.S. Department of Energy.

they spend outdoors (or indoors for some indoor air pollutants). Concentrations are determined by the quantity of pollution emitted and the volume of air into which it is mixed. Here we describe the factors that determine the volume of air into which pollutants mix. Emissions are discussed in the section about policy.

Concentrations are measured two ways—by weight and by volume. Weight measures can be confusing because

TABLE 2　Sulfur Emissions to Generate Electricity (grams per kilowatt-hour)

Regions (States)	Coal	Oil	Natural Gas
New England (Connecticut, Maine, Massachusetts, New Hampshire, Rhode Island, Vermont)	4.44	1.34	0.00
Middle Atlantic (New Jersey, New York, Pennsylvania)	2.03	0.62	0.00
East North Central (Illinois, Indiana, Michigan, Ohio, Wisconsin)	7.18	0.02	0.00
West North Central (Iowa, Kansas, Minnesota, Missouri, Nebraska, North Dakota, South Dakota)	8.71	0.02	0.00
South Atlantic (Delaware, District of Columbia, Florida, Georgia, Maryland, North Carolina, South Carolina, Virginia, West Virginia)	3.74	0.26	0.00
East South Central (Alabama, Kentucky, Mississippi, Tennessee)	5.24	0.05	0.00
West South Central (Arkansas, Louisiana, Oklahoma, Texas)	6.31	0.05	0.00
Mountain (Arizona, Colorado, Idaho, Montana, Nevada, New Mexico, Utah, Wyoming)	6.75	0.01	0.00
Pacific Contiguous (California, Oregon, Washington)	0.18	0.00	0.00
Pacific Noncontiguous (Alaska, Hawaii)	0.28	3.11	0.00

TABLE 3　Nitrogen Oxide Emissions to Generate Electricity (grams per kilowatt-hour)

Regions (States)	Coal	Oil	Natural Gas
New England (Connecticut, Maine, Massachusetts, New Hampshire, Rhode Island, Vermont)	1.40	0.40	0.01
Middle Atlantic (New Jersey, New York, Pennsylvania)	0.64	0.19	0.05
East North Central (Illinois, Indiana, Michigan, Ohio, Wisconsin)	2.27	0.01	0.00
West North Central (Iowa, Kansas, Minnesota, Missouri, Nebraska, North Dakota, South Dakota)	2.75	0.01	0.01
South Atlantic (Delaware, District of Columbia, Florida, Georgia, Maryland, North Carolina, South Carolina, Virginia, West Virginia)	1.18	0.08	0.05
East South Central (Alabama, Kentucky, Mississippi, Tennessee)	1.65	0.01	0.02
West South Central (Arkansas, Louisiana, Oklahoma, Texas)	1.99	0.01	0.13
Mountain (Arizona, Colorado, Idaho, Montana, Nevada, New Mexico, Utah, Wyoming)	2.13	0.00	0.03
Pacific Contiguous (California, Oregon, Washington)	0.06	0.00	0.04
Pacific Noncontiguous (Alaska, Hawaii)	0.09	0.93	0.15

weight is determined by the number of particles. A cubic meter of air at sea level (such as in Seattle) has more particles, and therefore weighs more than a cubic meter of air a mile above sea level (such as in Denver). To avoid the effects of altitude, scientists also measure pollutants relative to the total number of particles, such as parts per million (ppm) or parts per billion (ppb). The number of particles in a cubic meter of air is kept constant by measuring concentrations at standard temperature, which is 25°C, and standard pressure, which is 1 atmosphere.

Vertical and Horizontal Mixing

For a given level of emissions, concentrations depend on the volume of air into which the pollutant is mixed. This volume is determined by vertical and horizontal mixing. **Vertical mixing** refers to the altitude to which a pollutant is mixed; **horizontal mixing** refers to how far the pollutant is carried from its source. Increases in vertical and horizontal mixing enlarge the volume of air into which emissions are mixed, which reduces their concentration.

Rates of horizontal and vertical mixing are determined by advection and convection. **Advection** is the horizontal transfer of mass or energy as air masses move in response to pressure differences, such as winds. Horizontal movements tend to keep pollutants near the surface; therefore, their ability to reduce concentrations in air that people breathe is limited. Conversely, **convection** refers to transfer of energy and mass by motions in a liquid or gas. In the atmosphere convection often describes vertical interchange of air masses (Figure 19.11). Examples of convection include rising movements of warm surface air and sinking movements of cold air from upper levels of the atmosphere. These movements are powered by the same forces that power the "heat engine" shown in Figure 4.3 on page 59. Vertical movements, especially those that move pollutants away from the surface, can reduce concentrations in the air that people breathe.

The potential for vertical movements is influenced by the stability of the atmosphere, which is determined by the rate at which temperature declines with altitude (recall from Figure 14.1 on page 295 that temperature declines with altitude in the troposphere). The rate at which temperature (or some other variable) changes with altitude is known as the **lapse rate.** Suppose we start with a cubic meter of air at sea level. As we raise this parcel of air, pressure on the parcel drops, which allows the

parcel to expand beyond its original cubic meter. As it expands, the air cools. Conversely, lowering a parcel of air will increase its pressure, cause it to contract, and warm it. If these parcels of air were not allowed to lose energy as they move up or gain energy as they move down, the resultant temperature change would correspond to the **adiabatic lapse rate.** This lapse rate corresponds to a change of about 9.8°C/km or about 1°C per 100 m. In the "real world," the observed environmental lapse rate often is faster or slower than the adiabatic lapse rate.

Atmospheric Stability

Differences between the observed environmental lapse rate and the adiabatic lapse rate indicate how a parcel of air will move following a change in altitude. Suppose a parcel of polluted air at sea level is raised to an altitude of 100 meters (328 ft); as it rises those 100 meters, it does not exchange heat with the atmosphere. The reduction in pressure allows the parcel to expand and cool by about 1°C (1.8°F). This temperature change relative to the observed environmental lapse rate determines whether that parcel of polluted air will continue to rise, stay at its new altitude (100 m), or sink back toward sea level (Figure 19.12).

Suppose the observed environmental lapse rate is faster than the adiabatic lapse rate, so the air temperature at 100 m is 2°C (3.6°F) cooler than at sea level (Figure 19.12(a)). In this case the polluted air is 1°C (1.8°F) warmer than the surrounding air (the air we raised has cooled by 1°C while the atmosphere has cooled by 2°C). The newly raised polluted air is now less dense than the surrounding air. Because it is less dense, the raised parcel of air will continue to rise away from its original location. Such an atmosphere is said to be **unstable.**

Suppose the observed environmental lapse rate is the same as the adiabatic lapse rate (Figure 19.12(b)). In this case the air at 100 meters (328 ft) is the same temperature as the parcel of the polluted air. Because the polluted air has the same temperature and density, it tends to stay at its new location. Such an atmosphere is said to be **neutrally stable.**

Conversely, suppose the observed environmental lapse rate is slower than the adiabatic lapse rate, so the air at 100 meters (328 ft) is only 0.5°C (0.9°F) cooler than the air at sea level (Figure 19.12(c)). In this case the polluted air is 0.5°C cooler than the surrounding air. The newly raised air is thus denser than the surrounding air and will sink back toward sea level. Such an atmosphere is said to be **stable.**

The last few paragraphs are rather technical. Here's why these details are important: The differences among unstable, stable, and neutrally stable atmospheres influence the degree to which pollutants mix vertically. An unstable atmosphere tends to promote vertical mixing, which reduces pollutant concentrations near the surface. A neutrally stable atmosphere neither promotes nor hinders vertical mixing. And a stable atmosphere hampers vertical mixing, which amplifies the concentration of air pollutants at ground level. Severe pollution episodes, such as those

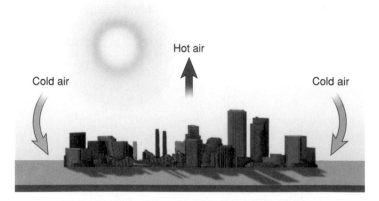

FIGURE 19.11 *Atmospheric Convection Solar energy is absorbed by a city's dark surfaces, such as roads and roofs. These materials warm the air above them. This warm air rises. As it does, cool air rushes in to take its place.*

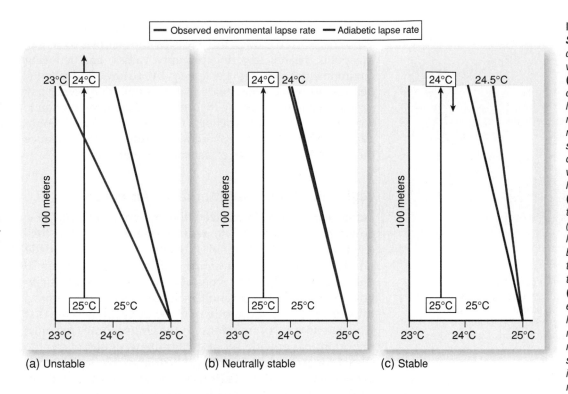

FIGURE 19.12 *Atmospheric Stability The atmosphere's stability determines the degree to which vertical mixing can dilute air pollutants.* **(a)** *In an unstable atmosphere, the observed environmental lapse rate (blue line) is steeper than the adiabatic lapse rate (red line). A parcel that has been raised 100 meters is warmer than the surrounding air, so the parcel of air continues to rise. These conditions favor vertical mixing, which reduces ground-level concentrations of pollutants.* **(b)** *In a neutrally stable atmosphere, the observed environmental lapse rate (blue line) is the same as the adiabatic lapse rate (red line). A parcel that has been raised 100 meters is the same temperature as the surrounding air, so the parcel of air remains at 100 meters.* **(c)** *In a stable atmosphere, the observed environmental lapse rate (blue line) is less steep than the adiabatic lapse rate (red line). A parcel that has been raised 100 meters is cooler than the surrounding air, so it sinks back to where it starts. These conditions slow vertical mixing, which increases ground-level concentrations of pollutants.*

that killed 4,000 people in London during 1952, are associated with an extreme form of a stable atmosphere called an **inversion.** As implied by its name, an inversion occurs when temperature rises with altitude (in the troposphere). Under these conditions there is little or no vertical mixing, and pollutants accumulate at ground level.

Radiation inversions occur daily due to radiational warming and cooling. As described in Figure 13.2 on page 271, the sun's rays warm Earth's surface, which emits long-wave radiation that is absorbed by greenhouse gases. On clear, calm nights the ground emits long-wave radiation faster than the lowest levels of the atmosphere can absorb it. Under these conditions, air just above the ground cools relative to the air above it. Such cooling can generate temperature inversions up to 100 meters (328 ft) above ground level. Radiation inversions occur frequently in desert cities such as Tucson and Phoenix. Despite their frequency, these inversions usually are short-lived. The next day the sun warms the ground and reestablishes a negative relationship between altitude and temperature.

Subsidence inversions occur where large air masses sink toward Earth. As they sink, pressure increases, which warms the air at ground level relative to the air above it. The altitude to which there is a positive relationship between temperature and altitude sets an upper bound on the vertical mixing, which is termed the **maximum mixing depth.** Subsidence inversions occur most frequently at about 30° north and south. Remember from Figure 4.7 on page 61 that this is where the air from the first Hadley cell tends to sink. Because of descending air masses, Los Angeles is frequently plagued by subsidence inversions, which is one reason for its high concentration of pollutants.

You can approximate horizontal and vertical mixing by observing smoke after it leaves a stack. If advection is the only mechanism diluting emissions, pollutants spread from a stack in the shape of a cone. The degree to which this cone spreads vertically reflects the stability of the atmosphere. A stable atmosphere causes the plume to move downwind with relatively little vertical movement—a shape known as fanning. If the atmosphere has an inversion, the plume will not rise above the maximum vertical mixing depth (escaping this maximum is one reason that some stacks are built very tall). Instead most of the mixing will occur in the downward direction. Because this pattern can generate very high ground-level concentrations, this pattern is known as **fumigation** (Figure 19.13(a)). In an unstable atmosphere the plume rises and falls, which is known as **looping** (Figure 19.13(b)). Looping dilutes pollutants effectively.

Observed Concentrations

Over the last two decades atmospheric concentrations for the five discussed pollutants have declined (Figure 19.14(a–f)). These data are calculated by averaging observations from hundreds of locations. Such measurements are used to determine the degree to which pollutants remain below the National Ambient Air Quality Standards, which are known as **NAAQS** (we will use this acronym because repeating National Ambient Air Quality Standards will make your eyes glaze over). These standards were set up by federal regulations that are described in the next section. NAAQS establish two standards. **Primary standards** are designed

(a) Fumigation   (b) Looping

FIGURE 19.13 *Plume Dispersion* *The potential for vertical mixing influences the degree to which a plume disperses.* **(a)** *A fanning plume can rise only as high as the maximum mixing depth, which then forces pollutants back towards the ground. This is known as fumigation.* **(b)** *A looping plume in an unstable atmosphere reduces ground-level concentrations.*

to protect public health, including the health of sensitive populations such as asthmatics, children, and the elderly. **Secondary standards** are designed to protect public welfare, including protection against decreased visibility and damage to animals, crops, vegetation, and buildings.

Despite the general decline in atmospheric pollutants, concentrations have not declined in all regions. Locations where concentrations remain above the NAAQS are termed **nonattainment areas.** Nonattainment frequently is associated with excess levels of ground-level ozone. In 1999 thirty-two areas had ozone concentrations exceeding the one-hour standard for two or more days. Despite these localized problems, the number of nonattainment areas declined from 293 in the early 1990s to 168 in 1999.

Although concentrations have not declined in all locations, the air that we breathe in the United States has become cleaner. Furthermore, as shown in Figure 19.1, concentrations are significantly lower in industrialized countries than in developing nations. Why has the air become cleaner while population and affluence have increased? Over the last two decades there has been little long-term change in vertical and horizontal mixing. Instead the improvement in air quality is due mainly to emission reductions.

WHY HAVE EMISSIONS DECLINED?

Air Pollution as an Externality

The single greatest cause for the decline in emissions is government efforts to reduce air pollution. Government has been forced to step in because air pollution is an externality (see Chapter 11). No one can own the atmosphere because individuals cannot stop others from using it. Under these conditions,

firms and households dump their wastes into the atmosphere at no charge even though these wastes impose costs by reducing public health, lowering property values, and reducing environmental services generated by natural ecosystems. Because firms and households that emit pollution do not pay for the damages caused by their wastes, there is no economic incentive for firms or households to reduce emissions. To remedy this externality, the government forces people to reduce their emissions in a series of state and federal laws.

Legislative Efforts to Internalize Air Pollution

Before economists formalized the notion of externalities, governments attempted to reduce emissions. In the thirteenth century the king of England, Edward I, forbade London residents from burning "dirty" coals (Chapter 20 will discuss the different types of coal). In the late 1800s city governments in Chicago and Cincinnati regulated emissions from factories and locomotives. In 1952 Oregon was the first U.S. state to control emissions. In the United States the most important nationwide set of government regulations is the Clean Air Act Amendments of 1970 and this legislation's subsequent amendments (it was amended most recently in 1990), which we refer to collectively as the Clean Air Act.

The Clean Air Act changed how the U.S. government approached air pollution. Before 1970 federal regulations required states to set up air quality control regions and support research on ways to reduce pollution. The Clean Air Act instructed the federal government, through the Environmental Protection Agency, to set nationwide standards for maximum concentrations of pollutants. These national standards (the NAAQS) provide the public with "an adequate margin of safety . . . from any known or anticipated adverse effects associated with air pollutant[s] in the ambient air."

Although the text of the Clean Air Act includes several hundred pages, the NAAQS are responsible for both its success and much disappointment. Most important for the economic well-being of individual states and cities, the NAAQS establish a lowest common denominator for all air pollution regulations. No state or city can allow a higher level of pollution and thereby give local firms an economic advantage by lowering the cost of compliance relative to cities or states that impose stricter environmental standards. Thus the NAAQS prevent states and cities from competing for businesses by weakening their environmental regulations. This so-called race to the bottom is an important component of the debate about globalization. Some people assert that developing nations attract firms (and their jobs) by weakening environmental regulations relative to developed nations.

By setting a lowest common denominator, the NAAQS pose the following dilemma: Many regions have pollution levels that are far below the NAAQS. Should air quality in these regions be allowed to deteriorate to the lowest levels set by the NAAQS? To settle this issue, the 1977 amendments set up areas where air pollutants must remain well below levels

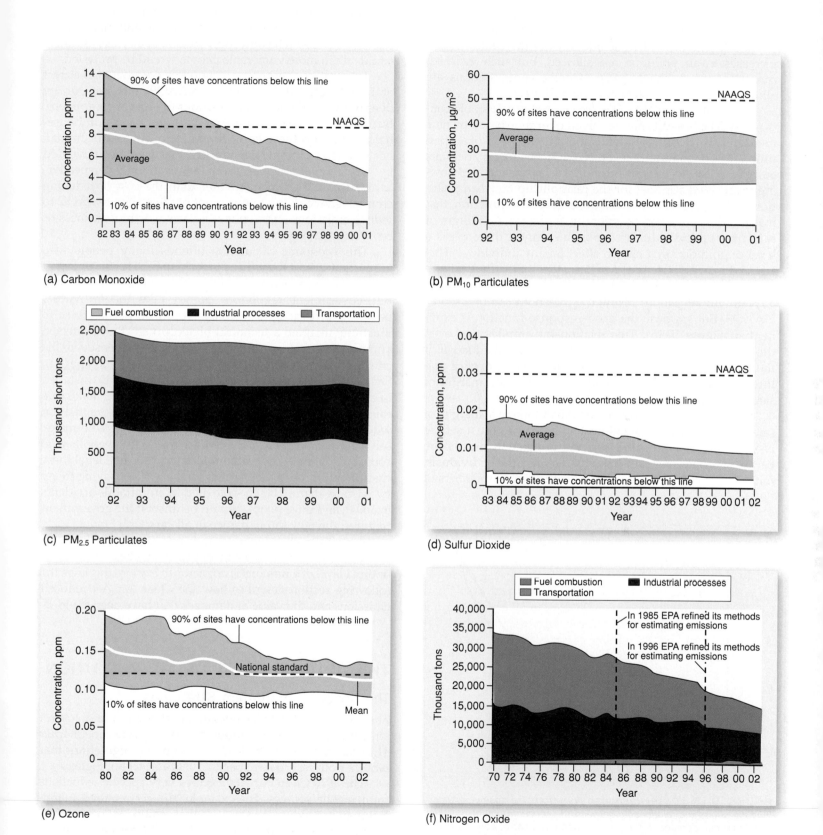

(a) Carbon Monoxide

(b) PM$_{10}$ Particulates

(c) PM$_{2.5}$ Particulates

(d) Sulfur Dioxide

(e) Ozone

(f) Nitrogen Oxide

FIGURE 19.14 *Declining U.S. Pollutants* Over the last twenty years the concentration and emission of five pollutants in the United States have declined. **(a)** *Carbon monoxide concentrations.* **(b)** *PM$_{10}$ particulates.* **(c)** *PM$_{2.5}$ particulates.* **(d)** *Sulfur dioxide concentrations.* **(e)** *Ozone (concentrations).* **(f)** *Nitrogen oxide emissions.* (*Source: Data from U.S. Environmental Protection Agency.*)

allowed by the NAAQS. In class I areas, such as national parks, no significant increases in air pollution are allowed. In class II regions, which include most of the United States, increases in air pollution are allowed, but such increases cannot approach the NAAQS. Finally, concentrations are allowed to reach the NAAQS in class III areas.

Establishing the NAAQS also created considerable controversy. The NAAQS have evolved over time based on new medical and scientific evidence about the impacts of air pollution. By mandating that the NAAQS ensure an adequate margin of safety for all individuals, lawmakers unknowingly specified a model for the relationship between pollution and damage, known as a **dose–response function,** that may not be supported by scientific evidence. A "margin of safety for all individuals" is possible only if there exists a level of pollution that has no effect on any individual. That is, the dose–response function has a threshold effect (Figure 19.15). Below the threshold, air pollution has no effect on any individual. Values below the threshold can be used as NAAQS. But suppose the dose–response function is exponential (Figure 19.15). This relationship implies that even the slightest pollution has an adverse effect, albeit small. If the dose–response function is exponential, no level of pollution, even the tiniest, provides "an adequate margin of safety for all individuals."

But what is the alternative? Abandoning a margin of safety for all creates difficulties for those who must set the NAAQS. For what segment of the population must there be an adequate margin of safety? For healthy individuals, for children or senior citizens, or for the sickest members of

society? Clearly the group chosen will be controversial. For example, increasing the NAAQS so that the pollution permitted has no effect on healthy individuals but increases sickness and death in more vulnerable people would be protested.

The cost of compliance is another controversial aspect of the NAAQS. When setting the NAAQS, the Clean Air Act does not allow regulators to consider the cost of compliance. But reducing emissions imposes costs on firms and households. How do these costs compare to the loss of human health and environmental services that have been avoided by reducing emissions? From an economic perspective, avoided costs should be greater than the cost of reducing emissions. According to this point of view, why spend $2 to reduce emissions and avoid a dollar's worth of sickness or ecosystem damage?

This economic calculation offends many people who argue that a cost–benefit approach makes sense unless it is your loved one who becomes ill because of pollutants. But government regulators cannot take the individual perspective because no activity would be permissible. To illustrate, imagine a technical breakthrough that protected automobile passengers from all injuries, however severe the crash. But suppose this safety device cost $1 million per car. Should the government require that the device be installed in all automobiles? Requiring installation would make cars unaffordable to most consumers; therefore, this regulation would cause a great deal of hardship for those who make, sell, and service automobiles and the many people who could no longer afford automobiles. At a societal level these costs would probably outweigh the benefits of saving many lives (more than 40,000 people died in automobile accidents in 2003), and most people would be upset if the government required the new safety device in all cars.

As with many economic decisions, decisions about the NAAQS come down to balancing costs and benefits—so the optimal level of emissions is not zero. To explore this idea, the following section describes how the Clean Air Act reduced emissions and the value of damages that have been avoided.

THE OPTIMAL LEVEL OF AIR POLLUTION

The Costs of Abatement Strategies

The Clean Air Act relies heavily on command and control strategies for reducing emissions. As described in Chapter 11, command and control strategies impose regulations that require polluters to use specific fuels and technologies.

The 1970 Clean Air Act requires very specific reductions from mobile sources. For example, the act required manufacturers to produce cars in 1975 that emitted 90 percent less hydrocarbons and carbon monoxide and 82 percent less nitrogen oxides. Subsequent revisions required further reductions and expanded the vehicles regulated. The 1990 amendments gave the Environmental Protection Agency the power to regu-

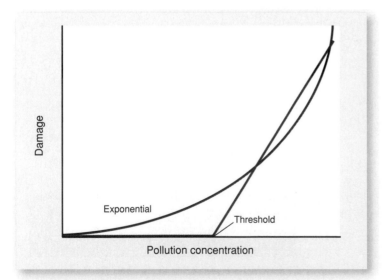

FIGURE 19.15 *Dose–Response Function* The dose–response function shows the relationship between the concentration of pollutants and damage. If the function has a threshold, concentrations below this threshold provide an adequate margin of safety for all individuals as required by the NAAQS. But if the function is exponential, no concentration greater than zero provides an adequate margin of safety for all individuals.

late emissions from diesel buses, motorboats, and even lawn mowers (burning a gallon of gasoline in a lawn mower built in 2006 emits ninety-three times more smog-forming emissions than burning that gallon in an average 2006 automobile).

One of the most effective command and control measures has been the catalytic converter. The three-way catalytic converter oxidizes hydrocarbons and carbon monoxide to form carbon dioxide. In addition, it converts nitrogen oxides to molecular nitrogen. By treating pollutants after they are formed, rather than changing the mixture of air and fuel in the engine as pollutants are formed, catalytic converters allow engines to run at nearly optimal conditions. This maintains both power, which makes drivers happy, and fuel efficiency, which makes drivers happy and lowers emissions.

The 1990 amendments focus on special blends of gasoline that produce fewer emissions. Reformulated gasoline reduces hydrocarbon emissions by including MTBE (the use of which is being reconsidered because it contaminates groundwater) or ethanol, which is derived primarily from corn. For example, Denver residents are required to burn reformulated gasoline between November 1 and February 29 to reduce nonattainment in carbon monoxide, which is caused by Denver's geography. (Denver is the "Mile High City": At high altitudes low oxygen concentrations favor the formation of carbon monoxide relative to carbon dioxide.) Regional gasoline blends are one reason why gasoline cannot be reallocated when one portion of the United States suffers a gasoline shortage.

Stationary sources turn over less rapidly than mobile sources (the average car lasts less than a decade, whereas most industrial facilities last several decades); therefore, the Clean Air Act differentiates between old and new plants. New facilities are subject to **new source performance standards,** which mandate that new facilities use the "best technological system of continuous emissions reduction"; these vary by region. In nonattainment areas new plants have to use technology with the lowest achievable emission rate. In areas that are designated to prevent significant deterioration, new plants must use the best available control technology. The nebulous meanings of these classifications of technology are still being worked out by the courts.

Existing sources (those in operation as of 1970) are subject to a different set of standards that are set and administered by the states. States set these standards as part of their **state implementation plans,** which describe how they will satisfy the NAAQS. In general, state implementation plans have not been able to satisfy the NAAQS due to the high cost of retrofitting existing facilities with abatement technologies and the reluctance of states to punish local employers and thereby threaten jobs. Because of these difficulties, the 1990 amendments to the Clean Air Act delay the deadline for state implementation plans to demonstrate compliance with the NAAQS until 2010 for some regions.

Recent efforts to reduce emissions from existing and new stationary sources focus on market-based mechanisms. Most successful are efforts to reduce sulfur emissions from coal-fired electric power plants. The 1990 amendments mandate that sulfur emissions be reduced by 9.1 million metric tons (10 million tons) relative to 1980 emissions. During the first phase of these reductions, the Environmental Protection Agency assigned emission allowances for each of 263 coal-fired units at 110 power plants. Starting on January 1, 2000, firms were allowed to emit sulfur based on their number of allowances. Emissions beyond this limit cost the utilities $2,000 per ton. The ability to sell excess allowances (and the need to buy extra allowances to avoid fines) generated an active market for sulfur dioxide allowances. See *Policy in Action: A Market for Sulfur Emissions.* Firms that could reduce emissions at low cost did so and sold their extra allowances to firms that had a high cost of emission reduction. Such trades reduced the total cost of reducing emissions by about $1 billion per year.

Despite these savings, the total cost of reducing emissions as mandated by the Clean Air Act is considerable. Between 1973 and 1990, surveys show that firms spent $628 billion. This estimate does not include indirect costs (and benefits). For example, waiting in line to have your car's emissions tested costs time, which has an opportunity cost. When these costs are included, the total costs of compliance dropped to $523 billion. Yes, *dropped*—because the automobile standards reduced expenditures on tune-ups and maintenance. Another estimate indicates that the regulations reduced consumer benefits by about $569 billion. These three estimates are similar and demonstrate that the Clean Air Act imposed a considerable cost on the U.S. economy.

These costs may be exacerbated by a loss in economic competitiveness. Remember, one reason for the Clean Air Act was to prevent states from attracting industry by weakening their environmental standards. Some people claim that the Clean Air Act has encouraged U.S. industries to move to nations where pollution laws are relatively lax. The size of this effect is thought to be small because complying with environmental legislation usually is a small component of total costs. For firms to move, the environmental savings must exceed the costs associated with moving, such as setting up a new factory and the increased cost of transporting goods to consumers. Because these conditions rarely are satisfied, empirical analyses find little support for the notion that the Clean Air Act reduced the competitiveness of U.S. industry or caused the loss of domestic jobs. Given that these indirect costs are relatively small, the vital question is how the costs of complying with the Clean Air Act compare with the benefits of cleaner air.

The Benefits of Cleaner Air

It is relatively easy to measure the costs of abatement by tracing the money spent by firms and households. But there are no such easily tabulated expenditures to measure the benefits of cleaner air. Many pieces of information are required: the degree to which the Clean Air Act reduced emissions; the reduction in concentrations associated with

As described in Chapter 11, economic theory says that society can lower the cost of reducing emissions using market-based mechanisms. The ability to translate theory to practice has been tested by the 1990 Clear Act Amendments. These amendments set up a market in which permits for emitting sulfur dioxide can be bought and sold. How successful is this market? Here we review information regarding the efficiency of the market and how the cost of reducing emissions has evolved.

Title IV of the 1990 Clean Act Amendments focuses on sulfur emissions by the largest polluters, which include electricity-generating plants and some larger manufacturers. Phase I of Title IV covers the 263 largest polluters (Figure 19.6). For 2000 and years thereafter, the regulations include nearly all electricity-generating plants. Title IV aimed to reduce sulfur emissions by about 9.1 million metric tons (10 million tons) per year in 2000 relative to 1980, which represents a reduction of about 50 percent.

This reduction is accomplished by controlling the number of allowances, which are permits that entitle the holder to emit one ton of sulfur in a given year. The year in which the owner can emit this ton is given by the permit's vintage. The holder can emit the ton of sulfur in the allowance's vintage year or can bank the allowance for later use. As such, allowances create a property right that allows the holder to emit sulfur.

This property right is enforced by comparing the number of allowances held by a firm to its total emissions. Yearly emissions are totaled by continuous monitoring equipment on each stack. At the end of each year, firms must have enough allowances to cover emissions. If a firm has extra allowances, it can bank them or sell them to firms that need additional allowances.

These sales are the basis for lowering costs. Firms that can reduce their emissions at low cost cut their emissions and sell their extra allowances to firms that have higher abatement costs. To use this mechanism firms must know their abatement costs and must be able to buy and sell their allowances in a market. In addition, the market must have enough buyers and sellers so that the cost of allowances reflects the marginal cost of reducing a unit of sulfur emissions.

Before trading started among units covered by Phase I, most analysts believed that allowances would cost $250–350. This price was expected to reach $500–700 as the number of units covered expanded and the number of allowances declined in Phase II. As predicted, allowances sold for $300 and $265 during trades in 1992 and early 1993. These sales were the only pieces of information available when trading started in March 1993.

Consistent with this lack of information, early trades on the market were highly erratic. Between March 1993 and mid-1994 the price of allowances varied over a wide range (Figure 1). Furthermore, prices set by the first EPA auction differed from prices set by private markets. These markets were set up by several financial companies such as Cantor Fitzgerald and the Emission Exchange Corporation. The large differences implied that there was no consistent price; therefore, the market did not lower the cost of reducing emissions.

But that changed quickly. Starting with the Environmental Protection Agency's second auction in mid-1994, prices set by the auction were consistent with private trades (Figure 1). In addition, the differences between the bid (the price initially offered by buyers) and ask (the price initially requested by sellers) declined from $20 per allowance in 1994 to $1.50 in 1997. Private markets also became more competitive: Commissions per trade declined from $3.50 in 1994 to $1.50 in 1996. Private markets now dominate the sale of allowances. Between 1993 and 1997 the number of allowances auctioned by the EPA doubled from 150,000 to 300,000. At the same time the number of allowances sold in private markets increased from 130,000 to 5.1 million.

In addition to becoming more efficient, the market also has reduced the cost of reducing emissions. Since 1992 the price of the allowances has declined. For a short time in 1996 allowances sold for less than $100. Since then allowances have traded between $200 and $300. This is much less than the cost antici-

lower emissions; the reduced effects on human health, agricultural productivity, environmental services, human structures, and visibility; and the economic benefits of these effects. Estimates for each step are uncertain. Nonetheless, the results described here show that the benefits of cleaner air between 1973 and 1990 greatly outweigh the costs.

Determining how emissions would have evolved after 1973 without the Clean Air Act is difficult. Using a complex version of the IPAT equation from Chapter 11, the Environmental Protection Agency estimates that emissions would have been considerably higher. For example, the Clean Air Act reduced 1990 emissions of sulfur dioxide by 40 percent, carbon monoxide by 50 percent, and volatile organic compounds by 45 percent.

To assess the impacts of lower emissions, we must translate them into concentrations. As described earlier, concentrations depend on the nature of emissions (such as the heights of stacks) and the local atmosphere. These effects on concentrations are simulated using computer models known as air quality models. Simulations show that lower emissions reduce concentration by varying amounts. For example, the 50 percent reduction in sulfur dioxide emissions reduced concentrations by 40 percent, whereas the 45 percent reduction in volatile organic compound emissions reduced ozone concentration by only 15 percent. The reduction in ozone concentrations is relatively low because natural sources for volatile organic compounds are unaffected by the Clean Air Act, and the reactions that generate ozone are highly nonlinear.

Next the Environmental Protection Agency estimates how lower concentrations ease the negative impacts of pollution by analyzing the relationships between concentrations and human health and ecosystem function. For example, the reduction in thoracic particulates (PM_{10}) caused 184,000 fewer people to die prematurely in 1990. The reduction in ozone

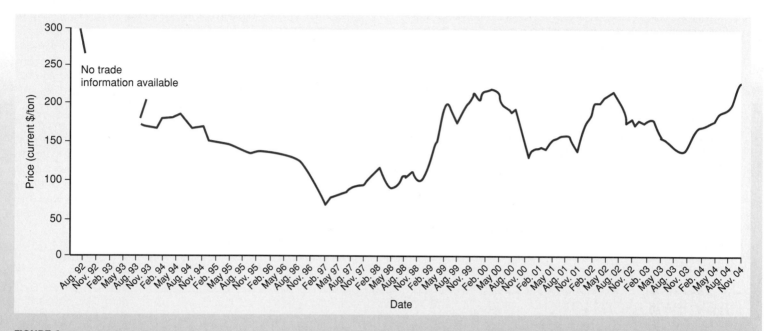

FIGURE 1 *The Price for Sulfur Emissions* The price for sulfur dioxide permits has declined since the market started even though the number of plants covered has increased and the total number of permits available has declined. (Source: Data from P.L. Joskow et al., "The Market for Sulfur Dioxide Emissions," American Economic Review 88: 669–685.)

pated by utilities when they lobbied against the 1990 Amendments (some utilities guessed that the reductions would cost $800–1,000 per ton) and the numbers anticipated by analysts at the start of the program. Based on the low costs and sizable reductions, many utilities and environmentalists agree: The market for sulfur emission allowances is a success!

ADDITIONAL READING

Joskow, P.L, R. Schmalensee, and E.M. Bailey. "The Market for Sulfur Dioxide Emissions." *American Economic Review* 88 (1998): 669–685.

STUDENT LEARNING OUTCOME

- Students will be able to evaluate whether the market for sulfur emission allowances was successful in reducing emissions.

avoided 850,000 asthma attacks. The reduction in all pollutants avoided an 0.08 percent reduction in winter wheat yields.

The final and perhaps most controversial aspect involves translating changes in human health and agricultural yields into dollar measures of benefits. The value of an avoided death is $4.8 million. The average cost of an asthma attack is $32 per episode. The economic value of agricultural impacts is relatively straightforward—it is the product of the reduction in yield (as measured in bushels per acre), acres planted, and price per bushel.

Summing these avoided costs indicates that the Clean Air Act lowered the negative impacts of air pollution by $22.5 trillion between 1973 and 1990. This comes to about $1.1 trillion annually. Clearly these benefits are significantly greater than the costs—a benefit–cost ratio of about 30:1. This high ratio is generated by the many avoided deaths that are attributed to the Clean Air Act and the high price associated with each of these deaths.

If the estimated number of deaths and the price per death are lowered, so too is the benefit–cost ratio. But even if significant reductions are made to the death rate and the value of life, the benefit–cost ratio remains large. The Clean Air Act is a net benefit to the U.S. economy despite some uncertainty about the size of its benefits.

The positive balance does not, however, eliminate the need to track the cost effectiveness of further amendments. Although the benefits of new regulations are subject to considerable debate, many argue that the cost of new regulations can be lowered by emphasizing market-based policies instead of command and control mechanisms. Empirical analyses of previous results show that the cost of command and control strategies is between two and ten times greater than achieving the same reduction through market-based mechanisms. These ratios imply that the compliance costs of the Clean Air Act could be reduced by 50 to 90 percent.

SUMMARY OF KEY CONCEPTS

- Air pollutants are gases, solids, or liquids that humans add to the atmosphere and that have the potential to harm life and property. Primary pollutants, such as carbon monoxide, hydrocarbons, particulate matter, sulfur dioxide, and nitrogen oxides, enter the atmosphere in the form in which they do their damage. When pollutants change form to become harmful, such as ground-level ozone, they are known as secondary pollutants.

- Damages caused by air pollution are determined by the duration of exposure and concentration. Concentrations are measured by both weight and volume and are determined by the degree of vertical and horizontal mixing. Horizontal movements generally do not move pollutants away from people, so they are less effective at easing exposure. Exposure is reduced by convection, which powers vertical movements in air masses. The rate of vertical mixing is determined by the atmosphere's stability.

- In the United States air pollution is mainly regulated by the Clean Air Act Amendments of 1970 and their subsequent amendments, which established National Ambient Air Quality Standards (NAAQS). These standards are based on a dose–response function that assumes a margin of safety for even the most sensitive segments of the population.

- The Clean Air Act depends heavily on command and control mechanisms. These strategies include mandating the use of catalytic converters, changing the composition of motor gasoline, and requiring use of the best available technology. Complying with the Clean Air Act cost the economy $500–700 billion between 1973 and 1990.

- The benefits of complying with the Clean Air Act include improved health, lower death rates, increased agricultural production, and enhanced environmental services. Although these benefits are difficult to measure, estimates indicate that the Clean Air Act cut the negative impacts of air pollution by about $22.5 trillion between 1973 and 1990. These immense gains are reduced, but not eliminated, if the estimated costs of poor health and premature deaths are lowered significantly.

REVIEW QUESTIONS

1. Explain why carbon monoxide, particulates, sulfur dioxide, nitrogen oxides, and hydrocarbons are either primary pollutants or secondary pollutants. For each, explain whether the relationships among emissions, concentration, and impacts are linear or nonlinear.

2. Explain how the difference between the observed lapse rate and the adiabatic lapse rate causes air parcels to move toward or away from their original location.

3. Explain why a competitive market will not solve the problem of air pollution on its own.

4. Why are the methods used to reduce emissions from mobile sources different from those used to reduce emissions from stationary sources?

5. Explain why stratospheric ozone is "good" but tropospheric ozone is "bad."

KEY TERMS

acid deposition	lapse rate	secondary standards
adiabatic lapse rate	looping	smoke
advection	maximum mixing depth	soot
aerosol	mist	stable
air pollutant	mobile sources	state implementation plan
carcinogens	NAAQS	stationary sources
coarse fraction	neutrally stable	subsidence inversions
convection	new source performance standards	thermal NO_x
dose–response function	nonattainment areas	thoracic particles
dust	primary air pollutants	ultrafine particles
fuel NO_x	primary particles	unstable
fumes	primary standards	vertical mixing
fumigation	radiation inversions	weekend effect
horizontal mixing	secondary air pollutants	wet deposition
inversion	secondary particles	

FOSSIL FUELS

The Lifeblood of the Global Economy

STUDENT LEARNING OUTCOMES

After reading this chapter, students will be able to

- Compare and contrast the ways in which the U.S. economy obtains and uses coal, oil, and natural gas.

- Compare and contrast the biological and geological processes that form coal, crude oil, and natural gas.

- Explain why at some point global oil production will decline over time rather than simply disappearing overnight.

- Describe why it is difficult to estimate the quantity of oil that remains.

- Explain the costs and benefits of efforts to reduce U.S. dependence on imported oil.

- Explain why crude prices tend to fluctuate and why they are so high now.

Automotive Status Symbols, Then and Now *Large cars such as the Cadillac were the status symbol during the 1950s and 1960s when motor gasoline was inexpensive. When gas prices rose in the 1970s and early 1980s, large cars declined in popularity. Will the same thing happen to the current automotive status symbol, the SUV, now that gasoline prices have risen?*

Déjà-vu is the feeling that one has seen something before; the redundant "all over again" was added by Yogi Berra, a catcher for the New York Yankees, who is famous for silly statements known as "Yogi-isms." But those of us who lived through the 1970s and 1980s find the first decade of the twenty-first century distressingly familiar.

Ask your parents about the "bad old days." The periods of 1973–1974 and 1979–1981 saw global reductions in oil production. During these shortages motorists waited in line to buy gasoline—on good days. On odd days of the month you could buy gasoline only if the last digit of your vehicle's license plate was an odd number, and vice versa. When taking a long trip, you had to worry about whether a gas station would be open. It was hard to find an open gas station on weekends or after 6:00 p.m.

These difficulties were forgotten with the 1986 price collapse. After that, for a while, motor gasoline was inexpensive; gas stations have become places where you can buy gas and food at nearly any hour. Another change has been the size of cars. Gone are most small, fuel-efficient cars. Many of you have driven and ridden in large vehicles such as minivans or SUVs.

But higher oil prices are back. After the hurricane season of 2005, motorists lined up to purchase $3-a-gallon gas. Although prices declined for a while and lines have disappeared, many are wondering what's next. Will smaller, more energy-efficient cars reappear? If so, the demographic change may look like what happened to the Cadillac and other big cars in the 1970s and 1980s.

In the 1950s and 1960s large cars such as the Cadillac were the automotive status symbol (opening photograph). Cadillacs were among the most expensive cars to own and operate. The income of the average Cadillac owner was well above the national average. But the rapid increase in gasoline prices changed the demographics of Cadillac owners. By the late 1970s Cadillacs and other large cars no longer were a status symbol. Although new Cadillacs were still expensive, pre-owned Cadillacs could be purchased at relatively low prices because their owners sold them so quickly that they created a glut.

Will SUVs suffer a similar fate? SUVs are among the least energy-efficient vehicles on the road. Most get fewer than 17 miles per gallon. As the price of motor gasoline rises, the desirability of owning an SUV is dropping. In 2006 sales of new SUVs slowed. If many SUV owners sell their cars, a glut of used SUVs could cut their prices. ■

THE PAST AND PRESENT OF FOSSIL FUEL USE

Fossil fuels include coal, oil, natural gas, oil shales, and tar sands. Of these coal, oil, and natural gas are termed **conventional fossil fuels,** which are the least expensive fuels to produce and therefore supply nearly all of the energy provided by fossil fuels. Oil shales and tar sands are **unconventional fossil fuels,** which are more expensive to produce than coal, oil, or natural gas; but they may replace these fuels as supplies of conventional fossil fuels are depleted.

Of the conventional fossil fuels, coal was the first to be used in large quantities (Figure 20.1). Coal is a solid and often is measured by its weight, such as metric tons. Although it has a higher energy density than wood (more energy per unit weight), coal use grew slowly at first. Coal smoke is more noxious than smoke from wood, and this makes coal difficult to burn in confined spaces with poor ductwork. In addition, most coal is dug from the ground, which makes it more difficult to collect than wood. Finally, coal is harder to ignite than wood. These challenges were an important obstacle to the expanded use of coal until a wood shortage throughout Europe during the sixteenth century provided an incentive for innovation. During the seventeenth and eighteenth centuries inventors developed

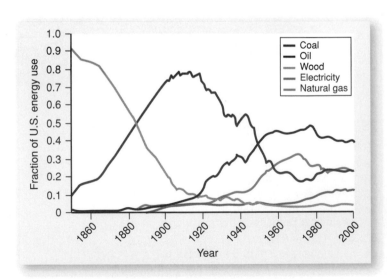

FIGURE 20.1 *Changes in U.S. Energy Use The Industrial Revolution started when coal displaced wood. The fraction of U.S. energy use from coal peaked early in the twentieth century as the use of oil and natural gas increased. Recently the fraction of energy use from these fuels declined due to greater use of coal and electricity generated from nuclear and hydroelectric sources.*

technologies that used coal to make steel and steam engines that used coal. These innovations expanded the use of coal and powered the first century of the Industrial Revolution (see Chapter 9).

The second century of the Industrial Revolution was powered primarily by oil. The first commercial oil well was drilled in Titusville, Pennsylvania, in 1859. Unlike coal, oil is a liquid and is often measured by barrels, which hold 42 gallons (159 liters). Under certain conditions crude oil flows spontaneously to the surface, which makes it easier to recover than coal. As was true for coal, the demand for oil was initiated by the diminishing supply of a renewable resource—whale oil, which was used for lighting. This use was replaced by kerosene, which is produced from crude oil.

The demand for oil increased rapidly during the first half of the twentieth century due to technical advances and oil's ease of use (Figure 20.1). Developing the internal combustion engine and producing cars via assembly lines increased the demand for oil. Internal combustion engines generate more power and are more efficient than the steam engines they replaced. Internal combustion engines also are easier to use. Gasoline flows from a car's gas tank to the engine—in contrast to the shoveling needed to move coal from a coal car to the boiler of a steam locomotive. Finally, because oil has a higher energy density than coal, oil requires less storage space than coal.

The use of natural gas increased significantly during the last half of the twentieth century because it has several advantages relative to both coal and oil (Figure 20.1). As a gas, it is usually measured by volume, such as cubic feet. Natural gas flows spontaneously to the surface and can be transported inexpensively via pipeline. Natural gas can be compressed into a small space, which makes it easy to store. Equally important, natural gas generates a very hot flame that emits relatively little pollution (Chapter 19).

In most places and at most times, coal is the least expensive conventional fossil fuel and natural gas is the most expensive. This ranking is due in part to differences in their end use characteristics. For many uses a Btu of oil and natural gas can do more work than coal. For example, a Btu of diesel fuel used to power a locomotive can move a train about three times farther than a Btu of coal. Similarly, household furnaces fired by natural gas are 75 percent efficient compared with furnaces that are fired by coal, which are 50 percent efficient. But not all differences among fossil fuels can be measured objectively. For instance, consumers want warm homes but prefer a fuel that requires relatively little storage space and does not have to be shoveled into a boiler. Finally, characteristics associated with the use of energy, such as pollution rates, also affect their prices. The price of natural gas is higher than that of coal because owners of gas-fired equipment do not need to buy expensive antipollution controls that the Clean Air Act requires for coal-fired equipment (Chapter 19).

The myriad characteristics associated with coal, oil, and natural gas, along with their prices, determine how society uses fossil fuels. To calculate how much energy you use, see *Your Ecological Footprint: How Much Energy Do You Use?* Coal is the least expensive fuel. But it has a low energy density (energy content per unit mass) and is highly polluting. Because of these characteristics, coal is used mainly as a stationary source of power in large boilers that are usually located far from population centers. For example, most coal is used to generate electricity and power industrial boilers (Figure 20.2(a)).

Oil is used primarily by the transportation sector (Figure 20.2(b)). Although oil is more expensive than coal, coal's low energy density and solid state make it an undesirable fuel for transportation. Both oil and natural gas can power transportation vehicles, but oil is preferred because it is a liquid, which makes it easier and faster to refill storage tanks. There are natural gas vehicles, but replacing natural gas canisters requires more time than refilling a tank with motor gasoline.

Natural gas is reserved for uses that require low emissions and high efficiency (Figure 20.2(c)). For example, the residential sector used about 22 percent of natural gas burned in the United States in 2003. The residential sector uses natural gas mostly to heat homes and water. Residential consumers are willing to pay more for natural gas because there are no bulky storage devices such as an oil tank. Homeowners also prefer natural gas because it burns cleanly, so there are no foul smells. These characteristics also offset the added expense of natural gas for many industrial users. Finally, the use of natural gas in electricity generation is increasing because recent advances have increased this technology's efficiency relative to coal-fired boilers.

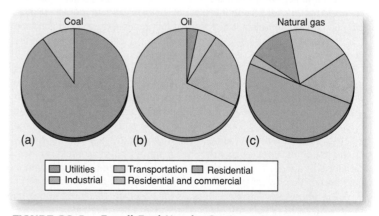

FIGURE 20.2 *Fossil Fuel Use by Sector* **(a)** *Coal use by sectors of the U.S. economy in 2003. Most coal is used to generate electricity.* **(b)** *Oil use by sectors of the U.S. economy in 2003. Most oil is used in the transportation sector.* **(c)** *Natural gas use by sectors of the U.S. economy in 2003. The residential sector is the largest user of natural gas.* (Source: Data from the Energy Information Administration, U.S. Environmental Protection Agency.)

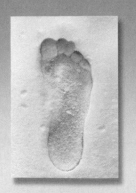

Your ECOLOGICAL *footprint*

How Much Energy Do You Use?

In 2003 the average U.S. citizen used about 25 barrels (146.2 million Btu) of oil, 76,000 cubic feet (78.1 million Btu) of natural gas, 3.4 metric tons (75.6 million Btu) of coal, and 3,500 kilowatt-hours (12 million Btu) of electricity generated from nuclear or hydroelectric power plants. These quantities are much greater than the average citizen of nearly every other nation. For example, the average German citizen used about 12 barrels (68 million Btu) of oil, 39,000 cubic feet (41.3 million Btu) of natural gas, 3.0 metric tons (67.2 million Btu) of coal, and 2,200 kilowatt-hours (7.4 million Btu) of electricity generated from nuclear or hydroelectric power plants, whereas the average Japanese citizen used about 16 barrels (92.5 million Btu) of oil, 24,000 cubic feet (24.6 million Btu) of natural gas, 1.3 metric tons (22 million Btu) of coal, and 2,700 kilowatt-hours (9.1 million Btu) of electricity generated from nuclear or hydroelectric power plants.

One reason for high rates of energy use in the United States is its high income. As indicated in Figure 10.12 on page 211, average income of U.S. citizens is among the highest. But income alone cannot explain the high rates of energy use. In 2003 average per capita income in Germany and Japan was only about 25 percent less than in the United States, but German per capita energy consumption was about 41 percent lower than in the United States, and Japanese per capita energy consumption was about 51 percent lower than in the United States.

Another cause for the high rates of U.S. energy use are relatively low rates of energy efficiency. As described in the start of this chapter, SUVs are the automotive status symbol of the first decade of the twenty-first century. But these automobiles average only 15 miles per gallon. As a result, the overall energy efficiency of the U.S. automobile fleet is low compared to that of Germany or Japan. Similarly, the efficiency of insulation, lighting fixtures, and heating and cooling equipment of U.S. buildings is low, so the United States uses more energy per square meter of floor space than Germany or Japan.

Here we provide information that will let you calculate how much energy you use. One aspect of this calculation is simple: Using the information you collected for Chapter 13's Your Ecological Footprint, add up your use of coal, oil, natural gas, and electricity. Following this procedure, you will conclude that you use little or no coal—much less than the 3.4 metric tons that are used by the average U.S. citizen. There seems to be a mistake.

There is no mistake. The fact that you seem to use no coal is caused by the difference between primary energy consumption and final energy consumption. Primary energy consumption includes energy that enters the economic system from the environment. In most developed nations, primary energy consumption includes coal, oil, natural gas, and electricity generated from nuclear and hydroelectric plants. Final energy consumption includes energy that is consumed but not resold. In most developed nations, final energy consumption includes coal, oil, natural gas, and electricity, regardless of how the electricity is generated.

The important difference between these two measures concerns the energy used to generate electricity. Power companies generate electricity by burning coal, oil, and natural gas, so these fuels are included in primary energy consumption. Electricity that is generated and resold to consumers is included in final energy consumption. Because generating electricity from fossil fuels is about 40 percent efficient, final energy consumption is always less than primary energy consumption. Separating primary energy consumption from final energy consumption prevents analysts from "double-counting" energy use. When totaling your energy

use, you should not count both the electricity you consume and the coal, oil, and natural gas that were used to generate it.

Calculating Your Footprint

To understand how the difference between primary and final energy consumption affects your measure of energy use, add up your direct uses of coal, oil, natural gas, and electricity. Table 1 allows you to translate physical measures of energy use, such as barrels, cubic feet, and kilowatt-hours, to energy units, such as Btus. For example, suppose you purchased 12 gallons of gasoline, 27 gallons of heating oil, 7,400 cubic feet of natural gas, and 270 kilowatt-hours of electricity. Your total energy use would be as follows:

Motor gasoline	1.5 million Btu = 12 gal × 125,000 Btu/gal
Heating oil	3.75 million Btu = 27 gal × 139,000 Btu/gal
Natural gas	7.5 million Btu = 7,400 ft³ × 1,030 Btu/ft³
Electricity	0.9 million Btu = 270 kilowatt-hours × 3,412 Btu/kilowatt-hour
Monthly total	13.65 million Btu

To calculate your primary energy consumption, use the information in Table 2 to translate your use of electricity back to the quantities of coal, oil, or natural gas that are used to generate it. Add these indirect uses to your direct uses of coal, oil, and natural gas to calculate your primary energy consumption. Continuing with the previous example, suppose you go to school in the Boston area. You would calculate the quantity of fossil fuels used to generate the 270 kilowatt-hours of electricity as follows:

Coal	1.6 million Btu = 270 kilowatt-hours × 5,990 Btu/kilowatt-hour
Oil	0.8 million Btu = 270 kilowatt-hours × 2,803 Btu/kilowatt-hour
Natural gas	0.05 million Btu = 270 kilowatt-hours × 184 Btu/kilowatt-hour
Monthly total	2.45 million Btu

TABLE 1   Energy Content of Fuels

Fuel	Units	Multiply to Convert to Thousand Btu
Coal	Pounds	11,474 Btu per pound
Motor gasoline	Gallons	125,000 Btu per gallon
Heating oil	Gallons	139,000 Btu per gallon
Natural gas	Cubic feet	1,030 Btu per cubic foot
Electricity	Kilowatt-hours	3,412 Btu per kilowatt-hour

But not all of your electricity is generated from fossil fuels, so you also have to determine the amount of electricity generated from fuels other than coal, oil, and natural gas. In the United States most of this other electricity is generated by nuclear or hydroelectric plants. Use the fractions in Table 3 on page 424 to calculate the quantity of electricity that you use generated by nuclear and hydroelectric sources. Continuing with our example, you would calculate these quantities as follows:

Electricity from nuclear power 0.2 million Btu = 270 kilowatt-hours × 3,412 Btu/kilowatt hour × 0.252 Nuclear power

Electricity from hydroelectricity 0.05 million Btu = 270 kilowatt-hours × 3,412 Btu/kilowatt-hour × 0.054 Hydroelectric power

To calculate your direct use of primary energy, you need to sum the quantity of coal, oil, and natural gas you use directly; the quantity of coal, oil, and natural gas used to generate the electricity you consume; and the electricity generated from nuclear and hydroelectric power plants. Finishing our example, this sum would be 15.45 million Btu (1.5 million Btu + 3.75 million Btu + 7.5 million Btu + 1.6 million Btu + 0.8 million Btu + 0.05 million Btu + 0.2 million Btu + 0.05 million Btu). Notice that this total is greater than the final consumption of energy, 13.65 million Btu.

Finally, your energy intake includes the energy used to produce the nonenergy goods and services you purchase. To estimate this quantity, refer to your purchases of goods and services that you used in Chapter 13's Your Ecological Footprint to calculate the quantity of carbon dioxide that you emit. Multiply the dollar values of your purchases by the energy intensities for those categories.

To illustrate, suppose you purchased a $30 pair of pants. Pants fall within the nondurable manufactures category. According to Table 4 on page 424, each dollar spent on goods from this category represents 4,970 Btu of coal, 6,110 Btu of oil, 2,520 Btu of natural gas, and 5,650 Btu of electricity. Based on these values, a $30 pair of pants requires 149,100 Btu of coal, 183,300 Btu of oil, 75,600 Btu of natural gas, and 169,500 Btu of electricity.

Interpreting Your Footprint

Now calculate your primary energy consumption over the year. To do so, sum your direct and indirect uses of primary energy. If you have calculated monthly or weekly values, be sure to increase them so they represent annual values.

Compare your total to the values in the opening paragraph. Is your total greater than or less than the national average? What is the reason for this difference? Try to answer using the relevant portion of the IPAT equation. In other words, is your income higher than or less than the national average? Similarly, do you tend to use technologies that have a high or low energy efficiency?

Finally, let's compare your energy use to that elsewhere. The opening paragraph has information about energy use in Japan and Germany. How could you change your energy use to match these nations?

Now let's compare your energy use to that in a developing nation, China. In 2003 the average Chinese citizen used about 1.5 barrels (9.0 million Btu) of oil, 933 cubic feet (0.9 million Btu) of natural gas, 1.0 metric tons (23.7 million Btu) of coal, and 244 kilowatt-hours (0.8 million Btu) of electricity generated from nuclear or hydroelectric power plants. Could you reduce your use of energy to this level? What if you consumed no energy directly and purchased only goods and services with a low energy intensity?

STUDENT LEARNING OUTCOME

- Students will be able to describe the differences between primary and final energy consumption.

TABLE 2 Btu's of Fuel Used to Generate a Kilowatt-hour of Electricity

Regions (States)	Coal	Oil	Natural Gas
New England (Connecticut, Maine, Massachusetts, New Hampshire, Rhode Island, Vermont)	5,990	2,803	184
Middle Atlantic (New Jersey, New York, Pennsylvania)	2,741	1,306	935
East North Central (Illinois, Indiana, Michigan, Ohio, Wisconsin)	8,631	37	77
West North Central (Iowa, Kansas, Minnesota, Missouri, Nebraska, North Dakota, South Dakota)	8,471	47	170
South Atlantic (Delaware, District of Columbia, Florida, Georgia, Maryland, North Carolina, South Carolina, Virginia, West Virginia)	5,239	557	953
East South Central (Alabama, Kentucky, Mississippi, Tennessee)	6,730	99	419
West South Central (Arkansas, Louisiana, Oklahoma, Texas)	5,895	98	2,520
Mountain (Arizona, Colorado, Idaho, Montana, Nevada, New Mexico, Utah, Wyoming)	7,509	11	658
Pacific Contiguous (California, Oregon, Washington)	177	5	702
Pacific Noncontiguous (Alaska, Hawaii)	358	6,111	2,747

Source: Calculations based on data from Electric Power Monthly, Energy Information Administration, U.S. Department of Energy.

Your Ecological Footprint (*continued*)

TABLE 3   Regional Share of Electricity Generated by Nuclear and Hydroelectric Plants

Regions (States)	Nuclear	Hydroelectric
New England (Connecticut, Maine, Massachusetts, New Hampshire, Rhode Island, Vermont)	.252	.054
Middle Atlantic (New Jersey, New York, Pennsylvania)	.349	.062
East North Central (Illinois, Indiana, Michigan, Ohio, Wisconsin)	.226	.007
West North Central (Iowa, Kansas, Minnesota, Missouri, Nebraska, North Dakota, South Dakota)	.0	.027
South Atlantic (Delaware, District of Columbia, Florida, Georgia, Maryland, North Carolina, South Carolina, Virginia, West Virginia)	.238	.020
East South Central (Alabama, Kentucky, Mississippi, Tennessee)	.184	.060
West South Central (Arkansas, Louisiana, Oklahoma, Texas)	.111	.014
Mountain (Arizona, Colorado, Idaho, Montana, Nevada, New Mexico, Utah, Wyoming)	.075	.084
Pacific Contiguous (California, Oregon, Washington)	.127	.409
Pacific Noncontiguous (Alaska, Hawaii)	.0	.086

Source: Calculations based on data from Electric Power Monthly, Energy Information Administration, U.S. Department of Energy.

TABLE 4   Energy Intensities of Nonenergy Goods and Services Measured in Thousand Btu per Dollar

	Coal	Oil	Natural Gas	Electricity
Biological Resources	2.24	6.77	1.40	1.83
Mineral Resources	16.63	8.47	7.77	10.50
Durable Manufacturers and Construction	3.65	3.61	1.68	2.85
Nondurable Manufacturers	4.97	6.11	2.52	5.65
Transportation Services	1.66	25.06	1.07	1.58
Other Services	2.05	3.10	1.35	1.68

Source: Data from R.W. England and S.D. Casler, "Fossil Fuel and Sustainable Development: Evidence from U.S. Input–Output Data, 1972–1985," *Advances in the Economics of Energy and Resources* 9: 21–44. Adjusted for the effects of inflation and gains in energy efficiency.

THE FORMATION OF FOSSIL FUELS

Fossil fuels are grouped together because they are derived from the net primary production of autotrophs that lived millions of years ago. This biological material did not follow the usual path in the global carbon cycle in which organic material decays back to carbon dioxide and flows back to the atmosphere (Figure 6.6 on page 105). Instead the organic material decayed partially and remained buried.

The geological processes that short-circuited the global carbon cycle had to satisfy several conditions. First large amounts of organic matter had to be available in a relatively small area. This implies that the organic precursors of fossil fuels originated in highly productive ecosystems. Next the organic matter had to be cut off from oxygen so that it would not decay aerobically according to Equation 5.3 on page 89. This partially decayed organic matter then had to be rearranged in a way that concentrated its energy content.

Finally, all these processes had to occur close to Earth's surface so that people could recover the energy in an economically feasible manner.

All of these conditions occurred together only rarely. Indeed, fossil fuel can be thought of as a geological accident. In sum, the fossil fuel resource base represents a bit more than a century of net primary production by terrestrial ecosystems (at current rates). (You can approximate this value by looking at the flows and storages in Figure 6.6 on page 105.) As a result, the supply of fossil fuels is limited.

Coal

Measuring the supply of fossil fuels relative to the net primary production of terrestrial ecosystems does not imply that all fossil fuels are derived from plants. But this comparison is relevant for coal. Coal is derived from terrestrial plants that lived millions of years ago. Originally people thought that coal beds came from ancient tropical ecosystems (see the discussion of continental drift in Chapter 4 on page 69). Although the theory of continental drift has been validated, this interpretation of coal beds is incorrect—most of the world's coal is derived from plants that lived in coastal and inland swamps that were located at latitudes greater than 40° north and south of the equator. For example, the coals in the Powder River basin in Montana and Wyoming were formed when this portion of North America was located at about 58°N (continental drift has moved it to about 43°N). As described in Chapter 7, swamps are highly productive ecosystems. This satisfies the first requirement for the formation of fossil fuels—a highly concentrated source of organic material.

The second requirement—that the organic material be cut off from oxygen—was satisfied by the chemical composition of the organic material and changes in sea level. Much of the organic material in swamps is derived from trees and other woody plants, which are made up mainly of cellulose and lignin. These materials are more resistant to decay than most other organic materials, and this resistance was enhanced by burial. As glaciers melted, sea levels rose and buried the organic material. When sea levels dropped, organic material would again accumulate. The melting and reforming of glaciers constituted a cycle that laid several layers of organic material atop each another.

As the layers of plant material accumulated, the pressure and temperature in the lower layers increased. The high temperature and pressure expelled water from the lower layers. Eliminating this water increased the energy content of coals. Equally important, the pressure and temperature changed the chemical composition of the coal. The plant material from which coal was formed had a carbon to hydrogen ratio of about 1:2. That is, there was about one atom of carbon for every two atoms of hydrogen. This ratio increased as the organic material was subjected to increasing temperature and pressure. Together these changes satisfied the third requirement—that the organic material be rearranged in a way that enhanced its energy content.

The increase in the carbon to hydrogen ratio improves the heat content of the coal, and these differences (and other characteristics such as sulfur content) are used to rank coals. The highest rank of coal includes anthracite, which is about 90 percent carbon (carbon to hydrogen = 9:1); in the United States anthracite contains 20–25.4 million Btu per metric ton (22–28 million Btu/ton). This rank is followed by bituminous coals, with 19–27.2 million Btu/metric ton (21–30 million Btu/ton), subbituminous coals at 21.8 million Btu/metric ton (24 million Btu per ton), and lignite at 11.8 million Btu/metric ton (13 million Btu/ton). Because lignite is relatively immature, you can sometimes see the woody plant material from which it is derived.

Finally, most of the processes that formed coal occurred relatively near the surface. As a result, much of the world's coal can be recovered economically.

Crude Oil and Natural Gas

Crude oil and natural gas are derived from aquatic organisms, mainly phytoplankton and zooplankton that lived millions of years ago in shallow marine environments such as estuaries. Like swamps, estuaries and shallow coastal seas have high rates of net primary production and so satisfy the first requirement for the formation of fossil fuels.

The second requirement, that the organic material be cut off from oxygen, happened as the organic material was buried by particles in a process known as **sedimentation.** Sedimentation is relatively rapid in estuaries. When a river reaches the sea, its speed decreases. As water speed decreases, the river no longer can carry its sediment load. The sediment settles to the bottom and forms layers that isolate the organic material from oxygen. In other areas known as nutrient traps, surface waters leave a shallow ocean basin and deep waters flow into the shallow ocean basin. As these deeper waters flow in, they often drop large amounts of sediment.

Once the organic material is cut off from oxygen, anaerobic bacteria use it as a source of energy. As they do so, they form an organic material known as **kerogen.** Kerogen is an insoluble organic material that is the main precursor of crude oil and natural gas. The formation of kerogen occurs over geological time as sedimentation continues. Burial is enhanced by the slow convective motion of the crust (see the discussion of plate tectonics in Chapter 4), which pulls the kerogen deeper.

Increasing depth subjects the kerogen to increasing temperature and pressure, which determine the type of fossil fuel that is formed. At relatively shallow depths, temperatures and pressures leave kerogen relatively unaffected. Under these conditions kerogen becomes trapped in materials that eventually become part of sedimentary rocks known as **oil shale.** Most petroleum occurs in the form of oil shale and tar sands.

The conversion of kerogen to crude oil or natural gas depends on the depth at which the kerogen and its derivatives are formed and trapped. Kerogen is converted mostly to crude oil at temperatures between 60°C and 150°C (140°F–302°F), which correspond to a depth of 1,500–4,500 meters (4,920–14,760 ft). This depth is known as the oil window because within this range most kerogen is converted to crude oil. **Crude oil** is a molecule made up of relatively long chains of carbon containing up to about twenty atoms. The number of carbon atoms in these chains is reduced by the high temperature and pressure that occur at greater depths. Here kerogen is split into molecules with fewer carbons than crude oil. The simplest of these carbon molecules has one carbon atom—methane (CH_4). Methane is the predominant component of natural gas. Most natural gas is recovered from fields that are within 2,500 meters (8,200 ft) of the surface.

Shortening the carbon chain reduces the viscosity of the kerogen, which allows the crude oil and natural gas to flow toward the surface. To reach their present position, crude oil and natural gas migrate toward the surface through porous *carrier beds.* If nothing lies in their path, oil and gas seep to the surface. The La Brea tar pits in downtown Los Angeles are just this type of surface seep. Occasionally an impermeable layer of rock traps the crude oil and natural gas. With its rise to the surface stopped, crude oil or natural gas accumulates below the trap and fills the spaces among the grains of the sedimentary reservoir rocks, just as water fills the spaces among soil particles (see Figure 18.2 on page 374). These accumulations are the reservoirs or fields from which society recovers crude oil and natural gas (Figure 20.3).

The sizes of oil fields vary greatly. The largest oil fields contain well over 50 billion barrels. Smaller oil fields often contain fewer than a million barrels. In general, the number of oil fields increases as their size decreases. There are very few giant oil fields. The number of midsized and smaller fields increases dramatically.

But don't be fooled by the large number of medium-sized and small fields. These fields contain only a small fraction of the total oil supply. Most oil is contained in a few giant oil fields—its distribution is highly skewed toward the largest fields. The ten largest oil fields in each of the world's twenty-five largest geological oil provinces contain 20 to 97 percent of the oil in that province. (A province is a large area whose rock structure, rock type, and/or rock age show common characteristics, especially origin in the same geological period.) In the United States the largest 81 fields contain about 40 percent of the oil that is known to exist in over 14,000 oil fields.

The concentration of oil in a few giant fields is reflected at the global level. There are 600 geological provinces with the potential to contain oil. Of these 600 provinces, 420 contain oil. Of these, only seven contain more than 25 billion barrels. These seven provinces account for two-thirds of the world's known oil supply. About half of the world's supply exists in the Arabian–Iranian province, which is the largest single province.

FIGURE 20.3   *Oil Field Close-Up Crude oil is stored among the particles of sedimentary rock. Notice that this figure looks like Figure 18.2 on page 374 which shows the distribution of water among soil particles.*

DISCOVERY, EXTRACTION, AND PROCESSING

Coal

In contrast to the highly skewed distribution of oil, commercially valuable coal deposits are found in most nations. Because coal is a solid, considerable effort is used to dig it from the ground and bring it to the surface. The technique used depends on the depth of the coal deposit. For the deepest deposits, some of which lie at depths comparable to oil fields, coal is extracted using **underground mining techniques** (Figure 20.4). To drill these mines, operators sink two or more shafts down to the coal deposit, which is known as a **seam.** One shaft is used to move miners, materials, and coal between the seam and the surface, while the other shafts are used to ventilate the mine. Tunnels are drilled between shafts to increase access to the seam. Underground mining is among the most hazardous occupations in the United States and throughout the world.

Coal is removed from the seams using a variety of techniques. **Room and pillar** techniques leave pillars of coal to prevent the ceiling from collapsing; therefore this technique leaves a significant fraction of the coal in place. The fraction of the coal removed, which is known as **extraction efficiency,** is increased by **continuous mining techniques.** Continuous mining machines allow the roof to cave in after the machine has removed all of the coal and "backs away" from the seam. Much of the coal dug from the eastern portion of the United States is recovered using underground mining techniques.

In other parts of the eastern United States, especially West Virginia, the seams are close enough to the surface to recover via **mountaintop removal.** Following this practice, the vegetation and soil are removed from the mountaintop using machines known as draglines. Next miners set off multiple explosions, which separate the coal from the rocks. Miners haul off the waste rock, which is dumped into a nearby valley.

In the western portion of the United States, where many seams are relatively shallow (within 60 meters—200 ft—of the surface) much coal is recovered using **surface mining techniques** (Figure 20.4). These techniques, also known as strip mining, remove the soil and rock above the coal seam, which is termed **overburden.** After this is removed, the seam is exposed and extracted by huge bulldozers and loaded into giant dump trucks. From here the coal is loaded onto trains or mixed with water and moved via pipeline to where it is needed. Both of these transport mechanisms are very expensive. To reduce transportation costs, coal is sometimes burned on-site at mine-mouth power plants, and the electricity is transported through the electricity grid. This strategy is cost-effective because most coal is used to generate electricity and little electricity is lost in transit relative to the great amounts of energy required to move coal. On the other hand, mine-

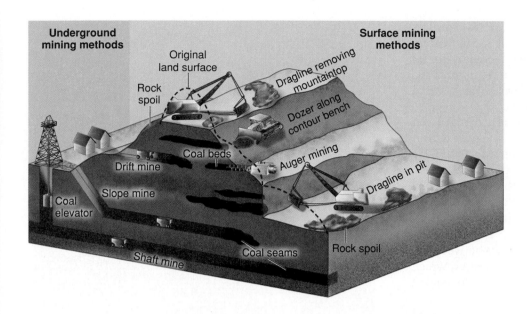

FIGURE 20.4   *Coal Mining Techniques*
Underground mining methods are used to extract coal when the seam is far from the surface. Where the coal is closer to the surface, surface mining techniques, such as mountaintop removal and strip mining, are **used.** *(Source: Redrawn from Kentucky Geological Survey at the University of Kentucky.)*

mouth plants can generate lots of pollution because the coal is not cleaned before being burned. This cleaning, which is known as **coal benefaction,** has the coal crushed, screened, and suspended in a liquid, where the solid impurities settle out. This process can remove up to 30 percent of the coal's sulfur, which allows coal to be burned in a way that complies with the Clean Air Act.

Crude Oil and Natural Gas

To find oil or natural gas, operators must drill wells. Rather than drilling wells at random, geologists use the organic theory of formation and plate tectonics to narrow the area where crude oil and natural gas might be found. Only sedimentary rocks can contain oil and gas fields. In addition, these fields must occur in association with some trapping structure, such as a salt dome or geological fault. Finally, sedimentary formations with the greatest potential for oil should overlay areas that have, or once had, high rates of net primary production. The large oil fields off the coast of Texas and Louisiana lie under highly productive estuaries, and the giant oil fields of the Middle East once were part of a highly productive shallow inland sea.

Given these criteria, oil firms have developed technology to identify promising locations. The earliest wells were drilled near surface seeps—areas where there is no geological structure to trap the migrating oil. As the seep sites were exhausted, firms looked for subsurface structures that would trap oil. These maps were produced using seismic equipment that shoots sound waves below the surface and listens for echoes. By measuring the speed and strength of these echoes, petroleum geologists can map the subsurface geology over wide areas. Recently these maps have been improved by three-dimensional seismic technology, which allows geologists to listen for echoes using a rectangular

grid of detectors rather than a single line of detectors. This technology improves the resolution of underground maps so that geologists now can identify relatively small features that are likely to hold crude oil or natural gas. In addition, these advances allow geologists to follow echoes from different types of waves. This is important because some types of waves go through liquids while others do not.

Once discovered, oil must be brought to the surface. An oil field is not a subterranean pool of free-flowing oil. Rather, it is an accumulation of oil, natural gas, and water among the grains of sedimentary rock. In the simplest terms, an oil field can be likened to an oil-soaked brick.

Within the sedimentary rock, the oil, gas, and water often exist in three layers (Figure 20.5). Natural gas is the least

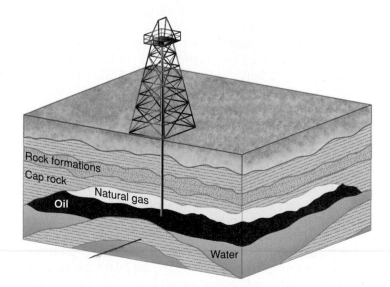

FIGURE 20.5   *Oil Reservoirs*   *Many oil fields consist of three layers: natural gas on top, a layer of crude oil, and a bottom layer of* water. *(Source: Redrawn from R.F. Flint and B.J. Skinner, Physical Geology.)*

dense. It rises toward the top of the reservoir and presses against the structural trap. Oil forms a middle layer, floating on a dense layer of water. At the micro level there is a similar separation. Within the oil layer, oil often exists as droplets suspended within water that coats the pores' walls. Natural gas often is dissolved within the oil droplets.

The amount of oil that can be recovered depends on the field's density, permeability, and porosity. Density measures the amount of mass per unit of volume. For oil fields density often describes the mass of rocks per unit volume of the field. High density indicates that there is relatively little space between particles, so there is little room for crude oil among the particles. As a result, the quantity of oil in a field has a negative relationship with the density of the sedimentary rock. **Porosity** measures the space between the particles that store the oil and provide pores through which oil can flow. These spaces generally account for 3–40 percent of total volume.

The rate at which oil flows through these spaces toward the well is influenced by permeability. For oil geologists, permeability depends on the viscosity of the liquids present, the size and shape of the formation, and the internal pressure. In general, permeability is negatively related to viscosity—molasses is more viscous than water and so pours slowly relative to water. Permeability also increases with the size of pores and with pressure. Think of how quickly gravity water drains from soil relative to capillary water, which can remain in the soil for weeks. High permeability is desirable because it increases the rate at which oil flows to the surface and increases the fraction of oil in the field that can be recovered.

Energy is required to move oil through the spaces of the sedimentary rock toward the well and up to the surface. When a well is first drilled into an oil or gas field, the pressure within the field is much greater than the atmosphere. This pressure gradient pushes the oil, gas, and water toward the well and up to the surface. Additional quantities of oil are pushed to the surface by the natural gas that bubbles out of oil in the same way that carbonation pushes soda out of a bottle when the bottle is first opened. Oil that is pushed to the surface by the pressure gradient in the field is termed **primary recovery.** Primary recovery is the least expensive way to recover oil and natural gas because society does not have to use additional energy to move oil and gas to the surface. In the United States, primary recovery typically recovers about 10 percent of the oil.

As primary recovery moves oil and natural gas to the surface, the pressure gradient weakens, which reduces the flow of oil or natural gas to the surface. When the pressure gradient no longer pushes economically significant quantities of oil to the surface, operators can increase production by repressurizing the field. To do so, they drill an injection well near the original producing well and inject water into the field (Figure 20.6). Water pushes additional quantities of oil toward the producing well and up to the surface.

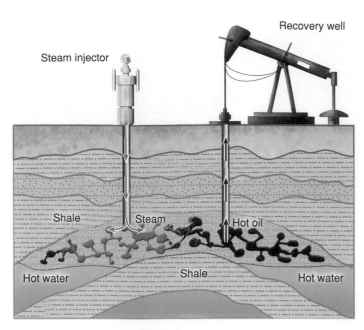

FIGURE 20.6   *Secondary Oil Recovery After the initial pressure gradient weakens, additional quantities of crude oil can be recovered by pumping steam, water, or carbon dioxide into the field through an injection well. These materials then push additional quantities of oil toward and up the recovery well. (Source: Redrawn figures from U.S. Geological Survey.)*

This effort is termed **secondary recovery.** In the United States such methods recover another 10–30 percent of the oil originally in place.

Eventually secondary recovery methods lose their effectiveness. The water cuts one or more paths toward the producing well, and these paths bypass the remaining oil. At this point **tertiary methods** can boost production. Tertiary methods include the injection of heat or materials that reduce the viscosity or surface tension of crude oil, which makes it flow more easily toward the surface. Tertiary methods can recover another 20 percent of the oil originally in place. On average primary, secondary, and tertiary methods extract 30–60 percent of the oil in a field, which is termed **extraction efficiency.** This implies that significant quantities of oil remain after a field is considered exhausted. The oil that remains is similar to hygroscopic water—the soil still has water, but it can't be removed by plants' roots.

Fields from which firms produce oil and natural gas are known as proved reserves. The American Petroleum Institute defines **proved reserves** as "volumes of crude oil which geological and engineering information indicate, beyond reasonable doubt, to be recoverable in the future from a reservoir under existing economic and operating conditions." Two phrases in this definition are critical. "Beyond a reasonable doubt" limits reserves to oil whose existence have been proved by a well that has been drilled into the field. "Under existing economic and operating conditions" limits proved reserves to the fraction of the oil that can be recovered with a profit using primary, secondary, and tertiary recovery. This definition was established by the Securities and Exchange Commission, which is the government body that oversees the stock market. Because the value of an oil company's stock is determined mainly by its proved reserves, a firm may be tempted to overstate its proved reserves. To prevent such fraud, the Securities and Exchange Commission established a conservative definition for oil companies that are traded on U.S. stock exchanges. To comply with this definition, British Petroleum had to reduce its estimate of proved reserves in 2004. For firms whose stock is not traded on U.S. exchanges, there is no agreed definition of proved reserves—so estimates for their proved reserves are an uncertain mix of geology, economics, and politics.

Efforts to increase proved reserves are termed exploration and development (E & D). Exploration programs can discover new quantities of oil. **Discoveries** is a broad term that includes two locations for potential finds. Firms can explore around the edge of known fields, which is termed developmental or delineation drilling. Successful developmental and delineation wells extend proved reserves by increasing the size of previously discovered fields. Or firms can drill for oil in formations not previously known to contain oil or natural gas. This type of drilling is termed **wildcatting.**

Alternatively, **revisions** are changes (either positive or negative) to proved reserves that are generated by new information other than an increase in acreage. For example, higher prices can increase recoverable oil supplies by making secondary or tertiary recovery techniques economically feasible. Similarly, technical advances such as horizontal drilling can increase extraction efficiency.

Crude oil cannot be used directly by society. Long carbon chains make crude oil viscous and limit its energy content. For example, crude oil cannot be ignited like motor gasoline and so cannot power automobiles.

To increase its usefulness, crude oil is transported by pipeline or tanker to an **oil refinery** (Figure 20.7). An oil refinery breaks and separates the long-chain carbon molecules into groups of shorter-chain molecules. These shorter-chain molecules are termed **refined petroleum products.**

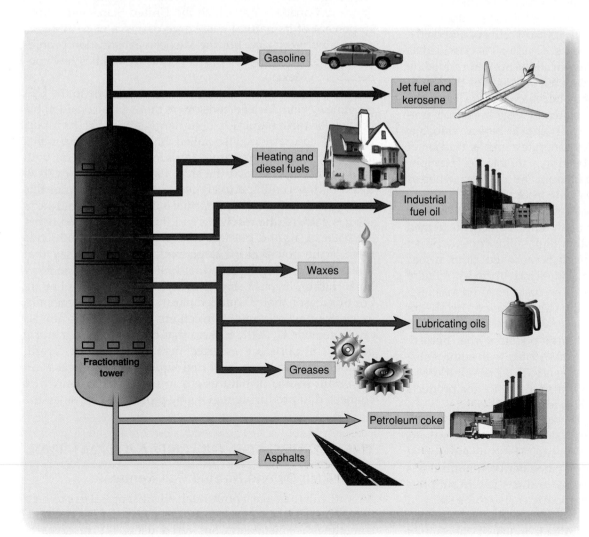

FIGURE 20.7 *Oil Refinery*
In an oil refinery, crude oil is heated, which breaks and rearranges the carbon chains in crude oil. The new molecules are refined petroleum products, which vary in the length of their carbon chains. Shorter-chain molecules, such as jet fuel and motor gasoline, are recovered near the top of the refinery, while longer-chain molecules, such as asphalt, are recovered near the bottom of the refinery.

Gasoline

Jet fuel and kerosene

Heating and diesel fuels

Industrial fuel oil

Waxes

Lubricating oils

Fractionating tower

Greases

Petroleum coke

Asphalts

The refinery works much like the condensation coil that you may use in your organic chemistry labs. Crude oil is heated, and the high temperatures cause the long-chain molecules to break into smaller-chain molecules. Once heated, the shorter-chain molecules evaporate and condense at different heights along the column, which depend on their length. Molecules with longer chains condense near the bottom. These bottom-of-the-barrel products include asphalt and residual fuel oil. Molecules with shorter chains condense near the top. These light products include jet fuel and motor gasoline. Motor gasoline consists of molecules with chains of eight to twelve carbon atoms. Molecules that have eight carbon atoms are known as octanes. You choose the octane content of your gasoline at the pump. The least expensive gasoline, regular, usually has an octane number of 87. The most expensive blend of gasoline, premium, has the highest octane number, usually 92 or 93. Octanes are ideal for a car's internal combustion engine and prevent knocking (poorly timed firing within cylinders). Most engines are designed to run on regular gasoline. Other than for high-performance cars, the expense of premium gasoline is not warranted.

The fractions of various refined petroleum products generated by a refinery depend on the type of refinery and on the type of crude oil used as feedstock. Crude oil is classified as either light or heavy. These classifications are determined by a crude oil's specific gravity, which is a proxy for the length of the carbon chains. Long-chain, heavy crudes are less desirable because they generate a relatively large fraction of bottom-of-the-barrel products, which tend to be less valuable. On the other hand, light crudes generate a greater fraction of light products, which tend to be more valuable. For example, Saudi light is a valuable crude oil from Saudi Arabia, whereas Maya is a less valuable heavy crude from Mexico.

The usefulness of crude oils also is determined by their sulfur content. Crude oils with a low sulfur content are termed sweet; sour crudes are oils with a high sulfur content. Sweet crudes generally are more valuable because the sulfur in sour crudes tends to form sulfuric acid, which corrodes refinery structures and increases maintenance costs. In addition, a high sulfur content tends to reduce the value of refined petroleum products. As described in Chapter 19, the sulfur in refined petroleum products is emitted as an air pollutant when burned. To reduce acid deposition that is associated with sulfur emissions, firms are required to burn fuels with a low sulfur count. This reduces the demand and lowers the price for products with a high sulfur content, such as residual fuel oil.

In contrast to crude oil, the quality of natural gas is similar, regardless of the deposit. Furthermore, natural gas can be used "as is" without refining. The consumption of natural gas is limited by the distance between natural gas fields and consumers. Because natural gas requires a pressurized container, it is usually too expensive to be shipped via railroad or truck. Instead most natural gas is transported to consumers via pipelines. Indeed, much of the United

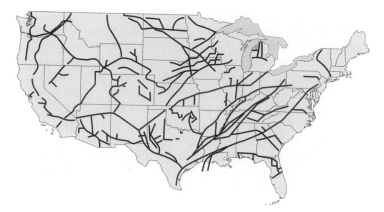

FIGURE 20.8   *U.S. Natural Gas Pipelines* Pipelines in the United States generally transport natural gas from producing areas, such as the U.S. Gulf Coast and the Texas–Oklahoma panhandle, to major consuming areas, such as the Northeast and upper Midwest.

States gets its natural gas via an extensive network of pipelines (Figure 20.8). The pipeline system connects the United States to Canada, from which the United States gets about 15 percent of its natural gas. Large deposits of natural gas are located in Mexico, but the Mexican constitution prohibits exports; therefore the U.S. pipeline system has few connections to Mexico.

New England has relatively few connections to the U.S. pipeline system. Instead consumers there get a considerable portion of their natural gas via large ships that transport liquefied natural gas (also called LNG). Such movements are costly because expensive facilities are needed on both ends of the trip. To load the ship, natural gas is converted to a liquid by cooling it to about –170°C (–274°F). After this cooling, it is pumped into special holding tanks that keep the gas cold. At the receiving end the gas is warmed using seawater or air and put into a local pipeline. At both ends there is the potential for large explosions—some say that if a tanker carrying liquefied natural gas exploded in New York Harbor, it could shatter every window in Manhattan. For this reason many liquefied natural gas facilities are built away from major population centers. Following the attacks on September 11, 2001, tankers that bring liquefied natural gas to U.S. ports are escorted by the U.S. Coast Guard. Despite these dangers, imported liquefied natural gas is an important source of natural gas in nations that cannot be connected to producing regions by pipeline, such as Japan.

THE FUTURE FOR OIL AND NATURAL GAS

How Much Oil and Natural Gas Remain?

It is difficult to forecast how much oil and natural gas remain to be found and how long supplies will last. This question sometimes is mistakenly answered using the **reserves to** production (R:P) ratio, which measures the quantity of oil in proved reserves relative to the current rate of oil production. For example, an R:P ratio of 10:1 implies that proved reserves can satisfy current rates of demand for ten years. In many nations this ratio varies between 10:1 and 25:1.

An R:P ratio of 10:1 does not imply that oil supplies will run out in ten years. At any time a firm can try to boost its proved reserves by increasing exploration and development. Even if supplies permitted, increasing the reserves to production ratio from 10:1 to 100:1 does not make economic sense. Looking for crude oil and natural gas costs money now, but a large reserve to production ratio implies that the oil or gas found now will not be produced for a long time. This lag means that it will take a long time for the firm to recoup its investment. To avoid costly delays, firms invest in exploration and development so that the rate at which they find oil is roughly equal to the rate at which they produce oil. Under these conditions, the reserve to production ratio remains fairly constant. For example, the reserves to production ratio in the United States has remained fairly constant over the last fifty years (Figure 20.9).

Firms can keep the reserves to production ratio constant only if they can find enough oil to replace the oil they produce. How much longer can firms do so? One way to estimate the rate at which oil or natural gas can be found is by measuring how much sedimentary rock remains to be explored and how much oil is found by wells drilled into that sedimentary rock. For nearly all of the 420 sedimentary provinces that contain oil, the amount of sedimentary rock is well documented. In the United States Alfred Zapp of the U.S. Geological Survey argued that there are about 5 million square miles of sedimentary formations. Exploring them fully requires about one well per two miles drilled to a depth of 20,000 feet. Given these numbers, about 5 billion feet of wells would be needed to explore all sedimentary rock formations in the lower 48 states.

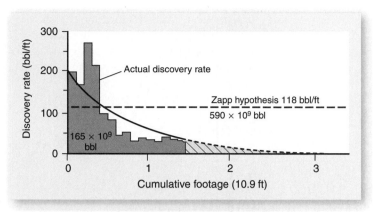

FIGURE 20.10   *Yield per Effort of Oil Exploration in the Lower 48 States* The dashed horizontal line represents the average amount of oil found per foot by the first billion feet drilled: 118 barrels per foot. The green bars represent the average yield per effort per 100 million feet of wells drilled. The smooth curve is the exponential decline curve estimated by M. King Hubbert. The dashed line represents the amount of oil to be found by future exploration efforts. *(Source: Redrawn from Hubbert.)*

But how much oil will these 5 billion feet find? This quantity can be estimated by looking at the success rate of drilling. The success rate can be measured by dividing the amount of oil found (in barrels) by the total feet of wells drilled to find it. This ratio measures the barrels of oil found per foot of well drilled and is termed **yield per effort.**

On average, the first billion feet drilled in the lower 48 states uncovered 118 barrels per foot. The Zapp hypothesis assumed that this average would remain constant over the next 4 billion feet. Based on this method, in the early 1960s the U.S. Geological Survey forecast that 590 billion barrels of oil would be found and produced in the lower 48 states. (At that time little drilling had been done in Alaska. Because only one giant field has been found in Alaska, most analyses still focus on the lower 48 states.)

Based on your understanding of the best first principle, you should recognize that the assumption made by the U.S. Geological Survey probably is incorrect. Closer examination of Figure 20.10 indicates that yield per effort is not constant over the first billion feet: Yield per effort declines as more wells are drilled. The explanation for this decline is simple—the first fields found tend to be the largest. A single well drilled into these giant fields can find great quantities of oil. After these giant fields are discovered, wells are drilled into ever smaller fields, and yield per effort declines.

M. King Hubbert, a petroleum geologist who worked at Shell Oil Company, was the first to recognize this pattern. To account for the best first principle, he fit the yield per effort data to an exponential decline curve, which is a curve with an ever-declining rate that eventually flattens out (Figure 20.10). Using this curve, Hubbert estimated that about 170 billion barrels of oil would be found in the

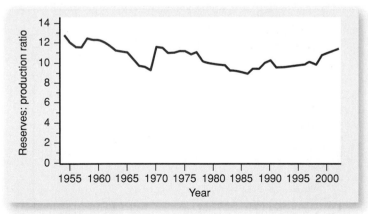

FIGURE 20.9   *U.S. Reserves to Production Ratio* The ratio of proved oil reserves to current rates of production in the United States has remained fairly constant over the last fifty years. Greatly increasing reserves relative to production does not make economic sense because a large ratio indicates that it will take a long time for firms to earn back the money they spend to find new reserves.

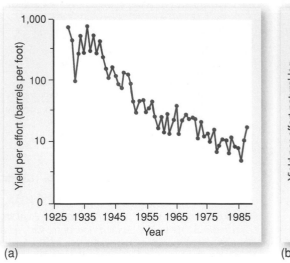

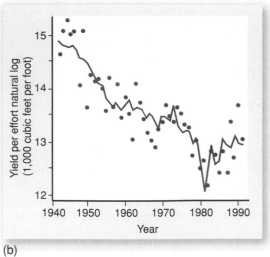

(a)   (b)

FIGURE 20.11   *Yield per Effort Explained*   (a) *Yield per effort for crude oil in the lower 48 states (circles) and the values indicated by a model (solid line) that uses cumulative drilling, annual rates of drilling, and real oil prices as explanatory variables. Yield per effort declines as cumulative drilling increases, annual rates of drilling increase, or real oil prices drop.* (Source: From C.J. Cleveland and R.K. Kaufmann, "Forecasting Ultimate Oil Recovery and Its Rate of Production: Incorporating Economic Forces into the Models of M. King Hubbert," The Energy Journal 12(2): 17–46.) **(b)** *Yield per effort for natural gas in the lower 48 states (circles) and the values indicated by a model that uses cumulative drilling, annual rates of drilling, and real oil prices as explanatory variables. Yield per effort declines as cumulative drilling increases, annual rates of drilling increase, or natural gas prices drop.* (Source: From C.J. Cleveland and R.K. Kaufmann, "Natural Gas in the US: How Far Can Technology Stretch the Resource Base?" The Energy Journal 18(2): 89–108.)

lower 48 states. With a similar technique he estimated that about 1,000 trillion cubic feet of natural gas would be found in the lower 48 states.

Recent analyses confirm the exponential decline in yield per effort and find that the decline also depends on the rates at which wells are drilled and the prices of oil and natural gas (Figure 20.11(a)–(b)). When firms drill many wells in a single year, some are drilled into formations that are less likely to contain oil or natural gas, and the low probability of success depresses yield per effort. When firms drill relatively few wells, most of these wells are drilled into formations that are likely to contain oil or natural gas, and the high probability of success increases yield per effort. Higher prices also tend to increase yield per effort because they allow firms to recover oil from fields that would be uneconomical at low prices. Remember, a field may contain oil or natural gas, but U.S. firms can add it to proved reserves only if they can make a profit in recovering it. If oil and gas are present but cannot be recovered with a profit, the well is considered a dry hole.

When Will We Run Out of Oil?

Although provocative, this question is misleading. You will not wake one day, turn on the television, and hear that the world has run out of oil. Oil production will not increase steadily and then one day drop to zero (Figure 20.12). As the world depletes its finite supply of oil, production will decline over time.

To get an idea of what this decline may look like, again we turn to M. King Hubbert, who generated a remarkably accurate forecast for one of the most important economic events in the twentieth century—the peak in U.S. oil production. In

the late 1950s Hubbert forecast that production in the lower 48 states would peak in 1970, and a total of 170 billion barrels of oil would be recovered. When issued, Hubbert's forecast was derided. At that time U.S. oil production had been increasing steadily since 1900. But production in the lower 48 states peaked in 1970 and declined thereafter. Hubbert is the only analyst who correctly anticipated the peak in U.S. production.

How could Hubbert forecast this peak? Hubbert assumed that cumulative rates of oil production would trace a logistic or S-shaped curve over time (Figure 20.13). He assumed

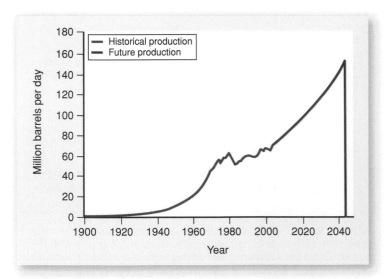

FIGURE 20.12   *When Will the World Run Out of Oil?* No one knows for sure, but it will **not** follow the path shown here in which production increases steadily and then suddenly collapses to zero.

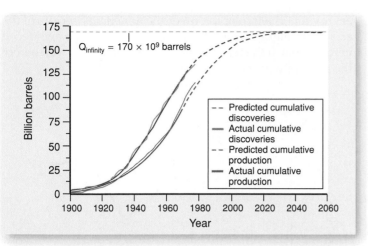

FIGURE 20.13 *Hubbert's Logistic Curve* M. King Hubbert fit data for cumulative oil discoveries and cumulative oil production to logistic curves. Notice that the logistic curve fits the observed values closely. By extending the logistic curves to the point at which they level off, Hubbert was able to estimate the ultimate quantity of crude oil that would be discovered and produced in the coterminous United States. He estimated these values to be 170 billion barrels for cumulative oil discoveries and production. *(Source: Redrawn from Hubbert.)*

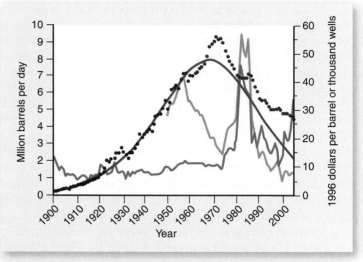

FIGURE 20.14 *Hubbert's Curve* Hubbert used the first derivative of the logistic curve in Figure 20.13 to model the annual rate of production. This first derivative traces a bell-shaped curve that has a peak in 1970. The black circles are observations that Hubbert had available when he made his forecast. Red circles are annual rates of production after Hubbert made his forecast. Notice that the peak given by the bell-shaped curve matches the peak indicated by the red circles. The green line gives the real price of crude oil corrected for inflation, and the orange line gives the number of wells drilled. Notice that production increased nearly tenfold between 1900 and 1970 even though prices were relatively stable. Between 1973 and 1985 oil prices increased greatly, as did the number of wells drilled; but production declined as forecasted by Hubbert's curve.

that cumulative discoveries and production would rise slowly at first when firms were not sure where to drill for oil and demand for oil was low. As exploration technologies improved and demand increased, cumulative discoveries and production would rise rapidly. As much of the oil was produced, exploration and development would become less successful, and cumulative discoveries would slow. Because of this prices would rise and demand would decline. Eventually cumulative discoveries and production would level off. This maximum is termed **Q infinity,** which is the total quantity of oil that will be discovered and produced.

Hubbert fit this logistic curve to the data available through the late 1950s. The first derivative of the logistic curve is a bell-shaped curve that traces the annual rate of production (Figure 20.14). This bell-shaped curve peaks in 1970, and this was the basis for Hubbert's forecast. The Q infinity is 172 billion barrels, which is very close to his estimate based on the analysis of yield per effort.

Because his forecast was so accurate, many believe that Hubbert was a genius. But subsequent analyses showed that Hubbert was both a genius and lucky. Specifically, economic factors, such as the price of oil, and institutional factors, such as the quantity of oil that the Texas Railroad Commission allowed producers in Texas to sell, affected the rate of U.S. oil production. These factors explain systematic errors in Hubbert's bell-shaped curve. The Texas Railroad Commission limited production in the 1950s and 1960s; therefore, Hubbert's curve overstates production during this period (Figure 20.14). After the production peak, oil prices rose, which allowed production to exceed Hubbert's bell-shaped curve during the late 1970s and early 1980s. Because of these effects, Hubbert's forecast for the peak in production would have been less accurate had

oil prices evolved along some other path, or had the Texas Railroad Commission controlled production using a criterion other than stabilizing oil prices.

We point to the importance of oil prices and the Texas Railroad Commission because economic and institutional factors probably are responsible for the perennial failure of Hubbert's method to generate an accurate forecast for the peak in global oil production. Using Hubbert's method, analysts have repeatedly forecast that the peak in world oil production is imminent, only to be proved wrong. These errors are probably caused by changes in oil prices and production decisions by OPEC.

Life after the Peak

Economic and institutional factors will affect the timing of the peak, but they do not change the basic insight that allowed Hubbert to forecast the general pattern of oil production. Oil production probably will follow a pattern that looks like a bell-shaped curve. The curve could be flatter or steeper—but at some point production will start to decline. And the timing of this decline is relatively insensitive to changes in the total quantity of oil that will be recovered. Even if oil supplies turn out to be larger than we currently think, this will delay the peak by perhaps a decade or two.

Many analysts believe that the peak in world oil production will occur within the next ten to twenty years. A decade or two seems far away, but the peak will be a watershed in the global economy. The world will go from an ever-increasing supply at relatively low prices to an ever-decreasing supply at ever-increasing prices. This change is illustrated by the seemingly contradictory relationship between real oil prices and oil production in the lower 48 states (Figure 20.14). Between 1900 and 1970 real oil prices were relatively stable, but production increased nearly tenfold. Between 1973 and 1985 production declined about 20 percent even though real oil prices and the number of wells drilled increased by a factor between two and three. In short, rising prices after the peak will not increase oil production.

Declining supplies of oil will have an important effect on the U.S. and world economy. Remember from Chapter 10 that energy use influences labor productivity, wages, and ultimately per capita GDP. Oil and natural gas are especially important for some sectors, such as agriculture. The ability of high-yield varieties to increase food supplies depends on fertilizer, which is made from natural gas via the Haber-Bosch process, and irrigation water, which is delivered by pumps powered by diesel fuel.

Furthermore, the peak in oil production portends a reduction in the supply of net energy. In the United States the energy return on investment for oil production parallels changes in production (Figure 20.15). The energy return on investment for oil production rises as production increases toward its peak. After the peak, the energy return on investment declines. This reduction means that more and more energy is required to produce each barrel of oil and each cubic foot of natural gas. As a result, less energy remains to power the nonenergy sectors of the economy. Net energy is the energy used to produce the nonenergy goods and services that we associate with a high material standard of living.

The decline in the energy return on investment implies that the end of oil may occur before the wells run dry. Before the last barrel of oil is pumped, the energy return on investment for oil will reach 1:1. This means that one Btu of energy is required to produce one Btu of oil. At this point oil no longer is a net source of energy to the economy, regardless of the quantity produced. Although it is possible to produce oil beyond the break-even point, the resource is exhausted, regardless of oil price. See *Policy in Action: The Energy Policy and Conservation Act of 1975.*

A decline in the energy return on investment for oil and natural gas as production peaks will cause difficulties for the transition to an alternative source of energy. As described in Chapters 21 and 22, potential replacements for oil and natural gas have a lower energy return on investment. This means that building the infrastructure for these replacements will require even more energy than that needed to produce oil and gas when the supply of oil and gas is declining. This would reduce greatly the amount of

energy that is left over to power the nonenergy sectors of the economy (Figure 20.16). Large reductions can be avoided if investments in the replacements for oil are made well before the peak in world oil production.

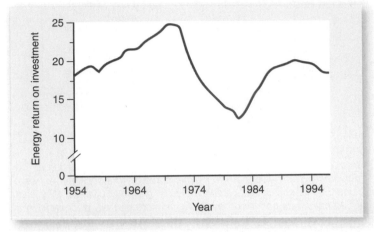

FIGURE 20.15   *Energy Return on Investment The energy return on investment for production of crude oil and natural gas in the United States. The energy return on investment increases through the early 1970s and declines relative to the peak thereafter.* (Source: From C.J. Cleveland, "Net Energy from Oil and Gas Extraction in the United States, 1954–1997," Energy: The International Journal, in press.)

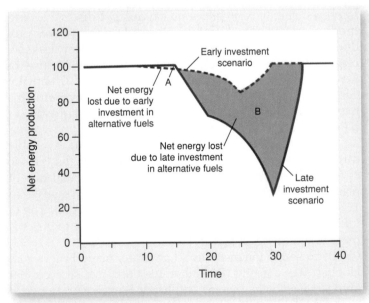

FIGURE 20.16   *Net Energy Supply The effect of a reduction in the energy return on investment on the supply of net energy. A model simulation indicates that this effect depends on the timing of efforts to produce alternative supplies of energy. If society starts to build alternatives before the energy return on investment for existing supplies declines in year 15 of the model simulation, there is a small reduction in net energy supply (area A) relative to the large reduction that occurs if society starts to build alternative energy sources after the energy return on investment of existing sources starts to decline (area B). Regardless of the strategy chosen, there is a reduction in net energy supplies in the short run. In the long run, net energy supplies return to their previous level once society increases its ability to produce energy.* (Source: Data from J. Gever et al,. Beyond Oil, Ballinger Press.)

POLICY IN ACTION

The Energy Policy and Conservation Act of 1975

The 1973–1974 oil shock presented a conundrum to U.S. politicians. On one hand, voters wanted protection from higher prices. Politicians also wanted to reduce U.S. imports, but they knew that U.S. output would not increase unless producers received higher prices. Policy makers hoped the Energy and Conservation Act of 1975 would achieve both, but the act accomplished neither goal.

Politicians hoped to protect consumers from higher oil prices because they cause economic hardship. Poor people and those on fixed incomes can be forced to choose among purchasing food, heating their homes, and being able to reach their jobs. For middle- and upper-income consumers, the extra expense of filling gas tanks and heating homes reduces the money available for nonenergy purchases. Fewer purchases of these other goods and services reduce consumers' income and reduce the income of businesses that produce them. To minimize these effects, the Energy Policy and Conservation Act of 1975 put an upper limit on oil prices, which is called a price ceiling.

But a price ceiling would interfere with the goal of reducing U.S. dependence on imported oil. One reason for the decline in U.S. production after 1970 was the depletion of low-cost sources of oil. To reverse this decline, U.S. producers needed to extract oil from fields that had higher production costs. The average cost of producing a barrel of oil in the United States increased by about 30 percent between 1965 and 1975. To cover these higher costs, U.S. producers needed higher oil prices.

The Energy Policy and Conservation Act of 1975 tried to balance these conflicting goals by creating a two-tiered pricing system. Under this act, the price for oil and natural gas produced in the United States was determined in part by the date at which it was discovered. Petroleum

discovered before 1978 was designated "old oil," and its price was controlled. Because "old oil" could be produced economically at the prices that prevailed before the 1973–1974 price increase, putting a ceiling on the price of "old oil" had a relatively small effect on production. But these controls damped the effect of higher prices on U.S. consumers because "old oil" accounted for much of the oil burned in the United States.

Petroleum discovered after 1978 (along with some wells that produce small quantities of oil and gas, which are termed stripper wells) was designated "new oil." The price of "new oil" was allowed to rise to levels charged in the international market. Higher prices induced domestic producers to increase their efforts to find "new oil" and produce "new oil" from fields previously considered uneconomical. Policy makers hoped these two changes would increase proved reserves and ultimately production.

The two-tiered pricing system was logically appealing but difficult to implement. There was no clear-cut difference between "old oil" and "new oil." For example, when developmental drilling after 1978 identified additional oil in a field discovered by a wildcat well drilled prior to 1978, were those additional supplies "new oil" or "old oil?" The big price differences gave oil producers a strong incentive to label "old oil" as "new oil."

But even when producers acted honestly and rationally, the Energy and Conservation Act of 1975 led to irrational outcomes. Martha Gilliland, at the University of Oklahoma, uncovered one of the more interesting irrationalities by examining the energy return on investment for natural gas production from a select group of fields. Some of the fields discovered after the enactment of the Energy and Conservation

Act of 1975 would have been uneconomical if prices were controlled at the level of old oil, but they were economical at the higher price of new oil. This was one of the legislation's goals.

But the legislators did not anticipate how the price difference between "old oil" and "new oil" made previously uneconomical fields economical. Gilliland identified several fields in which oil companies used large quantities of "old gas," which was priced at $1.40 per thousand cubic feet, to produce smaller quantities of new gas, which was priced at $3.00 per thousand cubic feet. Although the companies made an economic profit, they suffered a net energy loss—the amount of "old gas" used was greater than the amount of new gas produced. For these fields, the two-tiered pricing system did not increase usable supplies of energy to U.S. consumers. Because of these and other problems, President Reagan phased out the price difference between new oil and old oil, allowing all oil sold in the United States to reach levels set by the world market.

ADDITIONAL READING

Gilliland, M.W. "Energy Analysis and Public Policy." *Science* 189 (1975): 1051–1056.

U.S. General Accounting Office. *Net Energy Analysis Sees Little Progress and Many Problems.* GA1.13:EMD-77-57, 1977.

STUDENT LEARNING OUTCOME

- Students will be able to describe the advantages and disadvantages of putting a ceiling on oil prices.

WHAT DO YOU NEED TO KNOW ABOUT THE WORLD OIL MARKET?

Should the United States Reduce Its Dependence on Imported Oil?

The United States has been a net importer of crude oil since 1947 (Figure 20.17). Every time oil prices rise significantly, politicians seek ways to reduce U.S. dependence on imported oil. They argue against sending money out of the country (see Figure 1 in Chapter 17's Case Study on page 359). It would be better for national security and the economy if the United States could reduce or eliminate its imports. Failing that, the United States should reduce imports from politically unstable

nations or nations that have a different political agenda, such as Iran, Saudi Arabia, or Venezuela.

Although energy independence is politically popular, it probably would not enhance U.S. economic or military security. Even if the United States did not import oil from unfriendly or unstable governments, it would still be vulnerable because the world oil market can be envisioned as one big pool of oil. Any disruption to that pool ripples throughout the market, affecting everyone. Imagine that the United States imported all its oil from Canada. Now suppose Saudi Arabia stopped all oil exports. Saudi Arabia's customers would look for new suppliers, including Canada. This would increase oil prices for everyone, including the United States. Canada is not going to sell oil to the

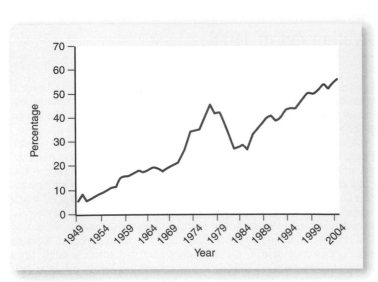

FIGURE 20.17   *U.S. Oil Imports* Net imports of oil by the United States. The United States became a net importer of oil in 1947, and that percentage has increased steadily except for a short period in the late 1970s and early 1980s. (Source: Data from U.S. Department of Energy.)

United States for a lower price than it could get from Saudi Arabia's former customers. Similarly, suppose the United States imported some of its oil from Iran, and Iran suddenly decided to stop selling oil to the United States. Iran would sell its oil to another nation, and that nation's previous supplier would now sell oil to the United States. Indeed, any effort to change the pattern of U.S. imports probably would increase prices to U.S. consumers because the market matches supply and demand in a way that minimizes the costs of transporting oil. This is why some of the oil that is produced in Alaska is exported overseas rather than being shipped to refineries on the U.S. West Coast.

Nor would it be economically attractive for the United States to be self-sufficient in oil. The United States imports oil because it is less expensive to get oil from overseas producers than it would be to replace that oil with domestic production. This cost saving, known as comparative advantage, is the basis for global trade. (See Chapter 8's *Your Ecological Footprint*.) Withdrawing from trade with the goal of economic self-sufficiency has created some of the world's poorest nations, such as North Korea and Albania.

The costs of trying to become self-sufficient in oil are illustrated by the economic effects of the oil boom of the late 1970s and early 1980s. During this period rising oil prices and government tax breaks spurred a tremendous increase in exploration and development, such as drilling exploratory wells. Paying for this effort used a significant fraction of total investment by the U.S. economy (Figure 20.18). In 1982 the U.S. oil and natural gas industry consumed nearly 7 percent of all U.S. investment, which was significantly greater than the fraction immediately before or after 1982. But by 1982 U.S. production was past its peak, and the energy return on investment was declining. So despite a large increase in exploration and development, relatively little oil was found. As a result, oil production continued to decline and the fraction of GDP produced by the oil and natural gas sector did not increase. In effect, all that extra capital was wasted. Had it been used in other sectors of the economy, such as the automobile or steel industry, U.S. citizens would have been better off.

Indeed, it is highly unlikely that the United States can increase its oil production significantly. Some people have argued that the United States should drill for oil in the Arctic National Wildlife Refuge. There are signs that the refuge could contain economically significant quantities of oil. But even if the most optimistic scenario came true, the

FIGURE 20.18   *Economic Effects of Efforts to Increase Oil Production* As oil prices rose in the late 1970s and early 1980s, the United States tried to boost domestic oil production by increasing exploration and development. This effort increased the fraction of national investment consumed by the oil and gas industry (red bars). Despite this effort, production continued to decline, so the fraction of GDP generated by the oil and gas industry (blue bars) also declined. This shows that the effort to increase domestic production wasted capital that could have been used more productively by other sectors of the economy. (Source: From R.K. Kaufmann, and C.J. Cleveland, "Policies to Increase U.S. Oil Production: Likely to Fail, Damage the Economy, and Damage the Environment," Annual Review of Energy 16: 379–400.)

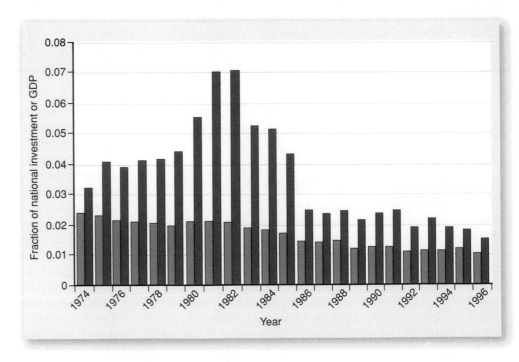

CASE STUDY

Changes in OPEC Pricing Strategy

The Organization of Petroleum Exporting Countries, which is known as OPEC, was formed on September 9, 1960. Originally OPEC included five nations: Iran, Iraq, Saudi Arabia, Venezuela, and Kuwait. These nations created OPEC in response to a steady decline in prices. Prices were dropping because oil supply was expanding faster than demand. To establish a balance between supply and demand, OPEC hoped to slow the increase in production by matching supply and demand as the Texas Railroad Commission had.

But even the modest goal of preventing further price declines was difficult. In 1960 OPEC nations produced 37 percent of world oil supply and controlled 54 percent of crude oil reserves. In spite of this dominance, OPEC had relatively little power over price. Foreign oil companies, not OPEC governments, controlled the oil fields, so OPEC could not restrict production. And even if it could reduce production, the United States could offset such reductions by opening spare capacity. In 1965 the Texas Railroad Commission shut in about 6.7 million barrels per day in production, which was about 22 percent of world oil supply.

Most of this spare capacity was reopened by the early 1970s. By 1973, when OPEC nations embargoed shipments to the United States and reduced supply to Europe and Asia, the United States could not offset this shortfall. This meant that OPEC could now have a significant effect on world oil supply; and this power gave OPEC considerable influence over prices. This ability changed OPEC's strategy. Rather than resist price reductions, OPEC would actively seek to raise oil prices.

This change in strategy is illustrated by OPEC's response to the reduction in Iranian oil production in 1978–1979. To offset the effect of this decline on world prices, member states could have increased production, or at worst done nothing. Instead they chose to push prices higher by reducing production. The reduction in supply raised prices to $40 per barrel in 1980.

The feedback effect of higher prices started another change in OPEC's pricing strategy. The large price increases in 1973–1974 and 1979–1980 caused hundreds of billions of dollars to flow from oil-consuming nations to oil-producing nations (see Figure 1 in the Chapter 17 Case Study on page 359). OPEC nations could not spend or invest this money domestically as fast as it came in. Rather than have all the money sit in local banks, OPEC nations invested some of the money in oil-consuming nations.

These investments changed the relationship between oil-consuming and oil-producing nations and ultimately OPEC's pricing strategy. OPEC's income from investments in oil-consuming nations depended in part on the economic health of those nations. If their economies grew, so did the income from OPEC's investments. But if oil-consuming economies went into recession, income from OPEC's overseas investments declined.

OPEC's ability to influence world oil prices allowed OPEC to influence the economies of oil-consuming nations. In the United States two of the worst recessions since the Great Depression (see Chapter 15, page 266) occurred after the oil price increases of 1973–1974 and 1979–1980. The effects of high oil prices on oil-consuming economies posed a conundrum for OPEC. OPEC could increase prices and thereby increase its member countries' revenues from selling oil. But higher prices reduced their income from overseas investments.

This trade-off forced OPEC to reconsider its pricing strategy. OPEC nations with significant overseas investments, such as Saudi Arabia, Kuwait, and the United Arab Emirates, favored moderate prices and became known as "price doves." On the other hand, "price hawks" urged OPEC to raise prices. In general, "price hawks" had few overseas investments. The views of the doves have generally prevailed because the doves are richer than the hawks and control a greater share of the oil market than the hawks.

But oil prices are high now—does that mean that the "price hawks" have won? Not really. It just means that the "price doves" like Saudi Arabia have lost their ability to keep oil prices low. Over the last couple of years Saudi Arabia opened spare capacity. But this extra production could not keep up with growing demand, so prices rose. In short, Saudi Arabia and other price doves cannot push oil prices down because they are pumping oil at capacity.

The story of OPEC pricing strategy highlights an important aspect of economic globalization. Globalization increases the connections among the parts of the economic system. These connections link the well-being of all parts. One part of the system cannot forever prosper at the expense of another, and this may induce decisions that benefit all members, not just a select group.

STUDENT LEARNING OUTCOME

- Students will be able to describe the factors that influence whether OPEC prefers high or low oil prices and OPEC's ability to get oil prices to their desired level.

---

refuge could produce about 2 million barrels per day starting twenty years after drilling begins (Figure 20.19). Two million barrels seems like a lot of oil, but that represents less than 10 percent of U.S. oil demand in 2005; by 2025 it would probably be much less than 10 percent. There seems to be no escape from the geological fact that most of the world's oil is deposited in a few very large fields. Of the oil that remains, most is located in OPEC nations, many of which are politically unstable (Figure 20.20).

A Competitive Oil Market?

Some people suggest that a competitive market will alleviate many difficulties associated with the supply and price for oil. In reality, it would be difficult to establish a competitive market because most of the world's oil is found in a few very large sedimentary provinces. This makes it relatively easy for a few big producers to cooperate and influence the supply of oil. Furthermore, when there was a competitive oil market in the United States, both consumers and producers were unhappy.

A cursory glance at the history of oil prices (Figure 20.21) shows three periods: an initial period of significant fluctuations (1900–1935), a period of relative price stability (1936–1973), and another period of wide price fluctuations (1973–present). During the first part of the twentieth century the U.S. oil market was as close as it has come to a competitive market. Although oil transportation was controlled by Rockefeller's Standard Oil, production was highly competitive—there were many firms, each with a small share of the market.

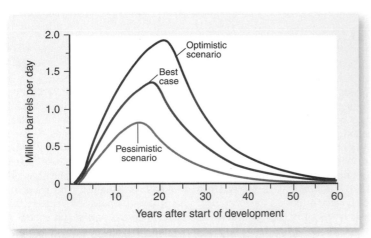

FIGURE 20.19   *Oil Production from the Arctic National Wildlife Refuge* *Scenarios for oil production from the Arctic National Wildlife Refuge, as forecast by the U.S. Geological Survey. Even if drilling started now, it would take nearly a decade for production to occur. And even under the most optimistic scenario, production would peak at less than 2 million barrels per day, which is less than 10 percent of current rates of U.S. consumption.* (Source: Data from U.S. Geological Survey, The Oil and Gas Resource Potential of the Arctic National Wildlife Refuge 1002 Area, Alaska, Open File Report 98–34.)

This competition led to cycles of boom and bust that caused big fluctuations in oil prices. Discovery of giant oil fields was followed by a glut of oil. This glut caused oil prices to drop and slowed the rate at which firms looked for new oil fields. Over time, low prices encouraged oil consumption, which allowed demand to catch up to the amount that could be produced from previously discovered fields. This caused prices to rise and thus firms to look for new fields. When new fields were discovered, the boom and bust cycle started anew.

Booms and busts hurt both producers and consumers. A bust can bankrupt producers, and a boom can bankrupt firms that use oil. These difficulties led to legislation aimed at damping price fluctuations. The Interstate Oil Compact Commission and the Connelly Hot Oil Act in 1934 and 1935 allowed state committees to control production. The Texas Railroad Commission was among the most important of these state committees because Texas accounted for about a third of U.S. production. When oil prices started to drop, the Texas Railroad Commission shut in production and reduced supply. Conversely, when prices started to rise, the Texas Railroad Commission allowed operators to reopen production. Following these behaviors, the Texas Railroad Commission operated as a benevolent cartel. As such, the Texas Railroad Commission damped the boom and bust cycle between 1947 and 1972; many people remember this period as the "Golden Age of Petroleum" when oil was abundant and inexpensive.

FIGURE 20.20   *Remaining Oil Supplies* *Estimates of the amount of oil that remains in various regions of the world. Notice that the greatest amounts are thought to be located in OPEC nations, especially those in the Middle East.* (Source: Data from National Geographic "The End of Cheap Oil," June, 2004, pp. 80–109.)

Despite this success, changes in geological and economic aspects of the oil industry eroded the ability of the Texas Railroad Commission to stabilize oil prices. Declining yield per effort for oil exploration and the large amount of oil held back by the Texas Railroad Commission reduced the profitability of U.S. production. In response, many U.S. firms directed much of their exploration and development efforts into foreign nations, some of which later formed the Organization of Petroleum Exporting Countries (OPEC).

The first signs that the United States was depleting its oil resources appeared in the late 1960s. The fraction of capacity allowed to operate by the Texas Railroad Commission increased from 29 percent in 1965 to just over 80 percent in 1970 (Figure 20.22). Despite the rapid reopening of capacity, U.S. production could not keep up with demand. In 1970 U.S. oil production peaked. By 1973 all U.S. fields were allowed to operate, and the United States imported about 35 percent of its oil consumption, almost half of which originated in OPEC countries. OPEC supplied an even greater fraction of oil to other developed nations.

The inability of the United States to increase production allowed OPEC to wrest control of the world oil market from the Texas Railroad Commission. By 1973 there was no spare capacity in the United States. That meant that the Texas Railroad Commission could not stabilize prices by opening oil wells that were previously closed. Instead OPEC was the only group of nations that had significant amounts of spare capacity. Spare capacity gave OPEC the power to stabilize oil prices. But OPEC wanted to raise prices. See *Case Study: Changes in OPEC Pricing Strategy.* But trying to do so exaggerated price increases—and later price decreases—which recreated the price volatility of the early 1900s.

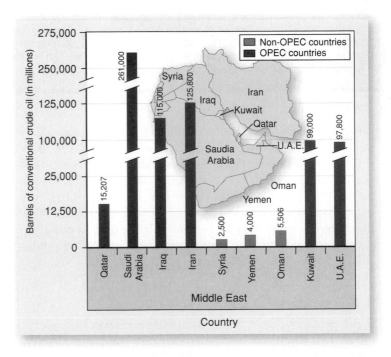

OPEC first demonstrated its power to increase prices in October 1973. Following the outbreak of war between Israel and its neighbors, OPEC cut exports by 5 percent and embargoed nations that supported Israel. Because no producer outside OPEC was able to offset this reduction by increasing production, oil prices increased dramatically.

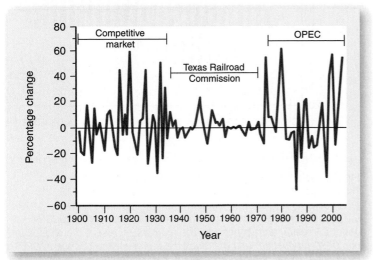

FIGURE 20.21 *Oil Price Regimes* *The annual rate of change in real oil prices shows three periods. During the first period, 1900–1935, a competitive market for production allowed prices to fluctuate greatly from one year to the next. During the next period of 1936–1973, the Texas Railroad Commission opened and shut production in Texas so that supply matched demand, and price fluctuations were greatly reduced. After 1973 OPEC was able to influence prices and caused prices to fluctuate greatly from one year to the next.* (Source: Redrawn from R.K. Kaufmann, "A Model of the World Oil Market for Project LINK: Integrating Economics, Geology, and Politics," Economic Modeling 12(2): 165–178.)

Oil prices started to rise again in December 1978, when the return of Ayatollah Khomeini plunged Iran into revolutionary chaos. During this period Iran stopped exporting 5 million barrels per day, which was slightly less than 10 percent of world production. This caused global oil prices to jump. To enhance this price increase, other OPEC nations also reduced production. By December 1980 a barrel of crude oil reached $40.

The great increase in real oil prices in the 1970s and early 1980s initiated the same forces that caused the bust phase in the oil price cycle earlier in the century (Figure 20.23). To reduce dependence on OPEC, oil companies increased exploration and development both in the United States and in other non-OPEC nations. Significant quantities of oil were found in the North Sea, Mexico, and the Canadian Arctic. Producing this oil became profitable after the 1979–1980 price increase.

The rapid increase in oil prices also reduced demand. Much of the reduction occurred in the residual fuel oil market. Residual fuel oil was used mainly to generate electricity. This use was replaced by coal, natural gas, or nuclear power. At the same time, the efficiency of energy-using devices increased. For example, the average fuel efficiency of the U.S. auto fleet increased from 13.4 mpg in 1973 to 17.5 mpg in 1985.

The combination of decreasing oil demand and increasing oil supply reduced the demand for oil from OPEC, which put downward pressure on prices. To keep prices high, OPEC shut in production as had the Texas Railroad Commission. But the steady increase in non-OPEC supply and reduction in demand forced OPEC to shut in ever-increasing quantities. By 1985 OPEC produced 50 percent less oil than it did in 1974. Much of this reduction was absorbed by Saudi Arabia, where

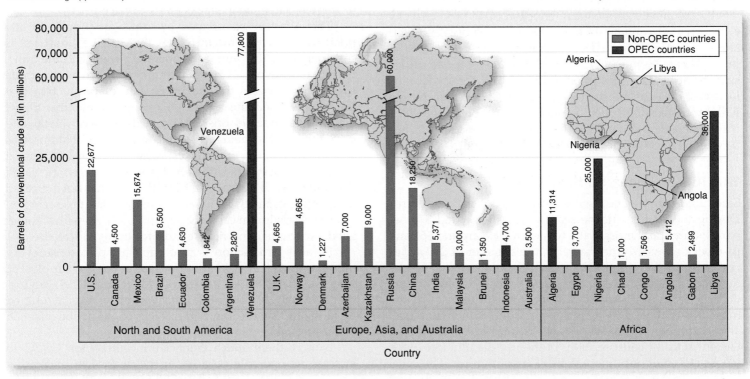

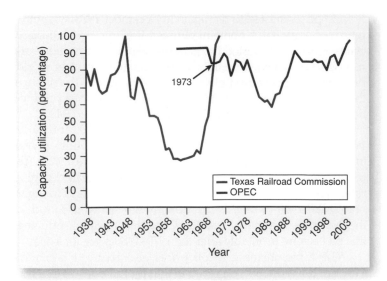

FIGURE 20.22   *Control over Marginal Supply* *Producers can affect oil prices by controlling the marginal supply of oil. That is, they can increase or decrease production, and no other producers can offset those changes. Prior to 1973 the Texas Railroad Commission shut in significant amounts of production so that firms operated at low rates of capacity. This allowed them to increase or decrease production. OPEC did not have enough spare capacity to offset these changes, so the Texas Railroad Commission was able to influence prices. After 1973 Texas operators produced near capacity and so had no spare capacity. This gave OPEC control over the marginal supply of oil and allowed them to influence price. Now OPEC has little spare capacity and so cannot reduce prices even if it wanted to.* (Source: Redrawn from R.K. Kaufmann, "A Model of the World Oil Market for Project LINK: Integrating Economics, Geology, and Politics," Economic Modeling 12(2): 165–178.)

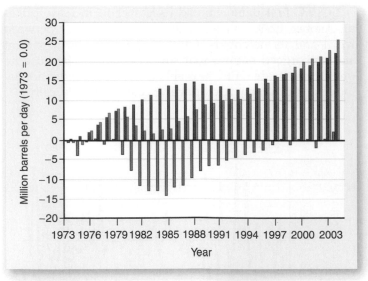

FIGURE 20.23   *Changes in Supply and Demand* *The changes in world oil supply and demand relative to 1973. The green bars represent the change in world oil demand relative to 1973; the final observation represents an increase in demand in 2003 of about 25 million barrels per day. The red bars represent the change in non-OPEC production, and the blue bars represent the change in OPEC production. Oil demand barely grew for nearly fifteen years after 1973, while non-OPEC production increased by more than 10 million barrels per day. As a result, OPEC had to reduce production significantly. Since the price collapse in 1986, demand has grown steadily but non-OPEC production has been relatively constant. As a result, OPEC production has increased significantly.* (Source: Redrawn from R.K. Kaufmann, "A Model of the World Oil Market for Project LINK: Integrating Economics, Geology, and Politics," Economic Modeling 12(2): 165–178.)

production fell from nearly 10 million barrels per day (mbd) in 1981 to 3.4 mbd in 1985.

By the mid-1980s the great reduction in oil production by OPEC became economically untenable. In December 1985 OPEC announced that it had changed strategy. Rather than shut in production to keep prices high, OPEC would recapture market share by expanding production. Early in 1986 OPEC production increased by 2 million barrels per day. This increase, coupled with continued production by non-OPEC nations, flooded the market and caused oil prices to collapse. Oil prices fluctuated between $15 and $25 per barrel between 1986 and 2002.

Why Are Oil Prices So High Now?

Since 2002 prices have risen steadily beyond $25 per barrel. During the 2005 hurricane season prices exceeded $70 per barrel; they spent most of the winter of 2006 around $60 and rose back above $70 per barrel in the summer of 2006. This rise was caused by reductions in stocks of crude oil held by developed nations and increases in capacity utilization by OPEC. Stocks represent oil in storage for future use and are measured in days of forward consumption, which is defined as the number of days that stocks could satisfy demand at current rates of consumption. In the middle and late 1980s stocks stored about ninety days of consumption. This level dropped to the low eighties and upper seventies by the first

decade of the twenty-first century. As with any other good or service, a decline in stocks tends to raise prices.

This effect is exacerbated by limits on supply. At any point in time, the amount of oil that can be produced is constrained by the number of operating wells and the transportation network. This maximum, known as **operable capacity,** represents the maximum rate of production that can be sustained during the following six months. OPEC did not increase its operable capacity between 1986 and 2005 because the demand for oil from OPEC nations was well below OPEC's operable capacity. Since 2002 increasing demand for oil by the United States and China has increased demand from OPEC nations, but OPEC still has not increased its operable capacity. As a result, suppliers are producing oil near their maximum capacity (Figure 20.22). As with all other goods, when demand rises relative to supply, prices rise.

Because production is near capacity, prices rise every time there is a real or imagined threat to current supply. For example, the hurricane season of 2005 reduced U.S. production by nearly 20 percent. Its effect on prices was magnified because stocks were low, and this loss could not be offset by increasing production elsewhere.

Nor is there much financial incentive for OPEC to increase its operable capacity in the near term. Increasing operable capacity would allow OPEC nations to sell more oil, which would increase their revenues. But this would also create spare capacity, which lower the price per barrel. Economic models predict that increasing capacity would reduce prices faster than it would increase sales. As a result, OPEC rev-

enues probably would decline if OPEC increased its operable capacity. Because most nations make policy based on self-interest, OPEC is unlikely to increase its operable capacity significantly unless it falls apart as an organization.

SUMMARY OF KEY CONCEPTS

- Fossil fuels powered the Industrial Revolution. Coal replaced dwindling supplies of wood and powered the first century of the Industrial Revolution. The second century was powered by oil, which replaced dwindling supplies of whale oil. Natural gas replaced both because it burns hotter and more cleanly.

- Fossil fuels are derived from the net primary production of autotrophs that lived millions of years ago. Their formation requires four conditions: (1) large amounts of organic matter (2) cut off from oxygen and (3) subjected to sufficient temperature and pressure to be rearranged in a way that concentrates its energy use (4) close enough to Earth's surface to be economically recoverable. Coal is derived from the net primary production of plants in highly productive ecosystems. Crude oil and natural gas are derived from phytoplankton and zooplankton.

- Coal is a solid that is extracted from the ground using underground mining techniques, mountaintop removal, or surface mining techniques. Coal is transported mostly by trains or pipelines, and these efforts constitute a large fraction of total costs. Oil and natural gas are found in sedimentary rock. Some of the oil in a field is brought to the surface via primary recovery, which is powered by pressure gradients between the field and the atmosphere. As this gradient

weakens, additional amounts can be brought to the surface via secondary or tertiary recovery.

- Coal is often cleaned to remove impurities before being used. Crude oil cannot be used as it is extracted. Oil refineries break long carbon chain molecules into smaller fragments. Short-chain molecules such as octanes tend to be more valuable; therefore, light crudes are more valuable than heavy crudes. Similarly, sulfur tends to damage refineries, so sour crudes are less valuable than sweet crudes. Natural gas can be used as it comes.

- The world will not run out of oil or natural gas suddenly. Rather, production will decline slowly, following a pattern that may look like a bell-shaped curve. M. King Hubbert used such a curve to generate an accurate forecast for the peak in U.S. oil production. The peak in global oil production will be a watershed event in the global economy.

- Despite the commonsense appeal of energy independence, efforts to increase domestic production will not enhance economic or military security. Because the world oil market is one big pool, disruptions in one segment of the market spread globally. Similarly, efforts to increase production from a declining resource base divert capital from other segments of the economy. Nor will a competitive market avoid problems in the world oil market.

REVIEW QUESTIONS

1. During the process of industrialization, why did coal replace wood? Why did crude oil and natural gas replace coal?

2. Explain why commercial valuable deposits of fossil fuels do not materialize if one of the four requirements for formation is not satisfied.

3. Explain how the solid nature of coal and the liquid state of crude oil influence the processes by which these fuels are extracted and distributed to consumers.

4. Explain why the definition of *proved reserves* tends to understate the amount of oil that remains to be recovered and why the ratio of proved reserves to production tends to remain constant over time.

5. Describe the costs and benefits associated with importing crude oil.

6. Describe why a competitive market for crude oil is unlikely and why such a market probably will not make oil producers or oil consumers happy.

KEY TERMS

coal benefaction	operable capacity	seam
continuous mining techniques	overburden	secondary recovery
conventional fossil fuels	porosity	sedimentation
crude oil	primary recovery	surface mining techniques
discoveries	proved reserves	tertiary methods
extraction efficiency	Q infinity	unconventional fossil fuels
kerogen	refined petroleum products	underground mining techniques
mountaintop removal	reserves to production (R:P)	wildcatting
oil refinery	revisions	yield per effort
oil shale	room and pillar	

CHAPTER

21

NUCLEAR POWER

STUDENT LEARNING OUTCOMES

After reading this chapter, students will be able to

- Distinguish between nuclear fission and fusion and explain why their differences are important in the economics of electricity generation.
- Draw a diagram of the nuclear fuel cycle.
- Explain why nuclear power plants are not currently being built in the United States.
- Describe the technological and political challenges of long-term nuclear waste disposal.
- Explain how human error contributed to accidents at Three Mile Island and Chernobyl.

The Cooling Towers of a Nuclear Power Plant
One of the most promising energy technologies, nuclear power, faces significant technical, economic, environmental, and public perception obstacles.

NUCLEAR POWER: A FAUSTIAN BARGAIN?

A popular legend in Western folklore and literature is based on Dr. Johannes Faust, who lived in early sixteenth-century Germany (Figure 21.1). Faust was a medical practitioner and had studied astrology and alchemy as well as philosophy. Rumors and legends about him began while he was still alive. He was reputed to be a charlatan who traveled from place to place in Germany, passing himself off as a physician, alchemist, astrologer, and magician. He also was rumored to be a necromancer (conjurer of the dead). After his death around 1540, the story set in that he had made a pact with the devil.

The legend of Faust has been retold in music, art, and literature. An English prose translation of 1592 inspired the play *The Tragicall History of D. Faustus* (1604) by Christopher Marlowe, Shakespeare's most important predecessor in English drama. According to the legend, Faust sought to acquire supernatural knowledge and power by a bargain with Satan. In this pact, signed with his own blood, Faust agreed that Mephistopheles, a devil, was to become his servant for 24 years. In return, Faust would surrender himself to Satan. Mephistopheles entertained his master with luxurious living, long intellectual conversations, and glimpses of the spirit world. After the agreed 24 years, during an earthquake, Faust was carried off to Hell.

Has society made a Faustian bargain by embracing nuclear power to generate electricity? Some people say yes. Like the promise of supernatural knowledge and power, the potential of nuclear power is seductive: The fission of an atom of uranium produces *millions* of times more energy than that produced by the combustion of a similar quantity of methane in natural gas. Nuclear power also generates small quantities of CO_2 and other air pollutants compared to electricity generation from fossil fuels. The Faustian nature of the issue is the potentially huge long-term costs that may be associated with the use of nuclear power today. Nuclear power generates very hazardous wastes that persist for millennia; and although the routine operations of a nuclear plant are safe, there is still a nonzero chance of a calamitous accident that would release harmful radiation to the environment.

FIGURE 21.1 *A Faustian Bargain?* Doctor Johannes Faust as portrayed in The Tragicall History of D. Faustus *(1604) by Christopher Marlowe.*

Supporters of nuclear power shun this analogy. They argue that the current generation of nuclear technology is extremely reliable, that the waste issue is political and not technological, and that the overall worker and public safety record of the industry is better relative to other energy industries. What's more, nuclear power is not a significant source of greenhouse gases—an important criterion in a world where decisions about future energy use must account for their impact on climate change.

This chapter provides information to help you sort this issue out by presenting an objective evaluation of the economic, technological, safety, and environmental characteristics of nuclear power. ■

THE NATURE, DISTRIBUTION, AND USE OF URANIUM RESOURCES

Cosmochemists believe that uranium was produced in one or more supernovae about 6.6 billion years ago, and that this material was inherited by the solar system of which Earth is a part. Natural uranium is radioactive, and it is present in most rocks and soils as well as in many rivers and in seawater. It is, for example, found in concentrations of about four parts per million in granite, which makes up 60 percent of Earth's crust (Figure 21.2). Uranium's radioactive decay provides the main source of heat in Earth's core, driving the motion of the crust and the rock cycle (chapter 4).

The two principal isotopes of natural uranium are ^{238}U and ^{235}U. The property of uranium important for nuclear weapons and nuclear power is its ability to

FIGURE 21.2 *Uranium Ore* A uranium mine with an exposed deposit of uranium ore. *(Courtesy of the U.S. Department of Energy.)*

fission, or split into two lighter fragments when bombarded with neutrons, releasing energy in the process. Of the naturally occurring uranium isotopes, only ^{235}U is **fissile,** meaning it can sustain a chain reaction—a reaction in which each fission produces enough neutrons to trigger another, so that the fission process is maintained without any external source of neutrons. The nucleus of a ^{235}U atom consists of 92 protons and 143 neutrons. When a ^{235}U nucleus absorbs a neutron it becomes unstable and will fission. The products of the splitting of the nucleus, the largest of which are called fission fragments, have a mass less than the mass of the nucleus of the original atom. (This conversion is described by Einstein's famous $E = mc^2$ equation, where E is energy, m is mass, and c is the speed of light.) The lost mass is converted into energy, mostly in the form of heat. Although this loss of mass is small, the energy released is large.

^{235}U is the only naturally occurring isotope that undergoes nuclear fission (Figure 21.3). The most common occurrence is when ^{235}U breaks down to ^{141}Ba, ^{92}Kr, two additional neutrons, and the one neutron that started the fission:

$$^{1}_{0}n + {}^{235}_{92}U \longrightarrow {}^{141}_{56}Ba + {}^{92}_{36}Kr + 3{}^{1}_{0}n \qquad (21.1)$$

The three neutrons released then may initiate three new fissions that release nine neutrons, and so on. Because the process generates more neutrons than are required to stimulate the reaction, a chain reaction is possible, and the process of fission can become self-sustaining.

In contrast, ^{238}U is a **fertile** isotope. It cannot sustain a chain reaction, but it can absorb neutrons and form ^{239}Pu (plutonium). Like ^{235}U, ^{239}Pu is fissionable and can split to release energy. In fact, up to half of the heat energy produced in a nuclear reactor can come from ^{239}Pu, even when none of this material is present in the initial fuel. ^{239}Pu, which is virtually nonexistent in nature, was used in the first atomic bomb tested on July 16, 1945, and in the one dropped on Nagasaki, Japan, on August 9, 1945.

Uranium Resources and Production

In some places geological processes have created uranium ore—rock containing uranium minerals in concentrations that can be mined economically (Figure 21.2). The most common mineral form is uranium oxide, U_3O_8. Typical ore grades mined today contain 0.45 to 1.8 kgs of U_3O_8 per metric ton of material extracted, or 0.05–0.20 percent U_3O_8.

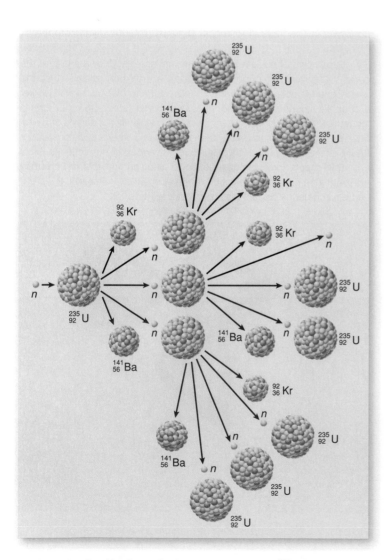

FIGURE 21.3   *Nuclear Fission* *The fission of a ^{235}U isotope.*

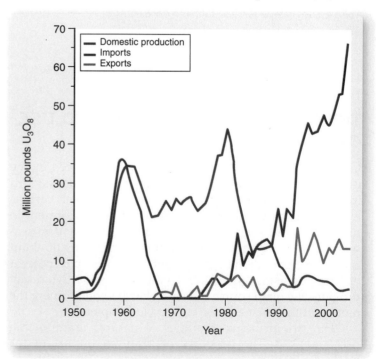

FIGURE 21.4   *U.S. Uranium Industry* *U.S. production, import, and export of uranium, 1950–2003.*

TABLE 21.1	Uranium Reserves and Resources, End of Year 2004 (Million Pounds U₃O₈)		
	FORWARD COST CATEGORY (DOLLARS PER POUND)		
Resource Category and State	$30 or Less	$50 or Less	$100 or Less
Reserves	265	890	1,414
New Mexico	84	341	566
Wyoming	106	363	582
Texas	6	23	38
Arizona, Colorado, Utah	45	123	170
Others	24	40	58
Potential resources			
Estimated additional resources	2,180	3,310	4,850
Speculative resources	1,310	2,230	3,480

Source: Data from U.S. Department of Energy.

FIGURE 21.5 *First Nuclear Electricity* Arco, Idaho, became the first U.S. town to use electricity produced from a nuclear reactor. *(Courtesy of the U.S. Department of Energy.)*

U.S. production of uranium declined from more than 40 million pounds of U_3O_8 in the early 1980s to about 2 million pounds in 2003 (Figure 21.4). At the same time imports of U_3O_8 increased substantially. Proved reserves of uranium are based on the cost of developing deposits of varying quality and location. As of the end of 2004, uranium reserves in the $30-per-pound and $50-per-pound categories were 265 and 890 million pounds, respectively. At current rates of production these proved reserves represent a 130- to 440-year domestic supply of uranium (Table 21.1).

THE PROMISE AND CURRENT STATUS OF NUCLEAR POWER

President Dwight Eisenhower's Atoms for Peace program proposed to divert fissionable materials from bombs to peaceful uses such as civilian nuclear power. In 1951 an experimental reactor sponsored by the U.S. Atomic Energy Commission generated the first electricity from nuclear power (Figure 21.5). The British completed the first operable commercial reactor in 1956. The U.S. Shippingport unit, a design based on power plants used in nuclear submarines, followed a year later. In cooperation with the U.S. electric utility industry, reactor manufacturers then built several demonstration plants and made commitments to build additional plants at fixed prices. This commitment helped launch commercial nuclear power in the United States.

The success of demonstration plants and the growing awareness of U.S. dependence on imported crude oil led to a wave of enthusiasm for nuclear electric power that sent orders for reactor units soaring between 1966 and 1974

(Figure 21.6). Many optimists saw this as the dawn of an era of unlimited energy that would be "too cheap to meter."

By 2005, however, the reality of nuclear power was far from these lofty ambitions. Nuclear power supplied 5 percent of the world's energy from about 450 commercial

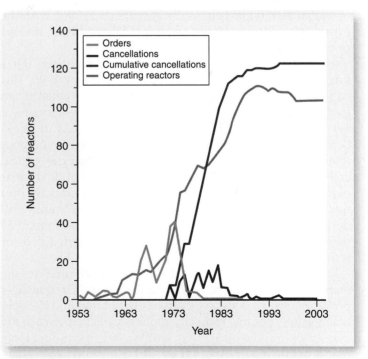

FIGURE 21.6 *U.S. Nuclear Power Industry* New orders, cancellations, and number of operating nuclear reactors in the United States, 1953–2003. Operating reactors continued to rise even when no new orders were made due to the long lead time of building a reactor. New plants are again being considered for new orders for the first time since 1978.

FIGURE 21.7 *World Nuclear Power Generation* *Nuclear power as a percentage of total electricity generation by nation in 2005.*

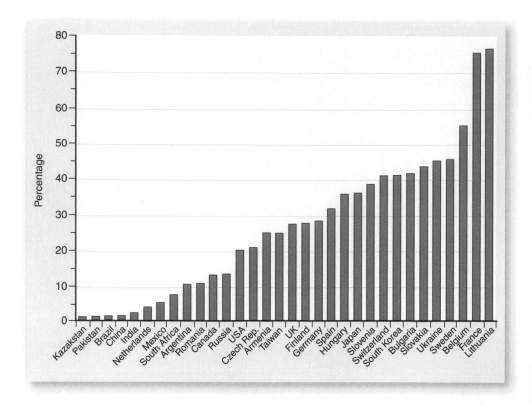

reactors in 32 countries; this represented about 15 percent of global electricity generation (Figure 21.7). But with the exception of France, Japan, and a few other nations, interest in nuclear power has dramatically declined.

Located near Spring City, Tennessee, the Watts Bar 1 nuclear plant connected to the grid in February 1996. This could be the last commercial nuclear reactor constructed in the United States, where no new nuclear reactors have been ordered since 1978. By the end of 2004, 124 units (48 percent of all ordered units) had been canceled, sometimes after billions of dollars had been spent on their design and construction (Figure 21.6). Currently about 100 nuclear plants generate about 20 percent of U.S. electricity. But nuclear generating capacity peaked in 1996 and will decline steadily unless new plants are built. More than 50 percent of current U.S. nuclear capacity is scheduled for retirement by 2020.

What happened? Several factors contributed to the demise of nuclear power in the United States. First, nuclear power plants have always been costly to build and, for several reasons, became radically more expensive between the mid-1960s and the mid-1970s. Utilities began building large plants before much experience had been gained with small ones. Expected economies of scale ("bigger is better") did not materialize. Many units were forced to undertake costly design changes and equipment retrofits, partially as a result of the Three Mile Island accident. Nuclear plants completed

in the 1980s cost $2–3 billion, with some plants costing up to $5 billion, and they typically took eight to fourteen years to construct. The cost of electricity from nuclear plants tripled between 1970 and 1990, making nuclear power unattractive relative to other methods of generating electricity.

Second, the optimistic predictions for the future of nuclear power were made in the 1960s when energy prices were low and the demand for electricity was growing rapidly. Most people—and especially electric utilities— expected these trends to continue, and as a result many new nuclear plants were ordered. But the energy price increases in 1973–74 and 1980–81 changed these expectations, slowing the increase in the demand for electricity for a period of time. As a result, utilities did not need to build new power plants in the 1980s as quickly as they had anticipated.

Third, accidents such as those at Three Mile Island and Chernobyl and unresolved problems such as the permanent disposal of radioactive wastes eroded public confidence in the safety and reliability of nuclear power. In some regions intense public opposition to nuclear power discouraged utilities from ordering new plants. Antinuclear protests reached a pinnacle on April 30, 1977, when 18,000 demonstrators under the banner of the Clamshell Alliance occupied the site of the proposed Seabrook reactor in New Hampshire (Figure 21.8). The size of the demonstration and the jailing of 1,400 nonviolent protestors galvanized antinuclear activists around the world.

FIGURE 21.8 *Nuclear Protests*
Protestors at the site of the proposed Seabrook nuclear reactor in New Hampshire in 1977. One of the two proposed reactors was eventually completed.

THE NUCLEAR FUEL CYCLE

The nuclear fuel cycle for typical light-water reactors is illustrated in Figure 21.9. The cycle consists of a "front end" that prepares uranium for use as fuel in reactor operations, as well as a "back end" that manages, prepares, and disposes of highly radioactive spent nuclear fuel.

The front end of the nuclear fuel cycle commonly is separated into the following steps. In **exploration,** a deposit of uranium, discovered by geophysical techniques, is evaluated and sampled to determine the amounts of uranium materials that are extractable from the deposit at specified costs and technology. In the **mining** stage, uranium ore is extracted by open pit and underground methods similar to those used for mining coal and other metals. In the **milling** stage, mined uranium ores are processed by grinding the ore materials to a uniform particle size and then extracting the uranium by chemical leaching. The milling process yields a dry powder called "yellowcake," which is sold on the uranium market as uranium oxide (U_3O_8). The U_3O_8 produced by a uranium mill is not directly usable as fuel for a nuclear reactor. In the **conversion** process, U_3O_8 is converted to uranium hexafluoride, UF_6, which is the form required by most commercial uranium enrichment facilities. A solid at room temperature, UF_6 can be changed to a gas at moderately higher temperatures.

The concentration of the fissionable isotope (^{235}U) in UF_6 is too low to sustain a nuclear chain reaction in light-water reactor cores. UF_6 must be "enriched" in the fissionable isotope to be used as nuclear fuel. The UF_6 is placed in special cylinders, solidified for transportation, and shipped to commercial **enrichment** plants (Figure 21.10). There are two enrichment processes in large-scale commercial use: gaseous diffusion and gas centrifugation. To understand how gaseous diffusion works, picture the UF_6 molecules as sand particles suspended in air. All the molecules, or "sand grains," are blown through thousands of sieves, one after another. Because the lighter ^{235}U particles travel faster than the heavier ^{238}U particles, more of them penetrate each sieve. As more sieves are passed, the concentration of ^{235}U increases. The process continues until the concentration of ^{235}U is raised to 3–5 percent (low-enriched uranium), which will sustain fission in a nuclear power plant. In contrast, nuclear weapons require highly enriched uranium with over 90 percent ^{235}U. In the gas centrifuge process, UF_6 gas is spun at high speed in a series of cylinders. This separates the ^{235}U and ^{238}U atoms based on their slightly different atomic masses.

In the **fabrication** stage, the enriched UF_6 is converted into uranium dioxide (UO_2) powder, which is then processed into pellet form. The pellets are fired in a high-temperature furnace to create hard, ceramic pellets of enriched uranium. The cylindrical pellets are ground into a uniform pellet size. Tubes are sealed to contain the fuel pellets, forming **fuel rods.** Fuel rods are grouped in **fuel assemblies** that are used to build up the nuclear fuel core of a power reactor.

The back end of the cycle deals with **spent nuclear fuel:** irradiated fuel that is permanently discharged from a reactor. Spent or irradiated fuel is usually discharged from reactors because of chemical, physical, and nuclear changes that make the fuel no longer efficient for the production of heat. Once a year, approximately one-third of the nuclear fuel inside a reactor is removed and replaced with fresh fuel. A typical nuclear power plant produces 18.1 to 22.7 metric tons of spent fuel each year.

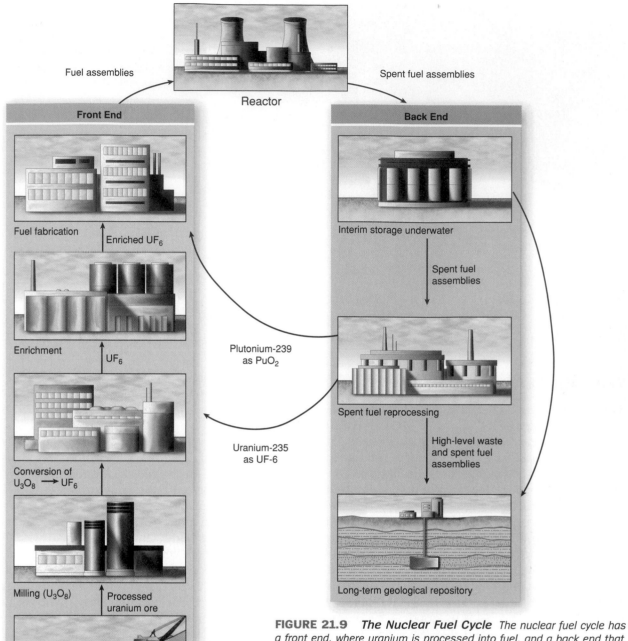

Fuel assemblies

Reactor

Spent fuel assemblies

Front End

Fuel fabrication

Enriched UF$_6$

Enrichment

UF$_6$

Conversion of
U$_3$O$_8$ → UF$_6$

Milling (U$_3$O$_8$)

Processed
uranium ore

Exploration and mining

Back End

Interim storage underwater

Spent fuel
assemblies

Spent fuel reprocessing

High-level waste
and spent fuel
assemblies

Long-term geological repository

Plutonium-239
as PuO$_2$

Uranium-235
as UF-6

FIGURE 21.9   *The Nuclear Fuel Cycle* *The nuclear fuel cycle has
a front end, where uranium is processed into fuel, and a back end that
deals with spent fuel wastes.*

The first step of the back end is **interim storage.** These
facilities are large pools of water immediately adjacent to
the reactor. The water acts as a shield against the radiation
and absorbs the released heat (Figure 21.11). Spent fuel is
generally held in such pools for a minimum of about five

months. As an alternative to storage in pools, some spent
fuel is stored aboveground at reactor sites in concrete or
steel containers called dry casks.

Ultimately spent fuel must either be reprocessed or
sent for permanent disposal. Spent fuel discharged from
light-water reactors contains appreciable quantities of fissile (^{235}U, ^{239}Pu), fertile (^{238}U), and other radioactive materials. These materials can be chemically separated and
recovered from the spent fuel in the **reprocessing** stage.
Reprocessing separates uranium and plutonium from
waste products by chopping up the fuel rods and dissolv-

FIGURE 21.10 *Nuclear Fuel Enrichment* *Gas centrifuges used to enrich nuclear fuel.* *(Courtesy of the U.S. Department of Energy.)*

FIGURE 21.11 *Interim Waste Storage* *An interim waste storage facility at a nuclear power plant.*

ing them in acid. Recovered uranium can then be used to produce new nuclear fuel. President Carter banned the reprocessing of spent fuel in 1977. His decision was based on the potential for plutonium to be captured by terrorists and on the abundance of natural uranium ore that could be used to produce enriched fuel more cheaply than reprocessing. Only five nations today engage in commercial reprocessing—Britain, France, India, Japan, and Russia.

Permanent waste disposal is the ultimate fate for spent fuel from reactors or, if the reprocessing option is used, wastes from reprocessing plants. These materials must be isolated from the biosphere until the radioactivity contained in them has diminished to a safe level. Under the Nuclear Waste Policy Act of 1982, as amended, the Department of Energy is responsible for developing a waste disposal system for spent nuclear fuel and high-level radioactive waste. Current plans call for the ultimate disposal of the wastes in solid form in licensed deep, stable geologic structures. We return to this issue later in this chapter.

INSIDE A NUCLEAR REACTOR

Nuclear energy is used to generate electricity in a nuclear reactor, an apparatus in which a nuclear fission reaction can be initiated, controlled, and sustained at a specific rate. A reactor includes fuel, moderating materials to control the rate of fission, a heavy-walled pressure vessel to house reactor components, shielding to protect personnel, a system to conduct heat away from the reactor, and instrumentation for monitoring and controlling the reactor's systems.

Nuclear reactors come in a variety of designs. Pressurized water reactors (PWR) account for about two-thirds of the nuclear power reactors in the United States (Figure 21.12). The **reactor core** holds the fuel assemblies and is where the fission chain reaction occurs. The core is housed in a **reactor vessel,** a large vessel of steel or concrete. This vessel is used to pressurize the coolant in gas–cooled and light water–cooled reactors. It is a cylindrical vessel whose top head is removable to allow for the refueling of the reactor. Between the fuel assemblies are **control rods** that precisely control the fission chain reaction. They do so by absorbing neutrons released by the fission events. Made of neutron-absorbing materials such as boron or cadmium, control rods can be moved in and out of holes in the core of a reactor and thereby slow or speed the chain reaction.

Each fission of a ^{235}U nucleus produces neutrons that move at a velocity of more than 3,500 meters (11,480 feet) per second. **Moderators** are used to slow down the neutrons and thereby increase their chance of being captured by another uranium atom. Deuterium (a form of hydrogen), graphite, and water are frequently used moderators.

A **primary coolant** is used to remove the heat generated by the fission process and maintain the temperature of the fuel within acceptable limits. Sometimes the moderator and the coolant are the same material (such as water). Powerful pumps force water through the core at approximately 378,541 liers (100,000 gallons) per minute. A **pressurizer** keeps the water flowing through the reactor vessel under very high pressure over 15.2 million pascals (2,200 pounds per square inch) to prevent it from boiling, even at operating temperatures over 315°C (600°F). The reactor coolant flows from the reactor to the steam generator. Inside the steam

FIGURE 21.12 *Inside a Nuclear Reactor* The design of a pressurized water nuclear reactor.

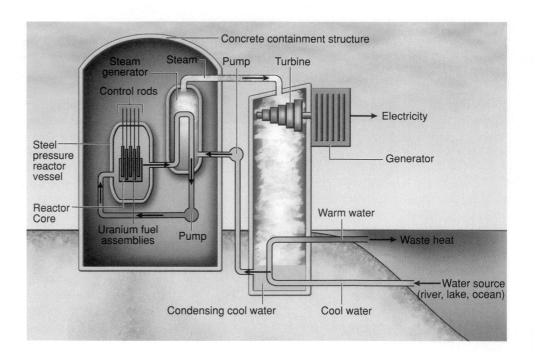

generator the hot reactor coolant flows through the many tubes. The **secondary coolant,** or feedwater, flows around the outside of the tubes, where it picks up heat from the primary coolant. When the feedwater absorbs sufficient heat, it starts to boil and form steam. The steam is transferred to the main **turbine generator,** where it is converted into electricity. After passing through the turbine, the steam is routed to the main condenser. Cool water, passing through the tubes in the condenser, removes excess heat from the steam, which allows the steam to condense. The water is then pumped back to the steam generator for reuse.

A **breeder reactor** is a particular sort of fission reactor that uses neutrons produced in the fission process to make new fuel. The most common design is the liquid metal fast breeder reactor, which uses liquid sodium as a coolant. The core contains a mixture of uranium dioxide and plutonium dioxide, surrounded by a blanket of nonfissile uranium-238. The U-238 captures neutrons from reactions in the core and is partially converted to fissile plutonium-239, which can then be reprocessed for use as nuclear fuel. In such a reactor more fissile plutonium nuclei can be produced than the number of fissile nuclei that undergo fission (hence the name "breeder").

Only a few countries have built breeder reactors because they are more expensive than conventional fission reactors. The largest fast breeder reactor to date, *SuperphÈnix,* entered service in France in 1984 but was shut down in 1997 due to high operating costs. The United States, the U.K., and Germany have drastically scaled back their breeder reactor programs. India remains the only country currently engaged in active development of breeder reactors.

Nuclear Reactor Safety

One of the first things people ask about nuclear power plants is whether a reactor can blow up like an atomic bomb. The answer is no; there is not enough fissionable material, and the material present is in the wrong arrangement to produce such an explosion. But there are other safety issues to consider.

Nuclear power plants have a complex system of controls designed to minimize the risk of radiation exposure to people or the environment. All of a plant's primary mechanical and electrical systems have secondary backup systems that are triggered if a primary system fails. For example, the emergency core cooling system combines a variety of safety measures designed to operate automatically to keep the core of a reactor covered with water if the primary circuit should spring a leak. Another system monitors temperatures in the core and automatically releases the control rods in the event of an unusual or sudden spike in temperature.

In addition to backup systems that monitor and regulate what goes on inside the nuclear reactor, U.S. nuclear power plants also use a series of physical barriers to prevent the escape of radioactive material. The first barrier is the nuclear fuel itself. The uranium fuel is in the form of solid ceramic pellets. Most of the radioactive by-products of the fission process remain locked inside the fuel pellets. The next barrier is the fuel rods that hold the fuel pellets. They are made of a zirconium alloy that is resistant to heat, radiation, and corrosion. The fuel rods are inside a large steel pressure vessel, with walls about eight inches thick. Finally, these barriers are enclosed in a massive reinforced concrete structure—called the **containment**—with walls that are about four feet thick (Figure 21.12). The containment's function is to prevent the release of radioactive material into the environment by enclosing it tightly.

THE DISPOSAL OF RADIOACTIVE WASTES

The Nature and Classification of Radioactive Waste

Nuclear wastes are generated by many activities, ranging from nuclear power to nuclear weapons facilities to hospitals and universities. In general, radioactive waste classes are based on the waste's origin, not on the physical and chemical properties of the waste that could determine how it should be managed (Table 21.2). By far the largest quantities—in terms of both radioactivity and volume—are generated by the commercial nuclear power and military nuclear weapons production industries and by nuclear fuel cycle activities to support these industries such as uranium mining and processing. A typical nuclear power plant produces 18.1–22.6 metric tons (20–25 tons) of spent fuel each year. All of the U.S. nuclear power plants together produce about 1,814 metric tons (2,000 tons) of used fuel every year; about 36,287 metric tons (40,000 tons) of used fuel are now stored at more than seventy nuclear plant sites around the country.

All parts of the nuclear fuel cycle—from uranium mining and enrichment, through use of the fuel to generate electricity, to management of used fuel and decommissioning—produce some radioactive waste. This waste includes materials with different physical, chemical, and radioactive characteristics that requires different types of management. The broad characteristics of radioactive waste that affect how it should be managed are how long the radioactivity lasts, the concentration of the radioactive material, and whether it generates heat (which is related to its concentration).

The half-lives of the radioactive elements determine how long the waste has to be managed. The concentration of the radionuclides and whether the waste generates heat dictates how it should be handled: how much, if any, shielding is needed. These considerations together determine which disposal methods are suitable.

There are several main categories of waste from the nuclear fuel cycle. Residues from processing uranium ore contain naturally occurring radioactive elements mined with the uranium, along with some chemicals used in the separation process. Their radioactivity is low-level but long-lived. The materials and equipment (contaminated rags, papers, filters, solidified liquids, tools, equipment, discarded protective clothing, dirt, construction rubble, concrete, and piping) that become contaminated during the operation of nuclear facilities are another form of low-level wastes. The spent nuclear fuel is high-level, long-lived waste. Wastes that result from dismantling nuclear reactors after the fuel has been removed and from fuel processing plants at the end of their operating lives are low-level wastes with a relatively short life.

High-level wastes in the spent fuel account for almost the entire radioactivity produced by nuclear electricity generation, but a very small proportion of the total volume of waste. The long-lived but low-level wastes from processing uranium ore account for most of the volume of all radioactive wastes—fifty to one hundred times more than the rest—but very little of the radioactivity.

TABLE 21.2	Classifications of Radioactive Waste
Category of Radioactive Waste	**Definition**
High-level waste (HLW)	Spent fuel: irradiated commercial reactor fuel. Reprocessing waste: liquid waste from solvent extraction cycles in reprocessing. Also the solids into which liquid wastes may have been converted.
Transuranic waste (TRU)	Waste containing elements with atomic numbers (number of protons) greater than 92, the atomic number of uranium (thus the term "transuranic" or "above uranium"). TRU includes only waste material that contains transuranic elements with half-lives greater than 20 years and concentrations greater than 100 nanocuries per gram. If the concentrations of the half-lives are below the limits, it is possible for waste to have transuranic elements but not be classified as TRU waste.
Low-level waste (LLW)	Defined by what it is not. This is radioactive waste not classified as high-level, spent fuel, transuranic, or by-product material such as uranium mill tailings. LLW has four subcategories: Classes A, B, C, and Greater-Than-Class-C (GTCC). On average, Class A is the least hazardous while GTCC is the most hazardous.
Class A	On average the least radioactive of the four LLW classes. Primarily contaminated with "short-lived" radionuclides (average concentration: 0.1 curies/cubic foot).
Class B	May be contaminated with a greater amount of "short-lived" radionuclides than Class A (average concentration: 2 curies/cubic foot).
Class C	May be contaminated with greater amounts of long-lived and short-lived radionuclides than Class A or B (average concentration: 7 curies/cubic foot).
GTCC	Greater-Than-Class-C radioactive waste. Much of this is irradiated material produced by the decommissioning of power plants.

The Long-Term Disposal of Radioactive Waste

The wastes from nuclear facilities pose one of the most daunting environmental challenges we face: How do we safely manage materials that remain potentially harmful for thousands of years? Scientists have debated many options, ranging from dropping containers of waste into deep ocean trenches to launching them into space. The concept of removing long-lived radioactive waste from the human environment by placing it in deep underground repositories—**geologic disposal**—was proposed over forty years ago. Geologic disposal involves packaging wastes inside long-lived containers that are placed deep underground and sealing these facilities with appropriate materials. The assumption is that conditions in these underground sites remain stable over the long periods needed to allow the radioactivity to decay to a sufficiently low level. Scientists now regard geologic disposal as the best way to deal with long-term nuclear wastes (Figure 21.9).

In geologic disposal, the waste is treated to achieve a suitable physical and chemical form. High-level solid wastes such as spent fuel rods are sealed in special stainless steel cylinders. High-level liquid wastes go through a process called **vitrification,** in which the liquid waste is heated to produce a dry powder that is incorporated into borosilicate (Pyrex) glass to immobilize the waste. The glass is then poured into stainless steel canisters, each holding about 400 kg (882 pounds) of glass.

Wastes must be transported by highway or rail from power plants to the site of final disposal, raising the issue of transportation safety. To minimize the potential for leakage in the event of an accident, a stainless steel cylinder holding the solid spent waste is encased in heavy metal shielding plus two more layers of steel. This cask measures about 5 feet in diameter and 5.2 meters (5.7 yards) in length. Casks shipped on trucks weigh less than 36.3 metric tons (40 tons), whereas those shipped by train weigh 63.5–90.7 metric tons (70–100 tons). The federal government subjects shipping casks to a variety of safety tests. In such tests, the casks have been broadsided by a 120-ton (108.8 metric ton) locomotive traveling at 80 mph (128.7 km/hr) (Figure 21.13).

The Yucca Mountain Site

The Nuclear Waste Policy Act of 1982 mandated that the U.S. Department of Energy (DOE) develop and manage a federal system for disposing of high-level nuclear wastes. The Nuclear Waste Policy Amendments Act of 1987 directed the DOE to study only Yucca Mountain, Nevada, to determine its suitability as a repository site for the disposal of spent nuclear fuel and high-level radioactive waste. The DOE must prove that a repository will be safe for 10,000 years. After 10,000 years of radioactive decay, according to Environmental Protection Agency standards, the spent fuel and high-level waste will no longer pose a threat to public health and safety. The deadline for completion of an operational high-level waste facility has been extended from 1985 to 1989 to 1998 to 2010 to 2015 due to delays in completing preliminary scientific investigation and due to political opposition by the state of Nevada and some antinuclear organizations.

Situated about 160 kilometers (100 miles) northwest of Las Vegas, Nevada, Yucca Mountain looks like many of the other outcroppings in that desert region (Figure 21.14). It was formed between 11 million and 13 million years ago by a series of volcanic eruptions some 32 kilometers (20 miles) away. The volcanic material in Yucca Mountain is more than 1,600 meters (5,000 feet) thick. Scientists and engineers believe that if radioactive wastes were stored in casks in a repository deep within the mountain, they would be unlikely to escape, even over long periods. They assume that over thousands of years, some of the

FIGURE 21.13 *Transportation of Nuclear Waste* *Casks that would be used to transport nuclear waste are tested for safety, in this case being broad-sided by a 120-ton locomotive traveling at 80 mph.*

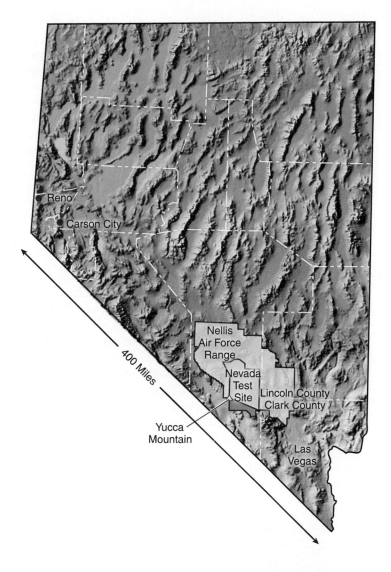

human-made barriers in a repository will break down. Once that happens, natural barriers will be counted on to stop or slow the movement of radioactive particles (Figure 21.15).

Following more than twenty years of study by the U.S. Department of Energy, in July 2002 President Bush signed legislation that recommended the Yucca Mountain site as the nation's first repository for the disposal of spent nuclear fuel and high-level radioactive waste. The next step in the repository's development is for the Department of Energy to submit a license application to the U.S. Nuclear Regulatory Commission (NRC). The Nuclear Regulatory Commission is the licensing and regulatory agency that will make the final decision on whether the Department of Energy is allowed to proceed with construction and subsequent licensing to operate the repository. The DOE's final environmental impact statement regarding the impacts of the repository will accompany that application.

Paying for Waste Disposal

The original plan for the Yucca Mountain Project called for its completion in 1985 at a cost of several billion dollars. The current deadline is 2015, and through 1998 the Yucca Mountain Project spent almost $3 billion. The government's estimate for the total cost to complete the design and to license, construct, operate, monitor, close, and decommission a monitored geologic repository at Yucca Mountain is $35 billion. Some critics charge that the cost is closer to $50 billion because the government has left out or underestimated some costs. As with most complex and controversial projects, the deadline and price tag will continue to rise.

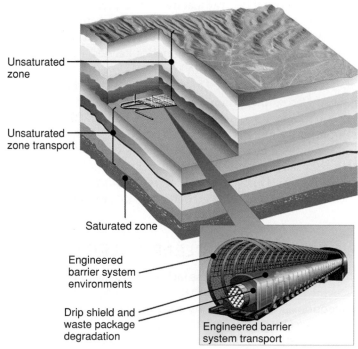

FIGURE 21.14 *Long-Term Waste Disposal* *Yucca Mountain of Nevada, site of the proposed long-term repository for high-level nuclear waste.* (Source: Courtesy of the U.S. Department of Energy.)

FIGURE 21.15 *The Yucca Mountain Repository* *The proposed system of underground burial of nuclear waste in the Yucca Mountain repository.*

Customers who use nuclear power pay for part of the cost of disposal of spent fuel. The federal government collects a fee of one mil (one-tenth of a cent) per kilowatt-hour of nuclear-generated electricity from utilities. In addition, the federal government will pay for disposal of high-level radioactive waste generated by Department of Defense programs.

Decommissioning

Nuclear power plants eventually cease operations, usually for financial reasons. A plant may require upgrades or repairs that are not economically justifiable, or the plant owners may find other sources of power that are less expensive than nuclear generation. After decades of absorbing the by-products of fission, many parts of the plant are classified as nuclear wastes. **Decommissioning** is the safe removal of a facility from service and reduction of residual radioactivity to an acceptable level. Decommissioning involves removing the spent fuel, dismantling any systems or components containing radioactive products, and cleaning up or dismantling contaminated materials. All radioactive materials have to be removed from the facility and shipped to a waste storage facility.

There are three methods of decommissioning. In the decontamination method, the equipment, structures, and portions of the facility and site that contain radioactive contaminants are removed or decontaminated immediately after the reactor shuts down. In the safe storage approach, the facility is placed in a safe, stable condition and maintained in that state while radioactive decay proceeds. It is subsequently decontaminated and dismantled. In the entombment method, radioactive structures, systems, and components are encased in a long-lived material such as concrete. The entombed structure is appropriately maintained, and continued surveillance is carried out until the radioactivity decays to a safe level.

Eleven reactors in the United States have been decommissioned. Decommissioning costs average slightly more than $400 million for a single-unit station and about $700 million for a two-unit station. A major variable in decommissioning cost and timing is the cost of low-level waste disposal, which has been increasing steadily over the past ten years with no slowdown in sight. During the next three decades more than 350 nuclear reactors will be taken out of service around the world.

HOW SAFE IS NUCLEAR ENERGY?

One of the principal problems facing the nuclear power industry is the public perception that nuclear energy is unsafe. This perception is exemplified by movies such as *The Horror of Party Beach* (1964), in which the dumping of radioactive waste off the coast of Connecticut turns a beautiful summer day into a nightmare when creepy monsters invade the beach. In *The China Syndrome* (1979) reporters inadvertently film an accident at a nuclear power plant and confront the industry's

cover-up. The title is nuclear jargon for the worst conceivable reactor accident in which the reactor core melts through the plant floor, all the way through the Earth to China (this can't actually happen). Regardless of how likely such disasters are, these stories reflected—and perhaps accelerated—growing public concern over nuclear power (Figure 21.16).

More damaging blows to the industry's safety record have come from real accidents (Table 21.3). Serious accidents such as those at Three Mile Island and Chernobyl, as well as less damaging accidents at many other facilities, indicate to some that nuclear power cannot be made safe enough to be a commercial energy source. A more detailed examination of the causes of these accidents will help us better understand safety issues.

Three Mile Island

The accident at the Three Mile Island Unit 2 nuclear power plant near Middletown, Pennsylvania, on March 28, 1979, was the most serious in U.S. commercial nuclear power plant history. Like most accidents at nuclear facilities, it was caused by a combination of equipment malfunctions, design-related problems, and worker errors.

The accident began about 4:00 a.m. on March 28, 1979, when the main feedwater pumps stopped running (due to either a mechanical or electrical failure), which prevented the steam generators from removing heat. The emergency feedwater system did not start as it should have because through either an administrative or human error, a valve was not reopened after a test several days earlier. Water flow to the core

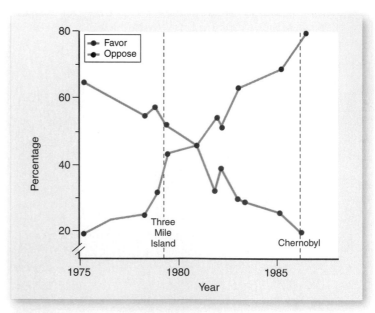

FIGURE 21.16 *Public Reaction to Nuclear Power* Results of *public opinion polls in the United States that asked, "Do you favor or oppose the construction of more nuclear power plants?"* (Source: Data from E.A. Rosa et al., "Prospects for Public Acceptance of a High-Level Nuclear Waste Repository in the United States: Summary and Implications," Public Reactions to Nuclear Waste, Duke University Press, pp. 291–328.)

TABLE 21.3	Accidents at Nuclear Power Plants and Their Release of One Form of Radioactive Material (^{131}I)		
Location and Year	Reactor Type	Accident Type	Iodine-131 Release (Curies)
Chalk River, Canada, 1952	Heavy water–moderated, light water–cooled, experimental reactor	Inadvertent supercriticality and partial meltdown	Release acknowledged; quantity unknown
Sellafield, Britain, 1957	Graphite-moderated, gas-cooled	Graphite fire	20,000
Chalk River, Canada, 1958	Heavy water–cooled and moderated reactor	Lack of coolant for a fuel element	Release contained within building
Near Idaho Falls, United States, 1961	Light-water reactor, BWR type	Accidental supercriticality; explosion, destruction of the reactor	80
Lagoona Beach (near Detroit), United States, 1966	Sodium-cooled fast breeder	Cooling system block, partial meltdown	Release confined to the secondary containment
Three Mile Island, near Harrisburg, PA, United States, 1979	Light-water reactor, PWR type	Cooling system failure, partial meltdown	13–17
Chernobyl, Ukraine, 1986	Graphite-moderated, water-cooled	Supercriticality, steam explosion, and graphite fire	7 million (minimum) (1,760 petabecquerels 10^{18} = peta)
Narora, Rajasthan, India, 1993	Heavy water–moderated and cooled, CANDU type	Turbine fire; emergency core cooling system operated to prevent meltdown system	Apparently no release of radioactivity
Monju, Japan, 1995	Sodium-cooled fast breeder	Major secondary sodium leak	Secondary sodium was not radioactive; extensive sodium contamination in plant

Source: Adapted from A. Makhijani, *The Nuclear Power Deception*, Institute for Energy and Environmental Research.

decreased. Within three hours, as much as two-thirds of the 3.6 meter (4 yard) high core was uncovered. Temperatures reached 1,927–2,204°C (3,500–4,000°F) or more in parts of the core during its maximum exposure. Although no "meltdown" occurred in the classic sense—fuel did not melt through the floor beneath the containment or through the steel reactor vessel—a significant amount of fuel did melt. Radioactivity in the reactor coolant increased dramatically, and there were small leaks in the reactor coolant system that caused high radiation levels in other parts of the plant and releases to the environment.

Shortly after the accident began, as much as 18.9 million liters (5 million gallons) of the water carrying fuel debris and fission products escaped from the reactor coolant system and flowed into the reactor building basement and into the nearby auxiliary building. There radioactive fission products were released to the environment through the building's ventilation system. This area of the plant became a major focus of the subsequent cleanup and decontamination.

On Friday, March 30, Governor Thornburgh of Pennsylvania ordered a precautionary evacuation of preschool children and pregnant women from a 5-mile zone nearest the plant, and suggested that people living within 10 miles of the plant stay inside and keep their windows closed. Most evacuees returned to their homes by April 4; by that time the situation at the reactor had been brought under control.

More than twenty years after the accident and after numerous scientific inquiries, debate remains over the human health impacts from the accident at Three Mile Island. Representing the federal government, the Nuclear Regulatory Commission maintains that there were no significant exposures to radiation and that no sicknesses or injuries can be attributed to the accident based on available scientific evidence. Scientists did find increased cancer rates in the area surrounding Three Mile Island, but claimed they were too small to be statistically significant and not likely to be associated with the nuclear accident. Based on these reports and the Nuclear Regulatory Commision's assessment, in 1993 a federal judge dismissed a lawsuit by 2,000 people who claimed that their health was damaged by the radiation released from the accident.

Some antinuclear groups and scientists disagree with the Nuclear Regulatory Commission's conclusion that Three Mile Island produced no harm. Their claims are bolstered by a 1997 study by researchers at the University of North Carolina School of Public Health. In that study scientists studied cancer cases from 1975 to 1985 in a population of 160,000 people living within 10 miles of Three Mile Island. These scientists claimed to use more sophisticated analytical and statistical techniques than the original cancer study. They concluded that following the accident, lung cancer and leukemia rates were two to ten times higher downwind of the Three Mile Island reactor than upwind. They

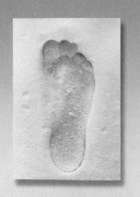

Your ECOLOGICAL *footprint*

How Much Nuclear Waste Do You Generate?

There are 104 commercial nuclear generating units licensed to operate in the United States (Figure 1). Every 12–24 months each plant is shut down, and the oldest fuel assemblies are removed and replaced. Those spent fuel assemblies become high-level nuclear waste. How much of that nuclear waste is associated with your lifestyle? The answer depends on how much of the electricity you use comes from nuclear power plants. In turn that depends on how much electricity you use and what fraction of that electricity comes from nuclear power.

The first step of calculating your nuclear waste footprint is to get a recent copy of your electric utility bill. If you live in a dorm or someplace where you don't see your bill, get a recent one from your parents. Your bill contains information about the quantity of electricity used in the past month (and perhaps year-to-date figures) in units of kilowatt-hours. From that information, estimate the total number of kilowatt-hours you use in a year (if you keep your old bills, you may be able to calculate this directly).

How many of those kilowatt-hours came from a nuclear facility? The answer depends on where you live. Table 1 lists the fraction of electricity generated from nuclear power by region in the United States for 2004. Note the strong regional differences and the fact that even if you live in a state with no nuclear reactors, you may still use power generated by a nuclear reactor. This is due to the fact that the power transmission system crosses state boundaries. Find your regional fraction from Table 1 and multiply it by your annual electricity use. This is the number of kilowatt-hours you use in a year that come from a nuclear facility.

Those 104 nuclear plants in the United States generate 532 million megawatt-hours of electricity each year. They also produce about 2,407 metric tons of uranium in the form of spent fuel each year. Use these two quantities to calculate a waste intensity for nuclear power in units of kilograms per kilowatt-hour. In the final step, multiply that intensity by

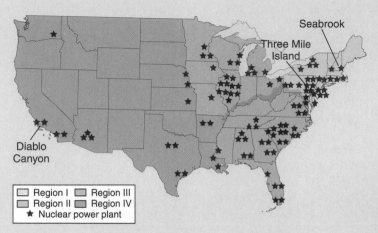

FIGURE 1 *The location of nuclear power plants licensed to operate in 2006. Note that a single nuclear facility can have more than one reactor. For example, the Diablo Canyon facility has in California two reactor units.* (Source: Data from U.S. Department of Energy.)

your annual electricity use to approximate nuclear waste generated by the electricity you use.

STUDENT LEARNING OUTCOME

- Students will be able to describe how their use of electricity is tied to the generation of nuclear wastes.

TABLE 1 Regional Share of Electricity Generated by Nuclear Plants	
Regions (States)	**Fraction of Total Electricity Use That Comes from Nuclear Power**
New England (Connecticut, Maine, Massachusetts, New Hampshire, Rhode Island, Vermont)	.25
Middle Atlantic (New Jersey, New York, Pennsylvania)	.35
East North Central (Illinois, Indiana, Michigan, Ohio, Wisconsin)	.23
West North Central (Iowa, Kansas, Minnesota, Missouri, Nebraska, North Dakota, South Dakota)	0
South Atlantic (Delaware, District of Columbia, Florida, Georgia, Maryland, North Carolina, South Carolina, Virginia, West Virginia)	.24
East South Central (Alabama, Kentucky, Mississippi, Tennessee)	.18
West South Central (Arkansas, Louisiana, Oklahoma, Texas)	.11
Mountain (Arizona, Colorado, Idaho, Montana, Nevada, New Mexico, Utah, Wyoming)	.08
Pacific Contiguous (California, Oregon, Washington)	.13
Pacific Noncontiguous (Alaska, Hawaii)	0

Source: Data from U.S. Department of Energy, *Electric Power Monthly*.

argued that the cancer findings, along with studies of animals, plants, and chromosomal damage in local residents, all point to much higher radiation levels than previously reported. Based in part on this new evidence, in 1999 a Federal Appeals court reversed part of the 1993 judgment against the 2,000 people who claimed that their health was damaged by the radiation released from the accident. New evidence will be allowed to be heard in this case.

Chernobyl

At the time of the Chernobyl accident, on April 26, 1986, the Soviet nuclear power program was based mainly on two types of reactors: a pressurized light-water reactor, and a graphite-moderated light-water reactor. The Chernobyl Power Complex, lying about 130 km (81 miles) north of Kiev, Ukraine (Figure 21.17), consisted of four nuclear reactors of the graphite–moderated design. Within a 30-kilometer radius of the power plant the total population was between 115,000 and 135,000.

The Unit 4 reactor was to be shut down for routine maintenance on April 25, 1986. Operators decided to take advantage of this shutdown to determine whether, in the event of a loss of power, the slowing turbine could provide enough electrical power to operate the emergency equipment and the core cooling water-circulating pumps until the diesel emergency power supply became operative. Unfortunately, this test was carried out without a proper exchange of information and coordination between the team in charge of the test and the personnel in charge of the operation and safety of the nuclear reactor. As a result, inadequate safety precautions were included in the design of the test.

The accident itself was caused by a destabilizing positive feedback loop unknowingly triggered by the plant operators. See *Case Study: The Chernobyl Disaster: Positive Feedback Run Amok.* A series of miscalculations by the opera-

tors produced an overwhelming power surge, estimated to be one hundred times the nominal power output. The sudden increase in heat production ruptured part of the fuel, and small hot fuel particles, reacting with water, caused a steam explosion that destroyed the reactor core. A second explosion added to the destruction two to three seconds later. The two explosions sent a shower with 7.3 of the 127 metric tons (8 of the 140 tons) of fuel one kilometer into the atmosphere—debris that contained plutonium and other highly radioactive materials. Ejected with the fuel was a portion of the graphite moderator, which was also radioactive. Cesium and iodine vapors were released both by the explosion and during the subsequent fire.

During the first ten days of the accident when important releases of radioactivity occurred, meteorological conditions changed frequently, causing significant variations in the direction of dispersion release (Figure 21.18). Major

FIGURE 21.17 *The Chernobyl Accident* The Chernobyl nuclear plant in Kiev, Ukraine, shortly after the accident in 1986.

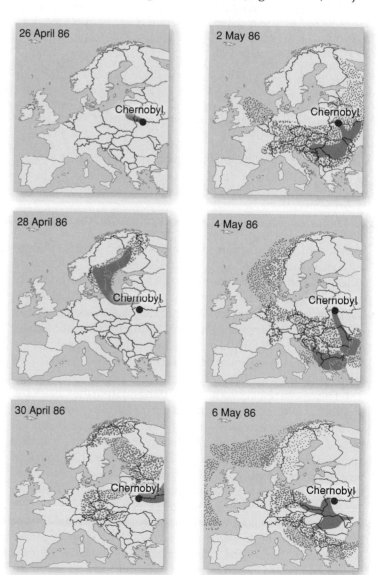

FIGURE 21.18 *Contamination from Chernobyl* Atmospheric dispersion of radioactive material from the Chernobyl accident. *(Source: Redrawn from National Atmospheric Release Advisory Center, University of California's Lawrence Livermore National Laboratory.)*

CASE STUDY

The Chernobyl Disaster: Positive Feedback Run Amok

Negative feedback stabilizes a system. Positive feedback destabilizes a system. These are fundamental principles of how systems work, and they go a long way toward explaining how and why the Chernobyl nuclear power plant exploded in April 1986.

A nuclear power plant is an extremely complex system. It has many parts that are connected to each other and to control systems that are designed to maintain the fission process within acceptable safety limits. Most of these controls are negative feedback loops. For example, an undesired increase in energy output will be sensed by the controls that operate the control rods. The control rods will be lowered into the fuel assembly, reducing the rate of fission and hence the output of heat. Similarly, an undesired increase in energy output will be sensed by the controls that operate the cooling system. More water would be pumped through the core, carrying away excess heat. These are examples of negative feedback: An initial increase in energy output is negated by the response of specific control systems.

The reactor at Chernobyl embodied what is now recognized as a serious design flaw that created the potential to trigger a dangerous positive feedback loop. Graphite-moderated reactors like the one at Chernobyl have what is called a positive void coefficient. Reactors that have a positive void coefficient can be unstable at low power and may experience a rapid, uncontrollable power increase.

In a water-cooled reactor, steam may accumulate to form pockets known as voids. If excess steam is produced, creating more voids than normal, the safe operation of the reactor is disturbed because (1) water is a more efficient coolant than steam and (2) the water acts as a moderator and neutron absorber, whereas steam does not. An increase in the quantity of steam relative to water can therefore diminish the effectiveness of the cooling system.

A reactor is said to have a positive void coefficient if excess steam voids lead to increased power generation, and a negative void coefficient if excess steam voids lead to a decrease in power. The coefficient is simply a measure of the speed of change of state of the reactor. When the void coefficient is negative, excess steam generation will tend to shut down the reactor. This is, of course, not a safety problem. When the void coefficient is positive, power can increase very rapidly because any power increase that occurs leads to increased steam generation, which in turn further increases power, and so on. In essence, a positive void coefficient can trigger uncontrollable positive feedback. Such increases are, therefore, very difficult to control.

This is exactly what happened at the Chernobyl plant. The positive void effect was triggered by the very low power output during the test. Human errors compounded the positive feedback effect. Key cooling systems had been shut off, and the control rods were improperly used. These decisions accelerated the power increase, which in turn enhanced the effects of the positive void coefficient. In a few short minutes positive feedback loops produced a huge surge in power that proved impossible to control. The rest, as they say, is history.

Since the Chernobyl disaster, designs have been altered and units have been retrofitted to protect them against the effects of the positive void coefficient. Reactors in the United States and most other nations do not use the Chernobyl reactor design.

STUDENT LEARNING OUTCOME

- Students will be able to trace the sequence of events that produced the explosion at the Chernobyl nuclear power plant.

---

releases of radionuclides continued for ten days and contaminated more than 200,000 square kilometers (77,220 square miles) of Europe. The largest particles, which were primarily fuel particles, were concentrated within 100 km (62.1 miles) of the reactor. Small particles were carried great distances by the wind and were deposited primarily with rainfall. Although all of the Northern Hemisphere was affected, only territories of the former Soviet Union and part of Europe experienced significant contamination.

Thirty-one rescue workers died outright from the Chernobyl accident, most of them from acute radiation poisoning. It is impossible to assess reliably the numbers of fatal cancers caused by radiation exposure due to the accident. Scientists and health officials predict that among the 600,000 persons receiving significant exposures in the most contaminated areas, the possible increase in cancer mortality due to this radiation exposure might be up to 4,000 additional fatal cancers.

From 1992 to 2002 in Belarus, Russia and Ukraine, more than 4,000 cases of thyroid cancer were diagnosed among those who were children and adolescents (0-18 years) at the time of the accident. Given the rarity of thyroid cancer in young people, it is likely that a large fraction of the thyroid cancers are attributable to radiation exposure from the accident. Scientists and health officials expect that the increase in thyroid cancer incidence from Chernobyl will continue for many more years, although the long term magnitude is difficult to quantify.

The economic costs of Chernobyl are immense. The local economy was the hardest hit, with 784,000 hectares (1.9 million acres) of agricultural land removed from production and 694,000 hectares (1.7 million acres) of forest in which timber production was halted. More than 350,000 people have been relocated away from the most severely contaminated areas. The total cost over the past two decades has been estimated at hundreds of billions of dollars.

Calculating the Risk

So just how safe are nuclear reactors? This is a very difficult question to answer precisely due to the uncertainty associated with the likelihood of mechanical or computer malfunctions, or the likelihood of human error in the operation of the plant. Scientists for the Nuclear Regulatory Commission in the United States have analyzed these possibilities and calculated the probability of reactor accidents of varying severity and the amount of radioactivity released by each (Table 21.4). The worst-case scenario is a so-called PWR 1 accident. This would

TABLE 21.4	Probabilities of Various Nuclear Reactor Accidents and Their Release of Radiation								
		Probability (reactor-yr^{-1})	Duration of Release (hours)	Warning Time for Evacuation (hours)	FRACTION OF CORE INVENTORY RELEASED				
Release	Category				Xe-Xr	Organic I	Cs-Rb I	Te-Sb	Ba-Sr
PWR 1	9×10^{-7}	0.5	1.0	90%	0.6%	70%	40%	40%	5%
PWR 3	4×10^{-6}	1.5	2.0	80%	0.6%	20%	20%	30%	2%
PWR 7	4×10^{-5}	10.0	1.0	0.6%	0.002%	0.002%	0.001%	0.0002%	0.00006%
PWR 9	4×10^{-4}	0.5	NA	0.0003%	7×10^{-7}%	1×10^{-5}%	6×10^{-5}%	1×10^{-7}%	1×10^{-9}%

PWR 1: Core meltdown followed by a steam explosion that ruptures reactor vessel and breaches the containment barrier; failure of heat removal systems.

PWR 3: Overpressure failure of the containment due to failure of heat removal; containment failure would occur prior to the commencement of core melting; core melting causes radioactive materials to be released through a ruptured containment barrier.

PWR 7: Core meltdown due to failure in the cooling systems; containment barrier would retain its integrity until molten core melts through the concrete containment base; radioactive materials released into the ground.

PWR 9: Large pipe break; core would not melt; minimum required engineered safeguards would function satisfactorily to remove heat from the core and containment.

NA = Not applicable.

Source: Data from U.S. Nuclear Regulatory Commission.

be a core meltdown followed by a steam explosion on contact of molten fuel, a failure of the heat removal systems, and a steam explosion that ruptures the top of the reactor vessel and breaches the containment barrier, resulting in a substantial release of radioactivity. Approximately 70 percent of the iodine radionuclides present in the core, a very hazardous group of materials, would be released. This type of accident also would release a large amount of heat energy.

The Nuclear Regulatory Commission estimates that the probability of this type of accident is 9×10^{-7} (.0000009) per reactor-year of operation. A reactor-year is defined as one reactor operating for one year. The United States has about a hundred nuclear plants. If we make the simplifying assumption that each reactor operates continuously each year, each year the industry accumulates about a hundred reactor-years of operation. Thus the probability of a PWR 1 accident occurring in any given year is about 0.00009, a very remote possibility.

These are *average* probabilities. Individual nuclear power plants vary greatly in their reliability, safety design, and operation. Some plants are much safer than others. In fact, the Nuclear Regulatory Commission has forced several plants in the United States to close before their scheduled retirement dates because they had significant safety problems, mostly related to reactor embrittlement and steam tube deterioration. Presumably the probability of accidents at these plants was greater—perhaps much greater—than those expressed in Table 21.4.

The accidents at Three Mile Island and Chernobyl caused sweeping changes in reactor design, emergency response planning, reactor operator training, human factors, engineering, radiation protection, and many other areas of nuclear power plant operations. These changes have increased the

safety of new plants and the operation of existing plants. It remains to be seen whether nuclear power is safe *enough* for the public and electric utilities to build new plants.

Proliferation and Diversion

The connection between civilian nuclear power and **nuclear proliferation** and **diversion** is a serious concern in a world rife with terrorism. Three critical diversion and proliferation risks confront civilian nuclear power. First, countries or terrorist groups could steal fissile or radioactive materials directly from the civilian nuclear fuel cycle and make them into nuclear explosives or so-called "dirty bombs" (conventional explosives surrounded by radioactive materials). Second, countries aspiring to obtain nuclear weapons could use civilian nuclear facilities to do so, and also to train scientists to produce weapons. Third, terrorists could release substantial amounts of radioactivity through attacks on reactors, spent fuel pools, or other nuclear facilities.

How serious is this threat? No one knows for sure, but these facts are not in dispute. A 1,000-megawatt nuclear power plant produces about 0.2 metric tons of plutonium annually. Virtually any combination of plutonium isotopes—including plutonium generated in commercial reactors—can be used to make a nuclear weapon, whether by unsophisticated proliferators or by advanced nuclear weapon states. Thus theft of plutonium would pose a grave security risk. In the long term, if nuclear power grows substantially as a counter to global climate change, proliferation and diversion issues will become urgent. New international agreements and cooperation will be needed (and probably demanded) that greatly exceed what is politically possible today.

THE ECONOMICS OF NUCLEAR POWER

One important issue surrounding nuclear power is the cost of its electricity compared to alternatives such as fossil fuel plants and renewable technologies. The easiest comparison is with power plants fired by coal, oil, and gas, for which substantial data are available.

The price that consumers pay for a kilowatt-hour of electricity has two components: capital cost and operating cost. The former is the cost to construct the plant; the latter is the cost to operate and maintain the plant over its lifetime. Costs include fuel costs (the cost of buying the enriched uranium, coal, oil, or natural gas) and the costs of operating the facility itself. The average operating cost for coal is about 2.1 cents per kilowatt-hour, for gas 3.5 cents, for oil 3.8 cents, and for nuclear fuel 2.3 cents. Thus in terms of operating costs alone nuclear power is competitive with fossil fuels.

However, nuclear power plants are more capital-intensive than fossil fuel plants. This means the cost of constructing a nuclear power plant is much greater than it is for fossil fuel plants. Nuclear plants require more expensive materials and systems than nonnuclear plants, and they take much longer to construct.

An important but mostly hidden cost of nuclear power is the large government subsidy received by the nuclear industry. See *Policy in Action: Should Taxpayers Subsidize Civilian Nuclear Power?* The government pays for nuclear research and for the enrichment of fuels. Perhaps the most important subsidy, however, is the limited insurance liability of the industry. In 1957 Congress passed the **Price-Anderson Act,** which indemnifies every utility operating a nuclear plant against damages greater than a certain amount. Current law caps accident liability for the *entire* nuclear industry at about $10 billion. The cost of a severe accident at a *single* reactor might exceed the cap for the entire industry. The government—that is, you the taxpayers—would pay any damages over the cap. Under intense pressure from the nuclear industry, Congress passed this law because private insurance companies were unwilling to provide the amount of insurance needed to cover worst-case accidents. The enormity of this subsidy suggests that the nuclear power industry would not exist without the subsidy provided by the Price-Anderson Act.

When capital costs and government subsidies are included, the average electricity production cost for nuclear power jumps to between 3 and 8 cents per kilowatt-hour. This is a threefold increase from the 1970s. As a result, the cost of electricity from nuclear power is greater than electricity from fossil fuels when subsidies are included.

FUSION

Another potential way to harness nuclear power is through **fusion**—the fusion of light elements into heavier ones, a process that releases energy. This process is similar to the one that powers the sun and other stars. A typical reaction fuses isotopes of hydrogen such as deuterium and tritium to produce helium and heat energy (Figure 21.19).

The energy released from fusion is enormous (Table 21.5). A kilogram of fusion fuel yields 160 times more than the energy from a kilogram of uranium oxide, the fuel in fission reactors, and more than 10 million times the energy from a kilogram of coal. The resource base for fusion is enormous. Deuterium occurs naturally at about 1 part in 6,000 in ordinary water. Tritium is a by-product of the decay of lithium. In theory, with this resource base fusion could provide centuries of electricity at current rates of use.

But the conditions necessary to liberate this energy pose considerable scientific and engineering barriers. To start a fusion reaction, gas from a combination of deuterium and tritium must be heated to 100 million °C (hotter than the core of the sun) and confined for at least one second. These conditions produce a state of matter known as plasma. Confinement of the plasma is essential to maintain a fusion reaction.

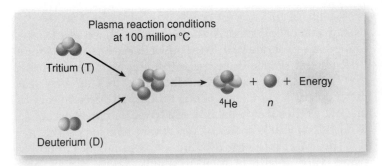

FIGURE 21.19   *Nuclear Fusion* The deuterium–tritium fusion process.

TABLE 21.5	Characteristics of Energy-Releasing Reactions		
	Chemical	**Fission**	**Fusion**
SAMPLE REACTION		$^{1}_{0}n + {}^{235}U \longrightarrow {}^{142}_{56}Ba + {}^{91}_{36}Kr + 3^{1}_{0}n$	H–2 + H–3 – > He–4 + n–1
Typical Inputs to Power Plant	Bituminous coal	UO_2 (3% ^{235}U + 97% ^{238}U)	Deuterium and lithium
Typical Reaction Temperature (K)	700	1,000	100,000,000
Energy Released per kg of Fuel (J/kg)	3.3×10^7	2.1×10^{12}	3.4×10^{14}

POLICY IN ACTION

Should Taxpayers Subsidize Civilian Nuclear Power?

Environmental economists and scientists agree that an important step toward sustainable use of the environment is the removal, to the fullest extent possible, of market externalities. An externality is a cost incurred in the production of a good or service that is not reflected in the price of that good or service. One common externality is a government subsidy: a payment to a firm or entire industry that lowers its cost of production.

The nuclear power industry receives considerable taxpayer support. Charles Komanoff, a leading analyst of the nuclear power industry, documented subsidies to the nuclear power industry from 1950 to the 1990s. The costs of generating nuclear power fall into two broad categories. Director reactor costs include the costs of plant construction and operation, the purchase of fuel, and decommissioning. Consumers pay these costs because they are included in the price of electricity generated by these plants. Indirect utility costs include money spent on plants that were canceled plus the fees utilities plan on charging in the future to pay for waste disposal and decommissioning. Part of the fee for electricity is a "fuel adjustment clause"—money that goes into a fund intended to pay for long-term waste disposal. Part of the fee also is set aside to pay for decommissioning nuclear plants. However, Komanoff and other industry analysts have found that these fees will not be sufficient to cover the escalating costs of waste disposal and decommissioning. Utilities eventually will have to raise the price of electricity to pay these full costs. All told, direct utility costs (6.3 cents/kilowatt-hour)

and indirect utility costs (1.0 cent/kilowatt-hour) produced a cost of 7.3 cents/kilowatt-hour for nuclear-generated electricity.

But society has borne additional costs of nuclear power that are not reflected in the prices that consumers currently pay, or will pay in the future, for electricity coming from nuclear power plants. The federal government has paid some of the costs of nuclear power through various subsidies:

1. The federal government has long paid for research and development of civilian nuclear technology because it is closely intertwined with military nuclear research.

2. The federal government, through agencies such as the Nuclear Regulatory Commission, has assumed responsibility for supervising the nuclear industry. Utilities pay for only part of this oversight through their licensing fees.

3. Utilities receive tax incentives through reduced corporate income taxes for their nuclear investments.

4. Although this is required by law, the Department of Energy does not charge utilities for the cost of enriching nuclear fuel.

5. The current fee consumers pay into the waste disposal fund (1 mil/kilowatt-hour) is insufficient to cover the escalating cost of building the Yucca Mountain geological repository. The actual cost will be closer to 3 mil/kilowatt-hour. The federal government (in other words, taxpayers) is likely to cover the shortfall.

These subsidies add up to about 1.6 cents/kilowatt-hour. Thus the real cost of nuclear power is closer to 9 cents/kilowatt-hour when subsidies are included. Note that this is an incomplete accounting of federal subsidies to the nuclear power industry. Excluded, for example, is the Price-Anderson Act, which shields the industry from full liability for costs to society from reactor accidents.

Like all subsidies, federal support for nuclear power distorts the actual cost of generating electricity from nuclear plants, making it appear cheaper than it really is. All other things equal, this would lead to overinvestment in nuclear energy relative to other energy sources. Proponents of renewable energy sources are concerned that subsidies to nuclear power make it more difficult to develop and commercialize alternative technologies such as wind and photovoltaic power. The cost of electricity from renewable technologies is higher than the market price of electricity from conventional sources such as nuclear power. But part of that gap is due to the subsidy to nuclear power. Removal of that subsidy would make renewable technologies more economically attractive, perhaps speeding their development.

STUDENT LEARNING OUTCOME

- Students will be able to discuss the role of subsidies in the economics of nuclear power.

There are three principal mechanisms for confining plasma: magnetic, inertial, and gravity. Magnetic confinement utilizes magnetic fields that are 100,000 times more powerful than Earth's magnetic field, arranged in a configuration to prevent the plasma from leaking out. Inertial confinement uses powerful lasers or high-energy particle beams to compress the fusion fuel.

The most promising fusion confinement device is the tokamak (from the Russian words *toroid-kamera-magnit-katushka*, meaning "the toroidal chamber and magnetic coil"); see Figure 21.20. A large electrical current flows through the plasma, which is heated to temperatures over a 100 million °C by high-energy particle beams or radio-frequency waves.

Despite decades of research and large sums of money invested by many governments in the development of

fusion technology, a commercially viable power plant remains a long way off. For this reason many governments have scaled back or abandoned their fusion programs.

Fusion does have potential environmental and safety benefits compared to conventional fission power. The fusion fuel cycle does not involve any radioactive material. It does generate radioactive wastes in the form of the intermediate fuel, tritium, and as radioactive structural materials generated by the absorption of neutrons. Over their lifetimes fusion reactors would generate, by component replacement and decommissioning, radioactive material similar in volume to that of fission reactors. However, the materials are less hazardous and have much shorter lifetimes than the high-level spent fuel wastes generated by fission. Due to the nature of the fusion reaction, there is no chance of a catastrophic runaway reaction as there is with fission.

FIGURE 21.20 *Inside a Tokamak Fusion Reactor* The image on the right was taken when the fusion reaction was actually underway and thus captures the plasma itself.

A NUCLEAR RENAISSANCE?

Can we expect to see a nuclear renaissance? Do we want one? The nuclear industry has worked hard to rebuild public confidence in the technology. A new generation of nuclear reactors embodies the lessons learned from the design, safety, management, and economic woes of the last quarter-century. They are cheaper, safer, and more reliable. We cannot overlook the fact that nuclear power emits fewer greenhouse gases and other air pollutants than do fossil fuels (Figure 21.21). Nuclear power's low carbon intensity suggests it merits consideration in the debate about energy policy.

But barriers remain, perhaps the most formidable of which is the high cost of electricity from nuclear power when subsidies are included. Even if the technological issues surrounding long-term waste disposal are resolved, many people have reservations about embracing a technology that generates wastes that will last for millennia. The industry still faces the reality that regardless of whether their reasoning is rational, many people fear a catastrophic accident.

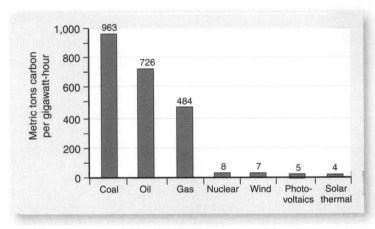

FIGURE 21.21 *Carbon Dioxide Released by Various Electricity Generation Technologies* Note the small amount released by nuclear power compared to fossil fuel technologies.

SUMMARY OF KEY CONCEPTS

- Energy from nuclear power today comes principally from the fission of uranium, which splits into two lighter fragments when bombarded with neutrons, releasing energy in the process.

- Although nuclear energy was once heralded with the promise of being "too cheap to meter," most nations no longer build new nuclear power plants. Cost issues, reduced electricity demand, accidents, and public opposition contributed to nuclear power's decline.

- The nuclear fuel cycle has a front end (mining, milling, conversion, enrichment, and fuel fabrication) and a back end (interim storage, reprocessing, and permanent waste disposal).

- Breeder reactors have the advantage of using neutrons produced in the fission process to make new fuel. However, high costs relative to conventional fission reactors have discouraged their widespread adoption.

- The plan for the long-term disposal of high-level radioactive wastes involves packaging wastes in long-lived containers that are placed deep underground and sealing these facilities with appropriate materials.

- The last stage in the life of a nuclear power plant is decommissioning, which involves removing the spent fuel, dismantling any systems or components containing radioactive products, and cleaning up or dismantling contaminated materials.

- The most serious nuclear accidents, Three Mile Island and Chernobyl, resulted from a combination of mechanical and technical failures and human error.
- In a world with significant terrorism, nuclear proliferation and diversion of plutonium from commercial reactors is a serious concern associated with any plan to build nuclear power plants.

- When government subsidies to nuclear power are taken into account, electricity from nuclear power is more expensive than electricity from many other sources.

REVIEW QUESTIONS

1. Describe the difference between a fertile and fissile isotope.
2. List and describe three reasons why no new nuclear power plants have been ordered in the United States since 1978.
3. Draw a diagram of the nuclear fuel cycle, distinguishing between the front and back ends.
4. What are the principal safety features of a nuclear reactor?
5. Should the government subsidize nuclear power? Why?

KEY TERMS

breeder reactor
containment
control rods
conversion
decommissioning
enrichment
exploration
fabrication
fertile
fissile
fission

fuel assemblies
fuel rods
fusion
geologic disposal
interim storage
milling
mining
moderators
nuclear proliferation and diversion
permanent waste disposal
pressurizer

Price-Anderson Act
primary coolant
reactor core
reactor vessel
reprocessing
secondary coolant
spent nuclear fuel
turbine generator
vitrification

22

RENEWABLE ENERGY AND ENERGY EFFICIENCY

A Concentrating Solar Thermal Collector in Golden, Colorado *Renewable energy technologies such as solar energy will eventually need to replace finite sources of fossil fuels.*

STUDENT LEARNING OUTCOMES

After reading this chapter, students will be able to

- Explain the quality/quantity paradox of solar energy.
- Distinguish between energy density and power density for renewable energy.
- Explain the strengths and weaknesses of renewable energy relative to fossil fuels.
- Describe how the federal government shapes U.S. energy policy.
- Distinguish among the major categories of energy end use, and identify where there is potential to improve efficiency in each.

SOLAR ENERGY: BACK TO THE FUTURE

Most people know from history books that the Industrial Revolution altered human existence in a fundamental way. You know from preceding chapters that the root of the Industrial Revolution was an unprecedented change in the type of energy resources humans tapped, and the accompanying changes in the types of machines used to convert that energy into useful work. You also know that the Industrial Revolution fundamentally altered the environment as well.

Many people assume that the lifestyles made possible by the changes set in motion in eighteenth-century Europe are more or less permanent, that "progress" is inevitable. A more cautious view of the future emerges when we look at these energy changes in the context of the history of human civilization. For the overwhelming majority of our existence, humans survived by using various forms of solar energy. First was the simple harvest of the products of photosynthesis, or animals sustained by the flow of solar energy through the food chain. Then came the use of fuel wood for heating and cooking and eventually for smelting metals and producing ceramics. Widespread use of fossil fuels did not begin until the late nineteenth century.

Let's assume that humans will extract and burn every last barrel of oil, every cubic foot of natural gas, and every ton of coal in Earth's crust. What would that profile of energy use look like in the long term? It would be a drop in the bucket, as indicated in Figure 22.1. At current and projected rates of energy consumption humans will exhaust fossil fuels in a few centuries—a very short period compared to how long civilization has existed.

It is clear that if humans are to survive as an industrial species, not too far in the future we will need new energy resources. In particular, once the fossil fuels are gone, we will again need to live on solar energy. There is no guarantee that the transition from fossil fuels to solar energy will be smooth, painless, or automatic. Solar energy offers some tremendous opportunities, but it also presents some formidable economic, technological, and political barriers. The purpose of this chapter is to introduce solar energy resources and technologies and to assess their potential to replace fossil fuels as the lifeblood of society. ■

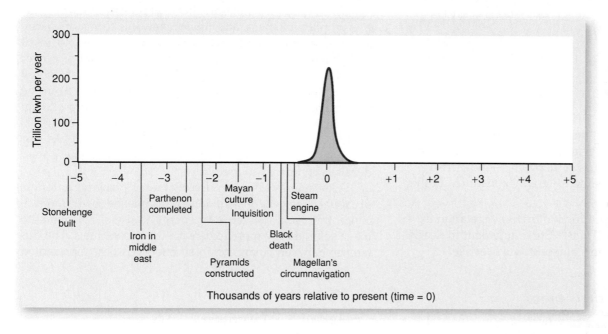

FIGURE 22.1 *The Fossil Fuel Era on a Timescale of Human Affairs* Note the relatively short time of the fossil fuel era relative to the history of human culture. *(Source: Modified from K. King Hubbert, Energy Resources—A Report to the Committee on Natural Resources: National Academy of Sciences—National Research Council. Publication 1000-D, p.141.)*

THE QUANTITY/QUALITY PARADOX OF RENEWABLE ENERGY

There is no shortage of energy on Earth (Table 22.1). Indeed the storages and flows of energy on the planet are staggering relative to human needs. Consider the following:

- The amount of solar energy intercepted by Earth every *minute* is greater than the amount of fossil fuel the world uses every *year*.
- Tropical oceans absorb 5.3×10^{20} Btu of solar energy each year, 1,600 times the world's annual energy use.

TABLE 22.1	Storages and Flows of Energy in the United States		
Energy Source	Storage (Btu)	Flow (Btu per year)	Multiple of Current U.S. Annual Energy Use
Coal	8.00×10^{19}		849.26
Oil	7.7×10^{18}		77.3
Oil shale	2.15×10^{19}		227.84
Tar sands	1.68×10^{17}		1.79
Gas	6.80×10^{18}		72.17
Unconventional gas	5.07×10^{18}		53.87
Uranium	2.70×10^{18}		28.68
Incident solar energy		4.67×10^{19}	495.75
Geothermal		3.97×10^{19}	421.28
Wind		1.55×10^{17}	1.64
Hydrogen		7.84×10^{16}	0.83
Biomass		4.68×10^{16}	0.50
Hydropower		2.09×10^{15}	0.02
Tides		7.24×10^{14}	0.01

- The potential energy in the winds that blow across the United States each year could produce more than 4.4 trillion kilowatt-hours of electricity—more than one and one-half times the electricity consumed in the United States in 2000.

- Annual net primary production by the vegetation in the United States is 4.7×10^{16} Btu, equivalent to nearly 60 percent of the nation's annual fossil fuel use.

In contrast to its vast *quantity*, the *quality* of solar energy is low relative to that of fossil fuels. Consider the energy flow in Earth's crust. The total heat loss from Earth's crust is 1.3×10^{18} Btu per year, equivalent to nearly four times the world's annual energy use. But this energy flow is spread over the entire 5.1×10^{14} square meters of Earth's surface. This means that the amount of energy flow per unit area is 2,400 Btu per square meter, an amount equivalent to just 1/100 of a gallon of gasoline.

Or consider incoming solar energy. The land area of the contiguous United States intercepts 4.7×10^{19} Btu per year, equivalent to 500 times the nation's annual energy use. But that energy is spread over nearly 7.8 million square kilometers (3 million square miles) of land area, so the energy absorbed per unit area is just 6.0×10^{12} Btu per square kilometer (1.5×10^{13} Btu per square mile) per year. Plants, on average, capture only about 0.1 percent of the solar energy reaching Earth. This means that actual plant biomass production in the United States is just 6.0×10^{9} Btu per square kilometer (1.6×10^{10} Btu per square mile) per year.

These examples illustrate that heat flow from Earth, solar energy, plant biomass, and other renewable forms of energy are diffuse, particularly when we compare them to fossil fuels (Figure 22.2). This is captured by the concept of **power density**. Power density combines two attributes of energy sources: the rate at which energy can be produced from the source and the geographic area covered by the source. A coal mine in Wyoming, for example, can produce about 10,000 watts per square meter of the mine (Figure 22.2). As our examples indicate, most solar technologies have low power densities compared to fossil fuels.

A low energy or power density means that large amounts of capital, labor, energy, and materials must be used to collect, concentrate, and deliver solar energy to users. This tends to make solar energy more expensive than fossil fuels. The difference between solar and fossil energy is best represented

FIGURE 22.2 *Power Density of Energy Sources and Energy End Uses* Fossil fuels have higher densities than renewable energy. *(Source: Modified from V. Smil, General Energetics: Energy in the Biosphere and Civilization, Wiley.)*

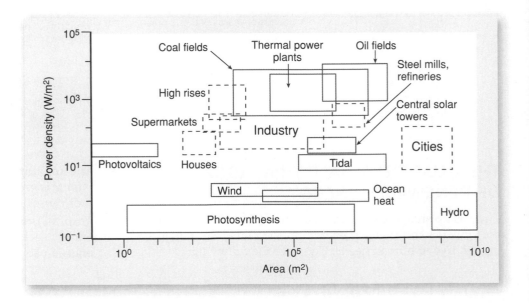

TABLE 22.2	Energy Densities of Fuels
Fuel	**Energy Density (Mj/kg)**
Peats, green wood, grasses	5.0–10.0
Crop residues, air-dried wood	12.0–15.0
Bituminous coals	18.0–25.0
Charcoal, anthracite coals	18.0–32.0
Crude oils	40.0–44.0

by their energy return on investment. The energy return on investment for fossil fuels tends to be large, whereas that for the various forms of solar energy tends to be low. This is, of course, one reason why humans aggressively developed fossil fuels.

Fossil fuels have allowed us to develop lifestyles that are quite energy-intensive. The places where we live, work, and shop have very high power densities. Supermarkets, office buildings, and factories use lots of energy for lighting, space conditioning, water heating, and other activities. This very energy-intensive way of living, working, and playing has been made possible by fossil fuel sources that are equally concentrated.

Another quality difference between renewable fuels and fossil fuels is their **energy density:** the quantity of energy contained per unit mass of a fuel (Table 22.2). Wood has a lower energy density than coal, which in turn has a lower energy density than oil. Higher energy densities also contribute to the higher energy return on investment for fossil fuels relative to many renewable fuels.

The challenge we face is to overcome the constraints imposed by the nature of solar energy and develop it in sufficient quantities to fuel both developed and developing nations. This is a formidable challenge. There is no guarantee that we will escape economic hardship in the transition from fossil to solar energy or that current lifestyles can be supported in an all-solar economy. But great strides are being made in many solar technologies. This progress is impelled by the growing awareness of the role that fossil fuels play in climate change and other environmental challenges.

THE DIRECT USE OF SOLAR ENERGY

The Solar Resource

The outer edge of Earth's atmosphere receives an average of 1,367 watts per square meter, a quantity known as the **solar constant.** Incoming solar energy is distributed unevenly among regions. Not surprisingly, solar radiation decreases with increasing latitude (Figure 22.3). Sunlight also varies with the seasons as the rotational axis of Earth shifts to lengthen and shorten days. The quantity of sunlight reaching any region also is affected by the time of day, the climate (especially cloud cover, which scatters the sun's rays), and local air pollution.

Solar Thermal Collectors

Solar thermal systems concentrate heat and transfer it to a fluid. The heat is then used to warm buildings, heat water, generate electricity, dry crops, or detoxify dangerous waste. Solar thermal collectors are divided into three categories. Low-temperature collectors provide low-grade heat (less than 43°C) for applications such as swimming pool heating and low-grade water and space heating. Medium-temperature collectors provide medium- to high-grade heat (usually 60°–82°C) either through glazed flat plate collectors that use air or liquid as the heat transfer medium, or through concentrator collectors that concentrate heat. These include evacuated tube collectors and are most commonly used for residential hot water heating (Figure 22.4).

High-temperature collectors rely on concentrators to collect and concentrate solar energy, convert it to heat, and transfer that heat to a fluid. In turn the fluid can be used for

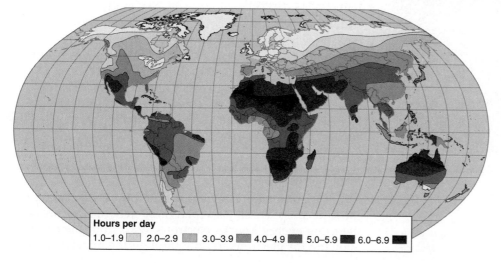

FIGURE 22.3 *Solar Energy Availability* Average solar energy as measured by the amount of solar energy received each day at Earth's surface during the least sunny month of the year. (*Source: Data from U.S. Department of Energy.*)

Hours per day
1.0–1.9   2.0–2.9   3.0–3.9   4.0–4.9   5.0–5.9   6.0–6.9

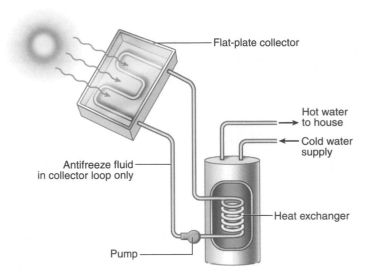

FIGURE 22.4 *Active Solar Energy* *An active solar hot water heating system.*

a variety of heating tasks or to generate electricity. These concentrating solar thermal systems use three different types of concentrators:

- Central receiver systems use heliostats (highly reflective mirrors) that track the sun and focus it on a central receiver (Figure 22.5).

- Parabolic dish systems use dish-shaped reflectors to concentrate sunlight on a receiver mounted above the dish at its focal point.

- Parabolic trough systems use parabolic reflectors in a trough configuration to focus sunlight on a tube running the length of the trough.

FIGURE 22.5 *Solar Power Tower* *The tower is a type of concentrating solar energy technology. The field of mirrors (called heliostats) reflects sunlight onto a central receiver, which transfers the energy to a working fluid. In turn the fluid is used to generate electricity.*

Solar thermal systems have several applications, the most familiar of which is water heating (Figure 22.4). Most solar water heating systems for buildings have two main parts: a solar collector and a storage tank. The most common system is a flat plate collector that transfers solar energy to a fluid—either water or another fluid, such as an antifreeze solution—to be heated. The storage tank holds the hot liquid. It can be as simple as a modified water heater, but it usually is larger and better insulated. Systems that use fluids other than water usually heat water by passing it through a coil of tubing in the tank, which is full of the hot fluid. Solar water heating systems can be either active or passive. Active systems rely on pumps to move the liquid between the collector and the storage tank, whereas passive systems rely on gravity and the tendency for water to naturally circulate as it is heated.

Solar heating systems are designed to provide hot water or space heating for nonresidential buildings. A typical system includes solar collectors that work with a pump, a heat exchanger, and one or more big storage tanks. The two main types of solar collectors used are flat plate and parabolic trough collectors.

High-temperature concentrating collectors can be used to generate electricity. The power tower system is designed specifically for this purpose (Figure 22.5). It uses a field of mirrors to concentrate sunlight on a reciever, which sits on the top of a tower. This heats molten salt flowing through the receiver. The salt's heat is used to generate electricity through a conventional steam generator. Molten salt has a high specific heat, so it can be stored for days before being converted into electricity. This means electricity can be produced on cloudy days or after sunset.

PHOTOVOLTAIC ENERGY

Photovoltaic (or PV) systems convert light energy into electricity. The term "photo" stems from the Greek photos, which means "light." "Volt" is a term honoring Alessandro Volta (1745–1827), a pioneer in the study of electricity. "Photo-voltaics," then, could literally mean "light-electricity." French physicist Edmond Becquerel first described the photovoltaic effect in 1839, but it remained a scientific curiosity for the next three-quarters of a century. At age 19 Becquerel found that certain materials would produce small amounts of electric current when exposed to light. Heinrich Hertz in the 1870s first studied the effect in solids such as selenium.

PV Technology

PV cells are made of semiconducting materials similar to those in computer chips. When these materials absorb sunlight, the energy knocks electrons loose from their atoms, allowing the electrons to flow through the material to pro-

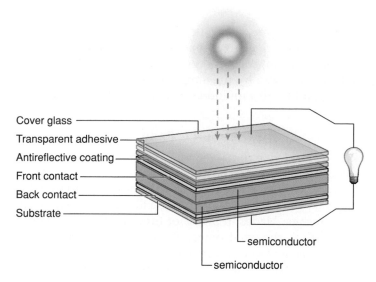

FIGURE 22.6 *Photovoltaic Effect* A schematic drawing of how the photovoltaic effect works.

FIGURE 22.7 *Photovoltaic Energy* A photovoltaic energy system at Natural Bridges National Monument in Utah. The unit provides electricity to run the Visitor Center and housing for the park rangers.

duce electricity. This process of converting light (photons) to electricity (voltage) is called the **photovoltaic effect.** A typical solar cell consists of a cover glass or other encapsulant, an antireflective layer, a front contact to allow the electrons to enter a circuit, a back contact to allow them to complete the circuit, and the semiconductor layers where the electrons begin and complete their voyages (Figure 22.6).

PV cells are typically combined into modules that hold about forty cells. The module is the smallest PV unit that can be used to generate substantial amounts of PV power. PV modules are manufactured with electrical outputs ranging from a few to more than a hundred watts of direct current (DC) electricity. About ten of these modules can be mounted in PV arrays that measure up to several meters on a side. These flat plate PV arrays can be mounted at a fixed angle facing south, or they can be mounted on a tracking device that follows the sun, allowing them to capture the most sunlight. About ten to twenty PV arrays can provide enough power for a household; for large electric utility or industrial applications, hundreds of arrays can be interconnected to form a single immense PV system (Figure 22.7).

In addition to PV modules, the components needed to complete a PV system may include a battery charge controller, batteries, an inverter or power control unit (for alternating current loads), safety disconnects and fuses, a grounding circuit, and wiring.

The performance of a PV cell is measured by its efficiency in converting sunlight into electricity. Only sunlight of certain energies will create electricity efficiently, and much sunlight is reflected or absorbed by the materials that make up the cell. Early PV cells in the 1880s converted light in the visible portion of the spectrum into electricity at an efficiency of 1–2 percent. In 1954 Bell Telephone Laboratories produced the first silicon PV cell with a 4 percent efficiency. Continued technological advances have boosted efficiencies to 15 percent, but that still means that

just one-sixth of the sunlight striking the cell is converted to electricity. Low efficiencies demand larger arrays, driving up the cost of electricity from PV systems.

The cost of PV systems has dropped sharply since the early 1980s due to the combined efforts of the PV industry and the Department of Energy, which sponsors some research and development (Figure 22.8). In the United States PV may be cost-effective for residential customers located farther than a

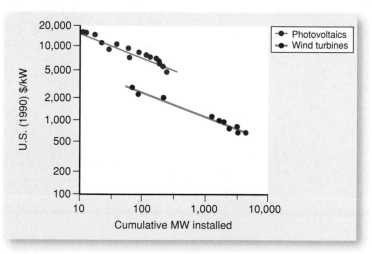

FIGURE 22.8 *Cost of Renewable Energy* The cost of photovoltaic and wind energy as a function of the cumulative amount of installed capacity.

quarter mile from the nearest utility line. If your site is already connected to a utility grid or is within a quarter mile of the grid, a PV system probably is not yet cost-effective.

Applications of PV

The most common application of PV is in consumer products that require small amounts of direct current. More than a billion handheld calculators and several million watches, portable lights, and battery chargers use PV cells for power. PV also is widely used in remote applications where connection to an existing electric grid is impossible or uneconomical. Examples of these applications include pumping water for small-scale remote irrigation, residential uses in remote villages, marine sump pumps, emergency radios, orbiting satellites, cellular telephones, and lighting billboards, security systems, highway signs, streets, and parking lots. A bigger challenge is to make PV an economical source for large-scale electricity generation.

WIND ENERGY

The terms **wind energy** and wind power describe the process by which wind is used to generate mechanical power or electricity. Wind turbines convert the kinetic energy in wind into mechanical power. This mechanical power can be used for specific tasks (such as grinding grain or pumping water), or it can drive a generator to produce electricity. Wind is a form of solar energy because the uneven heating of the atmosphere by the sun, the irregularities of Earth's surface, and Earth's rotation cause it (Chapter 4).

The Wind Energy Resource Base

Like water flowing in a river, wind contains energy that can be converted to electricity using wind turbines. The amount of electricity that wind turbines produce depends on the amount of energy in the wind passing through the area swept by the turbine blades in a unit of time. This energy flow is referred to as the wind power density. Power density typically is measured in watts per square meter. The wind power density at a given location is determined by the combined effects of several forces, including the cube of the wind speed, the frequency of various wind speeds, and the density of the air. Regions are grouped into wind power classes ranging from class 1 (the lowest) to class 7 (the highest); see Figure 22.9. Areas designated class 3 or greater are suitable for most wind turbine applications, whereas class 2 areas are marginal. Class 1 areas generally are not suitable.

Power densities reflect the energy physically available from the wind, but alone they do not determine the viability of a site. A variety of natural phenomena can impose additional costs. Too strong or off-peak winds, storms, freezing, lightning,

FIGURE 22.9 *Wind Power Density* Average annual wind power density for the United States (watts per square meter at 10 and 50 meters above the gound).

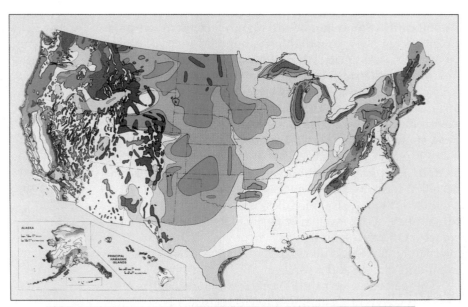

Wind power class	Classes of wind power density					
	10 m (33 ft)			50 m (164 ft)		
	Wind power	Speed		Wind power	Speed	
	W/m²	m/s	mph	W/m²	m/s	mph
	0–100	0–4.4	0–9.8	0–200	0–5.6	0–12.5
	100–150	4.4–5.1	9.8–11.5	200–300	5.6–6.4	12.5–14.3
	150–200	5.1–5.6	11.5–12.5	300–400	6.4–7.0	14.3–15.7
	200–250	5.6–6.0	12.5–13.4	400–500	7.0–7.5	15.7–16.8
	250–300	6.0–6.4	13.4–14.3	500–600	7.5–8.0	16.8–17.9
	300–400	6.4–7.0	14.3–15.7	600–800	8.0–8.8	17.9–19.7
	400–1000	7.0–9.4	15.7–21.1	800–2000	8.8–11.9	19.7–26.6

hail, vegetation, erosion, bird and animal habitat requirements, or other natural characteristics can raise costs, reduce output, or reduce wind's market value.

The cost of transmitting electricity can limit otherwise attractive wind energy sites. A site may be too far from an existing power grid to make connection economically viable. In other regions the grid may be at or near capacity and therefore be unable to accommodate much additional wind generation. Other sites may be in demand for alternative uses, which increase wind project costs and preclude otherwise excellent wind power projects—such as the Columbia Hills and Rattlesnake Hills (Washington), Burlington (Vermont), and Cape Cod (Massachusetts). In the West and Northwest environmental, scenic, cultural, and religious values restrict or preclude wind power development. Protection of birds, Native American cultural and religious values, and scenic vistas play prominently in the elimination of seemingly excellent wind sites.

Wind Turbine Technology

The blade or rotor of a wind turbine acts much like an airplane wing. When the wind blows, a pocket of low-pressure air forms on the downwind side of the blade. The low-pressure air pocket then pulls the blade toward it, causing the rotor to turn. This is called lift. The force of the lift is much stronger than the wind's force against the front side of the blade, which is called drag. The combination of lift and drag causes the rotor to spin like a propeller, and the turning shaft spins a generator to make electricity (Figure 22.10). Wind turbines can be grouped together into a single wind power plant, also known as a **wind farm,** that generates bulk electrical power. Electricity from wind farms is fed into the local utility grid just like electricity from conventional power plants.

Wind turbines are available in a variety of sizes and power output levels. The largest have blades that span more than the length of a football field, stand twenty stories high, and produce enough electricity to power 1,400 homes. A home-sized wind machine has blades between 2.4 and 7.6 meters (2.6 and 8.3 yards) in diameter, stands over 9 meters tall, and can supply about 50 kilowatts. Utility-scale turbines range in size from 50 kilowatts to several megawatts.

How Much Electricity Can Wind Provide?

This is estimated using the wind power densities shown in Figure 22.9 and by making assumptions about wind turbine technology (spacing between wind turbines, the efficiency of the machines, and so on). Current wisdom holds that good wind areas cover 6 percent of the lower forty-eight states' land area and have the potential to supply more than one and a half times the current electricity consumption of the nation. North Dakota alone has the potential to supply one-third of the total electricity consumed by the lower forty-eight states during 2000.

FIGURE 22.10 *Wind Energy Farm Facilities such as these still supply less than 1 percent of global electricity generation, but wind is the fastest-growing form of renewable energy use.*

The nation's rich wind potential must be tempered with economic and environmental realities. As a result, economic viability of wind power varies from region to region and from utility to utility. Estimates are based only on the availability of wind energy and characteristics of wind turbine technology. Estimates do not account for mismatches between supply and demand due to seasonal and daily fluctuations, constraints on the transmission of electricity, public acceptance, and other technological and institutional factors.

Current Status of Wind Energy

Wind energy is the fastest-growing energy technology in the world. In recent years worldwide capacity grew at average annual rates of more than 20 percent. Total wind energy capacity grew from 2,000 to 13,500 MW between 1990 and 2000. Wind energy's growth is concentrated in the Northern Hemisphere, particularly Europe, where renewable energy is a high priority. Of the ten countries with the highest installed capacity in 2000, only the United States, India, and China are outside Europe. Denmark uses wind power to generate approximately 10 percent of its electricity. At a smaller scale Germany's Schleswig-Holstein region uses wind for about 15 percent of its electricity; Spain's Navarra region gets more than 23 percent of its electricity from wind. But in Germany as a whole wind still provides less than 2 percent of the country's electricity.

The U.S. wind industry currently generates about 4 billion kilowatt-hours of electricity each year, which is less than 1 percent of the nation's electricity. The nation's leading wind energy generators are California, Texas, Minnesota, and Iowa. Wind energy is the fastest-growing form of new electricity generation capacity. A number of forces are driving this growth. Wind energy producers receive a generous subsidy in the form of a 1.5 cent per kilowatt-hour tax credit for utility-scale projects. This tax credit effectively lowers the cost of producing electricity from wind projects; it is intended to spur new investment. Increased competition in the electric utility industry has motivated a growing number of utilities to offer "green power" to their customers, and wind can be an attractive way to do this. States such as Minnesota and Iowa have adopted additional policies specifically designed to encourage investment in wind energy.

Environmental and Siting Issues

Generating electricity from wind produces very little air or water pollution compared to generating electricity from fossil fuels. For example, a gigawatt-hour of electricity from a coal-fired plant releases 963 metric tons of carbon; that same amount of electricity from a wind farm releases just 7 metric tons of carbon. Siting is another important issue for wind. Ecologically sensitive ecosystems can be disturbed by construction activities. For example, many attractive wind

energy sites are located in steep terrain where road and turbine construction can accelerate erosion. Careful siting review can minimize these problems.

The largest potential impact of wind energy is on migratory birds and bat populations. There are many documented accounts of significant numbers of birds and bats that have flown into wind turbines, or been sucked into the vortex behind the spinning blades, and been killed or injured. For example, wind facilities in the eastern United States are often located along natural corridors, such as mountain ridges, where some species of bats and birds concentrate during migration. Careful siting is essential to minimize these impacts.

BIOMASS ENERGY

The term **biomass** refers to all of Earth's vegetation and the many products that come from it. The biomass resource is continuously renewed through net primary production. We can harvest biomass directly (wood) or convert materials processed from biomass (the organic component of municipal wastes) into everyday forms of energy such as electricity and transportation fuels. Collectively these forms of energy are called **bioenergy** (Figure 22.11).

Bioenergy is the oldest form of energy, dating back thousands of years ago when people started burning wood to cook food or to keep warm. In addition to its renewable nature, biomass has a number of attractive features. Biomass is more evenly distributed over Earth's surface than fossil fuels, which are concentrated spatially. This provides the potential for more equitable access to bioenergy and the opportunity for local and regional control of energy supply. Some forms of biomass have fewer environmental impacts than fossil fuels. Certain forms of biomass

FIGURE 22.11 *Biomass Energy*
Pathways for converting biomass to useful forms of energy.

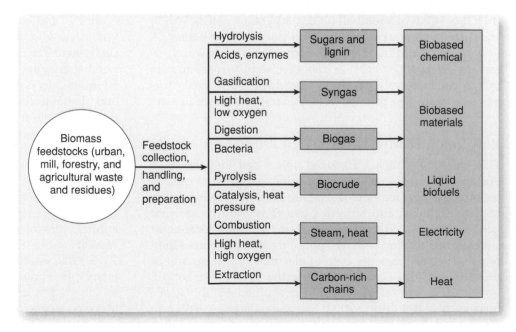

have substantial costs associated with their collection and conversion, as well as significant impacts on soil fertility if biomass is harvested improperly.

Biomass Resource Base

Terrestrial ecosystems occupy approximately 13.5 billion hectares. Net primary production for terrestrial ecosystem is about 4.6 tons per hectare per year. Of this, only a fraction can be harvested sustainably as an energy source for society. There may be as much as 255×10^{15} Btu of biomass available for human use, compared to total global energy use of about 425×10^{15} Btu.

The bioenergy resource base is defined by the availability of **biomass feedstocks** (Figure 22.11). These are the biological raw materials from which liquid fuel, gas, heat, or electricity can be produced. There are five feedstock categories. **Urban wood wastes** are the wood portions of wastes disposed of in municipal solid waste landfills, construction and demolition landfills, and yard waste taken directly to compost facilities. **Primary mill residues** are the wood wastes produced by sawmills and other wood processing facilities. **Forestry residues** are wood products in the form of logging residues, thinning cuts, and rough, rotten, and salvageable dead wood available from commercial forestry operations. This category does not include any harvest of growing stock. **Agricultural residues** are the organic materials that remain after harvest, such as wheat straw, corn stover (leaves, stalks, and cobs), and orchard trimmings. Energy crops are crops developed and grown specifically for fuel. These include fast-growing trees, shrubs, and grasses, such as hybrid poplars, willows, and switchgrass.

The quantity of biomass available for commercial use is sensitive to price. At $50 per metric ton, about 464 million metric tons (512 million tons) of biomass could be harvested annually. This represents about 8.1 quads of energy, or about 8 percent of the total energy used by the United States in 2000.

Current Status of Biomass

Biomass energy currently represents approximately 14 percent of the world's final energy consumption. The importance of biomass varies significantly across the world. In industrial Europe and the oil-rich Middle East the share of biomass averages 2–3 percent of total energy consumption. In the United States biomass accounts for about 3 percent of total energy use, with wood use by industry and households being the largest category. In developing countries biomass often is the main source of energy. Its share of total energy in Africa, Asia, and Latin America averages 30 percent but can be 80–90 percent in some of the poorest countries of Africa and Asia. Indeed, for some 2 billion people biomass is the only available and affordable source of energy for basic needs such as cooking and heating.

Biomass Technology

Biochemicals are produced when heat is used to chemically convert biomass into a fuel oil. The chemical conversion process is called **pyrolysis,** and it occurs when biomass is heated in the absence of oxygen. Pyrolysis converts biomass into a liquid that can be burned like oil to generate electricity. Pyrolysis oil is easier to transport and store than solid biomass material, and it can be refined in ways similar to crude oil. Chemicals extracted from pyrolysis oil are used to make wood adhesives, molded plastic, and foam insulation.

Biofuels are alcohols, ethers, and other chemicals made from biomass feedstocks such as fast-growing trees, grasses, aquatic plants, and agricultural and forestry residues (Figure 22.12). The two most common types of biofuels are **ethanol** and **biodiesel.** Ethanol is produced by converting the starch content of biomass feedstocks (such as corn, potatoes, beets, sugarcane, or wheat) into alcohol. The fermentation process is essentially the same process used to make alcoholic beverages: Yeast and heat are used to break down complex sugars into simpler sugars, creating ethanol. The United States currently produces more than 19 billion liters (5 billion gallons) of ethanol each year. Most of it is mixed with regular gasoline to cut down vehicles' carbon monoxide and other smog-causing emissions. But so-called flexible fuel vehicles are now available that run on mixtures of gasoline and up to 85 percent ethanol (Figure 22.13).

Ethanol is a controversial energy source. See *Case Study: Does Ethanol Have a Positive Energy Balance?* Proponents argue that it has important economic and environmental benefits. Ethanol production consumes nearly 600 million

FIGURE 22.12 *Forest Plantation* A forest plantation with fast-growing trees that are harvested for biomass energy.

CASE STUDY

Does Ethanol Have a Positive Energy Balance?

Ethanol is touted as a renewable replacement for conventional gasoline that will reduce both U.S. dependence on imported oil and the impact of transportation fuels on the environment. The potential for ethanol to live up to these expectations must be assessed with a systems perspective. Every energy system has many direct and indirect costs and benefits that must be accounted for in a consistent way before comparisons can be made.

David Pimentel, an agricultural and environmental scientist at Cornell University, is a long-time critic of ethanol. Pimentel argues that ethanol is not a viable alternative to conventional fuels because it has many negative economic and environmental costs. In particular, Pimentel finds that ethanol has an energy return on investment less than 1—that is, the quantity of energy in a gallon of ethanol is less than the energy required to produce it. The reason is the large indirect energy costs of ethanol production. The production of corn uses fertilizer, pesticides and other chemicals, machinery, labor, irrigation water, and other inputs.

All of these inputs require substantial amounts of fossil fuels to be manufactured and transported to the farm. More fossil fuels are used to transport the corn and to build, operate, and maintain the ethanol production facility. According to Pimentel, these direct and indirect energy costs amount to about 130,000 Btu per gallon of ethanol. A gallon of ethanol contains 76,000 Btu. Thus the energy return on investment for ethanol is 0.58 (76,0000 Btu ÷ 130,000 Btu).

Scientists at Argonne National Laboratory and the University of California at Berkeley disagree with Pimentel's results. Their research suggests that ethanol from corn has an energy return on investment of about 1.4, implying that ethanol yields 40 percent more energy than is required for its manufacture. These studies assume a higher yield of corn than Pimentel does, and they account for improvements in the efficiency of energy use on farms and in the manufacture of farm inputs such as fertilizers. Pimentel also includes some costs that the other studies exclude, such as the energy required to make the farm machinery that is used in corn production.

This debate raises an important question: How important is the energy return on investment for a fuel? Ethanol proponents argue that what really matters is that production of ethanol can achieve a net gain: a more desirable form of fuel. Abundant domestic supplies of natural gas and coal can effectively be used to convert corn into a premium liquid fuel that replaces imported petroleum, reduces emissions, and provides an important market for farmers. Others argue that it is ridiculous to tout a fuel as a viable long-term substitute when at best it barely breaks even in net energy terms. An economy cannot sustain itself, much less grow and develop, with energy systems that only break even in energy terms.

STUDENT LEARNING OUTCOME

- Students will be able to describe why ethanol is close to the energy break-even point.

FIGURE 22.13 *Ethanol* *Most ethanol is mixed with regular gasoline to cut down a vehicle's carbon monoxide and other smog-causing emissions.*

bushels of corn, provides a valuable source of income for farmers and reduces our dependence on imported oil. In addition, ethanol produces less air pollution, particularly less carbon dioxide, compared to regular gasoline. But critics charge that ethanol is an uneconomical energy source that would not be produced without a handsome subsidy from the federal government to farmers who grow crops for ethanol. Producers of ethanol receive a 5.4 cents/gallon tax credit. Refiners typically mix 1 gallon of pure alcohol with 9 gallons of gasoline to produce 10 gallons of ethanol. That translates into a 54-cent subsidy per gallon of alcohol.

In addition, critics question whether ethanol can be produced at a net energy gain—that is, with an energy return on investment greater than 1. Growing corn or another feedstock, transporting it to an ethanol facility, and converting it to a fuel requires a considerable amount of energy. Some analyses suggest that the energy return on investment for ethanol is less than 1. If correct, it is unlikely that ethanol can meet our long-term liquid fuel needs. Other analyses suggest that new fermentation technologies can produce ethanol with an energy return on investment greater than 1. See *Your Ecological Footprint: Your Renewable Energy Footprint*.

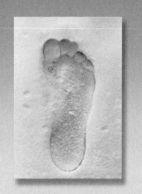

Your ECOLOGICAL *footprint*

Your Renewable Energy Footprint

In Chapter 20 you calculated your "fossil fuel footprint"—the quantity of fossil fuels you use directly and indirectly. In the long run we know that the use of oil and other fossil fuels is not sustainable because they are nonrenewable. Replacing fossil fuels with renewable energy sources has important implications for land use and biological productivity. What would your energy footprint look like if you replaced all the fossil fuels you use with renewable energy?

You can answer by calculating the land required to produce a renewable but technologically equivalent substitute for oil. Ethanol is one possible substitute for oil. The land equivalent of oil can therefore be represented as the productive land necessary to grow the biomass that is converted to the equivalent amount of ethanol. How much land would you need to replace your use of oil with a renewable substitute?

1. Find your consumption of oil from Chapter 20's Your Ecological Footprint. Convert that quantity of oil to Btu.

    Energy content of = _____ Btu oil consumption (Btu)

2. Convert your consumption of oil measured in Btu to gallons of oil (1 gallon oil = 130,000 Btu).

    Convert oil consumption = _____ Btu × 1 gallon/ 130,000 Btu to gasoline equivalent (gallons)

3. Convert your consumption of oil measured in gallons to ethanol (1 gal gasoline = 130,000 Btu; 1 gal ethanol = 76,000 Btu).

    Gallons ethanol needed = _____ gallons gasoline × (1.64 gals ethanol/gal gasoline) to replace 1 gallon gasoline

4. Convert your consumption of ethanol measured in gallons to the quantity of corn needed to produce that ethanol (1 bushel corn yields 2.5 gals ethanol).

    Quantity of corn needed = _____ gallons ethanol × (1 bushel corn/2.5 gal ethanol) to produce ethanol (bushels)

5. Convert the quantity of corn used to produce ethanol to the quantity of land needed to grow that corn (U.S. average corn yield = 110 bushels/acre).

    Quantity of cropland needed = _____ bushels corn × (0.0091 acres/bushel corn) to grow corn (acres)

    This last number is the amount of cropland that would be needed to grow the corn needed to replace your use of oil with ethanol.

STUDENT LEARNING OUTCOME

- Students will be able to describe the land requirements for biomass fuels.

Biopower is the use of biomass to generate electricity. Most biopower plants use direct-fire systems that burn biomass feedstocks directly to produce steam that is captured by a turbine and converted by a generator into electricity. In the lumber and paper industries, wood scraps are sometimes directly fed into boilers to produce steam for manufacturing processes or to heat buildings. Gasification systems use high temperatures and an oxygen-starved environment to convert biomass into a gas (a mixture of hydrogen, carbon monoxide, and methane).

Environmental Impacts of Biomass

Biomass currently supplies about 12% of the world's energy. This major utilization of biomass, in addition to habitat destruction and environmental pollution, contributes to the loss of biodiversity that we discussed in Chapter 12. Harvesting and processing biomass can have significant environmental impacts. Removing vegetation always has implications for soil quality, nutrient cycling, water availability, and biodiver-

sity. Harvesting crop residues, for example, can accelerate soil erosion and deplete the soil of nitrogen and other important nutrients.

One of the most important attributes of bioenergy is its effect on the carbon cycle. If the land from which biomass is harvested is replanted and managed sustainably, the growing trees and other plants remove carbon dioxide from the atmosphere during photosynthesis and store the carbon in their structures. When the biomass is burned, the carbon released back to the atmosphere will be recycled into the next generation of growing plants. When biomass is used for fuel in place of fossil fuels, the carbon in the displaced fossil fuel remains in the ground rather than being discharged to the atmosphere as carbon dioxide. The growth rate of the trees becomes an important consideration. Whereas slow-growing trees can take a long time to recapture released carbon, fast-growing trees can recycle carbon rapidly and will displace fossil fuel use with every cycle.

GEOTHERMAL ENERGY

The term "geothermal" comes from the Greek words *geo*, meaning Earth, and *therme*, meaning heat. Heat from radioactive decay deep within Earth has produced a huge warehouse of heat, storing over 7.6×10^{27} Btu of heat energy. This is 23 billion times the annual world energy use. But this energy moves slowly outward and is spread very diffusely over the entire Earth. If all the heat escaping from each square meter of the surface could be captured and used to heat a cup of water, it would take more than four days to bring it to a boil.

In special circumstances, however, geological cycles have concentrated this heat in **geothermal fields.** Geothermal fields are classified as low temperature (less than 90°C or 194°F), moderate temperature (90°–150°C or 194°–302°F), and high temperature (greater than 150°C or 302°F). The U.S. geothermal resource base is estimated to be 8,000 to 18,000 quads of energy (1 quad = 1×10^{15} Btu). By comparison, total energy use in the United States in 2006 was about 100 quads. Thus the resource base is vast, although these numbers reflect only physical availability, not economically or technologically recoverable resources.

Geothermal Technology

Three basic technologies convert geothermal resources into useful energy. The first is geothermal electricity generation. **Flash steam power plants** are the most common (Figure 22.14). Very hot water (greater than 182°C) flows up through production wells in the ground under its own pressure. As it flows upward, the pressure decreases and some of the hot water boils into steam. The steam is separated from the water and used to power a turbine generator. Any leftover water and condensed steam are injected back into the reservoir, where they can be reheated and reused. At The Geysers in California, the world's largest geothermal field, power is sold at about 3–3.5 cents per kilowatt-hour, making it competitive with conventional methods of electric power generation. Twenty-one countries generate about 8,000 megawatts of electricity from geothermal resources, accounting for less than 1 percent of the world's electricity.

The second major geothermal technology is **geothermal direct use.** As the name implies, this involves using the heat in the water directly (without first converting it to electricity) for space heat, industrial processes, greenhouses, aquaculture (growing of fish), and resorts. Direct use projects generally use resource temperatures between 38°C and 149°C. Geothermal direct use dates back thousands of years, when people began using hot springs for bathing, cooking food, and loosening feathers and skin from game. Today hot springs are still used as spas. In addition to these longstanding uses, wells are now drilled into geothermal reservoirs to provide a flow of hot water. The water is brought up

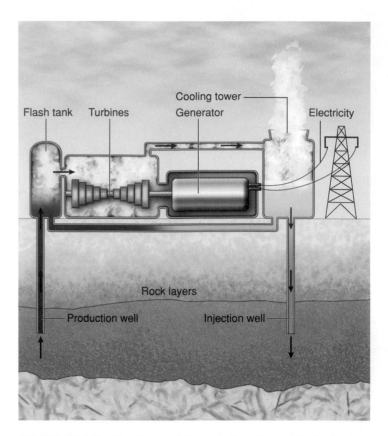

FIGURE 22.14 *Geothermal Energy* A geothermal flash steam plant that generates electricity.

through the well, and a system of piping, heat exchangers, and controls delivers the heat directly to its point of use. A disposal system then either injects the cooled water underground or disposes of it on the surface.

The third major geothermal technology is the **geothermal heat pump** (Figure 22.15). The shallow ground (the upper 3 meters of Earth) maintains a nearly constant temperature between 10°C and 16°C. This ground temperature is warmer than the air above it in the winter and cooler than the air in the summer. In the winter the heat pump removes heat from the heat exchanger in the soil and pumps it into the indoor air delivery system. In the summer the process is reversed, and the heat pump moves heat from indoor air into the soil. Heat removed from indoor air during the summer can also be used to heat water. Geothermal heat pumps tend to use less energy than conventional heating systems because they draw "free" heat from the ground.

Environmental Impacts of Geothermal Energy

Like photovoltaic and wind systems, the generation of electricity from geothermal systems releases significantly less air pollution than fossil fuel systems. For example,

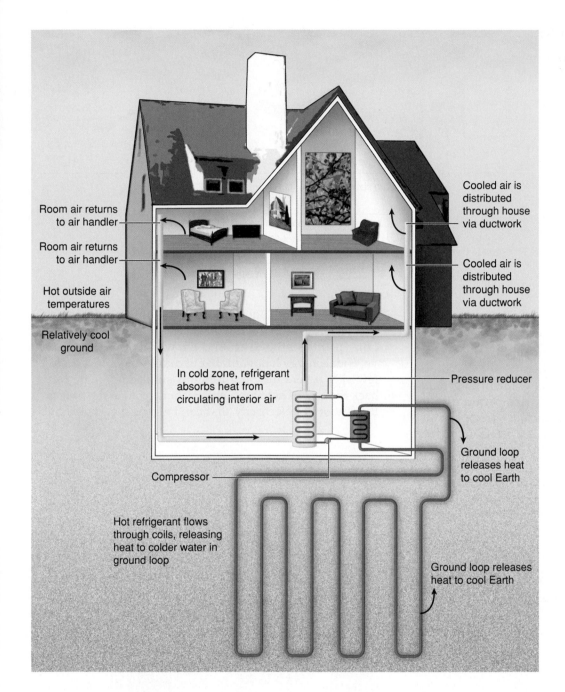

FIGURE 22.15   *Ground Source Heat Pump* *This energy system relies on temperature differences between the ground and a building to provide* *cooling.* *(Source: Redrawn from Geothermal Heat Pump Consortium, Inc.)*

Cooled air is distributed through house via ductwork

Room air returns to air handler

Room air returns to air handler

Cooled air is distributed through house via ductwork

Hot outside air temperatures

Relatively cool ground

In cold zone, refrigerant absorbs heat from circulating interior air

Pressure reducer

Ground loop releases heat to cool Earth

Compressor

Hot refrigerant flows through coils, releasing heat to colder water in ground loop

Ground loop releases heat to cool Earth

a geothermal power plant produces about one-sixth as much carbon dioxide as one powered by natural gas and even smaller amounts of nitrogen and sulfur-bearing gases. The land area required for geothermal power plants is smaller per megawatt than for almost every other type of power plant. However, geothermal plants process large amounts of water that contain salts and other potentially harmful substances. These wastewaters must be disposed of properly to avoid adverse environmental impacts.

OCEAN ENERGY SYSTEMS

Covering slightly more than 70 percent of Earth's surface, the oceans are the world's largest solar energy collector and storage system. On an average day the solar energy absorbed by 2.5 square kilometers of ocean is the equivalent of more than 7,000 barrels of oil. Each year tropical oceans absorb about 5.3×10^{20} Btu of solar energy, equivalent to 1,600 times the world's annual energy use. As we describe next, this is a vast but very diffuse energy source that requires a significant investment of time, energy, labor, and materials to capture.

FIGURE 22.16   *Ocean Thermal Energy*
Global distribution of the ocean thermal energy resource. The temperature difference (°C) is shown between the surface and 1,000 meters depth. (*Source: Data from U.S. Department of Energy, National Renewable Energy Laboratory.*)

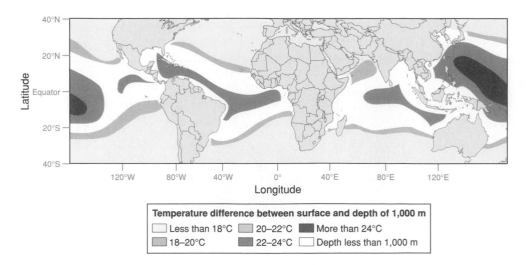

Temperature difference between surface and depth of 1,000 m
- ☐ Less than 18°C
- ☐ 18–20°C
- ☐ 20–22°C
- ☐ 22–24°C
- ■ More than 24°C
- ☐ Depth less than 1,000 m

Ocean Thermal Energy

In tropical regions the sun's energy heats surface waters to 25°C, establishing a strong temperature gradient with deep water at 5°C (Figure 22.16). **Ocean thermal energy conversion** (OTEC) systems use this thermal gradient to generate electricity (Figure 22.17). With a 20°C temperature difference, the best possible conversion efficiency is about 6 percent. The actual efficiency is more like 2–3 percent, and huge volumes of water must be pumped to capture the diffuse heat energy. An OTEC station generating 100 megawatts of electricity, for example, must pump 450 cubic meters of water through its system *every second*, making OTEC a very expensive technology. OTEC systems have a potential added bonus: In addition to electricity, they can generate a number of other valuable products such as freshwater, deep-water mariculture (the growth of commercially valuable fish and shellfish), and chilled water for air conditioning. Due to OTEC's high cost, tropical islands with growing power requirements, dependence on expensive imported oil, and shortages of freshwater are the most likely candidates for OTEC development.

Tidal Energy

The oceans yield another form of energy, but this one extracts energy from the kinetic energy of the Earth–moon–sun system. Tides are periodic variations in sea level caused by gravitational forces exerted by the moon. The usual technique (referred to as "barrage" technology) is to dam a tidally affected estuary or inlet, allowing the tidal flow to build up on the ocean side of the dam and then generating electricity during the two-hour high tide period. After the water level reaches maximum high tide, gate valves are closed and the water is impounded, awaiting low tide when it is released back through the dam where it again generates electricity. The largest **tidal energy** system is the Rance power plant in the Gulf of Saint-Malo, France, built

in the 1960s. The Rance plant has twenty-four power units that generate 10,000 kilowatts each. There are a handful of other plants scattered around the world.

Tidal energy has a number of drawbacks. First, there are few sites where the tidal range is sufficiently large to pay back the cost of development. Second, even at the best sites, the amount of electricity that can be generated is relatively

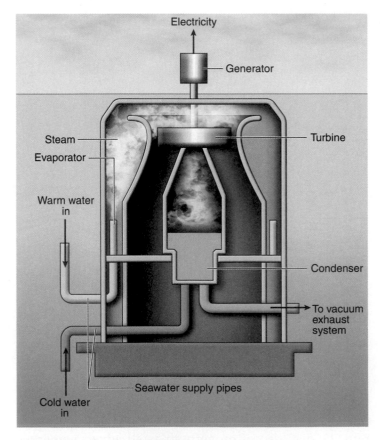

FIGURE 22.17   *Ocean Energy* An ocean thermal energy conversion (OTEC) energy system.

small. Third, tidal power is out of sync with peak electricity demand times because the cycle of tides doesn't always match the cycles of demand for electricity. Fourth, and perhaps most important, the best tidal sites are in estuarine and coastal ecosystems. Large-scale alteration of the hydrology of these systems could adversely affect the broad array of environmental services these ecosystems provide people and other species. All of these factors will limit the future role of tidal energy.

Wave Energy

Ocean waves are a concentrated form of wind energy. As wind blows across water, friction develops between air and water. As energy from wind flow is transferred to water, waves are produced. Wave energy is present in the form of potential energy due to the elevation of the wave above the still-water level, and kinetic energy due to the orbital motion of the component particles. Generating energy from waves is a matter of translating the motion of the waves into mechanical or electrical energy. The diffuse nature of this resource makes it prohibitively expensive. So far the only practical use of wave energy has been small self-powering navigational buoys.

HYDROPOWER

Hydroelectric energy (**hydropower**) is produced by the force of falling water. The most common type of hydropower plant uses a dam on a river to store water in a reservoir. Building up behind a high dam, water accumulates potential energy. This is transformed into mechanical energy when the water rushes down a tunnel or pipe called a penstock and strikes the blades of a turbine (Figure 22.18). The turbine's rotation spins electromagnets that generate current in stationary coils of wire. But hydropower doesn't necessarily require a large dam. Some hydropower plants use a small canal to channel the river water through a turbine.

In a **pumped storage plant,** the reverse is done. Power is sent from a power grid into the electric generators at a hydro facility. The generators then spin the turbines backward, which causes the turbines to pump water from a river or lower reservoir to an upper reservoir, creating a storage of potential energy. When needed, water is released from the upper reservoir back down into the river or lower reservoir, spinning the turbines forward and activating the generators to produce electricity. Pumped storage often is used as a reserve supply of energy that is tapped when demand for electricity is very high, such as during a stretch of hot days when many people use air conditioning systems.

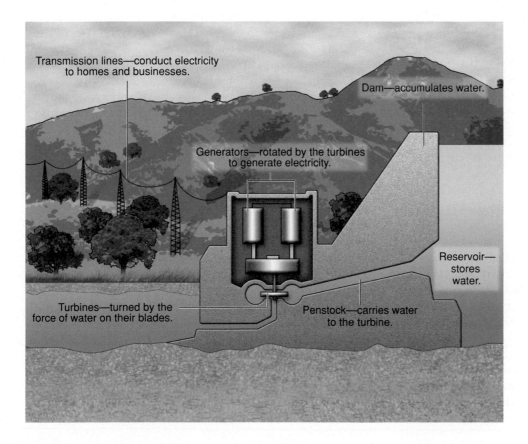

FIGURE 22.18 *Hydropower Facilities such as these generate about 7 percent of U.S. electricity needs.*

Transmission lines—conduct electricity to homes and businesses.

Dam—accumulates water.

Generators—rotated by the turbines to generate electricity.

Reservoir— stores water.

Turbines—turned by the force of water on their blades.

Penstock—carries water to the turbine.

Current Status of Hydropower

Hydropower currently accounts for about 2.5 percent of world energy use, making it the second largest renewable energy source after biomass. The largest producers of hydroelectric energy are the United States, Canada, and Brazil—immense nations with extensive river systems. In terms of hydropower's contribution to total electricity use, Norway, Brazil, Venezuela, and Canada all get more than half their electricity from water power. The United States supplies about 7 percent of its electricity from hydropower.

Environmental and Social Impacts of Hydropower

Hydropower has many desirable characteristics compared to electricity generated from fossil fuels. Because no fuels are burned, the release of air pollutants and solid wastes is minimal, and no resources are depleted. Yet we must weigh a host of environmental and social impacts unique to hydropower when comparing it to other forms of energy. These impacts generally relate to how a hydroelectric project affects a watershed ecosystem and its habitats. These changes include flooding of the original ecosystem, sedimentation behind the dam, damage to riparian habitat, and barriers to migratory fish.

In addition to the disruption of ecosystems, large hydropower projects can also disturb human communities. People living in the area that will be flooded by the reservoir are forced to relocate their homes and businesses; often entire communities must be moved. A good example of this is the James Bay Project (located in the Canadian province of Quebec), one of the world's largest power facilities. The Cree Indians lost their entire island home to the water diversion needed for the project. They also lost trap lines and fishing zones and faced difficulties in navigating the rivers and obtaining drinking water supplies due to the radically altered hydrology of the region. Roads and communication systems built to support the project accelerated the cross-cultural transplant of lifestyles and consumption patterns of Western society among the Cree.

HYDROGEN

Hydrogen (H_2) is a simple, abundant, and naturally occurring element that is found in materials such as natural gas, methanol, coal, biomass, and water. Hydrogen is not an energy source like coal or wind. Hydrogen must be manufactured from other materials such as methane or water, and sometimes stored and transported to its point of use. Hydrogen has many of the same attributes and uses as natural gas (CH_4), but it has one distinct disadvantage: a low energy density (Btu/cubic meter). Methane has an energy density of 35,070 Btu/cubic meter; for hydrogen this figure is 11,065 Btu/cubic meter.

Hydrogen Production, Storage, and Transport

The most widely used method to produce hydrogen is the steam reforming of natural gas:

$$CH_4 + 2H_2O \xrightarrow{\text{steam}} CO_2 + 4H_2 \qquad (22.1)$$

This process yields approximately 70–90 percent of the hydrogen in the methane input.

Splitting water into hydrogen and oxygen is the other principal method of producing hydrogen. The most widely used method of water splitting is electrolysis, in which an electric current is run through water, decomposing it into hydrogen at the negatively charged cathode and oxygen at the positive anode. Electric energy is used to split water into hydrogen and oxygen gas:

$$2H_2O \xrightarrow{\text{electricity}} 2H_2 + 2O_2 \qquad (22.2)$$

Renewable energy sources of electricity such as solar, wind, and hydropower can be used in this process.

Use of hydrogen as an energy carrier requires that it be stored and transported. The primary methods for hydrogen storage are compressed gas, liquefied hydrogen, metal hydride, and carbon-based systems. Compressing the gas allows a greater amount of energy to be contained in a given volume. New materials have permitted fabrication of storage tanks that can hold hydrogen at extremely high pressures, although at significant cost. Hydrogen also can be liquefied into denser forms. However, converting hydrogen gas to a liquid is costly and requires large inputs of energy.

Long-term (greater than 100 days), seasonal storage converts hydrogen to chemical hydrides. Various pure or alloyed metals can combine with hydrogen, producing stable metal hydrides. These hydrides decompose when heated, releasing the hydrogen. Hydrogen can be stored as a hydride at higher densities than simple compression.

A novel technology under development is the microsphere, a very small glass sphere that holds hydrogen at high pressures. The spheres are charged with gas at high temperatures where the gas can pass through the glass walls. At low temperatures the glass becomes impervious to hydrogen, thus locking in the gas.

Like conventional natural gas, hydrogen may be transported using several methods: pipeline, truck, rail, and ship. Long-distance transport of hydrogen is feasible only by pipeline. Most studies have focused on using existing natural gas pipelines to transport hydrogen alone or mixed with natural gas. Existing compressors and meters cannot be used, however, due to the low energy density of hydrogen. Hydrogen may also cause more physical damage to the pipeline. As a result, transmission costs of hydrogen are 50–80 percent higher than those of natural gas. Building new pipelines to carry hydrogen is prohibitively expensive, again due to its low energy density.

Fuel Cell Applications

Hydrogen can be used in stationary applications (to generate electricity) and to provide mobile power (to power a car or truck). In both stationary and mobile uses, **fuel cells** are envisioned to play a major role. Fuel cells generate electricity through an electrochemical process in which the energy stored in a fuel is converted directly into DC electricity. Because electrical energy is generated without burning fuel, fuel cells are extremely attractive from an environmental standpoint.

A fuel cell converts hydrogen fuel and oxygen from the atmosphere into electricity (Figure 22.19). Unlike an electrochemical battery, fuel cells do not store energy internally; rather fuel is stored externally. As long as there is a continuous supply of fuel, the fuel cell will continue to produce electricity. Fuel cells are small and efficient and produce little pollution. Individual fuel cells can be "stacked" until the desired power level is reached. Fuel cells presently are too expensive for large-scale use in transportation. However, new materials and engineering processes are being developed that offer the possibility of fuel cells with sufficiently high power, acceptable lifetimes, and affordable costs.

The Economic and Environmental Impacts of Hydrogen

Hydrogen's low energy density combined with the need to manufacture it from other materials creates significant technological and economic challenges. The cost of hydrogen from steam reforming, the most established technology, could reach $30/million Btu. In contrast, using natural gas directly as a fuel costs about $3/million Btu. Thus in the near term hydrogen remains very expensive compared to conventional fuels.

Steam reforming of methane has the disadvantage of venting carbon dioxide to the atmosphere, thus contributing to climage change. Research is focused on improving the economics of electrolysis using renewable sources of electricity such as photovoltaics, wind, and hydropower.

ENERGY EFFICIENCY

Thus far our discussion has focused on the supply side of the energy issue. The other side of the coin is the demand side: how and for what purposes is energy used. **Energy end use** refers to the desired physical function that is provided to a user of energy by a given energy service, such as heating, cooling, lighting, mechanical work, and so on. A closely related concept is **end use efficiency,** which is the ratio of the amount of energy service provided to the amount of energy consumed. For example, the efficiency of transportation is the quantity of gasoline needed to transport one person one kilometer. The efficiency of home heating is the quantity of natural gas burned in a furnace to heat a square meter of living space.

Energy end use typically is described in terms of its four broadest categories: transportation, commercial, residential, and industry and manufacturing. The distribution of energy use among these categories varies considerably among nations; Figure 22.20 shows energy end use for the United States in 2005.

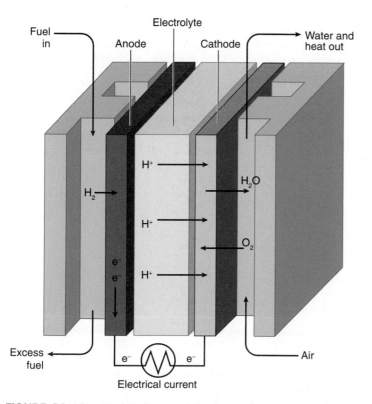

FIGURE 22.19   *Fuel Cell*   *The basic components of a fuel cell.*

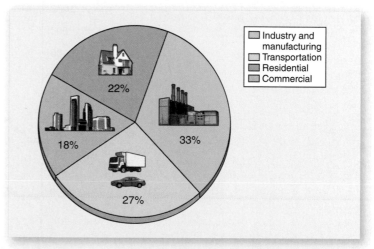

FIGURE 22.20   *Energy End Use*   *The major categories of energy end use for the United States in 2005.*

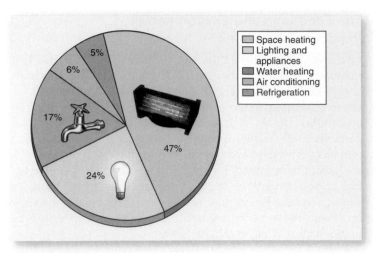

FIGURE 22.21 *Household Energy Use* A breakdown of typical activities that comprised household energy use in the United States in 2005.

The rapid transportation of people and goods is a hallmark of modern society. Transportation is the world's fastest-growing form of energy use, accounting for nearly 30 percent of global energy use and 95 percent of global oil use. Developing nations still rely heavily on walking, bicycles, and buses, whereas developed nations rely more on private cars and planes. Transportation choices are important because different modes have radically different demands for energy. The most significant drivers of rising energy consumption for transportation are the growing reliance on private cars in large developing nations such as China and the driving habits of the affluent in developed nations such as the United States, whose highways support about a fourth of the world's cars.

Energy use by households accounts for up to 20 percent of total energy use in developed nations and a much larger share for the rural poor in developing nations. Household energy use is determined by a complex array of factors including housing size, family size, income, the efficiency of energy-using devices, and the efficiency of the building itself. Space heating, water heating, lighting, and appliances dominate in developed nations, whereas cooking dominates in rural households in developing nations (Figure 22.21).

New technologies have dramatically reduced the energy used by appliances and electronic devices, due in large part to government support. See *Policy in Action: What Should the Role of Government Be in Shaping Our Energy Future?* A prominent example of this is the refrigerator (Figure 22.22). The size of refrigerators and energy use per unit increased together from 1950 to the early 1970s. At that time the size of refrigerators continued to increase, but energy use per unit dropped by nearly two-thirds by the turn of the century. The dramatic improvement in efficiency came from technological advances driven by government regulations requiring better energy performance.

Commercial energy use is increasingly important because more than half of the world's population lives in urban areas. The biggest energy users are retail, service, and office buildings, followed by schools, health care facilities, warehouses, hotels, and restaurants. More than half of energy use by the commercial sector is electricity. Major end uses include lighting, space conditioning, water heating, and office equipment.

Industry and manufacturing consume one-third of the world's energy. Petroleum refining and chemicals, paper, and metal manufacture consume most of the energy in this sector. Natural gas and electricity are the dominant forms of energy. Industry also uses some uncommon forms of energy such as wood wastes produced in papermaking, blast furnace gas produced in steel making, and still gas produced at petroleum refineries.

FIGURE 22.22 *Refrigerator Energy Efficiency* The trend over time in the size of refrigerators and the energy use per refrigerator. *(Source: Data from National Academy of Sciences.)*

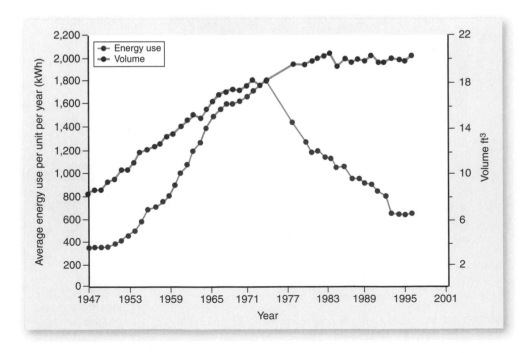

POLICY IN ACTION

What Should the Role of Government Be in Shaping Our Energy Future?

Should market forces determine what energy sources are developed, or should government policy shape our energy future? This is an enduring debate in energy and environmental circles. Federal and state governments influence every form of energy, but to greatly varying extents. One influence is the funding of research, development, and commercialization of energy technologies. Funding is allocated among different energy systems based on the priorities of the president and Congress, who propose and approve the federal budget. Not surprisingly, funding for specific technologies waxes and wanes with changes in the political winds. On the heels of the oil price shocks in the 1970s, President Carter won large increases in funding for renewable energy and energy efficiency. But Presidents Reagan and George H. Bush believed that research and

development should be done by the private sector, and they slashed funding for renewables. President Clinton increased support for renewables, although a Republican Congress elected in 1994 made this more difficult. President George W. Bush has emphasized support for fossil and nuclear fuels.

The U.S. Department of Energy (DOE) has spent more than $10 billion on the research and development (R&D) of energy-efficient technologies. Was it worth it? That is exactly what the National Academy of Science asked in a study of the benefits of the DOE's research. The study analyzed the costs and benefits of twenty-two specific case studies, including lighting, refrigerators, automobile engines, and energy-efficient glass used in new buildings. The committee found that the DOE's R&D programs have yielded significant benefits

(economic, environmental, and national security–related), important technological options for potential application in a different economic, political, or environmental settings, and vital additions to the stock of engineering and scientific knowledge in a number of fields (see Table 1). The National Academy stated, ". . . the benefits of these programs substantially exceed the programs' costs and contribute to improvements in the economy, the environment, and national security."

STUDENT LEARNING OUTCOME

- Students will be able to discuss the ways in which the federal government supports research and development of alternative energy technologies.

TABLE 1	Net Benefits for Selected Technologies Related to Energy Efficiency								
	Economic Benefits (Cumulative Net Energy Savings and Consumer Cost Savings)			Environmental Benefits (Cumulative Pollution Reduction)				Security Benefits (Oil Use or Outage Reduction)	
Technology	Cost of DOE and Private RD&D (billion $)	Electricity (Q of primary energy)	Net Cost Savings (billion $)	SO_2 (millions of metric tons)	NO_x (millions of metric tons)	Carbon (millions of metric tons)	Damage Reduction (billion $)	Oil and LPG (Q) Value (billion $)	Electricity Reliability
Advanced refrigerator/freezer compressors	≈0.002	1	7	0.4	0.2	20	1–5	0.04	0.02–0.1
Electronic ballast for fluorescent lamps	>0.006	2.5	15	0.7	0.4	40	1–10	0.1	0.05–0.3
Low-e glass	>0.004	0.5	8	0.3	0.2	20	0.5	0.2	0.1–0.7
Advanced lost foam casting	0.008	0.03	0.1	0.01	0.006	0.5	0.02–0.1		
Oxygen-fueled glass furnace	0.002		0.3		0.02	1	0.05–0.2		
Advanced turbine systems	≈0.356		≈0 by 2005		0.02	1	0.05–0.2		Yes
Total	≈0.4					0.2–1			

Q = quad = 1* 10^{15} Btu.

Source: National Research Council. *Energy Research at DOE: Was It Worth It? Energy Efficiency and Fossil Energy Research 1978 to 2000.* Washington, DC: National Academy Press, 2001.

AN ECONOMIC AND ENVIRONMENTAL COMPARISON OF SOLAR AND FOSSIL FUEL ENERGY

Proponents of solar energy argue that its time has come because it is now cost-competitive with fossil fuels, that government subsidies encourage fossil fuel use, and that fossil fuels have huge environmental externalities. Supporters of fossil fuel industries assert that conventional fuels are still the cheapest, that renewable fuels also receive generous government subsidies, and that solar proponents understate solar energy's costs.

To objectively compare the environmental impacts of renewable energy to those of fossil fuels we must use a systems approach. This evaluates the impacts of an energy system from "cradle to grave"—that is, from extraction to processing, use, and disposal. First let's look at the air, water, and solid waste impacts of various technologies to generate electricity (Table 22.3). Due to the diffuse nature of solar energy, photovoltaic, wind, and biomass power require significant quantities of land relative to coal and nuclear fuel cycles. However, the latter technologies use large quantities of fuel and water and generate lots of solid waste. Photovoltaic, wind, and geothermal power require no fuel inputs and thus use no water and release no solid waste. For similar reasons, renewable technologies also emit fewer air pollutants per kilowatt-hour of electricity than fossil fuels (Table 22.4), although some of the newer renewable technologies have not been fully assessed yet.

The numbers in Tables 22.3 and 22.4 probably *underestimate* the environmental impact of renewable technologies because they include only the direct impacts from the operation of a facility. Excluded are the indirect costs of the energy and materials used to build the power generating facilities. These costs are substantial for many renewable technologies that need to collect and concentrate diffuse solar energy. For example, wind technologies are more material-intensive than a gas-fired power plant. In fact, wind turbines use ten times more steel, concrete, and other materials to produce a unit of electricity than natural gas plants. Extracting and processing those materials has significant environmental costs, so solar energy is not as benign as the numbers in Tables 22.3 and 22.4 suggest.

There is no comprehensive study that accounts for all the costs and benefits of conventional and renewable energy systems on a common basis. The best information we have is for electric power generation. That information suggests that despite their environmental advantages and a decline in their cost due to technological improvements, most forms of solar energy remain more expensive than conventional fuels. The National Renewable Energy Laboratory, part of the U.S. Department of Energy, made such a comparison. Their study projected the future cost of various forms of electricity generation based on past and expected changes in technology. Only wind power was forecast to be as cheap as electricity from coal in the near future.

But the story is not this simple. These costs reflect only marketed costs; they exclude other costs such as subsidies and externalities. Both conventional fuels and renewable energy receive generous subsidies from local, state, and federal governments. The oil and gas industry, for example, receives subsidies in the form of defense of Persian Gulf oil supplies, provision of the Strategic Petroleum Reserve, tax breaks for domestic oil exploration and production, and support for oil-related exports and foreign production. These subsidies amount to billions of dollars each year. Environmental externalities of fossil fuels, such as global warming, air pollution, and oil spills, are not included in the price of fossil fuels. The subsidies and externalities for oil and other fossil fuels mean their market prices do not reflect these fuels' true cost to society.

Subsidies and externalities distort the costs for renewables as well. The federal government pays for much of the research and development costs for renewable energy technologies. Generators of wind energy and biopower receive a 1.5-cent per kilowatt-hour production tax credit for their electricity, some-

TABLE 22.3	Land, Water, and Solid Waste Impacts of Various Electricity Generating Technologies						
				Technology			
	PV	Wind	Geothermal	Biopower	Biopower	Coal	Nuclear
Resource	0.02–20 MW utility-scale, thin film	25 MW wind farm	50 MW flashed steam or binary	75 MW	50 MW direct-fired	360 MW with desulfurization	1000-MW light water
Land (ha/MW)	5	20–46 (nonexclusive)	0.2 (plant)	3.2 (steam field)	0.54 (plant) + 487 (crops)	0.69 (plant) + 2.18 (mining)	0.40 + mining
Fuel (metric tons/yr/MW)	0	0	0	3,560	5,420	3,140C	0.03E
Water (M³/Mwh)	0	0	0	0.07 (power plant)	0.81	1.81 (90.4)	1.79
Solid waste (metric tons/yr/MW)	0	0	0	269 (ash)	185 (ash)	475 (ash and sludge)	0.03 + low-level waste

TABLE 22.4	Atmospheric Emissions from Various Electricity Generating Technologies

	Coal		Biomass	PV		Wind	Geothermal	Nuclear
Emission	Average Current System	New Source Performance Standard	Gasification Combined Cycle	Grid-Tied Rooftop PV	Stand-Alone Rooftop PV with Battery		Flashed Steam (Reservoir Emissions Only)	Light-Water Reactor
Particulates	9.21	9.78	0.04	Unk.	Unk.	Unk.	0	0.09
SO_2	6.70	2.53	0.30	Unk.	Unk.	Unk.	0.03	0.16
No_x	16.1	14.6	0.69	Unk.	Unk.	Unk.	0	0.11
Carbon monoxide	1.3	1.5	0.08	Unk.	Unk.	Unk.	0	0.01
Non-CH_4 hydrocarbons	1.0	1.3	0.60	Unk.	Unk.	Unk.	0	Unk.
CO_2 from mining or cultivation	9	8	28	NA	NA	NA	00	Unk.
CO_2 from transportation	17	16	6	NA	NA	NA	0	Unk.
CO_2 from power generation	996	917	12	0	0	0	45–81	Unk.
Total CO_2	1,022	941	46	Unk.	Unk.	Unk.	45–81	36.6
Methane	4.4	5.2	0.005	Unk.	Unk.	Unk.	0.09–0.75	0.12
Total CO_2 equivalent	1,114	1,050	46	60–150	280–410	7–74	47–97	39.1

thing conventional fuels do not enjoy. In many states utilities must purchase electricity from renewable sources at higher prices than those paid for electricity from conventional sources.

Renewable energy sources also have environmental impacts that are not reflected in their price, although the size of this externality probably is much smaller than for fossil fuels.

SUMMARY OF KEY CONCEPTS

- Various forms of renewable energy are available in potentially large quantities, but their diffuse nature often requires an expensive system to exploit them. Thus the energy return on investment for fossil fuels tends to be large, whereas that for solar energy tends to be low.

- Solar thermal systems concentrate heat and transfer it to a fluid and can be used to deliver hot water, process heat in industry, or generate electricity.

- Photovoltaic (or PV) systems convert light energy to electricity, a potentially big advantage over conventional power generation in which two-thirds of a fuel such as coal is lost as waste heat.

- In good locations wind energy is cost-competitive with conventional power generation and releases few pollutants. The chief environmental concern with wind energy is its possible impact on migratory bird and bat populations.

- Biomass is a renewable form of energy that, when properly managed, does not produce a net increase in atmospheric carbon dioxide and has relatively small impacts on the land.

- Ocean energy is too diffuse and thus too expensive to develop on a large scale.

- Hydrogen is an attractive energy carrier because it can be produced from many sources and pathways. However, its low energy density on a volume basis poses significant technological and economic challenges for its transportation, storage, and use.

- Market imperfections such as subsidies and environmental externalities distort the price for all forms of energy, making it difficult to get a clear picture of the true costs and benefits of any energy system.

- Energy end use efficiency is the ratio of the amount of energy service provided to the amount of energy consumed. Efficiency has improved in many sectors of the economy due to the development of new technologies.

REVIEW QUESTIONS

1. Describe what is meant by the quality/quantity paradox of solar energy.
2. Define energy density and power density. How are they different?
3. Describe how the photovoltaic effect produces electricity.
4. What is the single greatest constraint to increased use of wind energy?
5. What technological barriers remain for hydrogen energy? How do these translate into economic barriers?
6. How is a fuel cell different than a battery?

KEY TERMS

agricultural residues
biochemicals
biodiesel
bioenergy
biofuels
biomass
biomass feedstocks
biopower
end use efficiency
energy density
energy end use

ethanol
flash steam power plants
forestry residues
fuel cell
geothermal direct use
geothermal fields
geothermal heat pump
hydropower
ocean thermal energy conversion
photovoltaic effect
power density

primary mill residues
pumped storage plant
pyrolysis
solar constant
solar thermal system
tidal energy
urban wood wastes
wind energy
wind farm

MATERIALS, SOCIETY, AND THE ENVIRONMENT

STUDENT LEARNING OUTCOMES

After reading this chapter, students will be able to

- Explain the different physical processes that create mineral resources.
- Explain the difference between a mineral resource and a mineral reserve.
- Understand why material prices have generally not increased over the past fifty years.
- Distinguish between economic and environmental energy costs of natural resource formation and extraction.
- Define and describe the major forms of source reduction.

Quarry Mining in Yorkshire, England *The mining of materials is the first step in the material life cycle that ultimately ends with recycling, reuse, or waste disposal.*

MATERIALS: THE STUFF OF LIFE

We seldom stop and think about where the stuff in our everyday lives comes from. Food seems to come from a supermarket, clothes from a department store, and automobiles from car dealers. Yet these goods are made from materials that are mined or harvested from the environment; their manufacture, use, and disposal returns those materials back to the environment, often in forms and in places that cause environmental challenges.

Consider the simple pencil (Figure 23.1). In 1565 the German–Swiss naturalist Conrad Gesner first described a writing instrument in which graphite, then thought to be a type of lead, was inserted into a wooden holder. But if you dig a little deeper, you will see that today's pencil is quite complex. The cedar wood is from forests in California and Oregon. The graphite (not lead) might come from Montana or Mexico, and it is reinforced with clays from Kentucky and Georgia. The eraser is made of soybean oil from Illinois and latex from trees in Brazil, reinforced with pumice from California or New Mexico, sulfur from Louisiana, calcium from Texas, and barium from Nevada. The metal band is aluminum or brass, made from copper and zinc, which were mined in one of thirteen states, nine Canadian provinces, or more than a half-dozen foreign countries. The paint to color the wood, the glue that holds the wood together, and the lacquer to make it shine are made from dozens of different materials.

This example illustrates the important role that materials play in our lives. Humans mine all ninety-two naturally occurring elements from Earth's crust and harvest hundreds of species of trees, plants, and animals for the materials they contain. Energy is used to refine and

What's in a pencil besides wood?

The cedar wood is from the forests in California and Oregon. The graphite (not lead) might come from Montana or Mexico, and is reinforced with clays from Kentucky and Georgia. The eraser is made from soybean oil and latex from trees in South America, reinforced with pumice from California or New Mexico, and also includes sulfur, calcium, and barium. The metal band is aluminum or brass, made from copper and zinc, mined in no fewer than thirteen states and nine Canadian provinces. The paint to color the wood and the lacquer to make it shine are made from a variety of different minerals and metals, as is the glue that holds the wood together.

FIGURE 23.1 *The Pencil* *The material composition of a pencil.*
(Source: Modified from Mineral Information Institute.)

transform those materials into goods and services. Over his or her lifetime the average American will consume more than 700,000 kilograms of minerals, 162,000 cubic meters of water, and 170 cubic meters of wood. Of course that flow of useful materials from mines, forests, and reservoirs is mirrored by a flow of wastes released to our atmosphere, water, and land. Over his or her lifetime the average American will generate more than 30,000 kilograms of municipal solid waste, 10,000,000 kilograms of wastewater, and 1,500,000 kilograms of air pollutants. Understanding the sources, uses, and ultimate fates of the material bases of our everyday lives is the purpose of this chapter. ■

THE MATERIALS CYCLE

The flow of materials is a principal connection between society and the environment. Figure 23.2 describes what is known as a **cradle-to-grave** or **life cycle analysis** of the flow of materials through the environment and the economy. The cycle begins with the extraction of a material—such as copper, timber, coal, or wool—from the environment by the extractive sectors of the economy. These materials are processed in various ways into raw materials that are converted into goods and services. Those goods and services are purchased and used by households, firms, and the government.

Life cycle analysis illustrates several important points about materials. First, their use obeys the law of conservation of matter. Every gram of every substance we extract from the environment is ultimately recycled or is released to the environment as waste. Second, every stage of the materials flow cycle generates waste that has potentially harmful

effects on the health of people and the environment. The recovery of wastes is a critical activity that reduces human impact on the environment and that can have positive economic benefits for the economy. See *Case Study: Tracing the Flows of Arsenic.*

MINERAL FORMATION, OCCURRENCE, AND ABUNDANCE

Formation and Occurrence

Mineral ore deposits are the result of work done in the environment that concentrates an element far above its average concentration. For example, Earth's crust contains an average of about 55 ppm (parts per million) of copper, whereas copper ores must contain about 5,000 ppm before they are economical to mine. Thus the planet's biogeochemical cycles need to concentrate copper at least a hundred times before

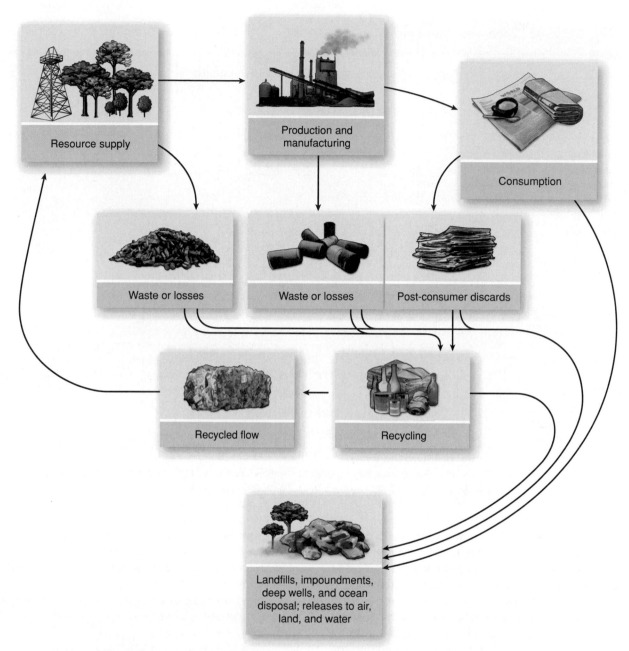

FIGURE 23.2   *Materials Flow Cycle* *The ultimate fate of materials is released to the environment or recycling.* *(Source: Redrawn from* Materials: A Report by the U.S. Interagency Working Group on Industrial Ecology, Material, and Energy Flows.*)*

we can use it. We then use fossil fuels to convert copper ore into pure copper metal, an additional two hundredfold increase in concentration.

Categories of mineral formation are broadly distinguished by whether they occur at or near the surface or in the subsurface (Table 23.1). Weathering (the chemical, biological, and physical decomposition of rock near the surface) concentrates some minerals in the soil. The most widespread resource is clay, which is mined for use in paints, brickmaking, and fine china. Under certain conditions intensive weathering of aluminum-rich rock concentrates aluminum

oxide. Sufficient concentrations of aluminum oxide produce bauxite, the main source for aluminum metal. Important deposits of nickel and cobalt also are produced from the weathering of iron- and magnesium-rich rocks.

Erosion by water and wind transports rock fragments produced by weathering. During the transport of rock fragments by water, there is a systematic sorting of the particles according to their size and density. Resources most commonly produced by weathering and erosion are sand and gravel—the basis for construction of many of the world's roads, bridges, and buildings. The same erosion processes

TABLE 23.1	Geologic Processes That Form Mineral Deposits, with Examples of Deposits Formed by Each Process and Elements Concentrated in Them
Type of Process	**Types of Deposits Formed/Minerals Concentrated**
Surface processes	
Weathering	Laterite deposits—nickel, bauxite, gold, clay
	Soil
Physical sedimentation	
Flowing water (stream or beach)	Placer deposits—gold, platinum, diamond, ilmenite, rutile, zircon, sand, gravel
Wind	Dune deposits—sand
Chemical sedimentation	
Precipitation from or in water	Evaporite deposits—halite, sylvite, borax
	Chemical deposits—iron, volcanogenic massive sulfide deposits, sedex deposits
Organic sedimentation	
Organic activity or accumulation	Hydrocarbon deposits—oil, natural gas, coal
	Other deposits—sulfur, phosphate
Subsurface processes	
Involving water	Groundwater and related deposits—uranium, sulfur
	Basinal brines—Mississippi Valley type
	Seawater—volcanogenic massive sulfide
	Magmatic water—porphyry copper, molybdenum
	Metamorphic water—gold, copper
Involving magmas	Crystal fractionation—chromium, vanadium
	Immiscible magma separation—nickel, copper, cobalt, platinum-group elements

Source: Modified from S. E. Kesler, *Mineral Resources, Economics, and the Environment*, Macmillan College Publishing Company.

also sort minerals such as gold, tin, and titanium. Minerals that make up these **placer deposits** tend to be heavy and resistant to weathering and abrasion as they are transported by water. Gold, for example, will settle out of water when it slows down over a rough spot or as it goes around a bend in a river. As a result, these minerals tend to form distinct deposits that are amenable to relatively simple mining methods. An example is the gold pan used by miners in the California Gold Rush of 1849, when thousands of fortune seekers panned and sluiced the rivers searching for placer deposits of gold (Figure 23.3).

Rivers often deliver huge volumes of dissolved materials to inland lakes or oceans. Over geologic time, lakes and shallow marine basins may become isolated and eventually dry up. In the evaporation process the dissolved materials precipitate and form concentrated accumulations called **evaporites.** Important examples include many types of salts, gypsum, calcium carbonate, and magnesium.

Igneous and metamorphic processes create many important metal deposits. The decay of radioactive material within Earth produces high temperatures that melt rocks and form magmas. When rocks melt they can move within Earth in response to pressure or density differences. When they eventually stop, they cool and crystallize. Cooling concentrates heavier metals at the bottom of the magma; lighter elements crystallize near the top. Chromium and platinum

FIGURE 23.3   *The Gold Rush* *American and Chinese gold miners working in California in 1852.*

CASE STUDY

Tracing the Flows of Arsenic

A systems perspective is needed to judge the sustainability of any particular material use because the flow of a material from the environment through the economy often is extremely complex. The case of arsenic demonstrates this point. For the first half of the twentieth century the greatest use of arsenic was in pesticides, owing to the ability of arsenic and arsenic compounds to kill insects, primarily termites and to a lesser extent grasshoppers, and its effectiveness as a fungicide. However, in the late 1940s the emergence of new types of organic pesticides reduced the demand for arsenic in agriculture. Decisions by the Environmental Protection Agency in the 1980s prohibited the use of arsenic in pesticides.

This ban was due to the fact that arsenic is an extremely toxic element. Exposure to arsenic can cause various health effects such as irritation of the stomach and intestines, decreased production of red and white blood cells, skin changes, and lung irritation. Very high exposure to inorganic arsenic can cause infertility, decreased resistance to infections, brain damage, and death. Arsenic bioaccumulates in freshwater food webs because plants easily absorb arsenic.

Yet despite the ban in agriculture, arsenic flows continued to rise in the United States (see Figure 1). Why? Look no farther than the wooden deck off the back of your house or apartment. Today arsenic is used mainly as

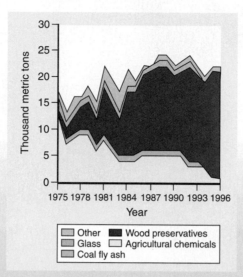

FIGURE 1 *Potential release of arsenic to the environment by source in the United States.*
(Source: Data from U.S. Environmental Protection Agency.)

a wood preservative, primarily in the form of chromated copper arsenate (CCA). Fueled by the building and real estate boom in the 1980s, arsenic use in pressure-treated wood skyrocketed. A residential deck measuring about 3.5 m (12 ft) on a side may contain over 750 g (1.65 lb) of inorganic arsenic. U.S. consumption of pressure-treated wood now accounts for 90 percent of worldwide arsenic demand.

Interestingly, domestic production of arsenic ended completely in 1985 when ASARCO, the sole remaining producer, closed its smelter in Tacoma, WA. Today U.S. demand is met entirely through imports—mostly from China, the world's largest producer, and Chile, a growing provider. Shifting arsenic production to foreign countries exemplifies the need for global material flow analyses that take into account all flows, including those crossing international borders.

This new use of arsenic has presented a new wave of health risks. Burning pressure-treated wood in open fires or in stoves, fireplaces, or residential boilers releases toxic chemicals in the smoke and ashes. Sawdust and construction debris constitute a serious health risk to those coming into contact with those materials.

As a result of these new health risks, the Environmental Protection Agency has worked with pesticide manufacturers to voluntarily phase out chromated copper arsenate use for wood products around the home and in children's play areas. Effective December 31, 2003, no wood treater or manufacturer could treat wood with chromated copper arsenate for residential uses, with certain exceptions.

STUDENT LEARNING OUTCOME

- Students will be able to describe the major ways in which arsenic is released into the environment.

---

are produced in this way. Diamonds are produced when magma containing small amounts of carbon crystallizes very slowly under high pressure.

Rocks undergo metamorphic change when they are subjected to great pressure and temperature from deep burial or contact with hot igneous rocks. The transformation of materials in metamorphism can produce valuable resources. Limestone metamorphoses into marble, and shale into slate. All are important building materials; metamorphosed materials are more valuable because they are harder, more durable, or more attractive (in the case of marble). Metamorphic zones also produce some of the world's most valuable accumulations of sapphires, rubies, and other gemstones.

Hydrothermal solutions are produced when water is heated by contact with magmas or hot rocks. Many hydrothermal solutions react with the rocks they move, chemically altering them to produce concentrations of particular

materials. Hydrothermal solutions can migrate away from their heat sources into folds or cracks to forms veins that may contain high concentrations of valuable minerals. Many of the world's richest deposits of copper, lead, and zinc were formed in this way.

Seawater hydrothermal systems form along midocean ridges at the divergent margins of continental plates. Seawater flows down into rock fractures, where it comes in contact with upwelling magma. The heated water emerges as spectacular hot springs that have been photographed by research submarines (Figure 23.4). Water leaves the vents at temperatures up to 350°C (662°F), but it doesn't boil due to the high pressures that prevail at the 5 kilometer (3.1 miles) depths where these vents are found. These hot springs are called **black smokers** because they emit water loaded with dissolved metals and hydrogen sulfide (H_2S) that forms rich metal sulfide deposits near the vents. The vents support unique biological communities that include white crabs,

FIGURE 23.4 *Black Smoker*
A "black smoker" hot spring in the Pacific Ocean. These smokers release dissolved metals and hydrogen sulfide (H_2S) that form rich metal sulfide deposits near the vents.

tube worms, and giant clams. These communities are based on chemosynthesis (as opposed to photosynthesis)—a process in which certain plants derive their energy from the breakdown of chemical compounds.

Currently it is too expensive to mine sulfide mineral deposits along oceanic ridges, although mining companies from several nations are exploring the possibility. The economic potential will have to be weighed against the potential impacts mining could have on these unique communities.

Classifying Mineral Resources

Earth is a vast warehouse of minerals. The mass of Earth's crust is 5.97×10^{24} kilograms, which is more than 2 *trillion* times the quantity of minerals the United States used in 2005. Dissolved in the world's oceans are additional huge amounts of many minerals. So in purely physical terms there is no shortage of minerals.

But economic, technological, political, and environmental constraints set practical limits on how much of this warehouse can—and should—be tapped. The U.S. Geological Survey has established a classification system that describes how much of a given mineral is currently and potentially available for human use. The system is based on two criteria: the extent of geologic knowledge and economic feasibility (Figure 23.5). A **mineral resource** is a concentration of naturally occurring solid, liquid, or gaseous material in or on Earth's crust in such form and amount that economic extraction of a commodity from the

concentration is currently or *potentially* feasible. In the geological sense, resources are classified somewhere between measured and speculative, reflecting an increasing amount of uncertainty about the abundance. In the economic sense, resources are classified as economic, marginally economic, or subeconomic.

The **reserve base** is the portion of an identified resource that can be extracted under current economic, technological, and legal conditions. These are the materials we mine for our everyday uses. The quantities of reserves are well known at any given time, but they change constantly. Extraction depletes reserves, but new discoveries and technological advances can increase reserves. Changes in the price of mineral products also increase or diminish the reserves.

There are many examples of resources becoming reserves. Soon after World War II the rich iron ore deposits in the Great Lakes region were in danger of running out. Technological advances in mining and processing allowed lower-quality and previously unusable deposits called taconites to be extracted. Taconites now supply most of the iron ore in the United States. The opposite also happens. In the 1960s vast quantities of natural gas were discovered in remote northern Alaska. They were added to reserves on the assumption that gas prices would remain high enough to finance a pipeline to the lower forty-eight states. In the 1980s, however, natural gas prices declined and political opposition to the construction of the pipeline mounted, so those deposits were moved from the reserve to resource classification.

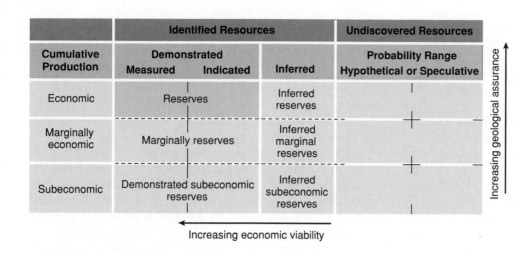

	Identified Resources			Undiscovered Resources
Cumulative Production	**Demonstrated**		**Inferred**	**Probability Range Hypothetical or Speculative**
	Measured	**Indicated**		
Economic	Reserves		Inferred reserves	
Marginally economic	Marginally reserves		Inferred marginal reserves	
Subeconomic	Demonstrated subeconomic reserves		Inferred subeconomic reserves	

Increasing geological assurance →

Increasing economic viability ←

FIGURE 23.5 *Resource Classification* *Major elements of a mineral resource classification system that is based on economic viability and geological assurance.* (Source: Redrawn from U.S. Geological Survey.)

Abundance and Distribution

Because the geologic processes that create mineral deposits are not random, it should come as no surprise that the distribution of resources and reserves also is not random. Distribution can be measured in two ways: the quantity of mineral in the crust and its geographic distribution. Recall that in Chapter 10 we discussed **crustal abundance:** the percentage of Earth's crust accounted for by metal (Figure 23.6). Abundant metals—iron, aluminum, chromium, manganese, titanium, and magnesium—exhibit a fairly wide distribution. Scarce metals—copper, lead, zinc, gold, silver, tin, platinum, uranium, mercury, and molybdenum—show much more limited distribution. This means, for example, that the number of iron deposits in the world is far greater than the number of gold deposits.

The geographic distribution of minerals is irregular due to the nonuniform distribution of the geologic processes that created them. Thus the larger the area covered by a nation, the greater the probability that it will contain significant amounts of any particular mineral. Five regions—the United States, Canada, Australia, South Africa, and the Commonwealth of Independent States—cover 34 percent of Earth's land area and possess the lion's share of many important mineral resources. Other notable examples includes China's holding of 44 percent of the world's tungsten reserves and Brazil's holding of 95 percent of the world's columbium reserves.

Mineral Exploration and Production

The search for and extraction of minerals is similar to oil and natural gas exploration, evaluation, mining, and processing. There are three basic types of mineral exploration. **Geologic exploration** relies on the mapping and analysis of geologic formations known to contain certain minerals. This works because many mineral deposits form in specific geologic

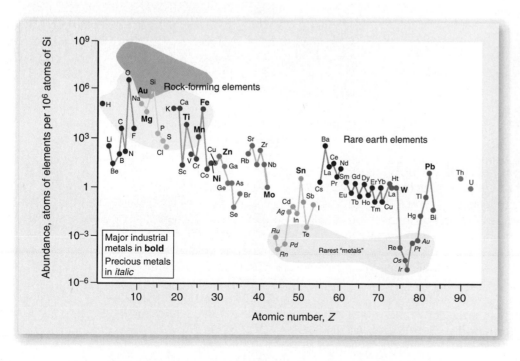

FIGURE 23.6 *Relative Abundance of the Chemical Elements in the Earth's Crust* *Some of the elements are classified into (partially overlapping) categories: (1) rock-forming elements (major elements in green field and minor elements in light green field); (2) rare earth elements in blue; (3) major industrial metals in bold; (4) precious metals in italics; and (5) the nine rarest metals.* (Source: Data from G.B. Haxel et al., "Rare Earth Elements—Critical Resources for High Technology," U.S. Geological Survey Fact Sheet 087-02.)

environments by processes that leave evidence in the rock. For example, hydrothermal processes frequently react with and change the composition of adjacent rocks. Porphyry copper deposits, the largest source of copper in the world, are formed in this way and sometimes can be identified by looking for a zone of altered rocks.

Geochemical exploration looks for mineral deposits formed by weathering of Earth's surfaces that disperses the components of the deposit into the surrounding water, soil, vegetation, and air to create chemically enriched zones. Weathering of rock that contains iron, zinc, and copper sulfides disperses those valuable elements to the environment, where they are transported and then deposited in high concentrations by wind and water. **Geophysical exploration** measures physical properties of the rock, such as its magnetic intensity, electrical conductivity, radioactivity, or the speed with which seismic waves pass through it. Under favorable conditions, measurements taken at the surface can "see through" younger sediments to ore-bearing rocks that lie hundreds of meters below.

Once a potentially attractive mineral deposit is identified through exploration, it must be evaluated by drilling. The drilling technology used to evaluate mineral deposits is similar to that used to evaluate potential oil and gas deposits. The purpose of drilling is to gain samples of buried rock for analysis that will determine whether it is worthwhile to mine the deposit.

A mineral deposit that has been identified and deemed economically attractive must then be extracted. This is done by mining in which the rock is physically removed from the ground, much as coal is mined. Deposits within a few hundred meters of the surface usually are recovered by open pit mining methods; deeper deposits are reached with underground methods (Figure 23.7). An open pit mine removes the ore and **overburden**—rock that overlies the ore.

Underground mines use horizontal and vertical shafts to transport people and equipment to the mineral deposit and to haul the mined material back to the surface for processing and transportation. In modern underground mines, vehicles use spiral ramps similar to those in parking garages to enter the mines. Mineral ores are removed in the same manner as underground coal: continuous mining, longwall mining, or a cyclic system of drilling, blasting, and hauling the ore.

How deep can mines go? The depth of surface mines is limited by the cost of removing ever-larger amounts of overburden and hauling ore to the surface. Most surface mines are less than 200 meters (656 feet) deep. The depth of underground mines is limited by the **geothermal gradient** (the increase in Earth's temperature with increasing depth), the increasing likelihood of cave-ins, and the greater costs associated with deeper mines. These forces limit the depth of underground mines to about 2,500 meters (8,202 feet).

FIGURE 23.7 *Surface Mining*
This open pit copper mine is an example of the technology used to extract many minerals from the crust.

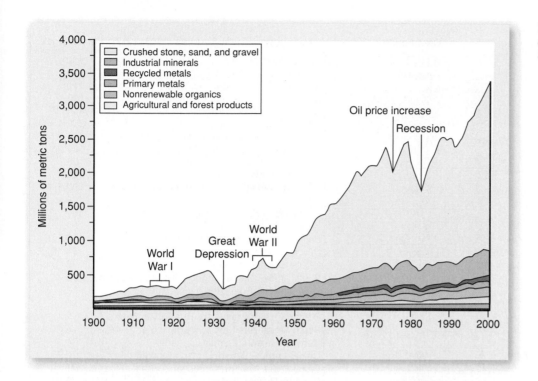

FIGURE 23.8 *Material Use* *The use of materials by broad category in the United States, 1900–2000.* (Source: Data from U.S. Geological Survey.)

MATERIALS AND THE ECONOMY

In Chapter 10 we discussed the close connection between economic activity and the use of energy and materials. In 1900 the United States used about 145 million metric tons (160 million tons) of materials, not including water (Figure 23.8). By 2005 material use had grown by a factor of 20, and the value of mineral products alone was nearly $450 billion. The nation exported another $40 billion of minerals.

In terms of sheer mass, crushed stone and construction sand and gravel make up as much as three-quarters of new resources used annually. These are used to make roads, bridges, ramps, and buildings. Other industrial mineral commodities account for the next largest share of materials used. These include cement for ready-mix concrete, potash and phosphate for fertilizer, gypsum for drywall and plaster, fluorspar for acid, soda ash for glass and chemicals, and sulfur, abrasives, asbestos, and various other materials for use in chemicals and industry.

The use of nonrenewable organics grew sharply after 1950 due to the introduction of **petrochemicals:** any chemical derived from crude oil, crude oil products, or natural gas. Petrochemicals are used in the manufacture of numerous products such as synthetic rubber, synthetic fibers (such as nylon and polyester), plastics, fertilizers, paints, detergents, and pesticides.

Like most developed nations the United States uses thousands of different materials that are derived from hundreds of natural resources. The United States has abundant supplies of many important minerals, including phosphate rock, iron ore, sulfur, and lead. However, significant amounts of other important minerals must be imported, including key metals such as zinc, cobalt, tin, and manganese (Figure 23.9). Some minerals are imported because domestic resources are inadequate to meet demand, others because foreign sources are cheaper than domestic ones. World trade in metals and minerals exceeds $300 billion each year.

Dematerialization

Intensity of use is a useful indicator of broad trends in our use of materials. It refers to the quantity of a material used to produce a unit of economic output. For example, the intensity of use of steel in the manufacture of automobiles is the kilograms of steel per car. The intensity of use of steel for the economy as a whole is the kilograms of steel per dollar of GDP.

The intensity of use for some important materials in the United States is shown in Figure 23.10. The quantity of timber, steel, copper, and lead used per dollar of GDP has decreased in the last half century. This is known as **dematerialization:** the absolute or relative reduction in the quantity of materials used or the quantity of waste generated in the production of a unit of economic output. Dematerialization is caused by a number of forces, including technological change, the substitution of new materials with more desirable properties, and government regulations.

In the case of timber, metals and then plastics gradually replaced wood, and wood was made more durable by creosote and other preservatives. A government ban on leaded gasoline produced the sharp decline in the intensity of use for lead, whereas aluminum was substituted for copper in electrical wiring because it was cheaper. These same forces produced increases in the intensity of use of other materials such as plastic, aluminum, potash, and phosphorus (the latter two materials are used to manufacture fertilizers).

FIGURE 23.9 *Mineral Imports* The *mineral import dependence of the United States.* (Source: Data from U.S. Geological Survey.)

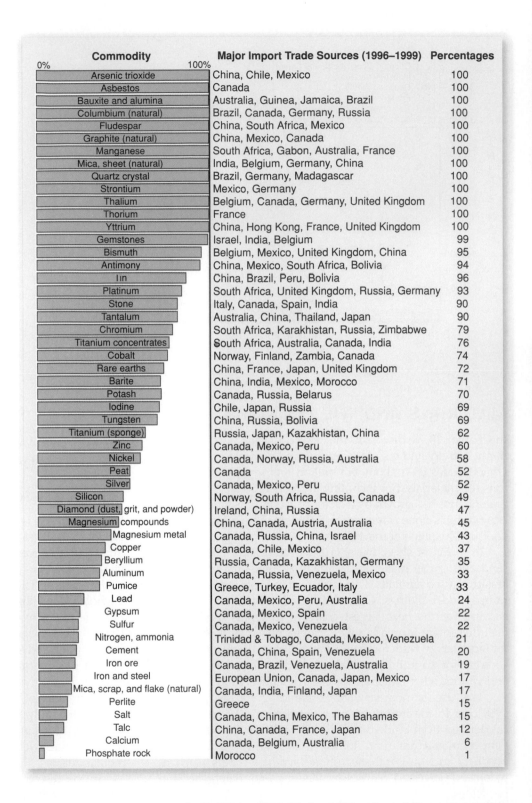

Commodity	Major Import Trade Sources (1996–1999)	Percentages
Arsenic trioxide	China, Chile, Mexico	100
Asbestos	Canada	100
Bauxite and alumina	Australia, Guinea, Jamaica, Brazil	100
Columbium (natural)	Brazil, Canada, Germany, Russia	100
Fludespar	China, South Africa, Mexico	100
Graphite (natural)	China, Mexico, Canada	100
Manganese	South Africa, Gabon, Australia, France	100
Mica, sheet (natural)	India, Belgium, Germany, China	100
Quartz crystal	Brazil, Germany, Madagascar	100
Strontium	Mexico, Germany	100
Thalium	Belgium, Canada, Germany, United Kingdom	100
Thorium	France	100
Yttrium	China, Hong Kong, France, United Kingdom	100
Gemstones	Israel, India, Belgium	99
Bismuth	Belgium, Mexico, United Kingdom, China	95
Antimony	China, Mexico, South Africa, Bolivia	94
Iin	China, Brazil, Peru, Bolivia	96
Platinum	South Africa, United Kingdom, Russia, Germany	93
Stone	Italy, Canada, Spain, India	90
Tantalum	Australia, China, Thailand, Japan	90
Chromium	South Africa, Karakhistan, Russia, Zimbabwe	79
Titanium concentrates	South Africa, Australia, Canada, India	76
Cobalt	Norway, Finland, Zambia, Canada	74
Rare earths	China, France, Japan, United Kingdom	72
Barite	China, India, Mexico, Morocco	71
Potash	Canada, Russia, Belarus	70
Iodine	Chile, Japan, Russia	69
Tungsten	China, Russia, Bolivia	69
Titanium (sponge)	Russia, Japan, Kazakhstan, China	62
Zinc	Canada, Mexico, Peru	60
Nickel	Canada, Norway, Russia, Australia	58
Peat	Canada	52
Silver	Canada, Mexico, Peru	52
Silicon	Norway, South Africa, Russia, Canada	49
Diamond (dust, grit, and powder)	Ireland, China, Russia	47
Magnesium compounds	China, Canada, Austria, Australia	45
Magnesium metal	Canada, Russia, China, Israel	43
Copper	Canada, Chile, Mexico	37
Beryllium	Russia, Canada, Kazakhstan, Germany	35
Aluminum	Canada, Russia, Venezuela, Mexico	33
Pumice	Greece, Turkey, Ecuador, Italy	33
Lead	Canada, Mexico, Peru, Australia	24
Gypsum	Canada, Mexico, Spain	22
Sulfur	Canada, Mexico, Venezuela	22
Nitrogen, ammonia	Trinidad & Tobago, Canada, Mexico, Venezuela	21
Cement	Canada, China, Spain, Venezuela	20
Iron ore	Canada, Brazil, Venezuela, Australia	19
Iron and steel	European Union, Canada, Japan, Mexico	17
Mica, scrap, and flake (natural)	Canada, India, Finland, Japan	17
Perlite	Greece	15
Salt	Canada, China, Mexico, The Bahamas	15
Talc	China, Canada, France, Japan	12
Calcium	Canada, Belgium, Australia	6
Phosphate rock	Morocco	1

The Price of Materials

In 1980 economist Julian Simon and biologist Paul Ehrlich made a famous bet. Ehrlich had long predicted massive shortages in various natural resources, whereas Simon argued natural resources were infinite. Simon offered Ehrlich a bet centered on the market price of metals. Ehrlich would pick a quantity of any five metals he liked worth $1,000 in 1980. If the 1990 price of the metals, after adjusting for inflation, was more than $1,000 (that is, if the metals became more scarce), Ehrlich would win. If, however, the value of the metals after inflation was less than $1,000 (if the metals became less scarce), Simon would win. The loser would mail the winner a check for the change in price. Ehrlich agreed to the bet and chose copper, chrome,

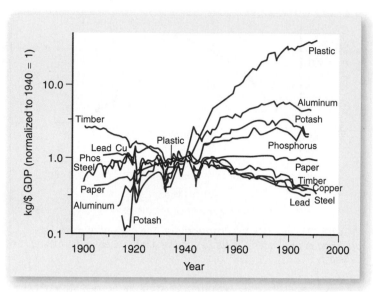

FIGURE 23.10   *U.S. Material Intensity of Use* *The intensity of use of various materials in the United States. Intensity of use is the ratio of material use to economic output. (Source: Modified from I.K. Wernick et al., "Materialization and Dematerialization: Measures and Trends." Daedalus 125(3): 171–198.)*

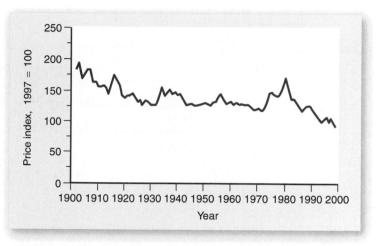

FIGURE 23.11   *U.S. Metal Prices* *The real price of metals in the United States. (Source: Modified from D.E. Sullivan et al., "20th Century U.S. Mineral Prices Decline in Constant Dollars." U.S. Geological Survey Open-File Report no. 00-389, Online version 1.0.)*

nickel, tin, and tungsten. By 1990 all five metals were below their inflation-adjusted price level in 1980. Ehrlich lost the bet and sent Simon a check for $576.07.

This story illustrates an important point: The price of many important minerals has not increased over time, even though the highest-quality deposits have been depleted. In fact, many mineral prices have declined over time. The U.S. Geological Survey calculates a composite price index that combines data for five metal commodities (copper, gold, iron ore, lead, zinc) and seven industrial mineral commodities (cement, clay, crushed stone, lime, phosphate rock, salt, and sand and gravel). Overall, inflation-adjusted prices shown in the composite index declined through the twentieth century (Figure 23.11).

The declining long-term price for minerals is the result of several forces. Technological change allows metals and other minerals to be extracted from lower-grade ores by new mining and processing methods. Product designs have also changed, such as the design of thin walls for metal beverage containers that results in using less material. This increases the supply of metal available for other products and helps hold down metal prices. The increased reuse and recycling of mineral materials also contribute to the declining trend in prices because they decrease the need to extract virgin materials. Finally, the United States imports large amounts of minerals, which helps keep domestic prices lower.

But not all natural resources are cheaper today than in the past. Consider the case of forest products. Consistent with the best first principle, people use the biggest trees closest to the sawmills before they harvest thinner, more distant trees. Over time the cost of producing lumber, veneer, furniture, and other forest products has risen due to this shift to lower-quality forests. In turn the price of those products has increased as well. From 1800 to the 1950s, the price of forest

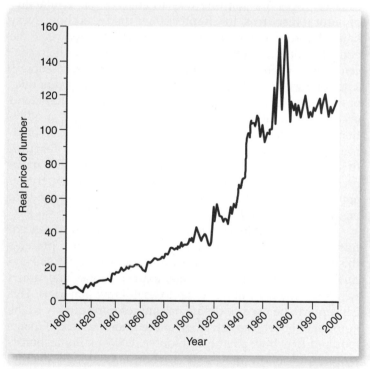

FIGURE 23.12   *U.S. Wood Prices* *The real price of forest products in the United States. (Source: Modified from C.J. Cleveland and D.I. Stern, "The Scarcity of Forest Products Revisited: An Empirical Comparison of Alternative Indicators." Canadian Journal of Forest Research, 23: 1537–1549.)*

products increased by a factor of ten due to the increasing scarcity of high-quality timber (Figure 23.12). Prices increased despite substantial technological improvements in forestry and sawmill operations that put downward pressure on prices. The price of forest products stabilized thereafter, largely due to the increased use of forest plantations that can produce a relatively uniform-quality product, albeit at a higher cost than the native forest ecosystems once provided.

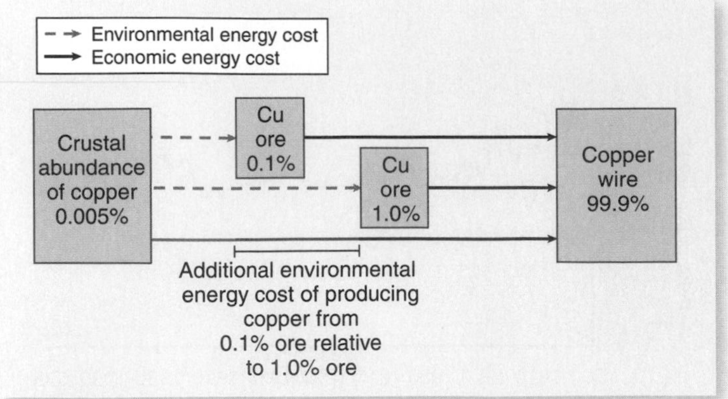

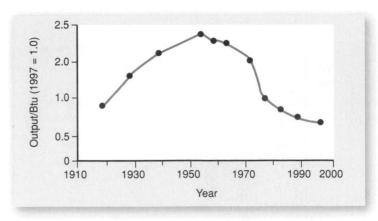

FIGURE 23.13 *Energy Cost of Resources* The relationship between the environmental energy cost and economic energy cost for natural resources.

FIGURE 23.14 *Metal Productivity* The amount of metal produced in the United States per unit of energy used in the mining industry.

ENERGY AND RESOURCE QUALITY

Market prices are just one way to analyze the availability of natural resources. In Chapter 10 we discussed how every natural resource has an **environmental energy cost:** the energy required to create a natural resource of a given concentration. There are two sources of environmental energy: solar energy and heat from Earth's interior. These energies power environmental processes—such as atmospheric and oceanic circulation, the rock cycle, and net primary production—that create and sustain natural resources. More environmental energy must be used to concentrate a resource further than its average concentration. Higher-grade metal ores have much higher concentrations than their crustal abundance and thus required significant environmental energy to be created.

An additional investment of energy is needed to concentrate, organize, or upgrade natural resources. This represents the **economic energy cost** of natural resources: the fossil and nuclear fuels required to locate, extract, refine, and otherwise upgrade natural resources to useful raw materials. This concept can be used to define natural resource quality as the amount of effort (capital, labor, and economic energy) that is required to extract a natural resource. Low-quality resources require more effort (more capital, labor, and economic energy) to extract and refine than do high-quality resources.

There is a trade-off between environmental energy cost and economic energy cost for many natural resources (Figure 23.13). Natural resources with high environmental energy costs have low economic energy costs. The more work that was done by biogeochemical cycles to create a natural resource, the less work the extractive sector must do to upgrade it to a useful form. For example, the crustal abundance of copper is 50 grams per metric ton. This amounts to about 0.005 percent by weight of the crust. The concentration of copper into a piece of metal wire that is 99.9 percent pure copper requires significantly more economic energy when it starts with a leaner ore.

The effects of depletion are evident in the metal mining industry. The amount of metal produced per unit of energy declined sharply in the last half century (Figure 23.14). Declining ore grades and increasing stripping ratios have driven up the energy costs of mining and concentration. For example, the average grade of copper ore was 2–3 percent a hundred years ago but now is just 0.5 percent. The increasing depth of mines increases the **stripping ratio:** the volume of overburden compared to the volume of ore. Consistent with the best first principle, stripping ratios tend to increase over time as higher-quality deposits are depleted, which raises energy costs. The potential for environmental harm also rises with the stripping ratio because more waste rock must be mined and processed per metric ton of useful mineral produced.

THE FATE OF MATERIALS

Materials have very different fates in the economy. A dam can last for centuries, a piece of tissue paper for a few months or years, some chemicals for only a few days or weeks. Pesticides are applied once and then dissipate into the environment, whereas the steel in a car can be recycled many times.

Materials that enter the economy have three potential fates: They become part of domestic stock, they are released to the environment as waste, or they are recycled (Figure 23.15). About 40 percent of material inputs become part of the built environment and industrial infrastructure. These terms refer to objects not consumed during normal use and designed to last for more than one year. Construction materials such as sand, gravel, stone, steel, asphalt, lumber, and cement used for roads, bridges, houses, and the like are the largest fraction of domestic stock. Smaller amounts of metals, forest products, and other materials are converted to industrial products, such as machines and tools, and consumer products, such as televisions, cars, furniture, and most clothing.

FIGURE 23.15 *Material Use in the United States, mid-1990s.* (Source: Data from I. Wernick and J. Ausubel, "National Materials Flows and the Environment," Annual Review of Energy and the Environment 20: 463–492.)

Legend:
- Domestic stock
- Atmospheric emissions
- Postconsumer waste
- Recycled materials
- Dissipation
- Other waste
- Processing waste

Material Wastes

The retrieval and processing of natural resources produces **extractive wastes,** which range from waste rock from mineral mining to biological residues generated in agriculture and forestry. Extractive wastes in the U.S. economy are enormous, amounting to more than 10 billion metric tons per year. These wastes seldom enter the economy, but they are an unavoidable by-product of resource extraction. Rock moved to expose desired minerals, fuels, and ores are the biggest component of extractive wastes. Here the majority of wastes are generated in the surface mining of coal, in which overburden must be removed to expose the coal seams. The amount of overburden mobilized varies from less than 2.3 cubic meters (3 cubic yards) per metric ton of coal in West Virginia to 36.7 cubic meters (48 cubic yards) per metric ton in Oklahoma.

Metal mining also mobilizes significant waste materials. On average 136 metric tons (150 tons) of overburden must be moved to mine 0.9 metric tons (1 ton) of copper metal. In the case of precious metals, a gold ring weighing a few grams generates more than 5.4 metric tons of extractive waste at a mine in Nevada or South Africa. Separating the metal from the ore generates more wastes called **tailings.** The total quantity of extractive waste is enormous: On a global scale, mining moves more of Earth's surface than natural erosion by rivers. This huge mobilization of material disturbs habitat, releases toxic metals, and pollutes surface water and groundwater.

The extraction of zinc typifies the environmental impact of materials extraction and processing. Zinc is used to make galvanized steel, brass, bronze, paint, batteries, and rubber products. For every metric ton of refined zinc metal produced, more than 50 metric tons of wastes are generated in the form of mining overburden, tailings from processing the ore, sulfur in the flue gas, and slag from the refining process (Figure 23.16). Note that some zinc is released to the environment in the wastes. This poses a significant health risk because zinc can be toxic to organisms. Much of the zinc mined in the United States is extracted in Alaska, where wastes disturb unique and delicate ecosystems. See *Your Ecological Footprint: Recycling Batteries.*

Processing and manufacture of raw materials into finished goods and services following extraction from the original natural resource generate **processing waste.** Water is a common processing waste because it is used for a variety of tasks, such as heating, cooling, washing, as a solvent, and as a raw material in the chemical industry. Steel in a car door is made from iron, limestone, and other materials. Sheets of raw steel are cut

Recycling Batteries

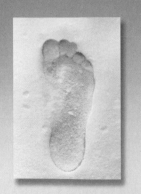

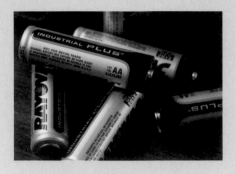

The alkaline battery was invented in 1959 by a Canadian chemical engineer named Lewis Urry, who was working for Eveready Battery. Alkaline batteries are a type of power cell; they derive power from a reaction between zinc and manganese dioxide (Zn/MnO_2). Since then the alkaline battery has become an ubiquitous feature of modern life. The common AA battery is used in hundreds of everyday products, including flashlights, TV remotes, cameras, toys of all sorts, clocks, radios, and so on (Figure 1). Like most consumer goods, used batteries usually end up in one of two places: a landfill or a recycling center. Batteries sent to landfills pose a number of environmental and human health risks because metals and chemicals in the batteries are released to the soil and water when the sealed batteries are broken.

Recycling batteries reduces these impacts and significantly reduces the amount of energy, wastes, chemicals, and pollutants associated with producing new batteries. To get a sense of these savings, let's look at your battery footprint, focusing only on AA batteries. The first step is to estimate the number of AA batteries you purchase in a year. You will need to inventory all the devices you use at home, in the car, in school, at the office, on vacation, and so forth. Be sure to check all the appliances and other electronic devices you use. Estimate how many you might use in a month, and then multiply by 12 to get an annual total.

A single AA battery contains about 0.8 ounces (23 grams) of zinc. Multiply that quantity times the number of batteries you use in a year.

This is the quantity of zinc you buy each year in the form of batteries. If you threw the batteries in the trash when they died, that also would be the amount of zinc you sent to a landfill.

Table 23.3 shows the reductions in the use of energy and materials due to the recycling of metals, including zinc. These data indicate that recycling reduces energy use by 1,096 Btu per ounce of zinc. Multiply that by the quantity of zinc you purchase to calculate the energy reduction associated with recycling your AA batteries. Table 23.3 also gives the reductions in water use, pollution, and land disturbed associated with recycling zinc. You can calculate your additional reduced footprint from recycling by using that information. (*Note:* Multiply tons by 32,000 to get ounces.)

STUDENT LEARNING OUTCOME

- Students will be able to discuss the environmental benefits of recycling batteries.

FIGURE 23.16 *Material Processing Wastes* *The relative quantities of wastes generated by the extraction and processing of zinc.* (Source: Data from R.U. Ayres, "Metals Recycling: Economic and Environmental Implications," Resource, Conservation, and Recycling 21: 145–173.)

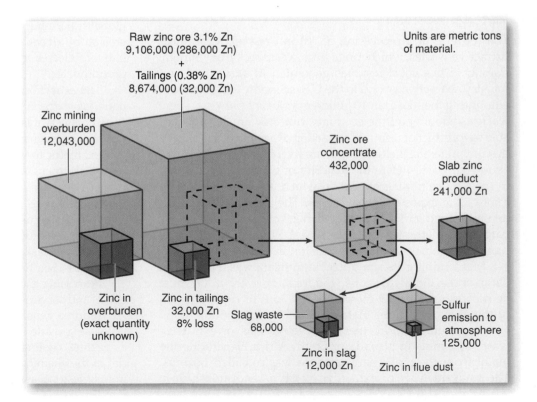

according to precise specifications for door size and shape. The pieces of steel discarded from the cutting process are treated as wastes, some of which are recycled. Much of the water used in industry and manufacturing is returned to the environment laden with materials that are hazardous to environment.

Atmospheric emissions result principally from the use of energy, especially fossil fuels. We usually associate the flow of energy through the economy with the provision of services such as lighting, mobility, shelter, warmth, and so forth. But energy flows are also material flows, and the combustion of fuels with ambient air generates a wide range of wastes. Figure 23.17 shows coal, oil, natural gas, biomass, and oxygen inputs to the combustion process for the entire U.S. economy as graphically scaled flows. This means that the length of each bar represents the quantity (mass) of each material flow. Note the wide range of atmospheric emissions. Carbon dioxide, nitrous oxide, and methane are greenhouse gases; sulfur dioxide contributes to acid deposition; and carbon monoxide and particulate matter are serious public health hazards (Chapter 19).

As we discussed in Chapter 10, **dissipation** refers to materials released directly to the environment, where no attempt to recover them is economically or technologically practical (Table 23.2). For example, road salt is used to reduce the hazard of winter driving in cold regions. Eventually the salt is crushed or dissolved and makes its way into soils and water bodies. Similarly, much of the fertilizer and chemicals applied to the soil or crops finds its way into soil or groundwater or runs off into lakes and rivers. Dissipation also covers materials released to the environment due to normal wear and tear: peeling paint, rusting bridges, worn fabric, and so on.

Postconsumer waste refers to a material discarded in a controlled fashion following use in product form. The largest component of this category is **municipal solid waste,** which is the paper, plastic, wood, and metal materials discarded by households and businesses. Solid waste is transported to a landfill or to an incinerator where it is burned, sometimes in the process of generating electricity.

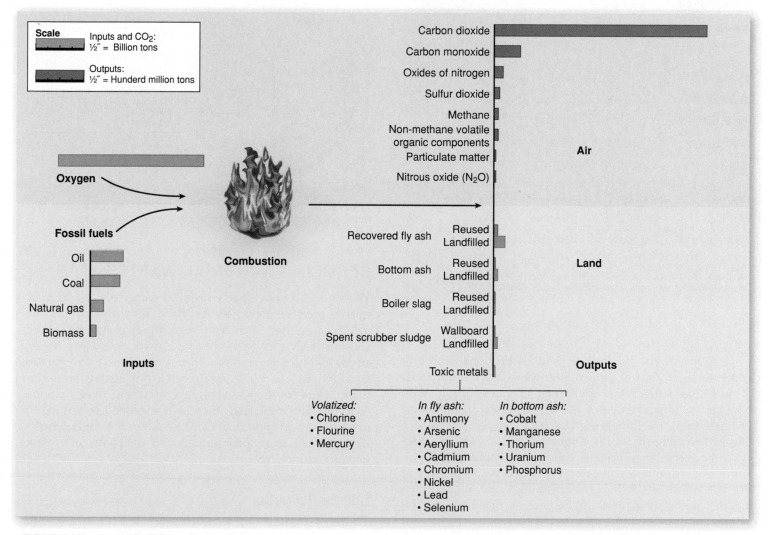

FIGURE 23.17   *Material Inputs to the Economy*   *The inputs of fossil fuels and biomass to the U.S. economy, and the outputs of wastes from those fuels, as graphically scaled flows. (Source: Redrawn from Materials: A Report by the U.S. Interagency Working Group on Industrial Ecology, Material, and Energy Flows.)*

TABLE 23.2	Examples of Dissipative Use of Materials (Global)	
	Quantity Released	
Substance	(Million metric tons)	Dissipative Uses
Chemicals		
Chlorine	25.9	Acid, bleach, water treatment, (PVC) solvents, pesticides, refrigerants
Sulfur	61.5	Acid (H_2SO_4), bleach, chemicals, fertilizers, rubber
Ammonia	93.6	Fertilizers, detergents, chemicals
Phosphoric acid	24.0	Fertilizers, nitric acid, chemicals (nylon, acrylics)
NaOH	35.8	Bleach, soap, chemicals
Na_2CO_3	29.9	Chemicals (glass)
Heavy metals		
Copper sulfate	0.10	Fungicide, algicide, wood preservative, catalyst
Sodium bichromate	0.26	Chromic acid (for metal plating), tanning, algicide
Lead oxides	0.24	Pigment (glass)
Lithopone (ZuS)	0.46	Pigment
Zinc oxides	0.42	Pigment (tires)
Titanium oxide (TiO_2)	1.90	Pigment
Arsenic	Unknown	Wood preservative, herbicide
Mercury	Unknown	Fungicide, catalyst
Tetraethyl lead	Unknown	Gasoline additive

Source: Data from R.U. Ayres, "Industrial Metabolism: Theory and Policy," eds. R.U. Ayres and U.E. Simonis, in *Industrial Metabolism, Restructuring for Sustainable Development*, United Nations University Press, pp. 119–162.

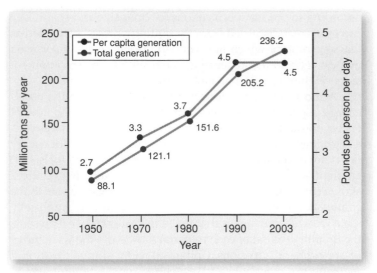

FIGURE 23.18   *U.S. Solid Waste* *The trend in total and per capita municipal solid waste generation in the United States.* (Source: Data from U.S. Environmental Protection Agency.)

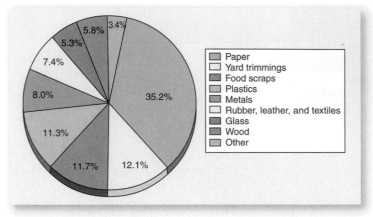

FIGURE 23.19   *U.S. Waste Composition* *The composition of municipal solid waste generation in the United States in 2005.* (Source: Data from U.S. Environmental Protection Agency.)

REDUCING MATERIAL WASTES

Wastes are an unavoidable result of using energy and materials to produce goods and services. They also are undesirable because they often have harmful environmental effects. Closing material cycles is, therefore, an important step toward sustainability. See *Policy in Action: Pay-as-You-Throw Programs for Municipal Solid Waste*. The technological strategy for reducing material waste includes source reduction, reuse, remanufacturing, and recycling. These activities tend to reduce the need to extract virgin materials and hence the environmental damage and energy consumption associated with such extraction and processing.

Source Reduction

The most effective way to reduce waste is to prevent it in the first place. **Source reduction** is the practice of designing, manufacturing, purchasing, or using materials (such as products

U.S. residents, businesses, and institutions currently generate about 240 million metric tons (265 million tons) of municipal solid waste each year. This is approximately 2 kilograms (4.5 pounds) of waste per person per day—an increase from 1.2 kilograms (2.7 pounds) per person per day in 1960 (Figure 23.18). Organic materials are the largest components of municipal solid waste based on weight. Paper and paperboard products account for 35 percent of the waste stream, with yard trimmings and food scraps together accounting for about 24 percent. Plastics comprise 11 percent; metals make up 8 percent; and rubber, leather, and textiles account for about 7 percent. Wood follows at 6 percent, and glass at 5 percent (Figure 23.19).

and packaging) in ways that reduce the amount or toxicity of waste created. Source reduction can be achieved through simple behavioral changes, such as buying products in bulk or larger containers rather than multiple small containers, making double-sided copies, removing yourself from junk mail lists, buying fresh produce without packaging, donating unwanted items (food, clothing, equipment, furniture, appliances) to charitable organizations that support your community, and avoiding single-use items such as disposable cups, razors, and lighters.

Source reduction also includes **reuse,** which involves using a product over and over in its original form. Refillable glass bottles accounted for 98 percent of all soft drink containers in 1958. Today they account for about 10 percent of the market, and only a handful of states allow returnable bottles. Refillable glass bottles can be used more than fifty times, which reduces the energy and environmental costs of a recycled glass bottle. An example of reuse is string or canvas shopping bags. Disposable bags have significant environmental effects. Plastic bags are made from nonrenewable fossil fuels and degrade very slowly in landfills. Many paper bags are made from virgin sources of timber, although they have the potential to be recycled.

There are more than 6,000 reuse centers in the United States, ranging from specialized programs for building materials, unneeded materials in schools, and used furniture to local programs such as Goodwill and the Salvation Army. In addition to its environmental benefits, reuse improves the well-being of communities by providing useful discarded products to people who can appreciate them. In many cases reuse supports local community and social programs while providing donating businesses with tax benefits and reduced waste disposal fees.

Business and industry also practice source reduction when they design, manufacture, or market a product with less waste. Such changes often are driven by a desire to reduce costs or to comply with government regulations that restrict the release of pollutants. For example, the weight of 2-liter plastic soft drink bottles has been reduced from 68 grams (2.4 ounces) in 1977 to 51 grams (1.8 ounces) today. That change reduces the amount of oil that needs to be extracted to make plastic, and it reduces the amount of plastic going into landfills. Glass jars also are 43 percent lighter than they were in 1970.

Recycling

Recycling is the collection and reprocessing of degraded materials so they can be made into new products. Because matter is not created or destroyed, it can cycle between firms and households many times. Most people associate recycling with personal efforts to recycle household materials such as glass, paper, and plastic. But recycling has a long history in industry: Recycled metal often is less expensive than metals extracted from the environment. Producers recycle metal scrap from demolition wastes, discarded motor vehicles, or other pieces of equipment. The process involves manual removal of valuable metal items such as stainless steel or copper, removal of nonmetallic items, shredding, sorting, and ultimately remelting and repurification.

Recycling has increased sharply in recent decades across most industrial nations. In the United States the fraction of municipal solid waste that gets recycled increased from 6 to 31 percent between 1960 and 2003 (Figure 23.20). Most of this gain is due to the increase in curbside recycling programs. Twenty years ago only one curbside recycling program existed in the United States. Today there are more than 9,000 curbside programs and 12,000 recyclable drop-off centers. In addition, nearly 500 materials recovery facilities have been established to process collected materials.

What gets recycled? Figure 23.21 shows the recycling rates for various materials in the United States. Materials

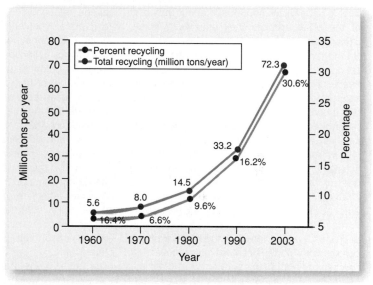

FIGURE 23.20 *U.S. Recycling Rates The trend in overall recycling rates of municipal solid waste in the United States. (Source: Data from U.S. Environmental Protection Agency.)*

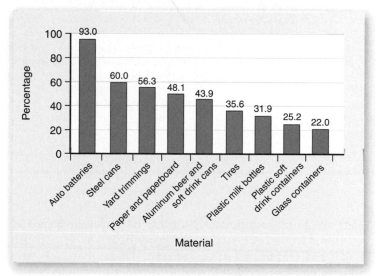

FIGURE 23.21 *The Recycling Rate for Selected Materials in the United States in 2005 The high rate of recycling for batteries is due to a law that requires car batteries to be recycled. (Source: Data from U.S. Environmental Protection Agency.)*

POLICY IN ACTION

Pay-as-You-Throw Programs for Municipal Solid Waste

Most U.S. communities charge for municipal solid waste (MSW) services by either levying a property tax or billing all residents an equal amount, regardless of the amount of waste they generate. In this system a household putting four trash barrels at the curb every week is charged the same as a household that generates one barrel of trash.

From an economic perspective, this system has some serious drawbacks because people usually lack information about the costs of the MSW services they use. As a result, citizens often act as if MSW services are free. No matter how much these individuals use the services, they incur no financial consequence. People typically respond to (apparently) free services by overusing them. In most communities this results in unnecessarily large amounts of garbage with excessive costs for which the local citizens must then pay.

Pay-as-you-throw (PAYT) is an economic incentive that encourages citizens to reduce waste. Under PAYT systems residents are charged for MSW services based on the amount of trash they discard. In doing so PAYT introduces price incentives into individuals' decisions, which then collectively determine a community's pattern of generation, collection, processing, and disposal of MSW. It does this by pricing MSW services in a way that reflects the resources (labor, equipment, fuel, and land) needed to dispose of household materials that are no longer useful—regardless of whether they were purchased (leftover food, discarded packaging, or broken products) or grown (grass clippings and tree trimmings).

This price signal gives residents *information* with which to choose the actions they can take to set out less trash and more recyclables. Pricing MSW services also gives individuals the *opportunity* to take actions that can make a financial difference. Both aspects of price—information and opportunity—induce individu-

als in PAYT communities to conserve MSW services just as they conserve any other service or product that has a price. PAYT provides a continuing motivation to residents to reduce their expenses for MSW services by managing their waste materials in a more environmentally sustainable fashion.

The biggest challenge in a PAYT system is determining how much to charge. The **proportional pricing system** creates the most direct relationship between trash amounts and price. Residents are charged the same amount of money for each unit of waste they set out for collection (for example, $1.50 for each 30-gallon bag). The price is based on the number of bags, tags, or stickers (usually sold at local retail stores or municipal offices) a resident uses. The advantage of this system is that it provides a strong, direct waste reduction incentive. Because residents must pay for the collection of each bag they place at the curb, they have a strong incentive to reduce waste and increase recycling and composting. A disadvantage of proportional pricing is that it may not accurately reflect the actual cost of a community's overall MSW system.

A **variable rate pricing system** charges different amounts per unit of garbage. Residents subscribe to one of several different container size options (typically these are 32- to 64-gallon sizes, although they can range from 10 to 96 gallons in capacity). The community bills residents based on their subscription level. For garbage that residents discard above their subscription level, they must pay an additional fee. The advantage of this system is that it gives a community increased control over the waste reduction incentive. MSW authorities can charge a price for additional containers that is higher than the subscription-level price if their goal is to create a strong incentive to reduce and recycle. A disadvantage of variable rate pricing is that such a system might be more

expensive to implement and administer (especially during the startup phase).

Currently over 2,000 communities in the nation practice some form of PAYT to finance their MSW programs (see Figure 1). For example, in 1994 the city of Gainesville, Florida, implemented a PAYT system based on a variable rate for residential collections: Residents pay $13.50, $15.96, or $19.75 per month according to whether they place 35, 64, or 96 gallons of solid waste at the curb for collection. In the first year alone the amount of solid waste collected decreased 18 percent, and the recyclables recovered increased 25 percent. The total disposal tonnage decreased from 22,120 to 18,116. This resulted in a savings of $186,200 to the residential sector, or $7.95 per household.

STUDENT LEARNING OUTCOME

- Students will be able to define the basic characteristics of a pay-as-you-throw program.

that are often recycled include car batteries, recycled at a rate of 93 percent; paper and paperboard at 48 percent; and yard trimmings at 56 percent. These materials and others may be recycled through curbside programs, drop-off centers, buy-back programs, and deposit systems.

Recycling now extends to materials that previously were never recycled. Recycled crushed concrete now is substituted for crushed stone, sand, and gravel to produce construction aggregates. These aggregates include natural mineral and rock materials used to produce concrete. Natural sources

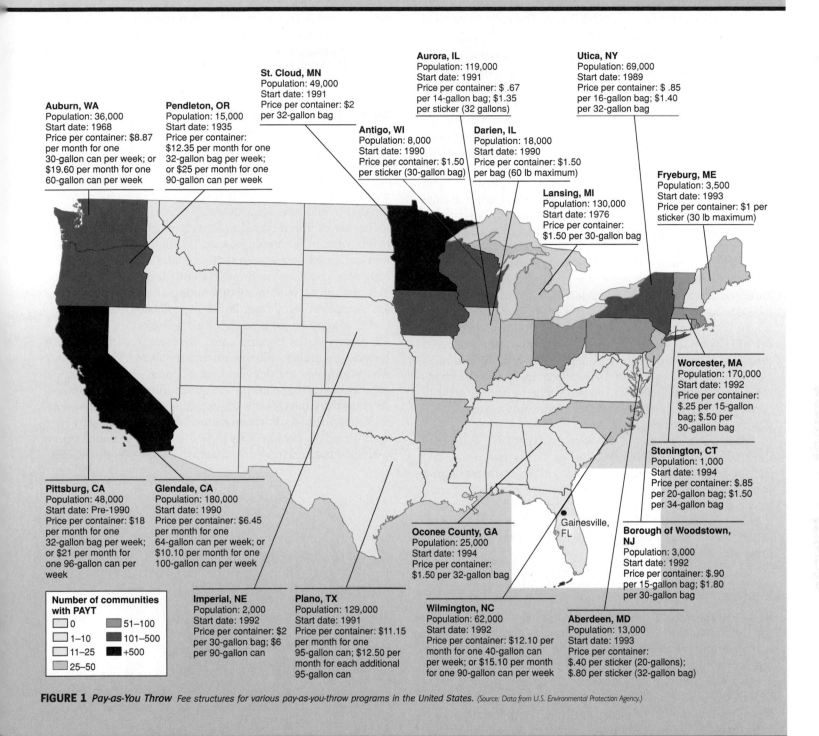

Auburn, WA
Population: 36,000
Start date: 1968
Price per container: $8.87 per month for one 30-gallon can per week; or $19.60 per month for one 60-gallon can per week

Pendleton, OR
Population: 15,000
Start date: 1935
Price per container: $12.35 per month for one 32-gallon bag per week; or $25 per month for one 90-gallon can per week

St. Cloud, MN
Population: 49,000
Start date: 1991
Price per container: $2 per 32-gallon bag

Aurora, IL
Population: 119,000
Start date: 1991
Price per container: $ .67 per 14-gallon bag; $1.35 per sticker (32 gallons)

Utica, NY
Population: 69,000
Start date: 1989
Price per container: $ .85 per 16-gallon bag; $1.40 per 32-gallon bag

Antigo, WI
Population: 8,000
Start date: 1990
Price per container: $1.50 per sticker (30-gallon bag)

Darien, IL
Population: 18,000
Start date: 1990
Price per container: $1.50 per bag (60 lb maximum)

Lansing, MI
Population: 130,000
Start date: 1976
Price per container: $1.50 per 30-gallon bag

Fryeburg, ME
Population: 3,500
Start date: 1993
Price per container: $1 per sticker (30 lb maximum)

Pittsburg, CA
Population: 48,000
Start date: Pre-1990
Price per container: $18 per month for one 32-gallon bag per week; or $21 per month for one 96-gallon can per week

Glendale, CA
Population: 180,000
Start date: 1990
Price per container: $6.45 per month for one 64-gallon can per week; or $10.10 per month for one 100-gallon can per week

Worcester, MA
Population: 170,000
Start date: 1992
Price per container: $.25 per 15-gallon bag; $.50 per 30-gallon bag

Stonington, CT
Population: 1,000
Start date: 1994
Price per container: $.85 per 20-gallon bag; $1.50 per 34-gallon bag

Oconee County, GA
Population: 25,000
Start date: 1994
Price per container: $1.50 per 32-gallon bag

Gainesville, FL

Borough of Woodstown, NJ
Population: 3,000
Start date: 1992
Price per container: $.90 per 15-gallon bag; $1.80 per 30-gallon bag

Number of communities with PAYT
- 0
- 1–10
- 11–25
- 25–50
- 51–100
- 101–500
- +500

Imperial, NE
Population: 2,000
Start date: 1992
Price per container: $2 per 30-gallon bag; $6 per 90-gallon can

Plano, TX
Population: 129,000
Start date: 1991
Price per container: $11.15 per month for one 95-gallon can; $12.50 per month for each additional 95-gallon can

Wilmington, NC
Population: 62,000
Start date: 1992
Price per container: $12.10 per month for one 40-gallon can per week; or $15.10 per month for one 90-gallon can per week

Aberdeen, MD
Population: 13,000
Start date: 1993
Price per container: $.40 per sticker (20-gallons); $.80 per sticker (32-gallon bag)

FIGURE 1 *Pay-as-You Throw* *Fee structures for various pay-as-you-throw programs in the United States.* (Source: Data from U.S. Environmental Protection Agency.)

of aggregate include crushed stone, sand, and gravel. These resources are abundant, but they are expensive to transport because they are so dense. As deposits of natural aggregates close to urban centers are depleted, costs rise steeply due to longer hauling distances. But cities contain "deposits" of aggre-

gates: demolished buildings, roads, and other concrete structures. The concrete in demolished structures—which often are closer than natural sources of stone, sand, and gravel—can be crushed on-site to produce aggregate and then transported to new construction sites.

There's Recyclable . . . and Then There's Recycled

There's more to recycling than setting out your recyclables at the curb. To make recycling economically feasible, people must buy recycled products and packaging. Doing so creates an economic incentive for recyclable materials to be collected, manufactured, and marketed as new products.

Product labels can be confusing to consumers interested in buying recycled goods because of the different recycling terminology used. **Recycled-content products** are made from materials that would otherwise have been discarded. Items in this category are made totally or partially from material destined for disposal or recovered from industrial activities—like aluminum soda cans or newspaper. Recycled-content products also can be items that are rebuilt or remanufactured from used products, such as toner cartridges or computers. **Postconsumer content** refers to material from products that were used by households or firms and would otherwise be discarded as waste. If a product is labeled "recycled content," the rest of the product material might have come from excess or damaged items generated during normal manufacturing processes—not collected through a local recycling program. **Recyclable products** can be collected and remanufactured into new products after they've been used. These products do not necessarily contain recycled materials and benefit the environment only if people recycle them after use.

Recyclable claims on labels and advertising mean that the manufacturer or seller of the products has proof that the products can be collected and used again or made into useful products. Some companies simply may say "Please Recycle" on their products. Such claims will be meaningful if these products are collected for recycling in your community, either through curbside pickup programs or drop-off programs.

Remanufacturing

Remanufacturing is the process of disassembling products, cleaning their parts, making any needed repairs, and then reassembling the products into sound working condition. A remanufacturer acquires discarded products from producers (air conditioning compressors, metal cutting machines) or consumers (tires, toner cartridges). Remanufacturers disassemble these products into their components, which are cleaned and examined for damage or wear; discard components that cannot be used or repaired, replacing them with others; refurbish all the useful original components; reassemble and test the products; and then sell them.

An example is an automobile engine that needs new gaskets and has worn-out bearings. The main industrial sectors with remanufacturing activities (typical remanufactured products in parentheses) are automotive (starter motors, engines, water pumps); compressors (air condition-

ers); machines (many types); office furniture (desks, office partitions); tires (truck and auto tires); and toner cartridges (laser printers). The U.S. remanufacturing industry in 2005 included 73,000 firms with total annual sales of $53 billion.

BENEFITS OF REDUCING MATERIAL WASTES

Source reduction, recycling, and remanufacturing produce significant environmental benefits (Table 23.3). Every ton of metal that is reused, recycled, or remanufactured replaces a ton that would otherwise have to be mined and purified. Each metric ton of iron recycled saves 11.3 metric tons (12.5 tons) of overburden in the mining sector, 2.5 metric tons (2.7 tons) of iron ore, nearly a metric ton of coal, and 72 metric tons (79.4 tons) of water. It also reduces CO_2 emissions by 0.4 metric tons (0.5 tons), along with many other pollutants. Recycling of each ton of paper saves 17 trees, 26,500 liters (7,000 gallons) of water, and 27 kilograms (60 pounds) of air pollution.

Many other benefits are associated with recycling. Recycling supplies valuable raw materials to industry, creates jobs, stimulates the development of more efficient technologies, and reduces the need for new landfills and waste incinerators. In the United States the recycling industry is comprised of 56,000 public and private sector facilities that employ more than a million people. This puts the recycling sector on par with the auto and truck manufacturing sector (Figure 23.22).

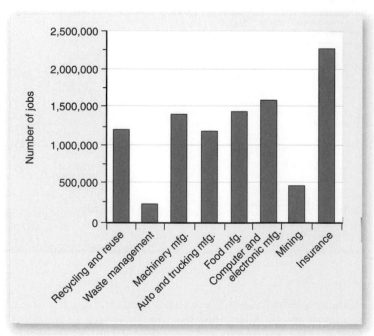

FIGURE 23.22 *Employment in Selected Sectors of the U.S. Economy* Recycling and reuse employed more than a million people in 2003. *(Source: Data from U.S. Environmental Protection Agency.)*

TABLE 23.3	Reductions in the Use of Energy and Materials Due to the Recycling of Metals*				
	Iron/Steel (Fe)	Aluminum (Al)	Copper (Cu)	Lead (Pb)	Zinc (Zn)
Ore grade	52.8%	17.5%	0.6%	9.3%	6.2%
Energy used (gJ/t)	22.4	256	120	30	37
Water flow in/out (t/t)	79.3	10.5	605.6	122.5	36.0
Material inputs					
Air	1.9	0.3	1.6	4.4	5.8
Solids	17.3	11.0	612.1	126.2	55.8
Total material inputs	19.2	11.2	613.7	130.5	61.6
Material outputs					
Product	0	0	0	0	0
By-products	0.2	0.1	1.0	6.7	4.9
Depleted air	1.5	0.2	1.3	1.2	2.4
CO_2	0.5	0.8	0.02	0.03	0.03
SO_x	0.01	0.06	1.47	0.005	0.01
Other gaseous material	1.18	0.002	0.15	0.28	0.0
Potential recycle	0.6		3.2	0.1	0.5
Overburden	12.5	0.6	395.4	72.5	37.3
Gangue	1.1	6.1	211.0	44.6	16.3
Other solids	1.5	1.4	0.1	2.5	0.1
Sludges, liquids	0.1	1.9	0.1	2.6	0.1
Total material outputs	19.2	11.2	613.7	130.5	61.6

*Unless specified otherwise, values are metric tons of material saved per metric ton of material recycled.

Source: Data from R.U. Ayres, "Metals Recycling: Economic and Environmental Implications," *Resource, Conservation, and Recycling* 21: 145–173.

SUMMARY OF KEY CONCEPTS

- A cradle-to-grave or life cycle analysis of materials traces their flow from the environment (mine, forest, agricultural field), through their processing, manufacture, and use by consumers and ultimately their disposal or recycling.

- Weathering concentrates some minerals in the soil.

- Hydrothermal solutions, produced when water is heated by contact with magmas or hot rocks, produce many of the richest deposits of copper, lead, and zinc.

- The mineral reserve base is the part of an identified resource that can be extracted under current economic, technological, and legal conditions.

- Dematerialization is caused by several forces, including technological change, the substitution of new materials with more desirable properties, and government regulations.

- The price of many important minerals has not increased over time even though the highest-quality deposits have been depleted. In fact, many mineral prices have declined over time.

- The energy cost of producing a unit of some mineral resources, especially some metals, has increased over time due to declining resource quality.

- The average American generates approximately 4.5 pounds of municipal solid waste per day. The main components of this waste are paper and paperboard products, yard trimmings and food scraps, and plastics.

- The most effective way to reduce waste is through source reduction, which is the practice of designing, manufacturing, purchasing, or using materials to prevent waste.

- Recycling has increased sharply in recent decades across most industrial nations, producing significant environmental benefits such as reduced energy use and less pollution.

REVIEW QUESTIONS

1. Distinguish between mineral resources and mineral reserves.
2. What are the major types of materials used in the United States? How have they changed over time?
3. Describe and give an example of dematerialization.
4. Why have most mineral prices declined over time?
5. Describe the relationship between economic and energy costs of natural resources.
6. What are the main components of the municipal solid waste stream in the United States?
7. List the principal environmental and economic benefits of recycling.

KEY TERMS

atmospheric emissions
black smokers
cradle-to-grave or life cycle analysis
crustal abundance
dematerialization
dissipation
economic energy cost
environmental energy cost
evaporites
extractive wastes
geochemical exploration
geologic exploration

geophysical exploration
geothermal gradient
hydrothermal solutions
intensity of use
mineral resource
municipal solid waste
overburden
petrochemicals
placer deposits
postconsumer content
postconsumer waste
processing waste

proportional pricing system
recyclable products
recycled-content products
recycling
remanufacturing
reserve base
reuse
source reduction
stripping ratio
tailings
variable rate pricing system

A SUSTAINABLE FUTURE

Will Business as Usual Get Us There?

24

STUDENT LEARNING OUTCOMES

After reading this chapter, students will be able to

- Compare and contrast the advantages and disadvantages of efforts to reduce environmental impacts by focusing on scale or efficiency.

- Compare and contrast the advantages and disadvantages of efforts to reduce environmental impacts using market-based mechanisms versus command and control strategies.

- Discuss the degree to which an environmental Kuznets curve implies that economic growth can increase affluence and reduce environmental impacts.

- Discuss whether slower population growth sets the stage for economic development or whether economic development sets the stage for slower population growth.

- Explain how biogeochemical cycles can be used to assess the sustainability of economic activities.

Arctic National Wildlife Refuge *The Arctic National Wildlife Refuge is a large area of tundra in Alaska. Preliminary results show that the refuge may sit atop a large oil field, but its status as a wildlife refuge prohibits oil production. Some people would like an exception to allow oil production, but others argue that the refuge should remain off-limits to drilling.*

College students are forced to analyze issues in depth; most voters are not. To simplify issues for voters, politicians often frame the discussion about environmental issues as a battle between "good guys" and "bad guys." Politicians who favor environmental protection often gloss over significant scientific uncertainties. They describe polluting industries as the "bad guys" who threaten the environment. On the other side, politicians who argue against environmental protection often exaggerate scientific uncertainties. They describe those in favor of environmental protection as "bad guys" who oppose economic growth and technical change. Consider the political debate about opening the Arctic National Wildlife Refuge to oil exploration. Those in favor of exploration label opponents as antigrowth preservationists who value caribou more than energy security or jobs. Those who argue against exploration label their opponents as tools of the oil industry who are unconcerned about the environment.

We hope you can see through this false dichotomy. There are no good guys or bad guys. Who would the good guys be? Who would the bad guys be? Yes, some industries use materials and energy in ways that degrade the environment. But are they the bad guys? After all, someone must purchase and use those goods or services. And that would make consumers the bad guys. But no one wants to think of himself or herself as a bad guy. Instead most consumers want to think of themselves as the good guys. And yes, many consumers conscientiously try to reduce their impact on the environment. Despite these efforts, consumers have a large impact on the environment—especially those who live in developed nations.

To ameliorate environmental challenges described in this book, political discussion must go beyond the notion of good guys and bad guys. Such a discussion requires additional information and nuance. Continuing with the example of drilling for oil in the Arctic National Wildlife Refuge, claims that oil from the refuge will reduce oil prices and U.S. dependence on imported oil are an exaggeration. As indicated in Figure 20.19 on page 438, at best the refuge will produce about 2 million barrels per day ten to twenty years after exploration starts. Such production would reduce U.S. dependence on imported oil by a few percentage points (Figure 24.1). Furthermore, it would probably leave prices unchanged because it would have little effect on OPEC production. On the other hand, claims that exploration and development will ruin the ecosystem probably also are an exaggeration. It is hard to imagine an accident that would have the consequences described by the worst-case scenario.

Why is drilling in the Arctic National Wildlife Refuge such a point of contention? For oil companies, efforts to open the refuge to drilling center on extending the economic life of the Trans-Alaska Pipeline System and the port of Valdez. These facilities were built for the giant oil field at Prudhoe Bay (Figure 24.2). Although this field is close to the Beaufort Sea, this sea cannot be navigated by oil tankers during much of the year. To avoid bottlenecks, oil from Prudhoe Bay is shipped via the pipeline to Valdez, where it is loaded on tankers. The economic life of the pipeline and the port facilities are limited by the lifetime of the field at Prudhoe Bay—once the field at Prudhoe Bay is exhausted, the pipeline and the port facilities will be very expensive junk (it cost more than $8 billion to build the pipeline).

Notice how close the Arctic National Wildlife Refuge is to the pipeline (Figure 24.2). If there is oil in the refuge, oil companies will build a short pipeline to the Trans-Alaska Pipeline and use the Trans-Alaska Pipeline System to ship oil from the wildlife refuge to Valdez. This will greatly extend the time during which the pipeline's owners can earn

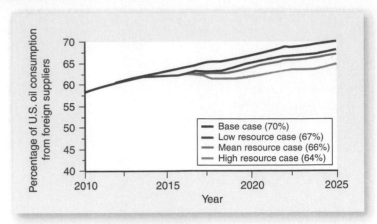

FIGURE 24.1 **U.S. Oil Imports** *Without oil production from the Arctic National Wildlife Refuge, U.S. oil imports are forecast to rise steadily to about 70 percent of consumption by 2025. This increase is relatively unaffected regardless of the quantity of oil produced from the refuge. The most optimistic scenario, the high resource case, reduces U.S. dependence to 64 percent; that figure is 67 percent under the pessimistic, low resource scenario.* (Source: Data from the U.S. Department of Energy.)

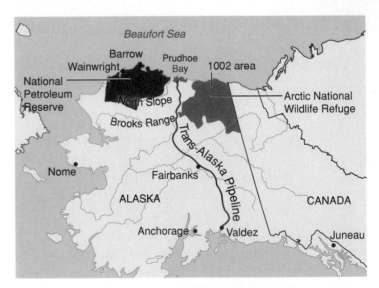

FIGURE 24.2 **Geography of Alaskan Oil** *To facilitate the transport of oil to refiners, oil from the giant Prudhoe Bay field in the North Slope is shipped via the Trans-Alaska Pipeline to Valdez, where it is loaded onto tankers. Should oil be produced from fields in the refuge, a pipeline would be built from the refuge to the Trans-Alaska Pipeline and the oil would be moved to Valdez. This would extend the economic lifetime of the pipeline and the loading facilities in Valdez.*

income. As such, the continued use of the pipeline and port would be an economic windfall to its owners, which include British Petroleum, Exxon Mobil, and several other oil companies, some of which may eventually drill in the refuge should it be opened for exploration. That is why they are so anxious to drill in the refuge.

But why the fierce opposition from environmentalists? There is legitimate concern about oil spills and accidents. But some of the opposition stems from the precedent that opening the refuge would represent. Should the refuge be opened, environmentalists fear that no national park or refuge would be beyond the reach of industry. If oil were found in Yellowstone or Yosemite, industry could point to the opening of the Arctic National Wildlife Refuge. In short, none of the protections gained by environmentalists over the last century would be permanent.

Getting the public to focus on these underlying issues would be difficult for both sides. Other than shareholders, the public is not going to be excited about increasing profits for large oil companies. Nor is the general public likely to be excited about protecting the legal precedent of refuge status. Doing so would require the public to educate themselves. Many people do not have the time to do so.

Nor would the need for education stop at the refuge. As described in the first twenty-three chapters, each environmental challenge is unique. The causes of and potential solutions to the reduction in stratospheric ozone are different from the causes of and potential solutions to global climate change. Educating the public about each of these issues would be expensive and time-consuming. It is much easier for industry and environmentalists to use the good guy versus bad guy approach and hope that you sign up for their side on all issues.

Differences among environmental challenges do not imply that the lessons learned from efforts to halt the reduction in stratospheric ozone are of no use to efforts to slow global climate change. Much has been learned from the last fifty years of scientific and political efforts to repair environmental degradation. This chapter describes some of the lessons learned and those yet to be fully understood. Lessons include the degree to which environmental policy should rely on the market; whether firms can be "bribed" to clean up pollution; and whether environmental challenges can be solved by slowing population growth, making poor people rich, affecting personal choices, and changing the risks that society is willing to incur. ∎

UNDERSTANDING POSSIBLE SOLUTIONS: THE PARABLE OF THE PLIMSOLL LINE

In general terms, potential solutions to environmental challenges address the efficiency or scale of economic activity. Efficiency gains reduce environmental degradation by lessening the effect of each unit of economic activity. Lowering the scale of economic activity also cuts environmental degradation.

Efficiency and scale can be illustrated by potential policies to curtail energy use. Total energy use can be described most simply by the following definition (a definition is an equation that is automatically true):

$$\text{Energy} = \frac{\text{Energy}}{\text{GDP}} \times \text{GDP} \qquad (24.1)$$

in which Energy is total energy use and GDP is real GDP. Policies that focus on efficiency seek to diminish the amount of energy used to produce an inflation-corrected dollar's worth of GDP ($\frac{\text{Energy}}{\text{GDP}}$). Policies that focus on scale seek to lower GDP.

Of the two, increasing efficiency seems more desirable. Any policy that reduces environmental degradation while maintaining economic well-being is a win–win solution. Returning to the example given by Equation 24.1, nobody cares about the energy/GDP ratio. (Do you know the U.S. energy/GDP ratio? It was 9,030 Btu per dollar in 2005.) This ratio has no clear connection to your material well-being. On the other hand, many people associate economic well-being with GDP. Even though you probably don't know the U.S. GDP (in 2005 U.S. GDP was $11.05 trillion as measured in inflation-corrected dollars), you know that it is "good news" when the government reports an increase in GDP. Clearly lessening energy use by cutting GDP would be unpopular.

Despite the appeal of increasing efficiency, focusing solely on economic efficiency ignores the possibility that there may be ecological limits on the scale of economic activity. There may be some upper bound to the sustainable size of the human population and its economic well-being. By ignoring this limit, policies that focus solely on efficiency may lead society beyond its carrying capacity.

Differences between policies that focus on efficiency versus scale and the dangers associated with ignoring the latter can be illustrated with a simple parable offered by ecological economist Herman Daly. He describes the problem of loading a ship (Figure 24.3). There are many ways to arrange cargo. All cargo can be loaded in the front or the back. But that arrangement would make the ship unstable and reduce the total amount of cargo the ship could carry. Alternatively, the cargo can be arranged evenly. This would keep the ship steady and would allow the ship to carry its maximum load.

But the most efficient pattern does not tell the cargo handlers when the ship has reached its maximum load. Even if the handlers load cargo efficiently, eventually the ship will sink because the load is too heavy. A ship with a rated capacity of 100 metric tons cannot carry 200 metric tons.

How is the capacity of a ship determined? Sinking is the most dramatic form of overloading. Before the ship sinks, it will sit so low in the water that even the smallest wave will

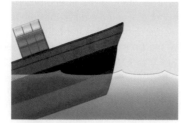
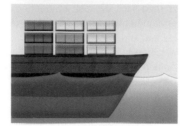

— Plimsoll line

FIGURE 24.3 *Efficiency, Scale, and the Plimsoll Line If all cargo is loaded on the front or back of the ship, the ship is unstable and is likely to sink. If the same amount of cargo is loaded evenly across the deck, the ship floats relatively high in the water, and the Plimsoll line remains above the water. In this case the ship is much less likely to sink. As the cargo increases, the ship sits lower in the water, and the Plimsoll line drops to the water's surface. Under these conditions even a small wave can wash across the deck and sink the ship. If too much cargo is loaded, the ship will sink even without a wave.*

wash over its deck, flooding the ship and possibly causing it to sink. To avoid this, ships are loaded so that their decks are well above any wave the ship is likely to encounter.

This safe height is identified by the Plimsoll line. The Plimsoll line usually is marked by a color change on the hull that differentiates the portion that is meant to be underwater from the portion that is meant to remain above the water. An empty ship floats high in the water. As cargo is loaded, the ship sits lower. Eventually only the portion that is supposed to remain above the water can be seen. Most prudent owners will not load cargo beyond this maximum.

It would be nice if there were an ecological or economic Plimsoll line that would tell us when the human population or its economic standard of living exceeds carrying capacity. See *Your Ecological Footprint: What Is Your Overall Impact?* Unfortunately there is no obvious equivalent to the Plimsoll line. Scientists, economists, and politicians cannot agree on an indicator. This inability does not mean that such limits do not exist or that they are unimportant. Rather, it implies that environmental policies must consider both efficiency and scale.

EFFICIENCY

Most efforts to protect the environment try to diminish the amount of environmental degradation that is generated by each unit of economic activity. These efforts can

ameliorate many environmental challenges by internalizing externalities, working with the market, eliminating subsidies, and encouraging personal choices that curtail environmental impacts.

Internalizing Externalities

Think about how you spend your income. Generally consumers spend their income on the goods and services that maximize their material well-being. Similarly, firms purchase the combination of capital, labor, materials, and energy that minimizes the cost of producing goods and services. These combinations are chosen based in part on the price of potential purchases.

According to economists, making decisions based on price is efficient *if* prices represent all costs associated with producing or consuming a good or service. But as described throughout this book, the price for many goods and services does not include the environmental damage that is associated with their production or consumption. For example, Chapter 13 describes the effects of carbon dioxide emissions on global climate and how these effects are not included in the price of fossil fuels.

Many chapters describe how environmental policy seeks to incorporate environmental damages in the price of a good or service. Internalizing externalities would allow you to evaluate how your purchases affect the environment: You could balance the benefits of purchasing a good or services against all costs associated with that good or service, including its effects on the environment. Under these conditions, internalizing externalities would generate the optimal level of pollution that is described in Chapter 11.

Despite these benefits, internalizing externalities is not politically popular. It is difficult to put a price on environmental degradation. For some environmental challenges, there is considerable uncertainty about the link between human actions and environmental harm. To illustrate, there is considerable uncertainty in the relationship between the emission of carbon dioxide and the size of the increase in global temperature. For other challenges, the link between environmental degradation and economic well-being is uncertain, as illustrated by an incomplete understanding of the link between the stratospheric concentration of ozone and economic well-being or human health.

Uncertainty often is used to delay efforts to internalize externalities. But doing nothing implies that human action has no effect on the environment—that is, the environmental damage has no economic value. This assumption is incorrect in nearly all cases. Many politicians do not want to impose carbon dioxide emission taxes until more is known about the effects of carbon dioxide emissions on global climate. This delay seems reasonable. But setting the carbon dioxide tax to zero implies that the economic damage associated with climate change also is zero. Clearly this is not correct.

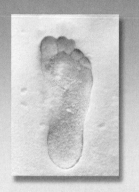

Your ECOLOGICAL footprint

What Is Your Overall Impact?

For this final chapter it would be nice if you could calculate a single measure that represents your overall impact on the planet. To do so, you would sum the uses and emissions of materials, energy, soil, and land that you calculated for previous chapters' Your Ecological Footprint.

But calculating a single measure is not easy. Think back to those exercises: Some had you track land use, whereas others asked you to trace energy use, water use, soil erosion, or the emissions of pollutants. Can such different impacts be combined?

Actually several measures of overall environmental impact are available. Some measure energy use; others measure the amount of land used. Here we help you translate calculations from previous Your Ecological Footprints to a land-based measure of your overall environmental impact.

You have already done part of this calculation. Chapter 8's Your Ecological Footprint had you calculate the amount of land associated with your use of net primary production. Chapter 16's Your Ecological Footprint had you calculate the amount of land used to grow your food. Chapter 18's Your Ecological Footprint had you calculate the land area used to capture the precipitation you use. It is easy to add these land areas. But how about your use and emissions of materials and energy? How can you convert energy use or carbon dioxide emissions to land area?

To convert oil and other forms of energy to land use, you can ask what would happen if you replaced all of your oil, coal, gas, and electricity with alternative fuels obtained from biomass (see Chapter 22). If you are willing to make this assumption, it is relatively simple to translate energy use to land area. Take the total energy use you calculated for Chapter 20's Your Ecological Footprint (in Btus) and convert it to kilocalories by multiplying by 0.25. Divide that product (in kilocalories) by 3.6 kcal per gram of carbon to get its carbon equivalent. Divide that carbon equivalent by the rate of net primary production in your local biome (see Table 1 in Chapter 8's Your Ecological Footprint). This quotient represents the amount of land required to generate the energy you use.

To illustrate, let's continue with the example given in Chapter 20's Your Ecological Footprint. In that chapter the example indicated a primary energy consumption of 13.65 million Btu per month. This is equivalent to 3.41 million kcal (13.65 million Btu × 0.25 kcal/Btu = 3.41 million kcal). These 3.41 million kcal are equivalent to 0.95 million grams of carbon (3.41 million kcal/3.41 g carbon/kcal = 0.95 million grams carbon). The example assumed that you went to school in the Boston area, which is located within the temperate forest biome, where net primary production is 701.92 grams of carbon per square meter per year. This implies the biomass from an area of 1,350 m² (0.95 million grams carbon/701.92 grams carbon/m²/year) would be needed to replace the energy used per month.

Now let's calculate the land area that would be required to remove your carbon dioxide emissions from the atmosphere. To do so, recall that net primary production represents the excess of photosynthesis (Equation 5.1) relative to respiration (Equation 5.3). According to these two equations, net primary production represents the amount of carbon dioxide removed from the atmosphere by autotrophs. If you divide the total amount of carbon dioxide emitted, which you calculated for Chapter 13's Your Ecological Footprint, by the rate of net primary production in your local biome (see Table 1 in Chapter 8's Your Ecological Footprint), you can determine the land area needed to remove the carbon dioxide you emit by burning fossil fuels.

To illustrate, the average U.S. citizen emitted about 20 metric tons of carbon dioxide per year in 2005. Of this carbon dioxide, about 27 percent is carbon (the atomic mass of carbon is about 12, the atomic mass of oxygen is about 16, and a molecule of carbon dioxide has an atomic mass of about 44 [12 + 16 + 16], of which 12 is carbon). This implies annual emissions of about 5.5 metric tons of carbon (5.5 metric tons = 20 metric tons carbon dioxide × 0.27). These 5.5 metric tons are equivalent to 5.5 million grams of carbon (1,000 grams per kilogram, 1,000 kg per metric ton). The temperate forests of the Boston area have a net primary production of 701.92 grams carbon/m²/year. At this rate, 7,836 m² (7,836 m² = 5.5 million grams carbon/701.92 grams carbon/m²/year) of temperate forests would be needed each year to take up the carbon emitted by fossil fuel use by the average U.S. citizen per year.

Interpreting Your Footprint

Summing the amount of land you use directly, in the form of net primary production, and the land equivalents of energy and carbon dioxide emissions gives you an idea of how much land you use. But does this land represent your overall environmental impact in a meaningful way?

We would answer no. You can convert many activities to their land equivalent, but many of these conversions have little ecological meaning. Converting fossil fuel use to land via net primary productivity is difficult. As described in Chapter 22, the conversion of biological energy to useful forms is relatively inefficient, so the conversion based on net primary production alone understates land use significantly. Similarly, converting water consumption to area over which the precipitation falls in Chapter 18 says nothing about whether collecting the precipitation disturbs the land. The land equivalent of your carbon dioxide emissions also contains an important simplification. Net primary production removes carbon dioxide emissions from fossil fuel use only if the plant material is prevented from decaying. Once a plant dies, the carbon it removed from the atmosphere via photosynthesis returns to the atmosphere via decomposition.

We hope you are not surprised by our pessimism about calculating an overall measure of your environmental impact. Reading this book and listening to your professor's lectures should have shown you that you are connected to the environment by many flows of energy and materials. These connections can be thought of as your niche, which any ecologist will tell you cannot be measured in a single unit.

The inability to convert your environmental impact to a single unit can be illustrated by the notion of the limiting nutrient. Remember from Chapter 6 that plants often are limited by a single nutrient. As such, measuring the environmental impact of a plant based on the combined weight of its nutrient use would have little ecological meaning.

ADDITIONAL READING

Wackernagel, M., and W. Rees. *Our Ecological Footprint: Reducing Human Impact on the Earth.* Gabriola Island, BC, and Philadelphia, PA: New Society Publishers, 1995.

STUDENT LEARNING OUTCOME

- Students will be able to explain why efforts to translate their use of energy, materials, and land to a single unit of measure may not be consistent with the ecological concept of a niche.

Even if scientists can eliminate uncertainty, politicians are reluctant to internalize externalities. In nearly all cases, internalizing externalities raises the price for the good or service. As such, internalizing externalities is akin to raising taxes. And raising taxes has been the political "kiss of death" in the United States during recent decades. But the analogy with taxes is incorrect. Internalizing an externality is not designed to raise money for the government. Raising prices is supposed to tell consumers that a good or service costs more than is indicated by the market price. And the higher prices are designed to discourage consumption of environmentally damaging goods and services.

The negative political connotation associated with raising government revenue can be reduced through creative policy. One approach offsets the revenue generated by internalizing an externality by lowering another tax. Such an approach is termed **revenue recycling.** Following this strategy, total revenue collected by the government remains constant.

The effectiveness of revenue recycling is being investigated for energy taxes and carbon dioxide taxes. Simulations generated by economic models show that imposing a carbon dioxide tax and lowering payroll taxes would reduce both emissions of carbon dioxide and the effect on economic growth (Figure 24.4). If correct, these simulations indicate that internalizing externalities and recycling the revenues could generate win–win solutions to environmental challenges. See *Case Study: Recycling Environmental Tax Revenues.*

Working with the Market

Environmental policies that are based on market mechanisms often are more cost-effective than policies based on command and control strategies. Their effectiveness is demonstrated by

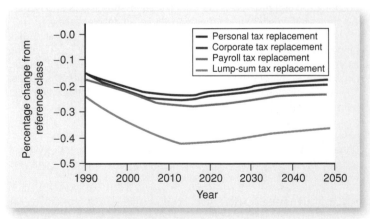

FIGURE 24.4 *Recycling Carbon Taxes Economic models indicate that a tax of $25 per ton of carbon dioxide emitted reduces GDP relative to the scenario in which there is no carbon dioxide tax. The reduction in GDP can be lessened (made less negative) if the carbon dioxide tax is offset by reducing another tax, such as personal, corporate, or payroll taxes.* (Source: Data from L.H. Goulder, "Effects of Carbon Taxes in an Economy with Prior Tax Distortions—An Intertemporal General Equilibrium Analysis," Journal of Environmental Economics and Management 29(3): 271–297.)

the market for sulfur emissions that is described in Chapter 19's Policy in Action. When the market was set up in the early 1990s, industry argued that the cost of reducing a ton of sulfur emissions would be $250–350. This price was expected to reach $500–700 when the number of power plants included in the market expanded and the number of permits in the market shrank. However, even after the number of permits was reduced in 2000, the cost of reducing a ton of emissions remained between $200 and $300.

Some of this lower-than-expected cost is associated with a continuous incentive to lessen environmental degradation. For example, the sulfur market provides a constant incentive to cut emissions because reductions allow firms to increase their revenues by selling more of their permits. This incentive contrasts with command and control strategies, which provide no incentive for further reductions once the original goals are reached. As described in Chapter 14, the Montreal Protocol is among the most successful environmental agreements. Nonetheless, it had an important failing. Because it was based on a command and control strategy, the protocol would not reduce the use of CFCs beyond its goal of a 50 percent reduction in production by 1997–1998. Once producers achieved that goal, the parties had to reconvene and establish new goals.

Market-based mechanisms also provide continuous incentives for research and development in pollution abatement technologies. So long as a firm can use a new technology to reduce its emissions at a cost lower than the price of a permit, firms will purchase new abatement technologies. Command and control approaches have no such incentives. As described in Chapters 18 and 19, portions of the Clean Water Act and the Clean Air Act require firms to install the best available technology, which is often described explicitly in the legislation. Once this is enshrined in legislation, there is no economic incentive to develop new abatement technologies because they would not fit the best available technology description and therefore could not satisfy legislative requirements.

Given the effectiveness of market-based mechanisms, why does so much environmental legislation still follow a command and control strategy? Command and control is popular with politicians because it allows them to take credit for environmental protection. For example, the Clean Air Act established national ambient air quality standards for specific air pollutants. Compliance with these standards allows politicians to rightfully claim they made the air cleaner. Making similar claims for market-based mechanisms is difficult. For example, politicians who imposed a carbon dioxide tax would not be able to identify the quantity of carbon dioxide emissions abated.

The use of taxes is another political argument against market-based mechanisms. No politician wants to raise taxes. Instead they impose "invisible taxes" by legislating command and control strategies. Requiring firms to purchase and install the best available technology imposes costs

CASE STUDY

Recycling Environmental Tax Revenues

Taxes are the "kiss of death" in American politics. Many view environmental taxes as an excuse to raise government revenues. Before accepting this cynicism, we need to differentiate two components of such efforts. Environmental taxes are designed to discourage the consumption of a good or service by raising its price. But the government does not have to keep the additional taxes. The additional revenues associated with environmental taxes can be offset by reducing other taxes so that total government revenues remain the same; this is known as revenue recycling.

If done correctly, imposing environmental taxes and reducing other taxes can protect the environment and boost incomes. Let's illustrate the win–win nature of this approach by describing the effect of a revenue-neutral carbon tax. By revenue-neutral we mean that the increase in government revenues associated with the carbon tax is offset by a corresponding reduction in some other tax.

The environmental benefits of a carbon tax are relatively straightforward. A carbon tax raises the price of energy based on the amount of carbon the fuel emits (see Table 13.2 on page 275). Because all fossil fuels emit carbon dioxide, a carbon tax increases the prices for all fossil fuels. But coal, oil, and natural gas emit carbon dioxide at different rates. A thousand Btus of coal emit more carbon dioxide than a thousand Btus of oil, which emit more carbon dioxide than a thousand Btus of natural gas. A carbon tax causes the price of coal to rise faster than the price of oil, which would rise faster than the price of natural gas.

How would these price changes reduce carbon emissions? Raising the price for all fossil fuels would encourage consumers to buy more efficient energy-using devices, such as fluorescent lightbulbs and hybrid cars. Rising energy prices also would encourage business owners to buy more efficient machinery, such as gas turbines. Together this more efficient machinery would lower energy use, which would reduce carbon emissions.

Carbon taxes also lower carbon emissions by stimulating interfuel substitution, in which one form of energy replaces another. Because a carbon tax raises the price of coal relative to oil and natural gas, consumers try to replace coal with either oil or natural gas. Such replacements also reduce carbon emissions because oil and natural gas emit less carbon than coal.

But if the government increases energy taxes by $1 and reduces other taxes by $1, why won't consumers purchase the same amount of energy? Energy use remains lower due to the higher price of energy relative to other goods and services. Even if consumers have the same amount of money to spend, they will spend more of their income on nonenergy goods and services because they are now less expensive than energy. Put simply, the tax on energy has reduced the bang from a dollar spent on energy relative to the bang from a dollar spent on non-energy goods and services. So to get the most happiness (in technical terms, utility) from their income, or to maximize their profits, consumers and business owners spend more on nonenergy goods than they did before the carbon tax.

But how can raising government revenues from a carbon tax and lowering revenues from other taxes speed economic growth? The other benefit of this solution comes from lowering taxes that slow economic growth. Such taxes are known as distortionary taxes. For example, taxes on labor, such as payroll taxes, curtail incentives to hire workers as well as people's incentive to work. Similarly, taxes on capital, such as corporate taxes, cut incentives to invest in new machinery.

Lowering distortionary taxes speeds economic growth. For example, dropping the payroll tax will increase the incentive to hire workers and will encourage laborers to work more. Based on this effect, policy makers try to lower distortionary taxes to match the increase in revenues associated with environmental taxes.

The degree to which recycling environmental tax revenues generates a win–win solution depends on the distortion caused by the nonenvironmental tax. If the government lowers a highly distortionary tax, the resultant increase in economic activity could exceed the negative effect associated with the environmental tax. Under these conditions the economy will grow faster than it would have previously; taxpayers will not pay any more to the government than they did previously; and environmental impacts will decline (the taxes paid by different segments of the economy would change). Such an outcome is a dream come true for both Republicans and Democrats!

ADDITIONAL READING

Bashmakov, I., C. Jepma, et al. "Policies, Measures, and Instruments." In *Climate Change 2001: Mitigation* (2001).

Goulder, L.H. "Effects of Carbon Taxes in an Economy with Prior Tax Distortions: An Intertemporal General Equilibrium Analysis." *Journal of Environmental Economics and Management* 29 (1995): 271–297.

STUDENT LEARNING OUTCOME

- Students will be able to explain how recycling environmental taxes may reduce environmental impacts and increase economic growth.

on firms. Because firms usually do not announce these costs, they remain invisible to voters.

Market-based mechanisms also have been tarred by descriptive phrases such as the "right to pollute." The notion that a firm has the right to pollute strikes a lot of people as morally wrong. Similarly, environmental taxes allow firms or individuals to choose between reducing emissions or paying taxes. Some people assume that rich firms will choose to pay the tax, thereby evading their moral responsibility to protect the environment.

Negative connotations associated with the "right to pollute" or the choice to pay the tax or pollute are based on the simplistic notion of good guys and bad guys. Polluters are not bad guys seeking to degrade the environment. Firms will exercise their right to pollute only if it is less expensive than reducing pollution and selling the permit. Similarly, firms will pollute and pay a tax only if paying the tax is less expensive than reducing emissions. And these are exactly the choices that society wants potential polluters to make. Remember, the optimal level for pollution is not zero—it is the level at which the benefits associated with the last good or service equal the environmental damage it causes.

Eliminating Subsidies

Subsidies are payments to firms or households that are designed to favor a particular action. They do so by lowering the cost of a particular good or service, which encourages its consumption. Politicians often talk about subsidizing firms for responsible behavior instead of taxing them for irresponsible behavior. As the saying goes, one can catch more flies with honey than with vinegar.

Goods or services that receive a subsidy are chosen by politicians and are written into the tax code. As a result, the tax code influences many decisions. Although many of these decisions do not involve the environment directly, many have the potential to exacerbate or to ameliorate environmental challenges.

Over the last century the United States has changed from an agricultural and extractive (mining, for example) economy in which environmental challenges were relatively unimportant to an industrial and service economy in which environmental challenges threaten long-term economic well-being. This change implies that many of the subsidies that were designed for the early twentieth century may no longer be appropriate for the twenty-first century. Notable examples include depletion allowances that are designed to encourage the production of fossil fuels, metals, and minerals. The U.S. federal government allows firms in these industries to reduce their tax payments as they deplete their in-ground resources. These reductions are worth several billion dollars per year.

Originally depletion allowances were designed to speed economic growth; but their logic no longer is clear. Does the government really want to speed the rate at which the United States depletes its low-cost supplies of energy and minerals? Clearly such a policy violates the best first principle. Furthermore, the energy and mining industries have major environmental impacts—so depletion allowances accelerate environmental degradation.

From an economic perspective, all subsidies should be eliminated. By definition, a subsidy hides the true cost of a particular action. This prevents consumers from weighing the costs and benefits of their purchases. And without this information, their decisions will not lead to the optimal level of pollution.

Given this logic, why do subsidies remain? They run counter to the Democratic Party's stated goal of cutting corporate welfare. Similarly, they run counter to the Republican Party's stated goal of getting the government out of the economy. But once enshrined in the tax code, subsidies are notoriously difficult to remove because doing so makes someone worse off. And that someone often is willing to spend a lot of time and money to maintain the subsidy.

To avoid this sort of fight, the government often will design a second set of subsidies to offset the effects of the original subsidy. Over time subsidies can get so complex that the government will sometimes end up on both sides of an issue.

One for the most absurd examples of this conflict concerns ongoing efforts to restore Everglades National Park in the southern tip of Florida (mentioned in Chapter 8). Much of the water in the Everglades is contaminated by phosphorus fertilizers that run off from sugar plantations in the area between Lake Okeechobee and the Everglades. The federal and Florida State governments have proposed actions to reduce phosphorus levels that would cost U.S. taxpayers and sugar growers billions of dollars. Growers object. They argue that the costs would put them out of business. Ironically, they are in business because government subsidies ensure that the price they receive is about double the world price. Without this subsidy U.S. sugar producers would not be able to compete with imports of abundant, low-priced sugar. In short, the government pays both to create the problem and to clean it up.

Nor should the government use subsidies to bribe firms to reduce their environmental impact. Subsidies may reduce environmental degradation by increasing the efficiency of individual firms in the short run. But in the long run, subsidies increase environmental degradation by increasing the scale of the industry. Subsidies reduce the average cost of producing a good; this increases the profitability of production, which attracts new firms and ultimately increases output. Although subsidies may appear to be a way to "catch more flies with honey," a more appropriate cliché may be "the road to hell is paved with good intentions."

Personal Choices

Policy makers will not eliminate all subsidies and internalize all externalities in the next few years; therefore prices will not convey complete information about costs (including environmental costs) and benefits associated with individual goods and services. Without that information, consumers cannot make optimal decisions. That leaves consumers and firms who wish to reduce their environmental impact in a quandary. Which goods and services should they purchase and which should they avoid?

Your Ecological Footprints illustrate several generalities that you can use to guide personal environmental decision making. Indirect impacts of an action can sometimes be as great as its direct impacts. As indicated in Chapter 20's Your Ecological Footprint, the energy used to render services often is similar to that used to provide goods. As a result, sound environmental decisions must consider the entire process of making a good or rendering a service.

Some choices are more important than others. The environmental impacts of paper bags are not too different from those associated with plastic bags. Similarly, the environmental impacts of producing a glass bottle from virgin materials are not much greater than those associated with producing the same bottle from recycled materials.

On the other hand, your choice of car or appliance can have a large environmental impact. The environmental impacts associated with extracting the materials required to make a 3,500 kg SUV are much greater than the impacts associated with those required to make a 1,500 kg sedan. Even a modestly efficient sedan uses about half the motor gasoline needed by the SUV and emits fewer air pollutants. Similarly, the most efficient refrigerator requires less than half of the electricity used by the least efficient model.

But which model should you choose? The purchase price of an energy-efficient refrigerator often is greater than a less efficient model. However, the cost of electricity used by the energy-inefficient model is greater than that for the energy-efficient model. For many appliances, including refrigerators, the lifetime costs of the more efficient models are less than those of the inefficient models. Under these conditions you can save both money and the environment.

Studies show that many consumers will purchase the less efficient model. Why do they make this "irrational" choice? Many explanations are offered. Sometimes the purchaser does not use the appliance. If you live in off-campus housing, you may have noticed that landlords often furnish apartments with low-cost, low-efficiency appliances and let you pay the higher operating costs. The landlords' behavior is rational, but it leads to a suboptimal outcome.

In other instances consumers are not rational. For example, MIT economist Jerry Hausman finds that consumers will put their money in a bank account with a low rate of interest. On the other hand, they will not pay extra for energy-efficient refrigerators and air conditioners even though savings earned from the extra efficiency are much greater than what they could have earned had they put that extra money in the bank.

Such irrational behavior poses an important challenge because many environmental policies are based on the assumption that consumers are rational. That is, policy makers try to establish conditions in which rational consumers and firms will make decisions that further the goals of environmental policy. By forcing manufacturers to label the energy efficiency of appliances and estimate their annual operating costs, policy makers hope consumers will buy models that have the lowest lifetime costs, which usually are the more expensive, energy-efficient models.

The prevalence of irrational behavior implies that environmental policy cannot depend solely on market-based mechanisms. The environmental consequences of decisions should be described in terms of material and energy flows, in addition to dollars. This will allow personal choice to play an important role in protecting the environment.

SCALE: AN UPPER LIMIT ON SIZE AND ECONOMIC WELL-BEING?

For reasons described throughout this book, there may be a limit on the size and economic well-being of the human population. These limits are highlighted by the difficulties the world may have in accommodating the rapidly growing Chinese economy. Over the last two decades the Chinese economy has grown nearly 10 percent per year (Figure 24.5). As a result, the economic standard of living for 1.3 billion Chinese consumers has risen sharply.

This rise also increased China's demand for energy and materials and its emission of pollutants. Between the early 1990s and the first decade of the twenty-first century, the rapid growth in oil demand changed China from an oil exporter to an oil importer (Figure 24.6). This change is partially responsible for the rapid increase in world oil prices since 2003. If Chinese consumers eventually own and drive cars like U.S. consumers, China would use 30–40 million barrels per day of motor gasoline, which is nearly half the world's current demand for all oil products. Similarly, as their incomes rose, Chinese consumers increased their

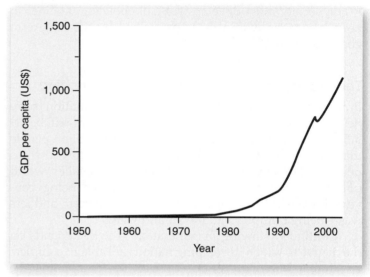

FIGURE 24.5   *Economic Growth in China* Per capita GDP grew slowly before the economic reforms of the early 1980s. After the economic reforms, per capita GDP grew rapidly. (Source: Data from International Financial Statistics.)

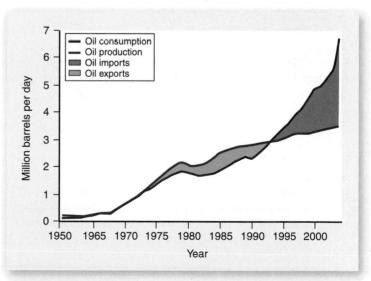

FIGURE 24.6   *Oil Balances in China* Until the early 1990s China produced more oil than it consumed and exported the surplus. After that oil consumption grew more rapidly than domestic production, and China increased imports to satisfy demand. (Source: Data from the International Energy Agency.)

consumption of meat (Figure 24.7). To satisfy this demand, large areas of natural ecosystems have been converted to agroecosystems, including many areas of Brazil that are now planted in soybeans. Similarly, hefty areas of forest in Indonesia are being harvested and replaced with tree plantations to satisfy demand for wood and palm oil.

Increases in material and energy flows and the conversion of natural ecosystems to human systems are critical to the issue of scale. Policy analysts must determine whether increases in efficiency can offset the increasing size and affluence of the human population. See *Policy in Action: When Is a Reduction Not a Reduction? Scale versus Efficiency.* If not, society must develop workable policy guidelines to keep its actions within those prescribed by the long-term goal of sustainability.

The Economics of the Demographic Transition

Projections for the size of the human population have declined. Recent U.N. forecasts project that there will be 8.9 billion people in 2050, as opposed to the 10.2 billion people that were projected in 1992. Many view this slowdown as positive. For environmentalists, having a smaller world population means fewer natural ecosystems being converted to agricultural land, less energy being used, and less waste being pumped into the air and water.

Continued slowing in population growth depends on the degree to which developing nations are able to complete the demographic transition (Chapter 9). Completing the demographic transition is more than making birth control available—limiting the number of children must fit families' needs. For parents in some developing nations, large fami-

lies take the place of social security systems that are available in developed nations. Parents who will depend on their children in old age will reduce the size of their family only if the parents are confident that most of their children will survive to adulthood or will earn high salaries. Similarly, parents will have fewer children if employment opportunities for women mean that families will forgo paychecks while women are pregnant and parents stay at home with the children.

Such decisions imply a link between the demographic transition and economic development. Economic development increases the opportunities and social infrastructure that support smaller families; smaller families increase funds available for economic and social development. So does population growth slow and generate the social infrastructure that is needed for economic development? Or does economic development occur first, with these income gains providing an incentive to slow population growth?

The success of the Chinese economy makes one wonder whether the market reforms that allowed the Chinese economy to grow so rapidly would have worked if they had not been accompanied by a rapid decline in birthrates (Figure 24.8). Did the reduction in the number of children born increase the amount of money that could be spent to educate each child and build the sewer systems, roads, and other social infrastructure that make China attractive to foreign investors? Or did the jobs and infrastructure associated

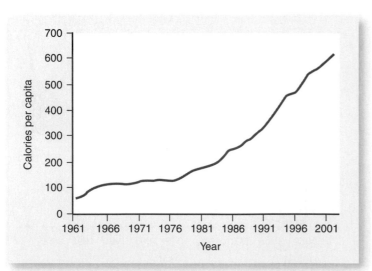

FIGURE 24.7 *Chinese Meat Consumption Prior to the early 1980s per capita Chinese meat consumption grew slowly. After economic reforms per capita GDP grew rapidly, and the increase in income allowed Chinese consumers to purchase more meat.* (Source: Data from United Nations Food and Agricultural Organization.)

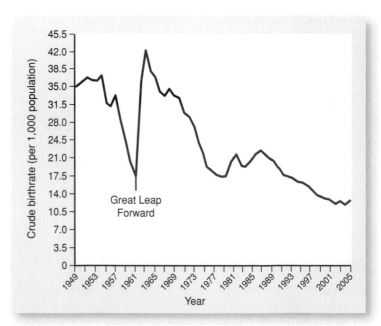

FIGURE 24.8 *China's Crude Birthrate Crude birthrates in China have declined over the last fifty years, with a rapid reduction between the early 1960s and late 1970s. Did these reductions make the economic growth in Figure 24.5 possible, or did the economic growth in Figure 24.5 make further reductions in birthrates possible?* (Source: Data from U.S. Census Bureau, International Data Base.)

POLICY IN ACTION

When Is a Reduction Not a Reduction? Scale versus Efficiency

One of President George W. Bush's first environmental actions was to reject the Kyoto Protocol. He promised a different approach. He announced that approach in 2002, when he proposed a voluntary goal of reducing the greenhouse gas intensity of the U.S. economy by 18 percent between 2002 and 2012. Supporters hailed the plan as ambitious. Critics argued that the plan would do little if anything. Is either side correct?

To understand, we must define the phrase *greenhouse gas intensity*. Greenhouse gas intensity measures the amount of greenhouse gases emitted by each unit of economic activity. The Bush plan measures greenhouse gas intensity by the amount of carbon dioxide that is released by each dollar of GDP. In 2002 greenhouse gas intensity was 575 metric tons of carbon dioxide per million dollars of GDP (base year 2000). If successful, the Bush plan would reduce intensity to 472 million metric tons per million dollars of GDP by 2012.

This reduction sounds impressive, but critics raise two objections. One group argues that an 18 percent reduction is little more than business as usual. They point out that the greenhouse gas intensity of the U.S. economy declined by about 17 percent from 1990 to 2002 (Figure 1). So the voluntary Bush plan does not require consumers to do anything beyond their ongoing efforts.

The other criticism relates to the difference between scale and efficiency. Greenhouse gas intensity is a measure of efficiency—it represents emissions per unit of economic activity. As such, it is similar to the energy/GDP ratio in Equation 24.1.

Recall from Chapter 13 that radiative forcing is determined by the total amount of carbon dioxide in the atmosphere. This implies that the U.S. contribution to global climate change is determined by its total emissions. As indicated by Equation 24.2, this total is determined by the product of greenhouse gas intensity and GDP.

As such, the reduction in greenhouse gas intensity is only part of the story. The effect of

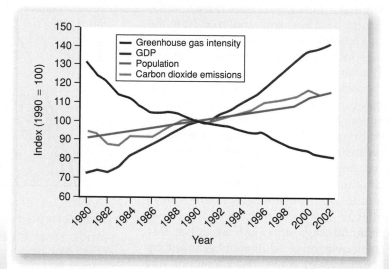

FIGURE 1
Population, Income, Efficiency, and Carbon Dioxide Emissions In the United States, efficiency as measured by the amount of greenhouse gases emitted per unit of GDP has increased over the last twenty years. During the same period carbon dioxide emissions increased because GDP (and population) increased.

reductions in greenhouse gas intensity on total emissions depends on the rate at which GDP grows. If GDP grows more slowly than greenhouse gas intensity declines, total emissions shrink. But if GDP grows faster than greenhouse gas intensity declines, total emissions increase.

We are not economic forecasters, but we can do some simple calculations to determine the GDP growth rate at which emissions will remain constant. A decline rate of 18 percent represents a change of 0.18 over twelve years. To determine the annual reduction, we can calculate $0.82^{1/12}$, which is 0.984. This implies that if the U.S. economy grows at 1.6 percent per year $(1 - 0.984) \times 100$, carbon emissions will remain constant.

What is the likelihood that the U.S. economy will grow faster than 1.6 percent per year? One way to assess this possibility is to examine the historical growth rate of the U.S. economy. Between 1990 and 2002 the U.S. economy grew at an average annual rate of about 3 percent, which would seem to imply that if greenhouse gas intensity declines by 18 percent, total emissions would rise. This conclusion is reinforced by looking at growth rates in the two components of GDP, population and per capita

GDP. Over the next decade the U.S. population is forecast to grow at an annual rate of about 1 percent. Unless per capita GDP grows very slowly (0.6 percent per year), GDP will grow faster than greenhouse gas intensity declines, which would cause total carbon emissions to rise. In this case the critics may be correct: A reduction in greenhouse gas intensity is not a reduction in total emissions.

How would such an outcome compare to the Kyoto Protocol? Recall from Chapter 13 that the Kyoto Protocol would obligate the United States to reduce its emissions below 1990 levels by the 2008–2012 period. Because emissions in 2002 are nearly 16 percent above 1990 levels, the Kyoto Protocol would require significant reductions in emissions compared to the likelihood of increased emissions under the Bush plan.

STUDENT LEARNING OUTCOME

- Students will be able to explain the effect of scale on efforts to reduce carbon emissions by increasing efficiency.

with market reforms create conditions that favored a birth-rate reduction? There is no definitive answer, but it is clear that (1) policy to slow population growth will not succeed unless limiting family size makes sense for individual families and (2) economic development will not succeed unless economies can generate funds that are needed to educate children and build social infrastructure.

Increasing Efficiency versus Scale: The Environmental Kuznets Curve

If income gains are needed to complete the demographic transition, could the solution to population growth create another problem—income-driven increases in environmental degradation? As discussed in Chapter 11, income gains generally increase the use of materials and energy and increase the emission of solid, liquid, and gaseous wastes.

To explore the relationship between income and environmental impacts in greater detail, scientists turned to the work of economist Simon Kuznets, who analyzed how the distribution of income changes with income. He found that when a nation's average income is low, income is distributed evenly because nearly everyone is poor. As average income rises, some people become better off while other people remain poor. Thus initial income gains increase the inequality of income distribution. Inequality continues to rise until some turning point. At this point further income gains reduce inequality through the formation of a large middle class. This inverted U-shaped relationship between income and the inequality of its distribution is termed a Kuznets curve.

Over the last decade environmental economists have made a similar argument about the relationship between income and environmental degradation. They argue that poor nations have little environmental impact because they use relatively small amounts of energy and material and therefore generate little waste. As income rises as measured by per capita GDP, so too does the use of energy and materials and the generation of wastes. Together these changes increase environmental degradation. Beyond some turning point, the relationship between income and environmental degradation changes so that further income gains reduce environmental degradation. This inverted U-shaped relationship between income and the use of materials or energy or the emissions of wastes is termed an environmental Kuznets curve (Figure 24.9).

The shape of the environmental Kuznets curve can be explained many ways. One explanation points to changes in the composition of production and consumption. Early in economic development, most people are farmers. Furthermore, most of these farmers use traditional techniques because they cannot afford the inputs associated with the Green Revolution. These agrarian economies have relatively little environmental impact. As economic development proceeds, the economy industrializes, which increases the consumption of energy and materials and the emissions of wastes. These changes increase environmental degradation. Continued economic development pushes the economy into a postindustrial service economy. This tends to reduce the use of energy and materials and waste emissions, which lessens environmental degradation.

Another explanation, though not necessarily contradictory, focuses on changes in consumer tastes and preferences and the availability of institutions to express them. Early in the process of economic development, people are poor and so must focus on day-to-day survival. Under these conditions, people are willing to sacrifice environmental quality for short-term economic gain. For example, poor farmers are not worried about the long-term effects of soil erosion—they need to feed their families today.

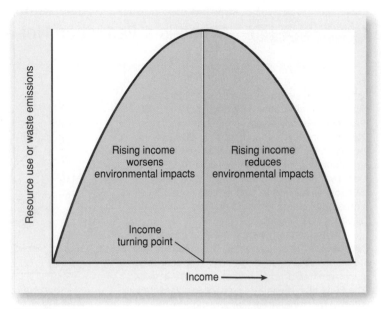

FIGURE 24.9 *Environmental Kuznets Curve* The environmental Kuznets curve represents the relationship between income and the use of natural resources or the emissions of wastes. The use of natural resources or the emission of wastes increases with income until a turning point. Beyond that turning point further increases in income reduce the use of resources or the emission of wastes.

As incomes rise beyond some threshold, confidence in day-to-day survival lets people take a longer view that includes amenities like a clean environment which allows healthy living and outdoor entertainment. This change in tastes and preferences prompts them to trade some economic well-being for environmental protection. A slightly more affluent farmer may have enough time and money to invest in soil conservation measures.

Concern for the environment can be expressed at higher income levels because income gains support the required institutions—such as the scientific workforce needed to identify environmental challenges and government bodies to develop and enforce environmental laws. A slightly more affluent farmer may get information about soil erosion and what to do about it from a government agricultural extension agent. Furthermore, the farmer may be willing to invest in soil conservation because the government will protect the farmer's property and allow it to be passed on to the farmer's children.

If correct, the environmental Kuznets curve suggests a win–win solution in which economic growth improves both living standards and environmental quality. This notion has been embraced by several important organizations. For example, the World Bank suggests that environmental degradation can be slowed by policies that protect the environment and promote economic development.

Promises of higher incomes and environmental protection have prompted many studies that examine the relationship between per capita GDP and a host of environmental indica-

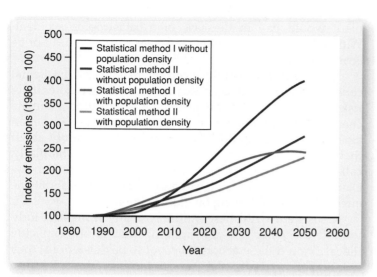

FIGURE 24.10 *Sulfur Emissions, Scale versus Efficiency* *The forecast for sulfur emissions generated by an environmental Kuznets curve that is driven by forecasts for income and population. Regardless of the statistical technique used to estimate the environmental Kuznets curve or the inclusion of population density, the large turning point implies that total sulfur emissions will increase throughout much of the forecast period if population and income grow at their current rates.* (Source: Data from T.M. Selden and D.Q. Song, "Environmental Quality and Development: Is There a Kuznets Curve for Air Pollution Emissions?" Journal of Environmental Economics and Management 27(2): 147–162.)

tors, such as sulfur emissions and energy use. Initial studies are consistent with the notion of an environmental Kuznets curve. Some studies show an inverted U-shaped relationship between per capita GDP and sulfur emissions, energy use, and carbon dioxide emissions. These results are contradicted by more rigorous studies, which find that the inverted U-shaped relationship disappears when the data are analyzed using more sophisticated statistical techniques or other factors are included in the model. For example, Amy Richmond at the U.S. Military Academy and her colleagues found that including energy prices and the share of total energy use from oil and gas relative to coal eliminates the turning point in the relationship between per capita GDP and energy use.

Even if there is an inverted U-shaped relationship between income and resource use or waste emissions, the ability to protect the environment by making people richer depends on the income level at which the turning point occurs. If the turning point occurs at a relatively low income level, rising incomes and increasing efficiency may improve environmental quality. On the other hand, if the turning point occurs at a relatively high income level, cumulative environmental damage may be too great and the scale of economic activity may be too large for higher incomes to fix environmental challenges.

Many estimates for the income turning points of several pollutants are too large to stop environmental degradation. For example, one study estimates that the income turning point for sulfur dioxide emissions is between $8,000 and $11,000. If this is correct, global sulfur emissions will rise

through 2040, regardless of the rate at which population and income grow (Figure 24.10). Similarly, an analysis found an inverted U-shaped relationship between per capita GDP and carbon dioxide emissions, but all scenarios for population growth overwhelmed the effect of income gains so that carbon dioxide emissions are forecast to rise through 2050.

In summary, it is highly unlikely that increasing efficiency will offset the increasing scale of economic activity. Policy aimed at increasing efficiency can slow environmental degradation; but in the long run, a different approach is needed to move society toward sustainability.

Living within Limits Imposed by Biogeochemical Cycles

Many environmental challenges are associated with activities that change storages and flows of biogeochemical cycles. Global climate change is caused by activities that transfer carbon dioxide to the atmosphere from storages as fossil fuels and biota. The anoxic zone in the Gulf of Mexico is caused by changes in the global nitrogen and phosphorus cycles that alleviate constraints associated with the limiting nutrient. Acid deposition is caused by activities that transfer sulfur from fossil fuels to the atmosphere.

Almost by definition, such changes are not sustainable. Storages and flows have evolved over long periods. This evolution has generated biogeochemical cycles that supply environmental goods and services at a fairly constant rate. But as humans have transformed themselves into the planet's dominant species, the ability of biogeochemical cycles to supply these goods and services may be undermined by significant alterations caused by increased human intervention.

Biogeochemical cycles can be used to judge the scale of human actions and their sustainability. Net primary production provides an upper bound on biological energy supplies. Humans cannot use more than 100 percent of net primary production—nor should society aspire toward this limit. As humans increase their use of net primary production, less remains for all other species. Indeed, increasing our share toward 33 percent is partially responsible for deforestation and the loss of biodiversity.

The nitrogen cycle imposes similar limits. The atmosphere stores large quantities of unavailable nitrogen, which can be converted to usable forms via the Haber-Bosch process. But there probably is a limit on the amount of available nitrogen the planet's ecosystems can handle. Quantities beyond this limit would change terrestrial and aquatic ecosystems in ways that society does not find useful. For example, the anoxic zone in the Gulf of Mexico and the phytoplankton blooms in the Gulf of California (Chapter 16) reduce food supplies by lowering the quantity of fish caught.

Given this importance, how can society use biogeochemical cycles to move toward sustainability? Policy makers could establish markets for nitrogen, phosphorus, and carbon dioxide similar to the sulfur market (see Chapter 19's Policy in Action). That is, governments could issue permits allowing holders to fix a given quantity of nitrogen or apply a given amount of phosphorus to their crops. Quantities could be set consistent with the underlying biogeochemical cycle.

Limiting human intervention in biogeochemical cycles would raise the price for the nutrients, which would force society to allocate its limited use of nitrogen or phosphorus toward activities that generate the greatest social well-being. For example, this policy would probably make it too expensive to fertilize your front lawn. In addition, higher prices would stimulate innovative ways to address environmental challenges. To illustrate, reforestation removes carbon dioxide from the atmosphere. A market for carbon dioxide permits would allow landowners to charge for the environmental services that they currently provide for free. Such economic returns could be a cost-effective means for preserving natural ecosystems and the diverse organisms that live there.

Operationalizing the Precautionary Principle in a World Filled with Risk

You are not safe in your bathroom or kitchen. People have been run over by cars while sleeping in their beds. Every action carries some risk. Society must decide how much risk it is willing to accept.

Often the willingness to accept risk is based on reward. The greater the reward, the greater the risk one is willing to assume. To understand, let's return to a ship's Plimsoll line. Loading more cargo increases profitability, but it also raises the probability that the ship will sink. To decide how much cargo to load, owners must gauge the sea, determine the risk of sinking, and compare that to the profit of loading an additional metric ton. Clearly an owner is less likely to load that last ton if the seas are rough.

What do the seas through which society is now sailing look like? If you believe that society is far from its carrying capacity or that technical progress is raising the Plimsoll line, the benefits of having another child or consuming more materials and energy are high relative to the associated risk. But if society is close to or beyond its carrying capacity, having another child or consuming more materials and resources carries a greater risk.

To decide, society must compare the benefits of a growing population and increases in material and energy use to the potential loss of environmental goods, services, and amenities. Comparisons that involve immense but highly unlikely losses are especially difficult to evaluate. The precautionary principle argues that society should err on the side of caution. That is, society should avoid actions that have the potential to cause significant and irreparable environmental damages.

Chapters 12–23 indicate that society acts in nearly the opposite way. In some cases society encourages activities that carry the potential for dramatic environmental consequences. For example, Chapter 21 describes how the Price-Anderson Act limits the damages that the owner of a nuclear power plant would pay in the case of a catastrophic accident. Without this protection nuclear power plants would not be economically viable.

In other cases scientific uncertainty about environmental consequences is used as an excuse to continue a potentially risky action. Consider the arguments about global climate change and the reduction in stratospheric ozone. No action against CFCs was politically possible until all doubt about their effects on the ozone layer was eliminated. Similarly, calls to limit carbon dioxide emissions are rejected on the claim that there still is considerable uncertainty about the effect of radiative forcing on global climate.

Consistent with these attitudes, many firms and individuals are willing to do anything that is not explicitly illegal. Should these actions turn out to be deleterious, they will be outlawed after the damage has been done. Because some environmental damages are not reversible, future generations may endure the costs of actions that generate benefits for the current generation. For example, the current generation enjoys the benefits associated with the use of fossil fuels. Because carbon dioxide emissions remain in the atmosphere for nearly a century, future generations will bear the brunt of their effect.

Obviously this allocation of costs and benefits does not generate behavior anything like the precautionary principle. To operationalize the precautionary principle, the current generation must pay the potential cost to future generations. One such method is **assurance bonding.** Firms or households undertaking actions that may damage the environment, now or in the future, could put up a bond to cover potential costs. This bond would be held by an agency that would return the bond (with interest) if the activity had no environmental impact. But if impacts occur, the agency would spend the funds to repair the damage.

Assurance bonds can be used to protect society against environmental challenges that can be best described as hit and run accidents. A firm profits from its actions. But should its actions harm the environment, a firm may declare bankruptcy. That leaves the government (and ultimately taxpayers) to pay for the cleanup. Assurance bonding would ensure that the responsible parties pay to clean up potential accidents. Furthermore, posting bonds would force firms and households to consider the potential environmental impacts of their actions.

This effect can be illustrated by returning to the controversy about opening the Arctic National Wildlife Refuge to oil exploration. If the refuge is opened, the government could codify the environmental claims made by the oil industry about the impacts of drilling by negotiating a specific list of what could go wrong. Industry would then post a bond to cover all possible damages. This would force the industry to "put its money where its mouth is." Refusing would show that the firm does not believe its own environmental claims or that the action is too risky if the firm has to pay for potential impacts.

SUMMARY OF KEY CONCEPTS

- Potential solutions to environmental challenges address the efficiency or scale of economic activity. Solutions that focus on efficiency reduce environmental impacts by reducing the effect of each unit of economic activity. Solutions that focus on scale reduce the level of economic activity. Solutions that focus on efficiency are politically desirable but ignore the possibility that there may be an upper bound on the size of the human population or its material standard of living.

- Internalizing externalities improves the efficiency of economic activity by including environmental damages in the prices of goods and services. Using market-based mechanisms is economically efficient and provides continuous incentives to reduce environmental impacts. But this tends to be unpopular because it is difficult to put a price on environmental degradation and because taxes are politically unpopular.

- The environmental impacts of scale are determined in part by population, which will depend on whether nations can complete the demographic transition. The ability to do so depends in part on economic development, but this link creates a dilemma. Does slowing population growth generate the social infrastructure needed for economic development? Or does economic development occur first, with income gains providing an incentive to slow population growth?

- The environmental Kuznets curve posits an inverted U-shaped relationship between income and resource use or between income and emissions. This implies that improving standards of living could reduce environmental impacts. Studies find that reductions occur after a relatively high turning point, which implies scale effects outweigh efficiency effects so that increasing GDP intensifies environmental impacts.

- Some environmental challenges involve impacts that could be very large but are highly unlikely to occur. Addressing these impacts may require society to use the precautionary principle, perhaps through instituting assurance bonds.

REVIEW QUESTIONS

1. Describe how the scale and efficiency of economic activity affect the use of natural resources or emissions. Which of these two effects has been more important in the United States over the last century?

2. Explain the following statement: "Maintaining economic efficiency will ensure the sustainability of human economic activities."

3. Explain why even the most conscientious environmentalist may choose options that have a greater impact than some other alternative.

4. Explain why an environmental impact may increase over time even if the relationship between per capita income and that impact can be described by an environmental Kuznets curve.

5. Describe an environmental challenge that is best approached using the precautionary principle. Explain whether this challenge can be solved using an assurance bond.

KEY TERMS

assurance bonding
revenue recycling
subsidies

Glossary

A

A horizon A soil layer that consists of organic matter mixed with mineral material, usually containing more mineral material than organic.

absolute water scarcity A ratio of annual water availability to population less than 1,700 m³ per person per year.

absorption The movement of nutrients to the interior portion of the body.

acid deposition A process by which sulfate particles (SO₄) are removed from the atmosphere by precipitation.

active range The span of body temperatures at which ectotherms can carry out their everyday activities.

adaptations Behavioral or physiological traits that allow a plant or animal to thrive in a particular environment.

adiabatic lapse rate The rate of temperature change experienced by a parcel of air when it is moved vertically in the atmosphere such that it cannot exchange heat with its environment.

advection The horizontal transfer of mass or energy as air masses move in response to pressure differences, such as winds.

aerobic respiration The release of energy from glucose or another organic substrate in the presence of oxygen.

aerosols Tiny solid particles or liquid droplets that remain suspended in the atmosphere for a long time.

affluence The average material standard of living in a nation, typically measured by per capita GDP or GNP.

afforestation Forest regrowth after a disturbance.

age classes The number of males and females in an age group.

age–dependency ratio The number of people under the age of 15 plus the number of people over age 65 divided by the number of people between those ages.

age structure The distribution of a population among age classes.

agricultural residues A potential source of biomass energy consisting of plant parts remaining in a field after the harvest of a crop, including stalks, stems, leaves, roots, and weeds.

agriculture The process by which a natural ecosystem is replaced with one that supports plants and animals of people's choosing.

agriculturists People who obtain food by changing natural ecosystems in a way that increases the amount of edible energy generated.

agroecosystems Simplified ecosystems that are set up and operated to produce food for people.

air pollutant Any gas or particle in the atmosphere that has the potential to harm life or property.

albedo The fraction of incoming solar radiation that is reflected back to space.

alien species Species that do not naturally occur within an area and that have usually arrived in the area as a result of human intervention (whether deliberate or accidental).

allelopathy Direct or indirect harmful effects of one plant on another through the production and release of chemical compounds.

allopatric speciation Evolution of a new species that occurs when a population becomes geographically isolated from its parent population and accumulates genetic or behavioral changes that differentiate it from the original population.

anaerobic respiration The release of energy from glucose or another organic substrate in the absence of oxygen.

annuals Plants that live for a single growing season.

anoxic Environments without oxygen.

anthropocentric The ethical position that the value of nonhuman species is determined by their value to people.

aquifer A soil layer in which all pore spaces are filled with water.

aridsols Soil in dry environments characterized by eluvial horizons that undergo relatively little leaching.

artesian well A well in which water rises to the surface due to internal pressure.

assurance bonding Bonds put up by firms or households to cover the potential damage of their actions now or in the future.

asthenosphere The upper part of the mantle, which has a consistency that is somewhere between a liquid and a solid and moves large quantities of heat from Earth's center toward the surface.

asymmetric information The failure of two parties to a transaction to have the same relevant information.

atmospheric emissions The release of gases or particulates into the atmosphere.

atom The smallest component of an element having the chemical properties of the element.

atomic mass The sum of the number of protons and neutrons in the nucleus of an atom.

atomic number The number of protons in a nucleus.

attribution The process of establishing a cause-and-effect relationship between human activity and an observed change in climate.

autotrophs Organisms that convert inorganic forms of energy to organic forms of energy.

available Chemical forms of nutrients that organisms can use.

avoided costs The costs of providing a service (such as chemical pest control) that are avoided by maintaining a natural service (natural pest control).

B

B horizon A soil layer known as *subsoil* that accumulates the minerals that wash out from the eluvial horizon.

baby boomers People born in the United States between 1946 and 1964.

background extinction Continuous, low-level rate of extinction not associated with human activities or catastrophic geological events.

basal metabolic rate The rate at which an organism uses energy while at rest.

benthic The area on or near the floor of a body of water.

best available technology The most effective, economically achievable, state-of-the-art technology currently in use for controlling pollution, as determined by the U.S. EPA.

best first principle The observation that humans use the highest-quality sources of natural resources before lower-quality resources.

best practicable control technology Sets uniform, industrywide effluent standards that approximate the average amount of control achieved from existing technology in the specific industry.

biocentric The ethical position that nonhuman species have value in and of themselves and have the right to exist independent of their usefulness to humans.

biochemical oxygen demand The amount of oxygen required for aerobic organisms to decompose organic material in wastewater over a five- to twenty-day period; the usual measure is five days.

biochemicals Chemicals produced by, or derived from, living organisms.

biodegradable waste Waste that is capable of undergoing anaerobic or aerobic decomposition, such as food and garden waste, paper, and paperboard.

biodiesel A less polluting fuel for most diesel internal combustion and turbine engines produced from a range of biomass-derived feedstocks including oilseeds, waste vegetable oils, cooking oil, animal fats, and trap grease.

biodiversity The number and variety of living organisms.

biodiversity hot spot An area with an especially large number of species.

bioenergy Energy derived from biological sources.

biofuels Fuels made from cellulosic biomass resources, such as ethanol, biodiesel, and methanol.

biogeochemical cycles The flow of matter among storages in the biological, geological, and chemical systems of Earth.

biomagnification Increased concentration of pesticides or other toxic materials in living organisms at higher trophic positions via the food chain.

biomass The mass of a species or group of species.

biomass feedstocks The sources for bioenergy and biofuels, including forest and agricultural biomass, as well as the organic portions of municipal solid wastes.

biome A major regional community of plants and animals with similar life forms and environmental conditions.

biopower Electricity generated from the combustion of biological sources.

biota All living organisms.

birthrate The number of new individuals produced by a population in a period.

black market A market in which goods or services are sold illegally.

black smoker A chimneylike structure on the seafloor made of metal sulfides, out of which flow hot (~350°C) fluids that look like black smoke. The black color of the fluid is due to mineral particles within it.

breeder reactor A nuclear reactor that manufactures more fissionable isotopes than it consumes.

C

C horizon A soil layer that consists of the parent material from which the soil formed and that shows little or no sign of soil formation.

cap and trade system A system of pollution control in which an environmental regulator establishes a "cap" that limits emissions from a designated group of polluters; emissions allowed under the new cap are divided into individual permits (usually equal to one ton of pollution) that represent the right to emit that amount. Companies are free to buy and sell permits to continue operating in the most profitable manner available to them.

capillary water Water that fills a soil's micropores and is held with moderate force.

capital Any asset or stock of assets, financial or physical, capable of producing income.

carbohydrates Organic molecules that contain only carbon, hydrogen, and oxygen.

carbonation and solution A form of decomposition in which carbon dioxide combines with water to form carbonic acid, which dissolves minerals within rocks and thereby opens spaces.

carcinogens Chemicals that cause cancer.

carnivores Animals that eat other animals.

carrying capacity The maximum number of individuals of a population that can be maintained indefinitely by the environmental goods and services that are generated by a given area of the environment.

cation exchange capacity A soil's ability to hold positively charged nutrients.

cations Positively charged ions.

centrally planned economies An economic system in which the government determines the quantities and prices of goods and services.

cereals Cultivated members of the grass family whose seeds (grains) are eaten by people or domesticated animals.

channel A path that is always filled with water and connects mudflats to the ocean.

chemical change A process in which a substance is transformed into a different substance by changing its chemical composition.

chemical energy Potential energy stored in chemical bonds of molecules.

chemical oxygen demand The amount of oxygen required to oxidize the organic material in a sample.

chemical reaction See *chemical change*.

clay Small soil particles that can be seen only using an electron microscope.

clean development mechanism A United Nations policy that allows a nation to earn credit for reducing emissions in another nation.

clear-cutting A harvest practice in which all commercially valuable trees are harvested at the same time.

climate Average weather conditions over a long period.

climate change A shift in the long-term average conditions of weather.

climate envelope Conditions under which populations of a species can persist in the face of competitors and natural enemies.

climax community Plant and animal species that tend to persist for an extended period.

coal benefaction A process in which coal is cleaned by being crushed, screened, and suspended in a liquid, where the solid impurities settle out.

coarse fraction Particles that are bigger than 2.5 microns but smaller than 10 microns.

coarse grains Cereals used to feed livestock.

coevolutionary hypothesis The notion that agriculture evolved via a positive feedback loop that includes the human population and the plants and animals humans eat.

combustion The complete oxidation of a substance through the use of air or O_2.

command and control A method of environmental regulation characterized by the application of direct regulatory controls to industry, such as by specifying allowable levels of pollution.

common property A resource owned by the public, such as fish in public waters, trees on public land, and the atmosphere.

community A group of interacting species.

competition Simultaneous demand by two or more individuals for limited environmental resources.

competitive exclusion principle The observation that no two species can share the same exact niche indefinitely unless at least one factor limits the density of the better competitor.

complexity The number of storages and flows and the number and strength of feedback loops in a system.

compound A chemical substance formed from two or more elements, with a fixed ratio determining the composition.

condense To change from a gas to a liquid as a result of being cooled.

cone of depression A drop in the water table around a well.

confined aquifer Groundwater that accumulates between two impermeable layers.

consolidation The process of joining particles to form sedimentary rocks.

consumption The purchase of goods and services by consumers.

containment The control of a nuclear fission reaction so that harmful radiation is not released to the environment.

Continental Divide The series of ridges through the Rocky Mountains that divides the country into two drainage basins.

contingent valuation A valuation technique that asks people directly how much they are willing to pay or to accept for improving or deteriorating environmental quality.

continuous mining techniques Recovery in which mining machines allow the roof to cave in after the machine has removed all of the coal and "backs away" from the seam.

contour plowing Sowing crops in rows that cut across the slope.

control rods Devices in the core of a reactor that absorb neutrons and are used to control the rate of fission and to stop the chain reaction.

convection Transfer of energy and mass by motions in a liquid or gas.

convection cell Regular movement in a gas or liquid due to the application of energy that creates gradients in temperature and pressure.

conventional fossil fuels Coal, oil, and natural gas that supply nearly all of the energy provided by fossil fuels due to their low cost.

convergent evolution Evolution of similar characteristics in unrelated species due to similar environmental stresses.

conversion The step in the nuclear fuel cycle in which solid uranium oxide (U_3O_8) is converted into the gas uranium hexafluoride (UF_6).

Coriolis effect The deflective effect of Earth's rotation on all freely moving objects.

correlative rights doctrine Rules that govern water use and force landowners to share water.

corruption Unlawful use of public office for private gain.

cradle-to-grave or **life cycle analysis** A comprehensive examination of the environmental and economic effects of a product at every stage of its existence, from production to disposal and beyond.

creep Movement of larger soil particles along the surface.

crop residues Plant parts remaining in a field after the harvest of a crop, which include stalks, stems, leaves, roots, and weeds.

crop yield potential The degree to which farmers' yields approach a theoretical maximum.

crude birthrate The number of births per 1,000 people per year.

crude death rate The number of deaths per 1,000 people per year.

crude oil Liquid that consists of molecules with relatively long chains of carbon, up to about 20 atoms.

crustal abundance The amount of an element in Earth's crust, measured as percentage by weight.

cultivars A shortened term for *cultivated varieties* that refers to plants people have bred for a specific trait or characteristic.

D

death rate The number of individuals that die in a given period.

deciduousness Plants that shed their leaves seasonally to avoid adverse weather conditions such as cold or aridity.

decommissioning The process of closing down and removing a nuclear reactor after its useful life has come to an end.

decomposers Organisms that get food energy from dead parts of other organisms.

decomposition Chemical changes in solid materials that lead to soil formation.

deforestation Reductions in forest areas caused by human activity.

dematerialization A reduction in the amount of material required to produce a good or service.

demographic transition The process of moving from high birthrates and high death rates to high birthrates and low death rates, and finally to low birthrates and low death rates.

denitrification The loss of gaseous nitrogen from soil by biological or chemical means.

density-dependent factors Variables that affect population growth but are not related to the size of the population.

density-independent factors Variables that affect population growth and are related to the size of the population.

dependent variable The variable that is affected by another variable, typically represented on the left side of an equation and along the *y*-axis of a graph.

depletion The movement from high-quality, low-cost resources to lower-quality, higher-cost resources.

desalinization Production of freshwater by removing the salt from salt water and brackish waters.

detection The process of determining whether climate is changing.

detritivores Organisms that eat decomposing organic material known as *detritus*.

detritus Dead or decaying organic matter.

diffuse chemical coevolution Natural selection favoring individuals that accumulate compounds effective against a wide variety of enemies.

digestion Breaking complex forms of organic molecules into smaller building blocks.

direct market valuation The use of the free market to place a value (price) on environmental goods and services.

discharge Water returned after use, frequently at or near its source.

discoveries A broad term that includes locations of new finds of resources.

disintegration Physical changes in solid material that lead to soil formation.

dispersal The distance a species can travel to find new environments.

dissipation The release of wastes by breaking up and scattering by dispersion.

dissipative waste Waste that is not technologically or economically feasible to collect and recycle.

dissolved oxygen deficit The difference between the amount of oxygen in water when it is fully saturated and the amount of oxygen actually present.

disturbances A discrete, punctuated killing, displacement, or damaging of one or more individuals (or colonies) that directly or indirectly creates an opportunity for new individuals (or colonies) to become established.

Dobson unit The unit used to measure the concentration of ozone in a column of the atmosphere.

domestication The process by which a species is modified relative to its wild ancestors by selective breeding.

dose–response function The relationship between pollution and damage.

doubling time The time required for a variable to double.

draft animal An animal used for pulling heavy loads.

drainage basin The area from which surface waters derive surface runoff and groundwater flows.

drip irrigation Slow, localized application of water just above the soil surface.

dust Solid material generated by crushing or grinding.

E

E horizon A soil layer from which minerals are leached as water percolates through the soil.

easterlies Surface winds that move from the poles toward the polar front.

ecocentric A philosophy that claims moral values and rights for both organisms and ecological systems and processes.

ecological efficiency The percentage of energy from one trophic level that is incorporated in the next level.

ecological footprint A calculation that estimates the area of Earth's productive land and water required to supply the resources an individual or group demands, as well as to absorb the wastes the individual or group produces.

ecological resistance The degree to which an ecosystem changes following a disturbance.

ecological restoration The process of reestablishing to the extent possible the structure, function, and integrity of indigenous ecosystems and sustaining the habitats they provide.

ecological stability The ability of an ecosystem to return to its original state following a disturbance.

economic efficiency A term that refers to the optimal production and consumption of goods and services. This generally occurs when prices of products and services reflect their marginal costs.

economic energy cost The quantity of energy required to extract and process a unit of natural resource.

economic growth An increase in a nation's production of goods and services.

economic system The collection of firms and households that produce and consume the goods and services people associate with material well-being.

economies of scale Reductions in the unit costs of production as the quantity produced increases.

ecosystem The community and the physical environment in which the community lives.

ecosystem diversity Variation between and within ecosystems with regard to species and function.

ecosystem health The degree to which ecosystems are stressed by human activities.

ecosystem management A harvest practice that balances human needs for wood as a raw material and an energy source with the forest's provision of environmental services.

ecosystem services The conditions and processes through which natural ecosystems, and the species that make them up, sustain and fulfill human life. Examples include provision of clean water, maintenance of livable climates (carbon sequestration), pollination of crops and native vegetation, and fulfillment of people's cultural, spiritual, and intellectual needs.

ectotherms Animals that obtain most of their body heat from the environment.

edge effects Habitat conditions created at or near the more or less well-defined boundary between ecosystems.

edges Boundaries between well-defined ecosystems.

efficiency In physics, *efficiency* refers to the amount of useful energy extracted from a system divided by the total energy put into a system. In economics, *efficiency* refers to the extent to which a given set of resources is allocated across uses or activities in a manner that maximizes whatever value it is intended to produce, such as output, market value, or utility.

Ekman transport The overall movement of a mass of water resulting from a balance between the Coriolis force and frictional stress at the bottom.

El Niño A change in ocean and atmospheric circulation associated with a weaker than normal zone of high pressure as measured at Tahiti.

electrical energy Energy of electrical charges as a result of their position or motion.

electromagnetic radiation Radiation consisting of electric and magnetic waves that travel at the speed of light. Examples include light, radio waves, gamma rays, and X-rays.

electrons Subatomic particles carrying a negative charge.

elements Any of the more than 100 known substances (of which 92 occur naturally) that cannot be separated into simpler substances and that singly or in combination constitute all matter.

eluvial horizon The soil layer from which minerals are leached as water percolates through the soil.

end use efficiency The efficiency with which energy is converted to useful work or heat at the point of end use.

endangered species A species that is in danger of extinction throughout all or a significant portion of its range.

endemic Species that are restricted to a certain geographic region and are thought to have originated there.

endotherms Animals that obtain most of their body heat from internal metabolic processes.

enemy release hypothesis The notion that the population of an alien species can grow rapidly (escape) if the number of pathogens it leaves behind in its native range exceeds the new pathogens it accumulates in its naturalized range.

energy The ability to do work.

energy converter A device that converts energy from one form into another, and in the process does useful work.

energy density The amount of energy stored per unit weight, volume, or space.

energy end use A set of devices, products, and systems that use energy for the same or similar purposes. Examples include transportation, cooking, lighting, heating, and refrigeration.

energy pyramid A diagram that compares the amount of energy available at each position, or level, in the feeding order.

enrichment The physical process of increasing the concentration of the uranium-235 isotope relative to the predominant uranium-238 isotope in natural uranium.

entropy A measure of the level of disorder or randomness in a closed system.

entropy law The general tendency for energy and materials to move from an ordered to a disordered state.

environmental accounting Any quantitative approach to linking financial and environmental performance.

environmental degradation Processes induced by human behavior and activities (sometimes combined with natural hazards) that damage the natural resource base or adversely alter natural processes or ecosystems.

environmental energy cost The amount of solar energy and heat from Earth's core that are used to produce a natural resource.

environmental gradients Changes in conditions from one region to the next.

environmental justice The fair treatment of people of all races, cultures, and incomes with respect to the development, adoption, implementation, and enforcement of environmental laws, regulations, and policies.

environmental performance bond A deposit that likely polluters and violators of environmental standards must pay to a certain environmental fund. These bonds are aimed at providing financial incentives to industry to adhere to environmental requirements.

environmental resistance The slowdown in population growth that occurs as a population approaches its carrying capacity.

environmental services See *ecosystem services*.

epifauna Organisms that are attached to or move about the surface of the bottom of a water body.

epilimnion The upper layer of a water body where sunlight may power high rates of net primary production.

epipelagic The oceanic zone extending from the surface to about 200 meters, where enough light penetrates to allow photosynthesis.

equatorial low A region of low pressure near the equator due to rapidly rising air.

equilibrium The state of a system in which there is no net change.

erodibility factor The ease with which soil particles can be detached and transported.

erosion The process of carrying away soil particles from their parent material by wind or water.

essential Molecules that an organism cannot synthesize from its constituents.

estuary A semi-enclosed coastal body of water that has a free connection with the open ocean, and within which seawater is measurably diluted with freshwater derived from land drainage.

ethanol A chemical formed by fermentation or synthesis used as a raw material in a wide range of industrial and chemical processes. An alternative automotive fuel that is usually blended with gasoline to form gasohol.

ethnobotany The study of how different groups of people, including indigenous cultures, use plants and animals.

euphotic zone The surface layer of the ocean that receives enough sunlight for photosynthesis.

eutrophic Water bodies that have high net primary production.

eutrophication A process in which water bodies receive excess nutrients that stimulate excessive growth of autotrophs.

evaporation Conversion of a liquid to a gas.

evaporites Salts deposited by evaporation of seawater.

ex situ Maintaining a species away from its natural habitat.

exhaustion The complete depletion of a natural resource.

existence value Environmental contribution to human well-being even when the environment is not used.

experiment A set of actions and observations to verify or falsify a hypothesis or research a causal relationship between phenomena.

exploration The search for and identification of new deposits of a natural resource.

exponential population growth Continuous population growth in proportion to the size of the population.

extensive land uses Practices that use large land areas relative to other factors of production.

externality In economics, a cost or benefit attributable to an economic activity that is not reflected in the price of the goods or services being produced. Often refers to the cost of pollution and other environmental impacts.

extinction The loss of living representatives of a given species either globally or locally.

extraction efficiency The fraction of the resource removed from its location in the environment.

extractive wastes A waste generated in the process of extracting a natural resource (mineral, energy, timber) from the environment.

extremophiles Organisms that live in extreme environments.

F

fabrication The process through which fissionable material is configured into precisely shaped fuel or target elements and made ready for use in a nuclear reactor.

facilitation A mechanism for succession in which species in early and middle successional communities change their microenvironments in ways that make them less hospitable for their own needs or more hospitable to species that inhabit later successional communities.

factors of production Inputs to the production of a good or service, such as labor, capital, technology, energy, and materials.

falsified Rejection of a hypothesis because observations or experimental results are inconsistent with expectations.

fats Organic molecules that contain only carbon, hydrogen, and oxygen and can store large amounts of energy per unit mass.

fecal coliform count Measures the number of coliform bacteria per 100 milliliters.

feedback loop Linkages that move through a system and ultimately connect back to itself.

feedlots Confined yards where livestock eat prepared or manufactured feed.

Ferrell cell A zonally symmetric pattern of atmospheric circulation located between 30° and 60° north and south of the equator.

fertile An atom or a collection of atoms that can produce fissile atoms under neutron irradiation. Fertile atoms or collections of atoms generally themselves do not undergo induced fission.

field capacity The quantity of both capillary and hygroscopic water stored by a fully wetted soil.

filter feeders Organisms that obtain food by separating it from passing water.

firms Establishments that produce goods and services for consumers.

first law of thermodynamics The principle that energy cannot be created or destroyed; it can only be converted from one form to another.

fishable A water body in which fish and shellfish can thrive and can be eaten safely by people.

fissile Any nucleus capable of undergoing fission by neutrons.

fission The splitting of a nucleus into at least two other nuclei and the release of a relatively large amount of energy.

fitness The number of offspring an individual leaves in the next generation.

flash steam power plants The most common type of geothermal power plant in which steam, once it has been separated from the water, is piped to a powerhouse where it is used to drive a steam turbine, which in turn generates electricity.

flows Movements of energy or materials between storages in a system.

food chain The simplest representation of energy flow in a community.

food web Interconnected feeding relationships in an ecosystem.

forestry residues Forms of biomass (bark, branches) left over from the harvesting of timber that are potential sources of energy.

fragmentation Breakup of a continuous habitat, ecosystem, or land use type into smaller areas.

frequency Number of events per unit time.

freshwater Water that contains less than 1,000 milligrams per liter of dissolved solids.

fuel Any material that can be burned to produce energy.

fuel assemblies Bundles of hollow metal rods containing uranium oxide pellets; used to fuel a nuclear reactor.

fuel cell An electrochemical engine (no moving parts) that converts the chemical energy of a fuel, such as hydrogen, and an oxidant, such as oxygen, directly to electricity.

fuel NO$_x$ Produced when the nitrogen in fossil fuels is oxidized.

fuel rods A long, slender tube that holds fuel (fissionable material) for nuclear reactor use.

fumes Condensed vapors.

fumigation Mixing in the downward direction that generates very high ground-level concentrations.

function A mathematical formula that relates one variable to another.

functional group Species that play a similar role in an ecosystem.

fungicides Chemicals designed to kill fungal diseases.

fusion The nuclear reaction whereby the nuclei of light isotopes, like hydrogen, are joined (fused) to form heavier elements, releasing large amounts of energy.

G

gap species Species whose range falls outside protected areas.

gathering The collection of edible plants from unaltered ecosystems.

general systems theory The study of relationships, structures, and interdependence of storages and flows.

genetic diversity Information in the DNA of plants, animals, and microorganisms.

genetically modified Cultivars with genes from other species.

geochemical exploration A method of mineral exploration that is based on the physical and chemical processes that have produced the observed distributions in Earth.

geologic disposal The long-term storage of nuclear waste in geologic formations in Earth's crust.

geologic exploration A method of oil and mineral exploration that employs basic data gathering and mapping skills. Surface data are used to project features to the subsurface and interpret the subsurface geology.

geometric population growth Growth in populations that produce a single batch of offspring in a year.

geophysical exploration A method of oil exploration that measures the physical properties of minerals and rocks to suggest the presence or absence of economic oil and gas concentrations.

geothermal direct use The use of geothermal energy for a variety of applications including space heating, agriculture, aquaculture, recreation, medical (balneology), and industrial use (process heating).

geothermal fields An accumulation of geothermal energy that can be developed for human use.

geothermal gradient The rate of change in temperature with depth.

geothermal heat pump A heat pump that uses Earth as a heat source and heat sink.

germinate The process by which a seed starts to grow and develop.

goods Physical, tangible products used to satisfy people's wants and needs.

governance Act of governing and exercising authority.

gradient A change in the entropy of energy or matter over a specific distance.

grains Seeds of cultivated members of the grass family that are eaten by people or domesticated animals.

gravitational water Water that moves into, through, or out of the soil by gravity within a day or two of a rainfall event.

grazers Organisms in the second trophic position that eat autotrophs.

green national accounts The incorporation of environmental benefits and costs into economic decision making. The phrase often refers specifically to incorporating the depreciation of natural resources and the environment into estimates of net domestic product or net national product.

greenhouse effect The atmosphere's ability to absorb reradiated energy with longer wavelengths and convert it to heat.

greenhouse gases Gases that absorb reradiated energy with longer wavelengths and convert it to heat.

gross domestic product (GDP) The total value of output, income, or expenditures produced within a country's physical borders.

gross national income (GNI) The total income earned by the citizens of a country.

gross national product (GNP) The total value of an output, income, or expenditures produced by a nation's citizens, even including those working abroad.

gross primary production The rate at which autotrophs convert inorganic forms of energy to organic forms of energy.

groundwater Water that fills the pore spaces of an aquifer.

groundwater discharge Locations where groundwater resurfaces.

growing season The number of consecutive days during which temperatures remain above 0°C.

gully erosion Water erosion in which rills are concentrated into deeper channels.

gyre A circular motion of water with a diameter of thousands of kilometers.

H

habitat Geographical locations and environmental conditions where a plant or animal lives.

habitat conversion Changes in the quality of land use or land cover associated with human activity.

hadal The deepest layer of the ocean, below 6,000 meters.

Hadley cell A pattern of atmospheric circulation driven by solar energy in which warm air rises near the equator, cools as it travels poleward at high altitude, sinks as cold air, and warms as it travels equatorward.

half-lives The time in which half the atoms of a radioactive substance will have changed, leaving half the original amount.

harvest index The ratio of grain to total crop biomass.

heat A form of energy that is transferred by a difference in temperature.

heat balance The difference between the amount of energy that enters the atmosphere and the amount of energy that leaves the atmosphere.

heat of fusion The amount of heat that must be supplied to change a unit mass of a substance at its melting point from solid to liquid. The heat of fusion of water is 80 calories per gram (80 kcal/kg).

heat of vaporization The amount of heat that must be supplied to change a unit mass of the substance at its boiling point from liquid to a gas or vapor state. For water this is 540 kcal/kg.

hedonic pricing The use of statistical techniques to determine, from the prices of goods with different measurable characteristics, the prices that are associated with those characteristics. The latter can then be used to construct what the comparable price of a good would be from its characteristics.

herbicides Chemicals designed to kill plants.

heterotrophs Organisms that obtain energy-containing molecules by eating other organisms.

hibernation A state in which the metabolic rate slows by as much as 99 percent.

high-grading The practice of harvesting only those trees that give the highest immediate economic return.

high-yield varieties Cultivars bred to produce more edible food per unit area.

homeostasis The ability of a system to maintain its behavior or set point when disturbed.

horizons Soil layers that are approximately parallel to the surface and have distinct characteristics that are related to the process of soil formation.

horizontal mixing Refers to how far a pollutant is carried from its source.

human development A complex concept of development, based on the priority of human well-being, and aimed at ensuring and enlarging human choices that lead to equal opportunities for all people in society and empowerment of people so that they participate in—and benefit from—the development process.

humus Partially decomposed plant or animal matter.

hunting Capture of wild animals from an ecosystem.

hydrologic cycle The flow of water from the ocean through the atmosphere to the land and back to the ocean.

hydrolysis Decomposition of a chemical compound in a soil's parent material by reaction with water.

hydropower The production of electricity by the action of moving water falling on a turbine generator.

hydrothermal solutions Hot water that concentrates, transports, or deposits minerals.

hygroscopic water Tightly held water that forms a thin film around individual soil particles.

hypolimnion The layer of water in a thermally stratified lake that lies below the thermocline, is noncirculating, remains perpetually cold, and is usually low in oxygen.

I

igneous rocks Rock formed when molten (melted) materials harden.

illuvial horizon A soil layer that accumulates the minerals that wash out from the eluvial horizon.

immigration Movement of a species into an area previously uninhabited by that species.

in situ Efforts to preserve species in functioning ecosystems.

income The amount of money received from employment (salary, wages, tips), profit from financial instruments (interest, dividends, capital gains), and other sources (welfare, disability, child support, Social Security, and pensions).

independent variable The variable that affects another variable, typically represented on the right side of an equation and along the x-axis of a graph.

industrial livestock production systems Feedlots that produce less than 10 percent of their own feed.

infauna Aquatic animals that live in the substrate of a body of water, especially in a soft sea bottom.

infiltration Vertical movements of water through the soil.

inhibition A mechanism of succession in which one species has a direct or indirect harmful effect on another species.

instream Uses of water that occur without the water being diverted or withdrawn from surface water or groundwater.

insurance The process of spreading the potential effects of risk.

integrated pest management The coordinated use of pest and environmental information along with available pest control methods to prevent unacceptable levels of pest damage by the most economical means, with the least possible hazard to people, property, and the environment.

integrated systems approach The use of information from many disciplines that is needed to understand and solve specific environmental problems and generate general policy that moves society toward sustainability.

intensification Reducing the use of land and labor while increasing the use of energy, materials, and water.

intensity of use The amount of energy or material used per unit of economic activity.

intensive land uses Practices that use small land areas relative to other factors of production.

interim storage Providing safe and secure storage in the near term to support continuing operations in the interim period until long-term storage or disposition actions are implemented.

intertidal zone The area between land that is wetted by the high tide but always covered by the low tide.

invalidation Rejection of a hypothesis because observations or experimental results are inconsistent with expectations.

invasive Species that displace indigenous species or spread into habitats where they were not previously common.

inversion An extreme form of a stable atmosphere in which temperature rises with altitude.

IPAT equation A method to describe the role of multiple factors in determining environmental degradation. It describes the multiplicative contribution of population (P), affluence (A), and technology (T) to environmental impact (I). Environmental impact (I) may be expressed in terms of resource depletion or waste accumulation; population (P) refers to the size of the human population; affluence (A) refers to the level of consumption per person; and technology (T) refers to the processes used to obtain resources and transform them into useful goods and wastes.

irrigation The controlled application of water to cropland, hayfields, and/or pasture to supplement natural precipitation.

isotopes An atom having the same number of protons but a different number of neutrons in its nucleus as other varieties of the element.

K

K selected An evolutionary strategy in which organisms allocate a relatively small fraction of their energy budget toward reproduction.

kerogen An insoluble organic material that is the main precursor of crude oil and natural gas.

keystone species Species whose presence and numbers control the integrity of a community or ecosystem and allow that system to persist within its natural range of environmental conditions.

kinetic energy Energy that a body has as a result of its motion. Mathematically, it is defined as one-half the product of a body's mass and the square of its speed.

kwashiorkor Severe malnutrition caused by a diet with insufficient protein that is found primarily in young children.

L

labor A measure of the work done by human beings, especially work done for wages.

labor productivity The rate of output of a worker or group of workers per unit of time.

land use change The practice of replacing natural ecosystems with others that meet human needs and wants.

lapse rate Change in temperature with altitude.

law of conservation of matter A fundamental principle of physics that matter cannot be created or destroyed.

legumes Plants that have symbiotic relationship with nitrogen fixers.

lentic Characterizing aquatic communities found in standing water.

Liebig's law of the minimum The growth rate of plants often is determined by the nutrient that is least abundant or least available relative to the needs of the plant.

life cycle or **cradle-to-grave analysis** A comprehensive examination of the environmental and economic effects of a product at every stage of its existence, from production to disposal and beyond.

light compensation point The light level at which photosynthesis generates energy equal to respiration.

light saturation point The light level at which further increases in light do not increase photosynthesis.

lignin Natural component of wood that binds the fibers in paper.

limiting factor The item that is in least supply relative to the needs of a population.

limiting nutrient The nutrient that is in least supply relative to the quantity required by an autotroph.

limnetic zone The well-lit, open-surface water area far from shore.

lithosphere The outermost layer of crust and uppermost mantle that consists of about two dozen major plates on which the continents ride.

littoral zone Shallow waters that are near the shore.

livestock Domestic animals raised for food and fiber.

loam soil Soil ideal for agriculture that consists of a roughly equal mixture of clay, sand, and silt.

logistic equation A function describing the idealized growth of a population subject to a density-dependent limiting factor.

looping A pattern in which a plume rises and falls in an unstable atmosphere.

lotic Of, relating to, or living in actively moving water.

M

macroenvironment Characteristics of a large geographical area.

macronutrients An element required in large proportion by plants and other life forms for survival and growth. Macronutrients include nitrogen (N), potassium (K), and phosphorus (P).

macropores Relatively larges spaces within soil.

maintenance respiration   The use of energy to maintain order in a living system.

marginal ice zone   Area along the Antarctic coast where melting ice forms a shallow layer of relatively fresh water on top of salt water that has abundant supplies of light and nutrients.

market   The world of commercial activity in which goods and services are bought and sold.

market-based incentives   The use of tradable permits and other economic instruments to internalize environmental costs into market decisions.

market failure   The result when the prices of goods and services do not reflect the true costs of producing and consuming those goods and services. In the context of environmental externalities, a market failure occurs when the price of goods and services does not reflect full societal costs, which are conventional financial costs plus environmental externalities.

market power   The ability of a firm to exercise control over industry prices or output.

mass extinctions   Periods when the extinction rate is much greater than the background extinction rate.

material waste   Any materials unused and rejected as worthless or unwanted.

maturation   A growth process in which juveniles increase in size and change in form to the point at which they are capable of reproduction.

maximum intrinsic growth rate   The largest per capita rate at which a population can grow under ideal conditions.

maximum mixing depth   The upper bound on vertical mixing set by the altitude at which there is a positive relationship between temperature and altitude.

mechanical energy   Energy that an object has because of its motion or position.

membrane separation   Physical separation of salt from water by pushing seawater through thin filters that do not allow minerals to pass.

mesopelagic   The middle layer of the ocean from 200 to 1,100 meters.

mesosphere   The layer of the atmosphere that extends from about 50 to 100 km above the surface, in which temperature declines with altitude.

metamorphic rocks   Rocks that have been physically altered by heat and/or pressure.

metamorphosis   A dramatic change in body form that occurs as juveniles change to adults.

microenvironment   Small-scale conditions at which an organism lives.

micropores   Small spaces within soil.

milling   Taking uranium ore extracted from Earth's crust and chemically processing it to prepare uranium concentrate (U_3O_8), sometimes called uranium octaoxide or "yellowcake."

mimicry   Species without protective chemicals that have the warning colors of animals with protective chemicals.

mineral resource   A concentration or occurrence of natural, solid, inorganic, or fossilized organic material in or on Earth's crust in such form and quantity and of such a grade or quality that it has reasonable prospects for economic extraction.

mineralization   Conversion of organic forms of nutrients to inorganic forms.

mining   The extraction of valuable minerals or other geological materials from Earth, usually (but not always) from an ore body, vein, or (coal) seam.

mist   An aerosol that consists of liquids.

mobile sources   Moving sources of pollution, such as cars or trucks.

moderators   Components of nuclear reactors that slow neutrons, thereby increasing their chances of being absorbed by fissile material. Natural water, heavy water, and nuclear-grade graphite are the most common moderators.

molecular nitrogen   Molecules of two nitrogen atoms that make up about 78 percent of the atmosphere.

molecular oxygen   Molecules of two oxygen atoms that make up about 21 percent of the atmosphere.

molecules   Particles consisting of two or more atoms held together by chemical bonds; the smallest unit of a compound that displays the properties of that compound.

mollisols   Soils with a deep A horizon that tend to form under grasslands.

monoculture   Growing a single crop over a large area.

mountaintop removal   A coal recovery practice in which vegetation and soil are removed from a mountaintop, explosives are used to separate coal from the rocks, and the rocks are subsequently dumped into a nearby valley.

mudflat   Large areas of mud in the intertidal zone that are exposed to the air during low tide.

municipal solid waste   Solid waste originating from homes, industries, businesses, demolition, land clearing, and construction.

mutation   Breaks and rearrangements of DNA molecules.

N

NAAQS   The National Ambient Air Quality Standards established by the U.S. federal government to limit air pollution.

native range   Areas where a species evolved or inhabits for a long period.

natural resource   Something from the natural environment (water, air, trees, fuels) that is used to meet people's needs and wants.

natural resource accounting   The process of adjusting national accounts such as GNP to reflect the environmental costs of economic production.

natural selection   The differential survival and reproduction of organisms with genetic characteristics that enable them to better utilize environmental resources.

naturalized range   An area that was previously uninhabited by a species.

negative feedback loop   Creating homeostasis by changing the effect of a disturbance after one complete loop so that the system is moved back toward its original state.

net primary production   The difference between gross primary production and maintenance in autotrophs.

neutrally stable   Environmental conditions in which the observed environmental lapse rate is the same as the adiabatic lapse rate.

neutron   A particle in the nucleus of an atom, which is without electrical charge and has approximately the same mass as a proton.

new source performance standards   Regulations that govern emissions of air pollutants from new plants.

niche   Totality of a species' environmental requirements.

nitrification   A process whereby ammonia is oxidized to nitrite and then to nitrate by bacterial or chemical reactions.

nitrogen fixation   A biological or chemical process by which molecular nitrogen, from the atmosphere, is converted to organic or available nitrogen.

nonattainment areas   Locations where concentrations remain above the NAAQS.

nondegradable wastes   Wastes that are incapable of being broken down into simple, less toxic compounds.

nonessential   Molecules an organism can synthesize from its constituents.

nongovernmental organizations   Private organizations that pursue activities to relieve suffering, promote the interests of the poor, protect the environment, provide basic social services, or undertake community development.

nonpoint pollutants   Pollutants that are not discharged or emitted from a specific point, such as a pipe or smokestack.

nonpoint source   Emissions that do not originate from a single location where the pollution can potentially be controlled.

nonrenewable resources   Resources that cannot be replaced in the environment (such as fossil fuels) because they form at a rate slower than their consumption.

nonspontaneous flow   Movement of energy or matter against the tendency toward a greater state of entropy.

nonspontaneous process   A process that cannot occur without an external input of energy.

nuclear decay   The set of various processes by which unstable atomic nuclei emit subatomic particles.

nuclear energy   Energy released when atomic nuclei undergo a nuclear reaction such as the spontaneous emission of radioactivity, nuclear fission, or nuclear fusion.

nuclear proliferation and diversion   The spread from nation to nation of nuclear technology, including nuclear power plants but especially nuclear weapons.

nucleus   The central part of an atom consisting of only protons and neutrons.

nutrients   Chemicals that are needed by living organisms.

O

O horizon   A soil layer that consists primarily of organic material, which serves as a precursor for soil formation.

ocean thermal energy conversion Electricity generation by making use of the temperature difference (as much as 20°C, or 68°F, in the tropics) between the top and bottom layers of the ocean to convert a fluid to vapor, which in turn powers a turbine generator.

offstream Water that is withdrawn or diverted from surface water or groundwater.

oil refinery An industrial installation that breaks and separates the long-chain carbon molecules of crude oil into groups of shorter-chain molecules known as refined petroleum products.

oil shale An unconventional fossil fuel in which kerogen is trapped in sedimentary rock.

oligotrophic Aquatic ecosystems with low rates of net primary production.

omnivores Heterotrophs that feed on both plants and animals.

operable capacity The maximum rate of oil production that can be sustained during the following six months.

opportunity costs The cost of using a resource based on what it could have earned if used for the next best alternative. For example, the opportunity cost of farming your own land is the amount you could have received by renting it to someone else.

ore A mineral deposit containing a metal or other valuable resource in economically viable concentrations.

orographic effect Precipitation that results from or is enhanced by mechanical lifting of an air mass over mountains.

overburden Soil and rock above a coal seam or other mineral source.

overdraft Removing water from an aquifer faster than it is recharged.

overshoot Population exceeding carrying capacity.

oxidation Occurs when oxygen combines with compounds in the parent material; this often weakens a material, which makes it more vulnerable to weathering.

oxisols Soils that often have a thin O layer and small amounts of nutrients.

ozone A molecule that consists of three oxygen atoms and is found largely in the stratosphere, where it absorbs a significant fraction of incoming UV-B radiation.

ozone depletion potential The ability of a chemical to destroy ozone.

P

parent material The mineral material from which a soil forms.

parenting Energy allocated toward reproduction in the form of caring for offspring.

passenger species Species whose loss would have little effect on ecosystem function.

pasture Land used to graze livestock or cultivated to produce animal feed.

pathogens Microorganisms that cause disease.

payback period The time it takes an energy investment to capture or save an amount of energy equivalent to the investment.

pelagic Open areas of the ocean away from the bottom.

perennials Plants that live for several growing seasons.

permafrost Soil material that remains below 0°C for two or more years.

permanent waste disposal The storage of nuclear waste in a form and location for a very long time.

persistent wastes Wastes that degrade very slowly over time.

personal consumption expenditures Money spent by households on the purchase of goods and services.

pesticides Chemicals designed to kill insect pests or plant diseases.

pests Species, such as insects, birds, or small mammals, that eat cultivated plants.

petrochemicals Chemicals obtained by refining crude oil. Petrochemicals are used as raw materials in the manufacture of most industrial chemicals, fertilizers, pesticides, plastics, synthetic fibers, paints, medicines, and many other products.

photodissociation The process by which solar energy splits a molecule.

photosphere The visible outer layer of the sun that reradiates energy absorbed from its interior.

photosynthesis The use of solar energy to break the bonds between the hydrogen and oxygen atoms in water molecules, and the incorporation of hydrogen atoms and carbon dioxide molecules to form glucose.

photovoltaic effect The generation of an electrical current in a circuit containing a photosensitive device when the device is illuminated by visible or nonvisible light.

phreatic zone The saturated soil below a stream.

physical change A change that affects the size, shape, or color of a substance but does not affect its composition.

phytoplankton Single-cell photosynthetic algae that live suspended in bodies of water and drift about.

pioneer community The first plants and animals to inhabit an area that was previously uninhabited.

placer deposits Naturally occurring localized concentrations of economically important ore minerals, formed as a result of physical processes at or near Earth's surface.

plate A large rigid slab of solid rock that makes up a portion of the lithosphere.

poaching Illegal killing or collecting of plants and animals.

polar cell A weak pattern of atmospheric circulation characterized by ascending motion in the subpolar latitudes (50°–70°), descending motion over the pole, poleward motion aloft, and equatorward motion near the surface.

polar front An area of low pressure at about 60° north and south of the equator.

polar stratospheric clouds Clouds that consist of very small droplets of water and nitric acid that are formed at very cold temperatures.

polar vortex A surface wind that blows in a circular pattern around the pole during the winter.

pollinators Animals that place pollen on the stigma of plants.

pollution The presence of a substance in the environment that, because of its chemical composition or quantity, prevents the functioning of natural processes and produces undesirable environmental and health effects.

pollution taxes A tax designed to reduce pollution by making the polluter pay the estimated amount of damages.

population Individuals of the same species that inhabit a specific area.

population density The number of individuals per unit area.

population growth An increase in population.

population histograms Diagrams that represent the number of males and females in various age groups.

population momentum Built-in potential for population growth due to a large number of individuals entering reproductive age.

pores Spaces between particles in a soil or an oil field.

porosity A measurement of space between soil particles that can hold water or oil.

positive feedback loop Destabilizes a system by reinforcing the effect of a disturbance so that the system is moved further away from its original state.

positive relationship Correlation between parts of a system such that an increase in one part of a system causes an increase in another part of the system.

postconsumer content A material or finished product that has served its intended use and has been diverted or recovered from waste destined for disposal, having completed its life as a consumer item.

postconsumer waste Waste collected after the consumer has used and disposed of it.

potential energy The energy that a body possesses by virtue of its position with respect to other bodies in the field of gravity.

potential evaporation The amount of water that would evaporate if water were available.

power The rate of doing work.

power density The rate of doing work per unit area or volume.

prairies The North American term for grasslands.

precipitation The flow of water from the atmosphere as a liquid (rain) or solid (snow or ice).

precision agriculture A system of agriculture that seeks to provide the inputs needed for crop growth (water and nutrients) and crop protection without deficiency or excess at each point in time during the growing season.

predator control Efforts to reduce the populations of species that compete with humans for crops or game.

predictability Variance in the average time between events.

pressurizer A high-strength tank containing steam and water used to control the pressure of the reactor coolant in the primary loop in a nuclear power plant.

Price–Anderson Act A U.S. law that indemnifies all nonmilitary nuclear facilities constructed in the United States against liability claims arising from nuclear incidents while still ensuring compensation for the general public.

primary air pollutants Pollutants that enter the atmosphere in a form that harms life or property.

primary consumers Organisms in the second trophic position that eat autotrophs (also known as *grazers*).

primary coolant The coolant in a nuclear power plant that first comes in contact with the core.

primary mill residues Biomass produced from the processing of timber (such as at sawmills) that is a potential source of energy.

primary particles Particulates that are emitted directly by economic activities.

primary recovery Oil that is pushed to the surface by the pressure gradient in the field.

primary standards Environmental regulations designed to protect public health, including the health of sensitive populations such as asthmatics, children, and the elderly.

primary succession Succession in areas where there was no soil or where a disturbance destroyed the soil.

primary treatment Treatment of municipal wastes that removes large solids by mechanical techniques such as screens and settling tanks.

principle of sustainability See *sustainability*.

prior appropriation doctrine A practice dictating that no one owns the water in a stream and that all people, corporations, and municipalities have the right to use water for beneficial purposes.

private costs Costs incurred by the individual who causes environmental degradation.

private property Ownership of property (or other assets) by individuals or corporations.

privatization Selling a state-owned business to private investors.

production In economics, manufacturing or mining or growing something (usually in large quantities) for sale.

property rights The rights of an individual to own property and keep the income earned from it.

proportional pricing system A method of charging for waste disposal in which residents pay the same amount of money for each unit of waste they set out for collection (perhaps $1.50 for each 30-gallon bag). The price is based on the number of bags, tags, or stickers (usually sold at local retail stores or municipal offices) the resident uses.

proteins Large complex molecules made up of one or more chains of amino acids.

proton A basic particle in an atom's nucleus that has a positive electrical charge.

proved oil reserves Volumes of crude oil that geological and engineering information shows, beyond reasonable doubt, to be recoverable in the future from a reservoir under existing economic and operating conditions.

public goods Goods and services that are supplied by the government because it is not sufficiently profitable for the private sector to do so. This term is also applied to resources that are said not to be diminished by their consumption by any single person.

public property Property owned by a government.

pumped storage plant A plant that usually generates electric energy during peak load periods by using water previously pumped into an elevated storage reservoir during off-peak periods when excess generating capacity is available to do so. When additional generating capacity is needed, the water can be released from the reservoir through a conduit to turbine generators located in a power plant at a lower level.

pyrolysis Decomposition of a chemical by extreme heat.

Q

Q infinity The total quantity of oil that will be discovered and produced.

R

R horizon A soil layer that consists of hard bedrock.

r selected An evolutionary strategy in which organisms allocate a large fraction of their energy budget toward reproduction.

radiant energy Energy traveling in the form of electromagnetic waves; measured in units of energy such as joules, ergs, or kilowatt-hours.

radiation balance The difference between incoming and outgoing radiation.

radiation inversions Short-lived inversions that occur daily due to radiational warming and cooling.

radiative forcing The total amount of energy (watts) that is absorbed by the gases that lie above an area of Earth's surface, from ground level to the top of the atmosphere.

radioactive isotopes An atom with an unstable nucleus that undergoes radioactive decay by emitting gamma rays and/or subatomic particles.

radioactivity The spontaneous emission of radiation from the nucleus of an atom.

radiocarbon dating A process that provides absolute dates by counting the radioactive decay of carbon in the remains of once-living plants and animals (charcoal, wood, bone, shell).

radioisotopes Radioactive isotopes.

random drift Accumulation of changes in the gene pool due to stochastic events.

range-fed meat Meat derived from animals that were raised on rangeland.

rangelands Relatively unaltered natural ecosystems where livestock graze natural vegetation that is predominantly native grasses, glasslike plants, forbs, or shrubs.

raw materials Unprocessed natural resources or products used in manufacturing.

reactants The substances that take part in a chemical reaction.

reactor core The core of a nuclear reactor, consisting of the fuel, moderator (in the case of thermal reactors), and coolant.

reactor vessel A cylindrical steel vessel in a nuclear power plant that contains the core, control rods, coolant, and structures that support the core.

reasonable use doctrine A doctrine that allows landowners to pump water for any beneficial use and does not recognize priority among users.

recharge area The area from which an aquifer receives its water.

recovered paper Old paper that is made available to the paper industry through recycling.

recyclable products Materials capable of being recycled.

recycled-content products Goods that contain material that has been recycled.

recycling The process by which materials that would otherwise become solid waste are collected, separated or processed, and returned to the economic mainstream to be reused in the form of raw materials or finished goods.

Redfield ratio The ratio of nitrogen atoms to phosphorus atoms that are needed by autotrophs.

reductionist approach A scientific methodology based on the premise that the best way to learn about something is to break it into its parts and study the parts separately.

redundancy The presence of more than one species in a functional group.

redundant species Species whose loss has little effect on ecosystem function.

refined petroleum products Carbon molecules derived from crude oil.

reflection The process whereby a surface turns back a portion of the radiation that strikes it.

regolith Unconsolidated material that lies above bedrock.

remanufacturing The process of restoring used durable products to "new" condition, to be used in their original function, by replacing worn or damaged parts.

renewable resources Resources that can be replenished at rates equal to or greater than their rates of depletion, such as solar, wind, geothermal, and biomass resources.

replacement level fertility The total fertility rate at which a population remains constant.

reprocessing The treatment of spent (irradiated) reactor fuel to separate plutonium from uranium and other fission products.

reproduction Allocation of energy to produce and care for offspring.

reproductive range The area where individuals have enough energy for reproduction.

reserve base That part of an identified natural resource that meets specified minimum physical and chemical criteria related to current mining and production practices, including those for grade, quality, thickness, and depth.

reserves to production (R:P) The quantity of oil in proved reserves relative to the current rate of oil production.

reservoir A natural or artificial pond or lake that is used for water storage or regulation.

residence time The time that an atom spends in a storage pool.

residues Materials that are left over from the normal operation of the timber industry, such as wood chips and scraps.

resilience The ability of a system to return to its set point following a disturbance.

resistance The ability of a system to withstand a disturbance.

resource depletion The "using up" of natural resources within a region; most commonly used in reference to the farming, fishing, mining, and timber industries.

resource depletion hypothesis The notion that agriculture is a response to population growth and the best first principle.

resource quality The amount of effort (capital, labor, energy) that is required to extract a natural resource.

respiration Biochemical pathways that convert food to energy.

reuse Using a product or component of municipal solid waste in its original form more than once—such as refilling a glass bottle that has been returned or using a coffee can to hold nuts and bolts.

revisions Changes (either positive or negative) to proved reserves that are generated by new information other than an increase in acreage.

rill erosion Water-driven soil erosion in which sheet erosion becomes concentrated in small channels.

riparian water rights Laws that allow a landowner to use a share of the water that flows naturally past his or her property but do not entitle a landowner to divert water for storage in a reservoir for use in the dry season or to use water on land outside the watershed.

riparian zone The transition zone between a waterway and the surrounding terrestrial environment.

risk management The process of making decisions without complete information due to the presence of a stochastic element.

rock cycle A series of processes through which a rock changes, over time, between igneous, sedimentary, and metamorphic forms.

room and pillar A coal extraction technique that leaves pillars of coal to prevent the ceiling from collapsing.

rooting depth or **rooting layer** The vertical distance from the soil surface that contains 95 percent of a plant's roots.

round wood Sources of fibers for papers that consist of whole trees.

rule of absolute ownership The principle that allows landowners to pump as much groundwater as they want.

runoff Water that flows horizontally across the land toward the nearest surface water.

S

salinity The concentration of mineral salts dissolved in water.

salt marsh The transition area from land to sea that is farthest from the sea.

salt water Water that contains more than 35,000 milligrams per liter of dissolved solids, most often salt.

saltation Wind-driven erosion in which particles bounce along the surface.

saltwater intrusion Occurs when overdrafts allow salt water to flow into aquifer pore spaces that were previously occupied by freshwater.

sand Soil particles that can be seen with the unaided eye.

scatter To disperse radiation in different directions.

scavengers Animals that eat portions of dead animals.

scenario analysis A modeling technique that involves entering different sets of data into a model and determining how changes in the input data affect the model's output.

seam A naturally occurring layer of coal usually thick enough to be mined for profit.

second best problem Implementing a suboptimal policy because real-world conditions are inconsistent with those required for the optimal policy.

second law of thermodynamics (1) Each time energy is converted from one form to another, some of the energy is always degraded to a lower-quality, more dispersed, less useful form. (2) No system can convert energy from one form to another useful form with 100 percent efficiency. (3) Energy cannot be spontaneously transferred from a cold body to a hot body. (4) The entropy of a system increases over time.

secondary air pollutants Pollutants that are formed when primary pollutants interact with sunlight and other gases in the atmosphere.

secondary consumers Organisms in the third trophic position that eat organisms in the second trophic position.

secondary coolant In a nuclear power plant, the part of the cooling system that absorbs heat from the primary coolant.

secondary particles Pollutants that are formed when particles emitted by human activity interact with natural components of the atmosphere.

secondary productivity The rate at which heterotrophs create new biomass per unit area in a given time period.

secondary recovery The injection of water into an oil field to push additional quantities of oil toward the producing well and up to the surface.

secondary standards Regulations designed to protect public welfare, including protection against decreased visibility and damage to animals, crops, vegetation, and buildings.

secondary succession Occurs when a disturbance destroys a climax or intermediate community without destroying the soil.

secondary treatment Treatment of municipal wastes that reduces the number of pathogens and accelerates the decomposition of organic wastes by enhancing the actions of aerobic and anaerobic bacteria.

sediment delivery ratio The fraction of eroded soil that ends up in sediments.

sedimentary rocks Rocks created by pressure and cementation of particles in a process known as *consolidation.*

sedimentation The burial of organic material by particles.

sediments Fine particles created from the weathering of rocks.

selective breeding The process in which only individuals with traits desired by agriculturists are allowed to reproduce.

selective harvesting The process of cutting individual trees or small groups of trees.

senescence Failure of body systems that decreases the probability of survival and reproduction.

services In economics, the nonmaterial equivalent of a good. Examples are airlines, banks and savings institutions, business services, financial services, food, lodging and travel services, information, entertainment and software, insurance, real estate, telecommunications, transportation, and utilities.

set point The level of a storage or flow that systems maintain via homeostasis.

sewage Waste and wastewater produced by residential and commercial users that is discharged into sewers.

sheet erosion Water-driven erosion in which a film of water moves across the soil surface.

shelterbelts Soil protection provided by rows of planted vegetation between strips.

shelterwood method A harvest method that removes trees in a series of two or three partial cuts that are designed to simulate a natural disturbance.

silt Intermediate-sized soil particles that can be seen under a microscope.

siltation Accumulation of eroded soil particles at the bottom of a water body.

slash and burn The process in which an agroecosystem is created by burning a patch of forest.

smoke Particles that form during the combustion of fossil fuels.

social costs The effects of environmental degradation on society.

soil The upper portion of the regolith that has been changed both chemically and biologically.

soil conservation techniques Techniques designed to slow soil erosion.

soil erosion The rate at which soil is moved.

soil profile The vertical arrangement of soil horizons.

solar constant The amount of solar radiation that reaches the upper layers of Earth's atmosphere: 1.97 calories per square centimeter per minute.

solar thermal system A system that uses radiation from the sun to produce heat energy.

soot Particles that form during the combustion of fossil fuels.

sorption A process in which soil organic carbon soaks up or attracts agricultural chemicals.

source reduction Reducing the amount and/or toxicity of an item before it is generated (examples include buying an item with less packaging or using a nontoxic cleaning compound).

spatial scale Geographic area.

specializations Adaptations well suited for a relatively narrow range of conditions.

speciation Evolution of a new species.

species diversity The total number of living species.

species evenness The distribution of individuals among species.

species richness The number of species in an area.

specific heat The amount of heat energy required to raise the temperature of a material of a particular mass.

spent nuclear fuel Fuel rods that no longer contain enough fissionable uranium to be efficiently used to produce power.

spodosols Soils formed under coniferous forests that have a relatively thin O horizon and a more thoroughly leached eluvial horizon.

spontaneous flow Movements of matter or energy that are consistent with the tendency toward a greater state of entropy.

spontaneous process A process that requires no influence from outside the system to proceed.

stability The ability of a system to return a storage or flow to a set point following a disturbance.

stable atmosphere An atmosphere in which the observed lapse rate is slower than the adiabatic lapse rate.

state implementation plan Describes how states will satisfy the NAAQS.

stationary sources Places or objects from which pollutants are released and that do not move around.

stochastic Containing uncertainty due to an element of chance.

storage A system part where energy or materials stay for an extended period.

stratosphere A layer of the atmosphere that extends from about 20 to 50 km above the surface, in which temperature rises by about 50°C.

stripping ratio The unit amount of overburden that must be removed to gain access to a similar unit amount of coal or mineral material.

strong interaction A trophic connection between species in which the predator eats only a few types of foods, so the likelihood of consumption of one species by another is high.

strong sustainability A criterion that prohibits human actions degrading the environment or using environmental goods or services faster than they are generated by the environment.

subsidence A drop in land level due to the weight of the overlying material compressing soil particles after water has been withdrawn.

subsidence inversion An increase in temperature with altitude produced by the adiabatic warming of a layer of subsiding air.

subsidies Payments to firms or households that are designed to favor a particular action.

subsoil A soil layer known as the *illuvial* or *B horizon*.

subsystem A system that is part of a larger system.

subtropical high A region of high pressure about 30° north and south of the equator due to the descending portion of the Hadley cell.

succession A change in plant and animal communities that follows a disturbance.

surface mining techniques Coal recovery techniques that remove the soil and rock above coal seams, thereby exposing the seams.

surface roughness Irregularities in the soil surface that retard soil erosion by slowing wind speed.

surface water Water that sits or flows above land, including lakes, oceans, rivers, and streams.

survival range The area where an individual can obtain enough energy and materials to survive.

suspension A wind-driven soil erosion process in which soil particles may be lifted high into the air and carried long distances.

sustainability A sustainable society meets the needs of the present generation without compromising the ability of future generations to meet their own needs.

sustainable development Economic activities that meet the needs of the present without compromising the ability of future generations to meet their own needs.

sustainable yield of timber Harvesting timber no faster than the rate at which trees produce new supplies.

swimmable A category of water quality in which recreation in and on the water will not threaten people's health.

symbiosis The intimate living together of two dissimilar organisms in a mutually beneficial relationship.

sympatric speciation The process of forming a new species in which individual traits isolate a subpopulation from the parent population and allows the populations to evolve separately.

system A collection of parts, which are known as *storages* and *flows*, that interact with each other to generate regular or predictable patterns or behaviors.

T

tailings Material that remains after all metals considered economic have been removed from ore during milling.

technical change hypothesis The notion that agriculture arose with increasing human technical capabilities.

technology The body of knowledge about the means and methods of producing goods and services.

temperature profile The change in temperature with depth.

temperature sensitivity The long-term change in temperature given a doubling in atmospheric concentration of carbon dioxide or radiative forcing.

terraces Ridges and channels that are constructed across a slope to prevent rainfall runoff from accumulating and causing serious erosion.

tertiary methods Injection of heat or materials that reduce the viscosity or surface tension of crude oil, which makes it flow more easily toward the surface.

tertiary treatment Treatment of municipal wastes in which undecomposed organic nutrients are separated from the wastewater, which is discharged back to the environment.

thermal desalting A process in which seawater is boiled or evaporated and the steam or evaporate is drawn off as pure water.

thermal NO$_x$ A process in which burning fossil fuels creates temperatures high enough to oxidize atmospheric nitrogen (N_2) and oxygen (O_2) to form nitrogen oxides.

thermocline The portion of the water column where temperature changes very rapidly.

thermohaline circulation Regular circulation of ocean waters between the surface and deep layers due to differences in temperature and salinity.

thermosphere The outer layer of the atmosphere that starts at about 100 km in which temperature reaches 1,200°C.

thoracic particles Particles smaller than 10 microns.

threatened species Species that are likely to become endangered in the foreseeable future throughout all or a significant portion of their ranges.

tidal energy Electricity generated by capture of the energy contained in moving water masses due to tides.

timber concession A contract that defines the rules for harvesting trees.

time lag The period that lapses between a cause and its effect.

tolerance model A mechanism for succession in which species replace one another based on their ability to withstand limiting factors.

topography The surface features of an area.

topsoil The A horizon.

total fertility rate The average number of children a woman will bear over her lifetime.

total saturation The maximum water capacity of a soil.

trace elements Elements essential for growth but required only in minute amounts.

trace gases A group of about twenty gases that make up about 1 percent of the atmosphere.

tradable permits An economic policy instrument under which rights to discharge pollution or exploit resources can be exchanged through either a free or a controlled permit market. Examples include individual transferable quotas in fisheries, tradable depletion rights to mineral concessions, and marketable discharge permits for waterborne effluents.

trade winds Ground-level winds associated with the pressure gradient that causes air to move from the subtropical high to the equatorial low.

transgenic cultivars New cultivars that are created by moving genes from other species into crop species.

transpiration The process by which water absorbed by plants, usually through the roots, evaporates from the plants' surface, principally from the leaves.

transportation costs The price paid for moving capital, labor, or other inputs into an area or moving agricultural goods, timber, or minerals out of an area.

transportation infrastructure Capital used in transportation, such as roads, railroad lines, and ports.

travel cost method A method used to estimate the economic use values associated with ecosystems or sites that are used for recreation based on the assumption that time and travel cost expenses that people incur to visit a site represent the "price" of access to the site.

trophic cascades Population changes transmitted via the food web from one trophic position to the next.

trophic position The position along the food chain or food web at which an organism obtains energy.

tropopause The end of the troposphere, where temperature starts to rise with altitude.

troposphere The lowest layer of the atmosphere extending from Earth's surface to an altitude of 10 to 20 kilometers, in which temperature declines with altitude at a rate of about 6.5°C per kilometer.

turbidity The degree to which light can penetrate water.

turbine generator A rotary engine driven by the pressure of water, air, or steam against the curved vanes of a wheel to transform heat, chemical energy, or water pressure into mechanical energy.

turbulent Winds that move across the surface and have an up-and-down component.

turnover rate The average time required to disturb an entire area.

U

ultrafine particles Particles smaller than 0.1 microns.

unavailable Forms of nutrients that cannot be used by an autotroph.

unconfined aquifer An aquifer that sits atop an impermeable layer.

unconventional fossil fuels Fossil fuels, such as oil shales and tar sands, that may eventually replace conventional fossil fuels.

underground mining techniques Coal recovery techniques in which shafts are drilled to the seam to give miners access.

universal soil loss equation The equation that represents the amount of soil moved by water.

unknown carbon sink An unknown mechanism that removes carbon from the atmosphere or a known mechanism that removes carbon faster than estimated by scientists.

unstable An atmosphere in which the observed environmental lapse rate is faster than the adiabatic lapse rate.

upwellings Areas where large quantities of deep ocean water rise back to the surface.

utility In economics, a measure of the happiness or satisfaction gained from a good or service.

V

validation Confirmation of a hypothesis because observations or experimental results are consistent with expectations.

variable rate pricing system A method of charging for waste disposal in which residents are charged different amounts per unit of garbage.

variance The degree of dispersion or scattering around a variable's expected value.

vegetable Edible seeds, roots, stems, leaves, bulbs, tubers, or nonsweet fruits of herbaceous plants.

vertical mixing The altitude to which a pollutant is mixed.

virgin materials Fibers used for paper production that come from trees.

vitrification A method of immobilizing nuclear waste that produces a glasslike solid that permanently captures the radioactive materials.

W

waste Any materials unused and rejected as worthless or unwanted.

waste assimilation The absorption of wastes by the environment.

waste heat Energy that must be dissipated to the environment from an energy conversion process.

waste recovery Any waste management operation that diverts a waste material from the waste stream and results in a product with a potential economic or ecological benefit.

water diversion Movement of water from surface water or groundwater over some distance to its point of use.

water pollution Purposeful or accidental addition of materials that contaminate water.

water table The top portion of the aquifer.

water use efficiency The amount of water transpired per unit of new biomass produced.

watershed The area from which surface waters derive surface runoff and groundwater flows.

waterworks Human systems for supplying water.

weak interaction A trophic linkage among species such that the likelihood of consumption of one species by another is small.

weak sustainability The principle that actions that degrade the environment or use a natural resource or a waste-processing service faster than it is generated by the environment can be sustainable if these losses are offset by an increase in either economic capital or social institutions.

wealth The total value of the accumulated assets owned by an individual, household, community, or country.

weather Atmospheric conditions at a particular place and point in time and how they change from day to day.

weathering The breakup of solid rock.

weekend effect Differences in climate variables on Saturday or Sunday relative to observations Monday through Friday.

westerlies Midlatitude surface winds that blow from the subtropical high toward the polar front.

wet deposition The process in which sulfate particles (SO_4) are removed from the atmosphere by precipitation.

wildcatting Drilling wells into formations not previously known to contain oil or natural gas.

wilting point The level of soil water at which plants lose the ability to support themselves because the soil water that remains is held more strongly than the plant's ability to absorb it.

wind energy Energy derived from the kinetic energy of the wind.

wind erosion equation The equation that represents the amount of soil particles that are moved by wind.

wind farm A collection of wind turbines all in the same location, used for the generation of electricity.

winds Horizontal motion of air caused by the uneven heating of the atmosphere combined with Earth's rotation.

withdrawals Water that is removed from its source.

X

xenobiotics Organic compounds that are synthesized by humans and therefore are relatively resistant to organic decay.

Y

yield The amount of edible food grown per land area.

yield per effort The ratio of resource obtained relative to the effort used to obtain it.

Z

zooplankton Small multicellular organisms that are among the most important primary consumers in aquatic environments.

Suggested Readings

We have compiled a list of suggested readings that presents a good overview of available resource materials and a starting point for more thorough investigations and in-depth research. For additional help in researching topics, visit our website at www.mhhe.com/kaufmann1e and click on **HOW TO RESEARCH A TOPIC.**

Chapter 1

Bahn P. and J. Flenley. 1992. *Easter Island, Earth Island*. London: Thames and Hudson.

The Corporation for Public Broadcasting. 1998. The secrets of Easter Island. www.pbs.org/wgbh/nova/easter/.

Van Tilburg, J. A. and J. Mack. 1995. *Easter Island: Archaeology, Ecology, and Culture*. Washington, D.C.: Smithsonian Institution Press.

Chapter 2

Graedel, T. E. and B. R. Allenby. 1995. *Industrial Ecology*. Englewood Cliffs, NJ: Prentice Hall.

Hall, C. A. S., C. J. Cleveland, and R. K. Kaufmann. 1986. *Energy and Resource Quality: The Ecology of the Economic Process*. Hoboken, NJ: John Wiley & Sons.

McFarland, E. L., J. L. Hunt, and J. L. Campbell. 2001. *Energy, Physics and the Environment*. Cincinnati: Thomson Custom Publishing.

Chapter 3

Bahn P. and J. Flenley. *Easter Island, Earth Island*. 1992. London: Thames and Hudson.

Redman, C. L. *Human Impact on Ancient Environments*. 1999. Tucson: University of Arizona Press.

Van Tilburg, J. A. *Easter Island*. 1994. Washington, D.C.: Smithsonian Institution Press.

Chapter 4

Bigg, G. R. 1990. El Niño and the southern oscillation. *Weather* 45:2–8.

Broecker, W. S. 1991. The great ocean conveyor. *Oceanography* 4:79–89.

Strahler, A. H. and A. Strahler. 2005. *Introducing Physical Geography*. Hoboken, NJ: John Wiley & Sons.

Chapter 5

Agrawal, A. A. 1998. Induced responses to herbivory and increased plant performance. *Science* 279:1201–1202.

Beatty, C. D., K. Beirinckx, and T. N. Sherratt. 2004. The evolution of Mullerian mimicry in multispecies communities. *Nature* 431:63–67.

Carbone, C., G. M. Mace, S. G. Roberts, and D. W. Macdonald. 1999. Energetic constraints on the diet of terrestrial carnivores. *Nature* 402:286–288.

Carranza, J. S. Alarcos, C. B. Sanchez-Prieto, J. Valencia, and C. Mateoa. 2004. Disposable-soma senescence mediated by sexual selection in an ungulate. *Nature* 432:215–218.

Chapter 6

Chameides, W. L. and E. M. Purdue. 1997. *Biogeochemical Cycles*. New York: Oxford University Press.

Prentice, I. C., G. D. Farquhar, M. J. R. Fasham, M. L. Goulden, M. Heimann, V. J. Jaramillo, H. S. Kheshgi, C. Le Quéré, R. J. Scholes, and D. W. R. Wallace. 2001. The carbon cycle and atmospheric carbon dioxide. In J. T. Houghton, Y. Ding, D. J. Griggs, M. Noguer, P. J. Van Der Linden, X. Dai, K. Maskell, and C. A. Johnson, eds. *Climate Change 2001: The Scientific Basis*. Cambridge, England: Cambridge University Press.

Smil, V. 2000. Phosphorus in the environment: Natural flows and human interferences. *Annual Review of Energy and the Environment* 25:53–88.

Chapter 7

Chabot, B. F. and D. J. Hicks. 1982. The ecology of leaf life spans. *Annual Review of Ecology and Systematics* 13:229–259.

Hall, C. A. S., J. A. Stanford, and F. R. Hauer. 1992. The distribution and abundance of organisms as a consequence of energy balances among multiple environmental gradients. *Oikos* 65:377–390.

Mclusky, D. S. 1989. *The Estuarine Ecosystem*. New York: Chapman and Hall.

Sumich, J. L. and J. F. Morrissey. 2004. *Introduction to the Biology of Marine Life*. Sudbury, MA: Jones and Bartlett.

Chapter 8

Bellwood, D. R., T. P. Hughes, C. Folke, and M. Nystrom. 2004. Confronting the coral reef crisis. *Nature* 429:827–833.

McCann, K. S. 2000. The diversity-stability debate. *Nature* 405:228–233.

Scheffer, M., S. Carpenter, J. A. Foley, C. Folke, and B. Walker. 2001. Catastrophic shifts in ecosystems. *Nature* 413:591–596.

Young, T. P. Restoration ecology and conservation biology. 2000. *Biological Conservation* 92:73–83.

Chapter 9

Cohen, J. 1995. *How Many People Can the Earth Support?* New York: W. W. Norton.

Diamond, J. 2005. *Collapse: How Societies Choose to Fail or Succeed*. New York: Viking Books.

Gever, J., R. K. Kaufmann, D. Skole, and C. Vorosmarty. 1991. *Beyond Oil: The Threat to Food and Fuel in the Coming Decades*. Boulder: University of Colorado Press.

Norgaard, R. B. 2002. Optimists, pessimists, and science. *Bioscience* 52:287–292.

Chapter 10

Costanza, R., J. C. Cumberland, H. E. Daly, R. Goodland, and R. Norgaard. 1997. *An Introduction to Ecological Economics*. Boca Raton: St. Lucie Press.

Daily, G. C., ed. 1997. *Nature's Services: Societal Dependence on Natural Ecosystems*. Washington, D.C.: Island Press.

Hall, C. A. S., C. J. Cleveland, and R. K. Kaufmann. 1986. *Energy and Resource Quality: The Ecology of the Economic Process*. Niwot, CO: University of Colorado Press.

Millennium Ecosystem Assessment. 2005. *Ecosystems and Human Well-Being: Synthesis Report*. Washington, D.C.: Island Press.

Chapter 11

Durning, A. 1992. *How Much Is Enough? The Consumer Society and the Future of the Earth*. New York: W. W. Norton.

Ehrlich, P. A. and A. H. Erhlich. 1990. *The Population Explosion*. New York: Simon and Schuster.

Smil, V. 2005. *Energy at the Crossroads: Global Perspectives and Uncertainties*. Cambridge, MA: MIT Press.

Willard, B. 2005. *The Next Sustainability Wave*. Gabriola Island, British Columbia: New Society Publishers.

Chapter 12

Burdick, A. 2005. *Out of Eden*. New York: Farrar, Straus, & Giroux.

Chapin, F. S. III, E. S. Zavaleta, V. T. Eviner, R. L. Naylor, P. M. Vitousek, H. L. Reynolds, and D. U. Hooper, et al. 2000. Consequences of changing biodiversity. *Nature* 405:234–242.

Gaston, K. J. 2000. Global patterns in biodiversity. *Nature* 405:220–227.

Metrick, A. and M. L. Weitzman. 1998. Conflicts and choices in biodiversity preservation. *Journal of Economic Perspectives* 12:21–34.

Purvis, A. and A. Hector. 2000. Getting the measure of biodiversity. *Nature* 405:212–219.

Chapter 13

Banuri, T., T. Barker, I. Bashmakov, K. Blok, D. Bouille, R. Christ, and O. Davidson, et al. 2001. *Climate Change 2001: Mitigation.* Cambridge: Cambridge University Press.

Harvey, L. D. 2000. *Climate and Global Environmental Change.* New York: Prentice Hall.

McCarthy, J., O. F. Canziani, N. A. Leary, D. J. Dokken, and S. White. 2001. *Climate Change 2001: Impacts, Adaptation, and Vulnerability.* Cambridge: Cambridge University Press.

Prentice, I. C., G. D. Farquhar, M. J. R. Fasham, M. L. Goulden, M. Heimann, V. J. Jaramillo, H. S. Kheshgi, C. Le Quéré, R. J. Scholes, and D. W. R. Wallace. 2001. The carbon cycle and atmospheric carbon dioxide. In J. T. Houghton, Y. Ding, D. J. Griggs, M. Noguer, P. J. Van Der Linden, X. Dai, K. Maskell, and C. A. Johnson, eds. *Climate Change 2001: The Scientific Basis.* Cambridge, England: Cambridge University Press.

Chapter 14

Appenzeller, C., J. Eberhard, N. R. P. Harris, and J. Staehelin. 2001. Ozone trends: A review. *Reviews of Geophysics* 39:231–290.

Day, T. A. and P. J. Neale. 2002. Effects of UV-B radiation on terrestrial and aquatic primary producers. *Annual Review of Ecology & Systematics* 33:371–396.

Parson, E. A. 2003. *Protecting the Ozone Layer: Science and Strategy.* New York: Oxford University Press.

Roan, S. 1990. *Ozone Crisis.* Hoboken, NJ: John Wiley & Sons.

Chapter 15

Ananda, J. and G. Herath. 2003. Soil erosion in developing countries: A socioeconomic appraisal. *Journal of Environmental Management* 68:343–353.

Brady, N. C. and R. R. Weil. 1999. *The Nature and Properties of Soils.* New York: Prentice Hall.

Hansen, Z. K. and G. D. Libecap. 2004. Small farms, externalities, and the Dust Bowl of the 1930s. *Journal of Political Economy* 112:665–694.

Chapter 16

Cassman, K. E. 1999. Ecological intensification of cereal production systems: yield potential, soil quality, and precision agriculture. *Proceeding National Academy of Sciences* 96:5952–5959.

Cronon, W. 1983. *Changes in the Land: Indians, Colonists and the Ecology of New England.* New York: Farrar, Straus, & Giroux.

Diamond, J. 2002. Evolution, consequences, and the future of plant and animal domestication. *Nature* 418:700–707.

Evenson, R. E. and D. Golin. 2003. Assessing the impact of the Green Revolution, 1960–2000. *Science* 300:758–762.

Huang, J., C. Pray, and S. Rozelle. 2002. Enhancing the crops to feed the poor. *Nature* 418:678–684.

Trewavas, A. J. 2001. The population/biodiversity paradox: agricultural efficiency to save wilderness. *Plant Physiology* 125:174–179.

Chapter 17

Cockburn, A. and S. Hecht. 1990. *The Fate of the Forest.* New York: Harper Perennial.

Perlin, J. 1989. *A Forest Journey: The Role of Wood in the Development of Civilization.* New York: W. W. Norton.

World Commission on Forests and Sustainable Development. 1999. *Our Forests: Our Future.* Cambridge: Cambridge University Press.

Chapter 18

Glennon, R. 2002. *Water Follies: Groundwater Pumping and the Fate of America's Fresh Waters.* Washington, D.C.: Island Press.

Jehl, D. and B. McDonald, eds. 2003. *Whose Water Is It? The Unquenchable Thirst of a Water-Hungry World.* Washington, D.C.: National Geographic.

Swain, S. 2004. *Managing Water Conflict: Asia, Africa, and the Middle East.* London: Routledge.

Ward, D. R. 2002. *Water Wars: Drought, Flood, Folly, and the Politics of Thirst.* New York: Riverhead Books.

Chapter 19

Boubel, R., D. Fox, A. Stern, and B. Turner. 2006. *Fundamentals of Air Pollution.* New York: Elsevier.

Prinn, R. G. 2003. The cleansing capacity of the atmosphere. *Annual Review of Environment and Resources* 28:9–57.

United States Environmental Protection Agency. 1997. *The Benefits and Costs of the Clean Air Act, 1970 to 1990.* Washington, D.C.

Chapter 20

Aleklett, K., R. Cavaney, C. Flavin, R. K. Kaufmann, and V. Smil. 2006. Peak oil. *World Watch Magazine* 19, no. 1 (January/February).

Encyclopedia of Energy. 6 vols. Oxford, UK: Elsevier.

Hall, C. A. S., C. J. Cleveland, and R. K. Kaufmann. 1986. *Energy and Resource Quality: The Ecology of the Economic Process.* Niwot, CO: University of Colorado Press.

Chapter 21

Domenici, P. V., B. J. Lyons, and J. J. Steyn. 2004. *A Brighter Tomorrow: Fulfilling the Promise of Nuclear Energy.* Lanham, MD: Rowman & Littlefield.

Medvedev, Z. A. 1992. *The Legacy of Chernobyl.* New York: W. W. Norton.

Rothwell, G. 2004. Nuclear power economics. In C. J. Cleveland, ed. *Encyclopedia of Energy.* Oxford, UK: Elsevier.

Chapter 22

Board on Energy and Environmental Systems, National Academy of Sciences. 2001. *Energy Research at DOE: Was It Worth It? Energy Efficiency and Fossil Energy Research 1978 to 2000.* Washington, D.C.: National Academy Press.

Boyle, G., ed. 2004. *Renewable Energy,* 2nd ed. New York: Oxford University Press.

Smil, V. 2005. *Energy at the Crossroads: Global Perspectives and Uncertainties.* Cambridge, MA: MIT Press.

Chapter 23

Ackerman, F. 1997. *Why Do We Recycle? Markets, Values, and Public Policy.* Washington, D.C.: Island Press.

United States Interagency Working Group on Industrial Ecology, Material and Energy Flows. 1998. *Materials.* Washington, D.C.

Wagner, L. A. 2002. Materials in the economy: Material flows, scarcity, and the environment. *Geological Survey Circular* 1221.

Wernick, I. K. and F. H. Irwin. 2005. *Material Flow Accounts: A Tool for Making Environmental Policy.* Washington, D.C.: The World Resources Institute.

Chapter 24

Daly, H. 1996. *Beyond Growth: The Economics of Sustainable Development.* Boston: Beacon Press.

Kutner, R. 1997. *Everything for Sale: The Virtues and Limits of Markets.* New York: Alfred Knopf.

Liu, J. and J. Diamond. 2005. China's environment in a globalizing world. *Nature* 435:1179–1186.

Credits

PHOTOS

Design Elements

(Saplings, Windmills): © CORBIS RF; (Pond Cleanup): © Digital Vision/Punchstock; (Solar Panels, Foot Print): © Getty Royalty Free;(Recycling Newspapers): © Masterfile; Case Study Icons(tree): © Artville; Dollar: © Eyewire; Factory: © Digital Vision; Policy Icons (building): © Eyewire; (Waterfall): © Digital Vision.

Chapter 1

Opener, 1.2: © B.S.P.I./CORBIS; 1.8: Courtesy KARRENA GmbH, Germany; 1.11: © Dan Lamont/CORBIS; 1.12: Courtesy Forest Stewardship Council (FSC); Figure 1, p. 16: © Easter Island Statue Project/Jo Anne Van Tilburg

Chapter 2

Opener: Courtesy NASA/JPL-Caltech; 2.1: © Paolo Koch/ Photo Researchers; 2.7: Courtesy Edgar Fahs Smith Collection, University of Pennsylvania

Chapter 3

Opener: © Angus McIntyre 1987

Chapter 4

Opener: © AP/Wide World Photos/20th Century Fox

Chapter 5

Opener: © Photofest; 5.2(left): Courtesy Donald Blanchard, Colorado Herpetological Society; 5.2(right): © Creatas/PunchStock; 5.3: © Dennis Prager; 5.5a: © Wim van Egmond / Visuals Unlimited; 5.5b: © Gerald & Buff Corsi/Visuals Unlimited; 5.6a: © Brand X Pictures/PunchStock; 5.6b: © Image 100/PunchStock; 5.9a: © Brian P. Kenney/ Animals Animals/Earth Scenes; 5.9b: U.S. Fish and Wildlife Service; 5.10a: © Gerold and Cynthia Merker/Visuals Unlimited; 5.10b: © Joe McDonald/CORBIS

Chapter 6

Opener: © David M. Dennis/Animals Animals/ Earth Scenes; 6.14a: © Torsten Blackwood/ AFP/Getty Images; 6.14b: © Tim Graham/Getty Images; p. 118 fig. 2: Courtesy Ocean Climate Laboratory/US National Oceanographic Data Center; p. 119 fig. 3(top): © NASA/GSFC and ORBIMAGE; p. 119 fig. 3(bottom): Courtesy Rudolph Husar from *Journal of Geophysical Research*, vol. 102, No. D14, 16889-16909, 1997; p. 119 fig. 4(left): Courtesy Goddard Earth Science DISC, the SeaWiFS Project at NASA Goddard Space Flight Center, and Orbimage Inc.; p. 119 fig. 4(right): © Wim van Egmond/Visuals Unlimited

Chapter 7

Opener(a): © Rachel Kaufmann; Opener (b): © Dr. Ellen K. Rudolph; 7.2a: © Vol. 19/ PhotoDisc/Getty Images; 7.2b: © Arthur C. Gibson; 7.3b: NASA image courtesy of JPL AIRS Project and JPL Visualization and Scientific Animation Group; 7.5(top & middle): © Rachel Kaufmann; 7.5(bottom): © Eve Kaufmann; 7.9(Borreal): © Brand X Pictures/ PunchStock; 7.9(Nambia Desert): © Stephen P. Lynch; 7.9(Mojave Desert): © Comstock/ Punchstock; 7.9(Ecuador): © Steve Holt/Aigrette Photography; 7.9(Pine needles): © Comstock Images/Alamy; 7.9(Oak leaves): © Vol. 13/Getty Images; 7.9(Oak tree): © Vol. 1/PhotoDisc/Getty Images; 7.9(Rainforest): © Sergio Pitamitz/ CORBIS; 7.10(left): © Brand X Pictures/ Punchstock; 7.10(right): © Vol. 6/PhotoDisc/ Getty Images; 7.11(left): © Eve Kaufmann; 7.11(right): © Patryck Vaucoulon/Peter Arnold; 7.12(left): © age fotostock/SuperStock; 7.12(right): © Vol. 13/Digital Vision/Getty Images; 7.13(left): © Rachel Kaufmann; 7.13b(right): © Ann Manner/PhotoDisc/Getty Images; 7.14(left): © Royalty-Free/CORBIS; 7.14(right): © Wardene Weisser/Bruce Coleman; 7.15(left): © Vol. 6/PhotoDisc/Getty Images; 7.15(right): © Michael Bader/Peter Arnold; 7.16(left): © Digital Vision/PunchStock; 7.16(right): © Brand X Pictures/PunchStock; 7.17(left): © Stephen P. Lynch; 7.17(right): © Creatas/PunchStock; 7.18(left): © Vol. 74/ PhotoDisc/Getty Images; 7.18(right): © Jaakko Putkonen, University of Washington; 7.19a,b: Courtesy SeaWiFS; 7.20a: © MedioImages/ Alamy Images; 7.20b: © Rachel Kaufmann; 7.20c: © Susan Speak Abruzzi; 7.21(top & bottom): Courtesy Leif Olmanson, Remote Sensing and Geospatial Analysis Laboratory, University of Minnesota; 7.22(left): Courtesy SouthWest Florida Water Management District; 7.22(right): © The McGraw-Hill Companies/John A. Karachewski, photographer; 7.23a: © Jim Zipp/Photo Researchers; 7.23b: © Creatas/PunchStock; 7.24: © DV319 Digital Vision/Getty Images; 7.25: © David Wrobel/Visuals Unlimited

Chapter 8

Opener(both): © U.S. Geological Survey; 8.1: © Adam Durant; 8.7b: Courtesy William Humphrey at rainforestinn.com; 8.7a: © Courtesy Ian and Genevieve Giddy, Cloudbridge Nature Reserve (www.cloudbridge.org); 8.13a: Great Barrier Reef Marine Park Authority, Townsville Australia; 8.13b: Courtesy David Bellwood, James Cook University, Australia, photo by Shaun Wilson; 8.13c: Courtesy David Bellwood, James Cook University, Australia; 8.13d: Courtesy David Bellwood, James Cook University,

Australia, photo by Ken Anthony; 8.13e: Great Barrier Reef Marine Park Authority, Townsville Australia; 8.13f: Courtesy David Bellwood, James Cook University, Australia, photo by Terry Hughes; 8.16(left): Painting by H.H. Nichols; 8.16(right): Courtesy Ken Moore, Arizona Strip District, Bureau of Land Management, St. George, UT; p. 147(mayfly): © Susan Speak Abruzzi; p. 147(trout): © Tom & Pat Leeson/Photo Researchers; p. 147(algae on oar): Courtesy G. Winfield Fairchild, Department of Biology, West Chester University; p. 147(caddis): Wayne Davis, USEPA Office of Environmental Information http://www.epa.gov/bioindicators.

Chapter 9

Opener: © Sylvaine Achernar/The Image Bank/ Getty Images

Chapter 10

Opener: © Reuters/CORBIS; 10.4: © Dr. Parvinder Sethi; 10.8: © Brand X/SuperStock RF; 10.9: Courtesy Lynn Betts, USDA, Natural Resources Conservation Center; 10.13: UN/ DPI Photo, photographer E. Huffman; 10.17: Courtesy Tom Kunz; 10.18: Courtesy Jason Horn; 10.24: Courtesy Massachusets Water Resource Authority/MWRA/Kerwin-RVA

Chapter 11

Opener: Courtesy Karen Seto, Stanford University; 11.2a-d: Courtesy Cornell University Library's New York State Historical Literature Collection from Orsamus Turner, *A Pioneer History of the Holland Purchase of Western New York* (Buffalo, Jewett Thomas, 1849); 11.4: © AP/ Wide World Photos; 11.9: © Brian J. McMorrow; 11.11: © Régis Bossu/Sygma/Corbis; 11.14: Courtesy Bill Rowland, Landfill of North Iowa; 11.15: © Bill Aron/Photo Edit; 11.18: © Gavriel Jecan/CORBIS

Chapter 12

Opener: © Creatas/PunchStock; 12.3(Horned lizards): Courtesy Texas Parks and Wildlife Department, photographer Chase A. Fountain; 12.3(Collared lizards): Courtesy Donald Blanchard, Colorado Herpetological Society; 12.3(Gila Monsters): Courtesy U.S. Fish and Wildlife Service, Jeff Servoss, photographer; 12.3(Desert Iguana): Photo by Mark Dimmitt © 2006 Arizona-Sonora Desert Museum.; 12.6(Six Allopatric species photos): © Arthur Anker and Denis Poddoubtchenko, Smithsonian Tropical Research Institute; 12.9: © Ashley Douthirt; p. 263(both): © Brand X Pictures/PunchStock

Chapter 13

Opener: © Rudy Brueggemann; 13.19: © Dr. Paul A. Zahl/Photo Researchers

Chapter 14

Opener: © European Space Agency/SPL/Photo Researchers

Chapter 15

Opener, 15.11a,c: Courtesy of USDA/National Resource Conservation Service; 15.11b,15.15: Courtesy of USDA, Natural Resources Conservation Service, photo by Lynn Betts; 15.16,15.17: Courtesy of USDA, National Resource Conservation Service, photo Tim McCabe

Chapter 16

Opener: From Andrew M.T. Moore and A. Legge, *Village on the Euphrates; From Foraging to Farming at Abu Hureyra,* (2000), fig 8.52, p. 234 Oxford University Press. Reprinted by permission of Oxford University Press, Inc., New York; 16.6-1: © Vol. 44/PhotoDisc/Getty Images; 16.6-2, 3: © Vol. 19/PhotoLink/Getty Images; 16.6-4: © Pixtal/age fotostock; 16.6-5: © Royalty-Free/CORBIS; 16.6-6: © Vol. 19/PhotoLink/Getty Images; 16.6-7: © Digital Vision/Getty Images

Chapter 17

Opener: Courtesy Dr. Graham Jones, University of Leicester; 17.4: Courtesy David R. Foster; 17.7: © Jacques Jangoux/Photo Researchers; 17.9a-c: Courtesy United Nations Environment Programme; 17.14: USGS, photographer Ed Harp; 17.15: © Joel W. Rogers/CORBIS

Chapter 18

Opener: Courtesy of Dr. Juris Zarins, Missouri State University; 18.5(top): © Ray Manley/ SuperStock; 18.5(bottom): Courtesy Central Arizona Project; 18.11a-d: USGS Earth Resources Observation and Science; 18.13 From Figure 2 in Fact Sheet 165-00 "Approximate location of maximum subsidence in the United States identified by research efforts of Dr. Joseph F. Poland(pictured); 18.15a: Courtesy Arizona Game & Fish/USGS; 18.15b: Courtesy Desert Laboratory Repeat Photography Collection/ Raymond M. Turner/USGS, Stake 937; 18.19: © Christi Carter/Grant Heilman

Chapter 19

Opener: © Mike Clarke/AFP/Getty Images

Chapter 20

Opener(top): © CORBIS Royalty Free; Opener(bottom): © The McGraw-Hill Companies, Inc/Christopher Kerrigan, photographer

Chapter 21

Opener: © Creatas/PunchStock; 21.1: © Mary Evans Picture Library/The Image Works; 21.2: © Robert Francis/Getty Images; 21.5, 21.8: © AP/Wide World Photos; 21.10: Courtesy U.S. Department of Energy; 21.11: © ODD Andersen/ AFP/Getty Images; 21.13a: © Sandia National Laboratories; 21.14(both), 21.15: Courtesy U.S. Department of Energy; 21.17 © ZUFAROV/ AFP/Getty Images; 21.20 Courtesy EFDA-JET

Chapter 22

Opener, 22.5: DOE/National Renewable Energy Laboratory; 22.7: © Rachel Kaufmann; 22.9: *Courtesy Wind Resource Atlas of the United States* prepared by D.D. Elliott, et al., Pacific Northwest Laboratory, for U.S. Dept. of Energy; 22.10, 22.12, 22.13: DOE/National Renewable Energy Laboratory

Chapter 23

Opener: © Digital Vision/PunchStock; 23.3: Collection of the Oakland Museum of California, Oakland Museum of California Donors Acquisition fund; 23.4: Photo provided by W. R. Normak, USGS and taken on the East Pacific Rise expedition sponsored by the Scripps Institution of Oceanography; 23.7: © Vol. 39/PhotoDisc/Getty Images; 23.15(tl): © Vol. 31/PhotoDisc/Getty Images; 23.15(tr): © Digital Vision/PunchStock; 23.15(br): © Digital Vision/PunchStock; 23.15(bl): © Vol. 39/PhotoDisc/Getty Images; p. 500: © McGraw-Hill Companies, John Thoeming photographer

Chapter 24

Opener: Courtesy U.S. Fish and Wilidlife Service, Alaska

Index

An emboldened page number signifies a definition. Figures and tables, indicated respectively with an f or t following a page number, are cited only when they appear outside the related text discussion.

A

A horizon, 313f, **314**, 315, 316f
Aberdeen PAYT program, 505f
absolute ownership, rule of, **393**
absolute water scarcity, **384**, 385f
absorption, **102**
absorptive capacity, 359
Abu Hureyra, 334
Acapulco climate graph, 137f
acid deposition/rain, 5, 9, **403**–5
acidity, 320
active range, **82**
adaptation(s), **124**
    environmental gradients and, 127–30
    leaf shape as, 125, 132f–133f
    niches and, 124–25, 127
    size as, 145–46, 147
Adelaide climate graph, 140f
adiabatic lapse rate, **410**
adult literacy rates in Canada *vs.*
        Kenya, 10t
advanced loss foam casting energy efficiency,
        483t
Advanced Very High Resolution Radiometer
        (AVHRR), 131
advection, **63**, **410**
aerobic respiration, **88**–89
aerosols, **114**, **270**, **400**
    as air pollutants, 400
    phosphorus cycle and, 114
    solar energy and, 270
    sulfate (*see* sulfate aerosols)
affluence, **210**
    automobile and, 226
    deforestation and, 358
    economic growth and, 210
    environmental impacts of, 231–33, 236–37
    international variations in, 211f
    materials use and, 213–14
    rise in, 210
afforestation, 356, **357**, 369
Africa. *See also specific country*
    land use determinants in, 126
    oil supplies by country in, 439f
African elephants, 262–63
age classes, **189**–91
age structure, **189**–91
age-dependency ratio, **190**–91
Agenda 21, 14
agricultural pollution, 391–92
    atmospheric emissions, 485t
    feedlots and, 236
    nitrogenous waste, 112, 120–21, 345–46,
        348
    nonpoint, 391
    phosphorus waste, 120–21, 345, 516
    policy, profit, and reduction of, 349
    pollinators and, 256
    wastewater and, 392
agricultural residues, **473**
agriculture, **194**, 333–52
    affluence and impact of, 231–32
    African determinants of, 126
    arsenic in, 491
    biodiversity insurance and, 255
    in Brazilian Amazon basin, 359
    carbon cycle and, 194
    carrying capacity and, 194
    climate change and, 287–88, 334
    coffee production, 265
    deforestation and, 228–29, 357–58, 367
    dietary land requirements footprint,
        346–47
    dust bowl, 312, 331
    ecology of, 338–39
    economics of, 338, 339–41, 344–48
    environmental Kuznets curve
        and, 520
    ethanol production and, 473–74

Everglades National Park and, 176
future of, 350–52
Green Revolution in, 341–48 (*see also* Green
        Revolution)
Mabira Reserve and, 264, 266
material inputs reduction in, 351–52
natural ecosystems and, 335
nitrogen cycle and, 111f, 112, 120–21
Norse and, 269
origins of, 335–37
oxisols and, 315, 317
pollution from (*see* agricultural
        pollution)
population growth and, 228–29, 336–37, 357
production increases, 350–51
residues of, 473
soil conservation programs, 330, 331
soil conservation techniques, 325–28
soil erosion and, 320, 324–25,
        327–29, 330
soil erosion calculations, 323
soil formation and type and, 312–13,
        315, 317
subsidies and, 329, 330, 367, 368
water table drops and, 387
water use in, 380–81, 382–83
agriculturists, **335**
agroecosystems, 338–**39**
    Chinese meat consumption and, 518
    Green Revolution and, 344
    temperature and, 365
air. *See* atmosphere
air pollutants, **399**
air pollution
    abatement costs, 414–15
    acid deposition, 5, 9, 403–5
    atmospheric stability and, 410–11, 412f
    carbon monoxide causing, 400 (*see also*
        carbon monoxide
        emissions)
    clean air benefits, 415–17 (*see also* Clean
        Air Act)
    coal benefaction and, 427
    electricity generating technologies and,
        484, 485t
    emissions footprint for, 408–9
    as externality, 412
    geothermal energy and, 476–77
    global climate change and, 402–3
    in Guangzhou, China, 399
    hydrocarbons causing, 407
    measurements of, 408–9
    nitrogen oxides causing, 405–7
    observed concentrations of, 411–12, 413f
    optimal level of, 414–17
    particulate matter causing, 400–401
    personal transportation and, 226
    pollutant concentrations in, 407–12
    primary *vs.* secondary, 399–400
    rate and impact of, 8
    reduction of, 411–15
    sulfur dioxide causing, 402–5 (*see also* sulfur
        dioxide emissions)
    vertical and horizontal mixing and, 410
air pressure gradients, 61–62
air quality
    clean air benefits, 415–17
    EPA index of, 408
    standards of, 411–12, 415
air travel, 227, 232f
Alaska
    oil exploration in (*see* Arctic National
        Wildlife Refuge oil exploration)
    pipeline system, 510
    tundra climate, 145f
Alaskan tussock, 124, 125f
albedo, **57**
alfisols, 315f
algae
    aquatic food chain disturbances and,
        170, 171f

coral reef communities of, 168, 289
    keystone fish species and, 254
alien species, **258**–59
alkaline battery footprints, 29, 500
allelopathy, **164**
allopatric speciation, **250**–52
alpha radiation, 26
aluminum, 12t, 507t
Amazon rain forest
    agricultural development in, 359
    climate in, 136f
    deforestation in, 356, 359, 361–62
    fires in, 360
    habitat conversion in, 257–58
    soybean production in, 359, 361
    transportation infrastructure in, 361–62
American bald eagle, 261
amino acids, 102
ammonia, 111f, 112, 502t
amphibians
    carrying capacity and, 183–87
    extinction of, 288
    Spadefoot toad, 140
    UV-B radiation and, 303–4
anaerobic respiration, **88**
andisols, 315f
animals. *See also specific animal*
    benthic, 150, 151
    domestication of, 340–41
    draft animals, 211, 213f
    in feedlots, 235–36, 343
    as heterotrophs, 77, 102
annuals, **159**
anoxic, **112**
Antarctica
    ozone reduction above, 294, 297, 299–302
    UV-B radiation effects on, 303
anthropocentric society, 240–41, **254**
antibiotic resistance, 351
Antigo PAYT program, 505f
Appalachian disturbances, 158
appliances, 482, 483t, 517
Aquas del Tunari, 395
aquatic biomes, 146–54. *See also specific biome*
    coral reefs, 151–52
    estuaries, 149
    lakes, 148–49
    limiting factors in, 146
    oil and natural gas formation in, 425
    open ocean, 152–54
    rivers and streams, 147–48
    temperate coastal seas, 150–51
    tropical coastal seas, 151–52
    upwellings, 152, 153f
aquatic ecosystems
    acid deposition and pH of, 405
    food chain disturbances in, 170–71
    succession in, 159, 164–65
aquifers, **374**, 378f
    confined *vs.* unconfined, 376
    Ogallala Aquifer, 375, 386f
Aral Sea, 385
Arco, Idaho, 445f
arctic foxes, 145
Arctic National Wildlife Refuge oil
        exploration
    assurance bonding and, 522
    cultural values and, 240, 241f
    good guy-bad guy dichotomy and, 510–11
    production scenarios, 436–37, 438f
Argentinean nuclear power, 446f
Argonne National Laboratory ethanol research,
        474
aridsols, 315f, 316f, **317**
Aristotle, 28
Arizona
    Colorado River water allotment for, 385
    Mt. Lemmon precipitation gradients, 128, 129f
    overdraft effects in, 388, 389f
    uranium reserves in, 445t
    Yuma climate graph, 139f

Armenian nuclear power, 446f
arsenic, 491, 502t
artesian wells, **376**, 378f
Asian oil supplies by country, 439f
assurance bonding, **522**
asthenosphere, **71**, 72f
asymmetric information, **238**
atmosphere
    of Biosphere, 247
    components of, 294
    ecosystem service of gas regulation
        in, 207t
    greenhouse gases in (*see* greenhouse gases)
    hydrologic cycle and, 373f, 374
    layers of, 294–95
    pollutants in (*see* air pollution)
    solar radiation and, 56–57, 270–71
atmospheric circulation, 59–64
    circulation cells, 61–62
    climate change and water supply and,
        378–79
    climate change simulations of, 280, 281f,
        286
    El Niño and, 68, 70
    environmental services of, 63–64
    gradients in, 59–62
    oceanic circulation and, 65–66
    ozone reduction and, 300, 301
    ozone reduction-climate change link
        and, 308
    precipitation patterns and, 62–63
    surface winds, 62
atmospheric emissions, **501**. *See also* green-
        house gases; tradable permits; *specific
        emission*
atmospheric storage
    of carbon, 106–9
    of nitrogen, 110–11, 112
    of phosphorus, 113f, 114
    of sulfur, 116f, 117
Atomic Energy Commission (U.S.), 445
atomic mass, **24**
atomic number, **24**
atomic radiation, 25–26
atoms, **24**–25
Atoms for Peace program, 445
attitudes, 240–41
attribution, 280–82
Auburn PAYT program, 505f
Aurora PAYT program, 505f
Australia
    climate graphs for, 137f, 138f, 140f
    Great Barrier Reef, 151, 167f
    Kyoto Protocol and, 291
    oil supplies of, 439f
automobiles. *See* gasoline consumption; motor
        vehicles
autotrophs, **77**. *See also* plants
    food chains and webs and, 89, 90f
    nutrient capture by, 100–101
available, **100**
Avanca Brazil, 362
AVHRR (Advanced Very High Resolution
        Radiometer), 131
avoided costs, **215**–16, 217f

B

B horizon, 313f, **314**, 316f
Babbitt, Bruce, 385
baby boomers, **190**–91
Bacillus thuringiensis (Bt), 351
background extinction, **256**
bacteria
    fecal coliform count, 389
    nitrogen-fixing, symbiosis and, 111–12
    UV-B radiation and, 303
badgers, 260
Baja California fires, 174–75

bald eagle, 261
banana plant leaves, 125
banking system
    debt for nature swaps and, 370–71
    oil price increases and, 359
    World Bank, 262, 395
basal metabolic rate, **78**, 80
bass, 254
bats
    avoided costs and, 216, 217f
    little brown bat reproductive
        range, 128
    wind energy and, 472
battery footprints, 29, 500
Becquerel, Edmond, 468
Becquerel, Henry, 25
beef
    feedlots and production of, 235–36, 343
    U.S. *vs.* Indian consumption of, 12t
beetles, 253, 348
Belem climate graph, 136f
Belgian nuclear power, 446f
beliefs, 240–41
bent grass, 159, 161
benthic animals, **150**, 151
Berra, Yogi, 420
best available technology, **396**
best first principle, **7–8**
    oil exploration and, 431–32
    opportunity costs and, 218–19, 220f
best practicable control technology, **396**
beta radiation, 26
beverages, 135t, 347t
Beyond the Limits, 198
bicycling, 226–27, 232f
big bluestem grass, 159, 160f, 161, 164
Bingham Canyon copper mine, 205f
biocentric, **241**, 254
biochemical oxygen demand, **390**
biochemicals, **473**
biodegradable wastes, **209**
biodiesel, **473**
biodiversity, **247**
    alien species and, 258–59
    bioenergy and, 475
    biogeochemical cycle changes and, 259–60
    Biosphere experiment in, 247
    coffee production and, 265
    decline of, 256–60
    ecosystem function and, 254
    environmental services and, 256
    extinction rate and, 256–57
    forests and, 363
    genetic knowledge and, 255–56
    habitat alteration and, 257–58
    hunting and harvesting and, 260
    importance of, 254–56
    as insurance, 255
    market forces and, 264–66
    measurements of, 247–49
    patterns and mechanisms for, 249–53
    preservation of (*see* biodiversity preserva-
        tion)
    of species, 254 (*see also* species diversity;
        specific species)
    use *vs.* replenishment of, 6t
biodiversity hot spot, **248**, 250–51
biodiversity preservation
    corruption and, 262–63
    criteria for, 260–61
    legal protections for, 261
    market-based incentives and, 264, 266
    phosphorus and, 260
    species and habitat in, 261–63, 264f
bioenergy, **472–73**
    biomass resource base of, 473
    biomass technology, 473–75
    biopower land, water, and solid waste
        impacts, 484t
    biopower tax credits, 484–85
    current status of, 473
    environmental impacts of, 475, 485t
    ethanol energy balance, 474
    U.S. consumption of, 31f, 466t
    U.S. flow of, 466t
    U.S. potential for electricity generation
        by, 466
biofuels, **473**
biogeochemical cycles, **102–21**
    biodiversity and changes in, 259–60
    carbon cycle, 104–9
    climate change and, 274–76
    human activity and, 117, 120–21
    interactions among, 117
    nitrogen cycle, 110–13
    phosphorus cycle, 113–15
    policy and, 118–19
    rules governing, 102–4
    sulfur cycle, 115–17
    sustainability and, 120, 521–22
    UV-B radiation and, 303

biological activity, 207t, 314. *See also specific
    activity*
biomagnification, 94–97, **95**
biomass, **89**, **472**
    maintenance respiration and, 91
    as material waste, 501
    nitrogen and phosphorus and, 259
    as renewable fuel (*see* bioenergy)
    species richness and, 259f
biomass feedstocks, **472f**, **473**
biome factor in soil erosion calculations, 322–23
biomes, **124**. *See also specific biome*
    aquatic, 146–54
    boreal forests, 143–44
    climate change and, 277, 280, 286–87
    coastal seas, 150–52
    deserts, 139–40
    environmental gradients and, 127–30
    estuaries, 149
    footprint on, 134, 135t
    habitats and niches in, 124, 125f
    human land use determinants in, 126
    lakes, 148–49
    Mediterranean, 140–41
    net primary production in, 172
    niche and adaptation in, 124–25, 127
    open ocean, 152–54
    rivers and streams, 147–48
    satellite remote sensing of, 131, 277, 280,
        356
    temperate forests, 142–43
    temperate grasslands, 141–42
    terrestrial, 130–46
    tropical dry forests, 136–37
    tropical rain forests, 134–36
    tropical savannas, 138–39
    tundra, 144–46
    upwellings, 152, 153f
biopower, **475**. *See also* bioenergy
Biosphere experiment, 247
biota, **102**
biotic storage
    of carbon, 104, 105–6
    of nitrogen, 111–12
    of phosphorus, 113–14
    of sulfur, 115–16
Birdman Cult, 4
birds
    biomagnification in, 94–95, 97
    Endangered Species Act and, 261
    energy use for reproduction in, 85
    latitude and species richness of, 249
    wind energy and, 472
birthrate, **179**, 188–89
black market, 262, **266**
black rhinoceroses, 262–63
Black Sunday, 312
blue baby syndrome, 112, 348
blue whales, 94
bluestem grasses, 159, 160f, 161, 164
body mass index (BMI) formula, 80
body temperature, 78, 79
Bolivian water privatization, 395
Bombay climate graph, 137f
bonds, assurance bonding, **522**
boreal forests
    as biomes, 143–44
    carbon storage in, 365
    climatic determinants in, 132t, 133t
    deforestation in, 355f
    net primary production in, 172t
Borlaug, Norman, 341
Bosch, Carl, 112
Boston Harbor, 220, 221f
brackish water, 379–80
Brazil
    agricultural development in, 359
    Belem climate graph, 136f
    deforestation in, 356, 359, 360, 361–62
    land policy in, 360
    nuclear power in, 446f
    policy responses to El Niño in, 70
    soybean production in, 359, 361
    transpiration infrastructure in, 361–62
Brazilian freetailed bats, 216, 217f
breeder reactors, **450**
Britain
    badger control in, 260
    deforestation of 1600s in, 200, 355
    heather field succession in, 159, 160
    nuclear accident in, 455t
    nuclear power in, 446f, 449
broadleaf forests, 133f
bromine, 305, 308
Bt (*Bacillus thuringiensis*), 351
Bulgarian nuclear power, 446f
Bureau of Economic Analysis, 222
burns from sun exposure, 306
bus travel, 226–27, 232f
Bush, George H., 483

Bush, George W.
    Kyoto Protocol and, 291, 519
    renewable energy and, 483
    Yucca Mountain Project and, 453

C

C horizon, 313f, **314**
cabbage root fly, 352
cacti, 132f, 140
caddisfly, 170, 171f
Cadillacs, 420
cadmium, 27–28
CAFE (corporate average fuel efficiency) stan-
    dards, 244
calcium, 23t
calcium carbonate, 405
California
    fires and fire suppression policy
        in, 174–75
    geothermal energy in, 476
    Headwaters Forest, 367
    PAYT programs in, 505f
    San Diego climate graph, 140f
    toxic waste exposure and income in, 233
    water allocation in, 379, 384–85, 393
Canada
    Chalk River nuclear accidents, 455t
    Dawson climate graph, 144f
    human development in, 10
    James Bay Project in, 480
    nuclear power in, 446f
cancer, 455, 457
cap and trade system, **242**, 243. *See also* trad-
    able permits
capillary water, **318**
capital, **203**
capital equipment
    energy efficiency of, carbon emissions
        and, 276
    energy efficiency of, policy and, 282, 290
    labor productivity and, 211–12,
        213f, 213t
carbohydrates, **100**
carbon
    climate change and, 274–76, 365
    climate change policy and, 118–19
    in coal, 425
    emissions of (*see* carbon dioxide emissions;
        carbon monoxide emissions)
    forests and, 117, 275, 365, 370
    net primary production use and, 106–7
    as nutrient, 22, 23t
    in oil and natural gas, 425
    radiocarbon dating, 26
carbon cycle, 104–9
    atmospheric storage in, 106–9
    bioenergy and, 475
    biotic storage in, 104, 105–6, 117, 365
    footprint in, 106–7
    fossil fuels in, 105, 275, 276
    human activity and, 105–6, 274–76
    ocean storage in, 106–8
    soil storage in, 104–5
    unknown carbon sink in, 108–9
carbon dioxide
    atmospheric concentrations of, 273, 275f, 276
    in carbon cycle, 104–9
    carbon monoxide and, 400
    characteristics of, 271t
    as greenhouse gas, 272
carbon dioxide emissions
    allowances/permits for, 290, 291, 521–22
    in Biosphere, 247
    climate change and water supply and,
        378–79
    by electricity generating technologies,
        462t, 485t
    energy efficiency benefits and, 483
    food supplies and, 287–88
    footprint of, 278, 279t
    forecast for, 275, 278, 279t, 477
    fossil fuel rates of, 275, 278, 279t, 477
    GDP and, 521
    geothermal rate of, 477
    Kyoto Protocol requirements for, 291
    land required to remove, calculating, 513
    as material waste, 501
    Mauna Loa Curve and, 110
    metal recycling and, 507t
    taxing of, 512, 514, 515
    temperature and, 275f, 285
    U.S. policy on reduction of, 519
carbon monoxide emissions
    by electricity generating technologies, 485t
    nitrogen oxides and, 405
    reduction of, 413f
    sources and health impact of, 400
carbonation and solution, **313**

carcinogens, **407**
Caribbean coral reefs, 167
carnivores, **89**
carpool, 227, 232f
carrier beds, 425
carrot fly, 352
carrying capacity, **179**
    feedback loop and fluctuations around,
        185–87
    fisheries and, 182
    given area and, 183–84
    maintained indefinitely, 184–85
    maximum number of individuals and,
        183, 184f
    Plimsoll line and, 511–12, 522
carrying capacity for humans
    debates over, identifying limits in, 196–97
    Easter Island and, 179
    ingenuity and, 198–201
    maintained indefinitely, 194, 196
    Malthusian theory and, 191–92
    maximum number and demographic tran-
        sition and, 192, 193f
    resource optimists and, 198–99
    resource pessimists and, 197–98
cars. *See* gasoline consumption; motor vehicles
Carson, Rachel, 352
Carter, Jimmy, 449, 483
catalytic converters, 415
cataracts, 307
catclaw acacia, 129f
caterpillars, 351
cation exchange capacity, **319–20**
cations, **319**
cattle ranches, 370
CCA (chromated copper arsenate), 491
Cedar Creek, Minnesota
    diversity and stability along, 166
    succession along, 159, 161, 163–64
center-of-accumulation hypothesis, 250
center-of-origin hypothesis, 250, 251
Central America
    deforestation in, 358, 360
    extinctions in, 288
Central Arizona Project, 379, 385
centrally planned economies, **238**
cereal yields, 343, 350
cereals, **335**
certification of sustainable forestry,
    13f, 369
CFCs. *See* chlorofluorocarbons
Chad, Faya Largeau climate graph, 139f
Chalk River nuclear accidents, 455t
channels, 149
chaparral, fire suppression and, 174–75
charcoal, 363
chemical changes, **25**
chemical elements of periodic table, 22f
chemical energy, **30**
chemical industry
    CFCs and, 294, 304, 305, 309–10
    ozone depletion studies and, 300
chemical oxygen demand, **390**
chemical reactions, **25**, 73
chemicals as dissipative wastes, 502t
chemicals, protective
    crop yields and, 346–48
    prey and, 86–88
Chernobyl Power Complex accident, 446, 455t,
    457–58, 459
children
    demographic transition and, 192, 193f,
        518–19
    impacts of having, 225
    lead poisoning in, 20, 21
    nuclear accidents and, 455, 457, 458
    waterborne disease in, 390
China
    air pollution in, 399
    demographic transition in, 518–19
    economic growth in, 517–19
    energy consumption by, 423
    Kyoto Protocol and, 291, 306
    meat consumption in, 518
    Montreal Protocol and, 306
    nuclear power in, 446f
    oil demand in, 517
    Taiyuan climate graph, 141f
    total fertility rate in, 189f
chlorine
    dissipative use of, 502t
    as nutrient, 23t
    ozone reduction and, 298–99, 301
    photodissociation of, 298–99
    stratospheric concentration of,
        305, 308
    technical breakthroughs and, 309
chlorofluorocarbons (CFCs)
    atmospheric concentrations of, 307
    characteristics of, 271t
    design and properties of, 298

INDEX   **541**

chlorofluorocarbons (CFCs) *(continued)*
  energy absorption by, 271–72
  leakage of, 199
  Montreal Protocol and, 305–7
  ozone reduction and, 294, 298–99, 309
  ozone reductionclimate change link and, 308
  policy debates over, 304–5
  technical redesign and replacement of, 309–10
chromated copper arsenate (CCA), 491
circular flow model, 204
circulation cells, 61–62
CITES (Convention on International Trade in Endangered Species), 260, 261
Clamshell Alliance, 446
Clapp, Nicholas, 373
clays, **318**, 320
Clean Air Act (U.S.)
  benefits of, 415–17
  command and control strategies of, 414–15
  economics of, 414
  emissions allowances/permits and, 415, 416–17
  industrial response to, 9, 415
  lead and, 21
  standards of, 412, 414
clean development mechanism, **290**
clean water. *See* sustainable clean water supply
Clean Water Act (U.S.), 21, 395–96
clear-cutting, **369**
climate, **269**
  biodiversity and, 256
  changes in (*see* climate change)
  ecosystem service of regulation of, 207t
  forests and, 363, 365
  human land use and, 126
  soil and, 314, 315, 316, 317
  of terrestrial biomes, 130, 132f, 133f (*see also specific biome*)
  weather *versus*, 269
climate change, **269**
  agriculture and, 287–88, 334
  air pollution and, 401–3, 404f
  analyses and modeling of, 280–86, 378–79
  attribution of, 280–82
  biogeochemical cycles in, 118–19, 274–76, 365
  biomes and, 286–87
  carbon dioxide emissions footprint and, 278, 279t
  detection of, 277, 280
  El Niño and (*see* El Niño)
  extinction and, 288–89
  food supplies and, 287–88
  future of, 284–86
  greenhouse effect and, 271–73
  greenhouse gas concentrations in, 275f (*see also* greenhouse gases)
  heat balance in, 270–73
  impacts of, 286–89
  IPCC and, 53, 283, 288, 290
  Kyoto Protocol and, 241, 290–91, 306, 309
  Norse settlements and, 269
  ozone levels and, 308, 309
  policy and, 118–19, 286, 289–91
  radiative forcing and human activity in, 273–76
  sea level and, 277, 288
  solar energy and, 270–73
  temperature in (*see* temperature change)
  unknown carbon sink and, 108
  U.S. water supply and, 378–79
  UV-B radiation and, 304
  vacillations in, 269–70
  weekend effect and, 401–3, 404f
climate envelope, **288–89**
climax community, **159**
Clinton, Bill, 483
clouds
  polar stratospheric, 301, 308
  solar radiation reflected by, 56–57
coal
  as conventional fossil fuel, 420
  discovery, extraction, and processing of, 426–27
  energy content of, 422t
  energy intensities for nonenergy goods and services produced with, 424t
  energy released by, 460t
  as extractive waste, 499, 501f
  footprint of use of, 422–23
  formation of, 424–25
  global consumption of, 12t, 422, 423
  iron production using, 200
  past and present use of, 420
  properties of, 420, 421
  sulfur cycle and, 117
  types of, 425
  U.S. consumption of, 12t, 31f, 420f, 466t
  U.S. storage of, 466t

coal benefaction, **427**
coal-fired power plants
  carbon emissions from, 275, 462f, 485t
  Clean Air Act and, 415
  energy and materials balance of, 35–36
  land, water, and solid waste impacts of, 484t
  methane emissions from, 485t
  mine-mouth plants, 426–27
  nitrogen oxide emissions from, 408t, 409t, 485t
  operating costs for, 460
  particulate emissions from, 485t
  sulfur emissions from, 275, 402, 404, 408t, 409t, 415, 485t
  taxing emissions of, 515
  U.S. electricity generated from, 423t
coarse fraction, **401**
coarse grains, **343**
coastal seas, 150–52, 153f
Cochabomba water privatization, 395
cod fisheries, 186–87
coevolutionary hypothesis, **336**
coffee production, 265
coin flips, 39, 40f
coliform bacteria, 389
Colorado River
  surface runoff and damming of, 376, 377f
  water diversion from, 379, 384–85
Colorado uranium reserves, 445t
combustion, **25**
command and control strategies, **242**
  of Clean Air Act, 414–15
  gasoline consumption reduction and, 244
  market-based mechanisms *versus*, 242–43, 514–15
  soil erosion and, 331
  U.S. legislation based on, 242
commercial energy end use, 481t, 482
common property, **215**
community, **124**
  climax community, 159
  pioneer community, 160
competition, **163–64**
  agroecosystems and, 339
  in global oil market, 437–41
competitive exclusion principle, **124**
complexity in systems, 47–48
compounds, **22**
Comprehensive Everglades Restoration Plan, 176
computer modeling, 52
  of climate change, 280–86, 378–79
concentrators (solar), 467–68
condense, **63**
condenses, 374
cone of depression, **377**, 378f
confined aquifers, 376
conflict, 368, 392
Connecticut
  New Haven harbor siltation, 365
  Stonington PAYT program, 505f
conservation
  of biodiversity (*see* biodiversity preservation)
  coffee production and, 265
  debt for nature swaps and, 370–71
  of energy, 244, 435
  of matter, 27–28, 35–36
  in situ and ex situ, 261
  of soil, 325–29, 330, 331
Conservation International, 371
Conservation Principles for Coffee Production, 265
Conservation Reserve Program, 330
consolidation, **72**
consumption, **204, 380**
containment, **450**
Continental Divide, **375**
continental drift, 69, 71, 72f
continents
  biodiversity and, 253
  deforestation across, 355, 356f
  freshwater availability across, 376
contingent valuation, **216**
continuous mining techniques, **426**
contour plowing, 325–26
control rods, 449
convection, **410**
convection cells, **59**, 61
Convention Concerning the Protection of the World Cultural and Natural Heritage, 262
Convention on International Trade in Endangered Species of Wild Fauna and Flora, 260, 261
Convention on Wetlands of International Importance, 261
conventional fossil fuels, **420**
convergent evolution, **124**

conversion, in nuclear fuel cycle, **447**
coolants (nuclear), 449, 450
Coon Creek soil erosion, 325
copper
  concentrations and mining of, 205
  energy costs for, 498
  ore-to-metal process of, 488–89
  recycling benefits for, 507t
  resource quality and opportunity costs and, 217–19, 220f
  U.S. *vs.* Indian consumption of, 12t
copper sulfate, 502t
coral reefs
  alternative states of, 168f
  as biomes, 151–52
  of Caribbean *vs.* Great Barrier reef, 167
  climate envelope of, 289
  fish biodiversity and, 250–51
  redundancy and, 167–68
  restoration of, 171
Coriolis effect, **62**
corn
  declining prices for, 344
  ethanol production and, 474
  trade footprint for, 134, 135t
  water table and, 387
corporate average fuel efficiency (CAFE) standards, 244
correlative rights doctrine, **393**
corruption, 262–63, 368
Costa Rica
  forestry practices in, 370
  genetic information rights in, 266
  soil erosion costs in, 325
  total fertility rate in, 189f
cotton
  genetically modified strains of, 351
  natural pest control for, 216, 217f
cottony cushion scale, 348
cradle-to-grave analysis, **488**
credit, 329, 370
Cree Indians, 480
creep, 321f, **322**
creosote bush, 125, 128, 129f
Crichton, Michael, 283
crop residues, **324**, 325
crop yield potential, **350**
crop yields
  energy and, 341–42
  future of, 350–52
  genetically modified crops and, 351
  herbicides/pesticides and, 346–48
  increases in, 343–44
  pollution reduction and, 349
crude birthrate, 188–89
crude death rate, 188–89, 192, 193f
crude oil, 425. *See also* oil
crude steel, 12t
crust
  creation and destruction of, 71, 72f
  elements in, 23, 205
  phosphorus storage in, 113
  plate tectonics and, 69, 71, 72f
  rock cycle and, 71–73
crustal abundance, **205, 493**
Crutzen, Paul, 294
cultivars, **341**, 351
cultural values, 226, 240–41
Curie, Marie, 25
cycling, 226–27, 232f
Czech Republic nuclear power, 446f

D

Daly, Herman, 511
damage functions for climate change, 289
dams
  on Colorado River, 377f
  environmental impact of, 480
  power generation by, 384, 479
  water storage by, 376
Darby, Abraham, 200
Darien PAYT program, 505f
Darwin, Charles, 40
Darwin climate graph, 137f
Dawson climate graph, 144f
Daza, Victor Hugo, 395
DDE, 95–96, 97
DDT, 95, 97, 348
death rate, **179**
  crude death rate, 188–89, 192, 193f
deaths
  from Chernobyl nuclear accident, 458
  human patterns of, 189
debt for nature swaps, **370–71**
decay, 313
  carbon cycle and, 105
  nuclear, 26, 27t

deciduousness, **143**
decommissioning of nuclear fuel plant, **454**, 461
decomposers, **91**
decomposition, **313**
decontamination of nuclear power plants, 454
deer, 84–85, 186
deforestation, **106, 355**
  agriculture and, 228–29, 357–58
  in Britain, 200, 355
  carbon cycle and, 105f, 106
  causes of, 357–62
  debt for nature swaps and, 370–71
  economic wellbeing and, 355, 367, 368
  environmental service pricing and, 368
  forests fires and, 358, 360
  mineral and energy industries and, 360–61
  oil prices and, 359
  policy and, 360, 362, 367–68
  property rights and, 360
  rate of, 355–56, 367–68
  slowing of, 368–71
  social structure and, 368
  sustainable forestry and, 368–69
  timber harvests and, 358, 360, 367–68
  transportation and, 361–62
  waterway siltation and, 365–66
degradation, **7, 216**. *See also* environmental degradation; *specific form of degradation*
dematerialization, **495**, 497f
demographic transition, **192**, 193f, 518–19
denitrification, **112**, 345, 348
Denmark wind energy, 472
Denno, Robert, 169
density-dependent factors, **180–81**
density-independent factors, **180**
deoxyribonucleic acid (DNA), 247, 302
Department of Energy (DOE)
  energy efficiency R&D costs of, 483
  NREL energy research in, 484
  Nuclear Waste Policy Act and, 449
  Yucca Mountain Project and, 452–53
dependent variable, **42**
depletion, **217**. *See also* resource depletion and degradation
depletion allowances, 516
depression, cone of, **377**, 378f
Derbyshire, England, 355
desalinization, **379–80**
desert iguana range, 82
deserts
  as biomes, 139–40
  climatic determinants in, 132f
  food chain/web from, 92f
  net primary production in, 172t
  precipitation patterns and, 63
detection, **277**
detritivores, **91**
detritus, **91**
deuterium, 460
developed nations
  agriculture in, 329, 340f, 350
  climate change policy and, 290
  debt for nature swaps and, 370–71
  deforestation in, 357
  demographic transition in, 192, 193f
  human development inequity and, 9–12
  labor productivity in, 210–11
  net primary production and, 194
  oil market and, 359
developing nations
  agriculture in, 329, 340f, 343, 350, 351–52
  air pollution in, 399
  bioenergy in, 473
  climate change policy and, 290
  debt for nature swaps and, 370–71
  deforestation in, 357
  demographic transition in, 192, 193f, 518–19
  energy end use in, 482
  food subsidies and soil conservation in, 329
  habitat conversion in, 257
  human development inequity and, 9–12
  Kyoto Protocol and, 291
  labor productivity in, 211
  oil market and, 359
  transportation and, 482
  wood use in, 363
diatoms, 119
diet
  energy pyramids for, 94, 95f
  environmental impact of, 231–32
  food chains/webs (*see* food chains; food webs)
  land requirements footprint for, 346–47
diffuse chemical coevolution, **255**
digestion, **102**
dinosaurs, 76, 93, 94f
direct market valuation, **215**

discharge, 375, **380**
discount rate, 324, 329
discoveries, **429**
disease/illness
  agricultural development and, 337
  air quality and, 408, 416–17
  arsenic and, 491
  biodiversity and medicines for, 255–56, 266
  carbon monoxide poisoning, 400
  genetically modified crops and, 351
  hydrocarbons and, 407
  industrial pollutants and, 391
  lead and, 20, 21
  monoculture and, 346–48
  nuclear accidents and, 455, 457, 458
  particulate matter and, 401
  photochemical smog and, 407
  poverty and environmental risk for, 233
  sulfur emissions and, 121, 402
  timber harvests and, 369
  waterborne, sewage treatment and, 388–90
disintegration, **313**
disodium carbonate, 502t
dispersal, **250**
  climate change and, 280, 286–87
  of reef fish, 250–51
dissipation, **501**
dissipative wastes, **208,** 501, 502t
dissolution, carbon cycle and, 105f, 108
dissolved oxygen deficit, **390,** 391f
distance effects, 50
distortionary taxes, 515
disturbances, **157**
  agents, effect, and role of, 157–59
  agricultural, 339
  in aquatic food chains, 170–71
  catastrophic ecosystem shifts and, 164–65
  ecosystem health and, 165–76
  ecosystem service of regulation of, 207t
  fire suppression policy and, 174–75
  humans causing, 165–76, 257–58
  succession following (*see* succession)
diversity, **247**–48. *See also specific type of diversity*
DNA (deoxyribonucleic acid), 247, 302
Dobson, G. M. B., 296
Dobson unit (DU), **296**
DOE. *See* Department of Energy
domestic sewage, 388–91
domestication, **335**–36, 340–41
dose-response function, **414**
doubling time, **188**
draft animals, **211,** 213f
drainage basin, **375**
drilling of mineral deposits, 494
drip irrigation, **394**
drug companies, 255–56, 266
dry forests, tropical, 132f, 136–37
dry shrublands, 132f
DU (Dobson unit), **296**
Dupont, 304
dust, **400**
Dust Bowl, 312, 331
dynamic uplift hypothesis of ozone depletion, 300–301
Dzamiin Uuded climate graph, 139f

E

$E = mc^2$, 56
E horizon, 313f, **314**
eagle, 261
Earth Summit, 14
earthquakes, 68
Earth's interior, 57, 71
Earth's surface. *See also* crust
  in climate change simulations, 281f
  interior heat and, 57
  radiant energy intercepts, 29, 30f
  rock cycle and, 71–73
  solar radiation and, 56–57, 270–73
  temperature of, 6, 108, 109f
East Pacific Rise, 68
Easter Island civilization, 2–3
  carrying capacity and, 179
  feedback loops and, 45
easterlies, 61f, **62**
ecocentric, **241**
eco-labeling, 195
ecological efficiency, **91**
ecological footprint, **15, 17**
  of battery use, 29, 500
  of biodiversity-friendly coffee, 265
  of biome geography, 134, 135t
  of carbon dioxide emissions, 278, 279t
  of eco-labeling, 195
  of energy flow and weight change, 80–81
  of fossil fuel emissions, 408–9
  of land disturbance, 172–73
  of land requirements for diet, 346–47
  of moai, 16
  of nations, 212
  of net primary production use, 106–7
  of nuclear waste generation, 456
  of overall energy use, 422–24
  of overall environmental impact, 513
  of renewable energy sources, 475
  of resource use and waste emission, 49
  of soil erosion, 322–23
  of solar energy use, 58
  of sun exposure, 306–7
  sustainability and, 15–17
  of transportation, 16, 17, 226–27
  of water use, 382–83
  of wood used for paper, 364
ecological resistance, **166**
ecological restoration, **169**–71, 176
ecological stability, **166**
ecology
  of agriculture, 338–39
  human land use patterns and, 126
economic activity
  affluence and environmental degradation and, 231
  alien species and, 258–59
  carbon cycle and, 105–6
  El Niño events and, 70
  environmental tax revenues and, 515
  growth in (*see* economic growth)
  IPAT equation and, 225
  materials use in, 495
  measurements of, 204
  oil production and, 434
  species preservation and, 261
  sustainability and, 12–13, 196, 197
  technology and environmental degradation and, 234
  waste assimilation and, 219–20, 221f
  water rights and, 393–94
  well-being and (*see* economic well-being)
economic energy cost, **498**
economic goods and services
  attitudes and beliefs and, 240–41
  eco-labeling of, 195
  economic system and, 203–4
  environmental tax revenues and, 515
  footprints for, 15–17, 134, 135t
  internalizing externalities and, 512, 514
  market and, 237–38
  market failures and, 238–40
  personal choices and, 516–17
  sources of, 196
  subsidies and, 516
  sustainability and, 12–13, 196
  as waste, 208, 499, 501
economic goods and services, production of, 206–8
  carbon dioxide emissions from, 278, 279t
  conservation of matter and, 27
  energy intensities for nonenergy items in, 424t
  food (*see* agriculture; food production; meat production)
  footprints for, 106–7, 423, 424t
  gunpowder, 112
  nitrogen-based, 112
  paper, 58, 363, 364
  technology and, 203–4, 234–37
  wastes from, 208
  water use in, 381, 382, 383t
economic growth, **209**
  affluence and materials use and, 213–14
  affluence levels and, 210, 211f
  energy and materials and, 214
  forces behind, 210
  GDP measuring, 204, 209–10
  technology and labor productivity and, 210–12, 213f, 213t
economic power, social structure and, 368
economic system, **40**
  agriculture in, 338, 339–41, 344–48
  circular flow model of, 204
  ecosystem services in, 206, 207t
  environment and, 204–9
  functions of, 203–4
  incentives in (*see* market-based incentives)
  Kyoto and Montreal Protocols and, 309
  market failures in, 238–40
  market in, 237–38
  natural resource creation in, 205–6
  oil price increases and, 359
  price-induced technological change and, 282
  production of goods, services, and wastes in, 206–8
  waste assimilation in, 208–9
economic well-being
  biogeochemical cycles and, 521–22
  deforestation and, 355, 367, 368
  demographic transition and, 518–19
  environmental degradation and, 216–20, 221f
  environmental Kuznets curve and, 519–21
  forests and, 362–63, 367
  GDP and, 511
  precautionary principle and, 522
  subsidence and, 387
  upper limits on, 517–22
economically efficient, **238**
economically excessive rates of soil erosion, **328**
economies of scale, **345**
economy. *See* economic system
ecosystem diversity, **247**–48
ecosystem health, **166**
  diversity and, 166–69
  measuring, 165–67
  restoration of, 169–71, 176
ecosystem management, **369**
ecosystem services, **206,** 207t, 216, 218f
ecosystems, **124**
  agroecosystems, 338–39, 344, 518
  biodiversity and, 254–56
  catastrophic shifts in, 164–65
  disturbances in (*see* disturbances)
  economic processes and, 206, 207t
  global, value of, 216, 218f
  health of (*see* ecosystem health)
  human activity and, 165–76, 257–58
  hunter-gatherers *vs.* agriculturalists and, 334–35
  species interaction and, 254
  successional species and, 161–62, 163f
  sustainable management of, 369
  wind energy and, 472
ecotourism, **264,** 266
ectotherms, **78**
  energy pyramids and, 91, 92f
  maintenance and, 78, 79f, 82
edge effects, **257**–58, 358, 360
edges, **257**
efficiency, **33**
  energy and (*see* energy efficiency)
  environmental degradation and, 511–17
  environmental Kuznets curve and, 519–21
Egyptian total fertility, 189f
Ehrlich, Paul, 496–97
Eisenhower, Dwight D., 445
Ekman transport, **66**
El Niño events, **67**
  coastal upwellings and, 152
  fish catch and, 198f
  oceanic circulation and, 67–69
  policy responses to, 70
  UV-B radiation and, 304
electrical energy, **30**
electricity, consumption of
  energy content of, 422t
  energy efficiency benefits and, 483t
  primary *versus* final, 422
  in U.S., 420f
electricity, generation of
  atmospheric emissions from, 485t (*see also specific emission*)
  biopower for, 472, 475
  carbon dioxide emissions footprint from, 278, 279t
  carbon dioxide emissions from, 462f, 485t
  coal-based (*see* coal-fired power plants)
  energy and materials balance in, 35–36
  energy intensities for nonenergy goods and services produced with, 424t
  fossil fuel emissions in, 408, 409t
  geothermal energy for, 476
  land, water, and solid waste impacts of, 484
  mine-mouth plants and, 426–27
  nuclear-based (*see* nuclear power plants)
  operating costs by fuel source for, 460
  photovoltaics for, 468–70
  solar energy for, 467–68
  subsidies and, 239t, 461
  U.S. sources of, 423t, 424t, 456t, 466, 472
  water used in, 381
  wind for, 470–72
electromagnetic radiation, **29,** 30f
electrons, **24**–25
elements, **21**
  compounds and, 22
  crustal, 23, 205
  essential nutrients, 22–23
  periodic table of, 22f
  types and symbols of, 21–22
  U.S. use of, 213
elephants, 262–63
eluvial horizon, 313f, **314,** 315, 316f
emissions. *See* greenhouse gases; tradable permits; *specific emission*
employment in selected U.S. sectors, 506f
end use efficiency, **481**–82, 483t
endangered species, **259**–60, 261
Endangered Species Act, 261
endemic species, **250**
endotherms, **78**–79, 91, 92f
enemy release hypothesis, **259**
energy, **28**
  biodiversity and, 253
  content in selected foods, 81t
  crop yields and, 341–42
  emissions from generation of, 408, 409t
  end use efficiency of, 481–82, 483t
  entropy and, 33–34
  net supply of, 434
  power and, 32
  resource quality and, 498
  subsidies and, 238, 239t
  types of, 28–30, 31f (*see also* renewable energy; *specific type or source*)
  use of (*see* energy use; United States energy consumption)
  water used in generation of, 381, 384
  work and, 30–32
energy and materials balance of electricity plant, 35–36
energy converters, **31**–32, 33
energy density, **467**
energy efficiency, 481–82
  carbon emissions and, 276
  of motor vehicles, 235, 244, 422
  policy and, 282, 290, 483
  technology and, 276, 282, 483t
energy end use, 466–67, **481**–82
energy flow in biological systems, 76–97
  dinosaurs and, 76, 93, 94f
  from environment to organisms, 76–77, 78f, 79f
  growth and, 82, 83f
  maintenance respiration and, 77–79, 82
  natural selection and, 84–86, 88–89
  between organisms (*see* food chains; food webs)
  protection and, 76f, 86–88
  reproduction and, 76f, 84–86
  storage and, 76f, 82–83
  weight change footprint, 80–81
Energy Policy and Conservation Act, 244, 435
energy pyramids, **91,** 92f, 93–94, 95f
energy return on investment (EROI)
  for ethanol production, 474
  individual organisms and, 76f, 78f
  for natural gas production, 434f, 435
  net energy supply and, 434
  for oil production, 434
  for solar energy *vs.* fossil fuels, 466–67
energy use
  for agriculture, 342, 345
  in biological systems (*see* energy flow in biological systems)
  deforestation and, 360
  economic processes and, 206–8, 214
  efficiency and scale and, 511
  environmental energy cost, 206, 498
  footprints of (*see* ecological footprint)
  labor productivity and technology and, 211–12, 213f, 213t
  meat production and, 231–32
  overall ecological impact and, 513
  primary *versus* final consumption, 422
  technology and efficiency in, 276, 282
  for transportation, 226, 421
  by U.S. (*see* United States energy consumption)
engines, 213t, 421
England. *See* Britain
enrichment of nuclear fuel, **447,** 449f
entisols, 315
entombment of nuclear power plants, 454
entropy, **33**–34
entropy law, **34,** 35–36
environmental accounting, **220**–22
environmental degradation, **216.** *See also specific form of degradation*
  accounting for, 220–22
  affluence and, 231–33, 236–37
  biogeochemical cycles and, 521–22
  demographic transition and, 518–19
  efficiency and, 511–17
  environmental Kuznets curve and, 519–21
  environmental performance bond and, 13
  environmental tax revenues and, 515
  footprint of, 15, 16
  human ingenuity and, 219
  impacts of, 216–20, 221f
  internalizing externalities and, 512, 514
  living standards and, 11–12
  market failures and, 238–40

environmental degradation (continued)
market-based mechanisms and, 514–15
personal choices and, 516–17
Plimsoll line parable and, 511–12, 522
policy and (see government policy; policy)
population and, 225, 228–31, 236–37
poverty and, 233
precautionary principle and, 522
scale and, 511, 517–22
subsidies and, 516
sustainability and, 3, 6–8, 196, 197
technology and, 234–37
transportation and, 226
environmental energy cost, **206, 498**
environmental footprint. See ecological footprint
environmental goods and services
carrying capacity and, 179, 183–87
deforestation and, 368
economic value of, 214–20
environmental degradation and, 216–20, 221f
GDP and, 214–15
sources of, 196
valuing of, 215–16, 217f, 218f
environmental gradients, **127–30**
environmental impact
IPAT equation determining, 225, 237
personal calculations of (see ecological footprint)
environmental justice, **233**
environmental Kuznets curve, 519–21
environmental performance bond, **13**
Environmental Protection Agency (EPA)
air pollutants monitored by, 400
air quality index of, 408
arsenic policy of, 491
Clean Air Act and, 412, 414–15
Clean Water Act and, 396
radioactive waste disposal and, 452
sulfur emissions allowances from, 416
environmental resistance, **181**
environmental services, **3**. See also environmental goods and services
of atmospheric circulation, 63–64
of biodiversity, 256
deforestation and pricing of, 368
of forests, 363, 365, 366, 367
of oceanic circulation, 67, 69
of plate tectonics, 73
sustainability and depletion and degradation of, 6–8
sustainable forestry and, 369
sustainable use of, 3
waste assimilation, 207t, 208–9, 219–20, 221f
of water, 381
environmental taxes, 15, 515
environmentalists, 510–11
EPA. See Environmental Protection Agency
Ephesus, 365
epifauna, **150**
epilimnion, **148**
epipelagic zone, **153**
equations. See formulas
equatorial low, **61**
equilibrium, **44**
equity, sustainability and, 4–5, 9–12
erodibility factor, **320**
EROI. See energy return on investment
erosion, **72**
as ecosystem service, 207t (see also soil erosion)
mineral deposits and, 489–90
essential, **102**
essential nutrients, 22–23
estuaries, **149**
ethanol, **473–74**
ethnobotany, **255**
euphotic zone, **146**
Europe
oil supplies in, 439f
wind energy in, 471–72
eutrophic lakes, 148f, **149**
eutrophication, **115**
fertilizers and, 345
organic wastes and, 390–91
phosphates and, 115
evaporation, **374**
climate change and, 378
hydrologic cycle and, 373f, 374
potential evaporation, 130
evaporites, **490**
Everglades National Park
ecological restoration in, 171, 176
government subsidies and, 516
evolution
coevolutionary hypothesis, 336
diffuse chemical coevolution, 255
plate tectonics and, 73
ex situ conservation, **261**

exhaustion, **216**
existence value, **366**
experiments, **51**
exploration, in nuclear fuel cycle, **447**
exponential population growth, **180**
extensive land uses, **360**
externalities, **12, 238**
air pollution as, 412
energy technologies and, 484–85
internalizing, 512, 514
nuclear power and, 461
extinction, **249**
biodiversity and rate of, 256–57
climate change and, 288–89
as environmental alarm, 6
human activity and, 247
extraction efficiency, **426, 428**
extractive wastes, **499, 501f**
extremophiles, **255**
eyes and sun exposure, 307

F

fabrication, in nuclear fuel cycle, **447**
facilitation, **163**
factors of production, **203–4**
economic growth and, 210
market and, 237–38
fairness, sustainability and, 4–5, 9–12
falsified, **51**
family size
demographic transition and, 192, 193f, 518–19
environmental impact of, 225
Farman, Joseph, 297
farming. See agriculture
fats, **100**
in body, 83
in food, 347t
Faust, Dr. Johannes, 443
Faya Largeau climate graph, 139f
fecal coliform count, **389**
feedback loops, **44–45**
agricultural, 336, 347
carrying capacity and, 185–87
in Chernobyl nuclear accident, 457, 458
complexity of, 47–48
deforestation and, 358, 360
fuel efficiency and, 235
human population growth and, 192
lentic ecosystem shifts and, 164–65
OPEC pricing strategy and, 437
ozone reduction-climate change link and, 308
pesticide use and development and, 347
population and density-dependent factors and, 181
temperature change and human activity and, 285–86
unknown carbon sink and, 108–9
feedlots, **235–36, 343**
Ferrell cells, 61f, **62**
fertile isotope, **444**
fertile soil, 6t
fertility rates, 189, 192
fertilizers
nitrogenous, 111f, 112, 348
phosphate, 114
as pollutants, 120–21, 345, 348
precision farming and, 351–52
soil conservation and, 329
fiber, 135t, 213–14
field capacity, **319**
filter feeders, **150**
final energy consumption, 422
financial system. See banking system
Finke, Deborah, 169
Finnish nuclear power, 446f
fire
in forests (see forest fires)
grasslands and, 142
savanna vegetation and, 138
shrubs and, 141
firms, **203–4**
first law of thermodynamics, **32–33**
fishable, **396**
fisheries
coastal upwellings and, 152, 205
coral reef habitats of, 152
depletion of, 7–8
eco-labeling benefits for, 195
harvest declines in, 93, 182, 198
home biomes for, 135t
logistic curve and management of, 186–87
marine food chain/web and, 93, 94, 96–97
market failures and, 238
production by ecosystem, 205t
species diversity in, 250–51
sustainable management of, 182, 195

fissile, **444**
fission, **26, 444**. See also nuclear fission
fitness, **40**
flash steam power plants, **476, 485t**
flies, 352
flooding
aquatic food chain disturbances and, 170, 171f
benefits and costs of controlling, 368
deforestation and, 366
Florida
Everglades National Park and, 171, 176, 516
Gainesville PAYT program, 504
flows, **39, 103**
of cadmium, 28
of carbon, 104–9, 275
of energy (see energy flow in biological systems)
of energy sources in U.S., 466t
interindustry, 49
of nitrogen, 110–13
of phosphorus, 113–15
rules governing and types of, 103–4
of sulfur, 115–17
turbulent, 323–24
fluorescent lamp energy efficiency, 483t
foam casting energy efficiency, 483t
food
energy content in, 81t
prices of, 344
protective chemicals in, 86, 87f
food chains, 89–91
aquatic, disturbances in, 170–71
fish catch and, 93, 96–97
pesticides and, 347–48
predators in, 91–94, 95f
protective strategies in, 86–88
toxin concentrations in, 94–97
food production
coffee, 265
deforestation and, 358
as ecosystem service, 207t
farms and (see agriculture)
increases in, 350–51
land requirement footprint for, 346, 347t
meat (see meat production)
wheat, 345, 349
food supply
climate change and, 287–88
fish harvests, 93, 96–97, 182, 198
of hunter-gatherers vs. agriculturists, 334–35
Malthusian theory and, 191–92
price subsidies and soil conservation and, 329
solar energy and, 58
water use in production of, 382–83
food webs, 89–91, 92f
ecosystem stability and, 168–69
fish catch and, 93, 96–97
lotic, 147–48
pelagic, 153–54
toxin concentrations in, 94–97
footprint. See ecological footprint
forest fires
deforestation and, 358, 360
as disturbance agent, 158, 159
ecological restoration and, 169
lodgepole pine and, 159
suppression policy and, 174–75
Forest Stewardship Council (FSC) certification, 13f, 369
forestry residues, **473**
forests, **354–71**. See also specific type of forest
bioenergy and, 475
biogeochemical cycles and, 117
carbon flow from, 275, 475
climate change and, 286–87
climatic determinants in, 132f–133f
direct contributions to humans, 362–63
ecological restoration in American Southwest, 169, 170f
habitat conversion in, 257–58
harvesting certification in, 13f, 369
indirect contributions to humans, 363, 365–66
loss of (see deforestation)
paper use footprint and, 364
policy and, 360, 367–68, 370
prices of products from, 497
regrowth of, 356, 357, 369
relative erosion rates for, 322t
residues from, 473
restoration and density changes in, 169, 170f
soil in, 313–14, 315–17
sustainable management of, 13f, 368–69, 370
tree loss in, 158

formulas
for aerobic respiration, 89
for anaerobic respiration, 88
for basal metabolic rate, 80
for body mass index, 80
for exponential population growth, 180
functions, 42, 43f
for hydrogen production, 480
IPAT equation, 225, 237
logistic equation, 181
for NDVI, 131
for nuclear fission, 444
for ozone formation, 405–6
photodissociation equations, 296, 298–99
for photosynthesis, 77
for population growth, 179
for soil erosion, 320, 322, 325
for sulfuric acid formation, 403
fossil fuels, **419–41**. See also specific fuel
burning temperatures and emissions of, 408
carbon cycle and, 105, 275, 276
carbon dioxide emissions footprint from, 278, 279t
carbon emissions and forests and, 365
carbon monoxide and, 400
climate change and, 275, 276, 365
conventional vs. unconventional, 420
discovery, extraction, and processing of, 426–30
Energy Policy and Conservation Act and, 435
energy use footprint for, 422–24
formation of, 73, 424–26
future of oil and natural gas, 430–34
human activity time scale of, 465f
hydrocarbon emissions and, 407
labor productivity and, 211–12
as material waste, 501
meat production and, 231–32
nitrogen fixation and, 111f, 112
nitrogen oxide emissions and, 405–7, 408, 409t
oil price fluctuations, 199, 200f, 420
past and present use of, 420–21
policy and reduction of use of, 309
solar energy versus, 466–67, 484–85
subsidies and, 484–85
sulfur content in, 402, 430
sulfur cycle and, 116f, 117
sulfur dioxide emissions and, 402–5, 408, 409t
taxing emissions of, 515
fossil record, 256
foxes, 145
fragmentation, **257–58**
France
nuclear power in, 446f, 449
Reims climate graph, 142f
tidal energy in, 478
frankincense, 373
freezer energy efficiency, 483t
freeze-thaw cycle, 313
frequency, **158**
freshwater, **374**
desalinization and, 379–80
hydrologic cycle and, 373f, 374
source and availability of, 376, 377f
frogs
carrying capacity for, 183–87
UV-B radiation and, 303–4
fruit, 135t
Fryeburg PAYT program, 505f
FSC (Forest Stewardship Council) certification, 13f, 369
fuel assemblies, **447**
fuel cells, **481**
fuel efficiency
affluence and, 232, 237
CAFE standards for, 244
gasoline consumption and, 235, 237
technology and, 234, 237
fuel NO_x, 405
fuel rods, **447**
fuels, **25**. See also specific fuel
energy densities of, 466–67
fossil-based (see fossil fuels)
renewable (see renewable energy sources)
fumes, **400**
fumigation, **411, 412f**
function(s), **42, 43f**
for climate change damage, 289
linear vs. nonlinear, 50
functional group, **167**
fungicides, **346–47**, 352
fusion, **26, 27f**
harnessing power from, 460–61, 462f
heat of fusion, 30

G

Gainesville PAYT program, 504
gaining stream, 389f
gamma radiation, 26
gangue, 507t
gap species, **263**, 264f
gas. *See* gasoline *entries*; natural gas *entries*
gas centrifugation in uranium enrichment, 447
gaseous diffusion in uranium enrichment, 447
gasoline
   carbon dioxide emissions from, 279t
   energy content of, 422t
   lead in, 20, 21
   octane levels of, 430
   reformulated, 415
gasoline consumption
   affluence and, 232, 237
   IPAT calculations for, 237
   policy and, 244
   population size and, 230, 231f, 237
   technology and, 234, 235, 237
gasoline prices
   fluctuations in, 420
   fuel efficiency and, 235, 244
   oil externalities and, 239–40
gathering, 193, 334. *See also* hunting and gathering
GDP. *See* gross domestic product
gelisols, 315f
gender inequality, 10
general systems theory, 51–52
genetic diversity, **247**, 255–56, 266
genetic resources, as ecosystem
     service, 207t
genetically modified crops, **351**
genetics
   genetically modified crops, 351
   particulate matter and, 401
   pesticides and, 347
geochemical exploration, **494**
geographic regions
   air quality index by, 408
   carbon dioxide emissions by, 278, 279t
   electricity from fossil fuels by, 423t
   electricity from nuclear power by, 456t
   sulfur dioxide and nitrogen oxide
     emissions by, 408, 409t
geologic disposal, **452**
geologic exploration, 493–94
Geological Survey oil reserve forecasts, 431
geology
   fossil fuel formation and, 73, 424–26
   locating fossil fuel reserves via, 427,
     431–33
geometric population growth, **179**
geophysical exploration, **494**
Georgia, Oconee County PAYT program, 505f
geothermal direct use, **476**
geothermal energy, 476–77
   atmospheric emissions from, 485t
   land, water, and solid waste impacts of,
     484t
   U.S. consumption of, 31f, 466t
   U.S. flow of, 466t
geothermal fields, **476**
geothermal gradient, **494**
geothermal heat pumps, **476**, 477f
Germany
   energy consumption by, 422
   nuclear power in, 446f
   total fertility rate in, 189
   wind energy in, 472
germinate, **136**
Gesner, Conrad, 488
The Geysers, 476
Gilliland, Martha, 435
given area, 183–84
Glacier Bay succession, 160
glaciers, 277, 425
glass bottles, 503
glass furnace energy efficiency, 483t
Glendale PAYT program, 505f
global warming. *See also* climate change;
     greenhouse gases
   forests and, 363
   IPCC statement on, 283
   potential effects of, 6
GNI (gross national income), 204
GNP (gross national product), 203, **204**
gold mining, 360, 490
golden toad, 288
good guybad guy dichotomy, 510–11
goods, **203**
   energy efficiency of, 482, 483t
   materials cycle of, 488, 489f (*see also*
     materials)
   photovoltaic applications for, 470
   services and (*see* economic goods and
     services; environmental goods
     and services)

as waste, 208, 232–33, 499–502
as waste, recovery/reduction of, 234–35,
   502–6, 507t
gopher tortoise, 254
governance, **262–63**
government policy. *See also specific*
   legislative act
   air quality standards, 412–17, 519
   biodiversity preservation and, 261, 262–63
   climate change and, 290–91, 309
   command and control strategies, 242–43,
     514–15
   debt for nature swap, 371
   deforestation and, 360, 362, 367–68
   El Niño and, 70
   energy funding and, 483
   energy pricing system, 435
   environmental accounting, 222
   fire suppression, 174–75
   gasoline consumption and, 244
   greenhouse gas intensity reduction, 519
   internalizing externalities and, 512, 514
   IPCC and, 53
   lead regulations, 21
   market failures and, 238, 241–42
   market-based incentives, 243
   nuclear power sector and, 461
   ozone reduction and, 304–7, 309
   Pay-as-You-Throw programs, 504, 505f
   revenue recycling, 514
   soil conservation and, 325, 329, 330
   subsidies (*see* subsidies)
   sulfur emissions reduction, 416–17
   sustainability and, 12–15
   systems perspective and, 9
   taxes (*see* taxes)
   water supply and, 393, 394–96
gradients, 40–41
   atmospheric, 59–62
   environmental, 127–30
grains, **335**
   coarse grains, 343
   diet based on, 231
   home biomes for, 135t
grasses
   agriculture and, 334
   as food chain detritus, 91
   succession and, 159, 160f, 161, 163–64
   in tropical savannas, 138–39
grasslands
   climate and deforestation and,
     363, 365
   climatic determinants in, 133f
   food production and, 228–29
   net primary production in, 172t
   relative erosion rates for, 322t
   soil in, 314, 315, 316f
   succession in, 159
   as temperate biomes, 141–42
grassy stunt virus, 255
gravitational water, **318**
grazers, **89**
   savanna vegetation and, 138–39
   of temperate grasslands, 142
Great Barrier Reef, 151, 167f
Great Britain. *See* Britain
Great Depression, Dust Bowl and, 312, 331
Great Plains of North America, 141, 375, 386f
Greece
   lead and, 20
   total fertility rate in, 189f
green national accounts, **221**, 222f
Green Revolution
   benefits of, 343–44
   costs of, 343, 344–48
   evolution of, 341–43
   future and, 350–52
greenhouse effect, **271–73**
greenhouse gas intensity, 519
greenhouse gases, **271**. *See also specific gas*
   characteristics of, 271–73
   concentrations of, 273–74, 275f, 284–85
   emissions forecast for, 276
   forests and, 365
   income and population effects on, 276
   as material wastes, 501
   nuclear power and, 462
   optimal emissions levels for, 289–90
   policy and reduction of, 289–91, 306,
     309, 519
   temperature change and, 274, 284–85
Greenland Norse settlements, 269
gross domestic product (GDP), **204**
   emissions and energy use and, 521
   environmental goods and services in,
     214–17
   increases in (*see* economic growth)
   inequities in, 10
   natural resource/environmental account-
     ing and, 220–22
   solid waste production and, 232–33

total energy use and, 511
gross national income (GNI), **204**
gross national product (GNP), **203, 204**
gross primary production, **89**
ground-level ozone. *See* ozone, ground-level
groundwater, **374**
   agricultural pollutants in, 392
   allocation rights for, 393
   diversion of, 379
   hydrologic cycle and, 373f, 374–75
   mining of, 386–88, 389f
   pumping of, 376–77, 378f
   sources of, 376, 378f
groundwater discharge, **375**
growing season, **130**. *See also*
   specific biome
growth
   of animals, 76f, 82, 83f
   of economy (*see* economic growth)
   of greenhouse gas emissions, 276
   of plants, 130, 317–19
   of populations (*see* population growth)
Guangzhou, China, 399
Gulf of California, 345
Gulf of Saint-Malo power plant, 478
gully erosion, **321**
gunpowder production, 112
gypsum, 116
gyre, **66**

H

H. J. Andrews Forest climate graph, 142f
Haber, Fritz, 112
Haber-Bosch nitrogen fixation, 111f, 112
habitat conversion, **257–58**
habitats, **124**
   biodiversity and alteration of, 257–58
   ecosystem services and, 206, 207t
   forests as, 363
   niches and, 124, 125f
   preservation of, 261–63, 264f
hadal zone, **153**
Hadley cells, **61**, 62
half-lives, **26**
   of agricultural chemicals, 392
   radioactivity and, 25–26, 27t
halogen depletion hypothesis, 298–99, 301
halon, 298t
Hardin, Garrett, 215
hardwood densities, 365f
harlequin frogs, 288
harvest index, **341**, 350
harvesting
   agroecosystems and, 339
   biodiversity and, 260
   black market and, 266
   of fish, 93, 96–97, 182, 198
   of timber, 367–69, 370
Hausman, Jerry, 517
hazardous waste. *See* toxins
HDI (Human Development Index), 11f
health. *See* disease/illness; ecosystem health
heat, **30**, 57
heat balance, **270–73**
heat equivalents of forms of energy, 30, 31f
heat of fusion, **30**
heat of vaporization, **30–31**
heat pumps, 476, 477f
heat transport, atmospheric, 60
heating oil
   emissions from, 279t, 408t, 409t
   energy content of, 422t
heating, solar systems for, 467–68
heavy metals, 391, 502t
hedonic pricing, **216**
heliostats, 468
helium, 24f
hemispheric patterns
   of radiative forcing, 272f, 273, 274
   of temperature change, 274
hemoglobin, 400
herbicides, **339**
   crop yields and, 346–48
   genetically modified crops and, 351
   integrated pest management and, 352
herbs, 86, 87f
herding, 340, 341f
Hertz, Heinrich, 468
heterotrophs, **77**, 102
hibernation, **83**, 128
hierarchy, 48
high development nations, 9–12
high-grading, **369**
high-level (radioactive) waste (HLW), 451
highways. *See* roads and highways
high-yield varieties, **341**, 350–51

histosols, 315f
HIV in Canada *vs.* Kenya, 10t
HLW (high-level (radioactive) waste), 451
homeostasis, **41**
   disturbance and, 41–42
   generating, 42–45
horizons, **313**
   soil profiles and, 315–17
   types and distribution of, 313–14
horizontal mixing, **410**
hot springs, minerals and, 491–92
household tasks, water intensities for, 382
housing, lead and, 21
Hubbard Brook Valley, NH, 162, 163f
Hubbert, M. King
   oil and natural gas estimates by, 431–32
   peak oil production forecast by, 432–33
human activity
   alien species and, 258–59
   attributing climate change to, 280–82
   biogeochemical cycles and, 117, 120–21
   carbon cycle and, 105–6, 274–76
   climate change and, 273–76, 280–86, 289–91
   ecosystem disturbances, 165–76, 257–58
   footprint of (*see* ecological footprint)
   fossil fuel time scale and, 465f
   ingenuity, environmental degradation
     and, 219
   land use determinants and, 126
   mass extinctions and, 247
   nitrogen cycle and, 112
   phosphorus cycle and, 113f, 114–15
   pollution from (*see* air pollution; water
     pollution)
   radiative forcing, 273–76
   sulfur cycle and, 115, 117
human body
   elements occurring in, 23t
   energy allocation in, 79f
   maintenance respiration in, 78
   temperature regulation in, 79
human development, 9–12
Human Development Index (HDI), 11f
humus, **314**
Hungarian nuclear power, 446f
hunting, **193, 334**
   biodiversity and, 260
   black market and, 266
hunting and gathering, **193, 334**
   development of agriculture *versus*, 335–37
   extant populations engaged in, 337
   natural ecosystems and, 334–35
hurricanes, 366, 401, 404f
hybrid plants, 351
hydrides, 480
hydrocarbons, 407
hydroelectric energy. *See* hydropower
hydrogen
   in coal, 425
   energy density of, 480
   as energy source, 466t, 480–81
   as nutrient, 22, 23t
hydrogen ions, 320
hydrologic cycle, 373–76, 378
hydrolysis, **313**
hydropower, 384, **479–80**
   footprint of use of, 423, 424t
   global consumption of, 422, 423
   U.S. consumption of, 31f, 466t
   U.S. electricity generated from, 424t
   U.S. flow of, 466t
hydrothermal solutions, **491**
hydroxyl radicals
   carbon monoxide and, 400
   ozone reduction-climate change link and, 308
   photochemical smog and, 407
hygroscopic water, **318**
hypolimnion, **148**
hypotheses, 50–51

I

ice ages, 269–70
ice core analysis, 273
Idaho nuclear energy, 445f
IEESA (integrated environmental and eco-
    nomic satellite accounts), 222
igneous processes, minerals and, 490–91
igneous rocks, **72**, 73f
Illinois PAYT programs, 505f
illness. *See* disease/illness
illuvial horizon, 313f, **314**, 316f
immigration, **249**
immune system, sun exposure and, 306
Imperial PAYT program, 505f
imported minerals, 495, 496f
in situ conservation, 261
incentives, sustainability and, 5, 12–13. *See also*
    market-based incentives

inceptisols, 315f
income, **220**
  energy use and, 422
  environmental impact of increases in, 225
  environmental Kuznets curve and, 519–21
  greenhouse gas emissions and, 276
  inequality in, 10–11
  internalizing externalities and, 512, 514
  transportation and, 226
independent variable, **42**
India
  Bombay climate graph, 137f
  GDP in, 10
  Green Revolution and, 342, 343
  Kyoto Protocol, 291, 306
  Montreal Protocol and, 306
  nuclear accident in, 455t
  nuclear power in, 446f, 449
  resource use in, 11, 12t
  soil erosion costs in, 325
  total fertility rate in, 189f
indigenous peoples
  Cree, 480
  deforestation and, 368
  Inuit, 269
Indonesia
  fish biodiversity in, 250–51
  net domestic product for, 221–22
industrial livestock production systems, **343**
industrial nations
  agriculture in, 340f
  Kyoto Protocol emission requirements
    for, 290–91
  population growth in, 192
Industrial Revolution
  fuel sources for, 421
  origins and impact of, 200, 465
industry
  Clean Air Act and, 9, 415
  Clean Water Act and, 396
  energy end use by, 481t, 482
  fossil fuel use by, 421f
  lead research funded by, 21
  precautionary principle and, 522
  recycling and reuse and, 503, 506
  remanufacturing sector in, 506
  subsidies and, 516
  sulfur emissions from, 275, 402, 404
  water pollutants from, 391
  water use by, 381, 382, 383t
inertial confinement of plasma, 461
infauna, **150**
infiltration, **374**
ingenuity, 198–201, 219
inhibition, **164**
insect pests, climate change and, 287–88
instream uses of water, **380**, 384
insurance, **255**, 460
integrated environmental and economic
    satellite accounts (IEESA), 222
integrated pest management, **352**
integrated systems theory, **52**
intensification, **342**
  benefits of, 343–44
  costs of, 344–48
intensities
  calculating, 49
  of energy for goods and services, 424t
  of energy for transportation, 232f
  of water for household tasks, 382
intensity of use, **495**, 497f
intensive land uses, **360**
Intergovernmental Panel on Climate Change
    (IPCC)
  composition and goal of, 53
  on policy issues, 290
  sea level forecast of, 288
  statement on warming, 283
interim storage of nuclear waste, **448**, 449f
interindustry flows, 49
internal combustion engine, 421
International Geosphere-Biosphere Program,
    322
International Rice Research Institute, 255
intertidal zone, **150**
Inuit, 269
invalidated, **51**
invasive species, **258**
inversions, **411**
investments
  OPEC pricing strategy and, 437
  return on (see energy return on investment)
  U.S. oil companies and proved oil reserves
    and, 429
iodine-131 releases from nuclear power
    plants, 455t
IPAT equation, **225**, 237
IPCC. See Intergovernmental Panel on Climate
    Change
iron
  climate change experiment with, 118–19

production of, 200
recycling benefits for, 507t
irrigation, **339**
  Aral Sea and, 385
  drip form of, 394
  drivers of, 380–81
  fertilizers and, 345f, 346, 349
  precision farming and, 351
islands
  Easter Island, 2–3, 45, 179
  mining on, 114, 115f
  species richness and, 249
  West Indian, 249
isolation, species richness and, 250
isotopes, **25**–26, 27t
Italy, Taranto climate graph, 140f

J

Jambato toad, 288
James Bay Project, 480
Japan
  energy consumption by, 422
  fish catch for, 8
  Monju nuclear accident, 455t
  nuclear power in, 446f, 449
Java soil conservation, 329

K

K selected, **86**
Kaibab Plateau, 186
Kansas, Manhattan climate graph, 141f
Kazakhstan nuclear power, 446f
Kenya
  human development in, 10
  total fertility rate in, 189f
kerogen, **425**
Keystone species, **254**
kinetic energy, **30**
Kinney, Clesson S., 393
Kisangani climate graph, 136f
Komanoff, Charles, 461
Komodo dragon, 77, 79f
Kraft process, 58
krill, 94
Kruger Park, 138
Kuala Lumpur climate graph, 136f
Kuznets, Simon, 520
kwashiorkor, **102**
Kyoto Protocol, 241, 290–91, 306, 309
  Bush, George W. and, 291, 519
  scale vs. efficiency and, 519

L

labor, **203**, 344
labor productivity, **210**
  materials use and, 213–14
  technology and, 210–12, 213f, 213t
Lagoona Beach nuclear accident, 455t
lakes
  as biomes, 148–49
  ecosystems of (see lentic ecosystems)
  evaporites from, 490
  hydrologic cycle and, 375–76
  recreational activity on, 384
land
  abandoned/degraded in Philippines and
    U.S., 173
  in climate change simulations, 280, 281f
  footprint of human impact on, 513
land cover, **172**
  relative erosion rates and, 322–23
  soil erosion and, 320, 324
land use, **172**
  biodiversity and habitat conversion and,
    257–58
  climatological and ecological determinants
    of, 126
  deforestation, property rights, fire and, 360
  electricity generating technologies and, 484
  extensive vs. intensive, 360
  feedlots and, 235–36
  footprints of, 172–73, 346–47
  population growth and, 228–29
  soil and, 312–13, 317
land use change, **275**
landfills, 229–30, 276
Lansing PAYT program, 505f
lapse rate, 295, **410**
large mouth bass, 254
latitude, biodiversity and, 249, 250–51, 252, 253
Lavoisier, Antoine, 27
law of conservation of matter, 27–28, 35–36
law of the minimum, **101**

laws of thermodynamics, **32**–34, 33t
  energy and materials balance and, 35–36
  waste heat and, 208
lead
  legislation on, 21
  pathways of, 24
  properties of, 20
  recycling benefits for, 507t
  uses and toxic effects of, 20, 21
lead oxides, 502t
Lead Poisoning Prevention Act (U.S.), **21**
lead-based battery footprint, 29
leaves
  adaptive shapes of, 125, 132f–133f
  of boreal forests, 143–44
  lotic food webs and, 148
  of temperate forests, 143
  of tropical dry forests, 132f, 137
  of tropical rain forests, 133f, 134–35
legislation. See government policy;
    specific legislative act
legumes, **112**
Lemmon, Mt., precipitation gradients, 128, 129f
lentic ecosystems, **148**
  as biomes, 148–49
  catastrophic shifts in, 164–65
  restoration of, 176
leukemia, 455, 457, 458
Liebig's law of the minimum, **101**
life cycle analysis, **488**
life expectancy in Canada vs. Kenya, 10t
light compensation point, **161**
light, photovoltaic effect and, 469
light saturation point, **161**
lightning, 111
lignin, **364**
limiting factors, **180**
  in aquatic biomes, 146
  for population growth, 180–83
  in terrestrial biomes, 130
limiting nutrient, **101**
The Limits to Growth, 197
limnetic zone, **148**
linear functions, 50
liquefied natural gas, 430
liquid metal fast breeder reactors
    (LMFBRs), 450
liquidating, 221
lithium, 460
lithophone, 502t
lithosphere, **71**, 72f. See also plate
    tectonics
Lithuanian nuclear power, 446f
little brown bat, 128
littoral zone, **148**
livestock, **340**
  feedlots and, 235–36, 343
  Green Revolution and, 342–43
  land use determinants in Africa and, 126
lizards
  active range of, 82
  energy use and food acquisition by, 77,
    78f, 79f
  environmental gradients and, 128–30
  habitats and niches of, 123f, 124
  species richness vs. evenness in, 248–49
LLW (low-level radioactive waste), 451
LMFBRs (liquid metal fast breeder reactors),
    450
loam soil, **319**
lodgepole pine, 159
logistic curve, 186–87
logistic equation, **181**
Longreach climate graph, 138f
looping, **411**, 412f
losing stream, 389f
lotic systems, **147**–48
low development nations, 9–12
low-e glass energy efficiency, 483t
low-level (radioactive) waste (LLW), 451
low-till agriculture, 327

M

Mabira Reserve, Uganda, 264, 266
MacArthur, Robert, 249
machinery. See capital equipment
macroenvironments, **160**
macronutrients, **22**
macropores, **318**
magnesium, 23t
magnetic confinement of plasma, 461
Magnitogorsk climate graph, 141f
Maine, Fryeburg PAYT program, 505f
maize, 351
Malaysia, Kuala Lumpur climate
    graph, 136f

Malthus, Thomas, 191
Malthusian theory of population limits, 191–92
  Green Revolution and, 344
  resource optimists vs. pessimists and, 197–98
Manhattan (Kansas) climate graph, 141f
mantle, 71, 72f
manufacturing sectors, 481t, 482, 506f
manure surplus, 236
marginal ice zone, **303**
marine ecosystems
  of Antarctica, UV-B radiation and, 303
  biomagnification in, 95–96
  food chains/webs in, 90f, 93, 94, 96–97
  travel cost methods and, 216
Marine Stewardship Council, 195
market, **237**–38. See also economic activity
market failures, **214**, 238
  causes of, 238
  environmental degradation and, 238–40
  government intervention and, 241–42
market power, **238**
market-based incentives, **242**
  air pollution reduction and, 415
  biodiversity preservation and, 264, 266
  command and control strategy
    versus, 242–43, 514–15
  deforestation and, 368
  gasoline consumption reduction and, 244
  motor vehicle use and, 226
  Pay-as-You-Throw programs, 504, 505f
  soil conservation and, 329–31
Marlowe, Christopher, 443
marriage, 225
Martin, Jon, 119
Maryland, Aberdeen PAYT program, 505f
mass balance, 103f
mass extinctions, 247, **256**–57
Massachusetts Water Resources Authority,
    220, 221f
Massachusetts, Worcester PAYT program,
    505f
material wastes, **208**
  reduction of, 502–6, 507t
  types and impact of, 208, 209, 499–502
materials
  affluence and, 213–14
  agricultural inputs, 342, 351–52
  dematerialization of, 495, 497f
  economy and, 214, 495–97
  energy and, 214, 498
  fate of, 498–502
  in feedlots, 235–36
  labor productivity and, 211–12, 213f, 213t
  life cycle of, 488, 489f
  price of, 496–97
  in production of goods and services,
    206–8
  recycling of, 502, 503–6, 507t
  remanufacturing of, 506
  source reduction for, 502–3
  U.S. use by category, 495f, 499f
  as waste products, 208, 209, 499–502
  waste reduction benefits for, 506, 507t
materials cycle, 488, 489f
matter, **21**–23
  chemical changes in, 25
  conservation of, 27–28
  entropy and, 33–34
  flow of (see biogeochemical cycles)
  nuclear changes in, 25–26, 27f, 27t
  nutrients (see nutrients)
  physical changes in, 25
maturation, **82**, 83f
Mauna Loa curve, 110
maximum intrinsic growth rate, **180**
maximum mixing depth, **411**
maximum number, 183, 184f, 192, 193f
maximum sustainable yield, 182
MAXXAM, 367
Meadows, Donella, 197–98
meat
  Chinese consumption of, 518
  home biomes for, 135t
meat production
  affluence and environmental degradation
    and, 231–32
  increased yields of, 343
  land requirements for, 347t
  technologies for, 235–36
  water used in, 232
mechanical energy, **30**
medicines, 255–56, 266
Mediterranean woodlands and scrublands,
    140–41
medium development nations, 9–12
membrane separation, **379**
Merck, Inc., 266
mercury, 360, 502t
mesopelagic zone, **153**
mesosphere, 71f, **295**
metals. See also mineral resources

as dissipative wastes, 502t
as extractive wastes, 499
as health threat, 391
plate tectonics and, 73
prices of, 497
productivity of, 498
waste reduction benefits for, 506, 507t
metamorphic processes, minerals and, 490–91
metamorphic rocks, **73**
metamorphosis, **82**, 83f
meteorites, 257
methane
  anaerobic sources of, 275–76
  atmospheric concentrations of, 273–74, 275f
  characteristics of, 271t
  emissions of, 275, 485t
  energy density of, 480
  as material waste, 501
  ozone reduction-climate change link and, 308
  temperature and, 275f
methehemoglobinemia, **112**, 348
methyl chloroform, 298t
Mexico
  Acapulco climate graph, 137f
  Colorado River flow and, 385
  demographic transition in, 192, 193f
  nuclear power in, 446f
Michigan
  Lagoona Beach nuclear accident, 455t
  Lansing PAYT program, 505f
microenvironments, **160–61**, 162f
microorganisms, 388, 389–90
micropores, **318**
microspheres, 480
Middle East
  development of agriculture in, 334
  oil supplies by country, 438f
Midgley, Thomas, Jr., 304
migratory animals, 303, 472
milling, in nuclear fuel cycle, **447**
mimicry, **88**
Minamata disease, 391
mine-mouth power plants, 426–27
mineral resources, **492**
  abundance and distribution of, 493
  classification of, 492, 493f
  deforestation and mining of, 360–61
  as dissipative wastes, 502t
  exploration for and mining of, 493–94
  formation and occurrence of, 488–92
  plate tectonics and, 73
  price of, 496–97
  U.S. imports of, 495, 496f
  waste reduction benefits for, 506, 507t
mineralization, **112**
minimum, Liebig's law of the, **101**
mining
  atmospheric emissions from, 485t
  of coal, 426–27, 499, 501f
  of copper, 205
  deforestation and, 360–61
  extractive wastes from, 499, 501f
  of forests, 370
  of gold, 360, 490
  of groundwater, 386–88, 389f
  of mineral deposits, 494
  of nitrogen and phosphorus, 113f, 114, 115f
  of oil and natural gas, 427–29
  sulfur cycle and, 117
  of uranium, 443f, 447
Minnesota
  Cedar Creek, 159, 161, 163–64, 166
  forest fires in, 158
  St. Cloud PAYT program, 505f
mist aerosols, **400**
Mitch (hurricane), 366
moai statues, 2, 16
mobile sources of air pollution, **400**, 414–15
mobility of organisms, 82, 83
Mobro (ship), 230
moderators, **449**
Mojave Desert, 139
molecular nitrogen, **294**
molecular oxygen, **294**
molecules, **24**
Molina, Mario, 294, 298
mollisols, **315**, 316f
Mongolian Desert, 139
Monju nuclear accident, 455t
monoculture, **346–48**
Montreal Protocol, 305–7, 309–10, 514
mortality rate, 10t, 416–17
moths, 216, 217f
motor vehicles
  CAFE standards for, 244
  carbon monoxide emissions of, 400, 405
  Clean Air Act and, 414–15

energy intensities of, 232f
footprint of use of, 226–27
gasoline consumption (*see* gasoline consumption)
impact of choice of, 516–17
internal combustion engine and oil use, 421
lead and, 20, 21
material composition of, 23
octane levels and, 430
ozone formation and, 405–7
price fluctuations and, 420
U.S. energy efficiency of, 422
motorcycle footprint, 226–27
mountains, precipitation and, 63
mountaintop removal, **426**, 427f
movies about nuclear energy, 454
mudflats, 149
municipal solid waste, **501**–2
  affluence and, 232–33
  electricity generating technologies and, 484
  Pay-as-You-Throw programs for, 504, 505f
  population factors and, 229–30
  technology and, 234–35
municipalities
  clean water technologies for, 396
  sewage generated by, effects of, 388–91
  water use by, 380
mutations, **302, 401**

N

NAAQS. *See* National Ambient Air Quality Standards
Narora nuclear accident, 455t
National Academy of Science, 483
National Aeronautics and Space Administration (NASA)
  CFC production blue book by, 305
  ozone monitoring by, 297
  Ubar imaging project, 373
National Ambient Air Quality Standards (NAAQS), **411**
  compliance with, 414
  margin of safety and, 412, 414
  primary *vs.* secondary standards, 411–12
  state implementation plans for, 415
national ecological footprints, 212
National Oceanic and Atmospheric Administration (NOAA) satellite imaging, 131
National Renewable Energy Laboratory (NREL) research, 484
National Research Council, 222
native range, **259**
Natural Bridges National Monument photovoltaic system, 469f
natural gas
  as conventional fossil fuel, 420
  discovery and extraction of, 427–28
  energy content of, 422t
  energy intensities for nonenergy goods and services produced with, 424t
  Energy Policy and Conservation Act and, 435
  energy return on investment for, 434f, 435
  formation of, 425–26
  future of, 430–34
  global consumption of, 422, 423
  industry subsidies and tax breaks for, 484
  liquefied, transport of, 430
  as material waste, 501
  properties of, 421
  remaining quantities of, 430–32
  transport of (*see* pipelines)
  U.S. consumption of, 12t, 31f, 420f, 466t
  U.S. storage of, 466t
  yield per effort for, 432
natural gas-fueled power plants
  carbon dioxide emissions from, 275, 279t, 462t
  sulfur dioxide and nitrogen oxides emissions from, 408t, 409t
  taxing emissions of, 515
  U.S. electricity generated from, 423t
natural resource accounting, **220–22**
natural resources, **3**. *See also specific resource*
  biogeochemical cycles and, 120
  depletion/degradation of (*see* environmental degradation)
  economic accounting of, 220–22
  economic growth and, 210
  economic processes and, 205–6
  energy costs of, 498
  as extractive wastes, 499, 501f
  footprint of use of, 49
  plate tectonics and, 73
  price of, 496–97
  resource quality and opportunity costs and, 217–19, 220f

sustainable use of, 3
use *vs.* replenishment of, 6t
wood as, 362, 367
natural selection, **46–47**
  energy storage and, 83
  energy use and, 88–89
  nitrogenous wastes and, 112
  reproduction and, 84–86
naturalized range, **259**
Nature Conservancy, 263, 371
Nauru (island), 114, 115f
NDVI. *See* normalized difference vegetation index
negative feedback loops, **44–45**. *See also* feedback loops
negative relationship, 43
nerve cells, 41
net primary production, **89**
  aquatic biomes and, 146
  average annual rate in select biomes, 172
  biodiversity and, 259–60
  crop yields and, 341
  footprint for, 106–7
  GDP and, 215
  in soil erosion calculations, 323
  succession and, 161–62
  terrestrial biomes and, 130f, 172t
  trade and, 194
  as upper limit, 521
  UV-B radiation and, 303
Netherlands
  animal waste in, 236
  nuclear power in, 446f
neutrally stable atmosphere, **410**
neutrons, 24–25
Nevada
  Colorado River water allotment for, 385
  Imperial PAYT program, 505f
  Yucca Mountain Project, 452–53, 461
New England
  afforestation in, 356
  deforestation in, 356, 357, 365, 366
  flooding in, 366
  liquefied natural gas and, 430
New Hampshire nuclear protest, 446, 447f
New Haven harbor siltation, 365
New Jersey, Woodstown PAYT program, 505f
New Mexico
  deer management in, 186
  uranium reserves in, 445t
new source performance standards of air quality, **415**
New York City
  consumption of goods and services in, 194
  solid waste in, 229–30
  waste assimilation in, 194
New York State, Utica PAYT program, 505f
NGOs (nongovernmental organizations), **262–63, 370–71**
niches, **124–25**, 127
Niger, Tahoua climate graph, 138f
nitric oxide
  as agricultural pollutant, 348, 349
  as air pollutant, 405, 406
  emissions of, by electricity generating technologies, 485t
  energy efficiency benefits and, 483t
nitrification, **112**
nitrogen
  as agricultural waste, 112, 345–46, 348
  allowances/permits for, 521–22
  biodiversity and, 259
  goods produced from, 112
  human populations and, 230–31
  inorganic forms of, 110
  molecular form of, 294
  as nutrient, 22, 23t
  oceanic abundance of, 118–19
  odd nitrogen hypothesis of ozone reduction, 299–300
  plant use of, 100
  Redfield ratio of, 101
  in sewage, treatment of, 208–9, 220
  successional change and, 163–64
  trees and, 117
nitrogen cycle, 110–13
  agricultural pollution via, 348
  atmospheric storage in, 110–11, 112
  biotic and soil storage in, 111–12
  human activity and, 112, 120–21
  limits of, 521
nitrogen dioxide
  as air pollutant, 405, 406
  emissions footprint for, 408, 409t
  emissions of, by electricity generating technologies, 485t
  energy efficiency benefits and, 483t
nitrogen fertilizers
  nitrogen cycle and, 111f, 112

as pollutants, 345–46, 348
precision farming and, 351–52
win-win solution in use of, 349
nitrogen fixation, **111–12**
nitrogen oxides
  as air pollutants, 405–7
  emissions footprint for, 408, 409t
  as material wastes, 501f
  reduction of, 413f
nitrous oxide
  as agricultural pollutant, 348, 349
  atmospheric concentrations of, 273f, 274
  characteristics of, 271t
  as material waste, 501
  in ozone reduction tests, 300–301
NOAA (National Oceanic and Atmospheric Administration) satellite imaging, 131
nomadic herding, 340, 341f
nonattainment areas, **412**
nondegradable wastes, **209**
nonessential, **102**
nongovernmental organizations (NGOs), **262–63, 370–71**
nonlinear functions, 50
nonpoint pollutants, **391**
nonrenewable resources, **6**. *See also specific resource*
  natural resource/environmental accounting and, 221
  nonrenewable materials from, 213–14
nonspontaneous flows, **104**
nonspontaneous process, **34**
normalized difference vegetation index (NDVI)
  calculating values in, 131
  as climate change indicator, 277, 280
  remote sensing and, 131
Norse settlements, 269
North American oil supplies, 439f
North Carolina
  Piedmont succession in, 159, 160, 163
  Wilmington PAYT program, 505f
North Dakota wind energy potential, 471
North Pole ozone levels, 301–2
North Slope climate graph, 145f
Northern Hemisphere radiative forcing, 272t, 273, 274
Norway, Vardo climate graph, 145f
no-till agriculture, 327
NREL (National Renewable Energy Laboratory) research, 484
nuclear changes in matter, 25–26, 27f, 27t
nuclear decay, **26**, 27t
nuclear energy, **30**
nuclear energy, consumption of
  footprint of, 423, 424t
  global rates for, 422, 423
nuclear fission, **26, 444**
  energy released by, 460t
  in liquid metal fast breeder reactor, 450
  nuclear fuel cycle and, 447
  in pressurized water reactor, 449
nuclear fusion, **26**, 27f, 460–61, 462f
nuclear power plants
  atmospheric emissions from, 485t
  carbon dioxide emissions from, 462t
  Chernobyl accident, 457–58
  decommissioning of, 454, 461
  economics of, 460, 461
  electricity generated by, 424t, 445–46, 456
  as Faustian bargain, 443
  fuel cycle in, 447–49
  government policy and, 461
  growth of, 198, 445–46, 462
  iodine-131 releases from, 455t
  land, water, and solid waste impacts of, 484t
  opposition to, 446, 447f
  reactor designs, 449–50, 459
  risk calculations for, 458–59
  safety issues in, 446, 450, 454–59
  subsidies for, 460, 461
  terrorism and, 459
  Three Mile Island accident, 454–55, 457
  waste disposal by, 451–54 (*see also* radioactive waste)
nuclear proliferation and diversion, **459**
Nuclear Regulatory Commission (NRC)
  accident risk calculations by, 458–59
  Three Mile Island nuclear accident and, 455
  Yucca Mountain Project and, 453
nuclear waste. *See* radioactive waste
Nuclear Waste Policy Act, 449, 452
Nuclear Waste Policy Act Amendments Act, 452
nuclear weapons, 447, 459
nucleus, **24**
nutrients, **22–23, 100**. *See also specific nutrient*

nutrients (continued)
  agroecosystems and, 339
  autotroph capture of, 100–101
  ecosystem service of cycling of, 207t
  heterotroph capture of, 102
  storage in soil, 319–20
  succession and, 162, 164–65
nutrition levels in Canada vs. Kenya, 10t

O

O horizon, 313–14, 315–17
ocean(s)
  as biome, 152–54
  climate change effects on, 277, 288
  climate change experiment in, 118–19
  fish production of, by ecosystem, 205t
  hydrologic cycle and, 373–74, 375–76
  Japanese fishing returns from, 8
  plate tectonics and chemical reactions
    in, 73
  temperature gradient in, 478
ocean energy systems
  ocean thermal energy conversion, 478
  tidal energy, 478–79
  wave energy, 479
ocean ridges, 491–92
ocean storage
  of carbon, 105f, 106–8
  of nitrogen, 111
  of phosphorus, 113f, 114
  of solar energy, 477
  of sulfur, 115–16
ocean thermal energy conversion (OTEC), 478
oceanic circulation, 64–69
  in climate change simulations, 280, 281f,
    286
  El Niño and, 67–69
  environmental services of, 67, 69
  patterns of, 65–67
  water's properties and, 64–65
Oconee County PAYT program, 505f
octanes, 430
odd nitrogen hypothesis of ozone depletion,
  299–300
offstream uses of water, 380–81
Ogallala Aquifer, 375, 386f
oil
  crude oil, 425
  discovery, extraction, and processing
    of, 427–30 (see also oil fields/
    reserves)
  energy intensities for nonenergy goods
    and services produced with, 424t
  energy return on investment for, 434
  formation of, 425–26
  as fossil fuel, 420
  industry subsidies and tax breaks for, 484
  as material waste, 501
  pricing of (see oil prices)
  production peak and decline projection
    for, 432–34
  properties of, 421
  storage in soil, 317–18, 319
  storage in U.S., 466t
  transport of (see pipelines)
oil consumption
  in China, 517
  energy efficiency benefits and, 483t
  Energy Policy and Conservation Act
    and, 435
  gasoline and, 230, 239–40 (see also gasoline
    consumption)
  global market and, 435–41
  global rates for, 422, 423t
  replenishment vs. rate of, 6t
  by transportation sector, 226
  in U.S., 12t, 31f, 420f, 466t
oil fields/reserves
  future of, 430–34
  quantity, size, and yield of, 426
  resource quality and opportunity costs
    and, 219, 220f
oil prices
  Amazon deforestation and, 359
  current, 440–41
  fluctuations in, 199, 200f, 420, 437–40
  global market competition and, 437–41
  oil production forecasts and, 433
  OPEC pricing strategies, 437
  U.S. Energy Policy and Conservation Act
    and, 435
oil refineries, 429–30
oil shales, 420, 425, 466t
oil-fueled power plants
  carbon dioxide emissions from, 275t,
    279t, 462t
  operating costs for, 460
  sulfur dioxide and nitrogen oxides emis-
    sions from, 408t, 409t

taxing emissions from, 515
  U.S. electricity generated from, 423t
Okeechobee, Lake, 176
oligotrophic, 148f, 149
omnivores, 91
OPEC (Organization of Petroleum Exporting
  Countries), 244, 437–41
operable capacity, 440
opportunity costs, 217–19, 220f
optimal level of pollution
  air pollution and, 414–17
  greenhouse gases and, 289–90
  policy on environmental degradation and,
    240, 241
orders of soil, 315–17
Oregon
  H. J. Andrews Forest climate graph, 142f
  Pendleton Pay-as-You-Throw program,
    505f
ores, 205, 206
organic wastes, 390–91
Organization of Petroleum Exporting
  Countries (OPEC), 244, 437–41
orographic precipitation, 63
Ortelius, Abraham, 69
OTEC (ocean thermal energy conversion), 478
overburden, 426, 494
  metal recycling and, 507t
  as waste material, 499
overdrafts, 386–88, 389f
overfishing, 7–8
overshoot, 186
oxidation, 313
oxisols, 315–17
oxygen
  atmospheric concentration of, 294
  in Biosphere, 247
  bodily, carbon monoxide and, 400
  eutrophication and, 390–91
  molecular form of, 294
  as nutrient, 22, 23t
  ozone formation and, 295–96, 298–99
oxygen sag curve, 390–91
oxygen-fueled glass furnace energy efficiency,
  483t
ozone, 295
ozone depletion cycle, 298–99
ozone depletion potential, 298t, 299
ozone, ground-level
  formation and concentrations of, 405–7
  reduction of, 413f, 416–17
  as secondary air pollutant, 400
  stratospheric ozone versus, 407
ozone, stratospheric, 295
  Antarctic marine ecosystems and, 303
  CFCs and (see chlorofluorocarbons)
  climate change and, 308, 309
  depletion effects, 302–4
  depletion hypotheses, 297–302
  depletion indicators, 296–97
  distribution of, 296
  fertilizers as source of, 348
  formation of, 295–96
  ground-level ozone versus, 407
  Montreal Protocol and, 305–7, 309–10
  North Pole and, 301–2
  policy deadlock over, 304–5
  restoration policies and, 304–10
  seasonal factors affecting, 301
  terrestrial organisms and, 303–4
  testing of, 300–301

P

Pacific Lumber Company, 367
Pakistan nuclear power, 446f
Panama Canal, 365–66
Panama land bridge, 251–52
paper goods, 135t
paper production
  paper use footprint and, 364
  solar energy and, 58
  wood and, 363
parabolic heating systems, 468
Paraná, Brazil, 359
parent material, 313, 314, 315
parenting, 85–86
particulate matter
  composition, size, and health impact of,
    400–401
  emissions of, by electricity generating
    technologies, 485t
  emissions of, reduction of, 413f, 416
  as material waste, 501
passenger species, 254
pasture, 342–43
pathogens, 388
  in enemy release hypothesis, 259
  sewage treatment and, 389–90
pay-as-you-throw (PAYT) programs, 504, 505f

payback period, 143
payroll taxes, 514, 515
pedalfer soils, 314
pedocal soils, 314
pelagic, 153
pelagic larval duration, 251
pencils, 488
Pendleton PAYT program, 505f
penguins, 303
Pennsylvania
  nuclear accident in, 446, 454–55, 457, 459
  Philadelphia climate graph, 142f
perennials, 159
periodic table of elements, 22f
permafrost, 145
permanent waste disposal, 449
permeability of oil field, 428
permits. See tradable permits
persistent wastes, 209
personal choices
  environmental impact of income and, 225
  footprints of, 15, 17 (see also ecological
    footprint)
  purchase considerations and, 516–17
personal consumption expenditures, 204
Peru
  debt for nature swap and, 370, 371
  policy responses to El Niño in, 70
pesticides, 339
  arsenic as, 491
  biomagnification of, 94–97
  crop yields and, 346–48
  genetically modified crops and, 351
  genetics and resistance to, 347
  integrated pest management and, 352
pests, 339
  climate change and food production and,
    287–88
  monoculture and, 346–48
  timber harvests and, 369
petagrams, 104, 105f
petrochemicals, 213–14, 495
petroleum. See oil entries
pH, 320, 404–5
pharmaceutical companies, 255–56, 266
Philadelphia climate graph, 142f
Philippines
  abandoned/degraded land in, 173
  fish biodiversity in, 250–51
phosphate fertilizers, 114, 120–21
phosphate rock, 12t
phosphoric acid, 502t
phosphorus
  as agricultural waste, 120–21, 345, 516
  allowances/permits for, 521–22
  biodiversity and, 259, 260
  biogeochemical cycle of, 113–15
  as nutrient, 22, 23t
  plant use of, 100
  Redfield ratio of, 101
phosphorus cycle, 113–15
photochemical smog, 348, 407
photodissociation, 296
  of CFC molecules, 298–99
  of nitrogen molecules, 299–300
photosphere, 56
photosynthesis, 77
  carbon cycle and, 104–5, 110
  crop yields and, 341
  general formula and mechanism of, 77
  Mauna Loa Curve and, 110
  soil moisture and, 128, 129f
  succession and, 161, 162f
  UV-B radiation and, 302
photosynthetically active radiation, 302
photovoltaic effect, 469
photovoltaic energy
  applications for, 470
  atmospheric emissions from, 485t
  carbon emissions from, 462f
  cost of, 469f
  land, water, and solid waste impacts
    of, 484t
  technology of, 468–70
phreatic zone, 147
physical activity, 80t, 82
physical change in matter, 25
phytoplankton, 89
  climate change experiment with, 118–19
  lentic ecosystem shifts and, 164–65
  pelagic food webs and, 153–54
  UV-B radiation and, 302–3
Piedmont succession, 159, 160, 163
Pimentel, David, 474
pioneer community, 160
pipelines
  energy sector and subsidies for, 239t
  hydrogen transport via, 480–81
  Trans-Alaska system of, 510
  U.S. system of, 430
piston engines, 213t
Pittsburg (California) PAYT program, 505f

placer deposits, 490
Plano PAYT program, 505f
plant cover, soil erosion and, 320, 322–23, 324
plants
  as autotrophs, 77
  biogeochemical changes and, 259–60
  coal formation from, 424–25
  in food chains/webs, 89, 90f
  genetically modified crops and, 351 (see
    also crop entries)
  growth determinants for, 130
  human consumption of, 334–35
  limiting factors for, 130
  nutrient capture by, 100–101
  protective defenses of, 86–88
  soil function and growth of, 317–19
  sulfur dioxide and, 402–3
  unknown carbon sink and, 108–9
  UV-B radiation and, 302
plasma, 460f, 461
plate, 69
plate tectonics
  El Niño and, 68
  natural resources and environmental
    services and, 73
  theory of, 69, 71, 72f
Plimsoll line, 5, 511–12, 522
plowing techniques, 325–27
plutonium, 26, 27t
  fissionability of, 444
  nuclear weapons and, 459
  reprocessing of, 449
poaching, 260
Point Barrow climate graph, 145f
polar cells, 61f, 62
polar fronts, 62
polar stratospheric clouds, 301, 308
polar vortex, 300, 301, 302, 308
policy issues
  agricultural pollution and profits, 349
  biodiversity preservation, 260–64, 266
  climate change, 118–19, 289–91, 309
  debt for nature swaps, 370–71
  energy-efficient capital, 282
  environmental degradation, 240
  fisheries, 96–97
  gasoline consumption, 244
  governmental response to (see
    government policy)
  internalizing externalities, 512
  land use, 126
  market-based incentives, 242–43
  ozone reduction, 304–7, 309–10
  policy formulation, 241–42
  renewable resources, 186–87
  short- vs. longterm objectives, 290
  sustainability, 12–13
  sustainable forestry, 370
political power, 368
politicians
  command and control strategies and, 514
  environmental accounting and, 222
  good guy-bad guy dichotomy and, 510–11
  internalizing of externalities by, 514
  taxes and, 514–15
pollination, 207t
pollinators, 256
pollution, 3
  agricultural (see agricultural pollution)
  of air (see air pollution)
  biogeochemical cycles and, 121
  energy efficiency benefits and, 483t
  gold mining and, 360
  nonpoint pollutants, 391
  optimal level of (see optimal level of pol-
    lution)
  poverty and exposure to, 233
  "right to pollute," 515
  of water (see water pollution)
pollution taxes, 242, 243
population, 179
population (human)
  age structure and population momentum,
    189–91
  birth and death patterns in, 189
  carrying capacity and (see carrying capac-
    ity for humans)
  demographic transition in, 518–19
  environmental impacts and, 225, 228–31,
    236–37
  gasoline consumption and, 230, 231f,
    237
population density, 180, 334–35
population growth, 179–80, 188–91
  agriculture and, 228–29, 336–37, 357
  carrying capacity and (see carrying capac-
    ity entries)
  deforestation and, 357
  economics of decline in, 518–19
  environmental impact of, 228–31
  environmental Kuznets curve and, 519–21
  exponential form of, 180

formula for, 179
greenhouse gas emissions and, 276
limiting factors for, 180–83
soil erosion and, 328–29
population histograms, **190**, 191
population momentum, **190**
pores, in soil, 317–18
porosity of oil field, **428**
positive feedback loops, **44**, 45. *See also* feedback loops
positive relationship, **42**, 43f
postconsumer content, **506**
postconsumer waste, **208**, **501**. *See also*
municipal solid waste
potable water, 58
potassium, 23t
potential energy, **30**
potential evaporation, **130**
poverty, 233, 329
power, **32**
power density, **466**
power plants, 421f, 422
atmospheric emissions by types of,
485t
coal-fired (*see* coal-fired power plants)
cost comparisons across types
of, 460
gas-fueled (*see* natural gas-fueled power
plants)
geothermal plants, 476–77
nuclear (*see* nuclear power plants)
oil-fueled (*see* oil-fueled power plants)
power tower system, 468
prairie dogs, 142
prairies, **142**
precautionary principle, 522
precipitation, **374**
agroecosystems and, 339
air pollution and, 401, 403–5
climate change and, 378
global patterns of, 62–63
gradients of, 127–28, 129f
hydrologic cycle and, 373f
as land use determinant, 126
soil and, 314, 315, 317
soil erosion and, 320, 323
in soil erosion calculations, 322
sulfate particle deposition by, 403–5
terrestrial biomes and, 130 (*see also*
specific biome)
U.S. annual average of, 383f
precision agriculture, **351**
predator control, 260
predator-prey relationship. *See* food chains;
food webs
predictability, 158–59
predictable behavior, 39–40
preservation. *See* biodiversity
preservation
President's Council on Sustainable
Development, 14–15
pressure
air pressure gradients, 61–62
fossil fuel formation and, 425
in oil field, 428
pressurized water reactors (PWRs), 449–50,
458–59
pressurizers, 449
prey. *See* food chains
price(s)
of environmental services, 368
of food, 344
of fossil fuels, 421
of gasoline (*see* gasoline)
hedonic pricing, 216
internalizing externalities and,
512, 514
market failures and, 238–40
of materials, 496–97
of oil (*see* oil prices)
proportional pricing system, 504
of radioactive waste disposal, 453–54
variable rate pricing system, 504
Price-Anderson Act, **460**
price-induced change
carbon emissions and, 515
deforestation and, 368
fuel efficiency and, 235, 244
market-based incentives and, 242
oil and (*see* oil prices)
Pay-as-You-Throw programs and, 504
technology and, 282
water supply and, 393–94
primary air pollutants, 399–400
primary consumers, 89
primary coolant, **449**
primary energy consumption, 422
primary mill residues, **473**
primary particles, **400**
primary recovery, **428**
primary standards of air quality, **411–12**
primary succession, 160

primary treatment, **396**
prior appropriation doctrine, **393**
private costs, **324**
private property, 214
privatization, **393–94**, 395
production, **203**
factors of, 203–4, 210, 237–38
of food (*see* agriculture; food production;
meat production)
of goods and services (*see* economic goods
and services, production of)
gross primary production, 89
net primary production (*see* net primary
production)
reserves to production, 430–31
of wastes, 208
propane consumption, 31f
property, common *vs.* public, 214–15
property rights, 214, 360
proportional pricing system, **504**
protection, 76f, 86–88
proteins, **100**
protons, 24–25
proved reserves, **428–29**
public goods, 238
public property, 214–15
public transport
energy intensity of, 232f
footprint of, 226–27
forms of (*see specific form*)
pulpwood, 12t
pumped storage plant, **479**
PV. *See* photovoltaic energy
PWRs (pressurized water reactors), 449–50,
458–59
pyrite, 116
pyrolysis, **473**

Q

Q infinity, **433**
Quebec, James Bay Project, 480

R

R horizon, 313f, **314**
r selected, **86**
radiant energy, **29**, 30f
radiation, atomic, 25–26. *See also* solar
radiation/energy
radiation balance, **60**
radiation exposure
NRC risk calculations for, 458–59
from nuclear accidents, 455, 457–58
nuclear reactor design and, 450
radiation inversions, **411**
radiative forcing, **271–73**
biogeochemical cycles and, 274–76
in climate change simulations, 280–82
future of emissions, 276
greenhouse gas concentrations and,
273–74, 275f
hemispheric temperature change and, 274
ozone reduction-climate change link
and, 308
radioactive isotopes, **25**
radioactive waste
disposal costs for, 453–54, 461
footprint of generation of, 456
fusion and, 461
long-term disposal of, 449, 452–53
nature and classification of, 451
plant decommissioning, 454, 461
public confidence and, 446, 447f, 454
reprocessing of, 448–49
waste disposal fund for, 461
radioactivity, 25–26, 27t
radiocarbon dating, **26**
radioisotopes, **25**
rail
energy intensity of, 232f
energy sector and subsidies for, 239t
footprint of use of, 227
rain forests. *See* Amazon rain forest; tropical
rain forests
rainfall. *See* precipitation
Rajasthan nuclear accident, 455t
Rance power plant, 478
random drift, **251**
range-fed meat, **235**
rangelands, **342**
ranges
climate change and, 280, 286–87
types of, 82, 128, 259
raptor biomagnification, 94–95, 97
raw materials, **31**, 207t
RBMK nuclear reactor accident, 457–58
reactants, 25

reactor core, **449**
reactor vessel, **449**
reactors, 449–50, 459. *See also* nuclear
power plants
Reagan, Ronald, 483
reasonable use doctrine, **393**
recharge area, **374**, 378f
rechargeable batteries, 29
recovered paper, 364
recovery
from disturbances, 158
primary *vs.* secondary, 428
of waste, 234–35, 503–6, 507t
recreation, 207t, 384
recyclable products, **506**
recycled-content products, **506**
recycling, **503**
of batteries, footprint of, 500
of revenue, 514
as U.S. employment sector, 506
of waste, 234–35, 503–6, 507t
red deer senescence, 84–85
Redfield ratio, **101**
reductionist approach, **51**
redundancy, 167–68
redundant species, **254**
redwood forest, 367
reefs. *See* coral reefs
refillable glass bottles, 503
refined petroleum products, **429**
refineries, **429–30**
reflect, 270
reflection of solar radiation, 56–57
reformulated gasoline, 415
refrigerators, 482, 483t
refugia, 207t
regolith, **313**
Reims climate graph, 142f
remanufacturing, **506**
remote sensing. *See* satellite remote sensing
renewable energy sources
biomass, 472–75 (*see also* bioenergy)
energy efficiency of, 481–82
footprint of use of, 475
geothermal fields, 476–77
government policy and, 483
hydrogen, 480–81
hydropower, 479–80
oceans, 477–79
photovoltaic systems, 468–70
quantity/quality paradox of, 465–67
solar radiation, 467–68
solar radiation *versus* fossil fuels, 484–85
tides, 478–79
waves, 479
wind, 470–72
renewable resources, **6**. *See also specific resource*
logistic curve in management of, 186–87
natural resource/environmental accounting and, 221
nonrenewable materials and, 213
soil as, 327
Repetto, Robert, 221
replacement level fertility, **189**
reprocessing of nuclear waste, **448–49**
reproduction, 76f, 84–86
reproductive range, **128**
research
on abatement technology, 514
on energy costs, 481
on energy efficiency, 483
environmental Kuznets curve and, 520–21
on ethanol, 474
on lead, 21
on nuclear technology, 461
on radiation exposure, 455, 457
reserve base, **492**
reserves
of oil and natural gas, 428–34
of uranium, 444f, 445
reserves to production (R:P), **430–31**
reservoirs, **376**
of oil and/or natural gas, 428–34
of water, 384, 480
residence time, **103**, 272–73, 274
residential energy use, 421, 481t, 482
residues, 324, 325, 364
resilience, **42**, 166, 168f
resistance, **42**
antibiotic, 351
ecological, 166
environmental, 181
resource depletion and degradation, **7**. *See also*
environmental degradation
resource depletion hypothesis, **336**
resource pessimists *vs.* optimists, 197–99
resource quality, **217–19**
respiration, **88**
carbon cycle and, 105, 110
formulas for, 88–89
maintenance respiration, 76f, 77–79, 82
Mauna Loa Curve and, 110

respiratory system, 401
restoration
ecological, 169–71, 176
of stratospheric ozone, 304–10
reuse, **503**
Revelle factor, 107–8
revenue recycling, **514**
revisions, **429**
Rhine River Basin cadmium flow, 28
rhinoceroses, 262–63
rice
cultivation of, 340
International Rice Research Institute, 255
precision farming and, 351–52
yields of, 342f, 350–51
Richmond, Amy, 521
Ricker Curves, 186, 187f
Ricker, Dr. William, 186
rideshare footprint, 227
"right to pollute," 515
rights doctrines for water allocation, 393
rill erosion, **321**
riparian water rights, **393**
riparian zone, **147**
risk
of disease, 233
precautionary principle and, 522
of radiation exposure, 458–59
risk management, **47**
rivers and streams
as biomes, 147–48
deforestation and, 365–66
hydrologic cycle and, 375–76
hydropower and, 384, 479–80 (*see
also* dams)
instream water usage, 380, 384
overdrafts and, 387–88, 389f
recreational activity on, 384
surface runoff and, 376, 377f
roads and highways
deforestation and, 358, 361–62, 367
ecotourism and, 266f
subsidies for, 239t
rock cycle, 71–73
rocket engines, 213t
Romanian nuclear power, 446f
Romans, lead and, 20
room and pillar techniques, **426**
rooting depth, **319**
rooting layer, **145**
round wood, 12t, 364
Roundup Ready© soybeans, 351
Rowland, Sherwood, 294, 298
rule of absolute ownership, **393**
runoff, **375**
hydrologic cycle and, 373f
of surface water, 376, 377f
Rural Electrification Administration, 376
Russia
climate graphs for, 141f, 144f, 145f
nuclear power in, 446f, 449

S

safe storage of nuclear power plants, 454
Saharan Desert, 139
Saint-Malo power plant, Gulf of, 478
salinity, **146**
salmon fisheries
eco-labeling benefits for, 195
logistic curve and management of, 186,
187f
restoration of, 171
salt marshes, 149
salt water, **373**
desalinization of, 379–80
hydrologic cycle and, 373–74
saltation, 321–22
saltwater intrusion, **387**, 388f
San Diego climate graph, 140f
San Fernando (Venezuela) climate graph,
138f
sands, **318**, 420
Santa Clara County, California, 233
Santa Cruz River, 388, 389f
satellite remote sensing
climate change and, 277, 280, 283, 284f
deforestation estimates and, 356
of ozone levels, 297
of terrestrial biomes, 131, 277, 280, 356
saturated zone, 374
savannas. *See* tropical savannas
scale
economic well-being and, 517–22
economies of, 345
efficiency *versus*, 519–21
spatial, 157–58
total energy use and, 511
scatter, 56, **270**
scavengers, **91**

scenario analysis, **52**
scientific evidence
    climate change policy and, 286
    ozone policy and, 304–5, 309
scientific method, 50–51
scrubland biomes, 140–41
sea level, climate change and, 277, 288
Seabrook nuclear protest, 446, 447f
seam, **426**, 427f
seasonal variations
    climatic, forests and, 363, 365
    ozone levels and, 301
    soil conservation and, 329
    tropical dry forests and, 137
second best problem, **329**
second law of thermodynamics, **33**, 208
secondary air pollutants, **400**
secondary consumers, **89**
secondary coolant, **450**
secondary particles, **400**
secondary productivity, **91**
secondary recovery, **428**
secondary standards of air quality, **412**
secondary succession, **160**
secondary treatment, **396**
Securities and Exchange Commission, 429
sediment, **72**, 325
sedimentary rock, **72**, 73f
    oil fields in, 427–28, 430
sedimentation, **425**, 490
seeds, 339
seismic activity, El Niño and, 68
seismic technology, oil and gas exploration
    and, 427
selective breeding, **335**
selective harvesting, **369**
Sellafield nuclear accident, 455t
senescence, **84–85**
Serengeti National Park, 138f
services, **203**. *See also economic goods and*
    services entries; environmental goods
    and services
set point, **41**
sewage, **388**. *See also water pollution*
    sustainable clean water and, 388–91
    treatment of, 208–9, 220
sheet erosion, **321**
shelterbelts, **326**, 327f
shelterwood method, **369**
shifting cultivation in tropical forests, 340f
shipbuilding, 362
Shisur, 373
shopping bags, 503
shrimp speciation, 251–52
shrublands
    climatic determinants in, 132f
    fire and, 141
    relative erosion rates for, 322t
Siberia, 144f
silt, **318**
siltation, **365–66**
Simon, Julian, 496–97
simulation models, **52**
size, as adaptation, 145–46, 147
skin and sun exposure, 306–7
skin cancer, 304, 307
skin protection factor, 307
slash and burn agriculture, **340**
slope, in soil erosion calculations, 322–23
Slovakian nuclear power, 446f
Slovenian nuclear power, 446f
smog, 348, 407
smoke, **400**
snakes, 79f
snapping shrimp speciation, 251–52
social costs, **324**
    of deforestation, 355, 365
    of hydropower, 480
    of soil erosion, 365
social incentives, 5
social institutions, sustainability and, 196, 197
social structure, deforestation and, 368
Soddy, Frederick, 25
sodium
    as nutrient, 23t
    in salt water, 374, 379–80
sodium bichromate, 502t
sodium hydroxide, 502t
softwood densities, 365f
soil, **313**
    acid deposition and pH of, 405
    agricultural pollutants in, 392
    agroecosystems and, 339
    biodiversity and, 256
    conservation techniques for, 325–27
    constraints on use of, 312–13
    ecosystem service of formation
        of, 207t
    erosion of (*see soil erosion*)
    fertile, use *vs.* replenishment of, 6t

formation of, 256, 313–15
    functions of, 317–20
    horizons of, 313–14
    hydrologic cycle and, 373f, 374
    orders/types of, 315–17
    storage by (*see soil storage*)
    subsidence of, 387
Soil Conservation Service districts, 331
soil conservation techniques, 325–27
soil erosion, **320**
    conservation techniques and, 325–29, 331
    deforestation and, 365–66
    dust bowl and, 312, 331
    equations/formulas measuring, 320,
        322, 325
    footprint for, 322–23
    impacts of, 324–25
    land cover and, 322–23
    policy and, 325, 329–31
    social costs of, 365
    steps in, 320
    U.S. rates of, 325
    water-driven, 320–21
    wind-driven, 312, 321–24
soil moisture
    hydrologic cycle and, 373f, 374
    photosynthesis and, 128, 129f
    soil formation and, 315
soil particles, 317–19
soil profile, **315–17**
soil storage
    of carbon, 104–5
    of nitrogen, 111–12
    of nutrients, 319–20
    of phosphorus, 113–14
    of water, 317–19
soil tension, 318
soil texture, 318, 319
solar activity in odd nitrogen hypothesis of
    ozone reduction, 299–300
solar constant, **56**, 467
solar radiation/energy (sunlight)
    atmospheric circulation and, 59–60, 61f
    climate change and, 270–73
    distribution of, 467
    earth's surface and, 56–57, 270–71
    El Niño and, 68
    footprint of use of, 58
    fossil fuels *versus*, 466–67, 484–85
    greenhouse effect and, 271–73
    oceanic collection and storage of, 477–78
    photodissociation of, 296
    photosynthesis and, 77
    production of, 56
    quantity *vs.* quality of, 466
    satellite remote sensing and, 131
    skin exposure to, 306–7
    subsidies and, 484–85
    succession and, 160, 161, 162f, 164
    terrestrial biomes and, 130 (*see also*
        specific biome)
    thermal collection of, 467–68
    U.S. consumption of, 31f, 466t
    U.S. flow of, 466t
    wind power as, 470
    work done by, 57
solar thermal systems, 462f, **467–68**
solid waste. *See municipal solid waste*
solution, 105f, 313
soot, **400**
sorption, **392**
source reduction, 502–3
South African nuclear power, 446f
South America
    deforestation in, 358, 360
    national oil supplies in, 439f
South Korean nuclear power, 446f
Southern Hemisphere
    radiative forcing in, 272t, 273, 274
    South Pole ozone levels, 294, 297, 299–302
    temperature change in, 274
Soviet Union nuclear accident, 446, 455t,
    457–58, 459
soybeans
    Brazilian production of, 359, 361
    Roundup Ready© variety of, 351
    sulfur dioxide and, 402–3
Spadefoot toads, 140
Spain
    nuclear power in, 446f
    wind energy in, 472
Spartina, 91
spatial scale, **157–58**
specializations, **125**
speciation, 249, 250–52
species dispersal, **250**
    during climate change, 280, 286–87
    of reef fish, 250–51
species diversity, **248**
    ecosystem health and, 166–69

preservation of, 261–63, 264f
    ratio of use *vs.* replenishment of, 6t
    succession and, 162
species evenness, **248–49**
species interaction, 254
species loss. *See extinction*
species richness, **248**
    in beetles, 253
    biomass and, 259f
    factors affecting, 249–53
    species evenness *versus*, 248–49
species-area relationships
    in rate of extinction forecasts, 288–89
    in West Indian Islands, 249
specific heat, **64**
spent nuclear fuel, **447–49**. *See also*
    radioactive waste
spices, 86, 87f
spodosols, **315**, 316f
spontaneous flows, 103–4
spontaneous process, **34**
sports utility vehicles (SUVs), 420, 422, 516–17
St. Cloud PAYT program, 505f
St. George's Bank, 186–87
St. Helens, Mount, 157
stability, 41
    of atmosphere, 410–11, 412f
    coral reef health and, 168f
stable atmosphere, 410–11
state implementation plans, **415**
The State of Fear, 283
stationary sources of air pollution, **400**, 415
steam
    bioenergy plants and, 475
    geothermal, 476, 485t
    nuclear reactors and, 450, 457, 458
steam engines, 213t
steam reforming, 480, 481
steel
    as material waste, 499, 501
    recycling benefits for, 507t
    U.S. *vs.* Indian use of, 12t
stochastic, **47**
stock market collapse of 1929, 312
stone walls, 356
Stonington PAYT program, 505f
storage, **39**
    of energy by organisms, 76f, 82–83
    of energy sources in U.S., 466t
    of hydrogen, 480–81
    of nuclear waste, 448–49, 452–54
storages, **102**
    of carbon, 104–9, 117, 365
    of nitrogen, 110–13
    of phosphorus, 113–15
    rules governing, 102–4
    of sulfur, 115–17
    sustainability and, 521–22
stratosphere, **295**
    in climate models, 283
    ozone in (*see ozone, stratospheric*)
    polar stratospheric clouds, 301
streams. *See rivers and streams*
stress, coral reefs and, 168
strip planting, 326, 327f
stripping ratio, **498**
strong interactions, **168**
strong sustainability, **196**
subatomic particles, 24–25
submerged plants, ecosystem shifts and,
    164–65
subsidence, **387**
subsidence inversions, **411**
subsidies, **238**, **516**
    agricultural, 329, 330, 367, 368
    deforestation and, 367, 368
    elimination of, 15, 516
    energy sector and, 238, 239t
    ethanol production and, 474
    market failures and, 238
    nuclear power and, 460, 461
    for renewable *versus* fossil fuel energy,
        484–85
    soil conservation and, 329, 330
    sustainability and, 12–13
    water prices and, 393
subsistence agriculture, 340f, 357–58
subsoil, 313f, **314**
subsystems, **48**
subtropical forests, 132f
subtropical highs, **61**
succession, **157**
    agroecosystems and, 339
    in aquatic food chains, 170–71
    disturbances preceding, 157–59
    ecological restoration and, 171
    ecosystem shifts and, 164–65
    fire suppression policy and, 175
    on Mount St. Helens, 157
    pattern of, 160–62, 163f

process of, 163–64
sugar maple trees, 83
sulfate aerosols, 117
    atmospheric concentrations of, 273f, 274, 276
    emissions forecast for, 276
    heat balance and, 270
    hemispheric temperature change and, 274
    residence time of, 272–73
sulfate particle deposition, 403–5
sulfur
    dissipative use of, 502t
    fossil fuel content of, 402, 430
    as nutrient, 22, 23t
sulfur cycle, 115–17
sulfur dioxide emissions
    allowances/permits for, 243, 415, 416–17, 514
    by electricity generating technologies, 485t
    energy efficiency benefits and, 483t
    footprint of, 408, 409t
    forecast for, 276
    GDP and, 521
    as material waste, 501
    metal recycling and, 507t
    reduction of, 413f, 415, 416
    sources, deposition, and health impact of,
        117, 121, 275, 402–5
sulfur trioxide, 403
sulfuric acid, 403
Sullivan, Louis, 317
sun/sunlight. *See solar radiation/energy*
suntan, 307
supply, energy-efficient capital and, 282
surface mining, **426**, 427f, 494
surface roughness, 324
surface seeps, 425–26, 427
surface water, **375**
    allocation rights for, 393
    diversion of, 379, 384–85
    hydrologic cycle and, 373f, 375–76
    source and availability of, 376, 377f
surface winds, 62
survival range, **128**
suspension, 321f, **322**
sustainability, **3**
    biogeochemical cycles and, 120, 521–22
    degradation and (*see environmental*
        degradation; specific form of
        degradation)
    eco-labeling footprint and, 195
    ecological footprint and, 15–17 (*see also*
        ecological footprint)
    efficiency and, 512–17
    equity and fairness and, 4–5, 9–12
    fish harvest and, 182, 195
    forestry and, 13f, 368–69, 370
    incentives for, 5, 12–13
    natural resources and environmental
        services and, 3, 6–8
    principles of, 3–5
    principles of, violation of, 6–13
    scale and, 517–22
    soil conservation and, 327–29
    strong *vs.* weak, 196
    systems perspective on, 3–4, 9
Sustainable America: A New Consensus for
    Prosperity, Opportunity, and a Healthy
    Environment for the Future, 14–15
sustainable clean water supply
    access to, 10t, 393–96
    agricultural pollutants and, 391–92
    groundwater mining and,
        386–88, 389f
    industrial pollutants and, 21, 391
    pollution control and, 21, 395–96
    privatization and, 395
    rights doctrines and marketing of, 393–94
    sewage and, 388–91
    surface water diversion and, 384–85
    threats to, 384–92
    user efficiency and, 394–95
sustainable development, 14, **194**
sustainable yield of timber, **368**
SUVs (sports utility vehicles), 420, 422, 516–17
Sweden
    demographic transition in, 193f
    nuclear power in, 446f
    travel modes in, 226t
    Umeå climate graph, 144f
swimmable, **396**
Swiss nuclear power, 446f
symbiosis, 111–12
sympatric speciation, **252**
synthetic fibers, 213–14
systems, 3–4, **39**
    complexity of, 47–48
    distance effects in, 50
    hierarchy of, 48
    homeostasis in, 41–45
    management of, 47–50
    policy and, 9

T

Tahoua climate graph, 138f
tailings, **117**, 499
Taiwanese nuclear power, 446f
Taiyuan climate graph, 141f
tar sands, 420, 466t
Taranto climate graph, 140f
taxes
    environmental tax reform, 15, 515
    internalizing externalities and, 512, 514
    market-based mechanisms and, 514–15
    on municipal waste services, 504
    nuclear power industry and, 461
    oil and gas industries and, 484
    on pollution, 242, 243
    soil erosion and, 329
    subsidies and, 516
technical change hypothesis, **336**
technology, **203**
    agriculture and, 336, 357
    anthropocentric society and, 240–41
    best available and best practicable, 396
    CFC reduction and, 309–10
    Clean Water Act standards and, 396
    economic growth and, 210–12,
        213f, 213t
    energy efficiency and, 276, 282
    environmental impacts of, 225, 234–37
    goods and services and, 203–4, 234–37
    Industrial Revolution and, 421
    market and, 237–38
    market failures and, 238–40
    paper production and, 363
Technology and Economics Assessment Panel,
    309–10
tectonics. *See* plate tectonics
temperate coastal seas, 150–51
temperate forests
    afforestation in, 356
    as biomes, 142–43
    climatic determinants in, 133f
    deforestation in, 355
    net primary production in, 172t
temperate grasslands, 141–42. *See also*
    grasslands
temperature
    air pollution and, 410–11
    atmospheric variations in, 294–95
    biomes and, 130 (*see also specific biome*)
    bodily, 78, 79
    change in (*see* temperature change)
    of Earth's interior, 57
    environmental gradients of, 127–28
    fossil fuel emissions and, 408
    fossil fuel formation and, 425
    oceanic gradient of, 478
    orographic precipitation and, 63
    soil and, 314, 315, 317
    solar radiation wavelength and, 270–71
temperature change, 6
    biomic shifts and, 286–87
    in climate change damage functions,
        289
    in climate change simulations,
        280–82, 283, 284f, 378–79
    detection of, 277
    extinction and, 288
    food supplies and, 287–88
    forests and, 363, 365
    global vacillations in, 270
    greenhouse gases and, 274, 275f,
        284–85
    hemispheric patterns of, 274
    models of, 285–86
    sea levels and, 288
    temperature sensitivity and, 285–86
    water supply and, 378–79
temperature profile, **64–65**
temperature sensitivity, **285–86**
tenant farmers, 329
terraces, **326**
terrestrial biomes, 130–46
    as bioenergy resource base, 473
    boreal forests, 143–44
    carbon reduction in, 275
    climate and soil formation and, 314
    climate change and, 277, 280, 286–87
    climatic determinants in, 130, 132f–133f
    deserts, 139–40
    Mediterranean woodlands and scrublands,
        140–41

ozone levels and, 303–4
satellite remote sensing of, 131, 277,
    280, 356
temperate forests, 142–43
temperate grasslands, 141–42
tropical dry forests, 136–37
tropical rain forests, 134–36
tropical savannas, 138–39
tundras, 144–46
terrorism, 459
tertiary methods, **428**
tertiary treatment, **396**
tetraethyl lead, 502t
Texas
    Plano PAYT program, 505f
    uranium reserves in, 445t
Texas horned lizards, 123f, 124
Texas Railroad Commission, 433, 438, 439f
TFR (total fertility rate), **189**, 192
Thames Blackwater herring fishery, 195
thermal desalting, **379**
thermal NO_x, 405
thermocline, 64f, **65**, 148
thermodynamics. *See* laws of thermo-
    dynamics
thermohaline circulation, **66–67**
thermosphere, **295**
Thomson, Allison, 378–79
thoracic particles, **401**
Thornburgh, Richard L., 455
threatened species, **261**
Three Mile Island nuclear accident, 446,
    454–55, 457, 459
tidal energy, 466t, **478–79**
Tiksi climate graph, 145f
Tilman, David, 164
timber concession, **368**
timber harvests
    deforestation and, 358, 360, 367–68
    paper use footprint, 364
    sustainable forestry and, 368–69
time lags, 48–50, 200–201
time, soil formation and, 315
titanium oxide, 502t
toads
    extinction of, 288
    Spadefoot life cycle, 140
    UV-B radiation and, 303–4
toilets, 394
tokamak fusion reactor, 461, 462f
tolerance model, **163**
topography, soil and, 315, 320
topsoil, 313f, **314**, 315, 316f
tortoise, 254
total fertility rate (TFR), **189**, 192
total saturation, **318–19**
toxins
    arsenic, 491
    biomagnification of, 94–97
    elements as, 23 (*see also specific element*)
    lead, 20, 21
    poverty and, 233
    waste assimilation of, 208–9
trace elements, **22–23**
trace gases, **294**
tradable permits, **242**
    biogeochemical cycles and, 521–22
    for carbon dioxide emissions, 290, 291,
        521–22
    as market-based incentive, 243
    for nitrogen and phosphorus, 521–22
    soil erosion and, 329
    for sulfur dioxide emissions, 243, 415,
        416–17, 514, 521
trade
    biomes and, 134, 135t
    net primary production consumption
        and, 194
    wildlife market, 260
trade winds, **61**, 62
"Tragedy of the Commons," 215
The Tragicall History of D. Faustus, 443
Trans-Alaska Pipeline System, 510
transgenic cultivars, **351**
transpiration, **101**
    climate change and, 378
    forests and, 363
    hydrologic cycle and, 373f
transportation
    of coal, 426
    deforestation and, 361–62
    emissions from, by electricity generating
        technologies, 485t
    energy end use and, 481t, 482
    energy intensities for modes of, 232f
    energy subsidies in, 239t
    energy use in, 226, 421
    footprint of use of, 16, 17, 226–27
    of hydrogen, 480–81
    of moai statues, 16

by motor vehicle (*see* gasoline consump-
    tion; motor vehicles)
    of natural gas, 430
    of radioactive waste, 452
    water modes of, 381, 384
    wind energy transmission, 471
transportation costs, **361**
    deforestation and, 361–62
    of personal transport, 226–27
transportation infrastructure, **361**
transuranic (radioactive) waste
    (TRU), 451t
travel cost methods, **216**
tree plantations, 368–69, 370
trees. *See* forests
tritium, 460
trophic cascades, **169**
trophic position, 89
tropical coastal seas, 151–52
tropical dry forests, 132f, 136–37
tropical forests
    climatic determinants in, 132f, 133f
    deforestation in, 355
    net primary production in, 172t
    shifting cultivation in, 340f
    use *vs.* replenishment of, 6t
tropical rain forests
    Amazonian (*see* Amazon rain forest)
    as biomes, 133f, 134–36
    carbon storage in, 365
    oxisol soils in, 317
tropical savannas
    as biomes, 138–39
    climatic determinants in, 133f
    net primary production in, 172t
    relative erosion rates for, 322t
tropics, precision farming and, 351–52
tropopause, **295**
troposphere, **295**
TRU (transuranic (radioactive) waste), 451t
trucks, 227, 232f. *See also* gasoline consumption;
    motor vehicles
Trumbull, Mt., 170f
tundras, 144–46, 172t
turbidity, **365**
turbine generators, **450**
turbines
    energy efficiency of, 483t
    nuclear, 450
    power output advances in, 213t
    wind, 471
turbulent flow, **323–24**
turnover rate, **158**
tussocks, 124, 125f

U

Ubar, 372f, 373
Uganda, 264, 266
Ukraine
    nuclear accident in, 446, 455t, 457–58, 459
    nuclear power in, 446f
ultisols, 315f
ultrafine particles, **400**
ultraviolet radiation. *See* UV radiation
Umeå climate graph, 144f
unavailable, **100**
unconfined aquifers, **376**
unconventional fossil fuels, **420**
underground mining techniques, **426**, 427f
United Nations, ecotourism defined by, 264
United Nations Framework Convention on
    Climate Change, Article 2, 290
United States
    abandoned/degraded land in, 173
    affluence in, 210
    air pollution regulation in, 412–17
    annual precipitation in, 383f
    CFC ban in, 304–5
    climate change and water supply of,
        378–79
    command and control legislation in, 242
    debt for nature swap by, 371
    electricity generation by source in, 423t,
        424t, 456t, 466, 472
    energy stores and flows, 466t
    environmental accounting by, 222
    footprint of carbon dioxide emissions in,
        278, 279t
    forest conversions in, 357
    GDP in, 10
    global environment and, 15
    GNP of 2004 in, 203
    Green Revolution in, 341–42
    Kyoto Protocol and, 291, 519
    legislation on (*see* government policy; *specific
        legislative act*)
    metal and mineral imports of, 495, 496f

nuclear power industry in, 445–46, 449,
    456f
    oil reserves in, 431 (*see also* Arctic National
        Wildlife Refuge oil exploration)
    ozone reduction and, 300–301, 309
    recycling in, 504–5
    resource use in, 11, 12t, 213, 495f, 499f
    soil erosion in, 324, 325, 330
    sustainability council, 14–15
    total fertility rate in, 189
    uranium production, reserves, and
        resources, 444f, 445
    water pollution regulation in, 395–96
    water rights allocations in, 393
United States energy consumption
    changes in, 420f
    Chinese consumption *versus*, 423
    efficiency benefits for, 483t
    footprint of, 422–24
    German and Japanese consumption
        versus, 422
    global oil market and, 435–41
    Indian consumption *versus,* 12t
    materials use and, 206–7, 208f
    by source and/or sector, 31f, 420f, 421,
        466t, 481, 482f
universal soil loss equation, **320**, 325
University of California at Berkeley ethanol
    research, 474
University of North Carolina cancer research,
    455, 457
unknown carbon sink, **108–9**
unpredictability of systems, 47
unstable atmosphere, **410**
upwellings, **67**, **152**, 153f, 205
uranium
    fissionability of, 444
    isotopes of, 25, 443–44
    nuclear decay and, 26, 27t
    in nuclear fuel cycle, 447 (*see also* radio-
        active waste)
    radioactivity of, 443
    resources and production of, 444–45
    U.S. consumption of, 31f, 466t
    U.S. storage of, 466t
urban air pollution, 399, 406
urban wood wastes, **473**
Urry, Lewis, 500
U.S. DOE. *See* Department of Energy
U.S. Geological Survey
    metal and mineral commodities price
        index of, 497
    mineral classification system of, 492, 493f
    oil reserve forecasts of, 431
U.S. NRC. *See* Nuclear Regulatory
    Commision
Utah uranium reserves, 445t
Utica PAYT program, 505f
utilities, 421f, 422. *See also* coal-fired power
    plants
utility, **204**
UV index, 306, 307
UV radiation
    absorption by ozone, 296
    skin exposure to, 306–7
    UV-A, 302
    UV-B, effects of, 302–4

V

vaccinations, sun exposure and, 306
validated, **51**
van Tilburg, Jo Anne, 16
vaporization, heat of, **30–31**
Vardo climate graph, 145f
variable rate pricing system, **504**
variables, **42**
variance, **47**, 126
variation, natural selection and,
    46–47
veal, 12t
vegetables, 135t, **334**
vegetarian diet, 94, 95f
Venezuela, San Fernando climate
    graph, 138f
Verkhoyansk climate graph, 144f
vertical mixing, **410**
vertisols, 315f
Videlia beetle, 348
vines of Biosphere, 247
virgin materials, **364**
vision and sun exposure, 307
vitrification of nuclear waste, **452**
volatile organic compounds,
    407, 501F
volcanoes, 68, 116f
von Bertlanfy, Frederick, 51

W

Walker, Daniel A., 68
walking, 226–27, 232f
Washington, Auburn PAYT program, 505f
waste(s), **208**
  assimilation of (*see* waste assimilation)
  biogeochemical cycles and, 120–21
  emissions footprint for, 49
  forms of, 208, 209, 499, 501 (*see also* munici-
    pal solid waste; radioactive
    waste; *specific waste*)
  production of, 208
  in water supply (*see* water pollution)
waste assimilation, **3**, 207t, **208**
  economic activity and, 219–20, 221f
  in New York City, 194
  sewage treatment as, 208–9
waste heat, **208**
waste recovery, **234**–35
wastewater. *See* water pollution
water
  agroecosystems and, 339
  aquatic biomes, 146–54 (*see also*
    specific biome)
  ecosystem service of regulation of, 207t
  electricity generating technologies
    and, 484
  geothermal, 476
  gravitational, 318
  hydrologic cycle of, 373–76
  hygroscopic, 318
  meat production and, 232
  mineral deposits and, 489–90
  nuclear power plant use of, 448,
    449–50, 458
  oceanic (*see* oceanic circulation)
  oil extraction and, 428
  physical properties of, 64–65
  as precipitate (*see* precipitation)
  scarcity of, 384, 385f
  soil erosion by, 320–21
  soil formation and, 313, 314, 315
  solar heating of, 467–68
  storage in soil, 317–19
  supply of (*see* water supply)
  terrestrial biomes and, 130
water diversion, **379**, 384–85
water fleas, 165
water heating, 467–68
water pollution, **384**
  agricultural, 348, 391–92
  control mechanisms for, 395–96
  domestic and municipal, 388–91
  geothermal source of, 477

industrial, 391
  population density and, 230–31
  treatment of, 208–9
water power, 213t
water quality, 185–87, 396
water supply
  conflict and, 392
  desalinization technologies, 379–80
  efficient use of, 394–95
  feedlots and, 236
  footprint of use of, 382–83
  groundwater, 376–79
  groundwater mining and, 386–88, 389f
  human use of, 373, 380–84
  instream use of, 381, 384
  marketing of, 393–94
  offstream use of, 380–81
  per capita availability of, 385f
  per capita use of, 380
  pollutants in (*see* water pollution)
  potable, solar energy and, 58
  privatization of, 393–94, 395
  sewage and, 388–91
  sources of, 376–80
  surface water, 376, 377f
  sustainable access to, 393–96
  threats to sustainability of, 384–92
  of Ubar, 372f, 373
  in U.S., climate change and, 378–79
  water diversion and, 379, 384–85
water table, **374**
  cone of depression and, 377, 378f
  overdrafts and, 386–88, 389f
water transport, 239t, 381, 384
water use efficiency, **287**
water vapor
  as greenhouse gas, 285–86
  ozone reduction-climate change link and, 308
  tropopause and, 295
waterborne disease, 388, 389–90
watershed, **375**
  conflict over, 392
  hydropower and, 480
waterworks, **380**
wave energy, 479
wavelength of solar radiation, 270–71, 302
weak interactions, **168**
weak sustainability, **196**
wealth, **220**
weather, **269**
weathering, **72**, **313**
  mineral deposits and, 489–90
  rock cycle and, 72
  soil formation and, 313
weekend effect, **401**–3, 404f
Wegener, Alfred, 69

weight, calculating energy flows and changes
  in, 80–81
wells
  oil and natural gas, 427–29, 430
  water, 376–77, 378f
West Indian Islands, 249
westerlies, 61f, **62**
wet deposition, **403**
wetlands, 172t, 261
whales, 94
wheat production in Mexico, 345, 349
wildcatting, **429**
wildlife market, 260
Wilmington PAYT program, 505f
Wilson, Edmund O., 249, 258
wilting point, **319**
wind energy, **470**
  atmospheric emissions from, 485t
  carbon dioxide emissions from, 462t
  cost of, 469f
  current status of, 471–72
  electricity production capacity
    of, 471
  environmental and siting issues
    with, 472
  land, water, and solid waste impacts
    of, 484t
  materials use in, 484
  output advances in, 213t
  resource base and classes of, 470–71
  tax credits for, 484–85
  turbine technology for, 471
  U.S. consumption of, 31f, 466t
  U.S. flow of, 466t
  U.S. potential for generating
    electricity, 466
  wave energy as, 479
wind erosion equation, **322**
wind farm, **471**
winds, **61**–62
  air pollution and, 401, 404f
  soil erosion by, 312, 321–24
  surface winds, 62
Wisconsin
  Antigo PAYT program, 505f
  soil erosion in, 325
withdrawals, **380**
women, gender inequality and, 10
wood
  arsenic in, 491
  biomass residues, 473
  coal *versus*, 420
  consumption of, 12t, 420f
  densities of, 364t
  home biomes for, 135t
  importance and use of, 362–63

natural resource benefits *vs.* loss of
  environmental services of, 367
paper use footprint and, 364
products, prices of, 497
trade footprint for, 134
wood crisis of 1600s, 200
woodlands, 140–41
Woodstown PAYT program, 505f
Worcester PAYT program, 505f
work, **30**
  by Earth's interior heat, 57
  energy and, 30–32
  labor productivity and technology and,
    211–12, 213f, 213t
  by solar energy, 57
World Bank, 262, 395
World Commission on Environment and
  Development, 194
World Conservation Union, 262
World Heritage Sites, 262
World Meteorological Organization, 305
World Wildlife Federation, 371
World Wildlife Fund, 371
Wyoming uranium reserves, 445t

X

xenobiotics, **391**

Y

Yaqui Valley, Mexico, 345, 349
yield per effort, **431**–32
yields, **339**. *See also* crop yields
Yucca Mountain Project, 452–53, 461
Yuma climate graph, 139f

Z

Zaire, Kisangani climate graph, 136f
Zapp, Alfred, 431
Zarins, Juris, 373
zinc
  battery recycling and, 500
  extraction and processing impact
    of, 499, 500f
  recycling benefits for, 507t
zinc oxides, 502t
zooplankton, **89**, 154